Refereed and selected contributions from

The 2nd International Conference on Nuclear Physics in Astrophysics

Debrecen, Hungary

May 16-20, 2005

edited by

Zs. Fülöp, Gy. Gyürky and E. Somorjai

Società Italiana di Fisica

Springer

Zsolt Fülöp
ATOMKI
P.O.B. 51
H-4001, Debrecen, Hungary
fulop@atomki.hu

György Gyürky
ATOMKI
P.O.B. 51
H-4001, Debrecen, Hungary
gyurky@atomki.hu

Endre Somorjai
ATOMKI
P.O.B. 51
H-4001, Debrecen, Hungary
somorjai@atomki.hu

The articles in this book originally appeared on the internet (www.eurphysj.org) as open access publication of the journal
The European Physical Journal A – Hadrons and Nuclei
Volume 27, Supplement 1
ISSN 1434-601X

Cataloging-in-Publication Data applied for

Bibliographic information published by Die Deutsche Bibliothek. Die Deutsche Bibliothek lists this publication in the Deutsche Nationalbibliografie; detailed bibliographic data is available in the Internet at http://dnb.ddb.de

ISBN 978-3-642-06940-6 e-ISBN 978-3-540-32843-8

Springer is a part of Springer Science+Business Media
springer.com

Printed in Italy

NPA-II Organization

The *2nd International Conference on Nuclear Physics in Astrophysics, NPA-II*, has been organized by the Board of the Nuclear Physics Division of the European Physical Society and the Institute of Nuclear Research (ATOMKI), Debrecen, Hungary.

International Advisory Committee	M. ARNOULD	(Brussels)
	R.C. JOHNSON	(Guildford)
	A. KUGLER	(Řež)
	R.G. LOVAS	(Debrecen)
	Yu.N. NOVIKOV	(Gatchina)
	B. RUBIO	(Valencia)
	Z. SUJKOWSKI	(Swierk)
	F. TERRASI	(Naples)
	N. VAN GIAI	(Orsay)
	R. WYSS	(Stockholm)

International Organizing Committee	C. ANGULO	(Louvain-la-Neuve)
	N. AUERBACH	(Tel-Aviv)
	E. GROSSE	(Dresden)
	K. LANGANKE	(Aarhus)
	M. LEWITOWICZ	(Caen)
	C. ROLFS (Chairman)	(Bochum)
	O. SCHOLTEN	(Groningen)
	E. SOMORJAI	(Debrecen)
	C. SPITALERI	(Catania)
	K. SÜMMERER	(Darmstadt)

Local Organizing Committee	Zs. FÜLÖP (Chairman)
	Z. ELEKES
	Gy. GYÜRKY
	G.G. KISS
	E. SOMORJAI

NPA-II Sponsors

HUNGARIAN ACADEMY OF SCIENCES
EPS YOUNG PHYSICISTS FUND
EPS EAST WEST TASK FORCE
PAKS NUCLEAR POWER PLANT

The European Physical Journal A

Volume 27 • Supplement 1 • 2006

■ Cross-Section Measurements and Nuclear Data for Astrophysics

List of participants

N.H. Allal
Laboratoire de Physique
Theorique Faculte de Physique USTHB
Algeria
allaln@yahoo.com

N. Auerbach
Tel Aviv University
School of Physics and Astronomy
Tel Aviv, 69978 Israel
auerbach@post.tau.ac.il

J. Äystö
Department of Physics
University of Jyväskylä
P.O.B. 35 (YFL), FIN-40014 , Finland
juha.aysto@phys.jyu.fi

D. Bardayan
Oak Ridge National Laboratory
Physics Division P.O.B. 2008 Bldg 6025,
MS-6354 Oak Ridge, TN 37831-6354, USA
bardayan@mail.phy.ornl.gov

W. Beiglböck
Springer-Verlag Heidelberg, Physics Editorial
Tiergartenstr. 17, D-69121 Heidelberg, Germany
beiglboeck@springer.de

D. Bemmerer
INFN Sezione di Padova
Via Marzolo 8, I-35131 Padova, Italy
bemmerer@pd.infn.it

P.F. Bortignon
Dipartimento di Fisica, Università di Milano,
and Istituto Nazionale di Fisica Nucleare
Via Celoria 16, I-20133 Milano, Italy
pierfrancesco.bortignon@mi.infn.it

R. Cerulli
INFN - Laboratori Nazionali del Gran Sasso
S.S. 17 bis km. 18.910, I-67010
Assergi, L'Aquila, Italy
riccardo.cerulli@lngs.infn.it

D. Chmielewska
Institute for Nuclear Studies (IPJ)
05-400 Świerk, Poland
danka@ipj.gov.pl

H. Costantini
Università di Genova and
INFN, Department of Physics
Via Dodecaneso 33, I-16142 Genova, Italy
costant@ge.infn.it

J. Csikai
Department of Experimental Physics
University of Debrecen
Bem tér 18/a, Debrecen, Hungary
csikai@delfin.klte.hu

K. Czerski
Hahn-Meitner-Institut Berlin
Glienicker Str. 100
D-14109 Berlin, Germany
czerski@hmi.de

J. Dilling
TRIUMF National Laboratory
4004 Wesbrook Mall Vancouver, BC
V6T 2A3 Canada
JDilling@triumf.ca

I. Dillmann
Institut für Kernphysik
Forschungszentrum Karlsruhe
and Universität Basel
Departement für Physik und Astronomie
Postfach 3640, D-76021 Karlsruhe, Germany
iris.dillmann@ik.fzk.de

Z. Elekes
Institute of Nuclear Research (ATOMKI)
4028 Debrecen Bem tér 18/c, Hungary
elekes@atomki.hu

G. Endrődi
Eötvös University Budapest
Pázmány P. sétány. 1/A
1117 Budapest, Hungary
endrodigergo@freemail.hu

B.D. Fields
Departments of Physics and Astronomy
University of Illinois, Urbana
IL 61801, USA
bdfields@uiuc.edu

A. Formicola
INFN - Laboratori Nazionali del Gran Sasso
S.S. 17 bis km. 18.910, I-67010
Assergi, L'Aquila, Italy
formicola@na.infn.it

Zs. Fülöp
Institute of Nuclear Research (ATOMKI)
4028 Debrecen Bem tér 18/c, Hungary
fulop@atomki.hu

L. Gaudefroy
Institut de Physique Nucleaire d'Orsay
15 rue G. Clemenceau
IPNO - Groupe NESTER
91400 Orsay, France
gaudefro@ipno.in2p3.fr

A. Gójska
Institute for Nuclear Studies (IPJ)
05-400 Świerk, Poland
gojska@ipj.gov.pl

H. Grawe
Gesellschaft für Schwerionenforschung
Planckstr. 1, D-64291 Darmstadt, Germany
h.grawe@gsi.de

E. Grosse
Institut für Kern- und Hadronenphysik
Forschungszentrum Rossendorf
Postfach 510119, D-01314 Dresden, Germany
E.Grosse@fz-rossendorf.de

C. Gustavino
INFN - Laboratori Nazionali del Gran Sasso
S.S. 17 bis km. 18.910, I-67010
Assergi, L'Aquila, Italy
gustavino@lngs.infn.it

R. T. Güray
Department of Physics, Kocaeli University
41380 Umuttepe, Kocaeli, Turkey
tguray@kou.edu.tr

Gy. Gyürky
Institute of Nuclear Research (ATOMKI)
4028 Debrecen Bem tér 18/c, Hungary
gyurky@atomki.hu

M. Hass
Weizmann Institute of Science
Department of Particle Physics
Rehovot 76100, Israel
michael.hass@weizmann.ac.il

T. Hayakawa
Japan Atomic Energy Research Institute
Advanced Photon Secience Reserch Center
619-0215 Kizu, Umemidai 8-1, Kyoto, Japan
hayakawa@jball4.tokai.jaeri.go.jp

Á. Horváth
Eötvös University Budapest
Pázmány P. sétány. 1/A
1117 Budapest, Hungary
akos@ludens.elte.hu

N. Jachowicz
Department of Subatomic and Radiation Physics
Ghent University
Proeftuinstraat 86, B-9000 Gent, Belgium
natalie.jachowicz@ugent.be

D. Jenkins
Department of Physics
University of York
York, YO10 5DD, UK
dj4@york.ac.uk

R.C. Johnson
Department of Physics, University of Surrey
Guildford, Surrey GU2 7XH, UK
R.Johnson@Surrey.ac.uk

J. José
Institut d'Estudis Espacials de Catalunya
C. Gran Capita 2-4, Ed. Nexus 201
E-08034 Barcelona, Spain
jjose@ieec.fcr.es

A. Junghans
Institut für Kern- und Hadronenphysik
Forschungszentrum Rossendorf
Postfach 510119, D-01314 Dresden, Germany
a.junghans@fz-rossendorf.de

T. Kajino
National Astronomical Observatory
University of Tokyo 2-21-1 Osawa
Mitaka Tokyo 181-8588, Japan
kajino@nao.ac.jp

F. Käppeler
Forschungszentrum Karlsruhe
Institut für Kernphysik
Postfach 3640, D-76021 Karlsruhe, Germany
franz.kaeppeler@ik.fzk.de

K.-U. Kettner
Ruhr-Universität Bochum
Institut für Experimentalphysik III
Universitätsstr. 150
D-44780 Bochum, Germany
kettner@warpserv.fh-bielefeld.de

S. Kimura
INFN-LNS
Via Santa Sofia, 62, I-95123 Catania, Italy
kimura@lns.infn.it

G.G. Kiss
Institute of Nuclear Research (ATOMKI)
4028 Debrecen Bem tér 18/c, Hungary
ggkiss@atomki.hu

J. Klug
Ruhr-Universität Bochum
Institut für Experimentalphysik III
Universitätsstr. 150
D-44780 Bochum, Germany
klug@ep3.rub.de

S. Kubono
Center for Nuclear Study, University of Tokyo
Hirosawa 2-1, Wako, Saitama, 351-0198 Japan
kubono@cns.s.u-tokyo.ac.jp

Y.K. Kwon
Department of Physics, Chung-Ang University
221 Huksuk-Dong, Dongjak-ku Seoul
156-756 Korea
ykkwon11@hanmail.net

G. Lévai
Institute of Nuclear Research (ATOMKI)
4028 Debrecen Bem tér 18/c, Hungary
levai@atomki.hu

B-A. Li
Department of Physics
Arkansas State University
P.O.B. 419, AR 72467-0419 USA
Bali@astate.edu

B.N. Limata
Università Federico II di Napoli
and INFN Sezione di Napoli
Complesso Universitario Monte Sant'Angelo
Via Cintia, I-80126 Napoli, Italy
limata@na.infn.it

R.G. Lovas
Institute of Nuclear Research (ATOMKI)
4028 Debrecen Bem tér 18/c, Hungary
rgl@atomki.hu

G. Martinez-Pinedo
ICREA and
Institut d'Estudis Espacials de Catalunya
Universitat Autonoma de Barcelona
Torre C-5 parell, planta 2
E-08193 Bellaterra, Spain
martinez@ieec.uab.es

A. Mengoni
CERN
CH-1211 Geneve, 23 Switzerland
alberto.mengoni@cern.ch

J.Y. Moon
Department of Physics, Chung-Ang University
221 Huksuk-Dong, Dongjak-ku Seoul
156-756 Korea
jymoon70@hanmail.net

L. Mornas
Departamento de Fisica, Universidad de Oviedo
Avda Calvo Sotelo, 18
E-33007 Oviedo (Asturias), Spain
lysiane@pinon.ccu.uniovi.es

A. Mukhamedzhanov
Cyclotron Institute, Texas A&M University
College Station, TX 77843-3366, USA
akram@comp.tamu.edu

C. Nociforo
GSI, Darmstadt
Planckstr. 1, 64291 Darmstadt, Germany
c.nociforo@gsi.de

F. Nozzoli
Dipartimento di Fisica
Università di Roma "Tor Vergata"
and INFN Sezione di Roma 2
Via della Ricerca Scientifica 1
I-00133 Rome, Italy
Francesco.Nozzoli@roma2.infn.it

N. Özkan
Department of Physics, Kocaeli University
41380 Umuttepe, Kocaeli, Turkey
nozkan@kou.edu.tr

G. Pagliara
Dipartimento di Fisica, Politecnico di Torino
and INFN Sezione di Ferrara
Via Paradiso 12, I-44100 Ferrara, Italy
pagliara@fe.infn.it

S.J.M. Peeters
University of Oxford
The Denys Wilkinson Building, Keble Road
Oxford OX1 3RH, UK
s.peeters1@physics.ox.ac.uk

S. Pirro
INFN Sezione di Milano
P.zza della Scienza 3, I-20126 Milano, Italy
Stefano.Pirro@mib.infn.it

R.G. Pizzone
Università di Catania and LNS-INFN
Via S. Sofia 62, I-95123, Catania, Italy
rgpizzone@lns.infn.it

F. Raiola
Ruhr-Universität Bochum
Institut für Experimentalphysik III
Universitätsstr. 150
D-44780 Bochum, Germany
raiola@ep3.rub.de

C. Rolfs
Ruhr-Universität Bochum
Institut für Experimentalphysik III
Universitätsstr. 150
D-44780 Bochum, Germany
rolfs@ep3.rub.de

S. Romano
Università di Catania and LNS-INFN
Via S. Sofia 62, I-95123, Catania, Italy
romano@lns.infn.it

G. Ruprecht
TRIUMF National Laboratory
4004 Wesbrook Mall Vancouver, BC
V6T 2A3 Canada
ruprecht@triumf.ca

M.M. Sharma
Physics Department, Kuwait University
13060 Kuwait
sharma@kuc01.kuniv.edu.kw

E. Somorjai
Institute of Nuclear Research (ATOMKI)
4028 Debrecen Bem tér 18/c, Hungary
somorjai@atomki.hu

K. Sonnabend
Institut für Kernphysik, TU Darmstadt
Schlossgartenstr. 9 D-64289, Darmstadt
Germany
kerstin@ikp.tu-darmstadt.de

C. Spitaleri
Università di Catania and LNS-INFN
Via S. Sofia 62, I-95123, Catania, Italy
spitaleri@lns.infn.it

F. Strieder
Ruhr-Universität Bochum
Institut für Experimentalphysik III
Universitätsstr. 150
D-44780 Bochum, Germany
strieder@ep3.rub.de

J. Stroth
Institut für Kernphysik
Max von Laue-Strasse 1
D-60438 Frankfurt, Germany
j.stroth@gsi.de

Z. Sujkowski
The Andrzej Soltan Institute
for Nuclear Studies (IPJ)
05-400 Świerk, Poland
sujkowsk@ipj.gov.pl

K. Sümmerer
GSI, Darmstadt
Planckstr. 1, D-64291 Darmstadt, Germany
k.suemmerer@gsi.de

A. Szanto de Toledo
Departamento de Física Nuclear
Instituto de Física da Universidade de Sao Paulo
C.P. 66318, 5315-970 Sao Paulo, S.P., Brasil
szanto@dfn.if.usp.br

N. Timofeyuk
Department of Physics, University of Surrey
Guildford, Surrey GU2 7XH, UK
N.Timofeyuk@surrey.ac.uk

N. Todorovic
Department of Physics, University of Novi Sad
Trg Dositeja Obradovica 4 21000 Novi Sad
Serbia and Montenegro
zikic@im.ns.ac.yu

Y. Togano
Nuclear and Radiation Physics Laboratory
Department of Physics
Rikkyo University 3-34-1 Nishi-Ikebukuro,
Toshima, Tokyo 171-8501, Japan
toga@ne.rikkyo.ac.jp

T. Toró
Department of Physics
Western University of Timisoara
str. Beethoven nr. 4
RO-1900, Timisoara, Romania
toro@qusar.uvt.ro

L. Trache
Cyclotron Institute, Texas A&M University
College Station, TX 77843-3366, USA
livius_trache@tamu.edu

E. Truhlík
Institute of Nuclear Physics
Řež, CZ 250 68, Czech Republic
truhlik@ujf.cas.cz

H. Utsunomiya
Department of Physics, Konan University
8-9-1 Okamoto, Higashinada
Kobe 658-8501 Japan
hiro@konan-u.ac.jp

E. Vigezzi
INFN Sezione di Milano
Via Celoria 16, I-20133 Milan, Italy
Enrico.vigezzi@mi.infn.it

A. Wallner
VERA Laboratory
Institut für Isotopenforschung und Kernphysik
Universität Wien
Währinger Strasse 17, A-1090 Wien, Austria
anton.wallner@univie.ac.at

J.L. Weil
Institute of Isotopes
Chemical Research Center of the
Hungarian Academy of Sciences
P.O.B. 77 1525 Budapest, Hungary
weil@alpha0.iki.kfki.hu

M. Yalcinkaya
Department of Physics
Istanbul University
34459 Vezneciler, Istanbul, Turkey
yalcinm@istanbul.edu.tr

P. Zarubin
Joint Institute for Nuclear Research
Dubna, Moscow Region, 141980 Russia
zarubin@lhe.jinr.ru

Eur. Phys. J. A **27**, s01, XV–XVI (2006)
DOI: 10.1140/epja/i2006-08-053-2

EPJ A direct
electronic only

Foreword

The *Nuclear Physics in Astrophysics II Conference* was held on May 16-20, 2005, in Debrecen, Hungary, hosted by ATOMKI (the Institute of Nuclear Research of the Hungarian Academy of Sciences).

The first *Nuclear Physics in Astrophysics Conference* was the *17th Nuclear Physics Divisional Conference* of the European Physical Society in 2002. Based on the success of the event, the Board of the Nuclear Physics Division decided to launch a series of conferences called *Nuclear Physics in Astrophysics (NPA)* devoted to the interplay between nuclear physics and astrophysics. *NPA-II*, "a Europhysics Conference" was organized under the auspices of the Nuclear Physics Board of the European Physical Society as its 20th Divisional Conference.

Nuclear physics and astrophysics have been strongly linked ever since it was realised that nuclear reactions were a key source of energy in stars. There has been a recent resurgence of activity in the field because of technological developments which have raised the possibility of measuring rates for nuclear processes which are very relevant to the physics of stars and which were previously inaccessible to laboratory measurement. It is fair to say that the issues addressed in nuclear astrophysics are some of the most important and fascinating in the whole of science. Many of them were discussed at this Conference.

The program consisted of review talks on recent developments in nuclear astrophysics and selected oral and poster contributions on experimental and theoretical results in the following fields:

- Big-Bang Nucleosynthesis
- Neutrino Physics
- Stellar (non-explosive) Nucleosynthesis
- Explosive Nucleosynthesis
- Cross-Section Measurements and Nuclear Data for Astrophysics
- Nuclear Structure Far from Stability
- Rare-Ion-Beam Facilities and Experiments
- Perspectives of Nuclear Physics and Astrophysics

The Editor-in-Chief and the publishers kindly agreed to publish a Topical Volume in EPJAdirect consisting of original and refereed papers from the conference as electronic-only supplement to The European Physical Journal A.

Independently, this Topical Volume will also be made available in book form for the conference participants and the library book market. The electronic version of this Topical Volume in EPJAdirect will be open access to everyone worldwide without a time limit. Both oral contributions and posters were peer reviewed by a dedicated committee of referees. This rigorous refereeing process assured a high scientific standard for accepted papers.

Nuclear astrophysics is an outstanding example of state-of-the-art interdisciplinary research. The origin of elements studied by geologists is explored by astrophysicists using reactions measured by the nuclear-physics community. Low-energy reactions also have an impact on solid-state physics. The *Nuclear Physics in Astrophysics II Conference* provided a good opportunity to discuss the progress in various topics of nuclear astrophysics and stimulated new collaborations in the field.

We would like to thank all of those who attended the conference and the International Advisory and Organizing Committees for their time and effort.

The conference would not have been a success without the financial help of the host institute ATOMKI, the Hungarian Academy of Sciences and the European Physical Society.

At its meeting in Albena, Bulgaria on 1st October, 2005, the Nuclear Physics Board of EPS recommended that the next *Nuclear Physics in Astrophysics Conference (NPA-III)* be held in Dresden, Germany in the spring of 2007.

Zs. Fülöp
(on behalf of the NPA-II Local Organizing Committee)

R.C. Johnson
(on behalf of the EPS Nuclear Physics Board)

S. Kubono
(on behalf of the EPJ A Editorial Board)

1 Big-Bang Nucleosynthesis

Eur. Phys. J. A **27**, s01, 3–14 (2006)
DOI: 10.1140/epja/i2006-08-001-2

EPJ A direct
electronic only

Big bang nucleosynthesis in the new cosmology

B.D. Fields[a]

Center for Theoretical Astrophysics, Departments of Astronomy and Physics, University of Illinois, Urbana, IL 61801, USA

Received: 30 June 2005 /
Published online: 13 February 2006 –

Abstract. Big bang nucleosynthesis (BBN) describes the production of the lightest elements in the first minutes of cosmic time. We review the physics of cosmological element production, and the observations of the primordial element abundances. The comparison between theory and observation has heretofore provided our earliest probe of the universe, and given the best measure of the cosmic baryon content. However, BBN has now taken a new role in cosmology, in light of new precision measurements of the cosmic microwave background (CMB). Recent CMB anisotropy data yield a wealth of cosmological parameters; in particular, the baryon-to-photon ratio $\eta = n_B/n_\gamma$ is measured to high precision. The confrontation between the BBN and CMB "baryometers" poses a new and stringent test of the standard cosmology; the status of this test is discussed. Moreover, it is now possible to recast the role of BBN by using the CMB to fix the baryon density and even some light element abundances. This strategy sharpens BBN into a more powerful probe of early universe physics, and of galactic nucleosynthesis processes. The impact of the CMB results on particle physics beyond the Standard Model, and on non-standard cosmology, are illustrated. Prospects for improvement of these bounds via additional astronomical observations and nuclear experiments are discussed, as is the lingering "lithium problem."

PACS. 98.80.Ft Origin, formation, and abundances of the elements – 26.35.+c Big Bang nucleosynthesis

1 Introduction

Big bang nucleosynthesis (BBN) offers the deepest reliable probe of the early universe, being based on well-understood Standard Model physics. Predictions of the abundances of the light elements, D, ^{3}He, ^{4}He, and ^{7}Li, synthesized at the end of the "first three minutes" are in good overall agreement with the primordial abundances inferred from observational data, thus validating the standard hot big bang cosmology (see [1]). This is particularly impressive given that these abundances span nine orders of magnitude —from $^4\text{He/H} \sim 0.08$ down to $^7\text{Li/H} \sim 10^{-10}$ (ratios by number). Thus BBN provides powerful constraints on possible deviations from the standard cosmology [2], and on new physics beyond the Standard Model [3].

We are presently entering an age of a "new cosmology," in which BBN plays a changing but crucial role. The basic world model of a hot big bang has been resoundingly confirmed by a wealth of new observations with unprecedented precision. The fundamental parameters of this world model, including the abundances of the matter and energy component of the cosmos, are now known to a few percent.

The advent of "precision cosmology" is largely spurred by measurements of the cosmic microwave background radiation (CMB). The CMB and BBN are intimately connected. This link traces back to the work of Gamow, Alpher, and Herman [4,5], who determined the thermodynamic conditions needed for nucleosynthesis in the early universe, and used this to extrapolate a present radiation temperature of order $T \sim 5$ K. This historic BBN–CMB connection has recently been deepened with the advent of high-precision measurements of the CMB anisotropy [6]. The measurements have led to determinations of cosmological parameters with an unprecedented accuracy. These include the baryon density, which is the sole parameter in standard BBN.

With the cosmic baryon density measured independently via BBN and the CMB, several new analyses are possible. 1) First and foremost, one can pit BBN against the CMB, asking whether the results of the independent BBN and CMB "baryometers" agree. This comparison marks a fundamental and non-trivial test of the hot big bang. We will see, in fact, that cosmology passes this test, but then other strategies are possible when one *combines* BBN and the CMB, taking advantage of the precision of the CMB baryon density determination. 2) For standard BBN, a CMB-based baryon density fixes all parameters. Then BBN simply makes definite predictions for the

[a] e-mail: `bdfields@uiuc.edu`

abundances of the light elements which can then be contrasted with their observational determinations [7]. The results shed new light on the nucleosynthesis of light elements *after* the big bang.

Finally, 3) one can relax the assumptions of standard BBN (*e.g.*, an early universe populated only by known Standard Model particles and interactions). With the baryon density fixed by the CMB, for the first time, all light elements are available to probe early universe physics. This is perhaps the most exciting new possibility for BBN, and highlights the other aspect of the "new cosmology," namely that our view of the constituents, dynamics, and history of the cosmos has drastically changed of late. The universe appears to be accelerating today, due to a dark energy whose nature is unknown, but may provide clues to such far-reaching realms as quantum gravity. How and whether the dark energy is connected with another cosmic period of acceleration —inflation— remains to be clearly elucidated. Also, the need for and abundance of dark matter is ever more firmly established, but its nature and any connection with dark energy (or baryons!) is unclear. All of these open questions almost surely will find their answers in the early universe. And in this context it is crucial to appreciate that BBN provides our earliest reliable probe.

Thus the new cosmology presents both challenges and opportunities for BBN in general and for nuclear physics in particular. These will be the focus of this review, which draws heavily from [8] which adopts a very similar approach. For other recent reviews see [1].

2 Big bang nucleosynthesis theory

The theory of BBN consists of following the microphysics of weak and nuclear reactions in the cosmological context of an expanding, cooling universe. In fact, the essential aspects of BBN can be understood in terms of the competition between the cosmic expansion rate and particle reaction rates. The expansion rate $H = \dot{a}/a$, with $a(t)$ the cosmic scale factor, is given by the usual Friedmann equation:

$$H^2 = \frac{8\pi G}{3}\rho, \tag{1}$$

where we have dropped the curvature term and cosmological constant as both are negligible in the early universe compared with the energy density. In the early universe, the energy density was dominated by relativistic species ("radiation")

$$\rho_{\mathrm{rad}} = \frac{\pi^2}{30}\left(2 + \frac{7}{2} + \frac{7}{4}N_\nu\right)T^4 \equiv \frac{\pi^2}{30}g_* T^4, \tag{2}$$

which consists of photons, electrons and positrons, and N_ν neutrino flavors (at higher temperatures, other particle degrees of freedom should be included as well). Thus time and temperature scale as $t \sim 1/H \sim 1/T^2$, and in standard BBN (*i.e.*, with $N_\nu = 3$), we roughly have $t/1\,\mathrm{s} \simeq (2.4/g_*)(1\ \mathrm{MeV}/T)^2$.

The synthesis of the light elements is sensitive to physical conditions in the early radiation-dominated era at temperatures $T \lesssim 1$ MeV, corresponding to an age $t \gtrsim 1$ s. At these and higher temperatures, weak interactions rates $\Gamma_{\mathrm{weak}} \gg H$ were rapid compared to the expansion rate, and thus the weak interactions were in thermal equilibrium. In particular, the processes

$$\begin{aligned} n + e^+ &\leftrightarrow p + \bar{\nu}_e\,, \\ n + \nu_e &\leftrightarrow p + e^-\,, \\ n &\leftrightarrow p + e^- + \bar{\nu}_e\,, \end{aligned} \tag{3}$$

fix the ratio of the neutron and proton number densities to be $n/p = e^{-Q/T}$, where $Q = 1.293$ MeV is the neutron-proton mass difference. At $T \gg 1$ MeV, $(n/p) \simeq 1$. As the temperature dropped, the neutron-proton inter-conversion rate, $\Gamma_{np} \sim G_F^2 T^5$, fell faster than the Hubble expansion rate, $H \sim \sqrt{g_* G_N}\ T^2$. This resulted in breaking of chemical equilibrium ("freeze-out") at $T_{fr} \sim (g_* G_N/G_F^4)^{1/6} \simeq 0.8$ MeV. The neutron fraction at this time, $n/p = e^{-Q/T_{fr}} \simeq 1/6$ is thus sensitive to every known physical interaction, since Q is determined by both strong and electromagnetic interactions while T_{fr} depends on the weak as well as gravitational interactions. Moreover, the sensitivity to the Hubble expansion rate affords a probe of *e.g.* the number of relativistic neutrino species [9]. After freeze-out the neutrons were free to β-decay so the neutron fraction dropped to $\simeq 1/7$ by the time nuclear reactions began. A useful semi-analytic description of freeze-out has been given in [10,11].

The nucleosynthesis chain begins with the formation of deuterium in the process $p(n,\gamma)$D. However, the number density of photons is huge relative to the baryon density, *i.e.*, the baryon-to-photon ratio $\eta = n_B/n_\gamma \sim 10^{-9}$. Indeed, η is the sole parameter in the standard BBN model.

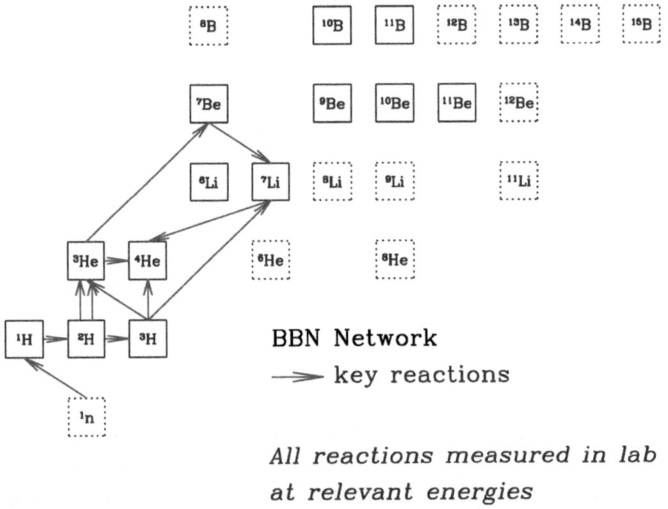

Fig. 1. An abbreviated nuclear network, showing the 12 most important reactions whose uncertainties dominate the theoretical error budget in BBN predictions.

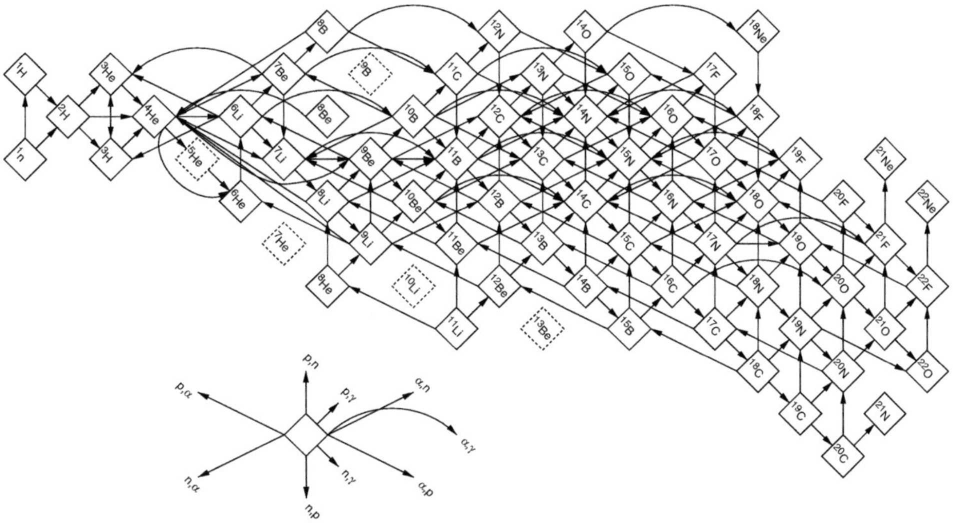

Fig. 2. The full nuclear network used in BBN calculations.

Because of the large number of photons per baryon, photodissociation delays production of deuterium (and other complex nuclei) well after T drops below the binding energy of deuterium, $B_{\mathrm{D}} = 2.23$ MeV. The degree to which deuterium production is delayed can be found by comparing the qualitative expressions for the deuterium production and destruction rates,

$$\begin{aligned} \Gamma_p &\approx n_B \sigma v\,, \\ \Gamma_d &\approx n_\gamma \sigma v e^{-B_{\mathrm{D}}/T}\,. \end{aligned} \tag{4}$$

When the quantity $\eta^{-1}\exp(-B_{\mathrm{D}}/T) \sim 1$, the rate for deuterium destruction ($\mathrm{D} + \gamma \rightarrow p + n$) finally falls below the deuterium production rate and the nuclear chain begins at a temperature $T \sim 0.1$ MeV. Only 2-body reactions are important because the density has become rather low by this time.

In addition to the $p(n,\gamma)$D reaction, there are 10 other major strong reactions (along with the neutron lifetime) leading to the production of the light elements. These reactions, illustrated in fig. 1, are

$$\begin{array}{ll} \mathrm{D(D,}\ p)\mathrm{T}, & {}^3\mathrm{He}(n,\ p)\mathrm{T}, \\ \mathrm{D(D,}\ n)^3\mathrm{He}, & \mathrm{D}(p,\gamma)^3\mathrm{He}. \end{array}$$

Followed by the reactions producing ^{4}He:

$$^3\mathrm{He(D,}\ p)^4\mathrm{He}, \qquad \mathrm{T(D,}\ n)^4\mathrm{He}.$$

The gap at $A = 5$ is overcome and the production and destruction of mass $A = 7$ are regulated by

$$\begin{array}{lll} {}^3\mathrm{He}({}^4\mathrm{He},\gamma)^7\mathrm{Be} & \rightarrow\ {}^7\mathrm{Li} + e^+ + \nu_e, & \\ \mathrm{T}({}^4\mathrm{He},\gamma)^7\mathrm{Li}, & {}^7\mathrm{Be(n,p)}^7\mathrm{Li}, & {}^7\mathrm{Li(p,}^4\mathrm{He)}^4\mathrm{He}. \end{array}$$

The gap at $A = 8$ prevents the production of other isotopes in any significant quantity. The nuclear chain in BBN calculations was extended [12] and is shown in fig. 2.

Nearly all the surviving neutrons when nucleosynthesis begins, end up bound in the most stable light element ^{4}He. Heavier nuclei do not form in any significant quantity both because of the absence of stable nuclei with mass number 5 or 8 (which impedes nucleosynthesis via $^4\mathrm{He}+n$, $^4\mathrm{He}+p$ or $^4\mathrm{He}+{}^4\mathrm{He}$ reactions) and of the large Coulomb barriers for reactions such as the $T(^4\mathrm{He},\gamma)^7$Li and $^3\mathrm{He}(^4\mathrm{He},\gamma)^7$Be reactions listed above. Hence the primordial mass fraction of ^{4}He, conventionally referred to as Y_p, can be estimated by the simple counting argument

$$Y_p = \frac{2(n/p)}{1+n/p} \simeq 0.25.$$

There is little sensitivity here to the actual nuclear reaction rates, which are however important in determining the other "left-over" abundances: D and ^{3}He at the level of a few times 10^{-5} by number relative to H, and ^{7}Li/H at the level of about 10^{-10} (when $\eta_{10} \equiv 10^{10}\eta$ is in the range 1–10). These values can be understood in terms of approximate analytic arguments [13,11]. The experimental parameter most important in determining Y_p is the neutron lifetime, τ_n, which normalizes (the inverse of) Γ_{np}. (This is not fully determined by G_F alone since neutrons and protons also have strong interactions, the effects of which cannot be calculated very precisely.) The experimental uncertainty in τ_n used to be a source of concern but has recently been reduced substantially. The Particle Data Group [14] world average is $\tau_n = 885.7 \pm 0.8$ s.

Historically, BBN as a theory explaining the observed element abundances was nearly abandoned due to its inability to explain *all* element abundances. Subsequently, stellar nucleosynthesis became the leading theory for element production [15]. However, two key questions persisted. 1) The abundance of ^{4}He as a function of metallicity is nearly flat and no abundances are observed to be below about 23% as exaggerated in fig. 3. In particular, even in systems in which an element such

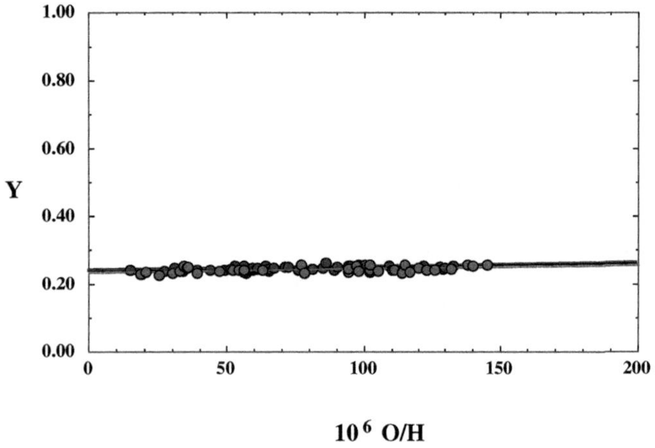

Fig. 3. The ^{4}He mass fraction as determined in extragalactic HII regions as a function of O/H.

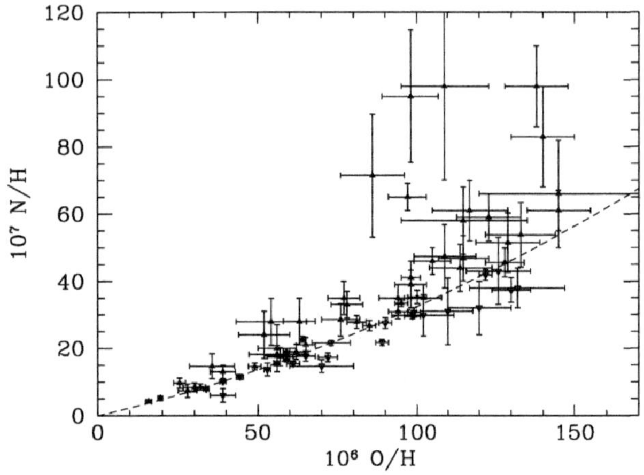

Fig. 4. The nitrogen and oxygen abundances in the same extragalactic HII regions with observed ^{4}He shown in fig. 3.

as oxygen, which traces stellar activity, is observed at extremely low values (compared with the solar value of $O/H \approx 8.5 \times 10^{-4}$), the ^{4}He abundance is nearly constant. This is very different from all other element abundances (with the exception of ^{7}Li as we will see below). For example, in fig. 4, the N/H *vs.* O/H correlation is shown [16]. As one can clearly see, the abundance of N/H goes to 0, as O/H goes to 0, indicating a stellar source for nitrogen. 2) Stellar sources cannot produce the observed abundance of D/H. Indeed, stars destroy deuterium and no astrophysical site is known for the production of significant amounts of deuterium [17]. Thus we are led back to BBN for the origins of D, ^{3}He, ^{4}He, and ^{7}Li.

Having sketched the basic physics of BBN, we now turn to the detailed predictions and the nuclear data on which they rely.

3 From nuclear data to primordial abundance predictions

The homogeneous nature of BBN, and the relatively small number of key reactions, makes the abundance evolution one of the most computationally simple in all of nuclear astrophysics. This relative simplicity presents us with the opportunity to calculate the abundances with a precision and statistical rigor that also is unique to nuclear astrophysics. Given the increasingly precise CMB measurements, it is fortunate that BBN calculations (and to some extent light element observations) can keep pace.

Because standard BBN theory rests upon the Standard Model of particle physics, the electroweak aspects of the calculation are very well determined and do not introduce an appreciable uncertainty. Instead, the major uncertainties come from the thermonuclear reaction rates. As noted above, there are 11 key strong rates (as well as the neutron lifetime) which dominate the uncertainty budget [18,19,20]. In contrast to the situation for much of stellar nucleosynthesis, BBN occurs at high enough temperatures (strong rates freeze out at $T \sim 0.1$ MeV) that laboratory data exist at and even below the relevant energies, so that no extrapolation is needed; this again places BBN in a unique position.

The goal of BBN theory is to derive primordial abundances which are as precise as the nuclear data allow, and to quantify the uncertainties in the predictions. This process takes several steps, going from the nuclear data to the final abundances. 1) First, the relevant nuclear data must be cataloged and evaluated. The data deemed reliable are used to infer both the best-fit cross section (in fact, the astrophysical S-factor), and to obtain an estimate of its uncertainty δS, both as functions of energy. 2) The cross section data are then averaged over a thermal Maxwell-Boltzmann distribution to obtain thermonuclear rates; similarly, one must propagate the cross section uncertainties to obtain thermonuclear errors. 3) Finally, the thermonuclear rates and uncertainties are placed in the BBN code. Monte Carlo techniques [18,19] are used to determine the best-fit abundances, and their uncertainties, at each η.

Recently the input nuclear data have been carefully reassessed [20,21,22], leading to improved precision in the abundance predictions. In addition, polynomial fits to the predicted abundances and the error correlation matrix have been given [23,22].

A treatment of the nuclear and statistical aspects of the BBN calculation was presented by Cyburt [22]. The nuclear data was updated, and the cross sections and their uncertainties fit as a function of energy in a systematic and statistically consistent manner. The error propagation from energy-dependent cross sections to thermal rates was structured to explicitly include the effect of systematic errors in the different dataset normalizations, and to respond to statistically discrepant datasets. As emphasized in [20], systematic errors of these kinds dominate over the statistical errors in the nuclear data. Cyburt explicitly presents a set of thermonuclear rates and their temperature-dependent uncertainties, tailored for use in BBN calculations. He finds that because the systematic errors are indeed large, the results are quite similar to previous work [18] which have adopted the approximation in which these errors dominate. A reanalysis [24] in

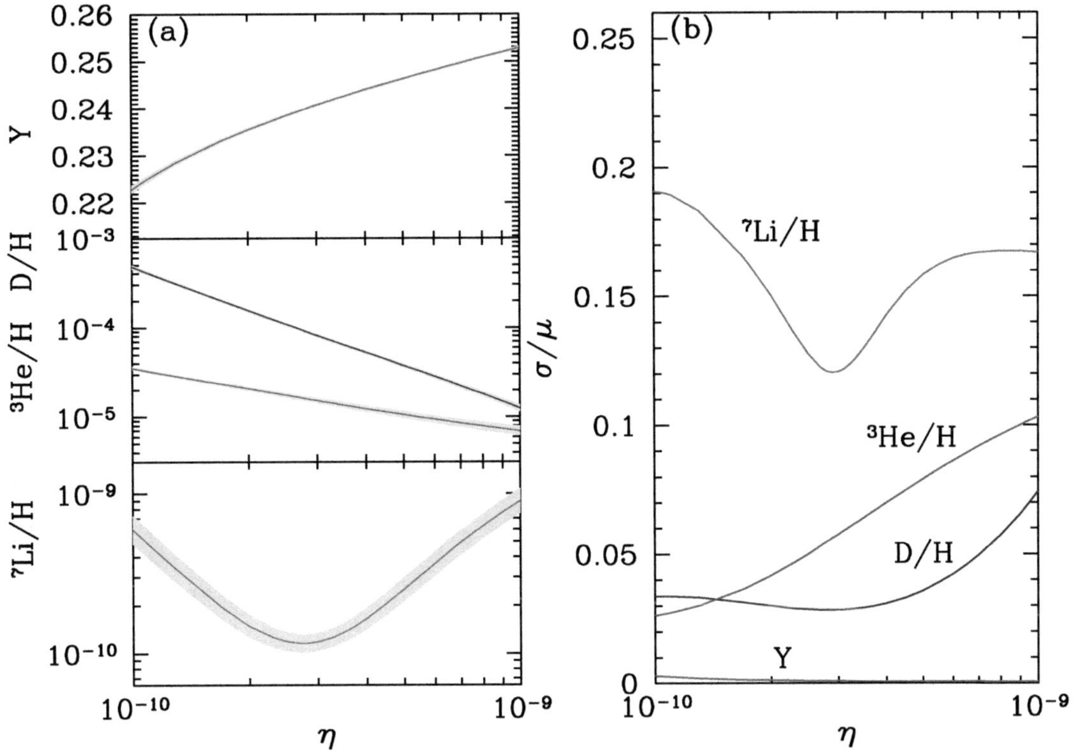

Fig. 5. The predictions of standard BBN [18], with thermonuclear rates based on the NACRE compilation [21]. (a) Primordial abundances as a function of the baryon-to-photon ratio η. Abundances are quantified as ratios to hydrogen, except for ^{4}He which is given in baryonic mass fraction $Y_p = \rho_{\rm He}/\rho_B$. The lines give the mean values, and the surrounding bands give the 1σ uncertainties. (b) The 1σ abundance uncertainties, expressed as a fraction of the mean value μ for each η.

the framework of R-matrix theory has recently been performed yielding comparable results.

The elemental abundances are shown in fig. 5 as a function of η_{10} [18]. The left plot shows the abundance of ^{4}He by mass, Y, and the abundances of the other three isotopes by number. The curves indicate the central predictions from BBN, while the bands correspond to the uncertainty in the predicted abundances. This theoretical uncertainty is shown explicitly in the right panel as a function of η_{10}. The uncertainty range in ^{4}He reflects primarily the 1σ uncertainty in the neutron lifetime.

We first consider standard BBN, which is based on Standard Model physics alone, so $N_\nu = 3$. (The implications of BBN for physics beyond the Standard Model will be considered briefly below, sect. 5). As noted above, in the standard case, the only free parameter is the density of baryons (strictly speaking, nucleons), which sets the rates of the strong reactions. The baryon density is usually expressed normalized to the blackbody photon density as $\eta \equiv n_B/n_\gamma$; while both density change with time and temperature, their ratio η remains constant from the end of BBN to the present. Because standard BBN is a one-parameter theory, any abundance measurement determines η, while additional measurements overconstrain the theory and thereby provide a consistency check. BBN has thus historically been the premier means of determining the cosmic baryon density.

Recently, however, a new "baryometer" has emerged in the form of the CMB. The release of the first-year WMAP results on the anisotropy spectrum of the CMB were a landmark event for all of cosmology, but particularly for BBN. As with other cosmological parameter determinations from CMB data, the derived $\eta_{\rm CMB}$ depends on the adopted priors [25], in particular the form assumed for the power spectrum of primordial density fluctuations. If this is taken to be a scale-free power law, the WMAP data implies $\eta_{10} = 6.58 \pm 0.27$, while allowing for a "running" spectral index lowers the value to

$$\eta_{10} = 6.14 \pm 0.25. \tag{5}$$

Equivalently, this can be stated as the allowed range for the baryon mass density today, $\rho_B = (4.2 \pm 0.2) \times 10^{-31}\ {\rm g\,cm^{-3}}$, or as the baryonic fraction of the critical density: $\Omega_B = \rho_B/\rho_{\rm crit} \simeq \eta_{10} h^{-2}/274 = (0.0224 \pm 0.0009)h^{-2}$, where $h \equiv H_0/100\ {\rm km\,s^{-1}\,Mpc^{-1}}$ is the present Hubble parameter.

The promise of CMB precision measurements of the baryon density suggests a new approach in which the CMB baryon density becomes an input to BBN. Thus, within the context of the Standard Model (*i.e.*, with $N_\nu = 3$), BBN becomes a zero-parameter theory, and the light element predictions are completely determined to within the uncertainties in $\eta_{\rm CMB}$ and the BBN theoretical errors. Comparison with light element observations then can be used to restate the test of BBN–CMB consistency, or to turn the problem around and test the astrophysics of post-BBN light element evolution [26]. Alternatively, one can consider possible physics beyond the Standard Model (*e.g.*, with $N_\nu \neq 3$) and then use all of the abundances to test such models; this is the subject of our final section.

4 Light element observations and comparison with theory

BBN theory predicts the universal abundances of D, ^{3}He, ^{4}He, and ^{7}Li, which are essentially determined by $t \sim 180$ s. Abundances are however observed at much later epochs, after stellar nucleosynthesis has commenced. The ejected remains of this stellar processing can alter the light element abundances from their primordial values, but also produce heavy elements such as C, N, O, and Fe ("metals"). Thus one seeks astrophysical sites with low metal abundances, in order to measure light element abundances which are closer to primordial. For all of the light elements, systematic errors are an important and often dominant limitation to the precision of the primordial abundances.

In recent years, high-resolution spectra have revealed the presence of D in high-redshift, low-metallicity quasar absorption systems (QAS), via its isotope-shifted Lyman-α absorption. These are the first measurements of light element abundances at cosmological distances. It is believed that there are no astrophysical sources of deuterium [27], so any measurement of D/H provides a lower limit to primordial D/H and thus an upper limit on η; for example, the local interstellar value of D/H $= (1.5 \pm 0.1) \times 10^{-5}$ [28] requires that $\eta_{10} \leq 9$. In fact, local interstellar D may have been depleted by a factor of 2 or more due to stellar processing. However, for the high-redshift systems, conventional models of galactic nucleosynthesis (chemical evolution) do not predict significant D/H depletion [29]; in this case, the high-redshift measurements recover the primordial deuterium abundance.

The five most precise observations of deuterium [30, 31,32,33] in QAS give D/H $= (2.78 \pm 0.29) \times 10^{-5}$, where the error is statistical only. These are shown in fig. 6 along with some other recent measurements [34,35,36]. Inspection of the data shown in the figure clearly indicates the need for concern over systematic errors. We thus conservatively bracket the observed values with a range D/H $= 2$–5×10^{-5} which corresponds to a range in η_{10} of 4–8 which easily brackets the CMB determined value. In principle, the steep decrease of D/H with η makes it a sensitive probe of the baryon density. We are optimistic that a larger sample of D/H in high-redshift, low-redshift, and local systems will bring down systematic errors, and thereby increase the precision with which η can be determined.

Using the WMAP value for the baryon density (5) the primordial D/H abundance is predicted to be [7]:

$$(\mathrm{D/H})_p = 2.75^{+0.24}_{-0.19} \times 10^{-5}. \qquad (6)$$

We note that the predicted value in eq. (6) is slightly higher than the value of D/H $= 2.62^{+0.18}_{-0.20} \times 10^{-5}$ quoted in [6], this is largely due to our use of the most recent nuclear rates as determined by the most recent nuclear data compiled by the NACRE Collaboration [21]; at higher values of η, this leads to 5–10% more D/H than older rates [26]. As one can see from fig. 7 (top left), this is in excellent agreement with the average of the 5 best determined quasar absorption system abundances noted above.

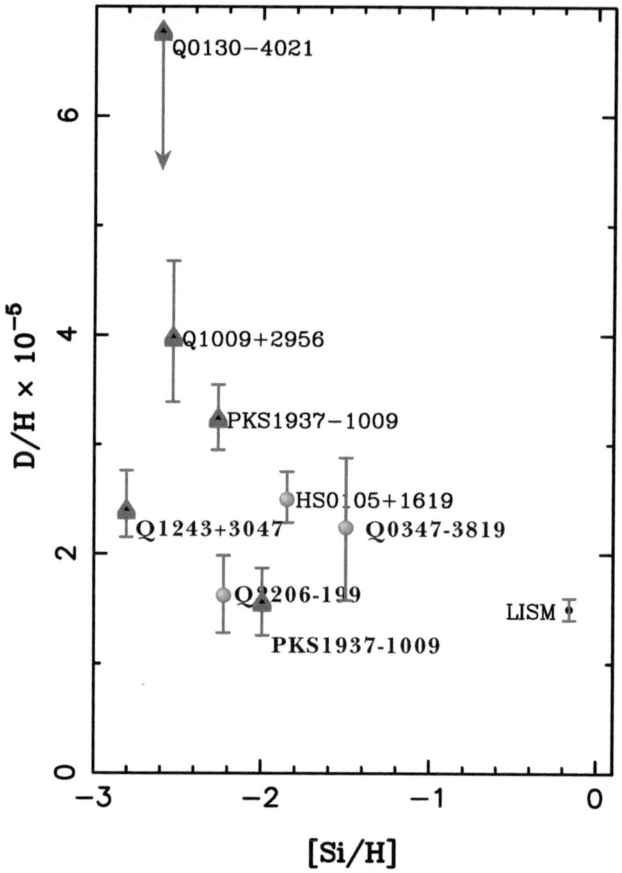

Fig. 6. D/H abundances shown as a function of [Si/H]. Labels denote the background QSO, except for the local interstellar value (LISM; [28]).

We observe ^{4}He in clouds of ionized hydrogen (HII regions), the most metal-poor of which are in dwarf galaxies. There is now a large body of data on ^{4}He and CNO in these systems [37]. These data confirm that the small stellar contribution to helium is positively correlated with metal production. ^{4}He abundance determinations depend on the overall intensity of the He emission line, but also on a number of physical parameters associated with the HII region. These include, the temperature, the electron density, optical depth and degree of underlying absorption. Previous extrapolations to zero metallicity have been based on various assumptions concerning these parameters and typically gave relatively low values for the primordial ^{4}He abundance. For example, a study using ^{4}He abundance where the electron density is derived from SII observations yielded [16,38] $Y_p = 0.238 \pm 0.002 \pm 0.005$. Here the latter error is an estimate of the systematic uncertainty; this dominates, and is based on the scatter in

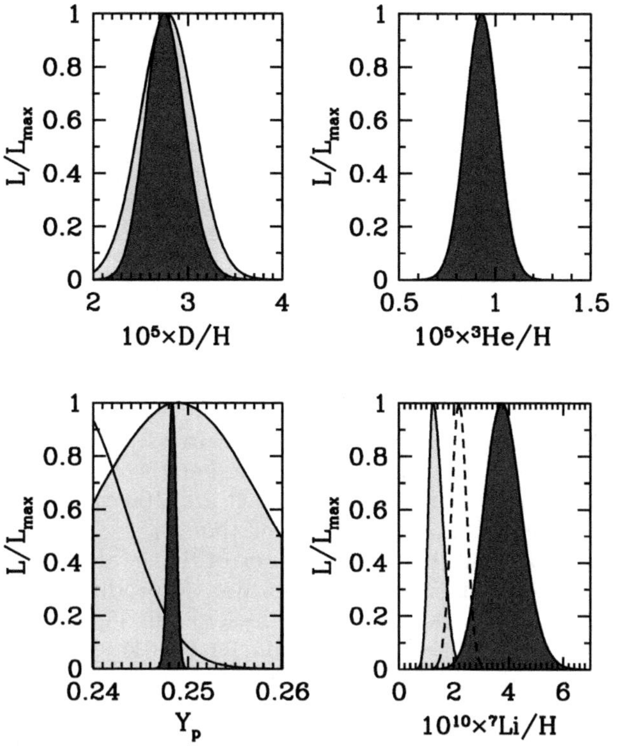

Fig. 7. Primordial light element abundances as predicted by BBN and WMAP (dark-shaded regions). Different observational assessments of primordial abundances are plotted as follows: (top left) the light-shaded region shows $D/H = (2.78 \pm 0.29) \times 10^{-5}$; (top right) no observations plotted; (bottom left) the solid curve shows $Y_p = 0.238 \pm 0.002 \pm 0.005$, and the light-shaded region shows $Y_p = 0.249 \pm 0.009$; (bottom right) the light-shaded region shows $^7Li/H = 1.23^{+0.34}_{-0.16} \times 10^{-10}$, while the dashed curve shows $^7Li/H = (2.19 \pm 0.28) \times 10^{-10}$.

different analyses of the physical properties of the HII regions [37,39]. Other extrapolations to zero metallicity give $Y_p = 0.2443 \pm 0.0015$ [37], and $Y_p = 0.2391 \pm 0.0020$ [40]. These are based on a self-consistent approach to determining the ^{4}He abundance. That is, several He lines are used to best fit the physical parameters. The effects of underlying absorption was not included in these analyses.

Recently a careful study of the systematic uncertainties in ^{4}He, particularly the role of underlying absorption, has led to a higher value for the primordial abundance of ^{4}He [41]. Using a subset of the highest quality from the data of Izotov and Thuan [37], all of the physical parameters listed above including the ^{4}He abundance were determined self-consistently with Monte Carlo methods [39]. Note that the ^{4}He abundances are systematically higher, and the uncertainties are several times larger than quoted in [37]. In fact this study has shown that the value determined for Y_p is highly sensitive to the method of analysis used. The extrapolated ^{4}He abundance was determined to be $Y_p = 0.249 \pm 0.009$. Conservatively, it would be difficult at this time to exclude any value of Y_p inside the range 0.232–0.258.

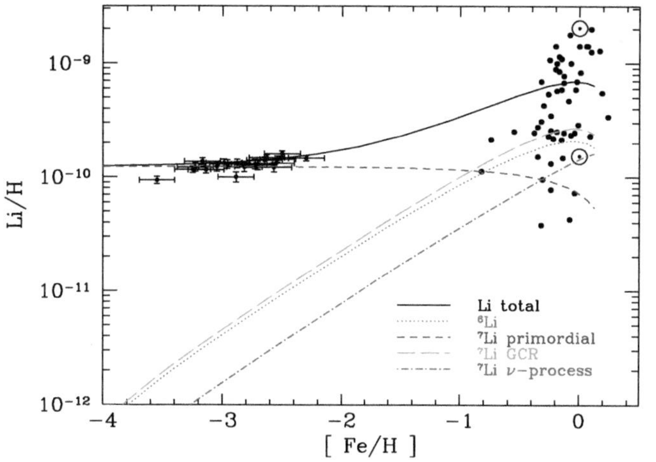

Fig. 8. Contributions to the total predicted lithium abundance from the adopted GCE model of [48], compared with low-metallicity stars and a sample of high-metallicity stars. The solid curve is the sum of all components.

At the WMAP value for η, the ^{4}He abundance is predicted to be [7]

$$Y_p = 0.2484^{+0.0004}_{-0.0005}. \tag{7}$$

While this value is considerably higher than any prior determination of the primordial ^{4}He abundance as can bee seen in fig. 7 (bottom left), it is in excellent agreement with the most recent analysis of the ^{4}He abundance [41]. Note also that the large uncertainty ascribed to this value indicates that while ^{4}He is certainly consistent with the WMAP determination of the baryon density, it does not provide for a highly discriminatory test of the theory at this time.

We also note that the CMB itself is sensitive to the helium abundance, which affects the mean number of electrons per baryon in the cosmic plasma. This opens the possibility that one can use the CMB to measure not only η but also ^{4}He, and thus test BBN entirely with the CMB, without reference to low-redshift measurements and prior to any stellar nucleosynthetic contamination [42,43]. The issues of Y_p degeneracy with other cosmological parameters remain to be sorted out, but we are optimistic that this approach may become an important new tool for BBN.

The systems best suited for Li observations are metal-poor stars in the spheroid (Pop II) of our Galaxy. Observations have long shown [44] that Li does not vary significantly in Pop II stars with metallicities $\lesssim 1/30$ of solar —the "Spite plateau". Recent precision data suggest a small but significant correlation between Li and Fe [45] which can be understood as the result of Li production from Galactic cosmic rays [46]. Extrapolating to zero metallicity one arrives at a primordial value [47] $Li/H|_p = (1.23 \pm 0.06) \times 10^{-10}$.

Figure 8 shows the different Li components for a model with $(^7Li/H)_p = 1.23 \times 10^{-10}$. The linear slope produced by the model is independent of the input primordial value. The model of ref. [48] includes in addition to primordial

^{7}Li, lithium produced in Galactic cosmic-ray nucleosynthesis (primarily $\alpha + \alpha$ fusion), and ^{7}Li produced by the ν-process during type-II supernovae. As one can see, these processes are not sufficient to reproduce the population I abundance of ^{7}Li, and additional production sources are needed.

The ^{7}Li abundance based on the WMAP baryon density is predicted to be [7]:

$$^7\text{Li/H} = 3.82^{+0.73}_{-0.60} \times 10^{-10} \qquad (8)$$

or in astronomical notation, $A(\text{Li}) \equiv \log_{10}(\text{Li/H}) + 12 = 2.58^{+0.08}_{-0.07}$ dex. This value is in clear contradiction with most estimates of the primordial Li abundance as can be seen from fig. 7 (bottom right).

The quoted value for the ^{7}Li abundance assumes that the Li abundance in the stellar sample reflects the initial abundance at the birth of the star; however, an important source of systematic uncertainty comes from the possible depletion of Li over the $\gtrsim 10$ Gyr age of the Pop II stars. The atmospheric Li abundance will suffer depletion if the outer layers of the stars have been transported deep enough into the interior, and/or mixed with material from the hot interior; this may occur due to convection, rotational mixing, or diffusion. Standard stellar evolution models predict Li depletion factors which are very small ($\sigma_{A(\text{Li})} < 0.05$ dex) in very metal-poor turnoff stars [49]. However, there is no reason to believe that such simple models incorporate all effects which lead to depletion such as rotationally induced mixing and/or diffusion. Current estimates for possible depletion factors are in the range ~ 0.2–0.4 dex [50]. As noted above, this data sample [45] shows a negligible intrinsic spread in Li leading to the conclusion that depletion in these stars is as low as 0.1 dex.

Another important source for potential systematic uncertainty stems from the fact that the Li abundance is not directly observed but rather, inferred from an absorption line strength and a model stellar atmosphere. Its determination depends on a set of physical parameters and a model-dependent analysis of a stellar spectrum. Among these parameters, are the metallicity characterized by the iron abundance (though this is a small effect), the surface gravity which for hot stars can lead to an underestimate of up to 0.09 dex if $\log g$ is overestimated by 0.5, though this effect is negligible in cooler stars. Typical uncertainties in $\log g$ are ± 0.1–0.3. The most important source for error is the surface temperature. Effective-temperature calibrations for stellar atmospheres can differ by up to 150–200 K, with higher temperatures resulting in estimated Li abundances which are higher by ~ 0.08 dex per 100 K. Thus accounting for a difference of 0.5 dex between BBN and the observations, would require a serious offset of the stellar parameters. We note however, that a recent study [51] with temperatures based on Hα lines (considered to give systematically high temperatures) yields $^7\text{Li/H} = (2.19^{+0.46}_{-0.38}) \times 10^{-10}$. These results are based on a globular cluster sample and do show considerable dispersion. This result is consistent with previous Li measurements of the same cluster which gave $^7\text{Li/H} = (1.91 \pm 0.44) \times 10^{-10}$ [52] and $^7\text{Li/H} = (1.69 \pm 0.27) \times 10^{-10}$ [53]. A related study (also of globular cluster stars) gives $^7\text{Li/H} = 2.29 \times 10^{-10}$ [54].

Finally another potential source for systematic uncertainty lies in the BBN calculation of the ^{7}Li abundance. As one can see from fig. 5, the predictions for ^{7}Li carry the largest uncertainty of the 4 light elements which stems from uncertainties in the nuclear rates. The effect of changing the yields of certain BBN reactions was recently considered by Coc *et al.* [55]. In particular, they concentrated on the set of cross sections which affect ^{7}Li and are poorly determined both experimentally and theoretically. In many cases however, the required change in cross section far exceeded any reasonable uncertainty. Nevertheless, it may be possible that certain cross sections have been poorly determined. In [55], it was found for example, that an increase of either the $^7\text{Li}(d,n)2^4\text{He}$ or $^7\text{Be}(d,p)2^4\text{He}$ reactions by a factor of 100 would reduce the ^{7}Li abundance by a factor of about 3.

The possibility of systematic errors in the $^3\text{He}(\alpha,\gamma)^7\text{Be}$ reaction, which is the only important ^{7}Li production channel in BBN, was considered in detail in [56]. The absolute value of the cross section for this key reaction is known relatively poorly both experimentally and theoretically. However, the agreement between the standard solar model and solar neutrino data thus provides additional constraints on variations in this cross section. Using the standard solar model of Bahcall [57], and recent solar neutrino data [58], one can exclude systematic variations of the magnitude needed to resolve the BBN ^{7}Li problem at the $\gtrsim 95\%$ CL [56]. Thus the "nuclear fix" to the ^{7}Li BBN problem is unlikely.

Finally, we turn to ^{3}He. Here, the only observations available are in the solar system and (high-metallicity) HII regions in our Galaxy [59]. This makes inference of the primordial abundance difficult, a problem compounded by the fact that stellar nucleosynthesis models for ^{3}He are in conflict with observations [60]. Consequently, it is no longer appropriate to use ^{3}He as a cosmological probe [61]; instead, one might hope to turn the problem around and constrain stellar astrophysics using the predicted primordial ^{3}He abundance [62]. For completeness, we note that the ^{3}He abundance is predicted to be [7]:

$$^3\text{He/H} = 9.28^{+0.55}_{-0.54} \times 10^{-6} \qquad (9)$$

at the WMAP value of η.

The overall concordance as evidenced in fig. 7 is encouraging and will be sharpened with further WMAP data, and subsequently from *Planck Surveyor* measurements. However, note that ^{7}Li is inconsistent with the CMB (as it is with D and ^{4}He) given the error budgets we have quoted. The question then becomes more pressing as to whether this mismatch comes from systematic errors in the observed abundance predictions, or whether there might be new physics at work.

5 Beyond the Standard Model

Given the simple physics underlying BBN, it is remarkable that it still provides the most effective test for the cosmological viability of ideas concerning physics beyond the Standard Model. Limits on particle physics beyond the Standard Model come mainly from the observational bounds on the ^{4}He abundance. As discussed earlier, the neutron-to-proton ratio is fixed by its equilibrium value at the freeze-out of the weak interaction rates at a temperature $T_f \sim 1$ MeV modulo the occasional free neutron decay. Furthermore, freeze-out is determined by the competition between the weak interaction rates and the expansion rate of the universe

$$G_F{}^2 T_f{}^5 \sim \Gamma_{\rm weak}(T_f) = H(T_f) \sim \sqrt{G_N g_*} T_f{}^2 . \qquad (10)$$

In the Standard Model, the number of relativistic particle species at 1 MeV is $g_* = 5.5 + \frac{7}{4} N_\nu$, where 5.5 accounts for photons and $e^\pm$, and N_ν is the number of (massless) neutrino flavors. The presence of additional neutrino flavors (or any other relativistic species) at the time of nucleosynthesis increases the overall energy density of the universe and hence the expansion rate leading to a larger value of T_f, (n/p), and ultimately Y_p. Because of the form of eq. (10) it is clear that just as one can place limits [9] on N_ν, any changes in the weak or gravitational coupling constants can be similarly constrained (for a discussion see ref. [63]).

The helium curves in fig. 5 were computed taking $N_\nu = 3$; the computed abundance scales as $\Delta Y_{\rm BBN} \simeq 0.013 \Delta N_\nu$ [10]. Clearly the central value for N_ν from BBN will depend on η, which is independently determined (with little sensitivity to N_ν). Here we have taken the WMAP value for η. As one can see from fig. 7 (bottom left), if the best value for the observed primordial ^{4}He abundance is 0.249, then, for $\eta_{10} \sim 6$, the central value for N_ν is very close to 3. The dependence of the light element abundances on N_ν is shown in fig. 9 [26]. Because of the relatively large error on the newly determined value for Y_p, the 2 σ upper limit to Y_p of 0.267 implies a conservative bound of

$$N_\nu \lesssim 4.5. \qquad (11)$$

Note that this bound is significantly weaker than previous bounds using smaller values for the primordial ^{4}He abundance.

It is also possible to use D to place an interesting limit on N_ν [7]. As seen in fig. 9, D is not as sensitive to N_ν as ^{4}He is, but nonetheless it does have a significant dependence. The relative error in the observed abundance of D/H ranges from 7–10%, depending on what systems are chosen for averaging. If the five most reliable systems are chosen, the peak of the N_ν likelihood distribution lies at $N_\nu \approx 3.0$, with a width of $\Delta N_\nu \approx 1.0$ as seen in fig. 10. However, if we limit our sample to the two D systems that have had multiple absorption features observed, then the peak shifts to $N_\nu \approx 2.2$, with a width of $\Delta N_\nu \approx 0.7$ as shown by the dashed curve. Adopting the five-system D average, D/H $= (2.78 \pm 0.29) \times 10^{-5}$, we get the following constraints on N_ν:

$$\hat{N}_\nu = 3.02 \,, \qquad (12)$$
$$1.26 < N_\nu < 5.22 \quad (95\% \text{ CCL}) .$$

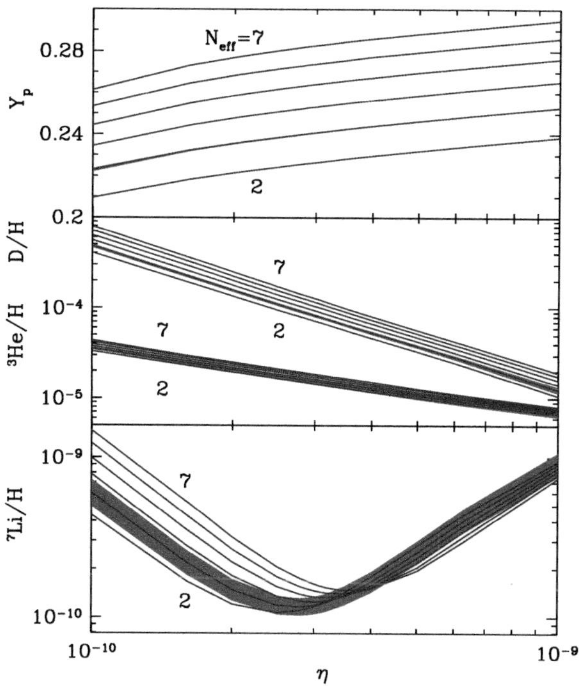

Fig. 9. BBN abundance predictions [26] as a function of the baryon-to-photon ratio η, for $N_\nu = 2$ to 7. The bands show the 1σ error bars. Note that for the isotopes other than Li, the error bands are comparable in width to the thickness of the abundance curve shown. All bands are centered on $N_\nu = 3$.

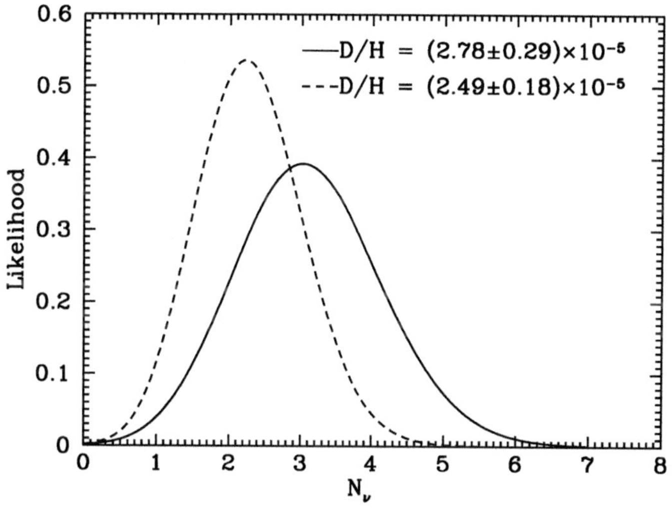

Fig. 10. Likelihoods for N_ν as predicted by the WMAP η (eq. (5)) and light element observations as in fig. 7.

While the limit from D/H remains weaker than the revised ^{4}He limit, it is certainly beginning to be competitive.

The limits on N_ν can be translated into limits on other types of particles or particle masses that would affect the expansion rate of the universe during nucleosynthesis. For example consider "sterile" neutrinos with only right-handed interactions of strength $G_R < G_F$. Such particles would decouple at higher temperature than (left-handed) neutrinos, so their number density ($\propto T^3$) relative to neutrinos would be reduced by any subsequent entropy release, *e.g.* due to annihilations of massive particles that become non-relativistic in between the two decoupling temperatures. Thus (relativistic) particles with less than full strength weak interactions contribute less to the energy density than particles that remain in equilibrium up to the time of nucleosynthesis [64]. If we impose $N_\nu < 4.5$ as an illustrative constraint, then the three right-handed neutrinos must have a temperature $3(T_{\nu_R}/T_{\nu_L})^4 < 1.5$. The temperature of the decoupled ν_R's is determined by entropy conservation, where $s = (4/3)\rho/T$ in a radiation-dominated Universe. In analogy with the determination of the ratio of the left-handed neutrino temperature to the photon temperature $(T_{\nu_L}/T_\gamma = (4/11)^{1/3}$, one finds $T_{\nu_R}/T_{\nu_L} = [(43/4)/g_*(T_d)]^{1/3} < 0.84$, where $(43/4)$ is the value of g_* at $T > 1$ MeV including $\gamma, e^\pm$, and 3 ν_L's and T_d is the decoupling temperature of the ν_R's. This requires $g_*(T_d) > 18$. Since g_* measures the number of relativistic species in thermal equilibrium, it increases with temperature as more particles are included in the radiation background. At the onset of BBN, $g_* = 10.75$; to achieve $g_* > 18$ requires the inclusion of quarks and gluons in the background, and thus one is forced to a temperature greater than the quark/hadron transition temperature or $T_d > 140$ MeV. The decoupling temperature is related to G_R through $(G_R/G_F)^2 \sim (T_d/3\,\mathrm{MeV})^{-3}$, where 3 MeV is the decoupling temperature for ν_Ls. This yields a limit $G_R \lesssim 10^{-2}G_F$. This limit becomes sensitive to the assumed upper limit to N_ν when T_d is constrained to be larger than the temperature of the quark-hadron transition.

6 Summary

As cosmology moves into a new and high-precision era, the utility of BBN is shifting from a test of the basic cosmological world model to one which is a sensitive probe of astrophysical processes which relate to the light element abundances as well as a sensitive probe of physics beyond the Standard Model. The parameters of the Standard Model as they relate to BBN are now well determined: $\eta_{10} = 6.14$ and $N_\nu = 3$. With these values, BBN is able to make very definite predictions for the primordial abundances of the light elements. These predictions are then contrasted with observational determinations of the abundances and reveal either our knowledge or ignorance of the astrophysical process which affect the primordial abundances. In the case of ^{4}He and D/H, within the large uncertainties in the observational data, the concordance is excellent. ^{7}Li remains problematic though several possible (if not perfect) astrophysical solutions are available. Finally, as the data on the light element abundances improves, BBN will sharpen its ability to probe physics beyond the Standard Model. As such it remains the only available tool for which to examine directly the very early universe.

The power and reach of BBN in the "new cosmology" rests upon the precision of its nuclear inputs, and of light element observations. Nuclear physicists can justly take pride that the cross sections used in BBN are now well measured, in the key cases to better than 10%. However, as cosmology develops ever higher requirements for precision, BBN must also improve. Thus there is a need to improve all nuclear reaction data. More specifically, some reactions stand out as particularly in need of further examination. Notable examples are $n(p,\gamma)d$, which are well measured but not in the crucial BBN regime near ~ 100 keV, and $^3\mathrm{He}(\alpha,\gamma)^7\mathrm{Be}$ which dominates production of ^{7}Li and thus lies at the heart of the "lithium problem." Active work on these reactions was reported in NPA-II, and we look forward to seeing these efforts bear fruit by the NPA-III meeting.

I am grateful to the organizers for a stimulating and enjoyable meeting, and for being especially gracious hosts. It is a pleasure to thank my longtime BBN collaborators, particularly Keith Olive and Richard Cyburt. The work of B.D.F. was supported by the National Science Foundation under grant AST-0092939.

References

1. T.P. Walker, G. Steigman, D.N. Schramm, K.A. Olive, K. Kang, Astrophys. J. **376**, 51 (1991); K.A. Olive, G. Steigman, T.P. Walker, Phys. Rep. **333**, 389 (2000); B.D. Fields, S. Sarkar, Phys. Rev. D **66**, 010001 (2002).
2. R.A. Malaney, G.J. Mathews, Phys. Rep. **229**, 145 (1993).
3. S. Sarkar, Rep. Prog. Phys. **59**, 1493 (1996).
4. G. Gamow, Nature **162**, 680 (1948).
5. R.A. Alpher, R. Herman, Nature **162**, 774 (1948).
6. C.L. Bennett *et al.*, Astrophys. J. Suppl. **148**, 1 (2003), arXiv:astro-ph/0302207; D.N. Spergel *et al.*, Astrophys. J. Suppl. **148**, 175 (2003), arXiv:astro-ph/0302209.
7. R.H. Cyburt, B.D. Fields, K.A. Olive, Phys. Lett. B **567**, 227 (2003), arXiv:astro-ph/0302431.
8. B.D. Fields, K.A. Olive, to be published in Nucl. Phys. A.
9. G. Steigman, D.N. Schramm, J. Gunn, Phys. Lett. B **66**, 202 (1977).
10. J. Bernstein, L.S. Brown, G. Feinberg, Rev. Mod. Phys. **61**, 25 (1989).
11. V. Mukhanov, arXiv:astro-ph/0303073.
12. D. Thomas, D. Schramm, K.A. Olive, B. Fields, Astrophys. J. **406**, 569 (1993).
13. R. Esmailzadeh, G.D. Starkman, S. Dimopoulos, Astrophys. J. **378**, 504 (1991).
14. S. Eidelman *et al.* Phys. Lett. B **592**, 1 (2004).
15. E.M. Burbidge, G.R. Burbidge, W.A. Fowler, F. Hoyle, Rev. Mod. Phys. **29**, 547 (1957).
16. B.D. Fields, K.A. Olive, Astrophys. J. **506**, 177 (1998).
17. H. Reeves, J. Audouze, W.A. Fowler, D.N. Schramm, Astrophys. J. **179**, 909 (1973); R.I. Epstein, J.M. Lattimer, D.N. Schramm, Nature **263**, 198 (1976); T. Prodanović, B.D. Fields, Astrophys. J. **597**, 48 (2003), arXiv:astro-ph/0307183.
18. R.H. Cyburt, B.D. Fields, K.A. Olive, New Astron. **6**, 215 (2001), arXiv:astro-ph/0102179.
19. L.M. Krauss, P. Romanelli, Astrophys. J. **358**, 47 (1990); M. Smith, L. Kawano, R.A. Malaney, Astrophys. J. Suppl. **85**, 219 (1993); N. Hata, R.J. Scherrer, G. Steigman, D. Thomas, T.P. Walker, Astrophys. J. **458**, 637 (1996); A. Coc, E. Vangioni-Flam, M. Casse, M. Rabiet, Phys. Rev. D **65**, 043510 (2002).

20. K.M. Nollett, S. Burles, Phys. Rev. D **61**, 123505 (2000), arXiv:astro-ph/0001440.
21. NACRE Collaboration (C. Angulo *et al.*), Nucl. Phys. A **656**, 3 (1999).
22. R.H. Cyburt, Phys. Rev. D **70**, 023505 (2004), arXiv:astro-ph/0401091.
23. G. Fiorentini, E. Lisi, S. Sarkar, F. L. Villante, Phys. Rev. D **58**, 063506 (1998), arXiv:astro-ph/9803177; S. Burles, K.M. Nollett, M.S. Turner, Astrophys. J. **552**, L1 (2001), arXiv:astro-ph/0010171.
24. P. Descouvemont, A. Adahchour, C. Angulo, A. Coc, E. Vangioni-Flam, arXiv:astro-ph/0407101.
25. M. Tegmark, M. Zaldarriaga, A.J.S. Hamilton, Phys. Rev. D **63**, 043007 (2001) arXiv:astro-ph/0008167.
26. R.H. Cyburt, B.D. Fields, K.A. Olive, Astropart. Phys. **17**, 87 (2002), arXiv:astro-ph/0105397.
27. R.I. Epstein, J.M. Lattimer, D.N. Schramm, Nature **263**, 198 (1976).
28. J. Linsky, Space Sci. Rev. **84**, 285 (1998).
29. D.D. Clayton, Astrophys. J. **290**, 42 (1985); B.D. Fields, Astrophys. J. **456**, 678 (1996).
30. S. Burles, D. Tytler, Astrophys. J. **499**, 699 (1998); **507**, 732 (1998).
31. J.M. O'Meara, D. Tytler, D. Kirkman, N. Suzuki, J.X. Prochaska, D. Lubin, A.M. Wolfe, Astrophys. J. **552**, 718 (2001), arXiv:astro-ph/0011179.
32. D. Kirkman, D. Tytler, N. Suzuki, J.M. O'Meara, D. Lubin, Astrophys. J. Suppl. **149**, 1 (2003), arXiv:astro-ph/0302006.
33. M. Pettini, D.V. Bowen, Astrophys. J. **560**, 41 (2001), arXiv:astro-ph/0104474.
34. D. Kirkman, D. Tytler, S. Burles, D. Lubin, J.M. O'Meara, Astrophys. J. **529**, 655 (1999).
35. S. D'Odorico, M. Dessauges-Zavadsky, P. Molaro, Astron. Astrophys., **368**, L21 (2001).
36. N.H. Creighton, J.K. Webb, A. Ortiz-Gill, A. Fernandez-Soto, astro-ph/0403512.
37. Y.I. Izotov, T.X. Thuan, V.A. Lipovetsky, Astrophys. J. **435**, 647 (1994); Astrophys. J. Suppl. **108**, 1 (1997); Y.I. Izotov, T.X. Thuan, Astrophys. J. **500**, 188 (1998).
38. K.A. Olive, G. Steigman, Astrophys. J. Suppl. **97**, 49 (1995), arXiv:astro-ph/9405022; K.A. Olive, E. Skillman, G. Steigman, Astrophys. J. **483**, 788 (1997), arXiv:astro-ph/9611166.
39. K.A. Olive, E.D. Skillman, New Astron. **6**, 119 (2001).
40. M. Peimbert, A. Peimbert, M.T. Ruiz, Astrophys. J. **541**, 688 (2000); A. Peimbert, M. Peimbert, V. Luridiana, Astrophys. J. **565**, 668 (2002).
41. E.D. Skillman, K.A. Olive, Astrophys. J. **617**, 29 (2004).
42. R. Trotta, S.H. Hansen, Phys. Rev. D **69**, 023509 (2004), arXiv:astro-ph/0306588.
43. G. Huey, R.H. Cyburt, B.D. Wandelt, Phys. Rev. D **69**, 103503 (2004), arXiv:astro-ph/0307080.
44. F. Spite, M. Spite, Astron. Astrophys. **115**, 357 (1982); P. Molaro, F. Primas, P. Bonifacio, Astron. Astrophys. **295** , L47 (1995); P. Bonifacio, P. Molaro, Mon. Not. R. Astron. Soc. **285**, 847 (1997).
45. S.G. Ryan, J.E. Norris, T.C. Beers, Astrophys. J. **523**, 654 (1999).
46. B.D. Fields, K.A. Olive, New Astron., **4**, 255 (1999); E. Vangioni-Flam, M. Cassé, R. Cayrel, J. Audouze, M. Spite, F. Spite, New Astron., **4**, 245 (1999).
47. S.G. Ryan, T.C. Beers, K.A. Olive, B.D. Fields, J.E. Norris, Astrophys. J. Lett. **530**, L57 (2000).
48. B.D. Fields, K.A. Olive, Astrophys. J. **516**, 797 (1999).
49. C.P. Deliyannis, P. Demarque, S.D. Kawaler, Astrophys. J. Suppl. **73**, 21 (1990).
50. S. Vauclair, C. Charbonnel, Astrophys. J. **502**, 372 (1998); M.H. Pinsonneault, T.P. Walker, G. Steigman, V.K. Narayanan, Astrophys. J. **527**, 180 (1998), arXiv:astro-ph/9803073; M.H. Pinsonneault, G. Steigman, T.P. Walker, V.K. Narayanan, Astrophys. J. **574**, 398 (2002), arXiv:astro-ph/0105439.
51. P. Bonifacio, *et al.*, Astron. Astrophys. **390**, 91 (2002).
52. L. Pasquini, P. Molaro, Astron. Astrophys. **307**, 761 (1996).
53. F. Thevenin *et al.*, Astron. Astrophys. **373**, 905 (2001).
54. P. Bonifacio, Astron. Astrophys. **395**, 515 (2002).
55. A. Coc, E. Vangioni-Flam, P. Descouvemont, A. Adahchour, C. Angulo, Astrophys. J. **600**, 544 (2004), arXiv:astro-ph/0309480.
56. R.H. Cyburt, B.D. Fields, K.A. Olive, Phys. Rev. D **69**, 123519 (2004).
57. J.N. Bahcall, M.H. Pinsonneault, S. Basu, Astrophys. J. **555**, 990 (2001).
58. SNO Collaboration (S.N. Ahmed *et al.*), Phys. Rev. Lett. **92**, 181301 (2004), arXiv:nucl-ex/0309004.
59. D.S. Balser, T.M. Bania, R.T. Rood, T.L. Wilson, Astrophys. J. **510**, 759 (1999).
60. K.A. Olive, D.N. Schramm, S.T. Scully, J.W. Truran, Astrophys. J. **479**, 752 (1997), arXiv:astro-ph/9610039.
61. T.M. Bania, R.T. Rood, D.S. Balser, Nature **415**, 54 (2002).
62. E. Vangioni-Flam, K.A. Olive, B.D. Fields, M. Casse, Astrophys. J. **585**, 611 (2003), arXiv:astro-ph/0207583.
63. E.W. Kolb, M.J. Perry, T.P. Walker, Phys. Rev. D **33**, 869 (1986); B.A. Campbell, K.A. Olive, Phys. Lett. B **345**, 429 (1995); L. Bergstrom, S. Iguri, H. Rubinstein, Phys. Rev. D **60**, 045005 (1999); C.M. Mueller, G. Schaefer, C. Wetterich, arXiv:astro-ph/0405373.
64. G. Steigman, K.A. Olive, D.N. Schramm, Phys. Rev. Lett. **43**, 239 (1979); K.A. Olive, D.N. Schramm, G. Steigman, Nucl. Phys. B **180**, 497 (1981).

2 Neutrino Physics

Eur. Phys. J. A **27**, s01, 17–23 (2006)
DOI: 10.1140/epja/i2006-08-002-1

EPJ A direct
electronic only

Salty neutrinos from the Sun

Results from the salt phase of the Sudbury Neutrino Observatory

S.J.M. Peeters[a]

For the SNO Collaboration
University of Oxford, Particle Physics, Denys Wilkinson Building, Keble Road, Oxford, OX1 3RH, UK

Received: 1 July 2005 /
Published online: 10 February 2006 –

Abstract. The Sudbury Neutrino Observatory is a 1 ktonne heavy-water Čerenkov detector. The first phase of this experiment has shown that neutrinos from the Sun change flavour. For the second phase of the experiment, approximately 2 tonnes of salt (NaCl) was added to the heavy water to enhance the neutral-current detection as well as the neutral-current and charged-current separability. Here the results are presented from the complete salt phase at the Sudbury Neutrino Observatory. The electron energy spectrum is presented for the first time. It is consistent with an undistorted ^{8}B spectral shape. It is also consistent with the Large Mixing Angle (LMA) parameters obtained through a global fit including the data presented in this paper. These parameters are found to be $\Delta m^2 = (8.0^{+0.6}_{-0.4}) \times 10^{-5}\,\text{eV}^2$ and $\theta = 33.9^{+2.4}_{-2.2}$ degrees. The total flux of active-flavour neutrinos from ^{8}B decay in the Sun is found to be $4.94^{+0.21}_{-0.21}(\text{stat})^{+0.38}_{-0.34}(\text{syst})$, whereas the integral flux of electron neutrinos for an undistorted ^{8}B spectrum is $1.68^{+0.0.6}_{-0.06}(\text{stat})^{+0.08}_{-0.09}(\text{syst})$, both in units of $10^6\,\text{cm}^{-2}\,\text{s}^{-1}$. The day-night asymmetry in the observed fluxes is consistent with both absence of matter-enhancement effects in the Earth and with the small level of these effects expected for the LMA model as stated above.

PACS. 26.65.+t Solar neutrinos – 14.60.Pq Neutrino mass and mixing

1 Introduction

Ray Davis' experiment [1] indicated over 35 years ago that the observed electron neutrino flux from the Sun was lower than expected from the Standard Solar Model [2]. As it was thought that the physics of Sun was well understood, this discrepancy was deemed "the solar neutrino problem".

In the first "D_2O" phase of the Sudbury Neutrino Observatory (SNO), the long-standing solar neutrino problem was solved. It demonstrated (see [3]) that the total active neutrino flux from the Sun, observed via neutral-current (NC) reactions, is consistent with Standard Solar Model (SSM), whereas the electron neutrino component of this flux, observed via charged-current (CC) reactions, is indeed too low. This shows that neutrinos change flavour on their way from the Sun to the Earth and thus solved the solar neutrino problem.

In the second "salt" phase of SNO, 2 tonnes of ultra-pure salt (NaCl) was dissolved into the heavy water. This enhanced the detection efficiency of the NC flux, but more importantly this increased the separability of the CC reactions and the NC reactions. The results of the first 254 days of data of this phase were published in [4]. Recently, the results of the full 391 days of data of this phase were published [5] and these results will be discussed in this article.

Finally, the third ("NCD") phase of SNO has started recently. In this phase 40 ^{3}He proportional counters (neutral-current detectors or "NCDs") have been placed inside the heavy-water volume. These absorb and detect the neutrons (produced by NC reactions) within the heavy-water volume and will allow an even better separation of NC events and CC events.

2 The Sudbury Neutrino Observatory

SNO [6] is a 1 ktonne water Čerenkov detector, located at a depth of 2092 m (6010 m of water equivalent) in the INCO Ltd. Creighton mine near Sudbury, Ontario in Canada. The detector (see fig. 1) consists of a 5.5 cm thick, 12 m diameter acrylic vessel (AV), holding the 1000 tonnes ultra-pure D_2O target, surrounded by 7 ktonnes of ultra-pure H_2O shielding. The AV is surrounded by a 17.8 m diameter geodesic sphere, holding 9456 inward-looking and 91 outward-looking 20 cm photomultiplier tubes (PMTs).

[a] e-mail: `s.peeters1@physics.ox.ac.uk`

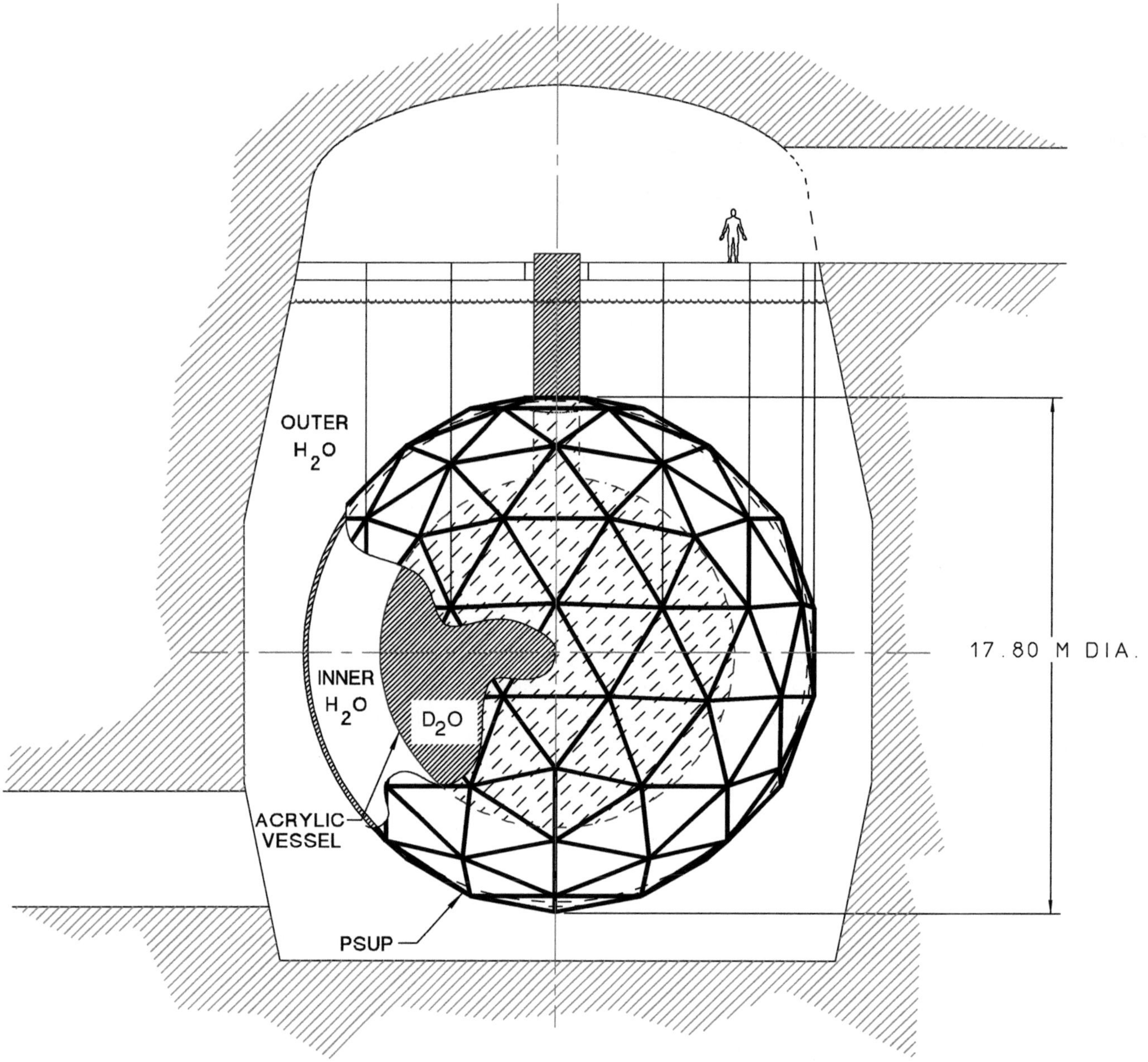

Fig. 1. The layout of the SNO detector. The PSUP holds the photomultiplier tubes.

The PMTs are surrounded by light concentrators, which increase the light collection efficiency. The effective coverage is 59%. The volume outside the geodesic support structure acts both as a cosmic-ray veto system and as a shield from naturally occurring radioactivity from the surrounding rock and the construction materials.

SNO detects neutrinos via elastic scattering (ES, $\nu_x + e^- \rightarrow \nu_x + e^-, x = e, \mu, \tau$), CC ($\nu_e + d \rightarrow p + p + e^-$) and NC ($\nu_x + d \rightarrow n + p + \nu_x$). The elastic scattering is sensitive to all neutrino flavours, but the electron neutrino reaction is enhanced by a factor of approximately 6.5 compared to the other flavours. The CC reaction is only sensitive to electron neutrinos. The NC reaction, however, is equally sensitive to all active neutrino flavours.

The ES and CC events are observed by the Čerenkov light generated by the electron produced in these reactions. The NC events are observed more indirectly. The neutron captures on either deuterium (first (pure D_2O) phase) or chlorine (second (salt) phase) creating an excited state which decays emitting γ's. These γ's primely Compton-scatter, producing relativistic electrons, which produce observable Čerenkov light.

2.1 Neutrino detection in the salt phase

Adding ultra-pure NaCl to the heavy-water target to a concentration of $(0.196 \pm 0.002)\%$ by weight increases the detection efficiency for neutrons, and thus the sensitivity

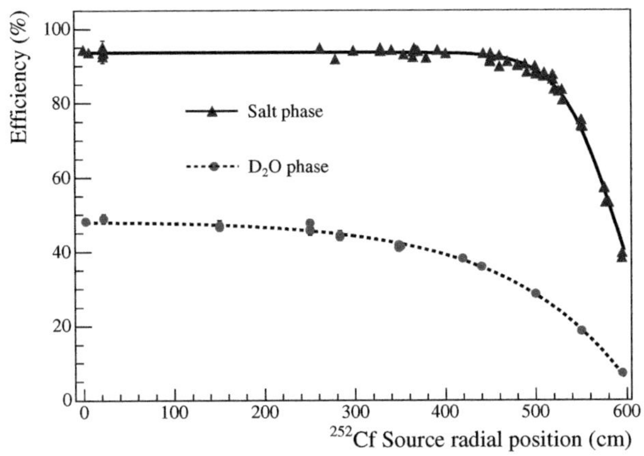

Fig. 2. Neutron capture efficiency *versus* radial position of the ^{252}Cf source for the pure D_2O and salt phase.

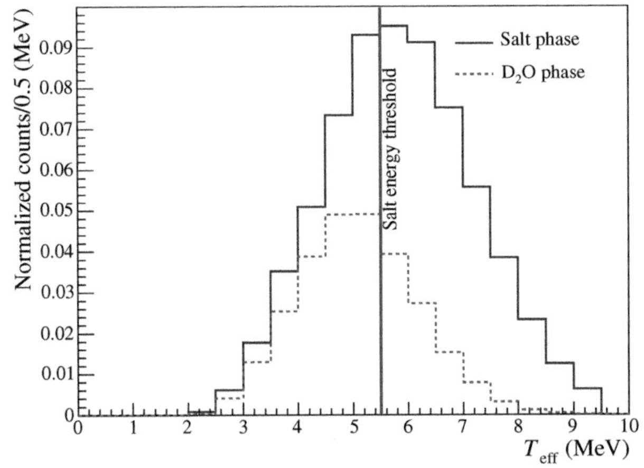

Fig. 3. Neutron energy response during the D_2O (dashed line) and salt (solid line) running periods. The vertical line represents the analysis energy threshold of $T_{\rm eff} = 5.5$ MeV. The distributions shown here are normalised to the neutron detection efficiency in the two phases for $R < 550$ cm.

to the NC reaction is enhanced. This increase of detection efficiency comes about from two effects.

Firstly, the neutron capture cross-section of deuterium is 0.5 mbarn, whereas the neutron capture cross-section of ^{35}Cl is 44 barn. Figure 2 shows a comparison of the neutron capture efficiency for the D_2O phase and the salt phase, as measured using calibration sources in SNO. It clearly shows a large increase neutron capture efficiency in the salt phase, compared to the D_2O phase.

Secondly, when a neutron captures on deuterium, a single γ of 6.25 MeV is emitted. In the salt phase, for a neutron produced within the AV, the probability of capture on ^{35}Cl is 90%. When a neutron captures on ^{35}Cl, multiple γ's are released with a total energy of 8.6 MeV. As shown in fig. 3, this enhances the detection probability, as more Čerenkov light is produced.

In the salt phase, neutron capture is observed by the Čerenkov light produced by relativistic electrons produced

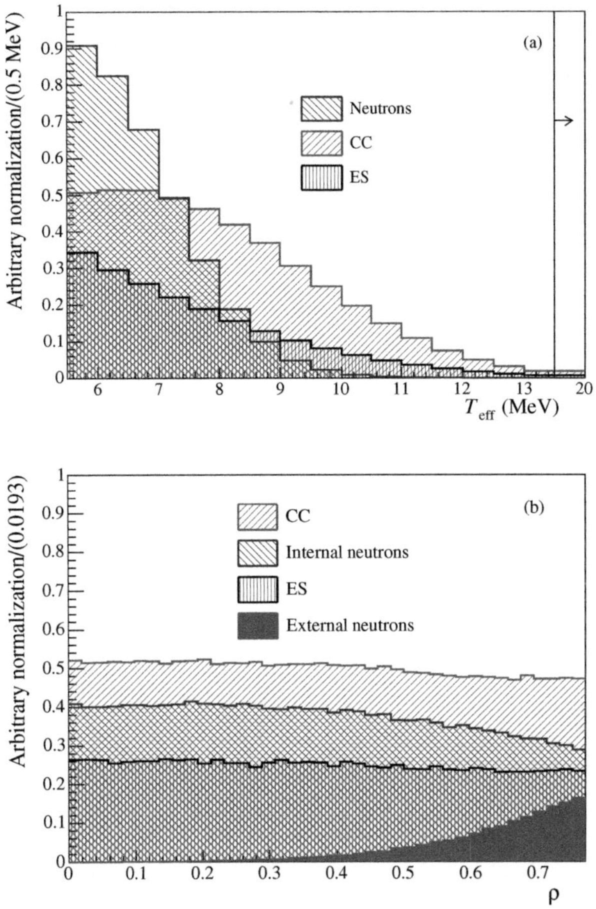

Fig. 4. $T_{\rm eff}$ (a) and ρ (b) distributions for NC, CC, ES, and external neutron events. Where internal and external distributions are identical the distribution is simply labeled neutrons. Note that the distribution normalizations are arbitrary and chosen to allow the shape differences to be seen clearly. The CC energy spectrum shape corresponds to an undistorted ^8B model.

by multiple γ's produced from the neutron capture on ^{35}Cl, instead of the single γ produced with neutron capture on D_2O. The isotropy of the hit distribution on the PMT array from the multiple γ-rays emitted from capture on ^{35}Cl is significantly different from that produced by a single relativistic electron (see fig. 5(a)). This extra information allows statistical separation of electrons created by the CC interactions and neutrons from NC interactions *without* any assumptions on the underlying neutrino energy distributions.

3 Data analysis

3.1 Signal extraction

Figures 4 and 5 show the four observables used to statistically separate the signals due to the three types of neutrino interaction. The variables used are derived from the

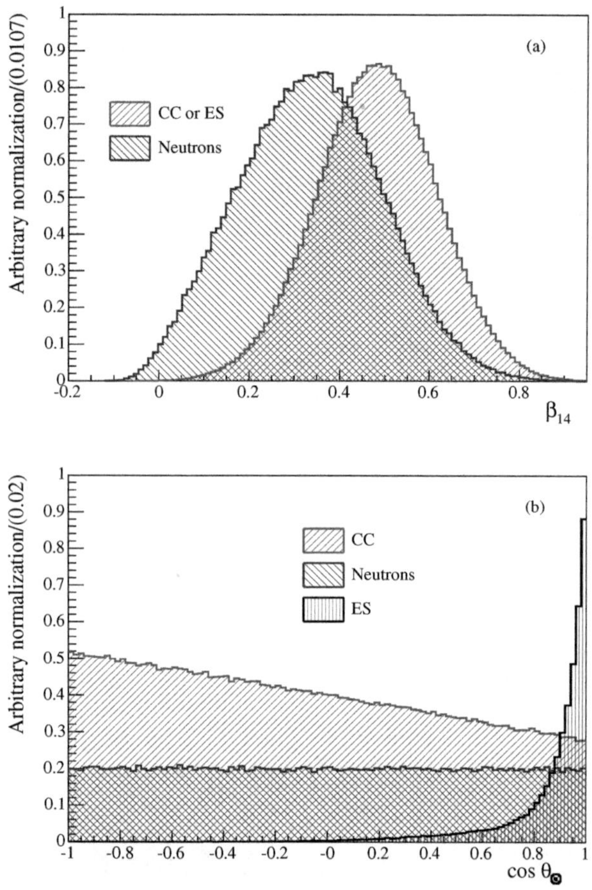

Fig. 5. β_{14} (a), and $\cos\theta_\odot$ (b) distribution for NC, CC, ES, and external neutron events. Where internal and external distributions are identical the distribution is simply labeled neutrons. Note that the distribution normalizations are arbitrary and chosen to allow the shape differences to be seen clearly.

number of triggered PMTs, the PMT trigger times and positions of an event. The variables are the effective electron kinetic energy, T_{eff}, a parameterisation of event isotropy, β_{14}, a parameterisation of the reconstructed radial position of the event ρ, and the cosine of the reconstructed event direction with respect to the Sun, $\cos\theta_\odot$.

The event position is parameterised by the volume-weighted radial position:

$$\rho = \left(\frac{R}{R_{\text{AV}}}\right)^3, \tag{1}$$

where R is the radius of the reconstructed event position and R_{AV} is the radius of the acrylic vessel (600.5 cm). The event isotropy is determined by the spatial distribution of triggered PMTs in an event and is parameterised as

$$\beta_{14} = \beta_1 + 4\beta_4 \tag{2}$$

with

$$\beta_l = \frac{2}{N(N-1)} \sum_{i=1}^{N-1} \sum_{j=i+1}^{N} P_l(\cos\theta_{ij}), \tag{3}$$

where P_l is the Legendre polynomial of order l, N is the total number of trigger PMTs in the event, and $\cos\theta_{ij}$ is the angle between PMTs i and j relative to the reconstructed event vertex.

There is a small background of external neutrons (EN) created by radiation outside the D_2O volume that can be separated using these observables. Background neutrons produced inside the D_2O volume (for example, produced by photodisintegration of deuterium by γ's from the low levels of natural radioactivity) are determined by separate analyses and extracted after the fit.

Detailed Monte Carlo simulations are used to generate probability density functions (PDFs) characterizing signal distributions and are shown in figs. 4 and 5. These PDFs are used in a extended maximum-likelihood fit to the 4722-event dataset to extract the electron neutrino energy spectrum, ^{8}B neutrino fluxes, and day-night asymmetries. As some neutrino oscillation scenarios can distort the electron neutrino spectrum, the energy distribution of the CC and ES interaction was not assumed. For these two signals PDFs were created separately for each 0.5 MeV interval in the range 5.5–13.5 MeV. For $T_{\text{eff}} > 13.5$ MeV a single bin was used due to the low statistics in this energy region. A single PDF was used, for both the NC and EN components whose T_{eff} spectra are determined by the energy release of neutron capture on ^{35}Cl are independent of neutrino energy. The normalisation for each of these PDFs was allowed to vary in the fit giving a model-independent measurement of the neutrino energy spectrum.

To take the correlations between the four variables, multi-dimensional PDFs were used. Ideally a 4-dimensional PDF should be used, which would take into account all correlations automatically. Unfortunately, statistical limitations made this unpractical. The two following factorisations were considered:

$$P(T_{\text{eff}}, \beta_{14}, \rho, \cos\theta_\odot) = P(T_{\text{eff}}, \beta_{14}) \times P(\cos\theta_\odot) \times P(\rho) \tag{4}$$

and

$$P(T_{\text{eff}}, \beta_{14}, \rho, \cos\theta_\odot) = P(T_{\text{eff}}, \beta_{14}, \rho) \times P(\cos\theta_\odot | T_{\text{eff}}, \rho). \tag{5}$$

Both factorisations were tested by applying the signal extraction procedure to 100 Monte Carlo datasets. It was found that the parameterisation in eq. (4) resulted in a small bias. Equation (5) showed no sign of bias. This reduction in bias can be understood by the inclusion of correlations with ρ and $\cos\theta_\odot$ in eq. (5). When the expected bias was corrected for, both approaches gave consistent results when applied to the data.

3.2 Systematic uncertainties

Systematic uncertainties in detector response are evaluated through comparisons of Monte Carlo simulations and calibration data. The primary calibration sources used to study systematic uncertainties are a ^{16}N 6.13 MeV γ-ray source and a ^{252}Cf fission neutron source. The ^{16}N source is used to study energy response, event reconstruction performance, and detector stability over time. The ^{252}Cf source is used to evaluate neutron response characteristics. Systematic uncertainties are propagated by perturbing the PDFs according to the estimated 1σ variation in each response parameter, and then repeating the signal extraction process. The dominant systematic uncertainties on the CC and NC extracted fluxes in the energy-unconstrained analysis are due to uncertainties in the β_{14} parameters. Uncertainties of less than a percent in the mean isotropy values translate to uncertainties of around 4% in the CC and NC fluxes. The energy scale uncertainty in the salt phase is estimated to be 1.15%, which contributes a 3.5% uncertainty in the NC flux, but has a smaller effect on the CC and ES fluxes. A radial scaling uncertainty of 1% is also one of the larger contributions to the overall systematic error, contributing roughly a 3% uncertainty to each flux. The ES flux uncertainty is dominated by a 5% systematic uncertainty due to the uncertainty in angular resolution. These are the major contributions to the systematic errors for the unconstrained analysis.

4 Results

4.1 Solar neutrino flux results

The energy unconstrained analysis as described in sect. 3.1 classified 2010 ± 85 events as NC, 2176 ± 78 as CC, and 279 ± 26 events as ES. The external neutron background is found to be 128 ± 28 events. Accounting for acceptance factors and detector lifetime, we can convert these extracted event numbers into equivalent ^{8}B solar neutrino fluxes:

$$\begin{aligned} \phi_{\mathrm{NC}} &= 4.94^{+0.21}_{-0.21}(\mathrm{stat})^{+0.38}_{-0.34}(\mathrm{syst}), \\ \phi_{\mathrm{CC}} &= 1.68^{+0.06}_{-0.06}(\mathrm{stat})^{+0.08}_{-0.09}(\mathrm{syst}), \\ \phi_{\mathrm{ES}} &= 2.35^{+0.22}_{-0.22}(\mathrm{stat})^{+0.15}_{-0.15}(\mathrm{syst}), \end{aligned} \tag{6}$$

in units of $10^6\,\mathrm{cm}^{-2}\,\mathrm{s}^{-1}$.

The flavour composition of the detected ^{8}B by SNO is illustrated in fig. 6.

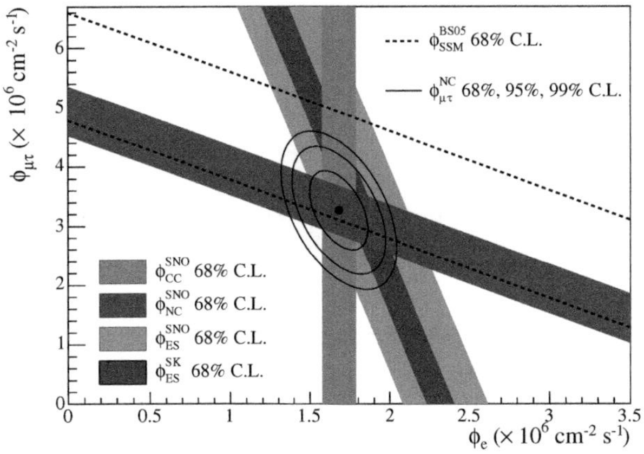

Fig. 6. Flux of μ and τ neutrinos *versus* flux of electron neutrinos. CC, NC, and ES flux measurements are indicated by the filled bands. The total ^{8}B solar neutrino predicted by the Standard Solar Model [2] is shown as dashed lines, and that measured with the NC channel is shown as the solid band parallel to the model predictions. The narrow band parallel to the SNO ES results corresponds to the Super-Kamiokande results in [7]. The intercepts of these with the axis represent the $\pm 1\sigma$ uncertainties. The non-zero value of $\phi_{\mu\tau}$ provides strong evidence for neutrino flavour transformation. The point represents ϕ_e for the CC flux and $\phi_{\mu\tau}$ from the NC-CC difference with 68%, 95% and 99% C.L. contours included.

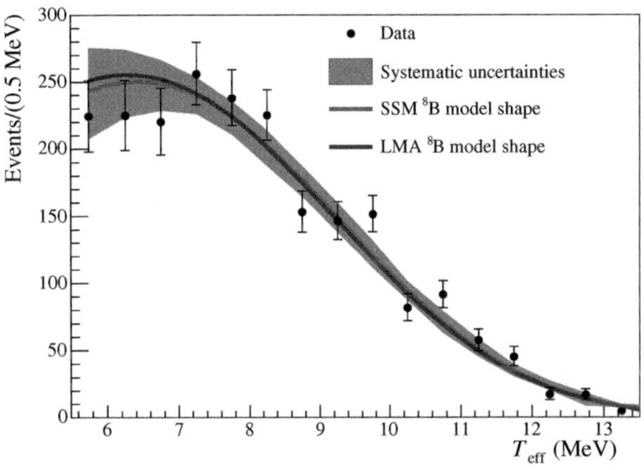

Fig. 7. The extracted CC T_{eff} spectrum with statistical error bars compared to the predictions for both an undistorted ^{8}B shape and the shape expected for the best-fit MSW model, corresponding to the LMA region, with combined systematic uncertainties, including both shape and acceptance components.

4.2 Charged-current spectrum

Figure 7 shows the extracted recoil electron energy spectrum from the CC reaction obtained from the extended maximum-likelihood fit. Statistical uncertainties are shown around the data points, whilst the systematic uncertainties are shown with respect to the prediction for an undistorted ^{8}B shape. While some uncertainties may change the PDF shapes leading to a change in the fitted number of events, other simply affect the overall acceptance of events. Both must be accounted for as the latter can lead to errors in the translation of differential event counts into differential neutrino fluxes.

The measured spectrum is consistent with no distortions, and also with the best-fit MSW model, corresponding to the Large Mixing Angle (LMA) region of the solar neutrino parameter space. The LMA spectrum does not differ significantly from the undistorted ^{8}B shape.

4.3 Day-night asymmetry measurement

For certain regions of mixing parameters, the MSW effect predicts a regeneration of solar electron neutrinos when the solar neutrino flux passes through the Earth.

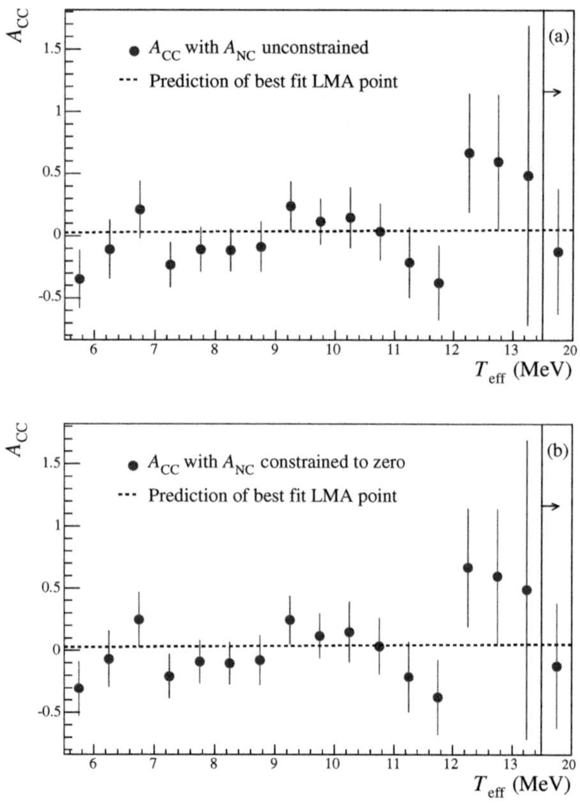

Fig. 8. Day-night asymmetries on each CC energy bin as a function of electron energy. Panel (a) shows the case in which no constraint is made on A_{NC}. Panel (b) shows the case in which A_{NC} is constrained to zero. The vertical lines show the expectation for $\Delta m^2 = 7 \times 10^{-5}\,\text{eV}^2$ and $\tan^2\theta = 0.40$.

The regeneration is measurable as an asymmetry of the solar flux:

$$A_{DN} = 2\frac{\phi_N - \phi_D}{\phi_N + \phi_D}, \tag{7}$$

where ϕ_N is the solar flux measured during the night and ϕ_D is the solar flux measured during the day. The energy-unconstrained analysis carried out as described in sect. 3.1 was carried out separately for the day and night neutrino candidate events. The resulting asymmetries in the fluxes are

$$\begin{aligned} A_{NC} &= +0.042 \pm 0.086(\text{stat}) \pm 0.072(\text{syst}), \\ A_{CC} &= -0.050 \pm 0.074(\text{stat}) \pm 0.053(\text{syst}), \\ A_{ES} &= +0.146 \pm 0.198(\text{stat}) \pm 0.033(\text{syst}). \end{aligned} \tag{8}$$

An asymmetry in the NC flux could indicate that the solar flux contains sterile neutrinos or that unexpected interactions inside the Earth take place. Figure 8 shows the day-night asymmetry of the CC flux, sensitive to the electron neutrino component of the solar flux only, as a function of energy for the assumption the A_{NC} equals zero or can be different from zero.

Within the uncertainties the asymmetries are compatible with both zero and with the LMA solution.

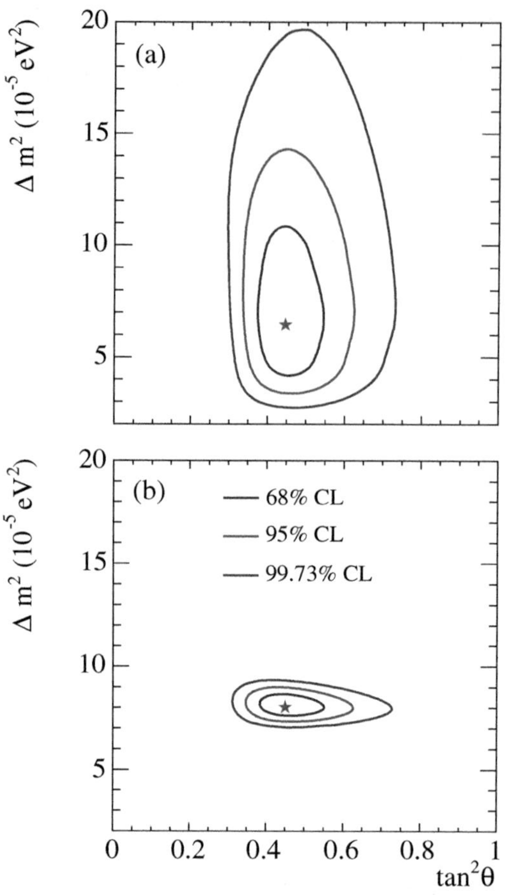

Fig. 9. Panel (a) shows the global neutrino analysis using only solar neutrino data, and panel (b) includes KamLAND's 766 ton-year data. The solar neutrino data included SNO's pure D_2O phase day and night spectra, SNO's salt phase extracted day and night CC spectra and ES and NC fluxes, the rate measurements from Cl, SAGE, Gallex/GNO, and SK-I zenith spectra. The ^8B flux was free in the fit, hep solar neutrinos were fixed at $9.3 \times 10^3\,\text{cm}^{-2}\,\text{s}^{-1}$. The stars are plotted at the best-fit parameters from the χ^2 analysis.

4.4 MSW parameter constraints

The salt phase results for the fluxes, spectra, and day-night asymmetries can be combined with SNO's previous results and the results of other solar neutrino experiments to produce constraints on the fundamental neutrino parameters in the MSW model. Figure 9(a) shows the results of a global χ^2 analysis. The best-fit point is $\Delta m^2 = (6.5^{+0.4}_{-2.3}) \times 10^{-5}\,\text{eV}^2$ and $\tan^2\theta = 0.45^{+0.09}_{-0.08}$.

Including the results for the KamLAND experiment [8, 9], the best-fit point is found to be

$$\begin{aligned} \Delta m^2 &= (8.0^{+0.6}_{-0.4}) \times 10^{-5}\,\text{eV}^2, \\ \tan^2\theta &= 0.45^{+0.09}_{-0.07}. \end{aligned} \tag{9}$$

The "survival probability" for solar neutrinos, which can be determined by SNO from the measurement of the ratio of the CC and the NC flux, places strong constraints on the value of $\tan^2 \theta$. Measurements of the NC rate in the third phase of the SNO experiment will further improve determinations of the fundamental neutrino oscillation parameters.

5 Conclusions

New results from the salt phase of the SNO experiment have been summarised, including a measurement of the flux of ^{8}B solar electron neutrinos through the CC reaction and a measurement of the ^{8}B solar neutrinos of all flavours through the neutral-current reaction. The use of the isotropy parameter as a tool to statistically separate the CC and NC events allows a model-independent measurement of the solar neutrino fluxes in the salt phase. The flux results confirm and improve previous results, demonstrating solar neutrino flavour change and contributing to evidence for solar neutrino oscillations.

Global analysis of solar neutrino and KamLAND data strongly favour LMA oscillations. The oscillation parameter space has now been tightly constrained to a region where the predicted distortion to the ^{8}B energy spectrum is small. The measured energy spectrum derived from the CC reaction is consistent with the expected spectrum assuming an undistorted ^{8}B shape, as well as with the predicted spectrum consistent with the best-fit LMA parameters. The day-night flux asymmetries predicted for the LMA scenario are also small and the day-night asymmetries measured here are consistent with these predictions, as well as with no day-night effect.

6 Outlook

Improved precision in the NC flux measurement is expected from the third phase of SNO. The salt was removed from the detector in October 2003, in preparation of the installation of the neutral-current detectors (NCDs). Forty string of ^{3}He proportional counters with a total length of approximately 350 m have been installed into the D_2O volume. This will provide an independent measurement of the NC flux on an event-by-event basis. The neutron capture mechanism of the counters is:

$$n + {}^3\mathrm{He} \rightarrow p + t. \tag{10}$$

The counters have been successfully installed into SNO and production data taking started at the beginning of 2005.

This research was supported by Canada: Natural Sciences and Engineering Research Council, Industry Canada, National Research Council, Northern Ontario Heritage Fund, Atomic Energy of Canada, Ltd., Ontario Power Generation, High Performance Computing Virtual Laboratory, Canada Foundation for Innovation; US: Department of Energy, National Energy Research Scientific Computing Center; UK: Particle and Physics and Astronomy Research Council.

References

1. Bruce T. Cleveland *et al.*, Astrophys. J. **496**, 505 (1998).
2. J.N. Bahcall, A.M. Serenelli, S. Basu, arXiv:astro-ph/0412440.
3. The SNO Collaboration, Phys. Rev. Lett. **87**, 011301 (2002).
4. The SNO Collaboration, Phys. Rev. Lett. **89**, 181301 (2004).
5. The SNO Collaboration, Phys. Rev. C **72**, 055502 (2005), arXiv:nucl-ex/0502021.
6. The SNO Collaboration, Nucl. Instrum. Methods A **449**, 172 (2005).
7. S. Fukuda *et al.*, Phys. Lett. B **539**, 179 (2002).
8. K. Eguchi *et al.*, Phys. Rev. Lett. **90**, 021802 (2003).
9. T. Araki *et al.*, arXiv:hep-ex/0406035.

Eur. Phys. J. A **27**, s01, 25–34 (2006)
DOI: 10.1140/epja/i2006-08-003-0

EPJ A direct
electronic only

Prospects in double beta decay searches

S. Pirro[a]

INFN - Sezione di Milano and Università di Milano Bicocca, Piazza della Scienza 3, I-20126 Milano, Italy

Received: 4 July 2005 /
Published online: 14 February 2006 – © Società Italiana di Fisica / Springer-Verlag 2006

Abstract. Neutrinos have recently provided us with the first tangible evidence of phenomena beyond the reach of our theory of the laws of particle physics, the remarkably predictive "Standard Model". The positive observation of oscillations in atmospheric and in solar neutrinos gives new motivations for more sensitive searches. Unfortunately, the oscillation experiments can only provide data on the squared mass differences of the neutrino mass eigenstates, while the absolute scale can only be obtained from direct mass measurements, ^{3}H end point measurements for example or, in the case of Majorana neutrinos, more sensitively by neutrinoless double beta decay. In fact, recently published constraints on the mixing angles of the neutrino mixing matrix make a strong case that if neutrinos are Majorana particles, there are many scenarios in which next-generation double beta decay experiments should be able to observe the phenomenon, trying to disentangle the mass scale of the neutrinos. The interest for next-generation double beta decay experiments is growing, for if the mass scale is below ~ 0.2 eV, double beta decay may be the only hope for measuring it.

PACS. 23.40.Bw Weak-interaction and lepton (including neutrino) aspects – 11.30.Fs Global symmetries (*e.g.*, baryon number, lepton number) – 14.60.Pq Neutrino mass and mixing

1 Introduction

All physical phenomena observed up to now in experiments performed with or without accelerators proved repeated confirmations of the validity of the Standard Model (SM). They enabled, moreover, the measurement of the set of fundamental constants which appear in their theoretical formalism. Many problems are still open, stimulating extensions of such a model: the high number of free parameters appearing in the SM (the three coupling constants, the masses of the fermions, the Cabibbo-Kobayashi-Maskawa matrix), the "running" coupling constants, the missing explanation of the hierarchy of the masses of the fermions. Recently, furthermore, with the discovery of neutrino oscillation the number of parameters increases with the need of introducing the analogue of the CKM matrix for the flavour mixing of neutrinos.

Following the pattern of what was done for the weak and electromagnetic interaction, physicists are looking for a "Unified Theory" of strong and electroweak interactions, named Grand Unified Theory (GUT). Such a theory should include the SM, in the same way as the weak theory includes electromagnetism and the theory of weak interactions. The fundamental idea is based on the assumption of the existence of an energy scale M_{GUT} ($\sim 10^{15}$ GeV) at which the three coupling constants, *i.e.* weak, electromagnetic and strong, converge toward a single coupling constant, which corresponds to a single interaction; such an interaction is based on a single gauge group. Many GUTs have been proposed, differing mainly in the choice of the gauge group on which they base themselves. The common characteristics of almost all GUTs are the violation of the lepton and baryon number. In order to have lepton number violation, the neutrino must be a Majorana particle.

As pointed out very recently by the Members of the APS Multidivisional Neutrino Study [1], Double Beta Decay (DBD) searches will play the central role in neutrino physics of the next decade.

2 Dirac neutrinos and Majorana neutrinos

Charged particles are, obviously, not their own antiparticles, due to the fact that they have an electric charge. In the same way, neutral particles, such as the neutron or the K^0, differ from their antiparticles since they have a baryon number. Neutrinos, however, could be equal to their own antiparticles: in this case the neutrino would be a Majorana particle. In the framework of the Standard Model, massive Dirac neutrinos consist of four different states: suppose the existence of a neutrino with negative helicity (left handed) ν_L; if CPT theorem holds, then there will

[a] e-mail: `Stefano.Pirro@mib.infn.it`; *Present address:* INFN - Laboratori Nazionali del Gran Sasso, S.S. 17 bis km 18+910, Assergi (AQ), Italy.

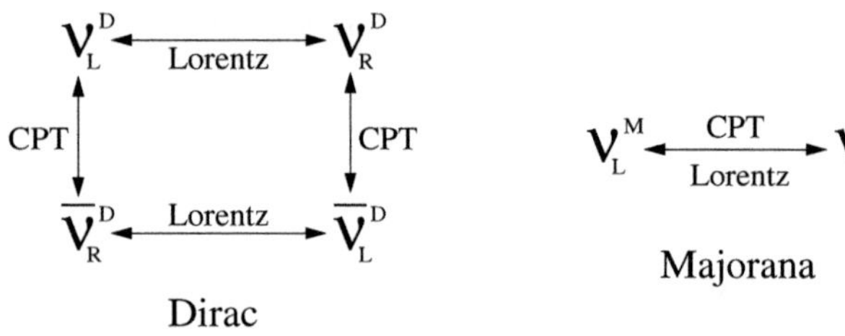

Fig. 1. Transformations among neutrinos: Dirac neutrinos (left); Majorana neutrinos (right).

be the corresponding CPT-transformed state, *i.e.* an antineutrino with positive helicity (right handed), $\bar{\nu}_R$. Now, if the neutrino has a mass then there will exist a Lorentz boost that permits the helicity flip. Thus if the neutrino has a charge (lepton number) and a mass, it consists of four different states and is called a Dirac neutrino. If, on the contrary, the neutrino does not have a charge, only the two helicity states are defined; this is called a Majorana neutrino (see fig. 1). Due to the $V-A$ structure of the Standard Model the right-handed neutrinos are sterile, thus only two neutrinos are able to interact, as in the Majorana case. The difference is that in the Standard Model they interact due to their charge while in the Majorana case they interact due to their helicity.

From a theoretical point of view, the possibility that neutrinos are Majorana particles is particularly appealing. The fact that neutrinos and charged leptons, belonging to the same weak isodoublet, have an extremely different mass (at least a factor 10^5) cannot be explained in the Dirac theory. Such an "anomaly" can be explained, without the insertion of an "*ad hoc*" symmetry, through the "see-saw" mechanism [2]: in the Majorana case, the mass of the neutrino naturally satisfies the relation $M_\nu\ M \approx M^2_{q,\ell}$, where $M_{q,\ell}$ represents the mass of a lepton or a quark, and M represents a mass scale.

Due to their tiny mass, the Dirac neutrinos in nature are produced, practically, always left handed, while anti-neutrinos are right handed. It is therefore impossible to discriminate whether they interact due to the lepton charge or due to their helicity. Different characteristics arise from the CPT transformation rules: it can be demonstrated [3] that if CPT is a conserved symmetry, then the Majorana neutrino cannot have an electric dipole or magnetic dipole moment. The magnetic moment for a Dirac neutrino can be evaluated [4] as $3.2 \cdot 10^{-19}\ m_\nu\ \mu_B$ (μ_B is the Bohr magneton and m_ν is expressed in eV). The present experimental limits are at least six orders of magnitude far away from the predicted value. There are other possibilities in order to search for lepton number non-conservation as in the case of pions and mesons decays [5]; the expected sensitivity, however, is very small compared to DBD. Thus the most favourable way to discriminate between Dirac and Majorana neutrino turns out to be the neutrinoless double beta decay (0νDBD).

After the discovery of neutrino oscillation, implying neutrino masses, the introduction of a mixing matrix between flavour eigenstates and mass eigenstates is straightforward. This matrix, analogous of the CKM matrix of the quark sector, is the Pontecorvo-Maki-Nakagawa-Sakata (PMNS) matrix [6]. The flavour eigenstates can then be written as $\nu_\alpha = \sum_i U_{\alpha i}\nu_i$. $U_{\alpha i}$ are the elements of the PMNS matrix, and are related to the observable mixing angles in the base where the charged lepton masses are diagonal. For Majorana neutrinos there is an additional matrix that takes into account the fact that in this case there are 2 more complex phases. Using the Chau and Keung [7] parameterization of the PMNS matrix, we have:

$$\begin{pmatrix}\nu_e\\ \nu_\mu\\ \nu_\tau\end{pmatrix} = UV\begin{pmatrix}\nu_1\\ \nu_2\\ \nu_3\end{pmatrix} =$$
$$\begin{pmatrix} c_{12}c_{13} & s_{12}c_{13} & s_{13}e^{-i\delta}\\ -s_{12}c_{23}-c_{12}s_{23}s_{13}e^{i\delta} & c_{12}c_{23}-s_{12}s_{23}s_{13}e^{i\delta} & s_{23}c_{13}\\ s_{12}s_{23}-c_{12}c_{23}s_{13}e^{i\delta} & -c_{12}s_{23}-s_{12}c_{23}s_{13}e^{i\delta} & c_{23}c_{13}\end{pmatrix}$$
$$\times\begin{pmatrix}1 & 0 & 0\\ 0 & e^{i\phi_2/2} & 0\\ 0 & 0 & e^{i(\phi_3/2+\delta)}\end{pmatrix}\begin{pmatrix}\nu_1\\ \nu_2\\ \nu_3\end{pmatrix}, \tag{1}$$

where $c_{ij} \equiv \cos\theta_{ij}$ and $s_{ij} \equiv \sin\theta_{ij}$; θ_{ij} are the mixing angles measured with the neutrino oscillations; δ is the Dirac CP phase. V is a diagonal matrix containing the Majorana CP phases (ϕ_2 end ϕ_3) that do not exist in the case of Dirac neutrinos and that, in any case, cancel in neutrino oscillations.

3 Double beta decay

The two-neutrino double beta decay mode (2νDBD) is expected to occur in the Standard Model as a second-order effect of the well-known beta decay Hamiltonian, and it imposes no special requirements on the properties of the neutrino. It will occur irrespective of whether the neutrino is a Majorana or a Dirac particle and irrespective of whether it has a mass or not. The possible 2ν decay modes are

$$\begin{array}{ll} (A,Z)\rightarrow(A,Z+2)+2e+2\bar{\nu}_e & \beta\beta^-\\ (A,Z)\rightarrow(A,Z-2)+2e^++2\nu_e & \beta\beta^+\\ (A,Z)+2e^-\rightarrow(A,Z-2)+2\nu_e & ECEC\\ (A,Z)+e^-\rightarrow(A,Z-2)+e^++2\nu_e & EC\beta^+ \end{array} \tag{2}$$

Nuclear transitions accompanied by positron emission or electron capture processes are, however, characterized by poorer experimental sensitivities and will not be discussed in the following. The first process of eq. (2) represents the analogue of the single beta decay mediated by the weak current in which a d quark transforms into an u with the emission of an e^- and an $\bar{\nu}$; even if mediated by the same weak interaction, DBD is a second-order transition, resulting in an extremely slow decay rate, namely the slowest process ever observed in nature. Using simple dimensional arguments it is straightforward to demonstrate that, as a first approximation, the 2νDBD is $\sim 10^{21}$ times slower

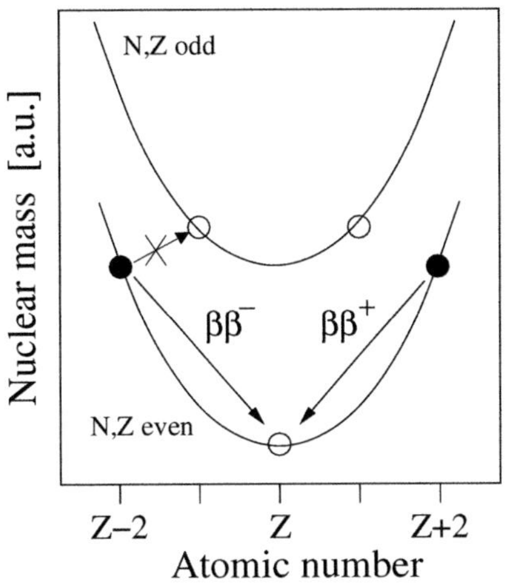

Fig. 2. Schematic picture of the atomic mass as a function of Z for isobar multiplets with even A.

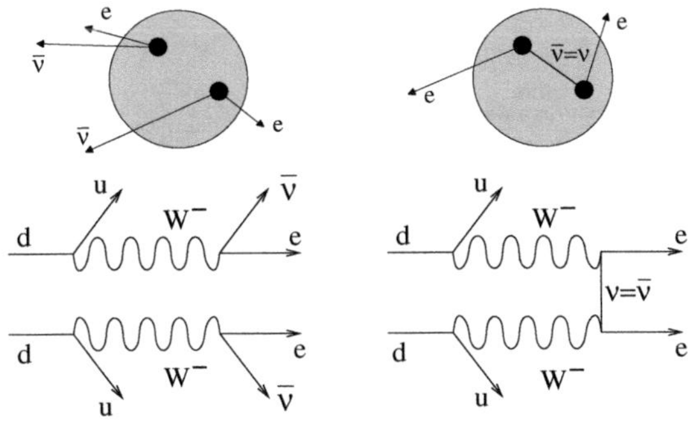

Fig. 3. Elementary scheme for 2νDBD (left) and 0νDBD (right).

with respect to the single beta decay, resulting in half-lives of the order of 10^{18}–10^{22} years. The experimental observation of 2νDBD is therefore possible only if the single beta decay is prohibited by energy conservation or, at least, strongly hindered by small transition energy and/or by large change of angular momentum. This happens, fortunately, with several nuclei in nature and is due to the "pairing" interaction [8] (see fig. 2). The first "direct" observation of the 2νDBD was obtained in 1987 [9] and is now observed in more than ten nuclei [10,11].

More interesting is the *neutrinoless* double beta decay (0νDBD), first proposed by Furry [12] in 1939. In this case there is the maximum lepton number violation ($\Delta L = 2$) and the decay is, therefore, not allowed by the Standard Model. The 0νDBD can occur only if two requirements are satisfied:

- the neutrino has to be a Majorana particle,
- the neutrino has to have a mass and/or the neutral current has to have a right-handed ($V+A$) component.

The second condition is needed because of the helicity of the neutrino. Due to the $V - A$ nature of the weak interaction, the neutrino emitted in the first vertex (see fig. 3) is right handed, while in order to be absorbed in the second one, it needs to change its helicity. Thanks to the finite mass this is possible, with a probability $\propto m_\nu/E_\nu$;

Table 1. Theoretically evaluated 0νDBD half-lives (units of 10^{28} years) for $|\langle m_\nu\rangle| = 10\,\text{meV}$. Only a few references are shown. The results of [13] are still considered by the authors as preliminary.

Isotope	[14]	[15]	[16]	[13]	[17]	[18]
^{48}Ca	8.83	–	–	–	2.5	–
^{76}Ge	17.7	14.0	2.33	6.0	3.6	3.7
^{82}Se	2.4	5.6	0.6	1.8	1.5	0.81
^{100}Mo	–	1.0	1.28	3.5	3.9	0.65
^{116}Cd	–	–	0.48	2.4	4.7	0.39
^{130}Te	5.8	0.7	0.5	3.0	0.85	0.52
^{136}Xe	12.1	3.3	2.2	7.3	1.8	0.27
^{150}Nd	–	–	0.025	> 0.3	–	–

from here it turns out that the amplitude of the decay is proportional to m_ν. Disregarding more unconventional contributions (SUSY or left-right symmetric models), the 0νDBD rate is usually expressed as

$$[T^{0\nu}_{1/2}]^{-1} = G^{0\nu}|M^{0\nu}|^2\frac{|\langle m_\nu\rangle|^2}{m_e^2}, \tag{3}$$

where $G^{0\nu}$ is the (exactly calculable) phase space integral $\propto Q^5_{\beta\beta}$ ($Q_{\beta\beta}$ represents the Q-value of the decay), $|M^{0\nu}|^2$ is the specific nuclear matrix element of the nucleus undergoing the decay and $|\langle m_\nu\rangle|$ (effective electron neutrino mass, often called $|m_{ee}|$) is the neutrino relevant parameter measured in 0νDBD. By using eq. (1) we have:

$$|\langle m_\nu\rangle| \equiv \left||U_{11}|^2 m_1 + |U_{12}|^2 m_2 e^{i\phi_2} + |U_{13}|^2 m_3 e^{i\phi_3}\right|, \tag{4}$$

where $e^{i\phi_2}$ and $e^{i\phi_3}$ are the Majorana CP phases ($= \pm 1$ in case of CP conservation), $m_{1,2,3}$ the mass eigenvalues and U_{1j} the matrix elements of the PMNS matrix. The presence of the ϕ_k phases implies that cancellations are, unfortunately, possible. Such cancellations are complete for a Dirac neutrino, since it is equivalent to two degenerate Majorana neutrinos with opposite CP phases. This stresses once more the fact that 0νDBD can occur only through the exchange of Majorana neutrinos.

From a Particle Physics point of view, 0νDBD represents a unique tool in order to measure the neutrino Majorana phases and to assess the absolute scale of the neutrino masses. As in evidence from eq. (3), the derivation of the crucial parameter $\langle m_\nu\rangle$ from the experimental results on 0νDBD lifetime requires a precise knowledge of the Nuclear Matrix Elements (NME) of the transition. Unfortunately, this is not an easy job, and a definite knowledge of NME values and uncertainties is still lacking in spite of the large attention attracted by this area of research. Many, often conflicting, evaluations are available in the literature and it is unfortunately not easy to judge their correctness or accuracy. Outstanding progress has been achieved over the last few years mainly due to the application of the QRPA method and its extensions. Renewed interest in Shell Model calculations has been boosted, on the other hand, by the fast development of computer technologies. Comparison with experimental 2νDBD rates has often been suggested as a possible way out (direct test of

Table 2. Summary of the present information on neutrino masses and mixings from oscillation data. For a recent review see [19].

Oscillation parameter	Central value	99% CL range
Solar mass splitting	$\Delta m^2_{12} = (8.0 \pm 0.3)\ 10^{-5}\,\mathrm{eV}^2$	(7.2–8.9) $10^{-5}\,\mathrm{eV}^2$
Atmospheric mass splitting	$\lvert\Delta m^2_{23}\rvert = (2.5 \pm 0.3)\ 10^{-3}\,\mathrm{eV}^2$	(1.7–3.3) $10^{-3}\,\mathrm{eV}^2$
Solar mixing angle	$\tan^2\theta_{12} = 0.45 \pm 0.05$	$30° < \theta_{12} < 38°$
Atmospheric mixing angle	$\sin^2 2\theta_{23} = 1.02 \pm 0.04$	$36° < \theta_{23} < 54°$
"CHOOZ" mixing angle	$\sin^2 2\theta_{13} = 0 \pm 0.05$	$\theta_{13} < 10°$

the calculation method). The evaluation methods for the two decay modes show, however, relevant differences (*e.g.* the neutrino propagator), and the effectiveness of such a comparison is still controversial [13,20,21]. A popular but doubtful attitude consists in considering the spread of the different evaluations as an estimate of their uncertainties. In such a way one obtains a spread of about one order of magnitude in the expected half-lives (table 1), corresponding to a factor of ~ 3 in $\langle m_\nu \rangle$. It is clear that a big improvement in the calculation of NME or, at least, in the estimate of their uncertainties would be welcomed. New calculation methods should be pursued while insisting on the comparison with dedicated measurements coming from various areas of nuclear physics [22]. On the other hand, an experimental effort to investigate as many $\beta\beta$ emitters as possible should be addressed.

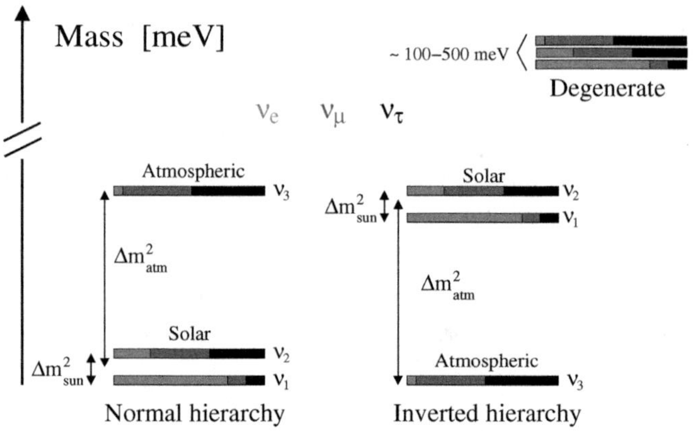

Fig. 4. Neutrino mass pattern based on the experimental relation $\Delta m^2_{\mathrm{sun}} \ll \Delta m^2_{\mathrm{atm}}$.

4 Prediction on the Majorana mass

The input parameters in order to figure out the possible mass pattern of the Majorana neutrinos are given by the neutrino oscillations. The most updated values [23,24,25,26,27,28] are given in table 2.

Given the two Δm^2 measured with the oscillations, and given the assumption of three neutrinos, the absolute mass scale is still missing. Nevertheless, three scenarios are possible, as shown in fig. 4. The previously quoted values can be accommodated in the framework of three neutrinos mixing, which describes the three flavour neutrinos (ν_e, ν_μ and ν_τ) as unitary linear combinations of the three massive neutrinos (ν_1, ν_2 and ν_3) having masses m_1, m_2, and m_3, respectively. The experimental measurements are compatible with three mass schemes:

1. Normal hierarchy: $m_1 < m_2 < m_3$, *i.e.* $\Delta m^2_{23} > 0$,
$$\Delta m^2_{12} \simeq \Delta m^2_{\mathrm{sun}} \simeq m^2_2\,, \quad \Delta m^2_{23} \simeq \lvert\Delta m^2_{\mathrm{atm}}\rvert \simeq m^2_3\,. \quad (5)$$
2. Inverted hierarchy: $m_3 < m_1 < m_2$, *i.e.* $\Delta m^2_{23} < 0$,
$$\Delta m^2_{12} \simeq \Delta m^2_{\mathrm{sun}}\,, \quad \Delta m^2_{23} \simeq -\lvert\Delta m^2_{\mathrm{atm}}\rvert \simeq -m^2_1\,. \quad (6)$$
3. Degenerate case: the values of Δm^2_{ij} are small when compared to each mass values. In this case the hierarchies are undistinguishable:
$$\lvert\Delta m^2_{ij}\rvert \ll m^2_1 \simeq m^2_2 \simeq m^2_3\,. \quad (7)$$

In order to extract the correct prediction, however, one has to use eq. (4); the steps are:

- use the matrix elements given by eq. (1) with the evaluated neutrino parameter from table 2,
- parameterize $m_2 = \sqrt{m^2_1 + \Delta m^2_{\mathrm{sun}}}$ and
$$m_3 = \sqrt{m^2_2 + \lvert\Delta m^2_{\mathrm{atm}}\rvert} \quad \text{for normal hierarchy,}$$
- parametrize $m_1 = \sqrt{m^2_3 + \lvert\Delta m^2_{\mathrm{atm}}\rvert}$ and
$$m_2 = \sqrt{m^2_1 + \Delta m^2_{\mathrm{sun}}} \quad \text{for inverted hierarchy.}$$

With this procedure the effective Majorana mass can be written as a function of the lightest neutrino mass ($m_{\mathrm{light}} \equiv m_1$ for normal hierarchy, $m_{\mathrm{light}} \equiv m_3$ for inverted hierarchy):

$$\lvert\langle m_\nu\rangle\rvert \equiv \lvert m_{ee}\rvert = f(m_{\mathrm{light}}, \phi_1, \phi_2, \text{observables}), \quad (8)$$

where *observables* are all the experimental data from neutrino oscillations. The plot of $\lvert\langle m_\nu\rangle\rvert \equiv \lvert m_{ee}\rvert$ is shown in fig. 5. The two disfavoured regions are given by the present limits on DBD experiments (see table 3) and by cosmological (Large-Scale Structures and anisotropies in the Cosmic Microwave Background) bounds [29].

The DBD-Experiments developed up to now, often called *First-Generation Experiments*, were designed to explore only the degenerate mass scenario. The proposed *Second-Generation Experiments* are designed to explore the inverted hierarchy scenario, with an expected sensitivity on $\lvert m_{ee}\rvert$ of the order of 10–50 meV.

Table 3. Best reported results on 0νDBD processes. Limits are at 90% CL except when noted. The effective neutrino mass limits and ranges are those deduced by the authors ($\langle m_\nu \rangle$) or according to table 1 ($\langle m_\nu^\dagger \rangle$). Only ‡ nuclei are presently investigated by high-sensitivity experiments.

Isotope	$Q_{\beta\beta}$ (keV)	*i.a.*	$T_{1/2}^{2\nu}$ (y)	$T_{1/2}^{0\nu}$ (y)	$\langle m_\nu \rangle$ (eV)	$\langle m_\nu^\dagger \rangle$ (eV)
^{48}Ca	4271	0.187	$(4.2 \pm 1.2) \times 10^{19}$	$> 9.5 \times 10^{21}$(76%) [30]	< 8.3	< 30
^{76}Ge	2039	7.8	$(1.3 \pm 0.1) \times 10^{21}$	$> 1.9 \times 10^{25}$ [31,32]	< 0.35	< 0.35–0.96
				$> 1.6 \times 10^{25}$ [33,34]	< 0.38–1.05	< 0.38–1.05
^{82}Se ‡	2995	9	$(9.6 \pm 1.0) \times 10^{19}$	$> 1.0 \times 10^{23}$ [35]	< 1.7–4.9	< 2.2–6.9
^{100}Mo ‡	3034	9.6	$(7.11 \pm 0.54) \times 10^{18}$	$> 4.6 \times 10^{23}$ [35]	< 0.7–2.8	< 1.36–3.3
^{116}Cd	2806	7.5	$(2.9 \pm 0.4) \times 10^{19}$	$> 1.7 \times 10^{23}$ [36]	< 1.7	< 1.5–5.3
^{130}Te ‡	2528	33.9	$(6.1 \pm 1.4) \times 10^{20}$	$> 1.8 \times 10^{24}$ [37,38]	< 0.2–1.1	< 0.5–1.8
^{136}Xe	2479	8.9	$> 1.6 \times 10^{22}$	$> 1.2 \times 10^{24}$ [39]	< 1.1–2.9	< 0.5–3.2
^{150}Nd	3367	5.6	$7.0^{+11.8}_{-0.3} \times 10^{18}$	$> 1.2 \times 10^{21}$ [40]	< 3	< 4.6–6.5

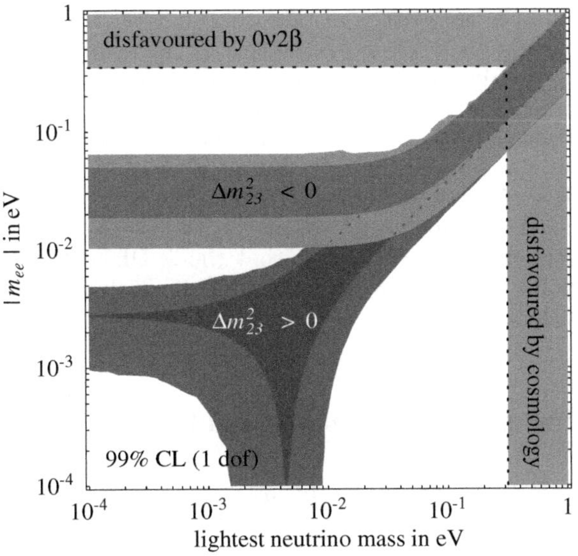

Fig. 5. 99% CL range for m_{ee} as function of the lightest neutrino mass. The darker regions show how the m_{ee} range would shrink if the best-fit values of oscillation parameters were confirmed with negligible errors (in this case the spread is due only to the Majorana CP phases). Picture given by courtesy of the authors of [19].

5 Experimental techniques

The experimental signatures of the nuclear double beta decays are in principle very clear: in the case of the 0νDBD, one should expect a peak (at the $Q_{\beta\beta}$ value) in the two-electrons summed energy spectrum, whereas a continuous spectrum (with a well-defined shape) will feature the 2νDBD.

In spite of such characteristic imprints, the rarity of the processes under consideration makes their identification very difficult. Such remotely probable signals have to be disentangled from a background due to natural radioactive decay chains, cosmogenic-induced activity, and man-made radioactivity, which deposit energy in the same region as the DBD, but at a faster rate. Consequently, the *main task* in 0νDBD searches is to diminish the background by using the state-of-the-art ultra-low background techniques and, hopefully, identifying the signal. Many are the experiments searching for DBD. They may be substantially subdivided into three categories:

- Geochemical determination of the total decay time through the measurement of the daughter nuclei $(A, Z + 2)$ produced by the parent nuclei (A, Z) in a sample of "old" geological rocks.
- Radiochemical measurement of the total decay time by the extraction of the radioactive daughter nuclei from the parent nuclei.
- Direct detection of the two electrons emitted in the transition.

The experiments belonging to the first two classes do not allow to distinguish among the two decay channels. They are, however, extremely sensitive to inclusive effects, since the so called "accumulation times" for the daughter isotope are very long. They gave, however, the first *indirect* prove [41] of the existence of the 2νDBD, but are no longer pursued nowadays. Direct experiments are the most interesting ones because they allow to distinguish the various modes of the double beta decay. It is therefore possible to search for the decay without neutrinos, which represents the most interesting process. The nuclear detector capable of revealing the two electrons emitted by the DBD-Emitter should have some basic properties:

- *High energy resolution*, since a peak must be identified over a background.
- *Low background*, which requires the use of extremely radiopure materials: natural radioactivity (mainly ^{232}Th and ^{238}U chains) exhibits decay times of the order of 10^9 years, extremely short if compared with the expected 0νDBD that should have a decay time larger than 10^{25} years. Furthermore it is absolutely necessary to operate the detector in underground laboratories in order to shield cosmic rays.
- *Large source* of DBD nuclei in order to have sensitivity to the decay time up to 10^{25}–10^{28} years.
- *Event reconstruction methods*, since the 0νDBD has a very characteristic decay with the two electrons that share the $Q_{\beta\beta}$ energy. Electron tracking can therefore help in rejecting background.

Unfortunately there are up to now no detectors that can fulfill these four requirements at the same time. Two are, substantially, the experimental approaches:

- homogenous detectors (or active source detectors), whose mean feature is to have the active $\beta\beta$ source inside the detector material,
- non-homogeneous detectors (or passive source detectors), in which the source and the detector are distinct.

Various conventional counters have been used so far in DBD direct searches: solid state devices (germanium spectrometers and silicon detector stacks), gas counters (time projection chambers, ionization and multiwire drift chambers) and scintillators (crystal scintillators and stacks of plastic scintillators). Techniques based on the use of low temperature *true* calorimeters have been, on the other hand, proposed and developed in order to improve the experimental sensitivity and enlarge the choice of suitable candidates for DBD searches investigable with an *active source* approach.

A common feature of all DBD experiments has been the constant work against backgrounds caused mainly by environmental radioactivity, cosmic radiation and residual radioactive contaminations of the detector setup elements. The further suppression of such backgrounds will be the *actual challenge* for future projects whose main goal will be to maximize the 0νDBD rate while minimizing background contributions.

In order to compare different experiments and in order to point out the advantages and the disadvantages of the two different detecting techniques, it is convenient to introduce a very important parameter, called "sensitivity", denoted by F. It is defined as the process half-life corresponding to the maximum number of signals (n) that could be hidden by the background fluctuations, at a given statistical CL. Let t be the measurement time, $N_{\beta\beta}$ be the number of atom candidates for $\beta\beta$ decay present in the source, B (expressed in number of counts per unit energy per unit time) the background counting rate in the energy region where the decay peak is expected and ΔE the FWHM energy resolution. The expected number of background counts in an energy interval equal to the FWHM energy resolution centered at the transition energy can then be written as $N_B = B\Delta E t$. The sensitivity factor F at 1σ level ($n = \sqrt{N_B}$) is defined as

$$F \equiv T_{1/2} = \frac{\ln 2\, N_{\beta\beta}\, t}{n} = \ln 2\, N_{\beta\beta}\sqrt{\frac{t}{B\Delta E}}\,. \quad (9)$$

If the background (measured as counts per unit energy per unit mass per unit time) is proportional to the detector mass the above formula can be rewritten as

$$F = \ln 2 \frac{i.a.\epsilon x}{A} N_A \sqrt{\frac{Mt}{B\Delta E}} \quad (68\%\ \mathrm{CL}), \quad (10)$$

where A is the compound molecular mass, x the number of $\beta\beta$ atoms per molecule, *i.a.* their isotopic abundance, M the source mass, N_A the Avogadro's number and ϵ the efficiency of the detector. In addition to its simplicity, eq. (10) has the advantage of emphasizing the role of the essential experimental parameters: mass, measuring time, isotopic abundance, background level, energy resolution and detection efficiency.

As far as the active source experiments are concerned, they can have detection efficiencies of the order of $\sim 90\%$, and energy resolution of the order of $\sim$ 0.2% FWHM (for Ge diodes and bolometers), while the background (mainly arising from the surrounding setup) cannot be easily rejected. As far as the passive source experiments are concerned, they are mostly performed with gas detectors (TPC, DC) in which the source is introduced into the volume of the detector as very thin sheets (of about $50\,\mu$m), to reduce the energy loss of the electrons emitted in the decay. The detection efficiency associated with this kind of measure is of the order of 30%. The great advantage of these experiments lies in the reduction of the background: the clear trace, which is peculiar in a drift chamber, for a 2 electrons event, guarantees a very good capability of background discrimination. The energy resolution, on the other hand, cannot be as good as $\sim$ 7–10% FWHM. As will be shown in sect. 7 the energy resolution plays an extremely crucial role in the second-generation experiments due to the fact that the 2νDBD close to the endpoint will result in an unavoidable/unrejectable source of background for the 0νDBD mode.

6 First-generation experiments

Impressive progress has been obtained during the last years in improving 0νDBD half-life limits for several isotopes and in systematically cataloguing 2νDBD rates (table 3). Although 2νDBD results are in some cases inconsistent, the effort to cover as many $\beta\beta$ nuclei as possible thus allowing a direct check for 2νDBD NME is evident. Optimal 0νDBD sensitivities have been reached in a series of experiments based on the active source approach. In particular, the best limit on 0νDBD comes from the Heidelberg-Moscow (HM) experiment [31] on ^{76}Ge even if similar results have been obtained also by the IGEX experiment [34] (table 3).

In both cases a large mass (several kg) of isotopically enriched (86%) germanium diodes, is installed deep underground under heavy shields for gamma and neutron environmental radiation. Extremely low background levels are then achieved thanks to a careful selection of the setup materials and further improved by the use of pulse shape discrimination (PSD) techniques. Both experiments quote similar background levels in the 0νDBD region of ≈ 0.2 (c/keV $\cdot$ kg $\cdot$ y) and ≈ 0.06 (c/keV $\cdot$ kg $\cdot$ y) before and after PSD. Taking into account the uncertainties in the NME calculations, such experiments indicate a limit of 0.3–1 eV for $\langle m_\nu \rangle$.

As will be discussed later, new ideas to improve such a successful technique characterize many of the proposed future projects. However, given the NME calculation problem, more $\beta\beta$ emitters than allowed by the use of conventional detectors (*e.g.* ^{76}Ge, ^{136}Xe, ^{116}Cd) should be investigated using the calorimetric approach. A solution to this problem, suggested [42] and developed [43] by the Milano group, is based on the use of low temperature calorimeters (*bolometers*). Besides providing very good energy resolutions they can in fact practically eliminate any constraint in the choice of the $\beta\beta$ emitter. Due to their very simple concept (a massive absorber in thermal contact with a suitable thermometer measuring the temperature increase following an energy deposition), they are in fact constrained only by the requirement of finding a compound allowing the growth of a diamagnetic and dielectric crystal. Extremely massive [44] detectors can then be built, by assembling large crystal arrays.

Thermal detectors have been pioneered by the Milano group for ^{130}Te (chosen, because of its favourable nuclear factor-of-merit and large natural isotopic abundance, from a large number of other successfully tested $\beta\beta$ emitters) in a series of constantly increasing mass experiments carried out at *Laboratori Nazionali del Gran Sasso* (LNGS), whose last extension, started in 2003, is the CUORICINO experiment [38]. Consisting of an array of 62 TeO_2 crystals totalling a mass of 40.7 kg and operating at a temperature of $\sim$ 8–9 mK, CUORICINO is characterized by a good energy resolution (7–8 keV on the average at the 0νDBD transition energy, 2528 keV) and a background level of ~ 0.18 (c/keV · kg · y). The quoted limit of 1.8×10^{24} y on the ^{130}Te 0νDBD half-life, corresponding to a limit of 0.2–1.1 eV on $\langle m_\nu \rangle$, represents the best limit after those reached by Ge diodes experiments (see table 3).

Half-way with next-generation experiments, NEMO III [45] is a passive source detector located in the Frejus underground laboratory at a depth of $\sim$ 4800 m.w.e. It consists of a tracking (wire chambers filled with an ethyl-alcohol mixture, operated in the Geiger mode) and a calorimetric (1940 plastic scintillators) system operated in a 30 gauss magnetic field. A well designed source system allows the simultaneous analysis of up to 10 kg of different 0νDBD active isotopes. Despite a relatively modest energy resolution (11% FWHM at 3 MeV), implying a non-negligible background contribution from 2νDBD, they achieved very good results on the study of the 2νDBD spectra of several $\beta\beta$ emitters (^{82}Se, ^{96}Zr, ^{116}Cd, ^{150}Nd). However, regarding the 0νDBD a good result was obtained with ^{100}Mo (see table 3).

In January 2002, a few members of the HM collaboration claimed evidence for ^{76}Ge 0νDBD [46] with $T^{0\nu}_{1/2} = 0.8\text{–}18.3 \times 10^{25}$ y (best value $T^{0\nu}_{1/2} = 1.5 \times 10^{25}$ y) corresponding to a $\langle m_\nu \rangle$ range of 0.11–0.56 eV (best value 0.39 eV). This claim is based on the identification of tiny peaks close to the 0νDBD region of ^{76}Ge, one of them at the energy of the Q-value of the DBD. However this announcement raised immediate scepticism [47]. Several reanalyses of the data were published by the claim's authors [48,49,50,51], while other authors [52,53,54] still criticize the claim. Probably a definite answer to the correctness of the claim will be given only by the very sensitive next-generation 0νDBD projects.

7 Towards second-generation experiments

We have seen that the field of 0νDBD searches is very active. The goal of the future experiments is to reach sensitivities capable of probing the inverted mass hierarchy, *i.e.* sensitivities in the decay time of the order of $\sim 10^{26}\text{–}10^{27}$ years. There are several possibilities in order to improve the sensitivities of the experiments. It is up to the experimentalist to choose the philosophy of the experiment and, consequently, to select the detector characteristics, privileging some properties with respect to others, having clearly in mind the final sensitivity of the set-up to half-life and, consequently, to $\langle m_\nu \rangle$. The question is: how to improve the experiments? Most of the criteria that need to be considered when optimizing the design of a new 0νDBD experiment follow directly from eq. (10) combined with eq. (3); the sensitivity at 68% CL on the neutrino mass, $F_{0\nu}$, can be written as

$$F_{0\nu} = \frac{m_e}{\sqrt{\ln 2 N_A}} \frac{1}{\sqrt{G^{0\nu}|M^{0\nu}|^2}} \sqrt{\frac{A}{i.a.\epsilon x}} \sqrt[4]{\frac{B \Delta E}{Mt}} . \quad (11)$$

7.1 Exposure time

The first consideration is that an "improvement" could be simply reached with the present experiments just by measuring for longer time; from eq. (11) we have

$$\langle m_\nu \rangle|_{t\to\infty} = \langle m_\nu \rangle_{\Delta T} \left(\frac{\Delta T}{\Delta T + t \to \infty} \right)^{1/4} , \quad (12)$$

where ΔT is the measurement lifetime that has allowed the present limit on $\langle m_\nu \rangle_{\Delta T}$. Now, almost all the experiments (except the CUORICINO and NEMO III experiments, that have, respectively, a lifetime of $\Delta T \sim 4$ months and of $\Delta T \sim 13$ months) have $\sim \Delta T \geq 2$ y. This implies that to have an improvement of only a factor 2 in the neutrino mass one has to measure $\sim$ 30 years! From this consideration it is also clear that, generally, DBD experiments have a, somewhat, "short" life.

7.2 Mass

The mass is one of the "parameters" with which one needs to deal. Regarding this point one has to consider also other variables, *i.e.* the isotopic abundance. From table 3 we see that all the interesting $\beta\beta$ emitters (except ^{130}Te) have isotopic abundances of the order of 5–10%. Let us consider, for example, two experiments with the same environmental natural background and with the same mass; let us suppose that one has the $\beta\beta$ emitter with natural *i.a.* (for example 5%) while the second has a 90% enrichment; then, from eq. (11), it turns out that the sensitivity of the enriched experiment is a factor $\sqrt{90/5} = 4.2$ better with respect to the other. This result can be seen also from another point of view: the time needed for the non-enriched experiment to reach the sensitivity of the enriched one will be $(90/5)^2 = 324$ times larger! This point is crucial and, in fact, all the most sensitive experiments carried out in the past (except the ones based on ^{130}Te) used enriched materials. This holds (see next section) also for the next planned experiments. It is also clear that enrichment is extremely expensive and raises tremendously the cost of an experiment.

Another crucial point comes directly from eq. (9). From table 1 it turns out that the decay time in order to reach a sensitivity on neutrino mass of the order of 10 meV (the goal of the most ambitious next-generation experiments) is of the order of 10^{28} y. Thus, assuming a measurement of three years, supposing to have *zero background* (it is clear that this is practically impossible) and considering at least three events for the 0νDBD discovery, then we need to have at least $N^*_{\beta\beta} = 10^{28}/\ln 2$ $\beta\beta$ nuclei, or $\approx 24 \cdot 10^3$ moles. It is therefore clear that the mass scale for the second-generation experiments is of the order of a ton. It is also clear that if no signal is detected with the next experiments it will be necessary to cover the normal hierarchy region, *i.e.* a few meV. Using the same argument it turns out that such *third-generation experiments* (if any) should be able to reach decay times of the order of $\sim 10^{30}$ y with a mass of ~ 100 ton.

7.3 Background

The background is a very delicate point and represent the main task of DBD experiments. There are several background sources that need to be taken into account. Cosmic-ray–induced background can, in principle, be made negligible operating deep underground and using, furthermore, active shielding to discard events correlated with the passage of a nearby muon. Neutron induced background (mainly (n,γ) reactions) can also be lowered to a negligible contribution by shielding the experimental setup with suitable moderators/neutron-catchers. It turns out, therefore, that the main source of background is the one arising from the natural radioactivity (mainly ^{232}Th and ^{238}U) overall present in traces in all kind of materials. Therefore the main work in order to build an experiment is the screening of materials used for the setup. The present techniques are based on High-Purity Ge diodes, neutron activation analysis (for solid and liquid samples), and High-Resolution Inductively Coupled Plasma Mass Spectrometry (HR-ICPMS) for liquid samples. The best limits that can be obtained with the first two techniques are around 10^{-11}–10^{-12} g/g (grams of contaminant/grams of sample) while 10^{-13} g/g can be reached on liquid samples. Both techniques are extremely delicate and, moreover, extremely long: for Ge measurement the time needed in order to reach such limits is of the order of several months.

Having screened all the *samples* of materials to be used for the experiment, one has to assemble the setup. *Assuming* that all the batches (also the ones that underwent mechanical machining) are as clean as the screened samples, extreme care has to be used in order to avoid possible re-contamination. Clean room operation is absolutely needed and operation in "synthetic" air (without Radon) or under clean atmosphere (pure nitrogen) is necessary. The background levels that need to be reached for the next experiments (see sect. 8) require reduction of the order of 10–100 (this means 10–1 c/(keV · ton · y)) with respect to the previous "pilot" experiments. These levels of radiopurity cannot be *directly* tested with the standard techniques [55], so that the only way to measure them will be the experiment itself. Many physicists consider this somehow as "bet".

Strictly connected with the background is also the choice of the $\beta\beta$ emitter. Apart some *extremely rare* high energy γ's from the ^{238}U chain, the highest natural γ line arising from natural radioactivity is the 2615 keV line of ^{208}Tl (from ^{232}Th chain). It would be, therefore, extremely useful to choose a $\beta\beta$ emitter with a $Q_{\beta\beta}$ value above this energy.

7.4 Energy resolution

Energy resolution will be the key point for some future experiments. Apart from the obvious role played in eq. (11), there is another fundamental aspect that has to be addressed. Assuming also the ability to reduce all the background sources, there is an intrinsic, unavoidable, undistinguishable "background": the 2νDBD. This source of background did not play a reasonable role in the present and past experiments due to the "low" sensitivity reached. As pointed out in [56] the fraction F of 2νDBD events that are contained in the ΔE^{FWHM} energy window centred at the $Q_{\beta\beta}$ value is given by

$$F \approx 7\frac{Q_{\beta\beta}\delta^6}{m_e}, \quad \delta = \frac{\Delta E^{\mathrm{FWHM}}}{Q_{\beta\beta}}. \tag{13}$$

An expression for the 0νDBD signal (S) to the 2νDBD background (B) ratio can be written

$$\frac{S}{B} \approx \frac{m_e}{7Q_{\beta\beta}\delta^6}\frac{T^{2\nu}_{1/2}}{T^{0\nu}_{1/2}}. \tag{14}$$

For example, looking at table 3, taking $T^{2\nu}_{1/2} \approx 10^{20}$ y and $Q_{\beta\beta} \approx 3$ MeV, in order to reach a sensitivity on $T^{0\nu}_{1/2}$ of the order of $10^{27}(10^{28})$ y with $S/B = 1$ the energy resolution should be less than 3.7 (2.5)% FWHM at the $Q_{\beta\beta}$ value. This energy resolution is extremely challenging for all detectors except Ge diodes and bolometers that have, normally, energy resolutions of the order of 0.2–0.4% FWHM. For the sake of completeness it has to be noted that the S/B ratio can be slightly enhanced by choosing an asymmetric analysis window defined as $Q_{\beta\beta} < E < Q_{\beta\beta} + \Delta E^{\mathrm{FWHM}}$.

8 Future experiments

So far, the best results have been obtained by exploiting the calorimetric approach (active source detectors) which characterizes therefore most of the future proposed

Table 4. Expected sensitivities of future projects. Last column evaluated from table 1.

Experiment	Isotope	kMoles ($\beta\beta$)	$T^{0\nu}_{1/2}$ (10^{26} y)	$\langle m_\nu \rangle$ (meV)
CUORE [57]	^{130}Te	1.6	7	(46–91)
EXO [58]	^{136}Xe	48	130	(5–30)
GERDA [59]	^{76}Ge	0.5	2	(105–300)
MAJORANA [60]	^{76}Ge	5.6	40	(24–66)
MOON III [61]	^{100}Mo	8.5	30	(15–36)
XMASS [62]	^{136}Xe	6.1	30	(9–63)
DCBA [63]	^{150}Nd	2.7	1	(16–55)
SUPERNEMO [64]	^{82}Se	1.1	2	(55–170)
CAMEO III [65]	^{116}Cd	2.7	10	(20–68)
CANDLES IV [66]	^{48}Ca	0.6	30	(55)

projects. Actually, a series of new proposals has been boosted by the recent renewed interest in 0νDBD following neutrino oscillation results. It is not so easy to classify them:

1. High energy resolution calorimetric experiments based on already consolidated techniques with improvements in background suppression/rejection (*e.g.* CUORE, GERDA, MAJORANA).
2. Calorimetric experiments based on consolidated techniques of scintillation light detection (CANDLES, CAMEO).
3. Calorimetric experiments with or without background identification techniques based on *non-standard* techniques that require further R&D (EXO, XMASS).
4. "Passive" experiments based on standard techniques that requires R&D (SUPERNEMO, MOON, DCBA).

Expected sensitivities on $T^{0\nu}_{1/2}$ of the proposed projects are compared in table 4. The sensitivities on $\langle m_\nu \rangle$ are evaluated using table 1. In many cases technical feasibility tests are requested, but the crucial issue will be the capability of each project to pursue the expected background suppression.

Many proposals have been recently suggested. However most of them are not officially approved or require further R&D to actually prove the feasibility. A complete report can be found in [67,68].

GERDA [59] (*GERmanium Detector Array*) is the only completely approved and funded experiment. It will be carried on in the INFN Gran Sasso National Laboratories. It is based on the technique already suggested by the HM Collaboration [69]: "naked" Ge diodes will be suspended in the centre of a very large liquid nitrogen container, which will act as a very effective shield. The experiment will consist of two phases: in the first one the *same* detectors of the HM Collaboration and the IGEX Collaboration will be "naked", removing all the components that are not needed for operating them in liquid nitrogen. The total mass will be ~ 17 kg of 86% enriched ^{76}Ge. The collaboration will probe the HM claim within the first 1–2 years. They plan to reach *zero* background in the 0νDBD region. Therefore, if the result of the HM Collaboration is true they will expect to confirm it at 5σ CL within the first year of operation. The second phase will consist of the addition of new enriched detectors for a total mass of 60 kg (0.7 kmol). For the second phase they will use background discrimination techniques and they, again, quote *zero background*. The first phase should start data taking in 2007.

CUORE [57] (*Cryogenic Underground Observatory for Rare Events*) will be a larger extension of CUORICINO. The experiment is scientifically approved and partially funded. CUORE will consists of 25 CUORICINO-like towers with 988 $5\times5\times5\,\mathrm{cm}^3$ TeO_2 *non-enriched* crystals. The number of $\beta\beta$ nuclei will be 1.6 kmol. The construction of a large custom cryostat, already funded, will start in 2006. Among all the proposed experiments it is the only one that needs a background suppression of "only" a factor 10 with respect to the "pilot" CUORICINO experiment. A factor of two was already achieved in 2004. The expected sensitivity will be $7 \cdot 10^{26}$ y. The experiment should start in 2009.

MAJORANA [60], which involves many of the IGEX collaborators, will consist of an array of 210 isotopically enriched Ge diodes for a total mass of 0.5 tons. As opposed to the GERDA design, the use of a very low activity conventional cryostat (extremely radiopure electroformed Cu) able to host simultaneously a large number of diodes is proposed. The driving principle behind the project is a strong reduction of the background by the application of very effective pulse-shape discrimination and the development of special segmented detectors. Despite the very promising R&D developed in the last years, the project is not yet funded.

9 Conclusions

A renewed interest in 0νDBD has been stimulated by recent neutrino oscillations results. Neutrinoless DBD is finally recognized as a unique tool to measure neutrino properties (nature, mass scale, intrinsic phases) unavailable for the successful experiments on neutrino oscillations. The attainability of such a goal strongly depends on the *true* capability of these projects to reach the required background levels in the 0νDBD region. An experimental confirmation of the (sometimes optimistic) background predictions of the various projects (even if extrapolated from the results of lower-scale successful experiments) is therefore worthwhile and the construction of preliminary test setups is absolutely needed. The ultra-low background understanding is a very complicated issue that needs careful investigations and experimental confirmation. The recently claimed evidence for a 0νDBD signal in the HM data seems still too weak but will be verified by the future experiments.

A strong effort to improve the NME evaluation should be encouraged while stressing the need of experiments addressed to different nuclei.

References

1. S.J. Freedman, B. Kayser (Co-chairs of the organizing committee), arXiv:physics/0411216.
2. R.N. Mohapatra, G. Senjanovic, Phys. Rev. D **23**, 165 (1981).
3. B. Kayser, *The Physics of Massive Neutrinos* (World Scientific, Singapore, 1989).
4. K. Fujikawa, R. Shrock, Phys. Rev. Lett. **45**, 963 (1980).
5. J. Kaulard *et al.*, Phys. Lett. B **422**, 334 (1998).
6. B. Pontecorvo, Zh. Eksp. Teor. Fiz. **33**, 549 (1957); **34**, 247 (1958); Z. Maki, M. Nakagawa, S. Sakata, Prog. Theor. Phys. **28**, 870 (1962); B. Pontecorvo, Zh. Eksp. Teor. Fiz. **53**, 1717 (1967).
7. Ling-Lie Chau, Wai-Yee Keung, Phys. Rev. Lett. **53**, 1802 (1984).
8. P.J. Siemens, A.S. Jensen, *Elements of Nuclei* (Addison-Wesley, 1987).
9. S.R. Elliot *et al.*, Phys. Rev. Lett. **59**, 2020 (1987).
10. V.I. Tretyak, Yu.G. Zdesenko, At. Data Nucl. Data Tables **80**, 83 (2002).
11. H. Ejiri, Prog. Part. Nucl. Phys. B **48**, 185 (2002).
12. W.H. Furry, Phys. Rev. **56**, 1184 (1939).
13. V.A. Rodin, A. Faessler, F. Simkovic, arXiv:nucl-th/0503063 (2005).
14. E. Caurier *et al.*, Nucl. Phys. A **654**, 973 (1999).
15. J. Engel *et al.*, Phys. Rev. C **37**, 731 (1988).
16. A. Staudt *et al.*, Europhys. Lett **13**, 31 (1990).
17. G. Pantis *et al.*, Phys. Rev. C **53**, 695 (1996).
18. O. Civitarese, J. Suhonen, Nucl. Phys. A **729**, 867 (2003).
19. A. Strumia, F. Vissani, arXiv:hep-ph/0503246.
20. J. Suhonen, Nucl. Phys. B (Proc. Suppl.) **143**, 240 (2005).
21. J. Suhonen, Phys. Lett. B **607**, 87 (2005).
22. H. Ejiri, Phys. Rep. **338**, 265 (2000).
23. B.T. Cleveland *et al.*, Astrophys. J. **496**, 505 (1998).
24. C. Cattadori, N. Ferrari, L. Pandola, Nucl. Phys. B (Proc. Suppl.) **143**, 3 (2005).
25. Super-Kamiokande Collaboration, hep-ex/0205075.
26. SNO Collaboration, nucl-ex/0502021.
27. KamLAND Collaboration, hep-ex/0406035.
28. S. Goswami, A. Bandyopadhyay, S. Choubey, Nucl. Phys. B (Proc. Suppl.) **143**, 121 (2005).
29. A. Lewis, S. Bridle Phys. Rev. D **66**, 103511 (2002); M. Tegmark *et al.*, Phys. Rev. D **69**, 103501 (2004); P. Crotty, J. Lesgourgues, S. Pastor, Phys. Rev. D **69**, 123007 (2004).
30. V.B. Brudanin *et al.*, Phys. Lett. B **495**, 63 (2000).
31. HM Collaboration (H.V. Klapdor-Kleingrothaus *et al.*), Eur. Phys. J. A **12**, 147 (2001).
32. C.E. Aalseth *et al.*, Nucl. Phys. B **48**, 223 (1996).
33. C.E. Aalseth *et al.*, Phys. Rev. C **59**, 2108 (1999).
34. IGEX Collaboration (C.E. Aalseth *et al.*), hep-ex/0202026.
35. R. Arnold *et al.*, Phys. Rev. Lett. **95**, 182302 (2005).
36. F. Danevich *et al.*, Phys. Rev. C **68**, 035501 (2003).
37. C. Arnaboldi *et al.*, Phys. Lett. B **557**, 167 (2003).
38. C. Arnaboldi *et al.*, Phys. Rev. Lett. **95**, 142501 (2005).
39. R. Bernabei *et al.*, Phys. Lett. B **546**, 23 (2002).
40. A. De Silva *et al.*, Phys. Rev. C **56**, 2451 (1997).
41. T. Kirsten, H. Richter, E. Jessberger, Phys. Rev. Lett. **50**, 474 (1983); A.L. Turkevich *et al.*, Phys. Rev. Lett. **67**, 3211 (1991).
42. E. Fiorini, T. Niinikoski, Nucl. Instrum. Methods **224**, 83 (1984).
43. A. Alessandrello *et al.*, Nucl. Instrum. Methods A **440**, 397 (1998).
44. C. Arnaboldi *et al.*, Nucl. Instrum. Methods A **554**, 300 (2005).
45. R. Arnold *et al.*, Nucl. Instrum. Methods A **536**, 79 (2005).
46. H.V. Klapdor-Kleingrothaus, A. Dietz, H.V. Harney, I.V. Krivosheina, Mod. Phys. Lett. A **16**, 2409 (2001).
47. C.E. Aalseth *et al.*, Mod. Phys. Lett A **17**, 1475 (2002).
48. H.V. Klapdor-Kleingrothaus *et al.*, arXiv:hep-ph/0205228.
49. H.V. Klapdor-Kleingrothaus *et al.*, Nucl. Instrum. Methods A **510**, 281 (2003).
50. H.V. Klapdor-Kleingrothaus *et al.*, Nucl. Instrum. Methods A **511**, 335 (2003).
51. H.V. Klapdor-Kleingrothaus *et al.*, Phys. Lett. B **578**, 54 (2004).
52. Yu.G. Zdesenko *et al.*, Phys. Lett. B **546**, 206 (2002).
53. F. Feruglio *et al.*, Nucl. Phys. B **637**, 345 (2002).
54. A. Giuliani, Nucl. Phys. B (Proc. Suppl.) **138**, 267 (2005).
55. G. Heusser, AIP Conf. Proc. **785**, 39 (2005).
56. S. Elliott, P. Vogel, Annu. Rev. Nucl. Part. Sci. **52**, 115 (2002).
57. C. Arnaboldi *et al.*, Nucl. Instrum. Methods A **518**, 775 (2004).
58. M. Danilov *et al.*, Phys. Lett. B **480**, 12 (2000).
59. GERDA Collaboration (Hardy Simgen), Nucl. Phys. B (Proc. Suppl.) **143**, 567 (2005).
60. The Majorana White Paper, 2003 Preprint, nucl-ex/0311013.
61. H. Ejiri *et al.*, Phys. Rev. Lett. **85**, 2917 (2000).
62. S. Moriyama *et al.*, *Proceedings of the XENON01 Workshop, December 2001, Tokyo, Japan* (2001).
63. Y. Kato *et al.*, Nucl. Instrum. Methods A **498**, 430 (2003).
64. NEMO Collaboration (F. Piquemal), talk presented at the IN2P3 scientific committee, 21 March 2005.
65. G. Bellini *et al.*, Eur. Phys. J. C **19**, 43 (2001).
66. T. Kishimoto *et al.*, Osaka University 2001 Annual Report.
67. F.T. Avignone III *et al.*, New J. Phys. **7**, 6 (2005).
68. S.R. Elliot, J. Engel, J. Phys. G **30**, R183 (2004).
69. H.V. Klapdor-Kleingrothaus, hep-ph/0103074.

Eur. Phys. J. A **27**, s01, 35–41 (2006)
DOI: 10.1140/epja/i2006-08-004-y

EPJ A direct
electronic only

Search for rare processes with DAMA/LXe experiment at Gran Sasso

R. Bernabei[1], P. Belli[1], F. Montecchia[1], F. Nozzoli[1], F. Cappella[2], A. Incicchitti[2], D. Prosperi[2], R. Cerulli[3,a], C.J. Dai[4], V.Yu. Denisov[5], and V.I. Tretyak[5]

[1] Dipartimento di Fisica, Università di Roma "Tor Vergata" and INFN, Sezione di Roma II, I-00133, Roma, Italy
[2] Dipartimento di Fisica, Università di Roma "La Sapienza" and INFN, Sezione di Roma, I-00185, Roma, Italy
[3] INFN - Laboratori Nazionali del Gran Sasso, I-67010 Assergi (AQ), Italy
[4] IHEP, Chinese Academy, P.O. Box 918/3, Beijing 100039, PRC
[5] Institute for Nuclear Research, MSP 03680 Kiev, Ukraine

Received: 20 June 2005 /
Published online: 22 February 2006 –

Abstract. After a short introduction on the low background liquid xenon DAMA set-up (DAMA/LXe) and its main previous results, we discuss the search for the nucleon, di-nucleon and tri-nucleon decays into invisible channels (disappearance or decay to neutrinos, Majorons, etc.) in the ^{136}Xe isotope. The obtained limits (90% C.L.) on the lifetimes are: $\tau_n > 3.3 \cdot 10^{23}$ yr, $\tau_p > 4.5 \cdot 10^{23}$ yr, $\tau_{np} > 3.2 \cdot 10^{23}$ yr, $\tau_{pp} > 1.9 \cdot 10^{24}$ yr, $\tau_{nnp} > 1.4 \cdot 10^{22}$ yr, $\tau_{npp} > 2.7 \cdot 10^{22}$ yr and $\tau_{ppp} > 3.6 \cdot 10^{22}$ yr. In particular, the tri-nucleon decay into invisible channels is investigated here for the first time.

PACS. 29.40.Mc Scintillation detectors – 95.35.+d Dark matter (stellar, interstellar, galactic, and cosmological) – 11.30.Fs Global symmetries (*e.g.*, baryon number, lepton number)

1 Introduction on previous results

The DAMA/LXe experiment has realized several prototype detectors and, then, has preliminarily put in measurement the set-up used in the data taking of refs. [1,2]; this set-up was firstly upgraded in fall 1995 as mentioned in refs. [3,4,5,6,7]. In particular, it has an inner vessel filled by $\simeq 6.5$ kg (*i.e.* $\simeq 2$ l) of liquid xenon. Firstly it used Kr-free xenon enriched in ^{129}Xe at 99.5% [8]; then, in 2000 the set-up was deeply modified reaching the configuration in fig. 5 of ref. [8] to handle also Kr-free xenon enriched in ^{136}Xe at 68.8% [9,10]. In this latter case, the interest has mainly been focused on the higher energy region. The main features of the set-up, details on the data acquisition, on the cryogenic and vacuum systems and on the running parameters control are described in [6,7,8,10,11]. In particular, the energy scale in the high energy region is determined with the help of external standard gamma sources, having identified the possibility to perform high energy calibrations with external sources near a flange located on the top of the detector, opening a limited upper part of the external shield. The energy resolution for ^{208}Tl line ($E_\gamma = 2614$ keV) is $\sigma \simeq 220$ keV, while for ^{137}Cs line ($E_\gamma = 662$ keV), which is the source usually exploited for the routine calibrations, σ is $\simeq 70$ keV [10,11,7].

We pointed out the interest in using liquid xenon as target-detector for particle Dark Matter (DM) investigation deep underground since long time [12]. After preliminary measurements both on elastic and inelastic DM particles-^{129}Xe scattering [1,3], the recoil/electron light ratio and pulse shape discrimination capability in a similar pure LXe scintillator have been measured both with Am-B neutron source and with 14 MeV ENEA-Frascati neutron generator [6]. Moreover, in 2000/2001 further measurements on the recoil/electron light ratio with 2.5 MeV ENEA-Frascati neutron generator have also been carried out; see ref. [5] for details and comparisons.

After upgrading of the LXe set-up, new results on the DM particles investigation have been obtained [4,6]. In particular, in ref. [6] pulse shape discrimination between recoils and electromagnetic background in the developed pure LXe scintillators has been exploited. Afterwards the inelastic excitation of ^{129}Xe by DM particles with spin-dependent coupling has further been investigated in ref. [4].

Several other rare processes have also been searched for by means of the detector filled with the Kr-free xenon gas enriched in ^{129}Xe. In particular, as regards the electron stability, limits on the lifetime of the electron decay in both the disappearance and the $\nu_e + \gamma$ channels were set in ref. [2]. The latter has been more recently improved up to $2.0(3.4) \cdot 10^{26}$ yr at 90% (68%) C.L. [13]. Furthermore,

[a] e-mail: riccardo.cerulli@lngs.infn.it

new lifetime limits on the charge non-conserving electron capture with excitation of ^{129}Xe nuclear levels have also been established to be in the range (1–4) $\cdot$ 10^{24} yr at 90% C.L. for the different excited levels of ^{129}Xe [14]. The stringent restrictions on the relative strengths of charge non-conserving (CNC) processes have been consequently derived: $\epsilon^2_W < 2.2 \cdot 10^{-26}$ and $\epsilon^2_\gamma < 1.3 \cdot 10^{-42}$ at 90% C.L. [14].

Moreover, we have searched for the nucleon and di-nucleon decay into invisible channels in the ^{129}Xe isotope by exploiting a new approach [15]. It consisted in a search (in the real-time experiment) for radioactive decay of unstable daughter nuclei which were created in result of the N or NN disappearance in parent nucleus. If the half-life of the daughter nucleus is of the order of 1 s or greater, such a decay will be time-resolved from prompt products if they were emitted and observed in a detector. This approach has the advantage of a branching ratio close to 1 and —if the parent and daughter nuclei are located in the detector itself— also of an efficiency close to 1. The obtained limits at 90% C.L. are: $\tau(p \rightarrow \textit{invisible channels}) > 1.9 \cdot 10^{24}$ yr; $\tau(pp \rightarrow \textit{invisible channels}) > 5.5 \cdot 10^{23}$ yr and $\tau(nn \rightarrow \textit{invisible channels}) > 1.2 \cdot 10^{25}$ yr. These limits are similar to or better than those previously available; moreover, they are valid for every possible disappearance channel [15] and the limits for the di-nucleon decay in $\nu_\tau \bar{\nu}_\tau$ are set there for the first time.

Afterwards, measurements have been carried out by using the Kr-free xenon gas containing 17.1% of ^{134}Xe and 68.8% of ^{136}Xe to investigate the ^{134}Xe and ^{136}Xe double beta decay modes. After the preliminary results of refs. [9, 16] a joint analysis of the $0\nu\beta\beta$ decay mode in ^{134}Xe and in ^{136}Xe (as suggested in ref. [17]) has been carried out. New lower limits on various $\beta\beta$ decay modes have been obtained: for the $0\nu\beta\beta(0^+ \rightarrow 0^+)$ decay mode in ^{134}Xe and in ^{136}Xe the limits at 90% C.L. are $T_{1/2} = 5.8 \cdot 10^{22}$ yr and $T_{1/2} = 1.2 \cdot 10^{24}$ yr, respectively; they correspond to a limit value on *effective light Majorana neutrino mass* ranging from 1.1 eV to 2.9 eV (90% C.L.), depending on the adopted theoretical model. For the neutrinoless double beta decay with Majoron (M) in the ^{136}Xe isotope the limit is $T_{1/2} = 5.0 \cdot 10^{23}$ yr (90% C.L.); for the $2\nu\beta\beta(0^+ \rightarrow 0^+)$ and the $2\nu\beta\beta(0^+ \rightarrow 2^+)$ decay modes in ^{136}Xe the limits at 90% C.L. are $1.0 \cdot 10^{22}$ yr and $9.4 \cdot 10^{21}$ yr, respectively. It is worthy to note that the experimental limit on the $2\nu\beta\beta(0^+ \rightarrow 0^+)$ decay mode is in the range of the theoretical estimate by [18] ($2.11 \cdot 10^{22}$ yr) and about a factor 5 higher than that of ref. [19][1].

A search for the charge non-conserving decay of ^{136}Xe into ^{136}Cs has also been performed for the first time [7], using the data collected during 8823.54 h and already published in ref. [10]. The used approach has been the investigation of the CNC processes by the search for the possible CNC decay firstly considered in [20]: if in a β decay $(A, Z) \rightarrow (A, Z+1) + e^- + \bar{\nu}_e$ some massless uncharged particle would be emitted instead of the electron (*e.g.*, ν_e or γ or Majoron), an additional 511 keV energy release would occur. Thus, usually forbidden decays to the ground state or to the excited levels of the daughter nuclei would become energetically possible. The presence of the $(A, Z+1)$ isotope or of its daughter products in a sample, initially free from them, would indicate the existence of the CNC decay searched for. Large advantages arise when the so-called "active-source" technique (source = detector) is considered as in the case described here. In particular, after the possible ^{136}Xe CNC decay, the daughter nucleus ^{136}Cs will be created. It is β unstable ($T_{1/2} = 13.16$ d) with quite high energy release ($Q_\beta = 2.548$ MeV) [7]. Comparing the experimental energy distribution with the expected response function, no evidence for the effect searched for has been found. Thus, the lifetime limit is: $\tau_{CNC}(^{136}\text{Xe} \rightarrow {}^{136}\text{Cs}) > 1.3 \cdot 10^{23}$ yr at 90% C.L. This limit is one of the highest available limits for similar processes [7]; however, the bound on the charge non-conserving admixture in the weak interactions, which can be derived according to ref. [21], is modest mainly due to the big change in nuclear spin in the considered CNC transition ($\Delta J^{\Delta\pi} = 5^+$).

[1] On the other hand, similar theoretical estimates suffer from the large uncertainties typically associated to the calculations of the nuclear matrix elements.

2 The search for nucleon instabilities into invisible channels in the ^{136}Xe isotope

The baryon number (B) conservation was introduced by Stuckelberg [22] and Wigner [23] more than fifty years ago and remains an accidental symmetry of the Standard Model, when this is seen as a renormalizable theory, not explained by deeper theoretical understanding. Modern theories of particle physics (GUTs, SUSY), unifying quarks and leptons into the same multiplets and predicting new interactions which transform quarks into leptons, naturally lead to the decay of the protons and the otherwise stable bound neutrons [24]. Many decay mechanisms, which violate B on one or two units, have been discussed [24,25]. In a recent work [26], a new process was examined in which two neutrons simultaneously decay to bulk Majoron, $nn \rightarrow M$, with typical lifetime 10^{32}–10^{39} yr; due to weak coupling of Majorons to normal matter they are not detected in an experiment, and such a process looks as a disappearance of two neutrons from a nucleus. Also mechanisms for the tri-nucleon decay have been proposed in the literature; in particular, very recently in ref. [27] also tri-nucleon decay processes with $\Delta B = 3$ have been considered. Moreover, disappearance of particles (electrons, e^-, or nucleons, N) are expected also in theories with extra dimensions [28,29,30]. No processes with baryon number violation were detected to date. We refer to our previous article [15] for a review of various ideas used in the search for the N and NN decays into invisible channels.

Here a search for the N, NN and NNN instabilities in the ^{136}Xe isotope is described, looking for decays of the unstable daughter nuclei. The experimental energy distribution collected during 8823.54 h by the LXe scintillator (enriched in ^{136}Xe at 68.8%) in the energy region 550–3550 keV (the same as in refs. [10] and [7]) is shown later.

Table 1. Daughter nuclei produced in N, NN and NNN decays in ^{136}Xe when the de-excitation of the daughter nucleus occurs by γ emission. The half-life times of the isotopes involved in the decay chains vary from 2.5 m (^{133}Sb) to 5.243 d (^{133}Xe) assuring that the chains are in equilibrium and that subsequent decays are well separated in time.

Decay	Daughter nucleus	Subsequent decays
n	^{135}Xe	$^{135}\text{Xe} \xrightarrow{\beta^-} {}^{135}\text{Cs}$ *
p	^{135}I	$^{135}\text{I} \xrightarrow{\beta^-} {}^{135}\text{Xe} \xrightarrow{\beta^-} {}^{135}\text{Cs}$ *
nn	^{134}Xe	Stable
np	^{134}I	$^{134}\text{I} \xrightarrow{\beta^-} {}^{134}\text{Xe}$
pp	^{134}Te	$^{134}\text{Te} \xrightarrow{\beta^-} {}^{134}\text{I} \xrightarrow{\beta^-} {}^{134}\text{Xe}$
nnn	^{133}Xe	$^{133}\text{Xe} \xrightarrow{\beta^-} {}^{133}\text{Cs}$
nnp	^{133}I	$^{133}\text{I} \xrightarrow{\beta^-} {}^{133}\text{Xe} \xrightarrow{\beta^-} {}^{133}\text{Cs}$
npp	^{133}Te	$^{133}\text{Te} \xrightarrow{\beta^-} {}^{133}\text{I} \xrightarrow{\beta^-} {}^{133}\text{Xe} \xrightarrow{\beta^-} {}^{133}\text{Cs}$
ppp	^{133}Sb	$^{133}\text{Sb} \xrightarrow{\beta^-} {}^{133}\text{Te} \xrightarrow{\beta^-} {}^{133}\text{I} \xrightarrow{\beta^-} {}^{133}\text{Xe} \xrightarrow{\beta^-} {}^{133}\text{Cs}$ **

* ^{135}Cs is not stable, but has $T_{1/2} = 2.3 \cdot 10^6$ yr and breaks the decay chain.
** Given here only the main part of the chain.

We remind that a cut has been applied to reject events whose charge is dominant in one PMT, that is to reject —as much as possible— background contribution from outside the inner vessel. As a consequence, a correction (which depends on the used hardware cut in high energy measurements) estimated by Monte Carlo program, properly considering the real geometry and the features of the detector, has been applied to the rate.

The isotopes given in table 1 are produced [31] after the disappearance of one, two or three nucleons in the parent $^{136}_{54}$Xe nucleus, when the daughter de-excitation occurs only by γ emission. In general, the created daughter nucleus —with one, two or three holes in nuclear shells due to disappeared nucleons— will be in an excited state, unless the nucleons were on the outermost shell. The holes will be filled in the subsequent de-excitation process in which different particles could be emitted. If the excitation energy is lower than the binding energy of the least bound nucleon, only γ quanta will be emitted; otherwise, heavy particles such as p, n or α will be ejected leading to daughter nuclei with lower atomic masses and numbers. In that follows we will take into account the N, NN and NNN disappearance from a few outermost shells in the parent nucleus, when only γ's would be emitted, avoiding in this way uncertainty in created daughter nuclide.

The lifetimes for the N, NN and NNN disappearance in ^{136}Xe can be calculated using the formula:

$$\tau = \frac{\epsilon_{\Delta E} \cdot N_{\text{nucl}} \cdot N^{\text{eff}}_{\text{obj}} \cdot T}{S_{\Delta E}}, \tag{1}$$

where $N_{\text{nucl}} = 2.00 \cdot 10^{25}$ is the number of ^{136}Xe nuclei; $T = 8823.54$ h is the time of measurement; $N^{\text{eff}}_{\text{obj}}$ is the "effective" number of objects (n, p, NN pairs or NNN groups) whose disappearance in the parent nucleus will result in the creation of the specific daughter nuclide; $S_{\Delta E}$ is the number of events which can be ascribed to the decay process searched for in the considered energy window, ΔE, while $\epsilon_{\Delta E}$ is the related detection efficiency.

Table 2. Values of $N^{\text{eff}}_{\text{obj}}$ used in the present data analysis to search for N, NN and NNN disappearance in ^{136}Xe. See ref. [15] and text.

Decay	n	p	np	pp	nnn	nnp	npp	ppp
$N^{\text{eff}}_{\text{obj}}$	32	26	2	7	1	1	1	1

Equation (1) requires the knowledge of the number of objects, $N^{\text{eff}}_{\text{obj}}$, that was calculated following the method already used in ref. [15]. As regards the NNN decays into invisible channels, there is no available information from previous search and, in particular, there are no reference criteria to evaluate $N^{\text{eff}}_{\text{obj}}$ in these cases. Thus, in the search we have carried out, we cautiously assumed $N^{\text{eff}}_{\text{obj}} = 1$ for all the NNN processes (nnn, nnp, npp and ppp), as done in the search for the NN decays into invisible channels in ref. [32]. The used values of $N^{\text{eff}}_{\text{obj}}$ are summarized in table 2. Referring to table 1, we note that, except for the ^{134}Xe nucleus which is stable and does not allow us to search for the nn disappearance, all other daughter nuclei are radioactive. In the subsequent decays (third column of table 1), if we exclude the ^{135}Cs which has $T_{1/2} = 2.3 \cdot 10^6$ yr and breaks the decay chain, the half-lives of the involved nuclides are relatively small. This ensures equilibrium in chains and thus equal number of decays for ^{135}I and ^{135}Xe (in case of the p disappearance), for ^{134}Te and ^{134}I (pp decay), etc. The expected response functions of the LXe detector for the β^- decays of the nuclei involved in the decay chains were simulated with the help of the EGS4 package [33]. The whole schemes of the decays, using the information from ref. [34] for the $A = 133$ isotopes, from ref. [35] for the $A = 134$ isotopes and from ref. [36] for the $A = 135$ isotopes, were implemented in an event generator which described the initial kinematics of the events. The response functions for the N, NN and NNN disappearance are given by a linear combination of the response functions obtained for the single decays of the generated decay chain. Comparison of the experimental spectrum with the calculated response functions gives no strong indication of the signals searched for. Thus, we limit ourselves to extract only the limits on the probability of these processes. We will see that, except for the np and pp channel, the calculated response function do not present any distinctive structure, as a peak, at energies above the experimental energy threshold (550 keV) but their shapes are similar to some extent to that of the measured spectrum. As a consequence, the limit on the amplitude of the expected response function, which can be hidden in the experimental data, was determined in a very cautious and simple way (used also in the investigation of other rare processes). In fact, it has been required that —in no energy interval, ΔE— the number of events which could be ascribed to the investigated process, $S_{\Delta E}$,

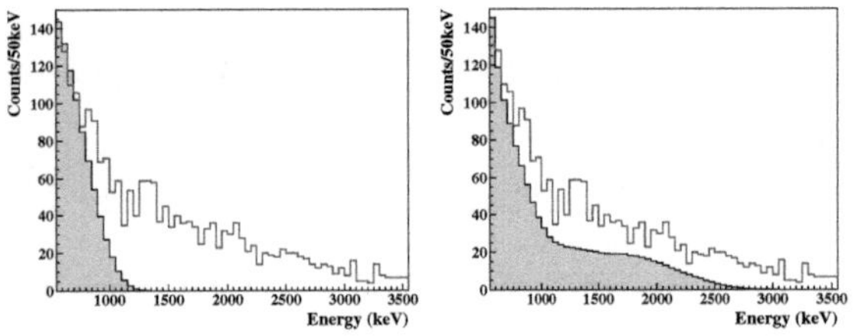

Fig. 1. Comparison between the experimental spectrum measured during 8823.54 h (thick histogram) and the expected signal (colored histogram) for: left) the n disappearance with $\tau_n = 3.3 \cdot 10^{23}$ yr excluded at 90% C.L.; right) the p disappearance with $\tau_p = 4.5 \cdot 10^{23}$ yr excluded at 90% C.L.

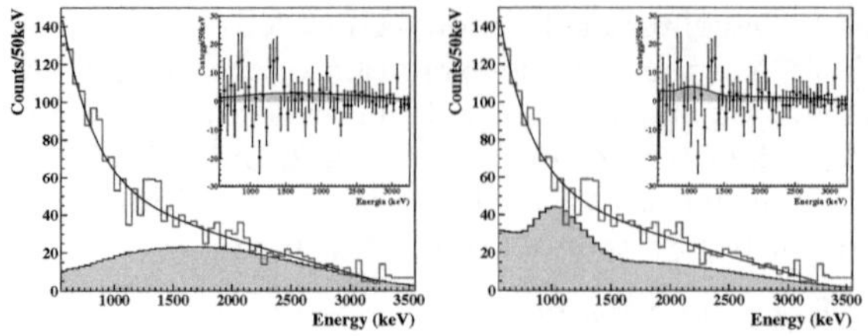

Fig. 2. Comparison between the experimental spectrum measured during 8823.54 h (thick histogram) and the expected signal (colored histogram) for: left) the np disappearance with $\tau_{np} = 4.0 \cdot 10^{22}$ yr excluded at 90% C.L.; right) the pp disappearance with $\tau_{pp} = 2.1 \cdot 10^{23}$ yr excluded at 90% C.L. In the insets the residuals between the experimental spectrum and the background fits are shown together with excluded distributions for np ($\tau_{np} = 3.2 \cdot 10^{23}$ yr) and pp ($\tau_{pp} = 1.9 \cdot 10^{24}$ yr) disappearances, respectively. See text.

can exceed the number of measured events plus $m \times \sigma$ (where m determines the C.L.). It is evident that the derived limits are very conservative because at least some events in the experimental spectrum can be induced by residual radioactive contaminations.

2.1 Results for N decays into invisible channel

A possible n disappearance in the ^{136}Xe nucleus is followed by the β^- decay $^{135}\text{Xe} \to {}^{135}\text{Cs}$ ($T_{1/2} = 9.14$ h, $Q_{\beta^-} = 1151$ keV) in the sensitive volume of DAMA/LXe. The expected energy distribution is mainly determined by the β^- decay with end-point ~ 900 keV (96% of the cases) followed by a 250 keV de-excitation γ (the β^- energy distribution is moved towards a higher energy of 250 keV) [11]. In the data analysis we considered the 550–700 keV energy region of the experimental spectrum (see fig. 1 left), where 369 events were accumulated. It gives $S_{\Delta E} < 369 + 1.29 \cdot \sqrt{369} = 393.6$ at 90% C.L. The related efficiency, calculated with the EGS4, is $\varepsilon_{\Delta E} = 20.2\%$. Substituting these values into eq. (1) together with the effective number of neutrons which should be taken into account ($N_n^{\text{eff}} = 32$), we obtain the following restriction on the neutron lifetime:

$$\tau_n > 3.3 \cdot 10^{23} \text{ yr at } 90\% \text{ C.L.} \tag{2}$$

The p disappearance in the ^{136}Xe nucleus is instead followed by a chain of two β^- decays: $^{135}\text{I} \to {}^{135}\text{Xe}$ ($T_{1/2} = 6.57$ h, $Q_{\beta^-} = 2648$ keV) and $^{135}\text{Xe} \to {}^{135}\text{Cs}$ (see above). The expected energy distribution for the ^{135}I β^- decay is characterized by a peak due to the ~ 527 keV metastable state of ^{135}Xe ($T_{1/2} = 15.29$ m), but below the energy threshold considered in this data analysis; the response function for the p disappearance is given by the $^{135}\text{I} + {}^{135}\text{Xe}$ distribution [11]. The most sensitive energy region of the experimental spectrum is 550–600 keV (see fig. 1 right). For 131 accumulated events, the 90% C.L. limit is $S_{\Delta E} < 145.7$; taking into account the calculated efficiency $\varepsilon_{\Delta E} = 12.7\%$ to detect the $^{135}\text{I} + {}^{135}\text{Xe}$ decays in the 550–600 keV interval and the effective number of protons $N_p^{\text{eff}} = 26$, the corresponding limit for the proton lifetime is:

$$\tau_p > 4.5 \cdot 10^{23} \text{ yr at } 90\% \text{ C.L.} \tag{3}$$

2.2 Results for NN decays into invisible channel

A possible np disappearance in the ^{136}Xe nucleus can be investigated by searching for the following β^- decay $^{134}\text{I} \to {}^{134}\text{Xe}$ ($T_{1/2} = 52.5$ m, $Q_{\beta^-} = 4175$ keV) in the sensitive volume of DAMA/LXe. Considering the expected energy distribution calculated for this decay, the most sensitive energy region of the experimental spectrum (see fig. 2 left) is 2250–2300 keV interval with 14 events collected: $S_{\Delta E} < 18.8$ events at 90% C.L. With the related efficiency $\varepsilon_{\Delta E} = 1.86\%$ and the effective number of pairs $N_{np}^{\text{eff}} = 2$, we obtain the restriction on the np lifetime: $\tau_{np} > 4.0 \cdot 10^{22}$ yr (90% C.L.).

The result of a pp disappearance in the ^{136}Xe nucleus is instead the creation of a ^{134}Te isotope and the subsequent β^- decays chain: $^{134}\text{Te} \to {}^{134}\text{I}$ ($T_{1/2} = 41.8$ m, $Q_{\beta^-} = 1550$ keV) and $^{134}\text{I} \to {}^{134}\text{Xe}$ (see above). The expected energy distribution for these processes in DAMA/LXe sensitive volume gives in particular for the ^{134}Te β^- decay a bump at ~ 1000 keV due to events where the de-excitation γ's, following the β^- ray with end-point ~ 730 keV (42% of the cases), ~ 610 keV (44%) and ~ 450 keV (14%), are fully contained in the detector, and a structure with energy below ~ 800 keV associated with events where one or more γ's escape the detector. For the pp disappearance the expected signal is given by the $^{134}\text{Te} + {}^{134}\text{I}$ distribution and the considered region of the experimental spectrum (see fig. 2 right) is 1100–1150 keV with 35 events. With the $S_{\Delta E} < 42.6$, $\varepsilon_{\Delta E} = 6.47\%$, and $N_{pp}^{\text{eff}} = 7$, it results in the value $\tau_{pp} > 2.1 \cdot 10^{23}$ yr at 90% C.L.

However, when the experimental data have a smooth behaviour and the expected response function of the effect has some peculiarities, as the peak-like structure near 1 MeV for the $^{134}\text{Te} + {}^{134}\text{I}$ decay (pp disappearance), it is justified to use an approach that allows to take into account the background contribution when estimating the limit on the lifetime. In particular, the experimental spectrum can be fitted by some appropriate background model together with the effect's response function with parameters of the model and the amplitude of the effect being the free parameters of the fit. The fit of the experimental

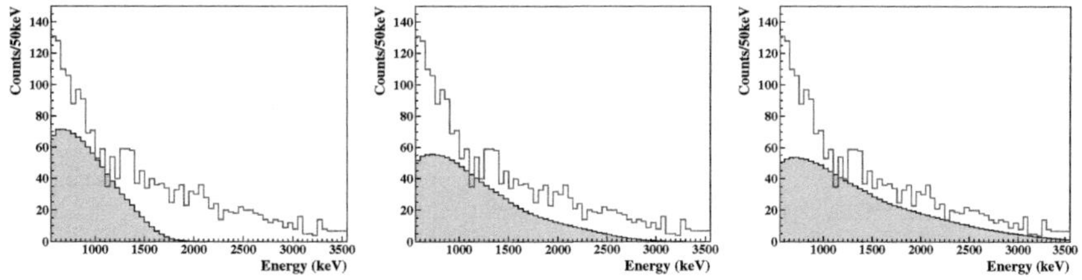

Fig. 3. Comparison between the experimental spectrum measured during 8823.54 h (thick histogram) and the expected signal (colored histogram) for: left) the *nnp* disappearance with $\tau_{nnp} = 1.4 \cdot 10^{22}$ yr excluded at 90% C.L.; center) the *npp* disappearance with $\tau_{npp} = 2.7 \cdot 10^{22}$ yr excluded at 90% C.L.; right) the *ppp* disappearance with $\tau_{ppp} = 3.6 \cdot 10^{22}$ yr excluded at 90% C.L.

spectrum in the 550–3200 keV region by the sum of an exponent and a straight line, as background model[2], and the ^{134}Te + ^{134}I decay response function gave for its area the value $S = 0.0 \pm 44.2$, providing no evidence for the effect being searched; the corresponding χ^2/n.d.f. value was equal to 1.1. According to the Feldman-Cousins procedure [37], the 90% C.L. limit on the number of observed ^{134}Te + ^{134}I events is $S < 72.5$. This value, together with the effective number of the *pp* pairs $N_{pp}^{\rm eff} = 7$, results in the *pp* lifetime limit:

$$\tau_{pp} > 1.9 \cdot 10^{24}\,\text{yr at 90\% C.L.} \quad (4)$$

In the same way the limit on the number of events under the ^{134}I distribution (*np* disappearance) was determined. The fit (χ^2/n.d.f. = 1.1) gave for the S value $S = 19.3 \pm 65.3$ that resulted in the bound $S < 126.4$ at 90% C.L. Taking into account that $N_{np}^{\rm eff} = 2$, we obtain the *np* lifetime limit:

$$\tau_{np} > 3.2 \cdot 10^{23}\,\text{yr at 90\% C.L.} \quad (5)$$

The distributions corresponding to a lifetime equal to these limit values for the *np* and *pp* disappearances in ^{136}Xe are shown together with the residuals between the experimental spectrum and the background model in the insets of fig. 2 left and right, respectively.

2.3 Results for NNN decays into invisible channel

The result of a *nnn* → *invisible channels* decay in the ^{136}Xe nucleus is the creation of the ^{133}Xe isotope, which is unstable and β^- decays to ^{133}Cs (stable) with $T_{1/2} =$ 5.243 d and $Q_{\beta^-} = 427.4$ keV. As evident, the maximum energy released in this process is below the 550 keV energy threshold of the present measurements; thus, this decay process cannot be investigated here.

As a consequence of a *nnp* disappearance in the ^{136}Xe nucleus, a ^{133}I nucleus is instead created in the sensitive volume of DAMA/LXe. This isotope is the parent of the β^- decay chain $^{133}\text{I} \rightarrow {}^{133}\text{Xe} \rightarrow {}^{133}\text{Cs}$. The energy distribution expected for the ^{133}I β^- decay ($T_{1/2} = 20.8$ h and $Q_{\beta^-} = 1770$ keV) in our liquid xenon set-up is mainly determined by the β^- decay with end-point ~ 1240 keV (83% of the cases) followed by a 530 keV de-excitation γ (the β^- energy distribution is moved towards higher energy of 530 keV); a 233 keV peak is instead due to the de-excitation of the ^{133}Xe metastable state ($T_{1/2} = 2.19$ d), which is reached by the $\sim 3\%$ of the decays [11]. The signal to be searched for in case of a *nnp* disappearance in ^{136}Xe is given by the ^{133}I + ^{133}Xe β^- decay distributions but the last one in the analysis does not play any role since it falls under the experimental energy threshold (550 keV).

The more selective region in the present case is $\Delta E =$ 1100–1150 keV (see fig. 3 left); it contains 35 events, which gives rise to the upper limit (90% C.L.) $S_{\Delta E} < 42.6$ events and being $\epsilon_{\Delta E} = 3.0\%$ gives:

$$\tau_{nnp} > 1.4 \cdot 10^{22}\,\text{yr at 90\% C.L.} \quad (6)$$

In the case of the process *npp* → *invisible channels* in the ^{136}Xe nucleus, a ^{133}Te nucleus is created. This isotope ($T_{1/2} = 12.5$ m and $Q_{\beta^-} = 2920$ keV) produces a β^- decay to ^{133}I followed by the decay chain $^{133}\text{I} \rightarrow {}^{133}\text{Xe} \rightarrow {}^{133}\text{Cs}$ already described above for the *nnp* channel. The simulation result for a ^{133}Te decay in the DAMA/LXe detector gives peculiar structures in the expected distribution determined by the γ's emitted in the ^{133}I de-excitation, whose energies are summed to the initial β^- of ^{133}Te. The more probable emitted γ's are those with energies 312 keV (62%) and 408 keV (27%), from the first two excited levels of ^{133}I. In fact, a *bump* at ~ 300 keV is due to the γ at 312 keV which, in most cases, releases all its energy in the detector, while a *bump* at ~ 700 keV is due to the sum of the energies released by the two γ's emitted in cascade. Each possible *npp* decay in ^{136}Xe is associated to a signal given by the sum of the energy distributions expected for the β^- decays of the ^{133}Te, of the ^{133}I and of the ^{133}Xe (this latter, as already mentioned, gives here a signal under the experimental energy threshold and, therefore, is not considered) [11]. Comparing the experimental spectrum measured (see fig. 3 center) and the expected signal, one gets that the more selective energy window is $\Delta E = 1100$–1150 keV which contains 35 events, giving rise to the upper limit: $S_{\Delta E} < 42.6$ events (90% C.L.). Taking into account the calculated efficiency $\epsilon_{\Delta E} = 5.7\%$, one can derive

$$\tau_{npp} > 2.7 \cdot 10^{22}\,\text{yr at 90\% C.L.} \quad (7)$$

Finally, a possible *ppp* → *invisible channels* decay in ^{136}Xe will create in the liquid xenon the ^{133}Sb nucleus which β^- decays with $T_{1/2} = 2.5$ m and $Q_{\beta^-} = 4003$ keV. As can be derived from the decay scheme of the ^{133}Sb, this

[2] Other parameterizations gave similar results.

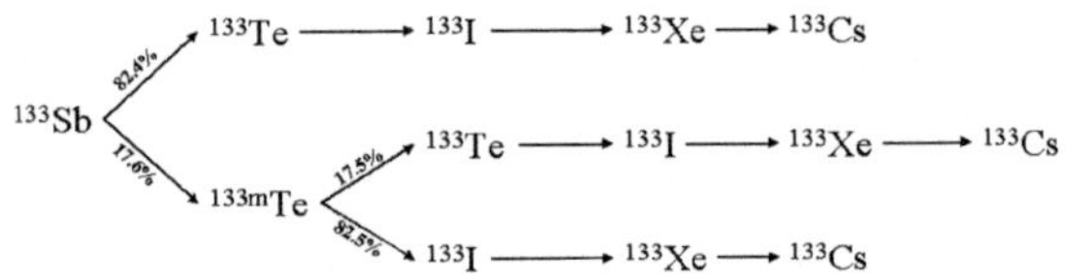

Fig. 4. The full decay chain generated by the ^{133}Sb.

process gives rise to the production of ^{133}Te nuclei in the ground state (82.4%) and in the $^{133\mathrm{m}}$Te metastable state at 334 keV (17.6%). The full decay chain generated by the ^{133}Sb is summarized in fig. 4. With simple calculation one can get that a possible *ppp* disappearance in ^{136}Xe is followed by a sequence of β^- decays weighted according to the following formula:

$$1 \times {}^{133}\mathrm{Sb} + 0.855 \times {}^{133}\mathrm{Te} + 0.176 \times {}^{133\mathrm{m}}\mathrm{Te} + 1 \times {}^{133}\mathrm{I} + 1 \times {}^{133}\mathrm{Xe}\,. \quad (8)$$

The energy distributions expected for the ^{133}Te, ^{133}I and ^{133}Xe isotopes have been already described above. As regards the response function of DAMA/LXe for the β^- decay of ^{133}Sb, it is more difficult to identify the structure due to the γ's emitted in the daughter nucleus de-excitation. In fact, the more probable γ's have energies of the order of MeV and, therefore, they can escape the detector releasing only a part of their energy. The last process in (8) is the $^{133\mathrm{m}}$Te decay ($T_{1/2} = 55.4$ m). The expected energy distribution for the $^{133\mathrm{m}}$Te decay process gives a 334 keV peak due to the de-excitation γ, which in the 17.5% of the cases move the $^{133\mathrm{m}}$Te into the ground state of ^{133}Te; the remaining part of the expected signal (branching ratio = 82.5%) is instead given by the $^{133\mathrm{m}}$Te β^- decay ($Q_{\beta^-} = 2920$ keV) to ^{133}I. In particular, the presence of the ^{133}I metastable state at 1634 keV with $T_{1/2} = 9$ s induces other peaks in the expected energy distribution (a 74 keV peak which can be well distinguished, but also peaks at 721 keV, 987 keV and 1634 keV) [11]. The signal expected for the *ppp* $\rightarrow$ *invisible channels* in ^{136}Xe can be obtained by summing according to (8) the obtained energy distributions [11] (we remind that the signal associated to the ^{133}Xe decay is below the experimental threshold 550 keV). Considering the expected energy distribution (see fig. 3 right), the most sensitive energy window is $\Delta E = 1100$–1150 keV which contains 35 events, giving rise to the upper limit: $S_{\Delta E} < 42.6$ events (90% C.L.); the detection efficiency is: $\epsilon_{\Delta E} = 7.6\%$. One obtains

$$\tau_{ppp} > 3.6 \cdot 10^{22}\,\mathrm{yr\ at\ 90\%\ C.L.} \quad (9)$$

3 Conclusion

The DAMA/LXe set-up, deeply improved several times, has allowed to achieve competitive results in the search for various rare processes as summarized in the first section of this paper.

Here in particular, the most recent results on nucleon decays into invisible channels in the ^{136}Xe isotope have been discussed. The considered experimental approach assures a high detection efficiency and a branching ratio ~ 1 with respect to other different approaches that used very large mass installations to compensate for the much lower values for those quantities. In particular, *NNN* decays into invisible channels have been investigated here for the first time. The obtained results are summarized in table 3. All the limits achieved here are valid for every invisible decay channel, including disappearance in extra-dimensions or decay into particles which weakly interact with matter.

Further data taking is foreseen.

Table 3. Experimental limits (90% C.L.) obtained by DAMA/LXe on the lifetime values of N, NN and NNN decays into invisible channels in the ^{136}Xe isotope.

Decay	τ_{limit} years (90% C.L.)
n	$3.3 \cdot 10^{23}$
p	$4.5 \cdot 10^{23}$
nn	–
np	$3.2 \cdot 10^{23}$
pp	$1.9 \cdot 10^{24}$
nnn	–
nnp	$1.4 \cdot 10^{22}$
npp	$2.7 \cdot 10^{22}$
ppp	$3.6 \cdot 10^{22}$

The authors thank F. Vissani for useful discussions.

References

1. P. Belli *et al.*, Nuovo Cimento C **19**, 537 (1996).
2. P. Belli *et al.*, Astropart. Phys. **5**, 217 (1996).
3. P. Belli *et al.*, Phys. Lett. B **387**, 222 (1996); **389**, 783 (1996)(E).
4. R. Bernabei *et al.*, New J. Phys. **2**, 15.1 (2000).
5. R. Bernabei *et al.*, EPJdirect C **11**, 1 (2001).
6. R. Bernabei *et al.*, Phys. Lett. B **436**, 379 (1998).
7. R. Bernabei *et al.*, in *Beyond the Desert 2003* (Springer, 2003) p. 365.
8. R. Bernabei *et al.*, Nucl. Instrum. Methods A **482**, 728 (2002).
9. R. Bernabei *et al.*, Phys. Lett. B **527**, 182 (2002).
10. R. Bernabei *et al.*, Phys. Lett. B **546**, 23 (2002).
11. F. Cappella, PhD Thesis, Università di Roma "Tor Vergata" (2005).
12. P. Belli *et al.*, Nuovo Cimento A **103**, 767 (1990).
13. P. Belli *et al.*, Phys. Rev. D **61**, 117301 (2000).
14. P. Belli *et al.*, Phys. Lett. B **465**, 315 (1999).
15. R. Bernabei *et al.*, Phys. Lett. B **493**, 12 (2000).
16. R. Bernabei *et al.*, in *Technique and Application of Xenon Detectors* (World Scientific Publ., 2002) p. 50.
17. F. Simkovic, P. Domin, A. Faessler, hep-ph/0204278.
18. A. Staudt *et al.*, Europhys. Lett. **13**, 31 (1990).
19. E. Caurier *et al.*, Nucl. Phys. A **654**, 973 (1999).
20. G. Feinberg, M. Goldhaber, Proc. Natl. Acad. Sci. U.S.A. **45**, 1301 (1959).

21. J.N. Bahcall, Rev. Mod. Phys. **50**, 881 (1978); *Neutrino Astrophysics* (Cambridge University Press, 1989) p. 359.
22. E.C.G. Stuckelberg, Helv. Phys. Acta **11**, 225 (1938).
23. E.P. Wigner, Proc. Am. Philos. Soc. **93**, 521 (1949).
24. P. Langacker, Phys. Rep. **71**, 185 (1981).
25. C.E. Carlson, C.D. Carone, Phys. Lett. B **512**, 121 (2001).
26. R.N. Mohapatra *et al.*, Phys. Lett. B **491**, 143 (2000).
27. K.S. Babu *et al.*, Phys. Lett. B **570**, 32 (2003).
28. F.J. Yndurain, Phys. Lett. B **256**, 15 (1991); N. Arkani-Hamed *et al.*, Phys. Lett. B **429**, 263 (1998); N. Arkani-Hamed *et al.*, Phys. Today **55**, February issue, 35 (2002).
29. S.L. Dubovsky *et al.*, Phys. Rev. D **62**, 105011 (2000); S.L. Dubovsky *et al.*, JHEP **08**, 041 (2000); V.A. Rubakov, Phys. Usp. **44**, 871 (2001).
30. S.L. Dubovsky, JHEP **01**, 012 (2002).
31. R.B. Firestone, V.S. Shirley *et al.* (Editors), *Table of Isotopes*, 8th ed. (John Wiley & Sons, New York, 1996).
32. C. Berger *et al.*, Phys. Lett. B **269**, 227 (1991).
33. W.R. Nelson *et al.*, SLAC-Report-265, Stanford, 1985.
34. S. Rab, Nucl. Data Sheets **75**, 491 (1995).
35. Yu.V. Sergeenkov, Nucl. Data Sheets **71**, 557 (1994).
36. Yu.V. Sergeenkov *et al.*, Nucl. Data Sheets **84**, 115 (1998).
37. G.J. Feldman, R.D. Cousins, Phys. Rev. D **57**, 3873 (1998).

Eur. Phys. J. A **27**, s01, 43–48 (2006)
DOI: 10.1140/epja/i2006-08-005-x

EPJ A direct
electronic only

On the importance of low-energy beta beams for supernova neutrino physics

N. Jachowicz[1,a] and G.C. McLaughlin[2,b]

[1] Department of Subatomic and Radiation Physics, Ghent University, Proeftuinstraat 86, B-9000 Gent, Belgium
[2] Department of Physics, North Carolina State University, Raleigh, NC 27695-8202, USA

Received: 17 June 2005 /
Published online: 21 February 2006 –

Abstract. We show that low-energy beta beams are very well suited to obtain information about supernova-neutrino interactions. Linear combinations of low-energy beta-beam spectra are fitted to supernova-neutrino energy distributions. The resulting synthetic spectra are able to reproduce the response of a nuclear target to an incoming supernova neutrino flux in a very accurate way. This can provide important information about the neutrino response in a terrestrial detector. We illustrate this technique using deuterium and ^{16}O as target material. The procedure provides an easy and straightforward way to apply the results of a beta-beam neutrino-nucleus measurement to a supernova neutrino detector, virtually eliminating potential uncertainties due to nuclear-structure calculations.

PACS. 25.30.Pt Neutrino scattering – 26.50.+x Nuclear physics aspects of novae, supernovae, and other explosive environments – 97.60.Bw Supernovae – 13.15.+g Neutrino interactions

1 Introduction

The supernova problem is a longstanding one. Since ancient times, literature has reported the observation of suddenly appearing bright new objects in the sky. While the name of the phenomenon reflects the confusion the origin of these events caused, today the general nature of core-collapse supernovae is understood. Notwithstanding the progress that has been made, the modeling of core-collapse supernova events and in particular the generation of successful explosions still seems to face major problems. The neutrinos from supernova 1987A confirmed that weak interactions play an important role in the explosion mechanism. Hence, neutrinos from the next galactic core-collapse supernova are much anticipated, since they are the only particles giving us the chance to see deep into the center of the event and shed light on the processes driving the explosion mechanism.

The terrestrial detection of supernova neutrinos can provide a broad variety of information [1,2]. The arrival times of the neutrinos are related to their mass and can moreover hint at the fate of the star [2,3]. Several reactions provide directional information, important for optical telescopes awaiting the photons in the wake of the supernova neutrino flux. The energy of the neutrinos can be inferred from the energy of the decay products. It indicates the decoupling site of the neutrinos and the temperature there. As mu and tau supernova neutrinos do not have enough energy to produce a massive lepton in a charged-current reaction, the flavor of the arriving neutrinos can be inferred from the frequency differences between the occurrence of neutral and charge-exchange processes. Whether a neutrino or an antineutrino entered the detector can be determined by looking at the charge of the outgoing lepton for electron (anti)neutrinos or by examining the spin of the outgoing nucleon in neutral-current nucleon knockout off nuclei [4,5]. When the signal in the detector is accurately resolved, the observed neutrino energies and flavors can help to disentangle the mixing scheme induced by oscillations [6,7,8].

Nuclei have relatively large cross-sections for neutrino reactions and are energy-sensitive in the range of interest, several particle-emission thresholds opening up with increasing incoming neutrino energies. This makes nuclear targets important as detecting material. Galactic supernova neutrinos could be detected by existing and proposed supernova neutrino detectors such as SNO [9], SuperKamiokande [10], KamLAND [1], LVD [11], MiniBooNe [12], OMNIS [13], LAND [14], and LENA [15]. Favored detection nuclei are ^{12}C, ^{16}O, ^{56}Fe, ^{208}Pb, and deuterium. However, the signal in the detector can only be interpreted as well as the relevant neutrino-nucleus cross-sections are understood. For most nuclei very little experimental neutrino data exists in the relevant energy

[a] e-mail: natalie.jachowicz@ugent.be
[b] e-mail: Gail_McLaughlin@ncsu.edu

Table 1. Percentage of the original ion flux that will be converted to neutrinos entering the target for the detector setup described in [22].

γ	%	γ	%
2	0.1	10	8
3	0.5	15	14
5	2		

region. This is due to the very small cross-sections for weak-interaction processes, and an additional limitation is caused by the fact that monochromatic neutrino beams are not available [16,17]. This has as a consequence that for most applications one has to rely on theoretical predictions, with their related uncertainties and model dependencies [18].

Beta beams, which are neutrino beams produced by the beta decay of nuclei that have been accelerated to high gamma factor, were original proposed for high-energy applications, such as the measurement of the third neutrino mixing angle θ_{13} [19]. Volpe [20,21,22] suggested that a beta beam run at lower gamma factor, would be useful for neutrino measurements in the tens of MeV range.

We suggest exploiting the flexibility these beta-beam facilities offer [23], combined with the fact that beta-beam neutrino energies overlap with supernova neutrino energies, to construct "synthetic" spectra that approximate an incoming supernova neutrino energy distribution. Using these constructed spectra we are able to reproduce total and differential folded supernova neutrino cross-sections very accurately.

2 Synthetic spectra

Traditionally, supernova neutrino energy distributions were parametrized using Fermi-Dirac distributions. The spectra are however not purely thermal, as the decoupling sites of the neutrinos are influenced by their flavor and energy, leading to the use of "effective temperatures" and "effective chemical potentials" in these distributions. Recent calculations showed that descriptions of a supernova neutrino spectrum are provided by a power law distribution [24]:

$$n_{SN[\langle\varepsilon\rangle,\alpha]}(\varepsilon) = \left(\frac{\varepsilon}{\langle\varepsilon\rangle}\right)^{\alpha} e^{-(\alpha+1)\frac{\varepsilon}{\langle\varepsilon\rangle}}, \quad (1)$$

where $\langle\varepsilon\rangle$ and α represent the average energy and the width of the spectrum, respectively. The average neutrino energy $\langle\varepsilon\rangle$ is related to the temperature at the decoupling site, and the effect of α is equivalent to that of the introduction of the effective chemical potential in the Fermi-Dirac distribution. Neutrino-nucleus reaction cross-sections depend on the square of the incoming energy, thus rising very fast with neutrino energies. Hence, the folded cross-sections reach their maximum at much higher energy values than the supernova neutrino energy spectrum does. Typically even neutrinos with energies more than twice

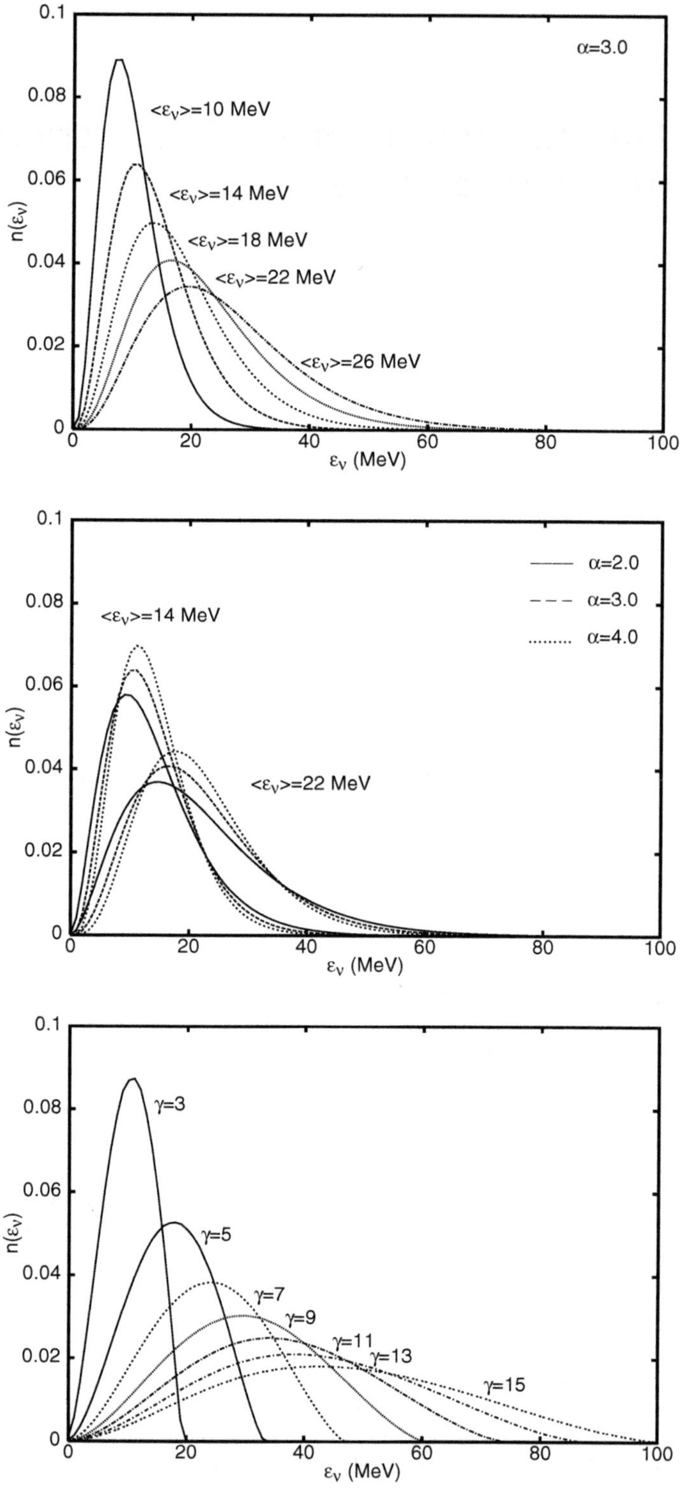

Fig. 1. Comparison between neutrino spectra. The top panel shows the power law parametrization for different values of the average neutrino energy. The width parameter is kept fixed at $\alpha = 3$. The middle panel compares spectra with different widths for average energies of 14 and 22 MeV. The bottom panel shows beta-beam spectra stemming from ^{18}Ne decay for different values of the boost parameter γ.

Table 2. Parameters for the best fit to the supernova neutrino spectrum, defined by average energy $\langle\varepsilon\rangle$ and width α as indicated in the first columns of the table. The left part of the table presents parameterizations $(a^{(3)}_{i=1,2,3}, \gamma^{(3)}_{i=1,2,3})$ for synthetic spectra constructed with three normalized gamma spectra, the right part of the table shows the parametrization $(a^{(5)}_{i=1,\ldots,5}, \gamma^{(5)}_{i=1,\ldots,5})$ for spectra obtained as the linear combination of five normalized gamma spectra stemming from ^{18}Ne decay.

$\langle E\rangle$	α	$a_1^{(3)}$	$\gamma_1^{(3)}$	$a_2^{(3)}$	$\gamma_2^{(3)}$	$a_3^{(3)}$	$\gamma_3^{(3)}$	$a_1^{(5)}$	$\gamma_1^{(5)}$	$a_2^{(5)}$	$\gamma_2^{(5)}$	$a_3^{(5)}$	$\gamma_3^{(5)}$	$a_4^{(5)}$	$\gamma_4^{(5)}$	$a_5^{(5)}$	$\gamma_5^{(5)}$
14	3	0.97	5	0.03	8			0.97	5	0.03	8						
18	2	0.77	5	0.18	8	0.05	11	0.73	5	0.12	7	0.06	8	0.06	9	0.03	12
18	4	0.86	5	0.05	7	0.09	8	0.86	5	0.07	7	0.05	8	0.01	9	0.01	10
22	3	0.77	6	0.19	9	0.04	12	0.56	5	0.16	7	0.15	8	0.11	10	0.02	13

the average energy of the distribution make sizable contributions to the folded cross-section, and integrated cross-sections only converge at energies above 60 MeV [25]. This makes the high-energy tail of the spectra very important for the determination of the nuclear response [26]. Figure 1 illustrates these distributions and compares them to beta-beam spectra at low gammas. The precise shape of the beta-beam spectra depends on the boost factor γ of the primary ion beam and the opening angle of the flux to the target, but is remarkably similar to the supernova neutrino energy distribution. Both classes of distributions are characterized by long tails. The range of the low-energy beta-beam spectra covers the energy region of interest for supernova neutrinos.

We exploit the flexibility offered by beta-beam facilities to construct linear combinations of beta-beam energy distributions,

$$n_{N\gamma}(\varepsilon_i) = \sum_{i=1}^{N} a_i n_{\gamma_i}(\varepsilon_i), \qquad (2)$$

where all distributions involved were normalized to 1:

$$\int \mathrm{d}\varepsilon_i n_{N\gamma}(\varepsilon_i) = 1\,, \qquad (3)$$

and

$$\int \mathrm{d}\varepsilon_i n_{\gamma_i}(\varepsilon_i) = 1, \quad \forall i\,. \qquad (4)$$

The constructed spectrum that represents the best fit to the original supernova spectrum is then obtained by minimizing the expression

$$\int_{\varepsilon_i} \mathrm{d}\varepsilon_i \left| n_{N\gamma}(\varepsilon_i) - n_{SN}(\varepsilon_i)\right|, \qquad (5)$$

where the similarities between beta-beam and supernova neutrino spectra assure that a good fit is easily obtained. In this way, the values for the expansion parameters a_i and the boost factors γ_i that yield a spectrum that is as close as possible to the original power law distribution are determined. Count rate considerations favor higher gammas, for example between γ-values between 5 and 15. Table 1 gives an overview of the count rates that can be expected in a detector 10 m away from a straight section in a beta-beam ring with a length of 90 m, while the target was taken to have an area of 4 m^2.

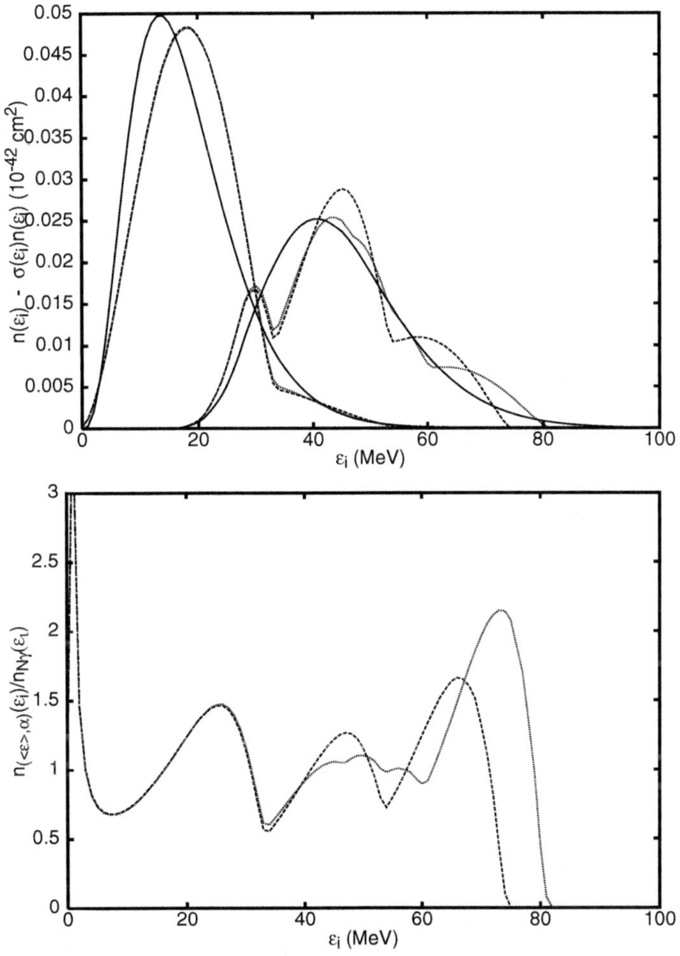

Fig. 2. Comparison between the original and the synthetic spectra for a neutrino energy distribution with average energy $\langle\varepsilon\rangle = 18$ MeV and $\alpha = 3$. In the top panel, the curves to the left represent the spectra, the curves to the right of the plot show total folded cross-sections for neutral-current neutrino scattering off ^{16}O. For each set of curves, the full line represents results obtained with the original spectrum, while the other shows the results for calculations with three (dashed) and five (dotted) beta-beam spectra in the fit. The ^{16}O cross-sections were obtained within a continuum random phase approximation calculation [27,28].

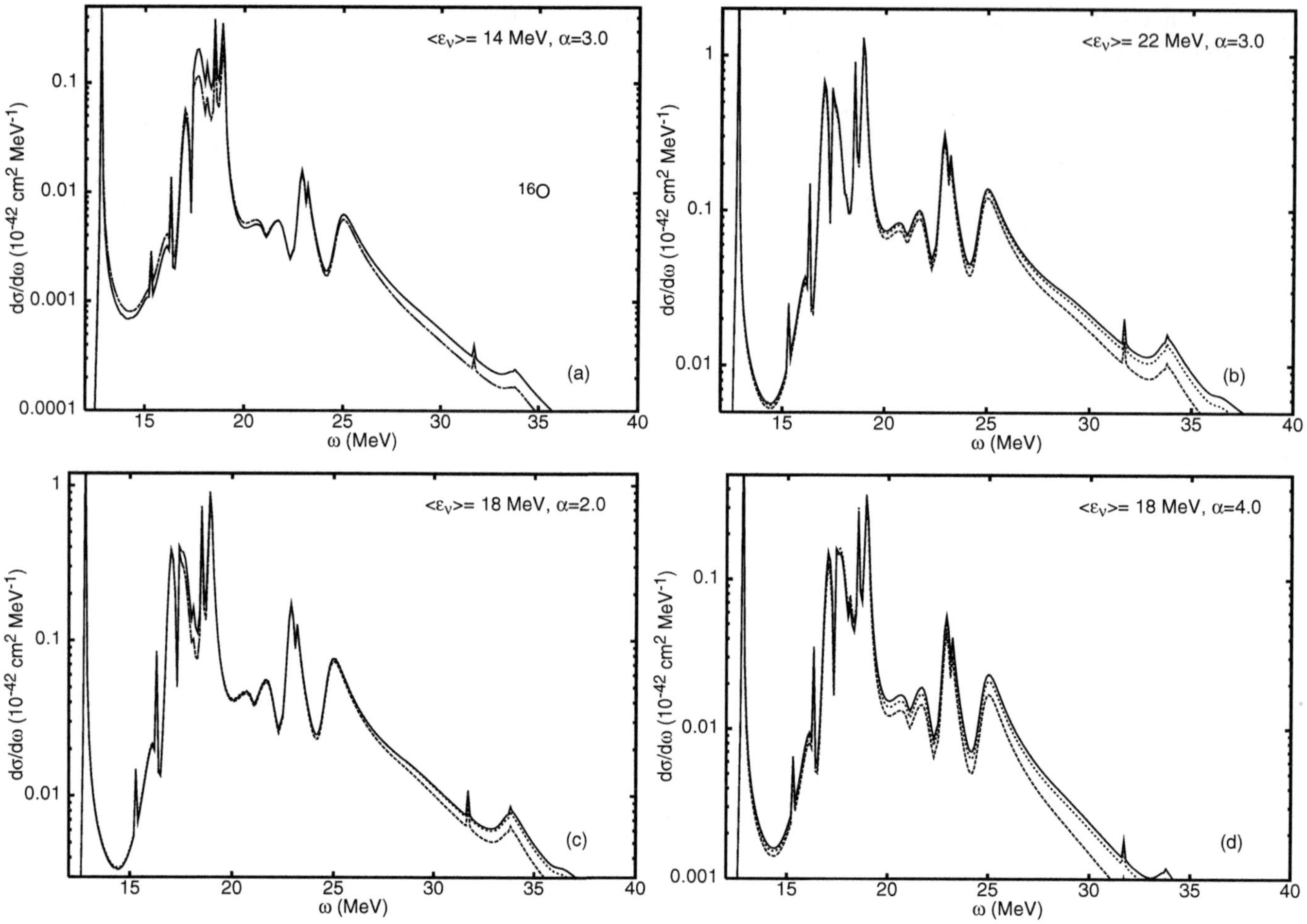

Fig. 3. Comparison between differential cross-sections for neutral-current scattering on ^{16}O, folded with a power law supernova neutrino spectrum (full line) and synthetic spectra with 3 (dashed line) and 5 components (dotted line) for different energy distributions: $\langle\varepsilon\rangle = 14$, $\alpha = 3$ (a), $\langle\varepsilon\rangle = 22$, $\alpha = 3$ (b), $\langle\varepsilon\rangle = 18$, $\alpha = 2$ (c), and $\langle\varepsilon\rangle = 18$, $\alpha = 4$ (d).

Table 2 shows the results of the fitting procedure for a number of relevant energies and widths of the supernova neutrino energy distribution. The parameters are shown for constructions with three and five beta-beam spectra in the fit of eq. (5). Whilst the main contribution stems from spectra at low gamma, contributions from higher gammas are very important for the reproduction of the spectrum's tail. This is illustrated in fig. 2. There we show the original power law spectra, together with the $N = 3$ and $N = 5$ fits, and the respective folded cross-sections for neutral-current scattering off ^{16}O as a function of incident neutrino energy. The figure clearly illustrates that the folded cross-section reaches its main strength for neutrinos in the high-energy part of the distribution. Therefore, it is important that the addition of some spectra at higher γ in the fit assures a good agreement between the constructed and the original spectrum in this energy region. The minimization procedure indeed fulfills this demand: whereas the peak of the fitted spectrum is slightly shifted to higher energies, the agreement between the folded cross-sections is rather satisfying. Including five beta-beam spectra in the construction clearly improves the performances of the minimization procedure at higher energies.

3 Detector response

Of course the spectrum as such is not an important observable. The information brought along by supernova neutrinos is encoded in the response of the detector to the incoming neutrino flux. This quantity is determined by the folded differential cross-section: the folded cross-section as a function of the excitation energy of the target indicates what the neutrino signal in the detector will look like.

In fig. 3, we show the differential cross-section for neutral-current neutrino scattering on an ^{16}O target for different energy distributions. The agreement between cross-sections folded with the power law supernova neutrino spectrum and those folded with the synthetic spectrum is remarkably good. The procedure is able to reproduce total strength, and the position and width of the resonances very accurately.

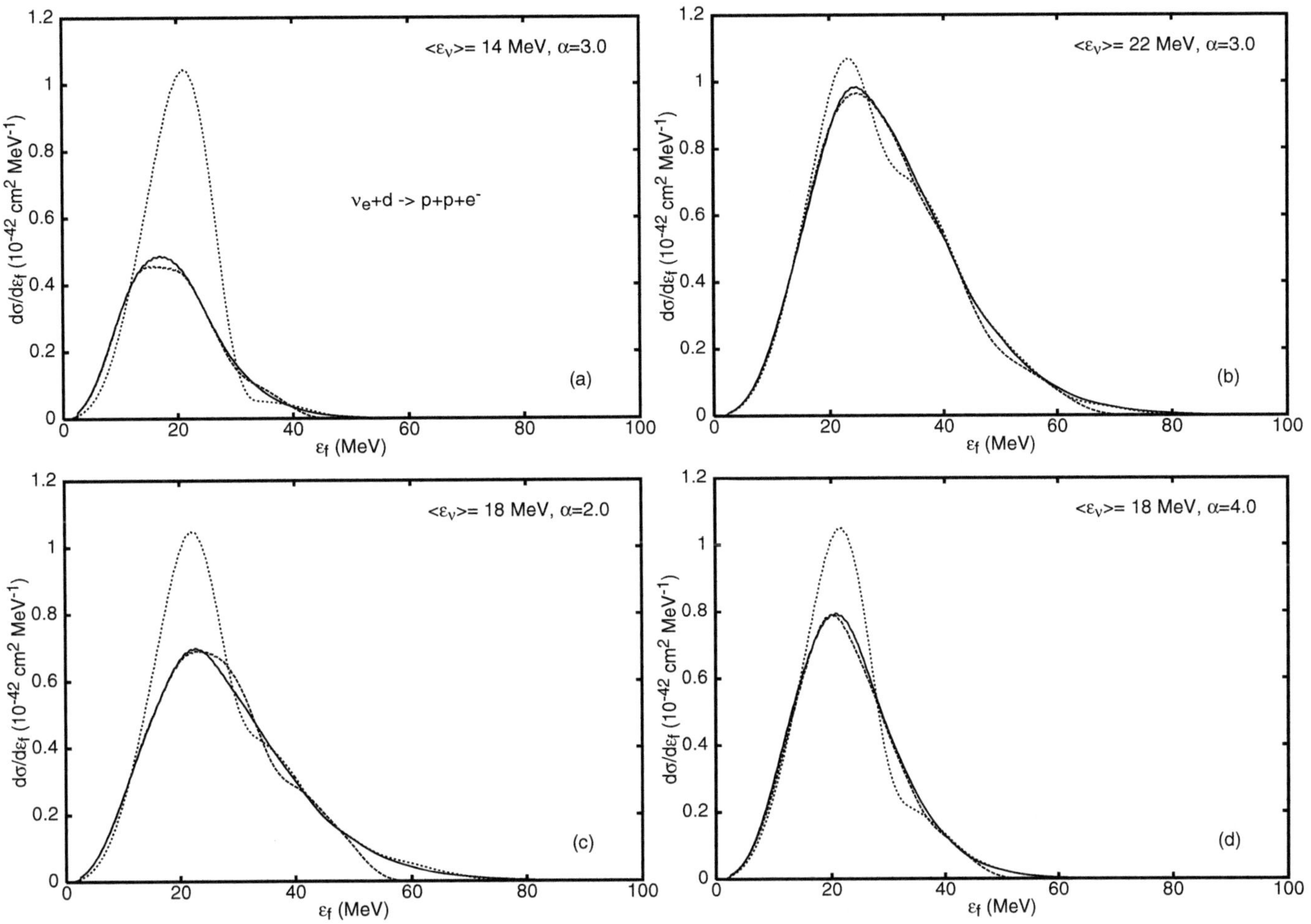

Fig. 4. Comparison between differential cross-sections for the charged-current reaction $\nu_e + d \rightarrow p + p + e^-$, folded with a power law supernova neutrino spectrum (full line) and synthetic spectra with 5 components, using gamma factors ranging from 5 to 15 (dotted line) and including lower energy beta-beam spectra with $\gamma = 2$ to 15 (dashed line), for different energy distributions: $\langle\varepsilon\rangle = 14\,\text{MeV}$, $\alpha = 3$ (a), $\langle\varepsilon\rangle = 22\,\text{MeV}$, $\alpha = 3$ (b), $\langle\varepsilon\rangle = 18\,\text{MeV}$, $\alpha = 2$ (c), and $\langle\varepsilon\rangle = 18\,\text{MeV}$, $\alpha = 4$ (d). The deuteron cross-sections were obtained from [29].

In principle, the formalism can be applied to any target material. Its efficiency, however, depends on the accuracy of the fitting procedure in the relevant energy range. Figure 4 illustrates differential cross-sections for the reaction $\nu_e + d \rightarrow p + p + e^-$ as a function of the energy of the outgoing electron. For these reactions, thresholds are situated at considerably smaller energy values than for ^{16}O and the cross-section peaks at smaller energies, typically around the maximum of the neutrino distribution. As a consequence, supernova spectra with lower average energies require smaller gamma components in the synthetic spectrum to obtain a good agreement. This is shown in the figure. For fits including beta-beam spectra only down to $\gamma = 5$, folded cross-sections at low average energies tend to be overestimated. Including spectra at lower gamma increases the agreement [30]. The introduction of a weight function in the fitting procedure can assure that the fit is optimal in the energy region dominating the nuclear response [30].

4 Conclusion

Low-energy beta beams are efficient tools to learn about the response of a neutrino detector to supernova neutrinos. Taking linear combinations of the detector response to beta-beam spectra provides a very accurate picture of the supernova neutrino signal in a terrestrial detector. This technique would work for almost any target nucleus. We have demonstrated theoretically its potential for deuterium and oxygen which are the nuclei relevant for SNO and for SuperKamiokande.

The authors would like to thank C. Volpe, M. Lindroos, and K. Heyde for interesting discussions. N.J. thanks the Fund for Scientific Research Flanders (FWO) for financial support. G.C.M. acknowledges support from the Department of Energy, under contract DE-FG02-02ER41216.

References

1. P. Vogel, Prog. Part. Nucl. Phys. **48**, 29 (2002); Phys. Rev. C **71**, 034604 (2005).
2. J.F. Beacom, R.N. Boyd, A. Mezzacappa, Phys. Rev. D **63**, 073011 (2001).
3. J.M. Soares, L. Wolfenstein, Phys. Rev. D **40**, 3666 (1989); J.F. Beacom, P. Vogel, Phys. Rev. D **58**, 053010 (1998) and references therein.
4. N. Jachowicz, K. Vantournhout, J. Ryckebusch, K. Heyde, Phys. Rev. Lett. **93**, 082501 (2004).
5. N. Jachowicz, K. Vantournhout, J. Ryckebusch, K. Heyde, Phys. Rev. C **71**, 034604 (2005).
6. G.M. Fuller, W.C. Haxton, G.C. McLaughlin, Phys. Rev. D **59**, 085005 (1999).
7. J. Engel, G.C. McLaughlin, C. Volpe, Phys. Rev. D **67**, 013005 (2003).
8. V. Barger, P. Huber, D. Marfatia, Phys. Lett. B **617**, 167 (2005).
9. The SNO Collaboration (C.J. Clarence), Nucl. Phys. Proc. Suppl. **100**, 326 (2001).
10. Y. Oyama *et al.*, Phys. Rev. Lett. **56**, 2604 (1987).
11. M. Aglietta, P. Antonioli, G. Bari, C. Castagnoli *et al.*, Nucl. Phys. Proc. Suppl. **138**, 115 (2005).
12. M.K. Sharp, J.F. Beacom, J.A. Formaggio, Phys. Rev. D **66**, 013012 (2002).
13. R.N. Boyd, A.St.J. Murphy, Nucl. Phys. A **688**, 386c (2001).
14. C.K. Hargrove *et al.*, Astropart. Phys. **5**, 183 (1996).
15. L. Oberauer, Mod. Phys. Lett. A **19**, 337 (2004).
16. B.E. Bodmann *et al.*, Phys. Lett. B **332**, 251 (1994).
17. `http://www.phy.ornl.gov/nusns`.
18. K. Langanke, G. Martínez-Pinedo, P. von Neumann-Cosel, A. Richter, Phys. Rev. Lett. **93**, 202501 (2004).
19. P. Zuchelli, Phys. Lett. B **532**, 166 (2002).
20. C. Volpe, J. Phys. G **30**, 1 (2004).
21. J. Serreau, C. Volpe, Phys. Rev. C **70**, 055502 (2004).
22. G.C. McLaughlin, Phys. Rev. C **70**, 045804 (2004).
23. `http://beta-beam.web.cern.ch/beta-beam`.
24. M. Keil, G.G. Raffelt, H.-T. Janka, Astrophys. J. **590**, 971 (2003).
25. N. Jachowicz, K. Vantournhout, J. Ryckebusch, K. Heyde, Nucl. Phys. A **758**, 51c (2005).
26. N. Jachowicz, K. Heyde, Phys. Rev. C **68**, 055502 (2003).
27. N. Jachowicz, S. Rombouts, K. Heyde, J. Ryckebusch, Phys. Rev. C **59**, 3246 (1999).
28. N. Jachowicz, K. Heyde, J. Ryckebusch, Phys. Rev. C **66**, 055501 (2002).
29. S. Nakamura *et al.*, Nucl. Phys. A **707**, (2002) 561; private communication.
30. N. Jachowicz, G.C. McLaughlin, submitted to Phys. Rev. Lett..

Eur. Phys. J. A **27**, s01, 49–55 (2006)
DOI: 10.1140/epja/i2006-08-006-9

EPJ A direct
electronic only

Neutrino-nucleon scattering rates in protoneutron stars and nuclear correlations in the spin S = 1 channel

L. Mornas[a]

Departamento de Física, Universidad de Oviedo, Avda Calvo Sotelo 18, 33007 Oviedo, Spain

Received: 20 June 2005 /
Published online: 27 February 2006 –

Abstract. The neutrino-nucleon cross-section is calculated in dense nuclear matter at finite temperature, in view of applications to supernovae and protoneutron stars. The main contribution to this parameter is provided by the axial response function. Nuclear correlations play an important role: while the ν-N cross-section is usually reduced by correlations, a collective mode in the spin $S = 1$ channel may be excited. In that case, the cross-section diverges and the neutrino mean free path would be drastically reduced. The predictions of various models of the nuclear force commonly used in the literature are compared in relation with a possible spin instability. While Skyrme or Gogny forces lead to attraction in this channel, variational or Brueckner-Hartree-Fock calculations predict a repulsive contribution. An interesting alternative appears to be the density-dependent M3Y interaction.

PACS. 26.60.+c Nuclear matter aspects of neutron stars – 25.30.Pt Neutrino scattering – 21.60.Jz Hartree-Fock and random-phase approximations

1 Introduction

With the development of more sensitive neutrino detectors, we would now be able to measure the neutrino burst produced in a supernova explosion with good statistics. At the same time, more powerful computers make it possible to solve numerically the general relativistic collapse together with neutrino transport described by the Boltzmann equations (see, *e.g.*, [1]). The shape of the neutrino burst could thus be predicted —even though this is still obscured in the early stages by the problem of shock stall. Of special interest is the tail of the signal, between 20 and 50 seconds after beginning of the collapse. The outer shells of the supernova should have then been expelled. This corresponds to the deleptonization and first cooling phase of the newly formed protoneutron star. The neutrinos, which were dynamically trapped in the interior by the high density and temperature reached during the collapse, are now gradually released out. During this phase, the neutrino transport can be described by a diffusion equation [2]. The main ingredient of this description, the mean free path, is affected by the state of the matter. Thus, the observation of the tail of the neutrino burst could bring some information on the properties of nuclear matter.

[a] e-mail: lysiane@fisi24.ciencias.uniovi.es

2 Neutrino-nucleon scattering rates

Since the purpose of this work is to investigate the reliability of the available descriptions of the nuclear interactions rather than to make accurate predictions for the neutrino mean free path, a number of simplifications will be made: i) among all possible processes involving neutrinos, only neutrino-nucleon scattering is considered; ii) the interaction potential in the p-h channel will be treated in the Landau-Fermi Liquid approximation. In the non-relativistic approximation, the differential cross-section for ν–N scattering is given as a function of the energy of the scattered neutrino E'_ν and of the scattering angle θ by

$$\frac{1}{V}\frac{\mathrm{d}\sigma}{\mathrm{d}\omega\mathrm{d}\Omega} = \frac{G_F^2}{8\pi^3}(E'_\nu)^2\left[1 - f(E'_\nu)\right] \times \left[(1+\cos\theta)\mathcal{S}^{(0)} + (3-\cos\theta)\mathcal{S}^{(1)}\right].$$

In this limit, the spin-0 and spin-1 responses $\mathcal{S}^{(0)}$, $\mathcal{S}^{(1)}$ correspond to the vector and axial couplings, respectively. The former couples to the density correlations/fluctuations while the latter couples to the spin-density correlator:

$$\begin{bmatrix}\mathcal{S}^{(0)}\\ \mathcal{S}^{(1)}\end{bmatrix} = \begin{bmatrix}c_V^2\mathcal{S}_V(k,\omega)\\ c_A^2\mathcal{S}_A(k,\omega)\end{bmatrix} \sim \int e^{-i\omega t+ik.x}\begin{bmatrix}\langle n(x,t)n(0)\rangle\\ \langle j^i(x,t)j^j(0)\rangle\end{bmatrix}.$$

The main contribution comes from the axial response function, for which reason the neutrino-nucleon mean free path will be specially sensitive to the properties of the nuclear interaction in the spin $S = 1$ channel.

2.1 Medium effects

It has long been known (see, *e.g.*, [3]) that the neutrino-nucleon cross-section is affected by the fact that the nucleons with which it is interacting are not free, but strongly correlated. The medium effects are of various types. A straightforward correction is to treat the nucleons in the mean-field approximation, where they acquire effective masses, momenta and chemical potentials due to their interactions with the other nucleons surrounding them. If one goes further to the random phase approximation (RPA), we are taking into account the screening of the nuclear interaction by particle-hole production. Usually, the RPA correlations are found to reduce the cross-section. If, however, a collective mode is excited, there is a sizable enhancement. For Skyrme parametrizations of the nuclear interaction, for example, a spin instability —which is identified in pure neutron matter with the onset of ferromagnetism— appears at high density. In this case, a pole is present in the cross-section and the mean free path consequently goes to zero [4,5,6,7]. Another effect, the broadening of the width of the nucleon, which is important for the Bremsstrahlung and modified URCA processes, will not be treated in this contribution.

The vector and axial structure functions are proportional to the imaginary part of the corresponding polarizations. In the random phase approximation, the polarizations obey Dyson equations:

$$\Pi^{\mathrm{RPA}}_{V/A} = \Pi_0 + \Pi^{\mathrm{RPA}}_{V/A}\, V_{(S=0/1)}\, \Pi_0 .$$

In asymmetric np matter in β equilibrium, these equations have a 2×2 matrix structure, *e.g.*,

$$\Pi_0 = \begin{bmatrix} \Pi^0_n & 0 \\ 0 & \Pi^0_p \end{bmatrix} \quad \text{(Linhardt function)}.$$

The interaction potential in the p-h channel in the Landau-Fermi Liquid approximation is

$$V_{(S=0)} = \begin{bmatrix} f_{nn} & f_{np} \\ f_{pn} & f_{pp} \end{bmatrix}, \qquad V_{(S=1)} = \begin{bmatrix} g_{nn} & g_{np} \\ g_{pn} & g_{pp} \end{bmatrix}.$$

The Landau parameters f_{ij}, g_{ij} describe the interaction between particles having momenta close to the Fermi momentum. They can be obtained, *e.g.*, from the second functional derivative of the polarized energy functional with respect to the density (or from the first derivative of the single-particle potential). In order to keep things as simple as possible we will work in the monopolar approximation $l = 0$ (no angular or momentum dependence). In this case, the Landau parameters are real functions of the partial densities of neutrons and protons.

2.2 Spin (in-)stability

The criterion for a spin instability to occur is when the determinant of the inverse magnetic susceptibility matrix (χ_{ij}, where $i, j \in \{n, p\}$) vanishes. In terms of the Landau parameters:

$$\mathrm{Det}\left(\frac{1}{\chi_{ij}}\right) = 0 \leftrightarrow \mathrm{Det}\begin{pmatrix} (1+G^{nn}_0) & G^{np}_0 \\ G^{pn}_0 & (1+G^{pp}_0) \end{pmatrix} = 0\,,$$

where $G^{ij}_0 = \sqrt{N^i_0 N^j_0}\, g^{ij}_0$ and $N^i_0 = m^*_i k_{Fi}/\pi^2$.

On the other hand, when solving the RPA equations, we obtain, *e.g.*,

$$\Pi^{\mathrm{RPA}}_{A\,nn} = \frac{(1 - g_{pp}\Pi^0_p)\Pi^0_n}{\mathrm{Det}[I - V_{(S=1)}\,\Pi_0]}\,.$$

In the limit where the temperature T and the energy transfer ω go to zero we find

$$\mathcal{R}e\,\Pi^i_0(\omega, k) \to -N^i_0 \quad \Rightarrow D_A = \mathrm{Det}[I - V_{(S=1)}\,\Pi_0] = 0.$$

A spin instability, therefore, gives rise to a peak in the structure function and in the neutrino cross-section. Such a spin instability has been observed to appear in non-relativistic (*e.g.* Skyrme) [8] as well as in relativistic ($\sigma\omega\rho$) models [9] of the nuclear interaction. It has been argued that it could be related to the exchange (Fock) part of the interaction. On the other hand, such an instability is not seen in recent non-relativistic Brueckner-Hartree-Fock calculations.

3 The effective nuclear interaction in the spin channel

In order to investigate this point more closely, we have tried a large variety of modelizations of the nuclear interaction commonly used to describe, on the one hand the properties of nuclei and of heavy-ion collisions, on the other hand the properties of neutron stars. Ideally, a single model should be able to accommodate all these facts before we can risk to perform extrapolations and make predictions in less accessible situations such as those encountered in protoneutron stars. Among the interactions probed were:

- The Skyrme interaction

$$\begin{aligned} V(r) &= t_0\,(1 + x_0 P_\sigma)\delta(r) \\ &+ \frac{1}{2} t_1\,(1 + x_1 P_\sigma)\left[k'^2\,\delta(r) + \delta(r)\,k^2\right] \\ &+ t_2\,(1 + x_2 P_\sigma) k'\,\delta(r)\,k + \frac{1}{6} t_3\,(1 + x_3 P_\sigma)\,\rho^\alpha_N \delta(r). \end{aligned}$$

The Skyrme interaction leads to a good description of properties of nuclei, nuclear matter and neutron stars. Recent parametrizations adapted to non-symmetric nuclear matter have been developed in the context of neutron stars, like the Skyrme-Lyon series [10] which was adjusted to reproduce microscopical neutron matter calculations. All Skyrme interactions (except the rather old and inaccurate SV parametrization [11]) lead to spin instabilities.

– The Gogny interaction

It has been argued that the reason for this behavior of the Skyrme interaction could be that it is written only as a combination of contact interactions. On the other hand, the Gogny interaction consists of a finite range part, to which a Skyrme-like contact term is added to describe correlations:

$$V(r) = \sum_{i=1,2} \left(t_i^W + t_i^B P_\sigma - t_i^H P_\tau - t_i^M P_\sigma P_\tau\right) \exp -(r/a_i)^2 + \sum_{i=1,2} t_{3i}(1 + x_{3i} P_\sigma)\rho^{\alpha_i}\delta(r).$$

The Gogny interaction yields a good description of the properties of nuclei and nuclear matter. Note that the original set D1S is not suitable for neutron matter, which would be found to be unstable and collapse. An improved parametrization D1P has been developed by Farine *et al.* [12]: it is adjusted to reproduce microscopical neutron matter calculations. It is indeed found [6] that the spin instability, which in the Skyrme models was found to occur around 1.5–3 times the saturation density $n_{\rm sat}$ (for some models, even *below* $n_{\rm sat}$!) is delayed in the D1P Gogny model to high densities, where other phenomena, like hyperon production and the transition to quark gluon plasma, are expected to occur anyway.

– The Modified Seyler-Blanchard interaction

In order to see whether the finite-range character of the interaction is the key to the suppression of spin instability, we next introduced several versions of the Modified Seyler-Blanchard interaction (MSB). This interaction, originally introduced by Myers and Swiatecki [13], consists of a finite-range (Yukawa) term and is momentum and density dependent:

$$V = -C_{ul}\left[1 - \frac{p^2}{b^2} - d^2(\rho_1 + \rho_2)^n\right]\frac{e^{-r/a}}{r/a}$$

with parameters C_{ul} for $\{n\uparrow, n\downarrow, p\uparrow, p\downarrow\}$ It yields a good description of the properties of nuclei, nuclear matter and neutron stars. It leads to spin instabilities at about twice the saturation density, and indeed has been used to describe the consequences of having polarized nuclear matter in the context of neutron stars [14]. The results denoted below by "MSB" will use the same values of the parameters as taken in [14].

– The density-dependent Michigan 3-range interaction

An interesting alternative to the Gogny interaction is the density-dependent Michigan 3-range interaction (M3Y). Whereas the Gogny force use Gaussians in the finite-range part, the M3Y uses three Yukawas with ranges corresponding to Compton lengths of σ, ω and π mesons. The phenomenological versions by Khoa *et al.* [15] or by Nakada [16] are based upon G-matrix elements of the Paris potential. The density dependence is introduced in 2 different ways: by multiplying the whole potential by a density-dependent scale factor $\mathcal{F}(\rho)$ [15], or by adding a density dependent Skyrme-like contact interaction [16]. A good description of the properties of nuclei is thus obtained. This interaction has been used to study neutron-rich nuclei and to describe low-energy nucleus-nucleus collisions [15]. For a study of the application of this model to the description of neutron star properties, see [17].

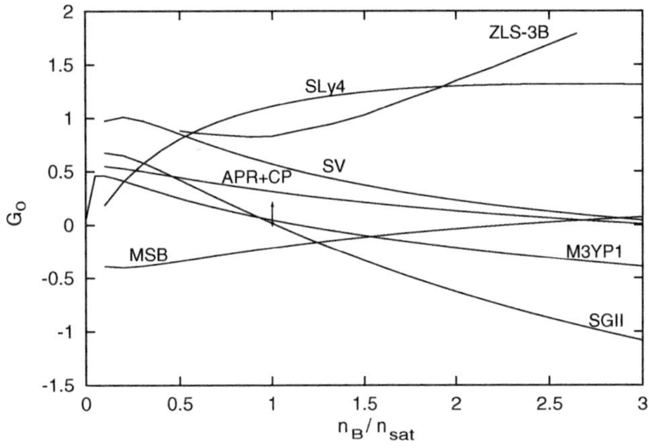

Fig. 1. Density dependence of spin Landau parameter G_0 in symmetric nuclear matter (meaning of the labels: see text).

– Parametrizations of microscopical calculations

We will compare these results to parametrizations of recent microscopical calculations which were explicitly performed in view of investigating the spin stability of nuclear matter and of applications to the neutrino-nucleon cross-section in protoneutron stars. On the one hand, we have the Brueckner-Hartree-Fock calculation of polarized nuclear matter performed by Vidaña and Bombaci [18] and the extraction of Landau parameters in pure neutron matter with application to the neutrino-nucleon cross-section published by Margueron *et al.* [19] (denoted here by the label "MVB"). A similar work was done by Zuo *et al.* and Shen *et al.* [20] in symmetric nuclear matter (denoted here by the label "ZLS"). On the other hand, we used a parametrization of variational calculations by Akmal, Pandharipande and Ravenhall [21] for spin saturated systems, supplemented by an extension by Cowell and Pandharipande [22] for spin-polarized systems (denoted here by the label "APR + CP"). A notable feature of the variational calculation is present around twice the saturation density and interpreted by the authors of the calculation as a transition to a pion condensed phase.

4 Density dependence of the Landau parameters

The Landau parameters have been calculated for each model and their density dependence investigated [17]. A sample of these results is displayed in figs. 1-4.

A first test that the Landau parameters have to pass is to reproduce the experimental constraints in symmetric nuclear matter at saturation density. Such a constraint

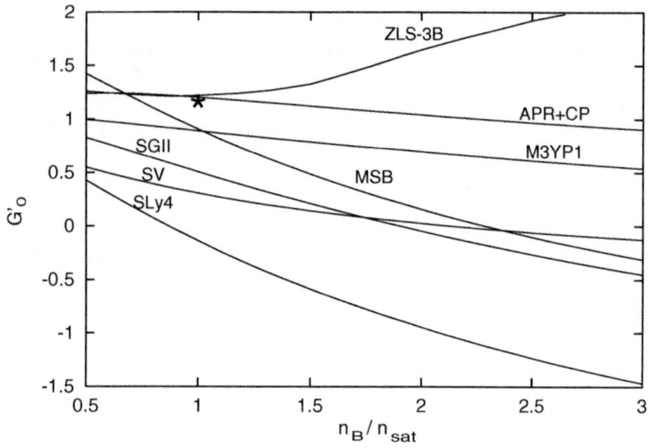

Fig. 2. Density dependence of spin-isospin Landau parameter G_0' in symmetric nuclear matter (labels: see text)

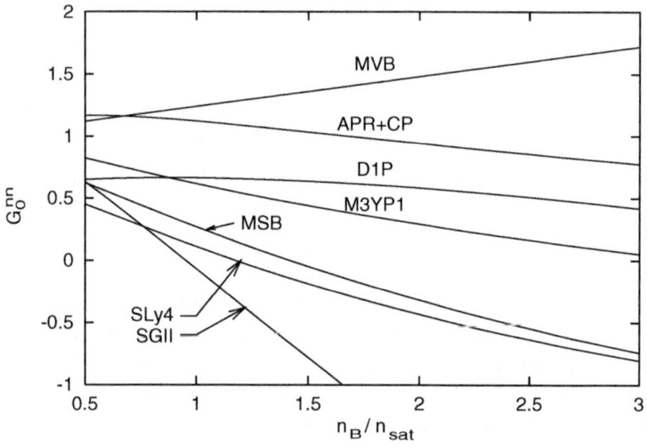

Fig. 3. Density dependence of spin Landau parameter G_0^{nn} in pure neutron matter (labels: see text).

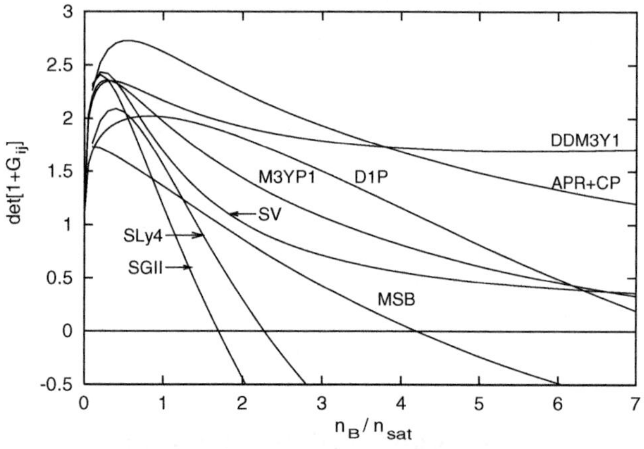

Fig. 4. Criterion for ferromagnetic instability in neutron star matter (labels: see text).

can be derived from the observation of the Gamow-Teller resonance in the spin-isospin channel. Zuo *et al.* [20] quote a value $G_0' \simeq 1.18$ referring to a work by Suzuki [23] (indicated by a star in fig. 2). In an older work by Bäckman *et al.*, [24] a higher value $G_0' \simeq 1.6$–1.7 is quoted. In the spin channel, the absence of a clear resonance sets a constraint on the value of G_0, namely it is compatible with zero within some error bars. From [24] we will take $G_0 = 0.1 \pm 0.1$ (indicated by a double arrow in fig. 1).

– In the spin-isospin channel ("Gamow Teller")

The microscopical calculations (Brueckner-Hartree-Fock from ZLS and variational from APR + CP) reproduce the experimental value at saturation density determined from observation of the Gamow-Teller giant resonance. They coincide for $n_B < n_{\rm sat}$ but differ strongly at high density. The M3Y interactions are also compatible with the experimental value at saturation density. On the other hand, the Skyrme interactions do not generally reproduce the quoted value for G_0'. An exception should be made for the SGII parametrization [25] of the Skyrme force, which was specially adjusted to reproduce the Gamow-Teller resonance. The value of G_0' obtained for this model is nevertheless rather low. A further discussion of the adequacy of the Skyrme force to describe the spin-isospin modes can be found in Bender *et al.* [26].

– In the spin channel ("Fermi")

The parametrization of variational results APR + CP also reproduces the recommended value $G_0 = 0.1 \pm 0.1$. The Landau parameters extracted from BHF calculations in contrast obtain too high values. The Skyrme results are in general too high at saturation density, except, again, the SGII parametrization. The MSB result is compatible with the available data. Finally, the M3Y models are again compatible with the experimental constraint.

– Pure neutron matter

In pure neutron matter there is only one relevant Landau parameter, G_0^{nn}, which is displayed in fig. 3. Its behavior is not very different from that of G_0' in symmetric nuclear matter. Let us note that the SGII parametrization leads to an instability at a low value of the density $\sim 1.6\ n_{\rm sat}$ in pure neutron matter.

– Neutron star matter in β equilibrium

In neutron star matter in β equilibrium there are three different Landau parameters, G_0^{nn}, G_0^{pp} and G_0^{np}. They can be combined to form the criterion for spin instability as explained in sect. 2.2. A sample of the results is displayed in fig. 4. We can see that the Skyrme parametrizations indeed lead to an instability at low density. In particular, this is also the case of the SGII parametrization: even though it was adjusted to give optimal results in symmetric nuclear matter, the stability criterion is here dominated by the interaction between neutrons. The MSB model presents a transition around $4\ n_{\rm sat}$, while the other models are stable up to at least $7\ n_{\rm sat}$.

5 Response functions

Parametrizations which are stable with respect to spin fluctuations give rise to response functions such as those

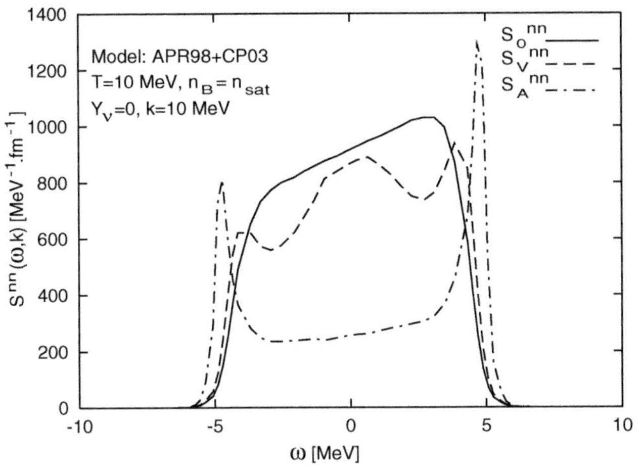

Fig. 5. Response function for the APR + CP model in neutron star matter in β equilibrium.

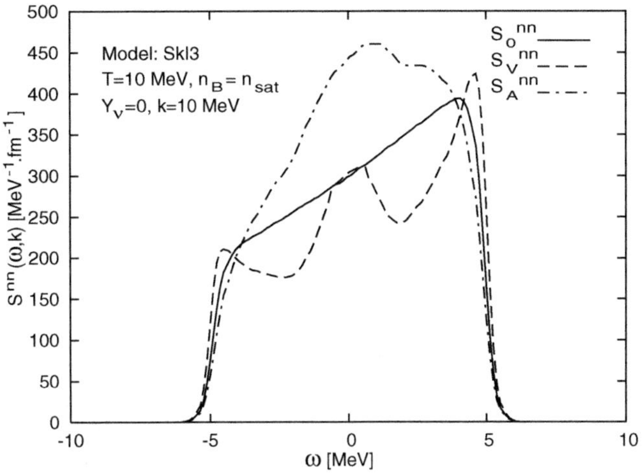

Fig. 7. Response function for the SkI3 Skyrme model in neutron star matter in β equilibrium.

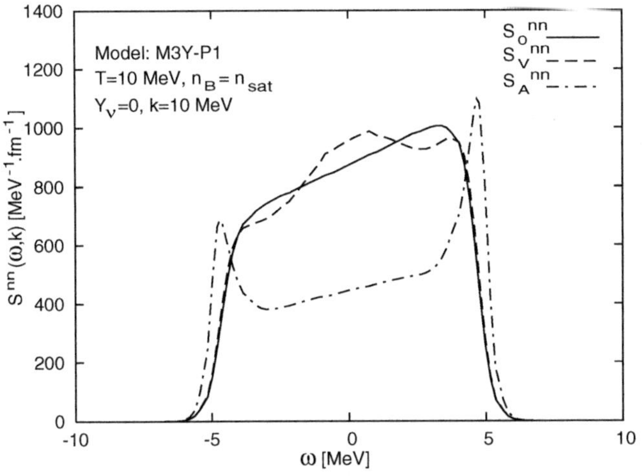

Fig. 6. Response function for the M3YP1 model in neutron star matter in β equilibrium.

displayed in figs. 5 and 6. This includes microscopical calculations (variational or Brueckner), the M3Y models as well as the Gogny parametrization at low density. In these figures, we have represented the contribution of the neutrons to the response function for the APR + CP [21, 22] (fig. 5) and M3YP1 [16] (fig. 6) parametrizations at saturation density, for neutrino-free matter in β equilibrium, at a temperature $T = 10\,\text{MeV}$ and for a momentum transfer $k = 10\,\text{MeV}$. The full line corresponds to the response S_0 in the mean-field approximation. Also represented are the vector S_V (dashed line) and axial S_A (dash-dotted line) response functions in the random phase approximation. It is seen that the RPA response for small energy transfer is reduced with respect to the mean-field approximation, in agreement with previous findings. The vector response displays a zero sound mode and the axial response a spin-zero sound mode, both appearing at the high energy edge of the strength distribution. The contribution of these modes to the neutrino-nucleon cross-section is strongly suppressed by phase-space as well as by Pauli-blocking factors. The global effect will be a reduction of the neutrino-nucleon cross-section with respect to the mean-field approximation.

On the other hand, parametrizations which lead to a spin instability will give rise to response functions such as those displayed in fig. 7. This is the case of the Skyrme or MSB interactions as well as of the Gogny interaction at high density. The model chosen for fig. 7 is the SkI3 Skyrme parametrization, which fulfills the criterion for spin instability for a density $\sim 3.1\ n_{\text{sat}}$, whereas the spin responses are plotted, as in figs. 5-6, at saturation density: even though we are still far from the transition, a precursor of the pole is already to be seen as an enhancement in the axial response function at zero energy $\omega = 0$. At higher densities, the response is completely dominated by the development of the instability.

6 Mean free path

Finally we represent in fig. 8 the RPA correction to the cross-section for various models as a function of density, *i.e.* we plotted the ratio of the total cross-sections (or equivalently the ratio of mean free paths) in the mean-field and random phase approximations $\sigma_{\text{RPA}}/\sigma_{\text{MF}} = \lambda_{\text{MF}}/\lambda_{\text{RPA}}$.

Again we see that Skyrme models (all except SV) or Modified Seyler-Blanchard (MSB) display an enhancement of the cross-section in the RPA with respect to the mean-field calculation and a divergence in the vicinity of spin instability. The Skyrme SV interaction, which is stable against spin excitations, still displays an enhancement (by a factor of 2 at 3 n_{sat})

The microscopical models (BHF or variational) predict a reduction of the cross-section from RPA correlations: at 3 n_{sat}, by a factor of 2 for ZLS, a factor of 3 for APR + CP, by a factor of 4 for MVB. A *caveat* is in order for model APR + CP: The parametrization performed by Cowell and Pandharipande [22] is only available up to $n_B = 0.24\ \text{fm}^{-3}$, *i.e.* below the transition observed

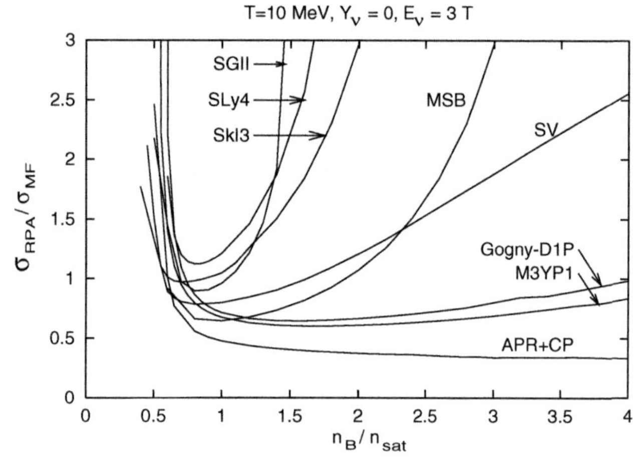

Fig. 8. RPA correction to the cross-section for various models of the nuclear interaction in neutron star matter in β equilibrium (labels: see text).

by Akmal *et al.* [21] around $\sim 2\ n_{\rm sat}$ and interpreted as pion condensation. It will be important to investigate the role of tensor force [27] (which was not taken into account here) in relation with the consequences of this transition on the neutrino mean free path.

The density-dependent Michigan three range models also predict a reduction of the cross-section from RPA corrections, but by a smaller factor (up to 40%). Very similar results are obtained for 8 different parametrizations of the M3Y forces. More detailed results for the M3Y forces will be published in [17].

7 Conclusion and outlook

The object of this contribution has been to investigate whether our knowledge of the nuclear interaction in the spin $S = 1$ is established on a sufficiently firm basis to allow to make predictions for the neutrino-nucleon cross-section in protoneutron star matter, and eventually to predict the shape of the neutrino burst. The answer is, unfortunately, not positive at the present time. This should not in fact come as a surprise. The interaction, which we have characterized here in terms of the Landau parameters, is immediately translatable in terms of "spin asymmetry" $E_{\rm s}$ and "spin-isospin asymmetry" $E_{\rm st}$ energies. A similar uncertainty as to the behavior of the asymmetry energy $E_{\rm t}$ (or, equivalently, of the Landau parameter F_0') as a function of density has long been the subject of intense research. Experiments involving neutron-rich nuclei or neutrino-nucleon scattering, in parallel with renewed efforts of groups who perform Brueckner as well as variational many-body calculations in the spin sector, will hopefully help reduce this uncertainty. In the meantime, some care has to be exerted when choosing an interaction to describe the neutrino-nucleon cross-section.

While the microscopical calculations are —up to a certain point— parameter free and more satisfactory from a conceptual point of view, their use is hampered by the heavy numerical work involved, so that a parametrization seems necessary. On the other hand, we have seen that the phenomenological interactions are not well under control. One could of course extract the Landau parameters from the microscopical calculations and parametrize them. The M3Y interactions can be a valuable alternative, since they have the interesting properties of reproducing without special adjustments the spin Landau parameters at saturation, and of being free of spin instabilities. Moreover, the 3-range Yukawa structure allow a more transparent connection to the meson-exchange models of the nucleon-nucleon potential: there have indeed been attempts to parametrize Brueckner calculations in terms of M3Y interactions [28].

This work was supported by the Spanish-European (FICYT/FEDER) grant No. PB02-076.

References

1. T. Janka, Astron. Astrophys. **368**, 527 (2001).
2. J. Pons *et al.*, Astrophys. J. **513**, 780 (1999).
3. N. Iwamoto, C.J. Pethick, Phys. Rev. D **25**, 313 (1982).
4. S. Reddy *et al.*, Phys. Rev. C **59**, 2888 (1999).
5. J. Navarro, E.S. Hernandez, D. Vautherin, Phys. Rev. C **60**, 045801 (1999).
6. J. Margueron, PhD Thesis, Orsay University, France (2001).
7. L. Mornas, Eur. Phys. J. A **23**, 365 (2005).
8. See, *e.g.*: A. Vidaurre, J. Navarro, J. Bernabeu, Astron. Astrophys. **135**, 361 (1984); M. Kutschera, W. Wojcik, Phys. Lett. B **325**, 271 (1994); J. Margueron, J. Navarro, Nguyen Van Giai, Phys. Rev. C **66**, 014303 (2002); A.A. Isayev, JETP **77**, 251 (2003); A.A. Isayev, J. Yang, Phys. Rev. C **69**, 025801 (2004).
9. P. Bernardos *et al.*, Phys. Lett. B **356** 175 (1995); T. Maruyama, T. Tatsumi, Nucl. Phys. A **693**, 710 (2001); **721**, 265c (2003).
10. E. Chabanat *et al.*, Nucl. Phys. A **635**, 231 (1998).
11. M. Beiner *et al.*, Nucl. Phys. A **238**, 29 (1975).
12. M. Farine *et al.*, J. Phys. G **25**, 863 (1999).
13. W.D. Myers, W.J. Swiatecki, Ann. Phys. (N.Y.) **55**, 395 (1969); **204**, 401 (1990).
14. V.S. Uma Maheswari *et al.*, Nucl. Phys. A **615**, 516 (1997); V.S. Uma Maheswari, J.N. De, S.K. Samaddar, Phys. Rev. D **57**, 3255 (1998).
15. D.T. Khoa, W. von Oertzen, A.A. Ogloblin, Nucl. Phys. A **602**, 98 (1996).
16. H. Nakada, Phys. Rev. C **68**, 014316 (2003).
17. L. Mornas, in preparation.
18. I. Vidaña, I. Bombaci, Phys. Rev. C **66**, 045801 (2002).
19. J. Margueron, I. Vidaña, I. Bombaci, Phys. Rev. C **68**, 055806 (2003).
20. W. Zuo, C. Shen, N. Van Giai, Phys. Rev. C **67**, 037301 (2003); C. Shen *et al.*, Phys. Rev. C **68**, 055802 (2003).
21. A. Akmal, V.R. Pandharipande, D.G. Ravenhall, Phys. Rev. C **58**, 1804 (1998).
22. S.T. Cowell, V.R. Pandharipande, Phys. Rev. C **67**, 035504 (2003).

23. T. Suzuki, H. Sakai, Phys. Lett. B **455**, 25 (1999); A. Kurasawa, T. Suzuki, N. Van Giai, Nucl. Phys. A **731**, 114 (2004).
24. S.O. Bäckmann, G.E. Brown, J.A. Niskanen, Phys. Rep. **124**, 1 (1985).
25. N. Van Giai, H. Sagawa, Phys. Lett. B **106**, 379 (1981).
26. M. Bender *et al.*, Phys. Rev. C **65**, 054322 (2002).
27. E. Olsson, C.J. Pethick, P. Haensel, Phys. Rev. C **70** 025804 (2004).
28. F. Hofmann, H. Lenske, Phys. Rev. C **57**, 2281 (1998); G. Bartnitzky *et al.*, Phys. Lett. B **386**, 7 (1996).

Eur. Phys. J. A **27**, s01, 57–62 (2006)
DOI: 10.1140/epja/i2006-08-007-8

EPJ A direct
electronic only

From DAMA/NaI to DAMA/LIBRA at LNGS

R. Bernabei[1], P. Belli[1], F. Montecchia[1], F. Nozzoli[1,a], F. Cappella[2], A. d'Angelo[2,5], A. Incicchitti[2], D. Prosperi[2], R. Cerulli[3], C.J. Dai[4], H.L. He[4], H.H. Kuang[4], J.M. Ma[4], and Z.P. Ye[4]

[1] Dipartimento di Fisica, Università di Roma "Tor Vergata" and INFN, Sezione di Roma II, I-00133, Roma, Italy
[2] Dipartimento di Fisica, Università di Roma "La Sapienza" and INFN, Sezione di Roma, I-00185, Roma, Italy
[3] INFN - Laboratori Nazionali del Gran Sasso, I-67010 Assergi (AQ), Italy
[4] IHEP, Chinese Academy, P.O. Box 918/3, Beijing 100039, PRC
[5] Scuola di Specializzazione in Fisica Sanitaria, Università di Roma "Tor Vergata", I-00133, Roma, Italy

Received: 20 June 2005 /
Published online: 22 February 2006 –

Abstract. DAMA is an observatory for rare processes based on the development and use of various kinds of radiopure scintillators; it is operative deep underground at the Gran Sasso National Laboratory of the INFN. Several low background setups have been realized and many rare processes have been investigated. In particular, the DAMA/NaI setup ($\simeq$ 100 kg highly radiopure NaI(Tl)) has effectively investigated the model-independent annual modulation signature; the data of seven annual cycles (total exposure of 107731 kg $\times$ d) have offered a 6.3σ C.L. model-independent evidence for the presence of a Dark Matter particle component in the galactic halo. Some of the many possible corollary model-dependent quests for the candidate particle have also been investigated. At present, the second-generation DAMA/LIBRA setup ($\simeq$ 250 kg highly radiopure NaI(Tl)) is in operation deep underground. This paper summarizes the main aspects.

PACS. 95.35.+d Dark matter (stellar, interstellar, galactic, and cosmological) – 29.40.Mc Scintillation detectors

1 Introduction

DAMA is an observatory for rare processes based on the development and use of various kinds of radiopure scintillators. Several low background setups have been realized; the main ones are: i) DAMA/NaI ($\simeq$ 100 kg of highly radiopure NaI(Tl)), which took data underground over seven annual cycles and was put out of operation in July 2002 [1,2,3,4,5,6,7,8,9,10,11,12,13,14]; ii) DAMA/LXe ($\simeq$ 6.5 kg liquid xenon) [15]; iii) DAMA/R&D, which is devoted to tests on prototypes and small-scale experiments [16]; iv) the new second-generation DAMA/LIBRA setup ($\simeq$ 250 kg highly radiopure NaI(Tl)) in operation since March 2003. Moreover, in the framework of devoted R&D for radiopure detectors and photomultipliers, sample measurements are regularly carried out by means of the low background DAMA/Ge detector, installed deep underground for more than 10 years and, in some cases, by means of Ispra facilities.

In the following, we will focus our attention only on the DAMA/NaI results, mainly recalling the investigation of the annual modulation signature due to a Dark Matter (DM) particle component in the galactic halo. At present, apart DAMA/LIBRA, no other experiment is sensitive, for mass and stability, to such a signature. The model-independent annual modulation signature (originally suggested in [17]) is very distinctive since it requires the simultaneous satisfaction of all the following requirements: the rate must contain a component modulated according to a cosine function (1) with one year period, T (2) and a phase, t_0, that peaks around $\simeq$ 2nd June (3); this modulation must only be found in a well-defined low-energy range, where the Dark Matter particle can induce signals (4); it must apply to those events in which just one detector of many actually "fires" (*single-hit* events), since the DM particle multi-scattering probability is negligible (5); the modulation amplitude in the region of maximal sensitivity is expected to be $\sim 7\%$ for usually adopted dark halo distributions (6), but it can be larger in case of some possible scenarios such as, *e.g.*, those in refs. [18, 19]. To mimic such a signature spurious effects or side reactions should be able both to account for the whole observed modulation amplitude and to contemporaneously satisfy all the requirements; none has been found [2,3].

Some of the main topics related to this argument on the DAMA/NaI results are shortly summarized in the following; for a detailed discussion see refs. [2,3] and references therein. For a description of the setup and its performances see refs. [1,2,3,10].

[a] e-mail: `Francesco.Nozzoli@roma2.infn.it`

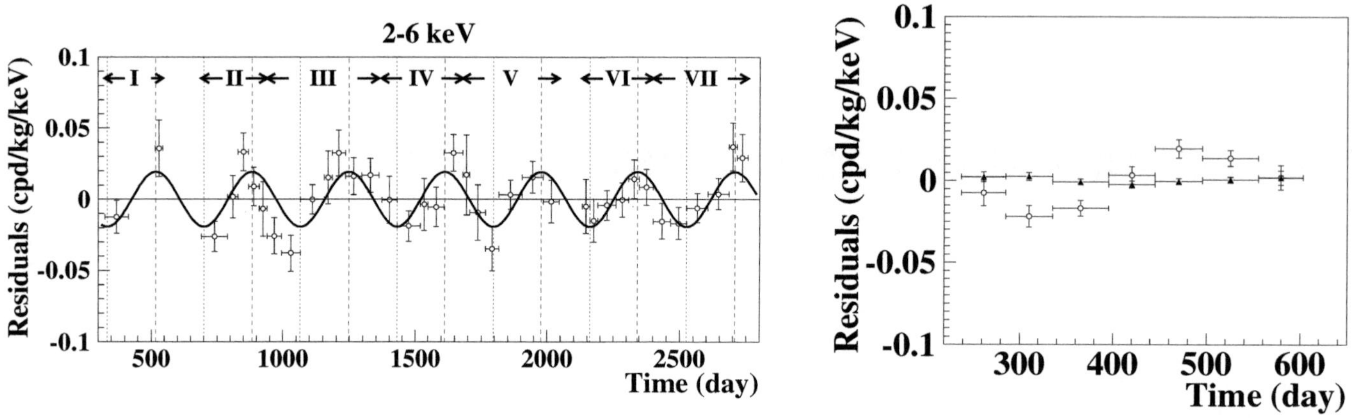

Fig. 1. On the left: experimental residual rate for *single-hit* events in the cumulative (2–6) keV energy interval as a function of the time over 7 annual cycles (total exposure 107731 kg × d); end of data taking, July 2002. The experimental points show the errors as vertical bars and the associated time bin width as horizontal bars. The superimposed curve represents the cosinusoidal function behaviour expected for a Dark Matter particle signal with a period equal to 1 y and phase exactly on 2nd June; the modulation amplitude has been obtained by best fit. On the right: experimental residual rates over seven annual cycles for *single-hit* events (open circles) —class of events to which Dark Matter particle events belong— and over the last two annual cycles for *multiple-hits* events (filled triangles) —class of events to which Dark Matter particle events do not belong— in the (2–6) keV cumulative energy interval. They have been obtained by considering for each class of events the data as collected in a single annual cycle and using in both cases the same identical hardware and the same identical software procedures. The initial time is taken on August 7th [2,3].

2 The model-independent result of DAMA/NaI

A model-independent approach on the data collected by DAMA/NaI over seven annual cycles offers an immediate evidence of the presence of an annual modulation of the measured rate of the *single-hit* events in the lowest-energy region. In particular, in fig. 1, left, the time behaviour of the residual rate of the *single-hit* events in the cumulative (2–6) keV energy interval is reported. The data favour the presence of a modulated cosine-like behaviour ($A \cdot \cos\omega(t-t_0)$) at 6.3σ C.L. and their fit for this cumulative energy interval offers modulation amplitude equal to (0.0200 ± 0.0032) cpd/kg/keV, $t_0 = (140 \pm 22)$ d and $T = \frac{2\pi}{\omega} = (1.00 \pm 0.01)$ y, all parameters kept free in the fit. The period and phase agree with those expected in the case of an effect induced by Dark Matter particles in the galactic halo ($T = 1$ y and t_0 roughly at $\simeq 152.5$th day of the year). The χ^2 test on the (2–6) keV residual rate disfavours the hypothesis of unmodulated behaviour giving a probability of $7 \cdot 10^{-4}$ (χ^2/d.o.f. = 71/37). The same data have also been investigated by a Fourier analysis: a clear peak corresponding to a period of $\simeq 1$ y is present [2,3]. Modulation is not observed above 6 keV[1]. Finally, a suitable statistical analysis has shown that the modulation amplitudes are statistically well distributed in all the crystals, in all the data taking periods and considered energy bins.

[1] We recall that DAMA/NaI took data up to the MeV energy region despite the optimization was done for the keV energy range.

As a further relevant investigation, the *multiple-hits* events collected during the DAMA/NaI-6 and 7 running periods (when each detector was equipped with its own Transient Digitizer with a dedicated renewed electronics) have been studied and analysed by using the same identical hardware and the same identical software procedures as for the case of the *single-hit* events (see fig. 1, right). The *multiple-hits* event class —on the contrary of the *single-hit* one— does not include events induced by Dark Matter particles since the probability that a Dark Matter particle interacts in more than one detector is negligible. The obtained modulation amplitudes are: $A = (0.0195 \pm 0.0031)$ cpd/kg/keV and $A = -(3.9 \pm 7.9) \cdot 10^{-4}$ cpd/kg/keV for *single-hit* and *multiple-hits* residual rates, respectively. Thus, evidence of annual modulation is present in the *single-hit* residuals (event class to which the Dark Matter particle induced events belong), while it is absent in the *multiple-hits* residual rate (event class to which only background events belong). Since the same identical hardware and the same identical software procedures have been used for the two classes of events, the obtained result offers an additional strong support for the presence of Dark Matter particles in the galactic halo further excluding any side effect either from hardware or from software procedures or from background.

Moreover, a careful investigation of all the known possible sources of systematics and side reactions has been regularly carried out and published at time of each data release and quantitative discussions can be found in refs. [2, 3,10]. No systematic effect or side reaction able to account for the observed modulation amplitude and to satisfy all the requirements of the signature has been found. Thus, very cautious upper limits (90% C.L.) on the possible con-

Table 1. Summary of the results obtained by investigating the possible sources of systematics or of side reactions [2,3]. No systematics or side reaction has been found able to give a modulation amplitude different from zero; thus very cautious upper limits (90% C.L.) have been calculated and are shown here in terms of the measured model-independent modulation amplitude, S_m^{obs}. As can be seen, they cannot account for the measured modulation amplitude and, in addition, cannot satisfy the peculiar requirements of the signature.

Source	Cautious upper limit (90%C.L.)
Radon	$< 0.2\% S_m^{obs}$
Temperature	$< 0.5\% S_m^{obs}$
Noise	$< 1\% S_m^{obs}$
Energy scale	$< 1\% S_m^{obs}$
Efficiencies	$< 1\% S_m^{obs}$
Background	$< 0.5\% S_m^{obs}$
Side reactions	$< 0.3\% S_m^{obs}$
In addition: no effect can mimic the signature	

tributions to the modulated amplitude have been calculated as summarized in table 1; for detailed and quantitative discussions see refs. [2,3,10].

In conclusion, the presence of an annual modulation in the residual rate of the *single-hit* events in the lowest-energy interval (2–6) keV, satisfying all the features expected for a Dark Matter particle component in the galactic halo is supported by the data of the seven annual cycles at 6.3σ C.L. No systematic effect or side reaction able to account for the observed effect has been found. This is the experimental result of DAMA/NaI; it is model-independent. No other experiment, whose result can be directly compared with this one in a model-independent way, is available so far in the field of Dark Matter investigation.

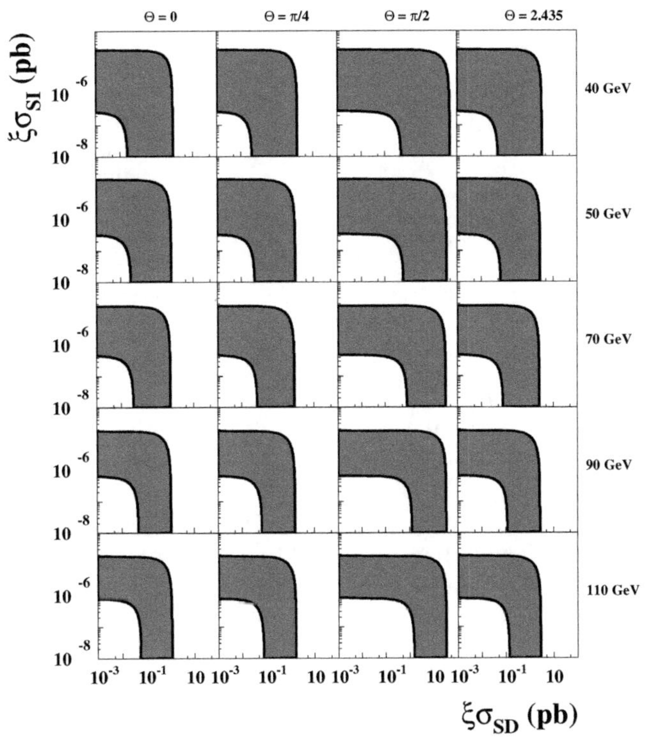

Fig. 2. Case of a Dark Matter particle with mixed Spin-Independent and Spin-Dependent interaction for the model frameworks given in refs. [2,3]. Coloured areas: example of slices (of the 4-dimensional allowed volume) in the plane $\xi\sigma_{SI}$ *vs.* $\xi\sigma_{SD}$ for some of the possible m_W and θ values. Here, ξ is the fractional amount of local density of Dark Matter particles, σ_{SI} and σ_{SD} are the point-like Spin-Independent and Spin-Dependent DM particle-nucleon cross-sections and θ, defined in the $[0, \pi)$ interval, is an angle whose tangent is the ratio between the effective DM particle-nucleon coupling strengths for the Spin-Dependent interaction.

3 Some corollary model-dependent quests for a candidate

On the basis of the obtained 6.3σ model-independent result, corollary investigations can also be pursued on the nature of the Dark Matter particle candidate. This latter investigation is instead model-dependent and —considering the large uncertainties which exist on the astrophysical, nuclear and particle physics assumptions and on the parameters needed in the calculations— has no general meaning (as is also the case of exclusion plots and of the Dark Matter particle parameters evaluated in indirect detection experiments). Thus, it should be handled in the most general way as we have pointed out with time passing [6,7,8,9,10,11,12,13,2,3].

Candidates, kinds of Dark Matter particle couplings with ordinary matter and implications, cross-sections, nuclear form factors, spin factors, scaling laws, halo models, etc., are discussed to some extent in refs. [2,3]. The reader can find in this latter paper and in references therein devoted discussions to correctly understand the results obtained in corollary quests and the real validity of any claimed model-dependent comparison in the field. Here, we just remind that the results briefly summarized here are not exhaustive of the many scenarios possible at the present level of knowledge, including those depicted in some more recent works such as, *e.g.*, refs. [19,20].

DAMA/NaI is intrinsically sensitive both to low and high Dark Matter particle mass having both a light (the ^{23}Na) and a heavy (the ^{127}I) target nucleus; in previous corollary quests for the candidate, Dark Matter particle masses above 30 GeV (25 GeV in ref. [6]) have been presented [7,9,11,12,13] for some model frameworks. Here, the present model-dependent lower bound quoted by LEP for the neutralino in the supersymmetric schemes based on GUT assumptions (37 GeV [21]) is just marked in fig. 3. This model-dependent LEP limit —when considered— selects the Dark Matter particle-iodine elastic scatterings as dominant. In any case also other scenarios have been considered[2].

[2] In fact, when the assumption on the gaugino mass unification at GUT scale is released, neutralino masses down to $\simeq$ 6 GeV are allowed [22,23].

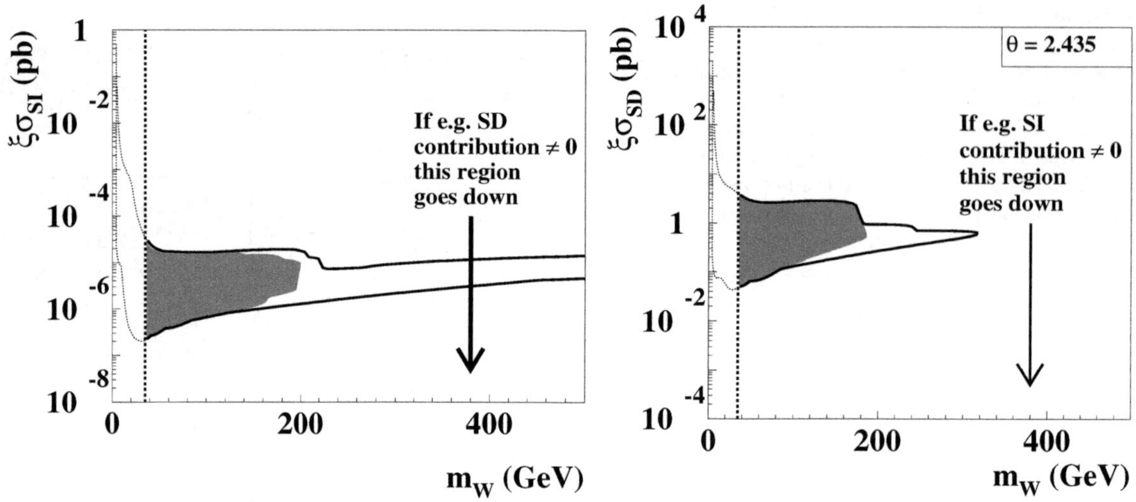

Fig. 3. On the left: case of a Dark Matter particle with dominant Spin-Independent interaction for the model frameworks given in refs. [2,3]. Region allowed in the plane (m_W, $\xi\sigma_{SI}$). The vertical dotted line represents a bound in case of a neutralino candidate when supersymmetric schemes based on GUT assumptions are adopted to analyse the LEP data; the low mass region is allowed for the neutralino when other schemes are considered (see text) and for every other Dark Matter particle candidate. While the area at Dark Matter particle masses above 200 GeV is allowed only for few configurations, the lower one is allowed by most configurations (the colored region gathers only those above the vertical line). Values of $\xi\sigma_{SI}$ lower than those corresponding to this allowed region are possible also, *e.g.*, in case of an even small Spin-Dependent contribution. On the right: case of a Dark Matter particle with dominant Spin-Dependent interaction in the model frameworks given in refs. [2,3]. Example of a slice (of the 3-dimensional allowed volume) in the plane (m_W, $\xi\sigma_{SD}$) at a given θ value (θ is defined in the $[0,\pi)$ range); here $\theta = 2.435$ (Z_0 coupling). Values of $\xi\sigma_{SD}$ lower than those corresponding to this allowed region are possible also, *e.g.*, in case of an even smaller Spin-Independent contribution.

For simplicity, here the results of these corollary quests for a candidate particle are presented in terms of allowed volumes/regions obtained as superposition of the configurations corresponding to likelihood function values *distant* more than 4σ from the null hypothesis (absence of modulation) in each of the possible model frameworks considered in refs. [2,3]. These allowed regions take into account the time and energy behaviours of the single-hit experimental data and have been obtained by a maximum-likelihood procedure which requires the agreement i) of the expectations for the modulated part of the signal with the measured modulated behaviour for each detector and for each energy bin; ii) of the expectations for the unmodulated component of the signal with the respect to the measured differential energy distribution and —since ref. [9] — also with the bound on recoils obtained by pulse shape discrimination from the devoted DAMA/NaI-0 data [4]. The latter one acts in the likelihood procedure as an experimental upper bound on the unmodulated component of the signal and —as a matter of fact— as an experimental lower bound on the estimate of the background levels. Thus, the C.L.'s, which we quote for the allowed volumes/regions, already account for compatibility with the measured differential energy spectrum and with the measured upper bound on recoils. Finally, it is worth noting that the best-fit values of cross-sections and Dark Matter particle mass span over a large range when varying the considered model framework.

Figures 2–4 show some of the obtained allowed slices/regions [2,3]. Here we only recall that tgθ is the ratio between the Dark Matter particle-neutron and the Dark Matter particle-proton effective Spin-Dependent coupling strengths and that θ is defined in the $[0,\pi)$ interval. Obviously, larger sensitivities than those reported in these figures would be reached when including the effect of other existing uncertainties on the astrophysical, nuclear and particle physics assumptions and related parameters; similarly, the set of the best-fit values would also be enlarged as well.

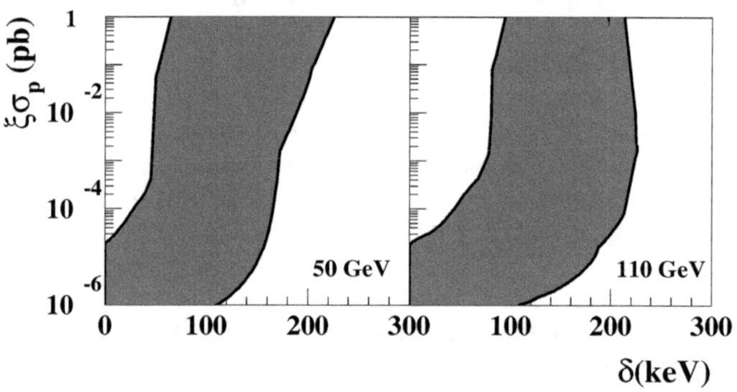

Fig. 4. Case of a Dark Matter particle with *preferred inelastic* interaction in the model frameworks given in refs. [2,3]. Examples of slices (coloured areas) of the 3-dimensional allowed volume ($\xi\sigma_p$, δ, m_W) for some m_W values. Here, δ is the mass splitting of the Dark Matter candidate.

In fig. 5 the theoretical expectations in the purely Spin-Independent coupling for the particular case of a neu-

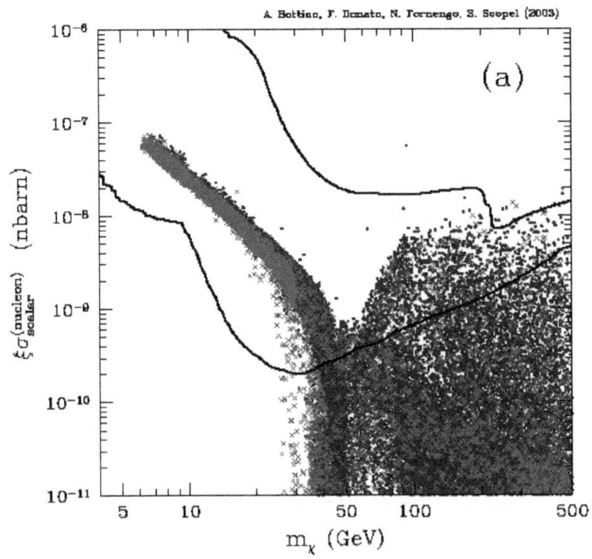

Fig. 5. Figure taken from ref. [23]: theoretical expectations of $\xi\sigma_{SI}$ *vs.* m_W in the purely SI coupling for the particular case of a neutralino candidate in MSSM with gaugino mass unification at GUT scale released; the curve is the same as in fig. 3, left.

tralino candidate in MSSM with gaugino mass unification at GUT scale released [23] are shown. The marked curve surrounds the DAMA/NaI purely SI allowed region as in fig. 3, left.

On the other hand, some positive hints are present in indirect detection experiments; in fact, an excess of positrons and of gammas in the space has been reported with the respect to a modelled background; they are not in contradiction with the DAMA/NaI result. Moreover, recently, it has been suggested [24] that these positive hints and the effect observed by DAMA/NaI can also be described in a scenario with multi-component Dark Matter in the galactic halo, made of a subdominant component of heavy neutrinos of the 4th family and of a sterile dominant component [24].

4 Towards the future: from DAMA/NaI to DAMA/LIBRA and beyond

The large merits of highly radiopure NaI(Tl) setup have been demonstrated in the practice by DAMA/NaI which has been the most radiopure setup available in this particular field. It has effectively pursued a model-independent approach to investigate Dark Matter particles in the galactic halo collecting an exposure several orders of magnitude larger than those available in the field and has obtained many other complementary or by-products results.

In 1996 DAMA proposed to realize a ton setup [25] and a new R&D project for highly radiopure NaI(Tl) detectors was funded at that time and carried out for several years in order to realize as an intermediate step the second-generation experiment, successor of DAMA/NaI, with an exposed mass of about 250 kg.

Fig. 6. The installation of the 25 NaI(Tl) crystals (9.70 kg each) of DAMA/LIBRA in high-purity nitrogen atmosphere. All the procedures as well as the photos have been carried out in high-purity nitrogen atmosphere.

Thus, new powders and other materials have been selected, new chemical/physical radiopurification procedures in NaI and TlI powders have been exploited, new growing/handling protocols have been developed and new prototypes have been built and tested. As a consequence of the results of this second-generation R&D, the new experimental setup DAMA/LIBRA (Large sodium Iodide Bulk for RAre processes), $\simeq$ 250 kg highly radiopure NaI(Tl) crystal scintillators (matrix of twenty-five $\simeq$ 9.70 kg NaI(Tl) crystals), was funded at the end of 1999 and realised. In fact, after the completion of the DAMA/NaI data taking in July 2002, the dismounting of DAMA/NaI occurred and the installation of DAMA/LIBRA started. The experimental site as well as many components of the installation itself have been implemented (environment, shield of the photomultipliers, wiring, high-purity nitrogen system, cooling water of air conditioner, electronics and data acquisition system, etc.). In particular, all the copper parts have been chemically etched before their installation following a new devoted protocol and maintained in high-purity nitrogen atmosphere until the installation. All the procedures performed during the dismounting of DAMA/NaI and the installation of DAMA/LIBRA detectors have been carried out in high-purity nitrogen atmosphere (see fig. 6).

DAMA/LIBRA has taken data since March 2003 and the first data release will, most probably, occur when an exposure larger than that of DAMA/NaI will have been collected and analysed in all the aspects. The highly radiopure DAMA/LIBRA setup is a powerful tool for further investigation on the Dark Matter particle component in the galactic halo having many intrinsic merits [1,2,3, 4,5,6,7,8,9,10,11,12,13,14] and a larger exposed mass, an higher overall radiopurity and improved performances with the respect to DAMA/NaI. Thus, DAMA/LIBRA will further investigate the 6.3σ C.L. model-independent evidence pointed out by DAMA/NaI with increased sensitivity in order to reach even higher C.L. Moreover, it will also offer an increased sensitivity to improve corollary quests on the nature of the candidate particle, trying to disentangle at least some of the many different possible astrophysical, nuclear and particle physics models as well as to investigate other new possible scenarios. As an example, we remind the effects induced on the Dark Matter particles distribution by the contributions from satellite galaxies tidal streams, by the possible existence of caustics and by the possible existence of "solar wakes" [26]. In particular, recently it has been pointed out [19] that contributions to the Dark Matter particles in the galactic halo should be expected from tidal streams from the Sagittarius Dwarf elliptical galaxy. Considering that this galaxy was undiscovered until 1994 and considering galaxy formation theories, one has to expect that also other satellite galaxies do exist and contribute as well. In particular, the Canis Major satellite Galaxy has been pointed out as reported in 2003 in ref. [27]; it can, in principle, play a very significant role being close to our galactic plane. At present, the best way to investigate the presence of a stream contribution is to determine more accurately the phase of the annual modulation, t_0, as a function of the energy; in fact, for a given halo model, t_0 would be expected to be (slightly) different from $\simeq 152.5$ d and to vary with energy.

Moreover, other interesting topics will be addressed by the highly radiopure DAMA/LIBRA, such as the study i) on the velocity and spatial distribution of the Dark Matter particles in the galactic halo [2,13]; ii) on possible structures as clumpiness with small scale size; iii) on the couplings of the Dark Matter particle with the ^{23}Na and ^{127}I target nuclei; iv) on the nature of the Dark Matter particles; v) on scaling laws and cross-sections (recently, it has been pointed out [20] that, even for the neutralino candidate, the usually adopted scaling laws could not hold); etc. A large work will be faced by DAMA/LIBRA, which will also further investigate with higher sensitivity several other rare processes.

Finally, at present a third-generation R&D effort towards a possible NaI(Tl) ton setup has been funded and related works have already been started.

5 Conclusion

DAMA/NaI has been a pioneer experiment investigating as first the Dark Matter particle annual modulation signature with suitable sensitivity and control of the running parameters. During seven independent experiments of one year each, it has pointed out at 6.3σ C.L. in a model-independent way the presence of a modulation satisfying the many peculiarities of an effect induced by Dark Matter particles in the galactic halo; no systematic effect or side reaction able to account for the observed effect has been found. As a corollary result, it has also pointed out the complexity of the quest for a candidate particle mainly because of the present poor knowledge on the many astrophysical, nuclear and particle physics related aspects. At present after a devoted R&D effort, the second-generation DAMA/LIBRA (a $\simeq 250$ kg more radiopure NaI(Tl) setup) has been realised and put in operation since March 2003 and a third-generation R&D toward a possible ton NaI(Tl) setup, which we proposed in 1996 [25], is also in progress.

References

1. R. Bernabei *et al.*, Nuovo Cimento A **112**, 545 (1999).
2. R. Bernabei *et al.*, Riv. Nuovo Cimento **26**, 1 (2003).
3. R. Bernabei *et al.*, Int. J. Mod. Phys. D **13**, 2127 (2004).
4. R. Bernabei *et al.*, Phys. Lett. B **389**, 757 (1996).
5. R. Bernabei *et al.*, Nuovo Cimento A **112**, 1541 (1999).
6. R. Bernabei *et al.*, Phys. Lett. B **424**, 195 (1998).
7. R. Bernabei *et al.*, Phys. Lett. B **450**, 448 (1999).
8. P. Belli *et al.*, Phys. Rev. D **61**, 023512 (2000).
9. R. Bernabei *et al.*, Phys. Lett. B **480**, 23 (2000).
10. R. Bernabei *et al.*, Eur. Phys. J. C **18**, 283 (2000).
11. R. Bernabei *et al.*, Phys. Lett. B **509**, 197 (2001).
12. R. Bernabei *et al.*, Eur. Phys. J. C **23**, 61 (2002).
13. P. Belli *et al.*, Phys. Rev. D **66**, 043503 (2002).
14. R. Bernabei *et al.*, Phys. Lett. B **408**, 439 (1997); P. Belli *et al.*, Phys. Lett. B **460**, 236 (1999); R. Bernabei *et al.*, Phys. Rev. Lett. **83**, 4918 (1999); P. Belli *et al.*, Phys. Rev. C **60**, 065501 (1999); R. Bernabei *et al.*, Phys. Lett. B **515**, 6 (2001); F. Cappella *et al.*, EPJdirect C **14**, 1 (2002); R. Bernabei *et al.*, Eur. Phys. J. A **23**, 7 (2005); R. Bernabei *et al.*, Eur. Phys. J. A **24**, 51 (2005).
15. R. Bernabei *et al.*, Nuovo Cimento A **103**, 767 (1990); Nuovo Cimento C **19**, 537 (1996); Astropart. Phys. **5**, 217 (1996); Phys. Lett. B **387**, 222 (1996); **389**, 783(E) (1996); **436**, 379 (1998); **465**, 315 (1999); **493**, 12 (2000); New J. Phys. **2**, 15.1 (2000); Phys. Rev. D **61**, 117301 (2000); EPJdirect C **11**, 1 (2001); Nucl. Instrum. Methods A **482**, 728 (2002); Phys. Lett. B **527**, 182 (2002); **546**, 23 (2002); in *Beyond the Desert 03* (Springer, 2004) p. 541.
16. R. Bernabei *et al.*, Astropart. Phys. **7**, 73 (1997); R. Bernabei *et al.*, Nuovo Cimento A **110**, 189 (1997); P. Belli *et al.*, Nucl. Phys. B **563**, 97 (1999); P. Belli *et al.*, Astropart. Phys. **10**, 115 (1999); R. Bernabei *et al.*, Nucl. Phys. A **705**, 29 (2002); P. Belli *et al.*, Nucl. Instrum. Methods A **498**, 352 (2003); R. Cerulli *et al.*, Nucl. Instrum. Methods A **525**, 535 (2004).
17. K.A. Drukier *et al.*, Phys. Rev. D **33**, 3495 (1986); K. Freese *et al.*, Phys. Rev. D **37**, 3388 (1988).
18. D. Smith, N. Weiner, Phys. Rev. D **64**, 043502 (2001).
19. K. Freese *et al.*, Phys. Rev. Lett. **92**, 11301 (2004).
20. G. Prezeau *et al.*, Phys. Rev. Lett. **91**, 231301 (2003).
21. K. Hagiwara *et al.*, Phys. Rev. D **66**, 010001 (2002).
22. A. Bottino *et al.*, Phys. Rev. D **67**, 063519 (2003); A. Bottino *et al.*, hep-ph/0304080; D. Hooper, T. Plehn, MADPH-02-1308; CERN-TH/2002-29; hep-ph/0212226; G. Bélanger, F. Boudjema, A. Pukhov, S. Rosier-Lees, hep-ph/0212227.
23. A. Bottino *et al.*, Phys. Rev. D **69**, 037302 (2004).
24. K. Belotsky *et al.*, hep-ph/0411093.
25. R. Bernabei *et al.*, Astropart. Phys. **4**, 45 (1995); R. Bernabei, *Competitiveness of a very low radioactive ton scintillator for particle Dark Matter search*, in *The Identification of Dark Matter* (World Scientific Publishers, 1997) p. 574.
26. F.S. Ling, P. Sikivie, S. Wick, astro-ph/0405231.
27. R.A. Ibata *et al.*, Mon. Not. R. Astron. Soc. **348**, 12 (2004).

Eur. Phys. J. A **27**, s01, 63–66 (2006)
DOI: 10.1140/epja/i2006-08-008-7

EPJ A direct
electronic only

Searching for Majorana neutrinos with double beta decay and with beta beams

L. Lukaszuk, Z. Sujkowski[a], and S. Wycech

The Andrzej Soltan Institute for Nuclear Studies, 05-400 Otwock - Świerk, Poland

Received: 11 July 2005 /
Published online: 14 March 2006 –

Abstract. The leading questions in today neutrino physics concern the transformation properties of neutrinos under charge conjugation, the lepton number conservation hypothesis and the values of neutrino masses. The answers can be obtained if the neutrinoless double beta decay, $0\nu 2\beta$, is observed and its rate is measured. The mere observation of the process proves the Majorana nature of neutrinos as well as the violation of the lepton number conservation, while the rate is a sensitive measure of the neutrino mass. The search for this effect at present concerns mainly the double β^- decay. This paper describes the recent proposal to search for the neutrinoless radiative double electron capture as an attractive alternative to the double β^- decay. It is shown, moreover, that the same information which is expected from the $0\nu 2\beta$ studies can in principle be also obtained from experiments using intense ν_e beams produced by radioactive ions decaying in flight (the "beta beams").

PACS. 13.15.+g Neutrino interactions – 23.40.Bw Weak-interaction and lepton (including neutrino) aspects – 14.60.Pq Neutrino mass and mixing

1 Introduction

The discovery of lepton flavour mixing for neutrinos, the so-called neutrino oscillations, has put the neutrino physics in the lime-light of interest of particle, astro-particle and nuclear physics. Although the number of answered questions is impressive, so is the shopping list of essential queries. Thus we know that there are three and only three kinds of weakly interacting neutrinos, we know that neutrinos are massive, we know the splittings between square masses. This paper considers some possibilities of getting answers to three of the remaining questions: the absolute mass values, the lepton number conservation and the neutrino-antineutrino identity.

The present limits on the mass values stem mainly from three sources: the cosmologic one, giving the sum of the three neutrino masses in a strongly model-dependent way (see, *e.g.*, [1]), the direct measurement [2] of the end point of the ^{3}H β^- spectrum ($m_{\nu_e} \leq 2.2\,\mathrm{eV}$) and the neutrinoless double β decay $0\nu 2\beta$ ($m_{\nu_e} < 0.3$–$1.0\,\mathrm{eV}$, where the range reflects the large nuclear matrix element uncertainty [3,4]; a claim for having observed the effect corresponding to (0.2–0.6) eV neutrino mass value is made in [5]; see [3] for comments). It should be emphasized that the mere observation of this decay will provide unambiguous answers to the next two questions: it will prove the Majorana nature of neutrinos as well as the violation of the lepton number conservation law.

After briefly reviewing the $0\nu 2\beta$ method in general we shall concentrate on the recent suggestion of studying the radiative neutrinoless double electron capture as an attractive alternative to the double β^- emission [6,7]. We shall also outline the possible experiments with beta-beam based neutrino factories. We shall show that the information obtainable from such experiments is at least in principle equivalent to that from the double beta decay.

2 The neutrinoless double beta processes

The double beta decay processes are very slow. They are observable only in cases for which the single $\beta^{\pm}$ or EC decays are energetically forbidden. We may distinguish the following decays:

- $\beta^-\beta^- : (Z,N) \rightarrow (Z+2, N-2) + \beta^- + \beta^- + \overline{\nu_e} + \overline{\nu_e}\,,$
- $\beta^+\beta^+ : (Z,N) \rightarrow (Z-2, N+2) + \beta^+ + \beta^+ + \nu_e + \nu_e\,,$
- $\beta^+\mathrm{EC} : (Z,N) \rightarrow (Z-2, N+2) + \beta^+ + \nu_e + \nu_e\,,$
- $\mathrm{ECEC} : (Z,N) \rightarrow (Z-2, N+2) + \nu_e + \nu_e.$

If neutrino is a massive Majorana particle then the first three processes can proceed without neutrino emission: the neutrino emitted in one weak interaction vertex is absorbed in the other one and only the β^- or β^+ electrons are emitted. These carry out the total energy excess. The double electron capture decay is not possible without

[a] e-mail: `sujkowsk@ipj.gov.pl`

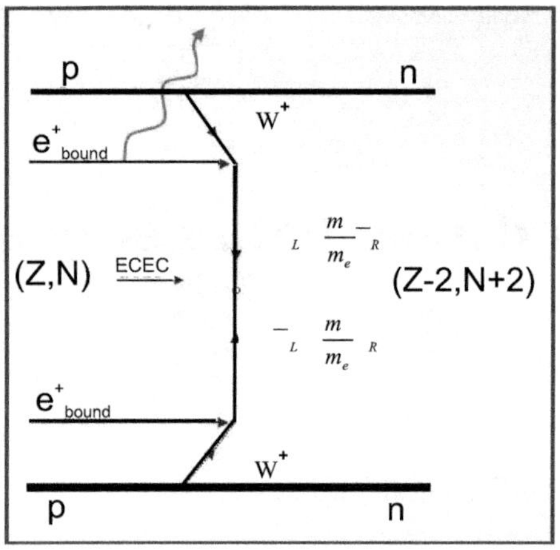

Fig. 1. Diagram for radiative neutrinoless double electron capture.

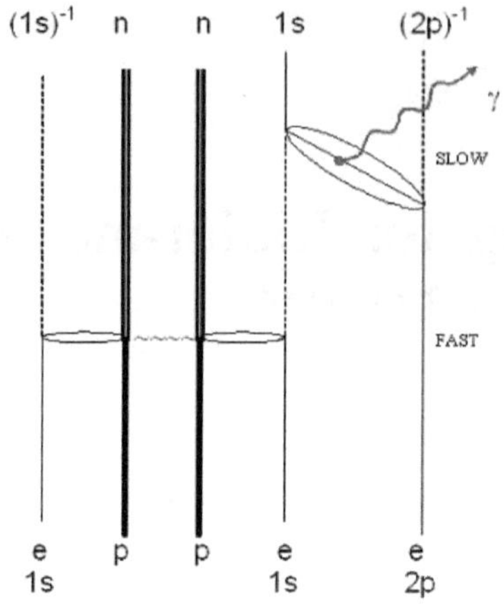

Fig. 2. Diagram for atomic resonance in radiative double electron capture.

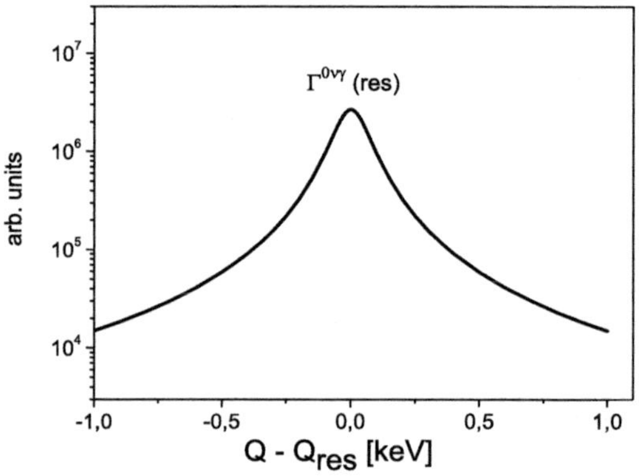

Fig. 3. The rate *vs.* energy dependence of 0νECECγ process in vicinity of the atomic resonance for a high Z atom.

an extra energy carrier. The most obvious one is a single photon radiated by one of the captured electrons. This is the radiative neutrinoless double electron capture, see fig. 1, discussed in some details below.

The value to be determined in any of the $0\nu2\beta$ processes is the rate. This is a sensitive measure of the neutrino mass. To a good approximation the rate can be factorized:

$$\Gamma = G(E,Z)M^2X^2, \tag{1}$$

where $G(E,Z)$ is the phase space factor which can be calculated rather accurately, M is the nuclear matrix element, not so easy to calculate, and X is proportional to the effective neutrino rest mass.

For a detailed discussion of the $0\nu2\beta$ processes, and especially of the best studied $0\nu\beta^-\beta^-$ one, we refer the reader to recent reviews [3,4]. Below, we return to the radiative neutrinoless double electron capture, 0νECECγ. The process is possible due to the admixture of neutrinos of opposite helicity. The amplitudes of these admixtures are proportional to the neutrino rest mass. It should be noted that because of the angular momentum conservation requirement the process precludes the capture of two $1S$ electrons. It is either the $1S+2S$ electrons (the magnetic type transition) or the $1S+2P$ electrons (the electric dipole transition). It is the second case which is of particular interest because of the possible atomic resonance enhancement of the rate.

The theory of radiative single electron capture has been given by Glauber and Martin in the fifties [8]. The continuous photon spectra of this internal bremsstrahlung process exhibit an interesting structure in the case of the $2P$ (or $3P$) electron capture: there is a singularity at the photon energy equal to the $1S$-$2P$ or $1S$-$3P$ atomic level energy difference. This prediction has been confirmed experimentally for several cases (cf., *e.g.*, [9]). The same approach can be applied to the 0νECECγ case. Figure 2 shows the Feynman diagram for such a resonance situation. The process can be visualised as follows: after the second $1S$ electron is virtually captured, the system lives long enough for one of the $1S$ holes to be filled by the $2P$ electron. The resonance condition is fulfilled if the energy of the resulting Kα X-ray transition coincides with the energy available for the internal bremsstrahlung photon. The rate can be expressed [6] as

$$\Gamma^{0\nu\gamma}(Q) = \frac{\Gamma^r(2P \to 2S)}{[Q - Q_{res}(Z-2)^2] + \left[\frac{\Gamma_r}{2}\right]^2} \left|R_{0\nu}\right|^2, \tag{2}$$

where Q is the available bremsstrahlung energy $Q_{res}(Z-2)$ is the Kα X-ray energy in the daughter atom and Γ_r is the natural width of the X-ray transition in the presence of an additional $1S$ hole (the width of a hypersatellite transition, of the order of 100 eV in heavy atoms).

Figure 3 shows the rate of the 0νECECγ process in vicinity of the resonance. The enhancement attainable in heavy emitters may reach in favourable cases several orders of magnitude. This more than compensates the usual retardate of a radiative process. For details of the theory of the 0ν radiation double electron capture we refer to [6,7,10,11]. Here we can mention that the process has several experimental advantages, the most important of which being the precious *coincidence trigger*. The trigger is due to the presence of characteristic Kα X-rays and, in cases of decays to excited states, of characteristic γ-rays de-exciting these states. The resonance conditions favour in many cases decays to such states, while in the $\beta^-\beta^-$ case the Q^5-dependence of the rate strongly inhibits the process. Without a coincidence trigger the requirements of shielding and purity of all the material, necessary in order to suppress the overwhelming random background, tend to the extremes. Other advantages such as, *e.g.*, the low physical background due to the competing two neutrino radiative process, are discussed in [6,10].

There are several candidates for 0νECEC decays which might meet the resonance condition. The rate estimates, however, suffer from large uncertainties due to low precision of mass measurements, typically a few keV to be compared with the resonance width of ~ 0.1 keV.

Mass measurements of the required accuracy are within the present state-of-the art, but have yet to be done. The nuclei to be considered are, *e.g.*, ^{112}Sn, ^{136}Ce, ^{152}Gd, ^{162}Er, ^{164}Er and ^{180}W. The corresponding $Q-Q_{res}$ energy values in keV are $-5.8(4.6)$, 27(13), 0.04(3.5), $-10.5(4)$, 5(4), 12(5), where Q is the decay energy available for the process [12] and Q_{res} is the resonance energy; see [6] for details.

3 Searching for Majorana neutrinos with beta beams

3.1 Beta beams

The present plans for developing high flux, high energy neutrino beams, the so-called "neutrino factories", are mainly motivated by the need for the next generation, high precision neutrino oscillation experiments. Once constructed, however, these facilities will inevitably be used for a number of other research projects. The principle is to produce a high intensity beam of a neutrino progenitor, be it muons or beta-decaying ions, accelerate these species to highly relativistic energies and store in a ring ([13,14, 15] and references therein). The "ring" will have a long straight-linear section. The neutrinos produced during the flight in this section will undergo a kinematical focusing and a relativistic energy boost. The former concentrates the neutrino beam at the detector (the "target"), the latter offers a dramatic increase of the interaction cross-section in the detector. The purpose of the present paper is to consider the use of the beta beam born neutrinos to study other fundamental properties of neutrinos: their Majorana or Dirac nature, the lepton number conservation hypothesis and the absolute mass scale.

3.2 The principle of experiment

The Majorana neutrinos produced in β^- or in β^+, EC decays differ only by their helicity. If ν is massive then in either kind of the decay there is an admixture of the helicity opposite to the dominating one. The magnitude of this admixture depends on the neutrino rest mass, m_ν, and its energy in the center of mass frame of the decaying species, $E_\nu^{\rm CM}$:

$$f\left(m_\nu, E_\nu^{\rm CM}\right) \sim \left(\frac{m_\nu}{2E_\nu^{\rm CM}}\right)^2 . \qquad (3)$$

We assume having

i) beams of either ν_e or $\overline{\nu_e}$ with 100% purity

ii) detectors with 100% selectivity to neutrinos of a given helicity.

We consider two situations:

a) single beam, two different detectors

b) single detector, two different beams.

For the purpose of the present argument these situations are equivalent.

Consider the case of the single ν beam and two different detectors labelled (det ν) and (det $\overline{\nu}$). (In the following we use the notation $\nu, \overline{\nu}$ for the neutrinos with dominating left or right handed helicity, respectively.) The corresponding rates observed by the two detectors are:

$$N_\nu(\det \nu) \sim N_\nu^0 \sigma_{\nu n}\left(E_\nu^{lab}\right),$$
$$N_\nu(\det \overline{\nu}) \sim N_\nu^0 \sigma_{\overline{\nu} p}\left(E_\nu^{lab}\right) f\left(m_\nu, E_\nu^{\rm CM}\right),$$

where N_ν^0 is the neutrino flux at the detector, E_ν^{lab} is the boosted neutrino energy in the laboratory frame and $\sigma_{\nu n}(E_\nu^{lab}), \sigma_{\overline{\nu} p}(E_\nu^{lab})$ are the cross-sections for the boosted neutrinos of the dominating and the admixed helicity, respectively. The admixed fraction, f, can then be expressed by the measured and/or calculated quantities:

$$f\left(m_\nu, E_\nu^{\rm CM}\right) = \frac{N_\nu(\det \overline{\nu})}{N_\nu(\det \nu)} \frac{\sigma_{\overline{\nu} p}(E_\nu^{lab})}{\sigma_{\nu n}(E_\nu^{lab})} \frac{\varepsilon_{\nu n}(E_\nu^{lab})}{\varepsilon_{\overline{\nu} p}(E_\nu^{lab})}, \qquad (4)$$

where $\varepsilon_{\nu n}$, $\varepsilon_{\overline{\nu} p}$ denote the detector efficiencies. The notation corresponds to the vector current interactions (ν-n) and ($\overline{\nu}$-p), respectively. Note that for any observation angle there is a continuous energy spectrum depending on energy and emission angle in the CMS.

The detection cross-sections can be calculated, albeit with sizeable nuclear structure uncertainties. Alternatively, they can be determined experimentally. In the ideal case of the experiment done with $\overline{\nu}$ and ν beams of the same energy and with the two different detectors, the $f(m_\nu, E_\nu^{\rm CM})$ function can be expressed in a simplified way. The terms of the detector efficiencies and the cross-sections (and thus of the nuclear structure dependence) cancel:

$$f\left(m_\nu, E_\nu^{\rm CM}\right) = \frac{N_\nu(\det \overline{\nu})}{N_\nu(\det \nu)} \frac{N_{\overline{\nu}}(\det \nu)}{N_{\overline{\nu}}(\det \overline{\nu})} . \qquad (5)$$

4 Helicity admixture estimates

4.1 Admixture in the CM system

Consider the ^{6}He decay (this is the prime candidate presently proposed for the beta beam, see the CERN feasibility study, [16]):

$$^6\text{He} \rightarrow {}^6\text{Li} + \overline{\nu} + \text{e}^- .$$

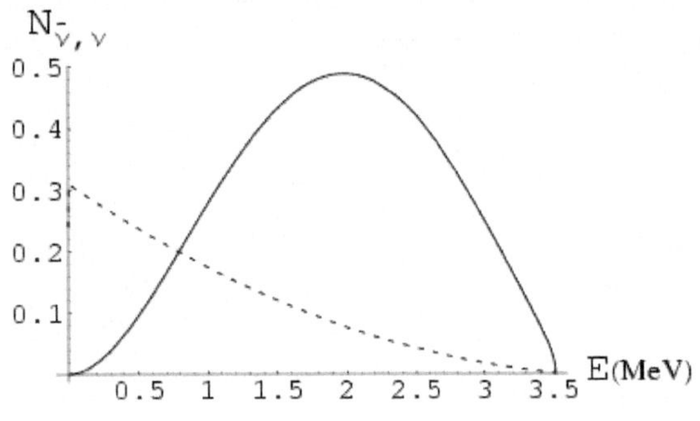

Fig. 4. The $\bar{\nu}$ energy spectrum (solid) and the opposite helicity admixture (dashed, $\times 10^{13}$) for ^{6}He decay in CM system. N_ν in arbitrary units.

The maximum $\bar{\nu}$ energy is about 3.5 MeV. The spectrum is shown in fig. 4. The spectrum of neutrinos with opposite helicity, calculated for $m_{\bar{\nu}} = 0.5\,\text{eV}$ and multiplied with 10^{13} is shown with the dashed line.

The neutrino CM energy dependence indicates the need to look for ions with decay energies as low as possible. A possible candidate is ^{178}W decaying via the electron capture decay:

$$^{178}_{74}\text{W}^{\text{EC}} \rightarrow {}^{178}_{74}\text{Ta}$$
with $Q(\text{EC}) = 91.3\,\text{keV}$.

The neutrino emitted in the CM system is mono-energetic, with the energy

$$E_\nu^{\text{CM}} = Q - B_K = 23.9\,\text{keV}$$

for neutral ^{178}W and $\simeq 22\,\text{keV}$ for the $^{178}_{74}\text{W}^{73+}$ ion. The gain has to be weighted, however, against the loss in intensity due to the elongated lifetime of the relativistic emitter.

4.2 Helicity admixture in the boosted system. The "helicity flip"

The helicity composition of the boosted beam of massive neutrinos will differ from that in the CM system. This comes about from the "helicity flip" of the neutrinos emitted in CM and reoriented due the boost in laboratory. Similar effect has been considered in [17] for pion decay. The very tiny rest mass of neutrinos, however, means that the gain in the helicity admixture is not very spectacular, at least not for the boosts presently attainable.

The ratio R of boosted admixture to that in the CMS can be very well approximated for a large range of laboratory energies by the formula [18]:

$$R = \frac{E_{max}}{E^{lab}} = 2\gamma E^{\text{CM}}/E^{lab}$$

Taking a mono-energetic neutrino, *e.g.*, 25 keV in CMS, and $\gamma = 1000$ one gets $E_{max} = 50\,\text{MeV}$ and thus at $E^{lab} = 1\,\text{MeV}$ the gain factor $R = 50$.

5 Conclusions and outlook

- Neutrino oscillations prove that at least one of the neutrino species has a finite rest mass and that the lepton flavour is not conserved. Information from other kind of data is needed to determine the values of neutrino masses, to establish the mass hierarchy and to test total lepton number conservation hypothesis.
- Neutrinoless double beta decay offers the most sensitive way to determine the effective neutrino mass. If discovered, it proves unambiguously the Majorana nature of neutrinos as well as the lepton number non-conservation.
- The radiative neutrinoless double electron capture is a promising alternative to the electron emission, with definite experimental advantages.
- Strong, resonance enhancement is predicted at photon energy equal to the energy difference of the $1S$-$2P$ atomic state.
- Experiments with beta beams can in principle provide information equivalent to that from the double beta decay. The rates are discouraging but improvements possible.

References

1. H. Ejiri, I. Ogawa (Editors), *Proceedings of the 1st Yamada Symposium on Neutrinos and Dark Matter in Nuclear Physics, Nara, Japan, 2003*, unpublished.
2. V. Lobashev *et al.*, Nucl. Phys. (Proc. Suppl.) **91**, 280 (2001); C. Weinheimer *et al.*, *International Conference on Neutrino Physics and Astrophysics Neutrino'02, May 2002, Munich*; N. Bohr, Mat.-Fys. Medd. K. Dan. Vidensk. Selsk. **18**, No. 8 (1948).
3. S. Elliott, P. Vogel, Annu. Rev. Nucl. Part. Sci. **52**, 15 (2002); S.R. Elliott, J. Engel, J. Phys. G **30**, R183 (2004).
4. Z. Sujkowski, Acta Phys. Pol. B **34**, 2207 (2003).
5. H.V. Klapdor-Kleingrothaus *et al.*, Part. Nucl. Lett. **1**, 20 (2001); Phys. Lett. B **586**, 198 (2004); H.V. Klapdor-Kleingrothaus, Nucl. Phys. B **143**, 229 (2005).
6. Z. Sujkowski, S. Wycech, Phys. Rev. C **70**, 052501(R) (2004).
7. Z. Sujkowski, S. Wycech, Nucl. Instrum. Methods B **235**, 81 (2005).
8. R.J. Glauber, P.C. Martin, Phys. Rev. **104**, 158 (1956); P.C. Martin, R.J. Glauber, Phys. Rev. **109**, 1307 (1958).
9. J. Żylicz *et al.*, Nucl. Phys. **42**, 330 (1963).
10. S. Wycech, Z. Sujkowski, Acta Phys. Pol. B **35**, 1223 (2004).
11. J. Bernabeu, A. de Rujula, C. Jarlskog, Nucl. Phys. B **223**, 15 (1983).
12. G. Audi, A.H. Wapstra, C. Thibault, Nucl. Phys. A **729**, 337 (2003).
13. F.T. Avignone *et al.*, Phys. At. Nucl. **63**, 1007 (2000); `http://www.phy.ornl.gov.orland`; see also `http://www.pas.rochester.edu/ksmcf/minerva`.
14. P. Zucchelli, Phys. Lett. B **532**, 166 (2002).
15. J. Serreau, C. Volpe, Phys. Rev. C **70**, 055502 (2004).
16. B. Autin *et al.*, J. Phys. G **29**, 1785 (2003); see also *CERN Workshop Physics with multi-MW proton source, CERN, May 2004*.
17. P. Langacker, J. Wong, Phys. Rev. D **58**, 093004 (1998).
18. L. Łukaszuk, Z. Sujkowski, S. Wycech, to be published.

Eur. Phys. J. A **27**, s01, 67–72 (2006)
DOI: 10.1140/epja/i2006-08-009-6

EPJ A direct
electronic only

Interactions of the solar neutrinos with the deuterons

B. Mosconi[1,a], P. Ricci[2,b], and E. Truhlík[3,c]

[1] Università di Firenze, Dipartimento di Fisica, and Istituto Nazionale di Fisica Nucleare, Sezione di Firenze, I-50019 Sesto Fiorentino (Firenze), Italy
[2] Istituto Nazionale di Fisica Nucleare, Sezione di Firenze, I-50019 Sesto Fiorentino (Firenze), Italy
[3] Institute of Nuclear Physics, Academy of Sciences of the Czech Republic, CZ-250 68 Řež, Czech Republic

Received: 15 June 2005 /
Published online: 23 February 2006 –

Abstract. Starting from chiral Lagrangians, possessing the $SU(2)_L \times SU(2)_R$ local chiral symmetry, we derive weak axial one-boson exchange currents in the leading order in the $1/M$ expansion (M is the nucleon mass), suitable for the nuclear physics calculations beyond the threshold energies and with the wave functions, obtained by solving the Schrödinger equation with the one-boson exchange potentials. The constructed currents obey the nuclear form of the partial conservation of the axial current. We apply the space component of these currents in calculations of the cross-sections for the disintegration of deuterons by the low-energy neutrinos. The deuteron and the $^1S_0$ final-state nucleon-nucleon wave functions are derived i) from a variant of the OBEPQB potential and ii) from the Nijmegen 93 and Nijmegen I nucleon-nucleon interactions. The extracted values of the constant $L_{1,A}$, entering the axial exchange currents of the pionless effective field theory (EFT), are in agreement with those predicted by the dimensional analysis. The comparison of our cross-sections with those obtained within the pionless EFT and other potential model calculations shows that the solar neutrino-deuteron cross-sections can be calculated within an accuracy of $\approx 3.3\%$.

PACS. 11.40.Ha Partially conserved axial-vector currents – 25.30.-c Nuclear reactions: specific reactions: Lepton-induced reactions

1 Introduction

The semileptonic weak nuclear interaction has been studied for half a century. The cornerstones of this field of research are i) the chiral symmetry, ii) the conserved vector current and iii) the partial conservation of the axial current (PCAC). In the formulation [1], the PCAC reads

$$q_\mu \langle \Psi_f | j^a_{5\mu}(q) | \Psi_i \rangle = i f_\pi m^2_\pi \Delta^\pi_F \left(q^2\right) \langle \Psi_f | m^a_\pi(q) | \Psi_i \rangle , \quad (1)$$

where $j^a_{5\mu}(q)$ is the total weak axial isovector hadron current, $m^a_\pi(q)$ is the pion source (the pion production/absorption amplitude) and $|\Psi_{i,f}\rangle$ is the wave function describing the initial (i) or final (f) nuclear state. It has been recognized [2] in studying the triton beta decay

$$^3\mathrm{H} \rightarrow {}^3\mathrm{He} + e^- + \bar{\nu} , \quad (2)$$

and the muon capture [3]

$$\mu^- + {}^3\mathrm{He} \rightarrow {}^3\mathrm{H} + \nu_\mu , \quad (3)$$

$$\mu^- + d \rightarrow n + n + \nu_\mu , \quad (4)$$

that in addition to the one-nucleon current, the effect of the space component of weak axial exchange currents (WAECs) enhances sensibly the Gamow-Teller matrix elements entering the transition rates. This suggests that the current $j^a_{5\mu}(q)$ can be understood for the system of A nucleons as the sum of the one- and two-nucleon components,

$$j^a_{5\mu}(q) = \sum_{i=1}^{A} j^a_{5\mu}(1, i, q_i) + \sum_{i<j}^{A} j^a_{5\mu}(2, ij, q) . \quad (5)$$

Let us describe the nuclear system by the Schrödinger equation

$$H|\Psi\rangle = E|\Psi\rangle , \quad H = T + V , \quad (6)$$

where H is the nuclear Hamiltonian, T is the kinetic energy and V is the nuclear potential describing the interaction between nucleon pairs. Taking for simplicity $A = 2$, we obtain from eq. (1) in the operator form and from eqs. (5) and (6) the following set of equations for the one-

[a] e-mail: mosconi@fi.infn.it
[b] e-mail: ricci@fi.infn.it
[c] e-mail: truhlik@ujf.cas.cz

and two-nucleon components of the total axial current:

$$\begin{aligned} \boldsymbol{q}_i \cdot \boldsymbol{j}_5^a(1,\boldsymbol{q}_i) &= [T_i\,,\,\rho_5^a(1,\boldsymbol{q}_i)] + if_\pi m_\pi^2 \Delta_F^\pi(q^2) \\ &\quad \times m_\pi^a(1,\boldsymbol{q}_i)\,, \quad i=1,2\,, \end{aligned} \tag{7}$$

$$\begin{aligned} \boldsymbol{q} \cdot \boldsymbol{j}_5^a(2,\boldsymbol{q}) &= [T_1+T_2\,,\,\rho_5^a(2,\boldsymbol{q})] + ([V\,,\,\rho_5^a(1,\boldsymbol{q})] \\ &\quad +(1\leftrightarrow 2)) + if_\pi m_\pi^2 \Delta_F^\pi(q^2)\, m_\pi^a(2,\boldsymbol{q}). \end{aligned} \tag{8}$$

If the WAECs are constructed so that they satisfy eq. (8), then the matrix element of the total current, sandwiched between solutions of the nuclear equation of motion (6), satisfies the PCAC (1).

It is known from the dimensional analysis [4], that the space component of the WAECs, $\boldsymbol{j}_5^a(2,\boldsymbol{q})$, is of the order $\mathcal{O}(1/M^3)$. Being of a relativistic origin, it is model dependent. This component of the WAECs was derived by several authors in various models. In the standard nuclear physics approach [5,6,7,8,9], the model systems of strongly interacting particles contain various particles (effective degrees of freedom), such as N, $\Delta(1236)$, π, ρ, ω and other baryons and mesons. Using these effective degrees of freedom and chiral Lagrangians, it was possible to describe reasonably nuclear electroweak phenomena in the whole region of intermediate energies. In particular, the existence of mesonic degrees of freedom in nuclei, manifesting themselves via meson exchange currents, was proven to a high degree of reliability [3].

One of the employed Lagrangians is the one [5] containing the heavy meson fields ρ and a_1, taken as the Yang-Mills gauge fields [10]. It reflects the $SU(2)_L \times SU(2)_R$ local chiral symmetry. Another used Lagrangian has been built up [11] within the concept of hidden local symmetries [12,13]. Besides possessing the chiral symmetry, our Lagrangians are characterized by the following properties: i) They respect the vector dominance model, reproduce universality, KSFRI, FSFR2, ii) they provide correct anomalous magnetic moment of the a_1-meson, iii) at the tree-level approximation, they correctly describe elementary processes in the whole region of intermediate energies ($E < 1\,\text{GeV}$) and iv) the current algebra prediction for the weak pion production amplitude is reproduced. Using such an approach, the exchange currents are constructed as follows. First, one derives the exchange amplitudes $J_{5\mu}^a(2)$ as Feynman tree graphs. These amplitudes satisfy the PCAC equation

$$q_\mu\, J_{5\mu}^a(2) = if_\pi\, m_\pi^2\, \Delta_F^\pi(q^2)\, M^a(2)\,, \tag{9}$$

where $M^a(2)$ are the associated pion absorption/production amplitudes[1]. The nuclear exchange currents are constructed from these amplitudes in conjunction with the equation, describing the nuclear states. Such exchange currents, combined with the one-nucleon currents, should satisfy eq. (1). In the present case, we describe the nuclear system by the Hamiltonian $H = T + V$ and the nuclear states by the Schrödinger equation (6). The nuclear exchange currents are constructed within the extended S-matrix method, in analogy with the electromagnetic meson exchange currents [15], as the difference between the

[1] We refer the reader for more details to ref. [14].

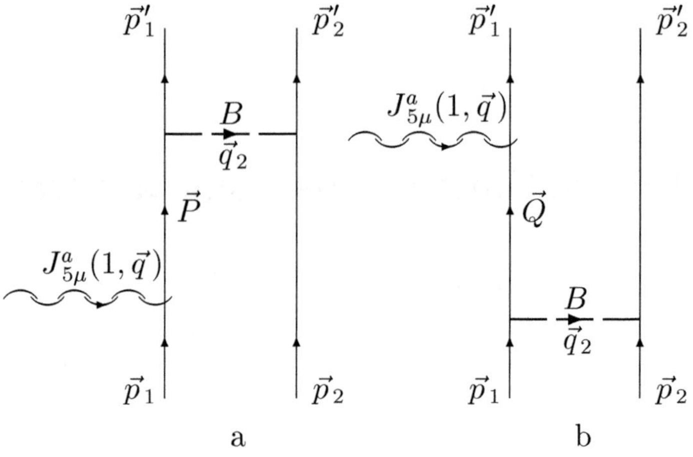

Fig. 1. The kinematics of the first Born iteration. The nucleon line in the intermediate state is on-shell.

relativistic amplitudes $J_{5\mu}^a(2)$ and the first Born iteration of the weak axial one-nucleon current contribution to the two-nucleon scattering amplitude, satisfying the Lippmann-Schwinger equation (see fig. 1). This method has already been applied [16,17] to construct the space component of the WAECs of the pion range.

On the other hand, EFTs are being developed since the early '90s. In this approach, one starts from a general chiral invariant Lagrangian with heavy-particle degrees of freedom integrated out and preserving N, $\Delta(1232)$ and π [18], or N and π [19,20], or only nucleons [21,22]. Such EFTs rely on systematic counting rules and on the existence of an expansion parameter, governing a perturbation scheme that converges reasonably fast. The expansion parameter is given as the ratio of the light and heavy scales.

In the pionless EFT [21,22], the heavy scale Λ is set to the pion mass m_π. This choice restricts the application of the scheme to the processes taking place at threshold energies, such as the interaction of solar neutrinos with the deuterons [23]. In the EFT with pions, the heavy scale is $\Lambda \approx 4\pi f_\pi \approx 1\,\text{GeV}$, restricting the application of the EFT to low energies.

The goal of this study is twofold: i) The construction of the WAECs of the heavy-meson range, suitable in the standard nuclear physics calculations beyond the long-wave limit, with the nuclear wave functions generated from the Schrödinger equation using the one-boson exchange potentials (OBEPs). ii) An application of the developed formalism to the description of the interaction of the low-energy neutrinos with the deuterons,

$$\nu_x + d \longrightarrow \nu_x' + n + p\,, \tag{10}$$

$$\overline{\nu}_x + d \longrightarrow \overline{\nu}_x' + n + p\,, \tag{11}$$

$$\nu_e + d \longrightarrow e^- + p + p\,, \tag{12}$$

$$\overline{\nu}_e + d \longrightarrow e^+ + n + n\,, \tag{13}$$

where ν_x refers to any active flavor of the neutrino. The reactions (10) and (12) are important for studying the solar neutrino oscillations, whereas the reactions (11) and (13) occur in experiments with reactor antineutrino beams.

The cross-sections for the reactions (10) and (12) are important for the analysis of the results obtained in the SNO detector [24,25,26]. The standard nuclear physics calculations [27,28] generally differ [23] by 5%–10%, which provides a good motivation to make independent calculations aiming to reduce this uncertainty.

In ref. [23], the effective cross-sections for the reactions (10)-(13) are presented in the form

$$\sigma_{\mathrm{EFT}}(E_\nu) = a(E_\nu) + L_{1,A}\, b(E_\nu)\,. \tag{14}$$

The amplitudes $a(E_\nu)$ and $b(E_\nu)$ are tabulated in [23] for each of the reactions (10)-(13) from the lowest possible (anti)neutrino energy up to 20 MeV, with 1 MeV step. The constant $L_{1,A}$ cannot be determined from reactions between elementary particles. Here we extract $L_{1,A}$ from our cross-sections calculated in the approximations of [23]: only the $^1S_0$ wave is taken into account in the nucleon-nucleon final state and the nucleon variables are treated non-relativistically. The knowledge of $L_{1,A}$ allows us to compare our cross-sections with $\sigma_{\mathrm{EFT}}(E_\nu)$.

2 Weak axial nuclear exchange currents

The starting quantities of our construction are the relativistic Feynman amplitudes $J^a_{5\mu,B}(2)$ of the range B ($B = \pi,\ \rho,\ \omega,\ a_1$). These amplitudes satisfy the PCAC constraint (9). The WAECs $j^a_{5\mu,\,B}(2)$ of the range B are defined as [14]

$$j^a_{5\mu,\,B}(2) = J^a_{5\mu,\,B}(2) - t^{a,\,\mathrm{FBI}}_{5\mu,\,B}\,, \tag{15}$$

where $t^{a,\,\mathrm{FBI}}_{5\mu,\,B}$ is the first Born iteration of the one-nucleon current contribution to the two-nucleon scattering amplitude, satisfying the Lippmann-Schwinger equation [15].

The PCAC for the WAECs, defined in eq. (15), is given by

$$\begin{aligned} q_\mu j^a_{5\mu,\,B}(2) = &\left([V_B, \rho^a_5(1)] + (1 \leftrightarrow 2)\right) \\ &+ i f_\pi m^2_\pi \Delta^\pi_F\left(\boldsymbol{q}^2\right) m^a_B(2)\,, \end{aligned} \tag{16}$$

where the nuclear pion production/absorption amplitude is given by

$$m^a_B(2) = M^a_B(2) - m^{a,\,\mathrm{FBI}}_B\,, \tag{17}$$

V_B is the potential of the range B and $\rho^a_5(1)$ is the one-nucleon axial charge density. We note here that the continuity equation (16) for our WAECs coincides with eq. (8).

It follows from eq. (16) that in order to make consistent calculations of the exchange current effects, one should use OBEPs for the generation of the nuclear wave functions and apply in the WAECs the same couplings and strong form factors as in the potentials. In our calculations, we employ the realistic OBE potentials OBEPQG [29], Nijmegen 93 (Nijm93) and Nijmegen I (NijmI) [30]. The potential OBEPQG is the potential OBEPQB [31], extended by including the a_1 exchange. The potential NijmI is the high-quality second-generation potential with the $\chi^2/\mathrm{data} = 1.03$.

Table 1. Values of the constant $L_{1,A}$ obtained by the fit to the cross-sections of the reactions (10)-(13) calculated using the NijmI, Nijm93 and OBEPQG potentials and by the fit (NSGK) to the cross-sections of table I of ref. [27].

Reaction		NijmI	Nijm93	OBEPQG	NSGK
(10)	$L_{1,A}$	4.6	5.2	4.8	5.4
	S	1.001	1.001	1.001	1.000
(11)	$L_{1,A}$	4.9	5.5	5.1	5.5
	S	1.001	1.001	1.001	1.000
(12)	$L_{1,A}$	4.1	5.0	–	6.0
	S	1.001	1.001	–	1.002
(13)	$L_{1,A}$	4.5	5.4	6.9	5.6
	S	1.001	1.000	0.9996	0.9997

In the next section, we use the WAECs, derived in the chiral invariant models [14,17,32], to calculate the cross-sections for the reactions (10)-(13). By comparing them with the EFT cross-sections (14), we extract the value of the constant $L_{1,A}$. We also compare our cross-sections with the cross-sections of refs. [27,28]. Our WAECs contain the following components [14]: the pair terms $\boldsymbol{j}^{\,a}_{5,\,B}(\mathrm{pair})$ ($B = \pi,\ \rho,\ \omega$), the non-potential exchange currents $\boldsymbol{j}_{5,\,\pi}^{\,a}(\rho\pi)$, $\boldsymbol{j}^{\,a}_{5,\,a_1\rho}(a_1)$ and the Δ excitation terms $\boldsymbol{j}^{\,a}_{5,\,B}(\Delta)$ ($B = \pi,\ \rho$).

The pion exchange part of our model WAECs is similar to the one employed in [27]. The representative cross-sections, presented in table I of ref. [27], are calculated using the AV18 potential [33], that is another high-quality second-generation potential[2] and the S- and P-waves are taken into account in the nucleon-nucleon final states.

We also compare our results with those reported in table I of ref. [28], where the calculations were performed i) with the Paris potential [34]; ii) with the currents taken in the impulse approximation; iii) with the S- and P-waves taken into account in the nucleon-nucleon final states.

3 Numerical results

Using the technique developed in refs. [35,36] one obtains the equations for the cross-sections $\sigma_{\mathrm{pot}}(E_\nu)$ that can be found in [14]. The equations are the same as those of ref. [27], but we treat the nucleon variables in the phase space non-relativistically. In ref. [23], the bounds on the phase space are defined in the neutral channel by

$$0 \le E'_\nu \le E_\nu - \nu - 2M_r + 2\left[M_r(M_r - |\epsilon_B|)\right]^{\frac{1}{2}}\,, \tag{18}$$

$$\mathrm{Max}\left[-1, \frac{E^2_\nu + E'^2_\nu + 4M_r(|\epsilon_B| - q_0)}{2E_\nu E'_\nu}\right] \le \cos\theta \le 1, \tag{19}$$

where M_r is the reduced mass of the neutron-proton system and $\epsilon_B = -2.2245\,\mathrm{MeV}$ is the deuteron binding energy. We have found that it is more effective to integrate

[2] However, it is not an OBEP.

Table 2. Scattering length and effective range (in fm) for the nucleon-nucleon system in the $^1S_0$ state, corresponding to the NijmI, Nijm93 [30], OBEPQG [29], AV18 [33] potentials and as used in the EFT calculations [23], and their experimental values.

	NijmI	Nijm93	OBEPQG	AV18	EFT	Exp.
a_{np}	−23.72	−23.74	−23.74	−23.73	−23.7	-23.740 ± 0.020^1
r_{np}	2.65	2.68	2.73	2.70	2.70	2.77 ± 0.05^1
a_{pp}	−7.80	−7.79	–	−7.82	−7.82	-7.8063 ± 0.0026^2
r_{pp}	2.74	2.71	–	2.79	2.79	2.794 ± 0.014^2
a_{nn}	−18.16	−18.11	−18.10	−18.49	−18.5	-18.59 ± 0.40^3
r_{nn}	2.80	2.78	2.77	2.84	2.80	2.80 ± 0.11^4

[1] Reference [37]. [2] Reference [38]. [3] Reference [39]. [4] Reference [40].

numerically within the bounds

$$-1 \leq \cos\theta \leq 1, \tag{20}$$

$$0 \leq E'_\nu \leq E_\nu \cos\theta - 2M + \left[4M_r^2 + 4M_r(E_\nu - |\epsilon_B|) - E_\nu^2(1-\cos^2\theta) - 4M_r E_\nu \cos\theta\right]^{\frac{1}{2}}. \tag{21}$$

For the charged channel, the momentum of the final lepton is restricted by

$$0 \leq p_l \leq p_{l,\max}, \tag{22}$$

where $p_{l,\max}$ is the solution of the equation

$$(E_\nu - p_l)^2 + 4M_r E(p_l) + 4M_r(\Delta - E_\nu) = 0. \tag{23}$$

Here $E(p_l) = (p_l^2 + m_e^2)^{\frac{1}{2}}$ and

$$\Delta = M_p - M_n + |\epsilon_B|, \quad M_r = M_p, \ \nu e^-, \tag{24}$$

$$\Delta = M_n - M_p + |\epsilon_B|, \quad M_r = M_n, \ \bar{\nu} e^+. \tag{25}$$

We extracted $L_{1,A}$ by comparing the cross-section $\sigma_{\rm EFT}(E_\nu)$ with our cross-sections $\sigma_{\rm pot}(E_\nu)$ using the least-square fit and also considering an average value of $L_{1,A}$

$$\bar{L}_{1,A} = \frac{\sum_{i=1}^N L_{1,A}(i)}{N}, \tag{26}$$

where

$$L_{1,A}(i) = \frac{\sigma_{{\rm pot},i} - a_i}{b_i}. \tag{27}$$

We estimated the quality of the fit by the quantity S defined as

$$S = \frac{1}{N}\sum_{i=1}^N \frac{\sigma_{{\rm EFT},i}}{\sigma_{{\rm pot},i}}. \tag{28}$$

It was found that the fit providing the average value (26) results in better agreement between $\sigma_{\rm EFT}(E_\nu)$ and $\sigma_{\rm pot}(E_\nu)$ and we present the results in table 1 only for this fitting procedure. We also applied this fit to the cross-sections of table I of [27] (cf. the column NSGK).

In table 2, we present the scattering lengths and the effective ranges, obtained from the NijmI, Nijm93, OBEPQG and AV18 potentials and also the values used in the EFT calculations [23]. For the generation of the final-state nucleon-nucleon wave functions from the NijmI and Nijm 93 potentials, we used the program COCHASE [41]. The program solves the Schrödinger equation using the fourth-order Runge-Kutta method. This can provide low-energy scattering parameters that slightly differ from those obtained by the Nijmegen group, employing the modified Numerov method [42]. Some refit was necessary, in order to get the correct low-energy scattering parameters in the neutron-proton and neutron-neutron $^1S_0$ states.

Table 3. Cross-section and the differences in % between cross-sections for the reaction (10). In the first column, E_ν (MeV) is the neutrino energy, in the second column, $\sigma_{\rm NijmI}$ (in $10^{-42} \times {\rm cm}^2$) is the cross-section calculated with the NijmI nuclear wave functions. Column 3 reports the differences between $\sigma_{\rm NijmI}$ (NijmI) and the EFT cross-section (14) $\sigma_{\rm EFT}$, calculated with the corresponding constant $\bar{L}_{1,A}$ from table 1. The differences between $\sigma_{\rm NSGK}$ ([27], table I) and $\sigma_{\rm EFT}$ are reported in column 4. Further, $\Delta_{1(2)}$ is the difference between the cross-sections $\sigma_{\rm NijmI}$ ($\sigma_{\rm Nijm93}$) and $\sigma_{\rm NSGK}$; Δ_3 is the difference between the cross-sections $\sigma_{\rm NijmI}$ and $\sigma_{\rm YHH}$, where the cross-section $\sigma_{\rm YHH}$ is from ([28], table I).

E_ν	$\sigma_{\rm NijmI}$	NijmI	NSGK	Δ_1	Δ_2	Δ_3
3	0.00335	1.2	0.4	−1.1	−0.5	–
4	0.0306	1.3	0.2	−0.8	−0.2	12.0
5	0.0948	1.3	0.2	−0.9	−0.2	5.0
6	0.201	1.1	0.1	−1.0	−0.3	10.2
7	0.353	1.0	0.1	−1.1	−0.4	8.1
8	0.551	1.0	0.2	−1.3	−0.5	10.1
9	0.798	1.0	0.4	−1.5	−0.7	8.9
10	1.093	0.4	−0.1	−1.6	−0.8	7.6
11	1.437	0.8	0.5	−1.6	−1.0	9.4
12	1.831	−0.1	−0.3	−2.1	−1.2	8.5
13	2.274	−0.1	0.0	−2.3	−1.4	9.9
14	2.767	−0.4	0.0	−2.6	−1.7	9.5
15	3.308	−0.8	−0.1	−2.9	−2.0	10.3
16	3.898	−1.2	−0.3	−3.2	−2.2	9.9
17	4.537	−1.6	−0.4	−3.5	−2.5	10.6
18	5.223	−1.9	−0.3	−3.9	−2.9	10.3
19	5.957	−2.3	−0.4	−4.2	−3.2	10.7
20	6.738	−2.9	−0.6	−4.6	−3.6	10.6

We shall now present the results for the reactions (10)-(13). In comparing our results with [23] we use in our calculations their values $G_F = 1.166 \times 10^{-5}\,{\rm GeV}^{-2}$ and $g_A = -1.26$. Instead we use the value $g_A = -1.254$, as employed in [27] and [28], when comparing our results with these works. In the cross-sections for the charged-channel reactions (12) and (13) the value $\cos\theta_{\rm C} = 0.975$ is taken for the Cabibbo angle.

Table 4. Cross-section and the differences in % between cross-sections for the reaction (11). For notations, see table 3, only instead of E_ν, now $E_{\bar{\nu}}$ is the antinetrino energy in MeV.

$E_{\bar{\nu}}$	$\sigma_{\rm NijmI}$	NijmI	NSGK	Δ_1	Δ_2	Δ_3
3	0.00332	0.6	0.1	−1.1	−0.5	–
4	0.0302	1.0	0.2	−0.8	−0.1	9.3
5	0.0928	1.0	0.1	−0.8	−0.1	0.9
6	0.196	1.1	0.3	−0.9	−0.1	5.7
7	0.342	0.8	0.1	−1.0	−0.2	2.0
8	0.531	1.4	0.8	−1.1	−0.3	3.1
9	0.765	0.8	0.2	−1.2	−0.4	0.9
10	1.043	0.6	0.2	−1.4	−0.5	−1.7
11	1.364	0.1	−0.2	−1.6	−0.7	−0.7
12	1.729	−0.2	−0.4	−1.7	−0.8	−2.8
13	2.136	−0.3	−0.2	−1.9	−1.0	−2.1
14	2.585	−0.5	−0.2	−2.1	−1.2	−3.9
15	3.074	−0.7	−0.2	−2.4	−1.4	−4.1
16	3.604	−0.9	−0.1	−2.6	−1.7	−5.6
17	4.173	−1.2	−0.2	−2.9	−1.9	−6.0
18	4.779	−1.6	−0.3	−3.3	−2.2	−7.6
19	5.422	−1.9	−0.3	−3.6	−2.5	−8.0
20	6.101	−2.2	−0.2	−3.9	−2.9	−9.4

3.1 Reaction $\nu_{\rm x} + {\rm d} \longrightarrow \nu'_{\rm x} + {\rm n} + {\rm p}$

In table 3, we present the difference in %, between the cross-sections, obtained with the NijmI and AV18 potentials models and the EFT cross-sections, calculated with the corresponding $\bar{L}_{1,A}$ from table 1. Besides, we give the differences between the cross-sections, computed with the wave functions of various potential models.

Comparing the columns NijmI and NSGK of table 3 we can see that the NSGK cross-section is closer to the EFT cross-section. This means that the standard approach and the pionless EFT differ, since the approximations, made in our calculations and in EFT, coincide: the nucleon-nucleon final state is restricted to the $^1S_0$ wave and the nucleon variables are treated non-relativistically. Besides, the inspection of columns Δ_1 and Δ_2 shows that our cross-sections closely follow the NSGK cross-section up to the energies when the P-waves in the nucleon-nucleon final state start to contribute. On the other hand, as follows from column Δ_3, it is difficult to understand the behavior of the cross-section $\sigma_{\rm YHH}$ in the whole interval of the considered neutrino energies.

3.2 Reaction $\bar{\nu}_{\rm x} + {\rm d} \longrightarrow \bar{\nu}'_{\rm x} + {\rm n} + {\rm p}$

In analogy with sect. 3.1, we present in table 4 a comparative analysis for the reaction (11).

Clearly, our cross-sections are closer to $\sigma_{\rm EFT}$ and also to the cross-section $\sigma_{\rm NSGK}$, than in the neutrino-deuteron case of table 3. The behavior of the cross-section $\sigma_{\rm YHH}$ is as little understandable as for the reaction (10).

One can also conclude from the differences given in the columns NijmI, NSGK, Δ_1 and Δ_2 of tables 3 and 4 that the cross-sections for the reactions (10) and (11) are described by both the potential models and the pionless EFT with an accuracy better than 3%.

Table 5. Cross-section and the differences in % between cross-sections for the reaction (12). For notations, see table 3.

E_ν	$\sigma_{\rm NijmI}$	NijmI	NSGK	Δ_1	Δ_2	Δ_3
2	0.00338	−5.5	−0.6	−7.6	−6.7	–
3	0.0455	−0.5	−0.3	−3.0	−2.0	–
4	0.153	0.5	−0.6	−1.9	−0.9	1.9
5	0.340	1.5	0.1	−1.6	−0.6	2.9
6	0.613	1.9	0.4	−1.6	−0.5	3.0
7	0.978	1.9	0.4	−1.6	−0.6	3.0
8	1.438	0.0	−2.4	−1.8	−0.7	3.1
9	1.997	−0.2	−2.3	−1.9	−0.8	2.9
10	2.655	0.1	−1.7	−2.1	−1.0	3.1
11	3.415	3.3	3.3	−2.4	−1.2	2.8
12	4.277	1.0	0.3	−2.6	−1.5	2.5
13	5.243	0.7	0.2	−2.9	−1.8	2.4
14	6.311	0.4	0.2	−3.2	−2.1	2.1
15	7.484	0.0	0.2	−3.6	−2.4	1.7
16	8.760	−0.5	−0.1	−4.0	−2.8	1.4
17	10.14	−0.9	−0.1	−4.4	−3.2	1.0
18	11.62	−1.3	−0.1	−4.8	−3.6	0.1
19	13.21	−1.7	−0.0	−5.3	−4.1	−0.1
20	14.89	−2.4	−0.3	−5.8	−4.5	−0.3

Table 6. Cross-section and the differences in % between cross-sections for the reaction (13). For notations, see table 3.

$E_{\bar{\nu}}$	$\sigma_{\rm NijmI}$	NijmI	NSGK	Δ_1	Δ_2	Δ_3
5	0.0274	−1.3	−0.9	−2.4	−1.5	9.0
6	0.116	0.1	−0.1	−2.1	−1.1	8.1
7	0.277	0.2	−0.2	−1.8	−0.7	7.4
8	0.514	0.5	−0.1	−1.7	−0.6	7.1
9	0.829	0.4	−0.2	−1.7	−0.6	6.9
10	1.224	0.9	0.4	−1.7	−0.6	6.8
11	1.697	0.7	0.2	−1.9	−0.7	6.0
12	2.249	0.6	0.1	−2.0	−0.8	6.1
13	2.876	0.4	0.0	−2.2	−1.0	5.5
14	3.578	0.4	0.2	−2.3	−1.1	5.2
15	4.353	0.0	0.0	−2.6	−1.3	4.9
16	5.200	−0.2	0.1	−2.8	−1.6	4.6
17	6.115	−0.3	0.2	−3.1	−1.9	3.5
18	7.097	−0.5	0.4	−3.4	−2.1	3.2
19	8.143	−0.9	0.2	−3.8	−2.5	2.8
20	9.251	−1.2	0.3	−4.1	−2.8	2.4

3.3 Reaction $\nu_{\rm e} + {\rm d} \longrightarrow {\rm e}^- + {\rm p} + {\rm p}$

The comparison of the columns NijmI, NSGK, Δ_1 and Δ_2 of table 5 shows that, disregarding the cross-sections for $E_\nu = 2\,$MeV, the cross-sections for the important reaction (12) are described with an accuracy of 3.3%. However, while our cross-sections and the cross-section [27] are smooth functions of the neutrino energy, the EFT cross-section exhibits abrupt changes in the region $7 < E_\nu < 12\,$MeV. In our opinion, the reason can be an incorrect treatment of the Coulomb interaction between protons in the EFT calculations. We have verified that the non-relativistic approximation for the Fermi function, employed in [23] is valid with a good accuracy in the whole interval of the solar neutrino energies.

Inspecting the difference of the cross-sections Δ_3 shows that the cross-section [28] is of the correct size in this case.

3.4 Reaction $\bar{\nu}_e + d \longrightarrow e^+ + n + n$

It follows from table 6 that our cross-sections for the reaction (13) are in a very good agreement with the EFT [23] and [27] calculations. This confirms our conjecture that the treatment of the Coulomb interaction between protons [23] in the reaction (12) is not correct.

The calculations [28] provide a too small cross-section. The most probable reason for this large difference is that the Paris potential does not describe the neutron-neutron interaction correctly.

4 Conclusions

We calculated here the cross-sections for the reactions of the solar neutrinos with the deuterons, (10)-(13), within the standard nuclear physics approach. We took into account the weak axial exchange currents of the OBE type, satisfying the nuclear continuity equation (8). These currents were constructed from the Lagrangians, possessing the chiral local $SU(2)_L \times SU(2)_R$ symmetry, in the tree approximation. Using the OBE potentials NijmI, Nijm93 and OBEPQG, we made consistent calculations. We took into account the nucleon-nucleon interaction in the $^1S_0$ final state and we treated non-relativistically the nucleon variables.

Our cross-sections for the reactions (10), (11) and (13) agree with the EFT cross-sections [23] and also with the cross-sections [27] within an accuracy better than 3%. The agreement for the reaction (12) is within 3.3%. In our opinion, the agreement for the reaction (12) can be improved by paying more attention to the treatment of the Coulomb interaction between the protons in the final state in the pionless EFT calculations.

This work is supported in part by the grant GA CR 202/03/0210 and by Ministero dell'Istruzione, dell'Università e della Ricerca of Italy (PRIN 2003). We thank M. Rentmeester for the correspondence.

References

1. S.L. Adler, Phys. Rev. **139**, B1638 (1965).
2. R.J. Blin-Stoyle, *Fundamental Interactions and the Nucleus* (North-Holland/American Elsevier, Amsterdam-London/New York, 1973).
3. D.F. Measday, Phys. Rep. **354**, 243 (2001).
4. K. Kubodera, J. Delorme, M. Rho, Phys. Rev. Lett. **40**, 755 (1978).
5. E. Ivanov, E. Truhlík, Nucl. Phys. A **316**, 437 (1979).
6. I.S. Towner, Phys. Rep. **155**, 263 (1987).
7. R. Schiavilla *et al.*, Phys. Rev. C **58**, 1263 (1999).
8. K. Tsushima, D.O. Riska, Nucl. Phys. A **549**, 313 (1992).
9. S.M. Ananyan, Phys. Rev. C **57**, 2669 (1998).
10. C.N. Yang, R.L. Mills, Phys. Rev. **96**, 191 (1954).
11. J. Smejkal, E. Truhlík, H. Göller, Nucl. Phys. A **624**, 655 (1997).
12. U.-G. Meissner, Phys. Rep. **161**, 213 (1988).
13. M. Bando, T. Kugo, K. Yamawaki, Phys. Rep. **164**, 217 (1988).
14. B. Mosconi, P. Ricci, E. Truhlík, nucl-th/0212042.
15. J. Adam jr., E. Truhlík, D. Adamová, Nucl. Phys. A **494**, 556 (1989).
16. J. Adam jr., Ch. Hajduk, H. Henning, P.U. Sauer, E. Truhlík, Nucl. Phys. A **531**, 623 (1991).
17. E. Truhlík, F.C. Khanna, Int. J. Mod. Phys. A **10**, 499 (1995).
18. T.R. Hemmert, B.R. Holstein, J. Kambor, J. Phys. G **24**, 1831 (1998).
19. T.-S. Park, D.-P. Min, M. Rho, Phys. Rep. **233**, 341 (1993).
20. T.-S. Park, K. Kubodera, D.-P. Min, M. Rho, Astrophys. J. **507**, 443 (1998).
21. D.B. Kaplan, M.J. Savage, M.B. Wise, Nucl. Phys. B **478**, 629 (1996); Phys. Lett. B **424**, 390 (1998).
22. J.W. Chen, G. Rupak, M.J. Savage, Nucl. Phys. A **653**, 386 (1999).
23. M. Butler, J.-W. Chen, X. Kong, Phys. Rev. C **63**, 035501 (2001).
24. SNO Collaboration, Phys. Rev. Lett. **87**, 071301 (2001); **89**, 011301 (2002).
25. S.N. Ahmed *et al.*, Phys. Rev. Lett. **92**, 181301 (2004).
26. SNO Collaboration (B. Aharmin *et al.*), nucl-ex/0502021.
27. S. Nakamura, T. Sato, V. Gudkov, K. Kubodera, Phys. Rev. C **63**, 034617 (2001).
28. S. Ying, W.C. Haxton, E.M. Henley, Phys. Rev. C **45**, 1982 (1992).
29. P. Obersteiner, W. Plessas, E. Truhlík, in *Proceedings of the XIII International Conference on Particles and Nuclei, Perugia, Italy, June 28-July 2, 1993*, edited by A. Pascolini (World Scientific, Singapore, 1994) p. 430.
30. V.G.J. Stoks, R.A.M. Klomp, C.P.F. Terheggen, J.J. de Swart, Phys. Rev. C **49**, 2950 (1994).
31. R. Machleidt, Adv. Nucl. Phys. **19**, 189 (1989).
32. J.G. Congleton, E. Truhlík, Phys. Rev. C **53**, 957 (1996).
33. R.B. Wiringa, V.G.J. Stoks, R. Schiavilla, Phys. Rev. C **51**, 38 (1995).
34. M. Lacombe *et al.*, Phys. Rev. C **21**, 861 (1980).
35. J.D. Walecka, *Semi-leptonic weak interactions in nuclei*, in *Muon Physics*, edited by V.W. Hughes, C.S. Wu (Academic Press, New York, 1972).
36. J.S. O'Connell, T.W. Donnelly, J.D. Walecka, Phys. Rev. C **6**, 719 (1972).
37. R. Machleidt, Phys. Rev. C **63**, 024001 (2001).
38. J.R. Bergervoet, P.C. van Campen, W.A. van der Sanden, J.J. de Swart, Phys. Rev. C **38**, 15 (1988).
39. R. Machleidt, I. Slaus, J. Phys. G **27**, R69 (2001).
40. G.F. de Téramond, B. Gabioud, Phys. Rev. C **36**, 691 (1987).
41. S. Hirschi, E. Lomon, N. Spencer, Comput. Phys. Commun. **9**, 11 (1975).
42. M. Rentmeester, private communication, 2005.

3 Non-explosive Nucleosynthesis

Eur. Phys. J. A **27**, s01, 75–78 (2006)
DOI: 10.1140/epja/i2006-08-010-1

EPJ A direct
electronic only

Relation between the $^{16}O(\alpha,\gamma)^{20}Ne$ reaction and its reverse $^{20}Ne(\gamma,\alpha)^{16}O$ reaction in stars and in the laboratory

P. Mohr[1,a], C. Angulo[2], P. Descouvemont[3], and H. Utsunomiya[4,b]

[1] Diakoniekrankenhaus, D-74523 Schwäbisch Hall, Germany
[2] Centre de Recherches du Cyclotron, UCL, Louvain-la-Neuve, Belgium
[3] PNTPM, Université Libre de Bruxelles, Brussels, Belgium
[4] Konan University, Kobe, Japan

Received: 24 May 2005 /
Published online: 23 February 2006 –

Abstract. The astrophysical reaction rates of the $^{16}O(\alpha,\gamma)^{20}Ne$ capture reaction and its inverse $^{20}Ne(\gamma,\alpha)^{16}O$ photodisintegration reaction are given by the sum of several narrow resonances and a small direct capture contribution at low temperatures. Although the thermal population of low-lying excited states in ^{16}O and ^{20}Ne is extremely small, the first excited state in ^{20}Ne plays a non-negligible role for the photodisintegration rate. Consequences for experiments with so-called quasi-thermal photon energy distributions are discussed.

PACS. 26.20.+f Hydrostatic stellar nucleosynthesis – 25.40.Lw Radiative capture – 25.20.-x Photonuclear reactions

1 Introduction

The small reaction rate of the $^{16}O(\alpha,\gamma)^{20}Ne$ capture reaction blocks the reaction chain $3\alpha \rightarrow {}^{12}C(\alpha,\gamma)^{16}O(\alpha,\gamma)^{20}Ne$ in helium burning at typical temperatures around $T_9 = 0.2$ (T_9 is the temperature in billion degrees K). Its inverse $^{20}Ne(\gamma,\alpha)^{16}O$ photodisintegration reaction is one of the key reactions in neon burning at higher temperatures around $T_9 = 1$–2 [1]. Only at very low temperatures below $T_9 = 0.2$ the $^{16}O(\alpha,\gamma)^{20}Ne$ reaction rate is dominated by direct capture. At higher temperatures the reaction rate is given by the sum of several resonances [2]. The contributions of several low-lying resonances to the reaction rate of the $^{16}O(\alpha,\gamma)^{20}Ne$ reaction are shown in fig. 1. The properties of three bound states in ^{20}Ne and four selected low-lying resonances are listed in table 1.

Usually, the reaction rates of inverse photodisintegration reactions are calculated from the capture rates using the detailed-balance theorem which is only valid if all nuclei involved are fully thermalized (see, *e.g.*, [2]). In the case of light nuclei where the first excited states are located at relatively high energies one finds very small occupation probabilities for these excited states. The scope of this paper is to analyze the relation between the $^{16}O(\alpha,\gamma)^{20}Ne$ and $^{20}Ne(\gamma,\alpha)^{16}O$ reaction rates

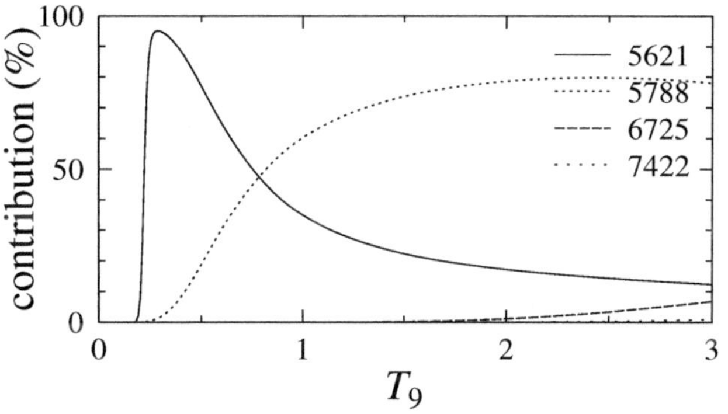

Fig. 1. Contribution of individual resonances to $^{16}O(\alpha,\gamma)^{20}Ne$ reaction rate. The resonances are labelled by their energies E_R in keV (see also table 1). At very low temperatures below $T_9 = 0.2$ direct capture is dominating.

Table 1. Properties of levels in ^{20}Ne below and above the ^{16}O-α threshold at $Q = 4730$ keV [3] (from [2,4]).

E_x (keV)	J^π	E_R^α (keV)	$(\omega\gamma)$ [4] (meV)	$(\omega\gamma)$ [2] (meV)	B^0 (%)	B^{1634} (%)
0	0^+	–	–	–	–	–
1634	2^+	–	–	–	100	–
4248	4^+	–	–	–	0	100
5621	3^-	892	1.7(3)	1.9(3)	7.6(10)	87.6(10)
5788	1^-	1058	17(3)	23(3)	18(5)	82(5)
6725	0^+	1995	71(12)	74(9)	0	100
7422	2^+	2692	146(19)	160(20)	$\leq 9.4(14)$	$\geq 90.6(14)$

[a] e-mail: WidmaierMohr@compuserve.de
[b] e-mail: hiro@konan-u.ac.jp

for the particular case of high-lying first excited states (^{16}O: $E_x = 6049$ keV, 0^+; ^{20}Ne: $E_x = 1634$ keV, 2^+). Furthermore, we discuss potential experiments using so-called quasi-thermal photon spectra [5,6] where the target nucleus is always in its ground state.

2 Resonant reaction rates

Because of the dominance of resonances in the rates of the $^{16}\mathrm{O}(\alpha,\gamma)^{20}\mathrm{Ne}$ and $^{20}\mathrm{Ne}(\gamma,\alpha)^{16}\mathrm{O}$ reactions we restrict ourselves to the discussion of resonances. The reaction rate $\langle\sigma\, v\rangle$ of a Breit-Wigner resonance with

$$\sigma_{\mathrm{BW}}(E) = \frac{\pi}{k_\alpha^2}\,\omega\,\frac{\Gamma_\alpha \Gamma_\gamma}{(E - E_R^\alpha)^2 + \Gamma^2/4} \tag{1}$$

is given by

$$\langle\sigma\, v\rangle = \hbar^2 \left(\frac{2\pi}{\mu kT}\right)^{3/2} (\omega\gamma)\, \exp\left(\frac{-E_R^\alpha}{kT}\right) \tag{2}$$

with the reduced mass μ, the resonance energy E_R^α in the c.m. system, and the resonance strength

$$(\omega\gamma) = \omega\,\frac{\Gamma_\alpha \Gamma_\gamma}{\Gamma} \tag{3}$$

and the statistical factor

$$\omega = \frac{(2J_R+1)}{(2J_1+1)(2J_2+1)}\,. \tag{4}$$

Γ_α is the α decay width of the resonance,

$$\Gamma_\gamma = \sum_b \Gamma_\gamma^b \tag{5}$$

the total radiation width (summed over all partial radiation widths Γ_γ^b to bound states b), $\Gamma = \Gamma_\alpha + \Gamma_\gamma$ the total decay width, and

$$B^b = \Gamma_\gamma^b / \Gamma_\gamma \tag{6}$$

is the γ-ray branching ratio to a bound state b.

The total astrophysical reaction rate is obtained by adding the contributions of all resonances and of a small direct contribution. In the case of the $^{16}\mathrm{O}(\alpha,\gamma)^{20}\mathrm{Ne}$ reaction, all contributions in the laboratory (with the target in the ground state) and under stellar conditions are identical up to high temperatures because of the high-lying first excited state in ^{16}O [2].

The cross section of the $^{20}\mathrm{Ne}(\gamma,\alpha)^{16}\mathrm{O}$ photodisintegration reaction is directly related to the capture cross section by time-reversal symmetry:

$$\frac{\sigma_{3\gamma}(E_\gamma)}{\sigma_{12}(E_\alpha)} = \frac{\lambda_\gamma^2}{\lambda_{12}^2}\,\frac{(2J_1+1)(2J_2+1)}{2\,(2J_3+1)}\,. \tag{7}$$

The cross section of a Breit-Wigner resonance for the capture reaction to a bound state b is given by

$$\sigma_{12}(E_\alpha) = B^b \times \sigma_{\mathrm{BW}}(E) \tag{8}$$

with the branching ratio B^b as defined in eq. (6). Using eq. (7), the thermal photon density from the Planck law

$$n_\gamma\,(E_\gamma,T)\,\mathrm{d}E_\gamma = \frac{1}{\pi^2}\,\frac{1}{(\hbar c)^3}\,\frac{E_\gamma^2}{\exp\left(E_\gamma/kT\right)-1}\,\mathrm{d}E_\gamma \tag{9}$$

and the definition of the reaction rate

$$\lambda = c \int n_\gamma(E_\gamma,T)\,\sigma(E_\gamma)\,\mathrm{d}E_\gamma\,, \tag{10}$$

one obtains a reaction rate for the target in a defined bound state b with spin J_3:

$$\lambda^b = \frac{1}{\hbar}\,\frac{(2J_1+1)(2J_2+1)}{(2J_3+1)}\,B^b\,(\omega\gamma)\,\exp\left(\frac{-E_R^\gamma}{kT}\right) \tag{11}$$

with the required photon energy

$$E_R^\gamma = E_R^\alpha + Q - E_x^b \tag{12}$$

and E_x^b is the excitation energy of the bound state b. Note that the required photon energy E_R^γ for the transition to a resonance at excitation energy E_R is reduced by the excitation energy E_x^b of the bound state b under consideration!

For simplicity we choose the example of the 1^- resonance at $E_R = 5788$ keV in ^{20}Ne in the following discussion. This resonance has a strength of $\omega\gamma = 23$ meV and branching ratios of $B^0 = 18\%$ to the ground state and $B^{1634} = 82\%$ to the first excited state in ^{20}Ne at $E_x = 1634$ keV with $J^\pi = 2^+$ [2,4] (see table 1).

For the ground-state rate λ^0 we obtain

$$\lambda^0 = \frac{1}{\hbar}\,B^0\,(\omega\gamma)\,\exp\left(\frac{-E_R}{kT}\right), \tag{13}$$

whereas the rate λ^{1634} for ^{20}Ne in its first excited state is

$$\lambda^{1634} = \frac{1}{\hbar}\,\frac{1}{5}\,B^{1634}\,(\omega\gamma)\,\exp\left(\frac{-(E_R - E_x^{1634})}{kT}\right). \tag{14}$$

Note the factor of 1/5 because of $J^\pi = 2^+$ for the first excited state at $E_x = 1634$ keV. Using the thermal occupation probability ratio

$$n^{1634}/n^0 = \frac{(2J^{1634}+1)}{(2J^0+1)}\,\exp\left(\frac{-E_x^{1634}}{kT}\right) \tag{15}$$

we can calculate the total astrophysical reaction rate $\lambda^\star$ for this single narrow 1^- resonance:

$$\begin{aligned}\lambda^\star &\approx \frac{1}{\hbar}\,(\omega\gamma)\left[B^0\,\exp\left(\frac{-E_R}{kT}\right)\right.\\ &\quad\left.+\,\frac{1}{5}B^{1634}\exp\left(\frac{-(E_R - E_x^{1634})}{kT}\right)5\,\exp\left(\frac{-E_x^{1634}}{kT}\right)\right]\\ &= \frac{1}{\hbar}\,(\omega\gamma)\,\left(B^0 + B^{1634}\right)\,\exp\left(\frac{-E_R}{kT}\right)\\ &= \frac{1}{\hbar}\,(\omega\gamma)\,\exp\left(\frac{-E_R^\alpha}{kT}\right)\,\exp\left(\frac{-Q}{kT}\right).\end{aligned} \tag{16}$$

Comparing eqs. (2) and (16), one finds the relation between the capture rate $\langle\sigma v\rangle$ and the photodisintegration rate $\lambda^\star$:

$$\frac{\lambda^\star}{\langle\sigma v\rangle} = \left(\frac{\mu\, kT}{2\pi\hbar^2}\right)^{3/2}\,\exp\left(\frac{-Q}{kT}\right). \tag{17}$$

This is identical to the result of the detailed-balance theorem for $J^\pi(\alpha) = J^\pi(^{16}\mathrm{O}) = J^\pi(^{20}\mathrm{Ne}) = 0^+$ and $G(\alpha) = G(^{16}\mathrm{O}) = G(^{20}\mathrm{Ne}) = 1$, where G are the temperature-dependent normalized partition functions as, *e.g.*, defined in [2]. Following [2], $G(T)$ do not deviate more than 1% from unity for α, ^{16}O, and ^{20}Ne up to $T_9 = 3$.

3 Discussion

There are several interesting aspects which arise from the calculations in sect. 2. As already pointed out, the first excited states of ^{16}O and ^{20}Ne have relatively high excitation energies. *E.g.*, at $T_9 = 1$ this leads to occupation probabilities of 3×10^{-31} for the 0^+ state in ^{16}O at $E_x = 6049$ keV and 3×10^{-8} for the 2^+ state in ^{20}Ne at $E_x = 1634$ keV. The normalized partition functions practically do not deviate from unity up to $T_9 = 3$ [2].

The reaction rate $\langle\sigma v\rangle$ for the $^{16}O(\alpha,\gamma)^{20}Ne$ capture reaction can be determined from laboratory experiments because the rates under stellar and laboratory conditions are identical for this reaction:

$$\langle\sigma v\rangle^\star = \langle\sigma v\rangle^{\rm lab}\,. \qquad (18)$$

However, this is not the case for the reaction rate of the $^{20}Ne(\gamma,\alpha)^{16}O$ photodisintegration reaction under stellar conditions and in the laboratory:

$$\lambda^\star \neq \lambda^{\rm lab}. \qquad (19)$$

The stellar reaction rate $\lambda^\star$ for a resonance is given by eq. (16); the full resonance strength $(\omega\gamma)$ of the capture reaction contributes to the stellar photodisintegration rate $\lambda^\star$. The laboratory reaction rate λ^0 is reduced by the branching ratio B^0 to the ground state as can be seen from eq. (13). It is interesting to note that for the chosen example of the 1^- resonance at $E_R = 5788$ keV the laboratory rate λ^0 is only 18% of the stellar rate $\lambda^\star$ because of the branching ratio B^0. The first excited state at $E_x = 1634$ keV in ^{20}Ne contributes with 82% to the stellar rate $\lambda^\star$ although the thermal occupation probability is only 3×10^{-8} at $T_9 = 1$! The reason for this surprising contribution can be understood from eqs. (9) and (14). The small thermal occupation probability at $E_x = 1634$ keV is exactly compensated by the much higher photon density at the required energy $E_R^\gamma = E_R^\alpha + Q - E_x^{1634}$.

The contribution of the laboratory reaction rate $\lambda^{\rm lab}$ to the stellar rate $\lambda^\star$ (taken from the detailed-balance calculation of [2]) is shown in fig. 2. The contributions of these resonances to the stellar $^{20}Ne(\gamma,\alpha)^{16}O$ reaction rate $\lambda^\star$ are identical to the contributions in the $^{16}O(\alpha,\gamma)^{20}Ne$ reaction rate which is shown in fig. 1.

The ratio $\lambda^\star/\langle\sigma v\rangle$ does not depend on individual properties of the respective resonance. Consequently, eq. (17) is also valid for the direct capture contribution which becomes relevant at relatively low temperatures. The detailed-balance theorem hence is also applicable to reactions between nuclei with high-lying first excited states as, *e.g.*, the $^{16}O(\alpha,\gamma)^{20}Ne$ and $^{20}Ne(\gamma,\alpha)^{16}O$ reactions. Further details on the direct capture contribution at low energies are given in [7].

A further consequence of the relation between capture and photodisintegration reactions is that the so-called Gamow windows of both reactions have a strict relation. The Gamow window —the energy region where a reaction mainly operates— of a capture reaction is characterized by its position at

$$E_{\rm G} = 1.22\left(Z_1^2 Z_2^2 A_{\rm red} T_6^2\right)^{1/3}\ {\rm keV} \qquad (20)$$

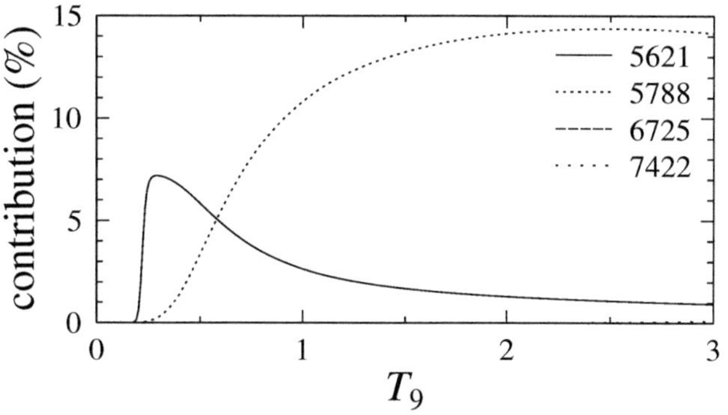

Fig. 2. Ground-state contribution $\lambda^{\rm lab}/\lambda^\star$ of individual resonances to the stellar reaction rate $\lambda^\star$ of the $^{20}Ne(\gamma,\alpha)^{16}O$ photodisintegration reaction as given by detailed balance [2]. The resonances are labelled by their energies E_R in keV. There is no contribution from the resonance at $E_R = 6725$ keV because of $B^0 = 0$ for this resonance. Note the different scale (compared to fig. 1).

and its width Δ

$$\Delta = 0.749\left(Z_1^2 Z_2^2 A_{\rm red} T_6^5\right)^{1/6}\ {\rm keV}. \qquad (21)$$

The corresponding Gamow window of the (γ,α) reaction is shifted by the Q value of the reaction

$$E_{\rm G}^\gamma = E_{\rm G}^\alpha + Q\,. \qquad (22)$$

The width Δ remains the same for the (α,γ) and (γ,α) reactions.

In the chosen example of the $^{16}O(\alpha,\gamma)^{20}Ne$ capture and $^{20}Ne(\gamma,\alpha)^{16}O$ photodisintegration reactions, one finds, *e.g.* at $T_9 = 1$, energies $E_{\rm G}^\alpha = 1141$ keV and $E_{\rm G}^\gamma = 5871$ keV and a width $\Delta = 725$ keV. As an obvious consequence the 1^- resonance at $E_R = 5788$ keV is dominating around $T_9 = 1$ (see also fig. 1). Properties of the Gamow window for (γ,α) reactions have been discussed in further detail in [8,9].

As shown above, the detailed-balance theorem holds for the $^{16}O(\alpha,\gamma)^{20}Ne$ and $^{20}Ne(\gamma,\alpha)^{16}O$ reactions provided that the population of excited states is thermal according to the Boltzmann statistics. How long does it take until thermal equilibrium is obtained in such cases where the first excited states are only weakly populated? In general, one can read from the formalism of photoactivation with a constant production rate (which is approximately fulfilled in the above case), that the time until equilibrium is reached is a few times the lifetime of the unstable product. In the case of the 2^+ state in ^{20}Ne at $E_x = 1634$ keV the mean lifetime is $\tau = 1.05$ ps [4]. Compared to the timescale of neon burning, which is of the order of years, thermal equilibrium is reached almost instantaneously.

4 Experiments with quasi-thermal photon spectra

Using a non-monochromatic photon spectrum, the experimental yield Y from a single resonance in the photodisintegration reaction $^{20}\mathrm{Ne}(\gamma,\alpha)^{16}\mathrm{O}$ is given by

$$Y = N_T n_\gamma(E_R) \left(\frac{\hbar c}{E_R}\right)^2 \pi^2 B^0 \frac{(2J_1+1)(2J_2+1)}{(2J_3+1)} (\omega\gamma). \tag{23}$$

Here N_T is the number of target nuclei, and $n_\gamma(E_R)$ is the number of incoming photons per energy interval and area at the energy of the resonance E_R. Using realistic numbers for present-day facilities and thin targets for direct detection of the α-particles, one obtains a relatively small yield. *E.g.*, $N_T \approx 10^{17}$, $n_\gamma \approx 10^5/\mathrm{keV\,cm^2\,s}$, and the weak 1^- resonance at $E_R = 5788\,\mathrm{keV}$ with the properties given in table 1, leads to $Y \approx 0.5/\mathrm{d}$. For stronger resonances the yields are higher by up to two orders of magnitude which makes experiments difficult but feasible with present-day facilities.

The astrophysical reaction rate λ^* for the $^{20}\mathrm{Ne}(\gamma,\alpha)^{16}\mathrm{O}$ photodisintegration reaction is well defined from experimental data of the $^{16}\mathrm{O}(\alpha,\gamma)^{20}\mathrm{Ne}$ capture reaction. Nevertheless, experimental yields for several resonances can be obtained simultaneously in one irradiation using the quasi-thermal spectrum which will become available at SPring-8 [6]. Consequently the ratios of all observed resonances will be measured in one irradiation —provided that the ground-state branching ratios are well known. This might help to resolve minor discrepancies between the adopted resonance strengths in [4] and [2].

There are further interesting experimental properties of the $^{20}\mathrm{Ne}(\gamma,\alpha)^{16}\mathrm{O}$ reaction which have to be discussed in the following. The analysis of experiments with non-monochromatic photons (see, *e.g.*, refs. [5,10,11,12]) requires the precise knowledge of the absolute number of incoming photons and their energy dependence. Using a non-monochromatic photon spectrum in combination with the $^{20}\mathrm{Ne}(\gamma,\alpha)^{16}\mathrm{O}$ reaction (or any other photodisintegration reaction which is dominated by narrow resonances) one finds emitted α-particles with discrete energies. Because the extremely high-lying first excited state in $^{16}\mathrm{O}$ is practically not populated in the $^{20}\mathrm{Ne}(\gamma,\alpha)^{16}\mathrm{O}$ reaction, the α energy is given by the difference between the resonance energy E_R and the Q value of the reaction. The experimental yield in each of the discrete lines is directly proportional to the resonance strength $\omega\gamma$, the ground-state branching B^0, and the number of incoming photons $n_\gamma(E_R)$ at resonance energy, as can be read from eq. (23). Provided that the resonance strengths and the branching ratios are well known, a $^{20}\mathrm{Ne}(\gamma,\alpha)^{16}\mathrm{O}$ experiment may help to determine the properties of the incoming-photon spectrum with good accuracy over a broad energy range. This is especially relevant for the measurement of (γ,α) reaction rates because the relevant energy region, the so-called Gamow window, is much broader than in the case of (γ,n) reactions which have mainly been analyzed in the last years. The intrinsic exponential decrease of photon intensity with energy for the photon source suggested in [6] may help to avoid the problems of reproducing the thermal photon distribution over a broad energy range which arise in the present technique using a superposition of bremsstrahlung spectra [5].

5 Conclusions

The relation between the $^{16}\mathrm{O}(\alpha,\gamma)^{20}\mathrm{Ne}$ capture reaction and the $^{20}\mathrm{Ne}(\gamma,\alpha)^{16}\mathrm{O}$ photodisintegration reaction has been discussed in detail. Whereas the stellar reaction rates of the $^{16}\mathrm{O}(\alpha,\gamma)^{20}\mathrm{Ne}$ reaction are identical to the laboratory rates [2], this is not the case for the photodisintegration rates under stellar and under laboratory conditions. Although the thermal population of the first excited state in $^{20}\mathrm{Ne}$ remains extremely small at typical temperatures of neon burning, it nevertheless provides an important contribution to the reaction rate under stellar conditions. The reason for this surprising behavior is that the increasing number of thermal photons at the relevant energy exactly compensates the small thermal occupation probability according to the Boltzmann statistics. The widely used detailed-balance theorem, which relates reaction rates of capture reactions to photodisintegration rates, remains valid for the case of the $^{16}\mathrm{O}(\alpha,\gamma)^{20}\mathrm{Ne}$ and $^{20}\mathrm{Ne}(\gamma,\alpha)^{16}\mathrm{O}$ reactions.

Additionally, it has been shown that the $^{20}\mathrm{Ne}(\gamma,\alpha)^{16}\mathrm{O}$ reaction may be helpful in calibrating new intense non-monochromatic photon sources as, *e.g.*, suggested in [6]. Such a calibration over a broad energy interval is especially relevant for (γ,α) experiments because of the broader Gamow window.

References

1. F.-K. Thielemann, W.D. Arnett, Astrophys. J. **295**, 604 (1985).
2. C. Angulo *et al.*, Nucl. Phys. A **656**, 1 (1999).
3. G. Audi, A.H. Wapstra, C. Thibault, Nucl. Phys. A **729**, 337 (2003).
4. D.R. Tilley *et al.*, Nucl. Phys. A **636**, 247 (1998); revised online version from 8 November 2000.
5. P. Mohr *et al.*, Phys. Lett. B **488**, 127 (2000).
6. H. Utsunomiya *et al.*, Nucl. Instrum. Methods Phys. Res. A **538**, 225 (2005).
7. P. Mohr, Phys. Rev. C **72**, 035803 (2005) and references therein.
8. P. Mohr *et al.*, Nucl. Phys. A **719**, 90c (2003).
9. P. Mohr, Proc. *Tours Symposium on Nuclear Physics V*, edited by M. Arnould *et al.*, AIP Conf. Proc. **704**, 532 (2004).
10. K. Vogt *et al.*, Phys. Rev. C **63**, 055802 (2001).
11. K. Vogt *et al.*, Nucl. Phys. A **707**, 241 (2002).
12. K. Sonnabend *et al.*, Phys. Rev. C **70**, 035802 (2004); **72**, 019901(E) (2005).

Eur. Phys. J. A **27**, s01, 79–82 (2006)
DOI: 10.1140/epja/i2006-08-011-0

EPJ A direct
electronic only

Enhanced d(d,p)t fusion reaction in metals

F. Raiola[1,a], B. Burchard[1], Zs. Fülöp[2], Gy. Gyürky[2], S. Zeng[3], J. Cruz[4], A. Di Leva[1], B. Limata[7], M. Fonseca[4], H. Luis[4], M. Aliotta[5], H.W. Becker[1], C. Broggini[9], A. D'Onofrio[6], L. Gialanella[7], G. Imbriani[7], A.P. Jesus[4], M. Junker[8], J.P. Ribeiro[4], V. Roca[7], C. Rolfs[1], M. Romano[7], E. Somorjai[2], F. Strieder[1], and F. Terrasi[6]

[1] Institut für Physik mit Ionenstrahlen, Ruhr-Universität Bochum, Germany
[2] Atomki, Debrecen, Hungary
[3] China Institute of Atomic Energy, Beijing, PRC
[4] Centro de Fisica Nuclear, Universidade de Lisboa, Portugal
[5] School of Physics, University of Edinburgh, UK
[6] Dipartimento di Scienze Ambientali, Seconda Università di Napoli, Caserta, Italy
[7] Dipartimento di Scienze Fisiche, Università Federico II and INFN, Napoli, Italy
[8] Laboratori Nazionali del Gran Sasso dell'INFN, Assergi, Italy
[9] INFN, Sezione di Padova, Padova, Italy

Received: 21 June 2005 /
Published online: 24 February 2006 –

Abstract. The electron screening in the d(d,p)t reaction has been studied for the deuterated metal Pt at a target temperature $T = 20$ °C to 340 °C, and for Co at $T = 20$ °C and 200 °C. The enhanced electron screening decreases with increasing temperature, where the data agree with the plasma model of Debye applied to the quasi-free metallic electrons. The data represent the first observation of a temperature dependence of a nuclear cross-section. We also measured the screening effect for the deuterated metal Ti (an element of group 4 of the periodic table) at $T = -10$ °C to 200 °C: above 50 °C the hydrogen solubility dropped to values far below unity and a large screening effect became observable. Similarly, all metals of groups 3 and 4 and the lanthanides showed a solubility of a few percent at $T = 200$ °C (compared to $T = 20$ °C) and a large screening became also observable. Within the Debye model the deduced number of valence electrons per metallic atom agrees with the corresponding number from the Hall coefficient, for all metals investigated.

PACS. 25.10.+s Nuclear reactions involving few-nucleon systems – 95.30.-k Fundamental aspects of astrophysics – 25.45.-z ^{2}H-induced reactions

1 Introduction

The cross-section of a charged-particle–induced nuclear reaction is enhanced at sub-Coulomb energies by the electron clouds surrounding the interacting nuclides, with an enhancement factor [1]

$$f_{lab}(E) = \frac{E}{E+U_e}\exp(-2\pi\eta(E+U_e)+2\pi\eta(E)), \quad (1)$$

for $S(E+U_e) \approx S(E)$, and where E is the center-of-mass energy, $\eta(E)$ the Sommerfeld parameter, and U_e the screening potential energy. The electron screening in d(d,p)t was studied previously for deuterated metals, insulators, and semiconductors, *i.e.* 58 samples in total [2,3] (see also [4,5]). As compared to measurements performed with a gaseous D_2 target ($U_e = 25\,\mathrm{eV}$ [6]), a large screening was observed in the metals (of order $U_e = 300\,\mathrm{eV}$), while a small (gaseous) screening was found for the insulators and semiconductors. An exception was found for the metals of groups 3 and 4 of the periodic table and the lanthanides, which showed a small screening; this is related to their high hydrogen solubility y ($= 1/x$ [2,3]), of the order of unity (see also [7]), that gives the deuterated targets of these metals the properties of insulators. Indeed, for the metals with high U_e values, the solubilities were small (a few percent) leaving the metallic character of the samples essentially unchanged. An explanation of the large screening was suggested [3] by the plasma screening of Debye applied to the quasi-free metallic electrons. The electron Debye radius around the deuterons in the lattice is given by

$$R_D = \sqrt{\frac{\varepsilon_o kT}{e^2 n_{eff}\rho_a}} = 69\sqrt{\frac{T}{n_{eff}\rho_a}} \quad [\mathrm{m}], \quad (2)$$

with the temperature T of the free electrons in units of K, n_{eff} the number of valence electrons per metallic

[a] e-mail: raiola@ep3.rub.de; for the LUNA Collaboration.

atom, and the atomic density ρ_a in units of atoms/m^3. With the Coulomb energy of the Debye electron cloud and a deuteron projectile at R_D set equal to $U_e = U_D$, one obtains

$$U_{e,D} = \frac{e^2}{(4\pi\varepsilon_o)R_D} = 2.09 \cdot 10^{-11}\sqrt{\frac{n_{eff}\rho_a}{T}} \text{ [eV]}. \quad (3)$$

A comparison of the calculated and observed U_e values led to n_{eff}, which was for most metals of the order of unity. The acceleration mechanism of the incident positive ions leading to the high observed U_e values is thus the Debye electron cloud at the small radius R_D, about one tenth of the Bohr radius. The n_{eff} values were compared with those deduced from the known Hall coefficient [8]: within 2 standard deviations the two quantities agreed for all metals. A critical test of the Debye model is the predicted temperature dependence $U_e(T) \propto T^{-1/2}$ (see also below); for deuterated Pt at $T = 20\,°\text{C}$ and $100\,°\text{C}$ the data agreed with prediction [3].

It is known [7] that the hydrogen solubility in metals decreases with increasing temperature. Thus, at higher temperatures the solubility of the metals of groups 3 and 4 and the lanthanides may be low enough (a few percent) that large U_e values can be observed and thus new n_{eff} values deduced. With the assumption that the temperature-dependent solubility $y(T)$ affects directly n_{eff}, one obtains

$$U_{e,D}(T) = \begin{cases} \frac{1}{4.78\cdot 10^{10}}\sqrt{\frac{(n_{eff}(T)(1-y(T))\rho_a}{T}} \text{ [eV]}, & y(T) \le 1, \\ 0, & y(T) > 1, \end{cases} \quad (4)$$

where a temperature dependence of $n_{eff}(T)$ is also taken into account. We report on the measurement of such temperature effects.

2 Setup and experimental procedure

The equipment, procedures, and data analysis have been described elsewhere [2,3]. Briefly, the surface of a given sample was —in a first step— cleaned *in situ* by Kr sputtering at 35 keV removing typically about 300 monolayers. The sample was —in a second step— deuterated at a given deuteron energy until a saturated yield was reached. This implantation procedure was repeated over the full energy range of the planned experiment taking typically about 4 days of running. Finally, the observed thick-target yield curve was differentiated to arrive at a thin-target yield curve, which was fitted using 2 free parameters, $y(T)$ and $U_e(T)$: the absolute yield provided information on the hydrogen solubility $y(T)$ and the energy dependence of the data gave the screening potential energy $U_e(T)$. We tested also the stability of the solubility against diffusion by switching the deuteron beam off for an extended period (typically 6 hours); the subsequent yield measurement was unchanged within experimental uncertainty indicating a stable solubility, both at room temperature and elevated temperatures. For the present measurements at elevated temperatures a new target holder was designed [9]. It consists of a diamond plate coated with a metallic layer (area $A = 20\times 20\,\text{mm}^2$, thickness $t = 1\,\text{mm}$) and heated by current flow. A given metal sheet ($A = 15\times 17\,\text{mm}^2$) is placed on top of the diamond plate with interim plates (from bottom to top) of MACOR ($t = 1\,\text{mm}$), Cu ($t = 3\,\text{mm}$), and MACOR ($t = 1\,\text{mm}$). At the center of the top MACOR plate there is a hole of $\varnothing = 5\,\text{mm}$ diameter, filled with another diamond of cylindrical shape and 2 mm height: it provides the thermal contact to the metal foil. The metal foil is electrically insulated for current measurement. Thermal elements measure the temperature at the diamond plate and the metal foil (near the area of the ion beam spot). The Si detectors in close geometry to the metal foil were cooled to $0\,°\text{C}$ using an Ultra Kryomat. The beam direction and spot on target were defined by 2 apertures, one of $\varnothing = 3\,\text{mm}$ at a distance $d = 62\,\text{cm}$ from the target and the other of $\varnothing = 6\,\text{mm}$ at $d = 280\,\text{cm}$; an electric quadrupole triplet placed between the 2 apertures was used to focus the beam. The beam current on target was kept below $2\,\mu\text{A}$ leading to a negligible influence on the target temperature (less than $2\,°\text{C}$ variation).

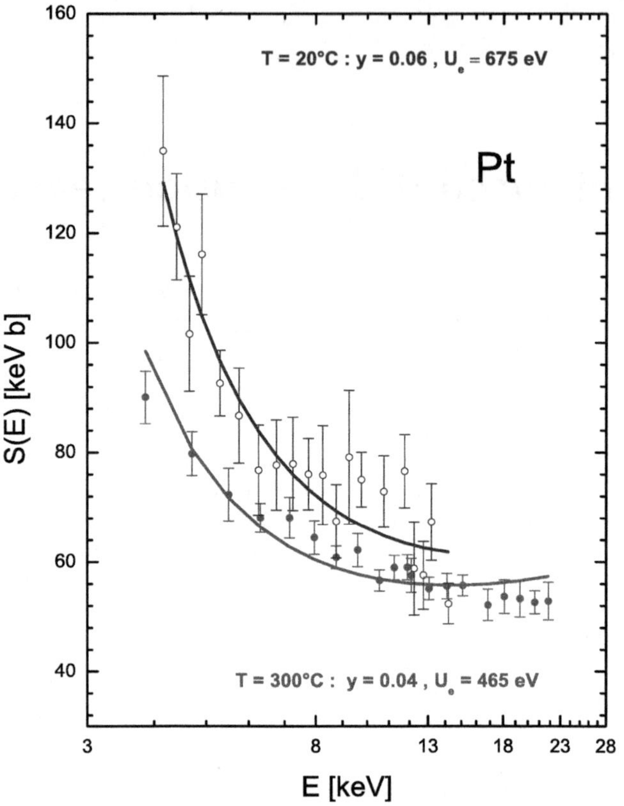

Fig. 1. $S(E)$ factor of d(d,p)t for Pt at $T = 20\,°\text{C}$ and $300\,°\text{C}$, with the deduced solubilities y. The curves through the data points include the bare $S(E)$ factor and the electron screening with the U_e values given.

3 Temperature dependence of Pt and Co

We measured the screening effect for the metal Pt at a target temperature between $T = 20\,°\text{C}$ and $340\,°\text{C}$, and

Table 1. Summary of the results.

Material[a]	T(°C)	U_e(eV)[b]	Solubility y[c]	n_{eff}[b]	n_{eff}(Hall)[d]
		Groups 3 and 4 and lanthanides			
Ce	200	200±50	0.11	1.5±0.7	(1.2±0.2)
Dy	200	340±70	0.09	4.9±2.0	1.5±0.3
Er	200	360±80	0.05	4.3±1.9	6±1
Eu	200	120±60	0.05	0.8±0.8	
Gd	200	340±85	0.08	4.2±2.1	2.2±0.4
Hf	200	370±70	0.04	4.0±1.5	(3.2±0.6)
Ho	200	165±50	0.07	0.9±0.5	
La	200	245±70	0.09	2.4±1.4	2.9±0.6
Lu	200	265±70	0.08	2.2±1.2	3.4±0.7
Nd	200	190±50	0.08	1.4±0.7	(2.2±0.4)
Sc	200	320±50	0.11	2.6±0.8	2.2±0.4
Sm	200	314±60	0.08	3.5±1.3	10±2
Tb	200	340±80	0.18	3.9±1.8	
Tm	200	260±80	0.05	2.2±1.4	1.0±0.2
Y	200	270±75	0.09	2.6±1.4	2.7±0.5
Yb	200	110±40	0.13	0.4±0.3	(0.6±0.1)
Zr	200	205±70	0.13	1.1±0.7	(1.1±0.2)
		Insulators			
C	200	≤ 50	0.15		
		T-dependence of Co and Pt			
Co	20	640±70	0.14		
	200	480±60	0.02		
Pt	20	675±50	0.06		
	100	530±40	0.06		
	200	530±40	0.05		
	300	465±38	0.04		
	340	480±70	0.04		
		T-dependence of Ti			
Ti	-10	≤ 30	2.1		
	50	≤ 50	1.1		
	100	250±40	0.26		
	150	295±40	0.23		
	200	290±65	0.20	1.7±0.7	4±1

[a] For details see ref. [9].
[b] Error contains no systematic uncertainty in energy dependence of stopping power.
[c] Estimated uncertainty is about 20%.
[d] From the observed Hall coefficient, with an assumed 20% error; the numbers in brackets are for hole carriers.

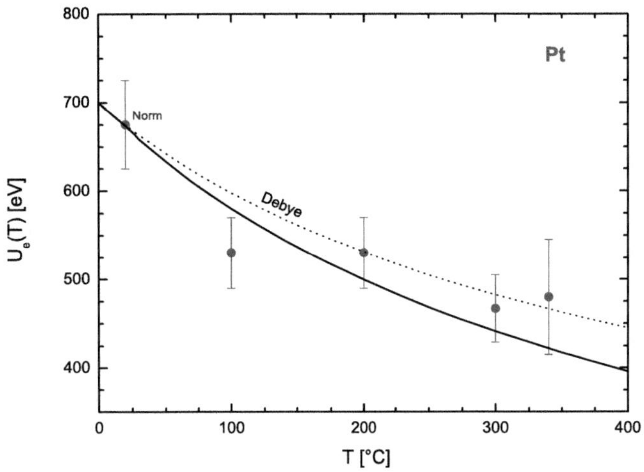

Fig. 2. The observed values $U_e(T)$ for Pt is shown as a function of sample temperature T. The dotted curve represents the prediction of the Debye model (eq. (5)) and the solid curve includes the observed T-dependence of the Hall coefficient [8, 10], *i.e.* $n_{eff}(T)$.

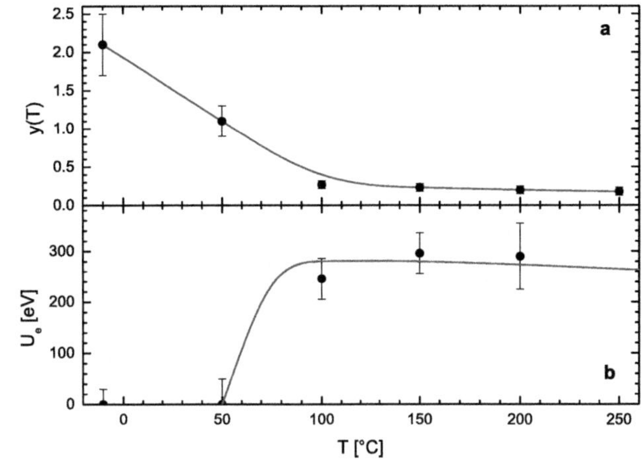

Fig. 3. a) Hydrogen solubility $y(T)$ in Ti as a function of sample temperature T. The curve through the data points is to guide the eye only. b) Observed $U_e(T)$ values as a function of T. The curve through the data points uses eq. (4) together with the observed $y(T)$ values and $n_{eff}(\mathrm{Ti}) = 1.7$.

for Co at $T = 20\,°\mathrm{C}$ and $200\,°\mathrm{C}$. Both metals have a solubility of a few percent at all T of the present work. The astrophysical $S(E)$ factor obtained at $T = 20\,°\mathrm{C}$ and $300\,°\mathrm{C}$ for Pt is shown in fig. 1. The results for Co and Pt are given in table 1 and the $U_e(T)$ values for Pt are plotted in fig. 2 together with the expected dependence $U_e(T) \propto T^{-1/2}$ (dotted curve). All data show a decrease of the screening, *i.e.* the U_e value, with increasing temperature: Over the present temperature range, the reported Hall coefficient for Pt increases by about 20% [8,10] leading to a corresponding decrease in n_{eff}, which we took into account (solid curve in fig. 2); there is good agreement between observation and expectation. The data represent the first observation of a temperature dependence of a nuclear cross-section.

4 Temperature dependence of metals with high solubility at T = 20 °C

We studied the electron screening effect for the deuterated metal Ti (group 4) at $T = -10\,°\mathrm{C}$ to $200\,°\mathrm{C}$, in steps of $50\,°\mathrm{C}$. The deduced solubility $y(T)$ is shown in fig. 3 and shows a sizable decrease with increasing temperature, where above $50\,°\mathrm{C}$ the solubility reaches values below unity and thus an enhanced screening should be observable at these higher temperatures. The observed U_e values (fig. 3b) verify this expectation, where the solid curve in fig. 3b uses eq. (4) including the observed $y(T)$-dependence and $n_{eff}(\mathrm{Ti}) = 1.7$ (table 1); the analysis indicates a maximum effect around $100\,°\mathrm{C}$. Note that a

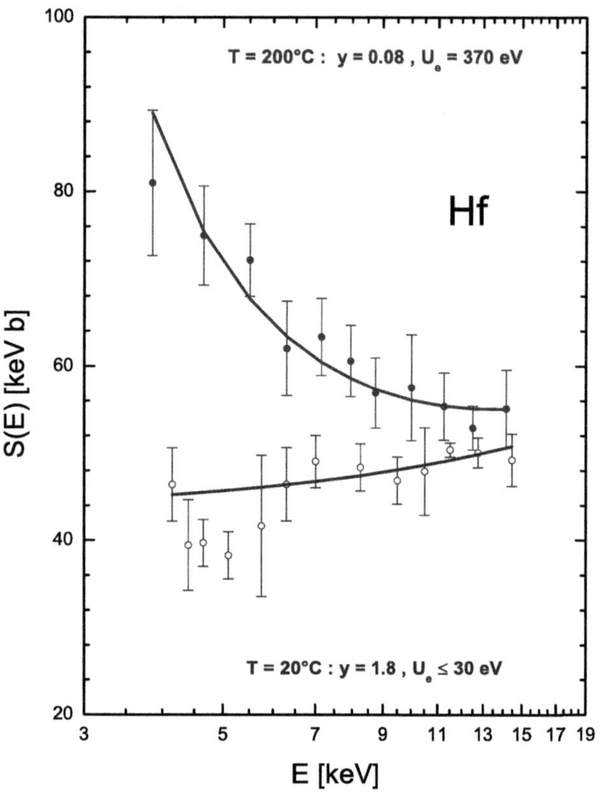

Fig. 4. $S(E)$ factor of d(d,p)t for Hf at $T = 200\,°\mathrm{C}$ and $T = 20\,°\mathrm{C}$, with the deduced solubilities y. The curve for $T = 20\,°\mathrm{C}$ represents well the bare $S(E)$ factor, while the curve for $T = 200\,°\mathrm{C}$ includes the electron screening with the U_e value given.

solubility of 10% leads to a 5% reduction in the maximum value of U_e, according to eq. (4).

Finally, all metals of groups 3 and 4 and the lanthanides have been studied at $T = 200\,°\mathrm{C}$. The astrophysical $S(E)$ factor obtained at $T = 200\,°\mathrm{C}$ for Hf is compared in fig. 4 with that obtained at $T = 20\,°\mathrm{C}$: at $T = 200\,°\mathrm{C}$ the solubility is reduced to a few percent and a large screening became observable, similarly as for Ti. In fact, all these metals exhibited a large reduction in solubility and thus showed a large screening, as expected according to eq. (4). The results for all metals are summarized in table 1, which also compares the deduced n_{eff} values with those from the Hall coefficient: there is again an agreement between both quantities within two standard deviations, for all metals of the present and previous work [3], *i.e.* 49 metals in total. As a consistency test we also studied the insulator C at $T = 200\,°\mathrm{C}$: the solubility decreased from 0.35 ($T = 20\,°\mathrm{C}$) to 0.15, but no enhanced screening was observed, as expected for an insulator with $n_{eff} = 0$ (eq. (4)).

5 Discussion

All data on the enhanced electron screening in deuterated metals can be explained quantitatively by the Debye model applied to the quasi-free metallic electrons. It was argued [3] that most of the conduction electrons are frozen by quantum effects and only electrons close to the Fermi energy (E_F) actually should contribute to screening, with

$$n_{eff}(T) = 0.67\frac{kT}{E_F} \propto T\,, \tag{5}$$

and thus there should be essentially no temperature dependence for $U_{e,D}$. However, this argument applies only to insulators and semiconductors with a finite energy gap, while for metals there is no energy gap and the Fermi energy lies within the conduction band. Note that the observed $n_{eff}(T)$ from the Hall coefficient decreases with increasing T, *e.g.* for Pt, contrary to eq. (5). Clearly, an improved theory is highly desirable to explain why the simple Debye model appears to work so well. Without such a theory, one may consider the Debye model as a powerful parameterization of the data.

This work was supported by BMBF (05CL1PC1), DFG (Ro429/31, 436Ung113), AvH (V-8100/B-ITA1066680), OTKA (T42733, T49245), China (2003CB716704), Portugal (FNU-45092-2002) and Dynamitron-Tandem-Laboratorium.

References

1. H.J. Assembaum, K. Langanke, C. Rolfs, Z. Phys. **327**, 461 (1987).
2. F. Raiola, *et al.*, Eur. Phys. J. A **13**, 377 (2002).
3. F. Raiola, *et al.*, Eur. Phys. J. A **19**, 283 (2004).
4. J. Kasagi *et al.*, J. Phys. Soc. Jpn. **71** 2881 (2002); J. Kasagi *et al.*, J. Phys. Soc. Jpn. **73**, 608 (2004).
5. K. Czerski, A. Huke, A. Biller, P. Heide, M. Hoeft, G. Ruprecht, Europhys. Lett. **54**, 449 (2001).
6. U. Greife, F. Gorris, M. Junker, C. Rolfs, D. Zahnow, Z. Phys. A **351**, 107 (1995).
7. A. Züttel, Naturwissenschaften **91**, 157 (2004).
8. C.M. Hurd, *The Hall Effect in Metals and Alloys* (Plenum Press, 1972).
9. F. Raiola, PhD Thesis, Ruhr-Universität Bochum (2005).
10. Landolt-Börnstein, Vol. **II.6** (Springer, Berlin, 1959).

Eur. Phys. J. A **27**, s01, 83–88 (2006)
DOI: 10.1140/epja/i2006-08-012-y

EPJ A direct
electronic only

Experimental and theoretical screening energies for the ^{2}H(d, p)^{3}H reaction in metallic environments

K. Czerski[1,2,a], A. Huke[1], P. Heide[1], and G. Ruprecht[1,3]

[1] Institut für Atomare Physik und Fachdidaktik, Technische Universität Berlin, Hardenbergstr. 36, 10623 Berlin, Germany
[2] Institute of Physics, University of Szczecin, Szczecin, Poland
[3] TRIUMF, Vancouver, Canada

Received: 13 July 2005 /
Published online: 23 February 2006 –

Abstract. The study of the ^{2}H(d, p)^{3}H reaction at very low energies in deuterized metallic targets provides a unique possibility to test models of the electron screening developed for dense astrophysical plasmas. Here, we compare the experimental screening energies obtained by our group as well as by other authors for different target materials with theoretical predictions based on an improved dielectric function theory. The calculations are performed within the self-consistent regime and include polarization of both quasi-free and bound electrons. Additionally, the cohesion screening, arising from different binding energies of deuterons and α-particles in crystal lattices, is taken into account. The proposed theory predicts only a weak material dependence of the screening energy in agreement with our experimental results but fails in the absolute strength of the effect by a factor of 2. The projectile-velocity dependence of the screening energy corresponding to the transition from the weak-screening regime to the strong-screening limit is discussed.

PACS. 25.45.Hi Transfer reactions – 95.30.Dr Atomic processes and interactions – 95.30.-k Fundamental aspects of astrophysics

1 Introduction

Electron screening of the Coulomb barrier between reacting nuclei leads to an enhancement of thermonuclear rates in dense astrophysical plasmas. For so-called weakly coupled plasmas (for example our Sun), where the kinetic energy of plasma particles is larger than the mean Coulomb repulsion energy, the electron screening contributes only to a few percent and can be described within the Debye-Hückel model [1] of the nearly perfect stellar gas. In the opposite limit of strongly coupled plasmas, at high densities and low temperatures the electron gas is degenerate and the ions undergo long-range correlation forces forming either a quantum liquid or a Coulomb lattice beyond a critical density. In such a case nuclear reaction rates can be increased by many orders of magnitude and are probably realized in White and Brown Dwarfs or Giant Planets. The study of d + d nuclear reactions at very low energies on deuterons embedded in metallic lattices makes it possible to test models of the electron screening developed for dense astrophysical plasmas in the terrestrial laboratory. The exponential-like increase of the reaction cross-section observed for decreasing projectile energies, as compared to the cross-section for bare nuclei, can be described by a screening energy. As could be shown in our first experiments [2,3], the screening energies determined for the d + d fusion reactions in metallic environments are by about a factor of 10 larger than that observed for the gas target [4] and up to a factor of 4 larger than the theoretical predictions [5]. This finding was also confirmed by results of other groups [6,7,8,9,10,11]. Especially, the data obtained by the LUNA Collaboration for almost 60 different target materials [11] enable us to compare the experimental results of different groups and to look for a theoretical description of the observed target material dependence as well as of the absolute screening energy values. The approach presented here is based on an improved dielectric function theory [12,13] that allows to derive a reliable deuteron-deuteron potential in the host metal including contributions not only from quasi-free valence electrons but also from polarized bound electrons. A special interest will be devoted to the dependence of the screening energy on projectile energies.

2 Experimental screening energy

Our experiments have been performed [3,13,14] using the D^+ and D_2^+ beams accelerated to energies between 5 and

[a] e-mail: czerski@kalium.physik.tu-berlin.de

60 keV and impinged on metallic targets (Li, Al, Zr, Pd and Ta) and amorphous graphite carbon foils. Most of the targets were implanted to large deuteron densities corresponding approximately to the chemically stable stoichiometry. In the case of Pd the implantation process was interrupted at a relatively small deuteron density (stoichiometric ratio $n_{\rm d}/n_{\rm Pd} \approx 0.3$) in order to study fusion reactions in the metallic environment possessing a small number of crystal-lattice defects and reducing the number of possible deuterium bubbles resulting from long-term irradiation. The deuteron density used for the Li target was even smaller ($n_{\rm d}/n_{\rm Li} \approx 0.03$) to prevent the target from oxidation. The reaction products (protons, tritons and ^{3}He particles) were detected by four Si detectors located in the reaction plane at backward angles.

The experimental determination of the electron screening energy is based on the assumption that the observed exponential-like enhancement of the reaction yield towards low projectile energies results from the reduction of the height of the Coulomb barrier. In the simple case of Bohr screening, the screened Coulomb potential energy between two reacting deuterons can be presented as follows:

$$V(r) = \frac{e^2}{r}\exp\left(-\frac{r}{a}\right) \approx \frac{e^2}{r} - \frac{e^2}{a}, \qquad (1)$$

where a is the screening length being of the order of the Bohr radius. For projectile energies used in accelerator experiments where $r \ll a$, the deuteron-deuteron potential can be simply described as the Coulomb potential reduced by a constant, the screening energy $U_e = e^2/a$. Thus, the "screened" cross-section, applying the transformation to the only weakly on energy dependent astrophysical S-factor, reads as follows:

$$\sigma_{scr}(E_{cm}) = \frac{1}{\sqrt{E_{cm}(E_{cm}+U_e)}} S(E_{cm}) \times \exp\left(-\sqrt{\frac{E_G}{E_{cm}+U_e}}\right). \qquad (2)$$

Here E_{cm} denotes the energy in the center-of-mass system and E_G is the Gamow energy. The screening energy U_e takes into account a drop of the Coulomb barrier in the expression for the penetration factor. In the experiment the strength of the screening effect is described by means of the thick-target enhancement factor $F(E)$ defined as the ratio between the angular integrated thick-target yields for screened and bare nuclei [3],

$$F(E) = \frac{Y_{scr}(E)}{Y_{bare}(E)} = \frac{\int_0^E \frac{\sigma_{scr}(E)}{\varepsilon(E)}\,\mathrm{d}E}{\int_0^E \frac{\sigma(E)}{\varepsilon(E)}\,\mathrm{d}E} = \frac{\int_0^E \frac{\sigma_{scr}(E)}{\sqrt{E}}\,\mathrm{d}E}{\int_0^E \frac{\sigma(E)}{\sqrt{E}}\,\mathrm{d}E}. \qquad (3)$$

Here, $\sigma(E)$ and $\varepsilon(E)$ are the cross-sections for bare nuclei and the stopping power taken at the beam energy E, respectively. The enhancement factor $F(E)$ is independent of the target deuteron density and of the absolute value of the stopping power; the only assumption used is that $\varepsilon(E) \propto E^{1/2}$ which agrees with the experimental data for all target materials investigated. The bare-nuclei cross-section is very well known from the precision measurements performed with the gas target [15]. From the exponential increase of $F(E)$ towards lower projectile energies the screening energy U_e can be determined. The U_e values obtained in our experiments for C, Li, Al, Zr, Pd and Ta targets are presented in fig. 2. For heavier metals the screening energy amounts to about 300 eV which is one order of magnitude larger than the value 25 ± 5 eV obtained in the gas target experiment [4].

3 Theoretical description of the electron screening effect

A charge point impurity embedded in a metallic environment leads to a polarization of surrounding degenerate electrons causing a cut-off of the screened Coulomb field at a characteristic distance of the inverse of the Fermi wave number. Additionally, the bound electrons of the host metal can also be polarized and contribute to the screening. Thus, using the standard Fourier representation of $1/r$, the screened Coulomb potential energy $V(r)$ between reacting deuterons within a static approximation can be expressed as follows [13]:

$$V(r) = \frac{e^2}{r}\Phi(r) = \frac{1}{(2\pi)^3}\int \frac{4\pi\,(e\varphi(q))^2}{\varepsilon_\nu(q)\,\varepsilon_c(q)\,q^2}\exp(i\boldsymbol{q}\boldsymbol{r})\,\mathrm{d}^3q, \qquad (4)$$

where $\varepsilon_\nu(q)$ and $\varepsilon_c(q)$ are the static wave-number–dependent dielectric functions resulting from quasi-free valence electrons and from bound metallic core electrons, respectively, and $\Phi(r)$ denotes the screening function. The elementary charge e is multiplied by a self-consistent charge form factor $\varphi(q)$ for deuterons with the screening electrons in the Thomas-Fermi approximation:

$$\varphi(q) = 1 - z + zq^2/\left(q^2 + k_{TF}^2\right). \qquad (5)$$

Here, the Thomas-Fermi wave number $k_{TF}^2 = 6\pi e^2 n/E_F$ has been used; n and E_F are the electron number density and the Fermi energy, respectively. The number z corresponds to the fraction of electrons bound to deuterons and can vary between 0 and 1. Since we are interested in the evaluation of the strongest possible screening effect, we set $z = 1$. In the absence of screening $\varepsilon_\nu \equiv \varepsilon_c \equiv 1$ and $z = 0$, $V(r)$ reduces to the bare Coulomb potential ($\Phi(r) \equiv 1$). The response of the valence electron gas to an external field is given by the dielectric function:

$$\varepsilon_\nu(q) = 1 - \frac{\nu(q)P(q)}{1+\nu(q)G(q)P(q)}, \qquad (6)$$

where $\nu(q) = 4\pi e^2/q^2$ and $P(q)$ is the static Lindhard RPA polarizability [16]. $G(q)$ is the static local field correction that takes into account the short-range electron correlation and the exchange interaction [17].

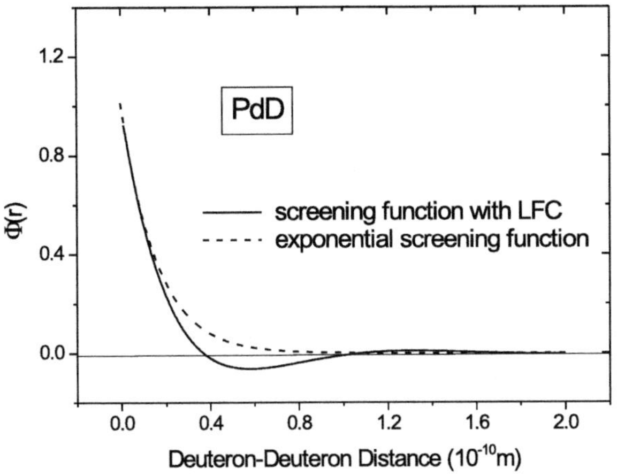

Fig. 1. Screening function calculated for PdD with the local field correction (LFC). For comparison the Bohr screening function with the same screening length is presented.

If we set $G(q) = 1$ and apply the long-wave approximation [12], the expression for the valence electron dielectric function (eq. (6)) reduces to the Thomas-Fermi form

$$\varepsilon_{TF}(q) = 1 + \frac{k_{TF}^2}{q^2}\,. \tag{7}$$

In this case the screening function can be described by the exponential function $\exp(-k_{TF}r)$ leading to the screening energy $e^2k_{TF} = e^2(4me^2/\pi\hbar^2)^{1/2}(3\pi 2n)^{1/6}$. Hence, the corresponding value depends only weakly on the electron density and amounts for Pd to 54 eV.

In the case of core electron polarization we applied the dielectric function proposed in [18]. Different from the valence electron polarization, ε_c takes a finite value at the limit $q = 0$. In the case of Ta the core-dielectric constant $\varepsilon_c(0) = 3.21$. The screening function $\Phi(r)$ calculated by a numeric integration of eq. (4) differs from the simple Bohr screening $\exp(-r/a)$ particularly for larger distances where the numeric potential becomes negative and shows characteristic Friedel oscillations. For smaller distances the potential becomes attractive reducing appropriately the screening length (see fig. 1).

In the metallic lattice, besides electrons also positive ions can contribute to the screening of the Coulomb barrier between reacting nuclei. This effect, called cohesion screening, can be calculated in analogy to the dense astrophysical plasmas within the ion-sphere model of Salpeter [1] providing in the case of the TaD target a screening energy of 18 eV. In our calculations we used a more realistic model based on the universal ion-ion potential introduced by Biersack [19]. This potential describes the interaction between light ions as well as between heavy ions at low energies with very good accuracy. Since the potential energy of two deuterons in the field of a host metal atom is larger than that of the helium atom produced in the fusion reaction, one obtains a gain in potential energy. For a rough estimation of the cohesion screening energy U_{coh}, we calculated the potential energy gain resulting from the surrounding 12 host atoms assuming the same fcc crystal structure for all target materials investigated [13].

The above description of the screening effect is limited to the charged particles with a velocity lower than the Fermi velocity v_F, for which the adiabatic approximation can be used. For higher velocities the electrons have not enough time to follow the ions and a wake wave [20] trails the ion through the electron gas. Thus, the electron screening gets weaker and depends on the velocity v of the ion. The screening length for the dynamic screening can be expressed by $a_d = v/\omega_p$, where ω_p is the plasmon frequency $\omega_p^2 = 4\pi ne^2/m$. Since the ion velocity v can be treated as velocity of electrons relative to the resting ion, the dynamic screening can also be applied to hot plasmas where the electron velocity arises from the plasma temperature T. Then $v^2 = k_BT/m$, where k_B is the Boltzmann constant. Consequently, the screening length in a hot plasma reads as follows:

$$a_{hp}^2 = \frac{k_BT}{4\pi e^2 n} \tag{8}$$

which corresponds to the Debye radius determining the electron screening within the Debye-Hückel model. In this sense, the velocity dependence of the screening length can describe the transition between a weak electron screening (hot plasmas) for $v > v_F$ and the strong electron screening (cold plasmas) for $v < v_F$. A corresponding formula has been proposed by Lifschitz and Arista [21],

$$\frac{1}{a^2(v)} = \gamma\frac{2v_F}{\pi}\left(1 + \frac{v_F^2 - v^2}{2v_F v}\ln\left|\frac{v + v_F}{v - v_F}\right|\right), \tag{9}$$

where γ is a number factor depending on the form of the screening function $\Phi(r)$. For small v the expression in parenthesis reaches the limit of 2 and the screening length its minimum value (strong screening).

Since the deuteron energies, for which an enhancement of the d + d reaction cross-sections due to the electron screening can be observed, are smaller than the corresponding deuteron Fermi energy (for Al $E_F(\mathrm{d}) = 46$ keV),

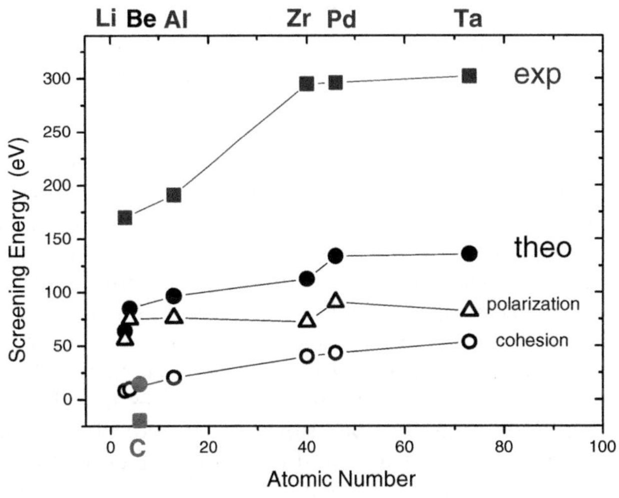

Fig. 2. Experimental and theoretical electron screening energies.

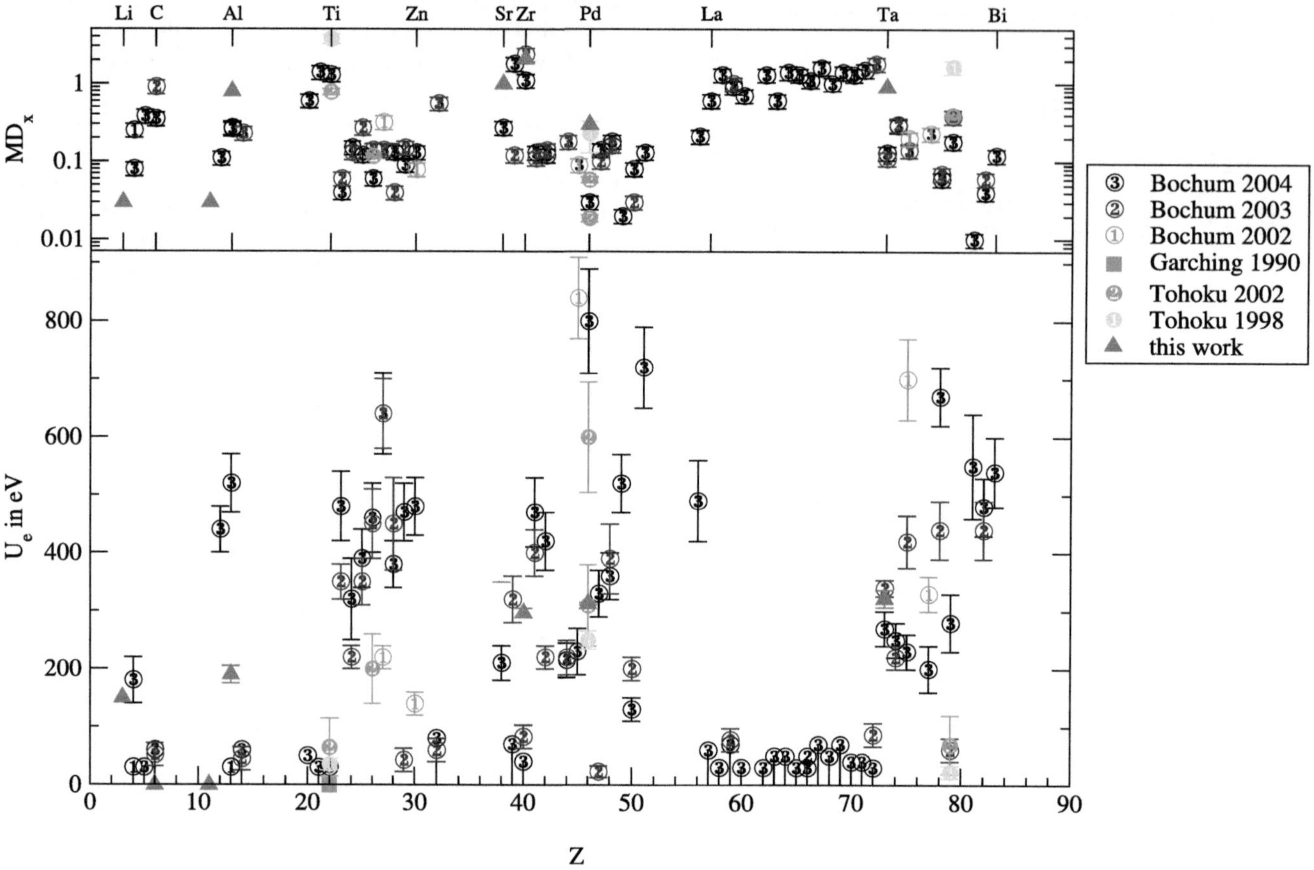

Fig. 3. Comparison of experimental screening energies. Also depicted are the corresponding densities of the implanted deuterons.

we consider the experimentally determined screening energy independent of velocity. Thus, the experimental values of U_e can be directly compared with the theoretical ones according to the prescription

$$U_{pol} = \lim_{r \to 0} \left(\frac{e^2}{r} - \frac{e^2}{r} \Phi(r) \right). \tag{10}$$

The theoretical value for the total electron screening energy U_e is a sum of the polarization and cohesion screening energies $U_{pol} + U_{coh}$. The theoretical and experimental U_e values determined for all target materials we have investigated are presented in fig. 2.

The theoretical calculations describe the observed material dependence of the screening energy qualitatively correctly. The main contribution to the theoretical values is provided by polarization of the free valence electrons [13], although the contribution of bound electrons (core polarization) cannot be neglected. In the case of TaD, the resulting core polarization energy amounts to about 1/3 of the valence electron screening energy. An increase of U_e with the atomic number arises mainly from the cohesion contribution. However, the absolute values of the theoretically calculated U_e fail by a factor of about two as compared to our experimental values. Including the self-consistent correction and the full wave number dependence of the dielectric function leads to screening energies lower than those determined within the simplified theory [12]. No reason for such a large discrepancy between theoretical and experimental values has been found so far. Even if a possible contribution of the channeling effect to the experimentally determined U_e values would be taken into account —in the case of Ta much smaller than 100 eV [22]— the difference between experiment and theory remains large.

4 Comparison with results of other authors

The screening energies measured by different groups together with the deuteron densities achieved in the experiments are presented in fig. 3. The largest part of data was obtained by the LUNA Collaboration (Bochum group) which measured the screening energies for more than 50 metals in three different experiments. Compared to our U_e values that show only a weak material dependence and a kind of saturation for heavier metals (Zr, Pd, Ta) with a screening energy of about 300 eV, there are partially large deviations.

The screening energy determined for Ta by the LUNA Collaboration [8] (309±12 eV) and for Pd by the Japanese group (Tohoku 1998 and 2002) [6,7] (310 ± 50 eV) are very close to our values 302 ± 13 eV and 296 ± 15 eV, respectively. On the other hand, the corresponding Pd value obtained by the LUNA Collaboration is much larger amounting to 800 ± 90 eV [11]. For Zr the LUNA Collaboration reported a significantly smaller U_e value than ours, whereas the values obtained for Al changed from a low value in the first experiment to a large one in the third experiment where the target surface was cleaned by Kr sputtering immediately before the deuteron incidence.

The strong variation of the experimental screening energies for different metals, as depicted in fig. 3, contradicts our results and cannot be explained within the proposed theory. For some metals, the experimental screening energies are even smaller than theoretical ones. In order to explain this, an application of the Debye-Hückel model was suggested [10,11,23]. The authors, setting room temperature into the expression for the Debye screening length (eq. (8)) and comparing with experimental U_e values, obtain charge carrier densities which are close to those determined from experimental Hall constants. Consequently, the experimental screening energies should be dependent on temperature and proportional to the density of charge carriers, *i.e.* electrons and holes. However, as shown in the previous section, the Debye-Hückel screening is applicable only for large temperature ($k_BT > E_F$) for which the electron degeneration vanishes and the Maxwell-Boltzmann statistics can be used. For low temperatures ($k_BT < E_F$) or correspondingly low projectile energies, the strong-screening limit should be applied. According to eq. (9), the screening length within this limit is smaller than the Debye length at the temperature $k_BT = E_F$ by a factor of $\sqrt{2}$. By the same factor the screening energy increases for low velocities. Additionally, no temperature dependence should be observed for the strong screening. Furthermore, the dominant contribution to the screening effect, the valence electron polarization, should only weakly depend on the electron density, in accordance with the Fermi-Dirac model $U_{FD} \propto n^{1/6}$ (see fig. 2).

Thus, the strong variation of U_e observed in some experiments probably arises from experimental problems with keeping constant a homogeneous deuteron density in the metallic targets. Contrary to our experiments, the large screening energies were measured using targets with relatively low deuteron densities. This can cause a temporal increase of the deuteron density in the surface region during the irradiation by projectile with lower energies and lead to an artificially large U_e values. In turn, too low screening energies can result from an oxidation layer on the target surface. Since the deuteron density in such a layer is much smaller than in the metallic bulk, the increase of the cross-section at low beam energy should be reduced [24]. Even cleaning of the target surface by a sputtering gun cannot help much under high-vacuum conditions, since in a vacuum of order 10^{-8} mbar the targets can re-oxidize within a few minutes (see, for example, [25]). This effect depends very strongly on the chemical reactivity of the target material and can be, on the other hand, reduced by the sputtering process of the target surface during the deuteron irradiation, which is, however, also target material dependent. Thus, the small value of U_e for some metals being significantly smaller than both our experimental and theoretical values, might be due to the re-oxidation process of the target. For a further detailed discussion of experimental results, we refer to our forthcoming paper [26].

5 Discussion and conclusions

In contradiction to results of the LUNA Collaboration, our experimental screening energies show only a weak target material dependence. As already stated above, discrepancies probably arise from an inhomogeneous depth distribution of deuterons within the irradiated targets. The situation can certainly be improved in the future experiments performed under ultra-high–vacuum conditions with an on-line monitoring of the deuteron density.

Clearly, the target material dependence of the screening energy is very important for the theoretical description of the effect. The improved dielectric function theory presented here supports a weak target material dependence of the screening energy. The theory provides, however, absolute values being by a factor of 2 smaller than the experimental ones. Therefore, one of the aims of future experiments remains to prove which screening contribution —valence electron polarization, core electron polarization or cohesion screening— is enhanced in the deuterized metals. Such a test is also very interesting for the physics of dense astrophysical plasmas. A large advantage of the presented theoretical approach is its ability to determine the deuteron-deuteron potential also for large distances (fig. 1). This enables to calculate the effective screening energies down to room temperature and consequently to compare the experimental results at higher energies with those achieved in the cold-fusion experiments by means of the heavy-water electrolysis. As shown in [13], the screening energy of order 300 eV determined in accelerator experiments can explain the neutron production rate observed by Jones *et al.* [27] at room temperature. Much larger U_e values of order 750 eV obtained in some accelerator experiments would increase the neutron production rate at room temperature by a factor of 10^7, which is, however, not observed.

The method proposed to include the velocity dependence for the dynamic screening allows to demonstrate the transition from the weak- to the strong-screening regime. Since the electron screening effect in the nuclear reactions is observable only at very low projectile energies, the theoretical description in the frame of the adiabatic dielectric function theory is well founded. On the contrary, the model based on the Debye-Hückel theory is for low temperatures and projectile velocities below the Fermi velocity not applicable.

References

1. E.E. Salpeter, Aust. J. Phys. **7**, 373 (1954).
2. K. Czerski, A. Huke, P. Heide, M. Hoeft, G. Ruprecht, in *Nuclei in the Cosmos V, Proceedings of the International Symposium on Nuclear Astrophysics, Volos, Greece, July 6-11 1998*, edited by N. Prantzos, S. Harissopulos (Editions Frontières, 1998) p. 152.

3. K. Czerski, A. Huke, A. Biller, P. Heide, M. Hoeft, G. Ruprecht, Europhys. Lett. **54**, 449 (2001).
4. U. Greife, F. Gorris, M. Junker, C. Rolfs, D. Zahnow, Z. Phys. A **351**, 107 (1995).
5. S. Ichimaru, Rev. Mod. Phys. **65**, 252 (1993).
6. H. Yuki, J. Kasagi, A.G. Lipson, T. Ohtsuki, T. Baba, T. Noda, B.F. Lyakhov, N. Asami, JETP Lett. **68**, 823 (1998).
7. J. Kasagi, H. Yuki, T. Baba, T. Noda, T. Ohtsuki, A.G. Lipson, J. Phys. Soc. Jpn. **71**, 2281 (2002).
8. F. Raiola *et al.*, Eur. Phys. J. A **13**, 377 (2002).
9. F. Raiola *et al.*, Phys. Lett. B **547**, 193 (2002).
10. C. Bonomo *et al.*, Nucl. Phys. A **719**, 37c (2003).
11. F. Raiola *et al.*, Eur. Phys. J. A **19**, 283 (2004).
12. K. Czerski, A. Huke, P. Heide, Nucl. Phys. A **719**, 52c (2003).
13. K. Czerski, A. Huke, P. Heide, G. Ruprecht, Europhys. Lett. **68**, 363 (2004).
14. A. Huke, *Die Deuteronen-Fusionsreaktionen in Metallen*, PhD Thesis, Technische Universität Berlin, (2002).
15. R.E. Brown, N. Jarmie, Phys. Rev. C **41**, 1391 (1990).
16. G. Grosso, G.P. Parravicini, *Solid State Physics* (Academic Press, 2000).
17. S. Moroni, D.M. Ceperley, G. Senatore, Phys. Rev. Lett. **75**, 689 (1995).
18. D.E. Penn, Phys. Rev. **128**, 2093 (1962).
19. J.F. Ziegler, J.P. Biersack, U. Littmark, *The Stopping and Ranges of Ions in Solids* (Pergamon Press, New York, 1985).
20. V.N. Neelevathi, R.H. Ritchie, W. Brandt, Phys. Rev. Lett. **302**, 302 (1974).
21. A.F. Lifschitz, N.R. Arista, Phys. Rev. A **57**, 200 (1998).
22. K. Czerski, A. Huke, P. Heide, G. Schiwietz, Nucl. Instrum. Methods B **193**, 183 (2002).
23. F. Raiola *et al.*, J. Phys. G **31**, 1141 (2005).
24. A. Huke, K. Czerski, P. Heide, Nucl. Phys. A **719**, 279c (2003).
25. K. Czerski *et al.*, Nucl. Instrum. Methods B **225**, 72 (2004).
26. A. Huke, K. Czerski, P. Heide, to be published.
27. S.E. Jones *et al.*, Nature **338**, 737 (1989).

Eur. Phys. J. A **27**, s01, 89–94 (2006)
DOI: 10.1140/epja/i2006-08-013-x

EPJ A direct
electronic only

Influence of chaos on the fusion enhancement by electron screening

S. Kimura[1,a], A. Bonasera[1], and S. Cavallaro[1,2]

[1] Laboratorio Nazionale del Sud, INFN, via Santa Sofia, 62, 95123 Catania, Italy
[2] Dipartimento di Fisica, Università degli Studi di Catania, via Santa Sofia, 64, 95123 Catania, Italy

Received: 21 June 2005 /
Published online: 24 February 2006 –

Abstract. We study the effect of screening by bound electrons in low energy nuclear reactions. We use molecular dynamics to simulate the reactions involving many electrons: D+d, D+D, ^{3}He+d, ^{3}He+D, ^{6}Li+d, ^{6}Li+D, ^{7}Li+p, ^{7}Li+H. Quantum effects corresponding to the Pauli and Heisenberg principles are enforced by constraints in terms of the phase space occupancy. In addition to the well-known adiabatic and sudden limits, we propose a new "dissipative limit" which is expected to be important not only at high energies but in the extremely low energy region. The dissipative limit is associated with the chaotic behavior of the electronic motion. It affects also the magnitude of the enhancement factor. We discuss also numerical experiments using polarized targets. The derived enhancement factors in our simulation are in agreement with those extracted within the R-matrix approach.

PACS. 25.45.-z ^{2}H-induced reactions – 34.10.+x General theories and models of atomic and molecular collisions and interactions (including statistical theories, transition state, stochastic and trajectory models, etc.)

1 Introduction

The relation between the tunneling process and dynamical chaos has been discussed with great interests in recent years [1,2]. Though the tunneling is a completely quantum-mechanical phenomenon, it is influenced by classical chaos. In the sense that the chaos causes the fluctuation of the classical action which essentially determines the tunneling probability. We study the phenomenon by examining the screening effect by bound electrons in the low energy fusion reaction. In the low energy region the experimental cross-sections with gas targets show an increasing enhancement with decreasing bombarding energy with respect to the values obtained by extrapolating from the data at high energies [3]. Many studies attempted to attribute the enhancement of the reaction rate to the screening effects by bound target electrons. In this context one often estimates the screening potential as a constant decrease of the barrier height in the tunneling region through a fit to the data. A puzzle has been that the screening potential obtained by this procedure exceeds the value of the so-called adiabatic limit, which is given by the difference of the binding energies of the united atoms and of the target atom and it is theoretically thought to provide the maximum screening potential [4]. For several years, the re-determination of the bare cross-sections has been proposed theoretically [5] and experimentally [6], using the Trojan Horse method [7,8,9]. The comparison between newly obtained bare cross-sections, *i.e.*, astrophysical S-factors, and the cross-sections by the direct measurements gives a variety of values for the screening potential. These values are often smaller than the sudden limit or larger than the adiabatic limit. Theoretical studies performed using the time-dependent Hartree-Fock (TDHF) scheme [10,11] suggest that the screening potential is between the sudden and the adiabatic limits.

One of the aims of this paper is to try to assess the effect of the screening quantitatively. Up to now, the dynamical effects of bound electrons have been studied only in some limited cases with a few bound electrons (the D+d with atomic target [10,11] and molecular D_2 target [12], the ^{3}He+d [10]) with the TDHF method. We investigate here the dynamical effects, including the tunneling region, for other systems with many bound electrons: D+D, ^{3}He+D, observing the effect of the electron capture of projectile. We consider also some reactions including Li isotopes: ^{6}Li+d, ^{6}Li+D, ^{7}Li+p and ^{7}Li+H.

To simulate the effects of many electrons, we use the constrained molecular dynamics (CoMD) model [2,13,14]. At very low energies fluctuations are anticipated to play a substantial role. Such fluctuations are beyond the TDHF

[a] e-mail: kimura@lns.infn.it

scheme. Not only are TDHF calculations, by construction, cylindrically symmetric around the beam axis. Such a limitation is not necessarily true in nature and the mean field dynamics could be not correct especially in the presence of large fluctuations. Molecular dynamics contains all possible correlations and fluctuations due to the initial conditions (events). For the purpose of treating quantum-mechanical systems like target atoms and molecules, we use classical equations of motion with constraints to satisfy the Heisenberg uncertainty principle and the Pauli exclusion principle for each event [13]. In extending the study to the lower incident energies, we would like to stress the connection between the motion of bound electrons and chaos. In fact, depending on the dynamics, the behavior of the electron(s) is unstable and influences the relative motion of the projectile and the target. The feature is caused by the non-integrability of the N-body system ($N \geq 3$) and it is well known that the tunneling probability can be modified by the existence of chaotic environment. We discuss the enhancement factor of the laboratory cross-section in connection with the integrability of the system by looking the inter-nuclear and electronic oscillational motion. More specifically we analyze the frequency shift of the target electron due to the projectile and the small oscillational motion induced by the electron to the relative motion between the target and the projectile. We show that the increase of chaoticity in the electron motion decreases the fusion probability.

The paper is organized as follows. In sect. 2 we introduce the enhancement factor f_e and describe the essence of the constrained molecular dynamics approach briefly. In sect. 3 we apply it to assess the effect of the bound electrons during the nuclear reactions. We discuss also the relation between the amplitudes of the inter-nuclear oscillational motion and the enhancement factor. We summarize the paper in sect. 4.

2 Formalism

2.1 Enhancement factor

We denote the reaction cross-section at incident energy in the center of mass E by $\sigma(E)$ and the cross-section obtained in the absence of electrons by $\sigma_0(E)$. The enhancement factor f_e is defined as

$$f_e \equiv \frac{\sigma(E)}{\sigma_0(E)}. \qquad (1)$$

If the effect of the electrons is well represented by the constant shift U_e of the potential barrier, following [15, 10], ($U_e \ll E$):

$$f_e \sim \exp\left[\pi\eta(E)\frac{U_e}{E}\right], \qquad (2)$$

where $\eta(E)$ is the Sommerfeld parameter [16].

2.2 Constrained molecular dynamics

We estimate the enhancement factor f_e numerically using molecular dynamics approach:

$$\frac{d\mathbf{r}_i}{dt} = \frac{\mathbf{p}_i c^2}{\mathcal{E}_i}, \qquad \frac{d\mathbf{p}_i}{dt} = -\nabla_{\mathbf{r}} U(\mathbf{r}_i), \qquad (3)$$

where $(\mathbf{r}_i, \mathbf{p}_i)$ are the position, momentum of the particle i at time t. $\mathcal{E}_i = \sqrt{\mathbf{p}_i^2 c^2 + m_i^2 c^4}$, $U(\mathbf{r}_i)$ and m_i are its energy, Coulomb potential and mass, respectively. We set the starting point of the reaction at 10 Å inter-nuclear separation. In eqs. (3) we do not take into account the quantum effect of Pauli exclusion principle and Heisenberg principle. In order to take the feature of the Pauli blocking into account in this framework, we use the Lagrange multiplier method for constraints and modify the classical equations of motion (3).

Our constraint which corresponds to the Pauli blocking is $\bar{f}_i \leq 1$ in terms of phase space density; note that the phase space density can be directly related to the distance of two particles, *i.e.*, $\mathbf{r}_{ij}\mathbf{p}_{ij}$, in the phase space. Here $\mathbf{r}_{ij} = |\mathbf{r}_i - \mathbf{r}_j|$ and $\mathbf{p}_{ij} = |\mathbf{p}_i - \mathbf{p}_j|$. The relation $\bar{f}_i \leq 1$ is fulfilled, if $\mathbf{r}_{ij}\mathbf{p}_{ij} \geq \xi_P \hbar \delta_{S_i,S_j}$, where $\xi_P = 2\pi(3/4\pi)^{2/3}$. i, j refer only to electrons and $S_i, S_j (= \pm 1/2)$ are their spin projection. For the Heisenberg principle $\mathbf{r}_{ij}\mathbf{p}_{ij} \geq \xi_H \hbar$, where $\xi_H = 1$, i and j refer not only to electrons but also to the nucleus. ξ_H is determined to reproduce the correct energy of hydrogenic atoms. Obviously the conditions $\mathbf{r}_{ij}\mathbf{p}_{ij} = \xi_{H(P)}\hbar$ must be fulfilled in the ground-state configuration rather than $\mathbf{r}_{ij}\mathbf{p}_{ij} > \xi_{H(P)}\hbar$.

Using these constraints, the Lagrangian of the system can be written down as

$$\mathcal{L} = \sum_i \frac{\mathbf{p}_i^2 c^2}{\mathcal{E}_i} - \sum_{i,j(\neq i)} U(\mathbf{r}_{ij}) + \sum_{i,j(\neq i)} \lambda_i^H \left(\frac{\mathbf{r}_{ij}\mathbf{p}_{ij}}{\hbar} - 1\right) + \sum_{i,j(\neq i)} \lambda_i^P \left(\frac{\mathbf{r}_{ij}\mathbf{p}_{ij}}{\xi_P \hbar}\delta_{S_i,S_j} - 1\right), \qquad (4)$$

where λ_i^P and λ_i^H are Lagrange multipliers. The variational calculus leads to

$$\frac{d\mathbf{r}_i}{dt} = \frac{\mathbf{p}_i c^2}{\mathcal{E}_i} + \frac{1}{\hbar}\sum_{j(\neq i)} \left(\frac{\lambda_i^H}{\xi_H} + \frac{\lambda_i^P}{\xi_P}\delta_{S_i,S_j}\right) \mathbf{r}_{ij}\frac{\partial \mathbf{p}_{ij}}{\partial \mathbf{p}_i}, \qquad (5)$$

$$\frac{d\mathbf{p}_i}{dt} = -\nabla_{\mathbf{r}} U(\mathbf{r}_i) - \frac{1}{\hbar}\sum_{j(\neq i)} \left(\frac{\lambda_i^H}{\xi_H} + \frac{\lambda_i^P}{\xi_P}\delta_{S_i,S_j}\right) \mathbf{p}_{ij}\frac{\partial \mathbf{r}_{ij}}{\partial \mathbf{r}_i}. \qquad (6)$$

In order to obtain the atomic ground-state configuration, we perform the time integration of eqs. (5) and (6). The value of λ_i^H and λ_i^P are determined depending on the magnitude of $\mathbf{r}_{ij}\mathbf{p}_{ij}$. If $\mathbf{r}_{ij}\mathbf{p}_{ij}$ is (smaller) larger than $\xi_{H(P)}\hbar$, λ has positive (negative) sign. Thus we change the phase space occupancy of the system. The constraints restrict us to variations $\Delta\mathcal{L} = 0$ that keep the constraints always true [14]. In this way we obtain many initial conditions which occupy different points in the phase space microscopically.

In order to treat the tunneling process, we define the collective coordinates $\mathbf{R}^{coll}$ and the collective momentum $\mathbf{P}^{coll}$ as

$$\mathbf{R}^{coll} \equiv \mathbf{r}_P - \mathbf{r}_T, \qquad \mathbf{P}^{coll} \equiv \mathbf{p}_P - \mathbf{p}_T, \tag{7}$$

where $\mathbf{r}_T, \mathbf{r}_P$ ($\mathbf{p}_T, \mathbf{p}_P$) are the coordinates (momenta) of the target and the projectile nuclei, respectively. When the collective momentum becomes zero, we switch on the collective force, which is determined by $\mathbf{F}_P^{coll} \equiv \dot{\mathbf{P}}^{coll}$ and $\mathbf{F}_T^{coll} \equiv -\dot{\mathbf{P}}^{coll}$, to enter into imaginary time [17]. We follow the time evolution in the tunneling region using the equations

$$\frac{\mathrm{d}\mathbf{r}^{\Im}_{T(P)}}{\mathrm{d}\tau} = \frac{\mathbf{p}^{\Im}_{T(P)}}{\mathcal{E}_{T(P)}}, \qquad \frac{d\mathbf{p}^{\Im}_{T(P)}}{d\tau} = -\nabla_{\mathbf{r}} U(\mathbf{r}^{\Im}_{T(P)}) - 2\mathbf{F}^{coll}_{T(P)}, \tag{8}$$

where τ is used for imaginary time to be distinguished from real time t. $\mathbf{r}^{\Im}_{T(P)}$ and $\mathbf{p}^{\Im}_{T(P)}$ are position and momentum of the target (the projectile) during the tunneling process, respectively. Adding the collective force corresponds to inverting the potential barrier which becomes attractive in the imaginary times. The penetrability of the barrier is given by [17]

$$\Pi(E) = (1 + \exp(2\mathcal{A}(E)/\hbar))^{-1}, \tag{9}$$

where the action integral $\mathcal{A}(E)$ is

$$\mathcal{A}(E) = \int_{r_b}^{r_a} \mathbf{P}^{coll}\, \mathrm{d}\mathbf{R}^{coll}, \tag{10}$$

r_a and r_b are the classical turning points. The internal classical turning point r_b is determined using the sum of the radii of the target and projectile nuclei. Similarly from the simulation without electron, we obtain the penetrability of the bare Coulomb barrier $\Pi_0(E)$.

Since nuclear reactions occur with small impact parameters on the atomic scale, we consider only head-on collisions. The enhancement factor is thus given by eq. (1),

$$f_e = \Pi(E)/\Pi_0(E) \tag{11}$$

for each event in our simulation. Thus we have an ensemble of f_e values at each incident energy.

3 Application to the electron screening problem

3.1 D+d and D+D reactions

Figure 1 shows the incident energy dependence of the enhancement factor for the reactions D+d and D+D, where the systems involve 1 and 2 electrons, respectively. The open and closed squares show the average enhancement factors $\bar{f}_e$ over events for the reactions D+d and D+D, respectively. The variances $\Sigma = \sqrt{\bar{f_e^2} - (\bar{f}_e)^2}$ are shown with error bars. The dotted and dash-dotted curves show

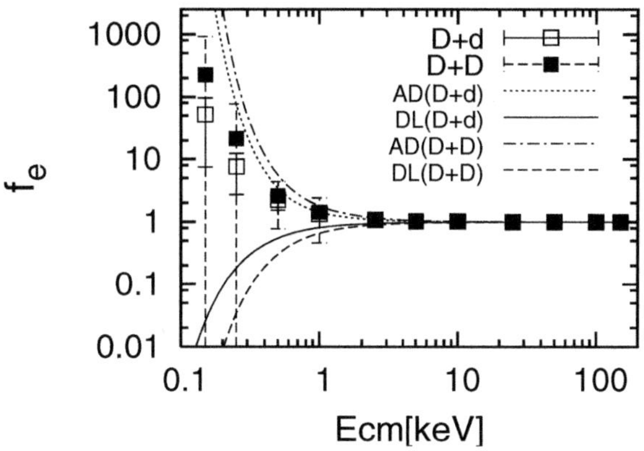

Fig. 1. Enhancement factor as a function of incident center-of-mass energy for the reactions D+d and D+D. Error-bars represent the variances obtained from the events generated for each beam energy.

the enhancement factors in the adiabatic limit $f_e^{(AD)}$ for an atomic deuterium target. For the reaction D+d $f_e^{(AD)}$ is obtained by assuming equally weighted linear combination of the lowest-energy gerade and ungerade wave function for the electron, reflecting the symmetry in the D+d, *i.e.*,

$$f_e^{(AD)} = \frac{1}{2}\left(e^{\pi\eta(E)\frac{U_e^{(g)}}{E}} + e^{\pi\eta(E)\frac{U_e^{(u)}}{E}}\right), \tag{12}$$

where $U_e^{(g)} = 40.7\,\text{eV}$ and $U_e^{(u)} = 0.0\,\text{eV}$ [11,10]. If we take into account the electron capture of the projectile, *i.e.*, in the case of D+D, the enhancement factor in the adiabatic limit is

$$f_e^{(AD)} = \frac{1}{4}e^{\pi\eta(E)\frac{U_e^{(g.s.)}}{E}} + \frac{3}{4}e^{\pi\eta(E)\frac{U_e^{(1es)}}{E}}, \tag{13}$$

where $U_e^{(g.s.)} = 51.7\,\text{eV}$ and $U_e^{(1es)} = 31.9\,\text{eV}$ [18]. The solid curve and dashed curve show the enhancement factors in the dissipative limit $f_e^{(DL)}$ for the reactions D+d and D+D, respectively. Notice how the calculated enhancement factor with their variances nicely ends up between the adiabatic and the dissipative limits. We performed also a fit of our data using eq. (2) including the very low energy region and obtained $U_e = 15.9 \pm 2.0\,\text{eV}$ for D+d case and $U_e = 21.6 \pm 0.3\,\text{eV}$ for D+D.

Now we look at the oscillational motions of the particle's coordinates as the projection on the z-axis (the reaction axis). We denote the z-component of $\mathbf{r}_T, \mathbf{r}_P$ and $\mathbf{r}_e$ as z_T, z_P and z_e, respectively. Practically, we examine the oscillational motion of the electron around the target, $z_{Te} = z_e - z_T$, and the oscillational motion of the inter-nuclear motion, *i.e.*, the motion between the target and the projectile, $z_s = z_T + z_P$, which essentially would be zero due to the symmetry of the system in the absence of the perturbation. In fig. 2 these two values are shown for 2 events, which have the enhancement factor $f_e = 170.8$ (ev. A), and $f_e = 6.5$ (ev. B), at the incident energy $E_{cm} = 0.15\,\text{keV}$. The panels show z_s, z_{Te} as a function of time. The asterisks indicate the time at

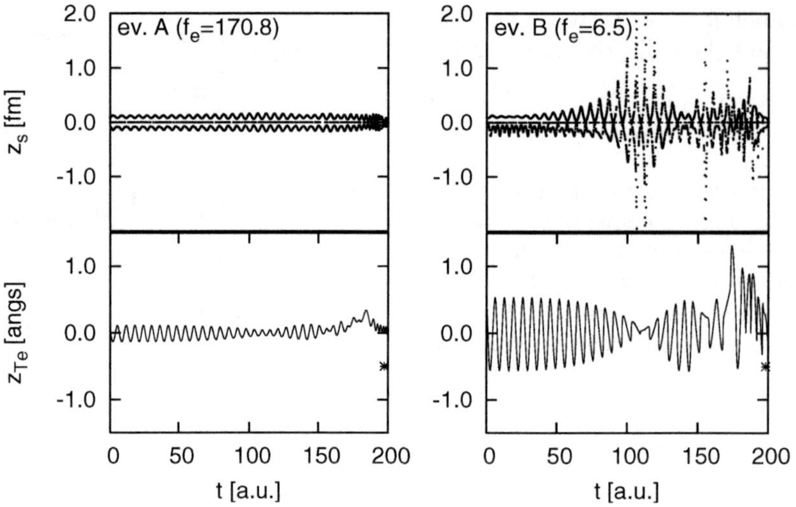

Fig. 2. The oscillational motion of the electron around the target (lower panels) and the inter-nuclear motion (upper panels) as a function of time, in atomic unit, for two events, with large f_e(ev. A) and small f_e (ev. B), for the D+d reaction at the incident energy 0.15 keV. The inter-nuclear separation is 10 Å at $t = 0$.

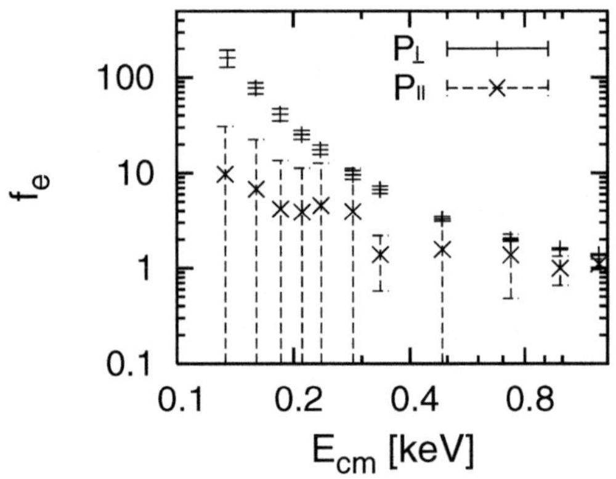

Fig. 3. Incident energy dependence of the enhancement factor for $P_\perp$ and $P_\parallel$ targets.

which the system reaches the classical turning point. It is clear that in the case of event B the orbit of the electron is much distorted from the unperturbed one than in event A. Characteristics of z_s are that 1) its value often becomes zero, as is expected in the unperturbed system, and 2) the component of the deviation from zero shows a periodical behavior. It is remarkable that the amplitude of the deviation becomes quite large at some points in the case of event B which shows the small enhancement factor. Note that in event B one observes clear beats, *i.e.*, resonances. Thus for two events, with the same macroscopic initial conditions, we have a completely different outcome, which is a definite proof of chaos in our 3-body system. We can understand these results in first approximation by considering the motion of the ions to be much slower than the rapidly oscillating motion of the electrons [2]. From fig. 2 we can deduce the following important fact. If the motion of the electron is initially in the plane perpendicular to the reaction axis, the enhancement factor is large, event A (notice $|z_{Te}| \ll R_B$, *i.e.*, the Bohr radius, at $t \sim 0$). On the other hand, if there is a substantial projection of the electron motion, as in event B (the amplitude of $|z_{Te}| \sim R_B$ at $t \sim 0$), on the reaction axis the enhancement factor is relatively small because of the increase of chaoticity. The fact suggests that if one performs experiments at very low bombarding energies with *polarized targets*, the enhancement factor can be controlled by changing the *polarization.* The largest enhancement would be gained with targets polarized perpendicularly to the beam axis.

In order to test this estimation, we prepared ensembles of target atoms which are polarized perpendicular ($P_\perp$) and parallel ($P_\parallel$) to the beam axis, numerically. In fig. 3 we show the incident energy dependence of the average enhancement factor for the $P_\perp$ and $P_\parallel$ targets with pluses and crosses, particularly in the low energy region. The enhancement factors from the $P_\perp$ targets are always larger than that from the $P_\parallel$ targets. In contrast to the average enhancement from the $P_\perp$ targets, which increases monotonically as the incident energy becomes smaller, the average enhancement from the $P_\parallel$ targets fluctuates. It has also large variances at low energies. A remarkable thing is that with the parallel targets there are many events in which the enhancement factor becomes less than 1. It means that in this case the bound electron gives the effect of hindrance to the tunneling probability.

3.2 ^{3}He+d and ^{3}He+D reactions

An excess of the screening potential was reported for the reactions ^{3}He+d with atomic gas ^{3}He target, and D_2+^{3}He with deuterium molecular gas target, for the first time in ref. [3]. Since then various experiments have been performed for these reactions. The incident energy covers from 5 keV to 50 keV for ^{3}He+d. Though once the problem of the discrepancy between experimental data and theoretical prediction seemed to be solved by considering the correct energy loss data [19], recent measurements using

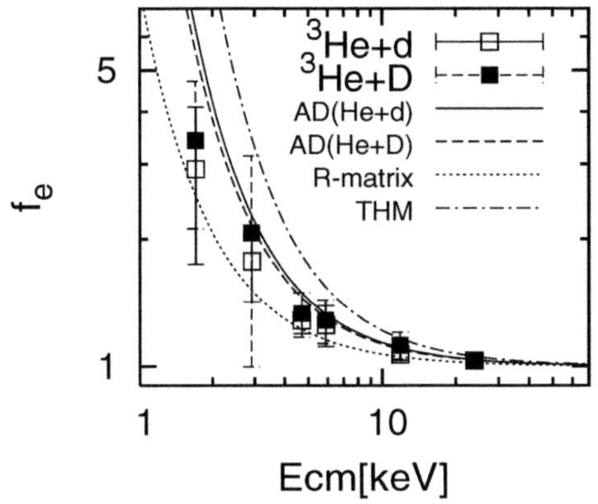

Fig. 4. Enhancement factor as a function of incident center-of-mass energy for the reactions ^{3}He+d and ^{3}He+D.

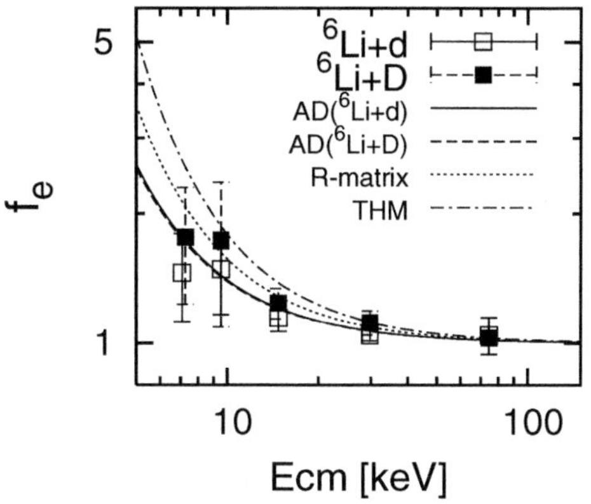

Fig. 5. Same as fig. 4 but for the reactions ^{6}Li+d and ^{6}Li+D.

measured energy loss data [20] report larger screening potentials than in the adiabatic limit for both reactions.

The electron capture by the projectile plays a minor role in the case of ^{3}He+d, since electrons are more bound in helium targets. However, the recent measurement by Aliotta *et al.* was performed using molecular D_2^+ and D_3^+ targets [20]. Thus we assess the contribution from the reaction ^{3}He+D, as well.

The enhancement factor in the adiabatic limit gives $U_e = 119\,\text{eV}$ for ^{3}He+d and $U_e = 110\,\text{eV}$ for ^{3}He+D, respectively. These are shown in fig. 4 with the solid curve for ^{3}He+d and with the dashed curve for ^{3}He+D. The comparison of these two adiabatic limits shows that the electron capture of projectile would give a hindrance compared with the case in the absence of the capture. Meanwhile the latest analysis of the experimental data using R-matrix two level fit [5] suggests the screening potential $U_e = 60\,\text{eV}$ (the corresponding enhancement factor is shown with the dotted curve). The comparison between direct measurement and an indirect method, the Trojan Horse method, suggests the screening potential $U_e = 180 \pm 40\,\text{eV}$ (the corresponding enhancement factor is shown with the dot-dashed curve) [9]. The average enhancement factors $\bar{f}_e$ over events in our simulations using the CoMD are shown with the open and closed squares for the reactions ^{3}He+d and ^{3}He+D, respectively. The enhancement factors of both reactions ^{3}He+d and ^{3}He+D are in agreement with the extracted values using the R-matrix approach within the variances over all the events. Notice that our calculated enhancement factors for the two systems display an opposite trend as compared to the adiabatic limits. The average enhancement factor of the reaction ^{3}He+D agrees with the estimation of the adiabatic limit and the reaction ^{3}He+d is below the corresponding adiabatic limit. The paradoxical feature comes from the fact that an electron between two ions is often kicked out during the reaction process, *i.e.*, the electron configuration seldom settles down the ^{5}Li$^+$ ground state in the reaction ^{3}He+d. It is known as autoionization in the context of the Classical Trajectory Monte Carlo method [21]. On the contrary, in the case of ^{3}He+D, the deuterium projectile brings its bound electron in a tight bound state around the unified nuclei of ^{3}He and d; practically it ends up with a ground-state configuration of the ^{5}Li atom. The fits of the obtained enhancement factors suggest the screening potentials $U_e = 82.4 \pm 1.9\,\text{eV}$ for the ^{3}He+d and $U_e = 102.8 \pm 3.0\,\text{eV}$ for the ^{3}He+D.

3.3 ^{6}Li+d, ^{6}Li+D, ^{7}Li+p and ^{7}Li+H

The S-factors for the reactions ^{6}Li+d, ^{6}Li+p and ^{7}Li+p were measured over the energy range $10\,\text{keV} < E_{cm} < 1450\,\text{keV}$ by Engstler *et al.* [22]. They used LiF solid targets and hydrogen projectiles as well as hydrogen molecular gas targets and Li projectiles. In the case of LiF target which is a large band gap insulator, one often approximates the electronic structure of the target ^{6}Li(^{7}Li) state by the ^{6}Li$^+$(^{7}Li$^+$) with only two innermost electrons. Thus for all three reactions one expects the screening potential in the adiabatic limit $U_e^{(AD)} = 371.8 - 198.2 \sim 174\,\text{eV}$. On the contrary, if one uses the ground state of the ^{6}Li(^{7}Li) atom and of the bare deuteron target as the initial state, $U_e^{(AD)} = 186\,\text{eV}$ [23], which is given by the solid curve in fig. 5. However one should be aware that the deuteron or hydrogen projectile plausibly moves with a bound electron in LiF solid insulator target [24]. Under such an assumption we could estimate the screening potential $U_e^{(AD)} = 389.9 - 198.2 \sim 192\,\text{eV}$. In the case of molecular D_2 or H_2 gas targets, as well, we should consider the electron capture by the lithium projectile.

The bare S-factors for the same reaction have been extracted using an indirect method, the Trojan-Horse method through the reaction ^{6}Li(^{6}Li, $\alpha\alpha$)^{4}He [8]. The comparison between direct and the indirect methods gives the screening potential $U_e = 320 \pm 50\,\text{eV}$. The corresponding enhancement factors are shown with the dash-dotted curve. The contrast between the direct measurement data and the theoretical estimation for the bare S-factor using the R-matrix theory gives $U_e = 240\,\text{eV}$. It is shown with the dotted line. The extracted U_e with the two different methods are larger than the adiabatic limit.

We simulate the reactions ^{6}Li+d, ^{6}Li+D, ^{7}Li+p and ^{7}Li+H. In fig. 5 (and 6) the open and closed squares show

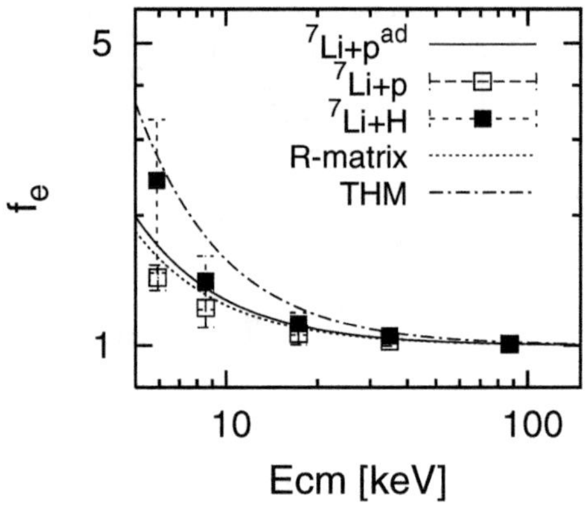

Fig. 6. Same as fig. 4 but for the reactions ^{7}Li+p and ^{7}Li+H.

the enhancement factor for the reactions ^{6}Li+d and ^{6}Li+D (and ^{7}Li+p and ^{7}Li+H), respectively. Again the average enhancement factors of the reaction ^{6}Li+D (^{7}Li+H) are larger than those of the ^{6}Li+d (^{7}Li+p). The enhancement factors of the reaction ^{6}Li+D are in agreement with the extracted values using the R-matrix approach within the variances over all the events. The fit of the obtained average enhancement factors suggests the screening potentials $U_e = 152.0 \pm 9.9$ eV for ^{6}Li+d and $U_e = 214.4 \pm 18.5$ for ^{6}Li+D. The screening potential for the reaction ^{6}Li+d in our simulation does not exceed the adiabatic limit nor the extracted values using the R-matrix theory and the THM, but that for ^{6}Li+D verges to the extracted values using the R-matrix approach.

4 Summary

We discussed the effect of the screening by the electrons in nuclear reactions at the astrophysical energies. We performed molecular dynamics simulations with constraints and imaginary time for the reactions D+d, D+D, ^{3}He+d, ^{3}He+D, ^{6}Li+d, ^{6}Li+D, ^{7}Li+p, ^{7}Li+H. For all the reactions it is shown that both the average enhancement factors and their variances increase as the incident energy becomes lower. Using bare projectiles we obtained the average screening potential smaller than the value in the adiabatic limit for all reactions. This is because of the excitation or emission of several bound electrons during the reactions. The comparison between the bare and atomic projectile cases for each reaction revealed that the electron capture of the projectile leads to larger enhancements. The derived enhancement factors in our simulation are in agreement with those extracted within the R-matrix approach including the variances over all the events.

We report also the results of the numerical experiments using polarized targets for the reaction D+d. Using $P_\perp$ targets, we obtained relatively large enhancements with small variances. While $P_\parallel$ targets give large variances of the enhancement factors and relatively small averaged enhancement factors. This is because with the $P_\parallel$ targets the force exerted from the electron to the relative motion of the nuclei is oscillational, in the direction of the beam axis, and the motion of the electron becomes often excited or unstable. It is the case where the chaoticity of the electron motion affects the tunneling probability and at the same time the enhancement factor of the cross-section. This suggests that if one performs experiments at very low bombarding energies with polarized targets, the enhancement factor can be controlled by changing the polarization. The largest enhancement will be obtained in reactions with targets polarized perpendicularly to the beam direction.

References

1. W.A. Lin, L.E. Ballentine, Phys. Rev. Lett. **65**, 2927 (1990); O. Bohigas, S. Tomosvic, D. Ullumo, Phys. Rev. Lett. **65**, 5 (1990); A. Shudo, K.S. Ikeda, Phys. Rev. Lett. **74**, 682 (1995).
2. S. Kimura, A. Bonasera, Phys. Rev. Lett. **93**, 262502 (2004); S. Kimura, A. Bonasera, Nucl. Phys. A **759**, 229 (2005).
3. A. Krauss *et al.*, Nucl. Phys. A **467**, 273 (1987); S. Engstler *et al.*, Phys. Lett. B **202**, 179 (1988).
4. C. Rolfs, E. Somorjai, Nucl. Instrum. Methods B **99**, 297 (1995).
5. F.C. Barker, Nucl. Phys. A **707**, 277 (2002).
6. M. Junker *et al.*, Phys. Rev. C **57**, 2700 (1998).
7. M. Lattuada *et al.*, Astrophys. J. **562**, 1076 (2001).
8. A. Musumarra *et al.*, Phys. Rev. C **64**, 068801 (2001).
9. M. La Cognata *et al.*, Phys. Rev. C **72**, 065802 (2005); M. La Cognata *et al.*, Nucl. Phys. A **758**, 98 (2005).
10. T.D. Shoppa, S.E. Koonin, K. Langanke, R. Seki, Phys. Rev. C **48**, 837 (1993).
11. S. Kimura, N. Takigawa, M. Abe, D.M. Brink, Phys. Rev. C **67**, 022801(R) (2003).
12. T.D. Shoppa *et al.*, Nucl. Phys. A **605**, 387 (1996).
13. M. Papa, T. Maruyama, A. Bonasera, Phys. Rev. C **64**, 024612 (2001); S. Terranova, A. Bonasera, Phys. Rev. C **70**, 024906 (2004).
14. S. Kimura, A. Bonasera, Phys. Rev. A **72**, 014703 (2005).
15. H.J. Assenbaum, K. Langanke, C. Rolfs, Z. Phys. A **327**, 461 (1987).
16. D.D. Clayton, *Principles of Stellar Evolution and Nucleosynthesis* (University of Chicago Press, 1983) Chapt. 4.
17. A. Bonasera, V.N. Kondratyev, Phys. Lett. B **339**, 207 (1994); T. Maruyama, A. Bonasera, S. Chiba, Phys. Rev. C **63**, 057601 (2001).
18. Y. Kato, N. Takigawa, nucl-th/0404075.
19. K. Langanke, T.D. Shoppa, C.A. Barnes, C. Rolfs, Phys. Lett. B **369**, 211 (1996).
20. M. Aliotta *et al.*, Nucl. Phys. A **690**, 790 (2001).
21. T. Geyer, J.M. Rost, J. Phys. B **36**, L107 (2003) and references therein.
22. S. Engstler *et al.*, Z. Phys. A **342**, 471 (1992).
23. L. Bracci, G. Fiorentini, V.S. Melezhik, G. Mezzorani, P. Quarati, Nucl. Phys. A **513**, 316 (1990).
24. K. Eder *et al.*, Phys. Rev. Lett. **79**, 4112 (1997); P. Roncin *et al.*, Phys. Rev. Lett. **83**, 864 (1999).

4 Explosive Nucleosynthesis

Eur. Phys. J. A **27**, s01, 97–106 (2006)
DOI: 10.1140/epja/i2006-08-014-9

EPJ A direct
electronic only

Recent astrophysical studies with exotic beams at ORNL

D.W. Bardayan[a]

Physics Division, Oak Ridge National Laboratory, Oak Ridge, TN 37831, USA

Received: 17 June 2005 /
Published online: 28 February 2006 –

Abstract. The availability of exotic beams has produced great opportunities for advances in our understanding of the nucleosynthesis occurring in stellar burning and stellar explosions such as novae, X-ray bursts, and supernovae. In these extreme environments, synthesized radioactive nuclei can undergo subsequent nuclear processing before they decay, and thus to understand these events, we must understand reaction rates involving radioactive nuclei. At the ORNL Holifield Radioactive Ion Beam Facility (HRIBF), we have made a number of measurements using proton-rich beams such as ^{18}F and ^{7}Be and neutron-rich beams such as ^{82}Ge and ^{84}Se that help clarify the structure of astrophysically-important nuclei. We are also poised to begin studies with doubly-magic ^{132}Sn. The experimental methods and results are discussed.

PACS. 25.60.-t Reactions induced by unstable nuclei – 26.30.+k Nucleosynthesis in novae, supernovae and other explosive environments

1 Introduction

Nuclear astrophysics addresses some of the most compelling questions in nature: What are the origins of the elements that make life on earth possible? How did the sun, the solar system, the stars, and our galaxy form, and how did they evolve? What is the total density of matter in the universe, and will the universe eventually collapse or expand forever? Astrophysical models that address these crucial questions require a considerable amount of nuclear physics information as input. The majority of this required information, however, is currently based on extrapolations or theoretical models and does not have a firm experimental basis. Nuclear data is also an important ingredient in the interpretation of new observations made by ground-based observatories such as the Keck and European Southern Observatory (ESO) Very Large Telescopes, by space-borne observatories such as the Hubble Space Telescope and the Chandra X-Ray Observatory, and by large subterranean detectors such as the Sudbury Neutrino Observatory and Super-Kamiokande. More complete and precise nuclear physics measurements are therefore needed to improve astrophysical models and to decipher the latest observations [1].

Because radioactive nuclei play an influential, and in some cases the dominant, role in many cosmic phenomena, information on these nuclei is particularly important to improve our understanding of the processes that shape our universe. In the explosive environments of novae and X-ray bursts, hydrogen and helium react violently with heavier seed nuclei to produce proton-rich nuclei via hot-CNO burning and the αp- and rp-processes. Neutron-induced nucleosynthesis may occur in the neutrino-wind–driven shock front of supernova explosions initiating the r-process and producing extremely neutron-rich nuclei. These cataclysmic stellar explosions produce reaction flows through nuclei far from the valley of beta stability. To understand and interpret observations of these events we must understand the nuclear reactions, the nuclear structure, and the decay mechanisms for unstable nuclei. Because the lifetimes of most of the nuclei of interest are too short for use as targets in experiments, the required information can best be obtained by producing energetic beams of radioactive ions.

2 Experimental details

At the ORNL Holifield Radioactive Ion Beam Facility (HRIBF), exotic beams are produced by the ion source on-line (ISOL) method [2]. Light ion beams, accelerated by the Oak Ridge Isochronous Cyclotron (ORIC), bombard thick, hot, refractory targets to produce radioactive atoms. The atoms diffusing from the target material are ionized and extracted by a close-coupled ion source. After undergoing two stages of mass analysis, the radioactive ions are then injected into the 25 MV tandem accelerator, accelerated to the appropriate energy for the experiment, and delivered to the experimental station. At the HRIBF, a full suite of developed beams is available including 120

[a] Representing the RIBENS Collaboration; e-mail: `bardayan@mail.phy.ornl.gov`

radioactive beams and 79 beams of stable species [3]. Available radioactive beams include proton-rich species such as ^{7}Be and ^{17,18}F and neutron-rich beams such as ^{82}Ge and doubly-magic ^{132}Sn. Energies up to ~ 10 MeV/u are possible for low mass beams and up to ~ 5 MeV/u for beams near ^{132}Sn.

While the availability of radioactive beams provides great opportunities for the better understanding of nuclei and nuclear reactions far from stability, there are also great challenges that must be overcome to utilize such beams. Radioactive beams are generally available only at a much lower intensity than stable beams ($\sim 10^{4-7}$ ions/s compared to 10^{9-12} ions/s for stable beams). High-efficiency large solid-angle detector arrays have been developed to deal with low beam intensity measurements. The Silicon Detector Array (SIDAR) [4] is a segmented array of silicon strip detectors that has been used for several of the nuclear astrophysics measurements at the HRIBF. Other detector arrays available at the HRIBF include the CLARION array of Clover germanium detectors and the HYBALL array of CsI detectors [5]. Another challenge is that radioactive beams are frequently contaminated with an unwanted isobar. While this contaminant can sometimes be removed with chemical techniques at the ion source or by fully stripping the beam (usually with accompanying drops in intensity), a general solution does not exist and many times an isobarically mixed beam will be delivered to the experimenter. A detector system with high selectivity is many times the best solution to the problem. Kinematically-complete measurements utilizing coincidence techniques is one method to identify the events of interest in such cases. We have frequently detected the beam-like recoil from a reaction in a gas-filled ionization counter, identifying the proton number (Z) of the recoil, and thus determining the reaction channel. Because the measurements of astrophysical interest are usually performed in inverse kinematics, the beam-like recoils are forward focused and can be efficiently detected in an ionization counter covering relatively small laboratory angles ($< 10°$).

While direct measurements of the reaction of interest are extremely important, it is not generally possible (or even desirable) to measure the reaction cross-section at all of the energies required to completely determine the stellar reaction rate. It is crucial to have complimentary measurements which may not directly determine the reaction rate but do help to elucidate the structure of the compound nuclear system involved. A good understanding of the properties of states in the compound nuclei can be used to calculate the stellar reaction rate over a large range of astrophysical temperatures. Direct measurements of reaction cross-sections should thus be complimented with other studies such as single-nucleon transfer reactions populating states of interest or particle-decay branching ratio studies from stable beam measurements. These points are discussed further in the following sections describing our studies with ^{18}F beams (sect. 3), ^{7}Be beams (sect. 4), and neutron-rich beams (sect. 5).

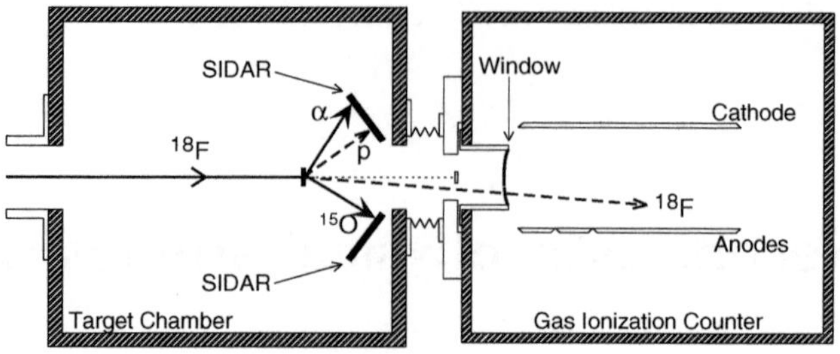

Fig. 1. Our experimental configuration is shown with the ^{18}F ions impinging on a polypropylene target. For the ^{1}H(^{18}F$,p)^{18}$F measurement, scattered protons were detected in the SIDAR array in coincidence with recoil ^{18}F ions detected by the ionization counter. For the ^{1}H(^{18}F$,\alpha)^{15}$O measurement, both the recoil ^{15}O ions and α particles were detected in coincidence in the SIDAR.

3 Studies with ^{18}F beams

Beams of ^{18}F have been used to study ^{18}F production in novae. ^{18}F is relatively long-lived ($t_{1/2} = 1.8$ h) and produced in large amounts in novae. It is therefore a target for γ-ray astronomy, and its observation in novae would provide a rather direct test of nova models [6]. To interpret observations such as these, however, we must know the rates of thermonuclear reactions that affect ^{18}F production. The ^{18}F$(p,\alpha)^{15}$O reaction is one of the most important reactions to understand as it destroys ^{18}F in the nova environment but is also one of the most uncertain [7]. We have made several measurements at the HRIBF to better determine the ^{18}F$(p,\alpha)^{15}$O reaction rate in novae.

^{18}F beams were produced by bombarding a fibrous HfO_2 target with an 85 MeV ^{4}He beam. The ^{16}O$(\alpha,pn)^{18}$F reaction produced ^{18}F atoms which were subsequently ionized, extracted, and accelerated by the HRIBF tandem electrostatic accelerator. Typical beam currents were 2×10^5 ^{18}F/s with an ^{18}O contamination at a ratio of ^{18}O/^{18}F $\sim 8/1$. This ^{18}O contamination can be removed by stripping the beam to $q = 9^+$ before the energy-analyzing magnet with some loss in ^{18}F beam intensity depending on the energy of the beam.

Our first measurement [8] was of the ^{1}H(^{18}F$,p)^{18}$F elastic-scattering cross-section as a function of energy. A ^{18}F beam was used to bombard a hydrogen-containing $(CH_2)_n$ target and elastically-scattered protons were detected by the Silicon Detector Array (SIDAR) [4]. Beam-like recoils were detected and identified in coincidence at forward angles in an isobutane-filled gas ionization counter (see fig. 1). This coincidence requirement was necessary to discriminate protons scattered by ^{18}F from protons scattered by ^{18}O. The ^{1}H(^{18}F$,p)^{18}$F events were then counted as a function of beam energy to produce the excitation function shown in fig. 2. This study resolved significant uncertainties from previous discrepant measurements [9,10] concerning the width and energy for an important $3/2^+$ resonance at $E_{c.m.} = 665$ keV.

This measurement has now been extended to a broader energy range (0.3–1.3 MeV) [11] using the thick-target technique [12]. A 24 MeV ^{18}F was stopped in a thick

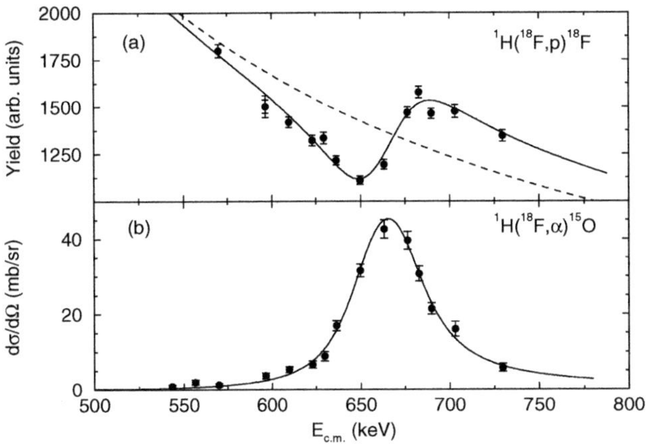

Fig. 2. (a) The ^{1}H(^{18}F$,p)^{18}$F excitation function is shown as a function of energy. The solid curve shows the best fit to the data, while the dashed curve shows the expected excitation function in the absence of resonances in this region. (b) The ^{1}H(^{18}F$,\alpha)^{15}$O center of mass differential cross-section is plotted as a function of energy. The cross-section was found to be isotropic.

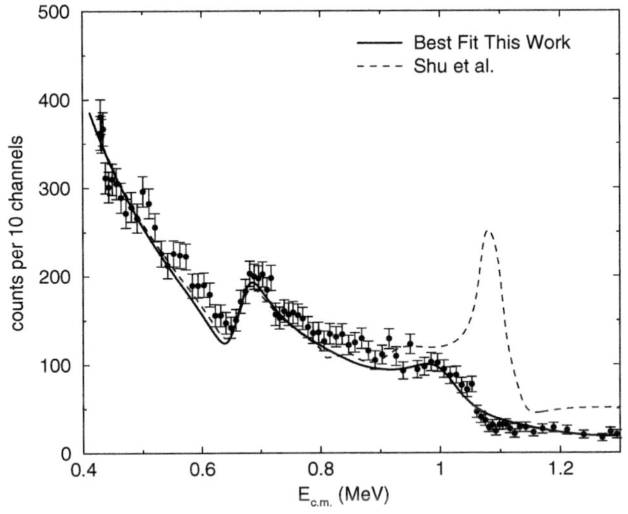

Fig. 3. The proton energy spectrum from the ^{1}H(^{18}F$,p)^{18}$F reaction at $\theta_{lab} = 12°$ is shown. The solid line shows the best fit assuming a $\frac{7}{2}^+$ resonance at $E_{c.m.} \simeq 1.01$ MeV. The dashed line shows the excitation function expected using the resonance parameters from ref. [13].

2.8 mg/cm^2 polypropylene target. The beam had been fully stripped to remove the ^{18}O contamination. Scattered protons from the ^{1}H(^{18}F$,p)^{18}$F reaction were detected at $\theta_{lab} = 8°$–$16°$ by a double-sided silicon strip detector (DSSD). Because the scattered protons lose relatively little energy in the target, measurements of the proton's energy and angle of scatter were sufficient to determine the center-of-mass energy at which the reaction occurred [12]. A measurement of the scattered proton energy spectrum at a fixed angle can thus be used to extract the excitation function for the ^{1}H(^{18}F$,p)^{18}$F reaction over a wide range of center-of-mass energies. Using this technique, we measured the excitation function shown

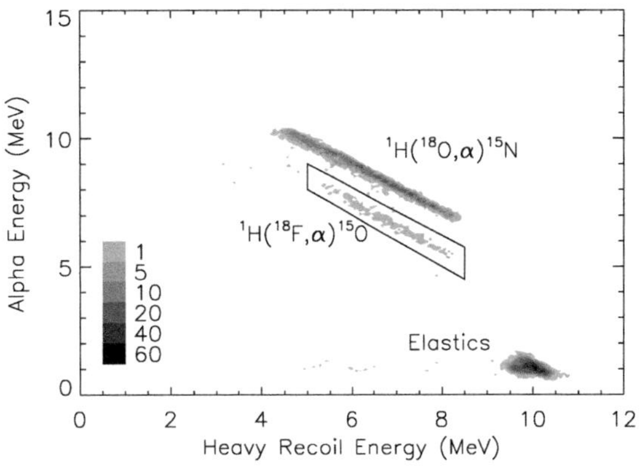

Fig. 4. The α particle energy is plotted *vs.* the heavy recoil energy. Owing to the different Q-values for the reactions, the ^{1}H(^{18}F$,\alpha)^{15}$O events are readily distinguished from the ^{1}H(^{18}O$,\alpha)^{15}$N events.

in fig. 3. The previously-observed resonance at $E_{c.m.} = 0.665$ MeV was clearly evident along with a new resonance at $E_{c.m.} = 1.01$ MeV. From an R-matrix fit to the data, we determined the most-likely properties of the latter to be $J^\pi = 7/2^+$, $E_x = 7.420 \pm 0.014$ MeV, $\Gamma_p = 27 \pm 4$ keV, and $\Gamma_\alpha = 71 \pm 11$ keV. Possibly more important for novae temperatures, we were able to set upper limits on the proton widths of unobserved resonances, several of which are expected based upon comparisons with the mirror nucleus, ^{19}F. Because the proton widths are typically much smaller than the alpha widths for ^{19}Ne energy levels near the proton threshold ($E_x = 6.411$ MeV), these upper limits directly constrain the resonance strengths and thus the contributions these unobserved levels could possibly make to the ^{18}F$(p,\alpha)^{15}$O rate.

In addition to the elastic-scattering measurements, we have also measured the ^{1}H(^{18}F$,\alpha)^{15}$O cross-section directly for astrophysically-important resonances at $E_{c.m.} =$ 330 and 665 keV. The 665 keV measurement was made simultaneously with the elastic-scattering measurement discussed above [14]. In this case, α particles and ^{15}O recoils were both detected in coincidence in the SIDAR array. The events of interest were identified by plotting (fig. 4) the detected α energy *vs.* the heavy recoil energy. Reactions for which both outgoing particles were detected appear as lines of constant total energy in fig. 4. Owing to the different Q values for the reactions, the ^{1}H(^{18}F$,\alpha)^{15}$O events were readily distinguished from the more intense ^{1}H(^{18}O$,\alpha)^{15}$N events. By counting the number of ^{1}H(^{18}F$,\alpha)^{15}$O events observed as a function of bombarding energy, the excitation function plotted in fig. 2(b) was produced. A simultaneous fit to the ^{1}H(^{18}F$,p)^{18}$F and ^{1}H(^{18}F$,\alpha)^{15}$O data sets yielded a precise value of the resonance strength, $\omega\gamma = 6.2 \pm 0.3$ keV [14]. This work also provided the first conclusive evidence for the $3/2^+$ spin assignment of this resonance.

We have extended this technique to measure the resonance strength of an important $3/2^-$ level at $E_{c.m.} =$ 330 keV [15]. Since the 1982 study by Wiescher and Ket-

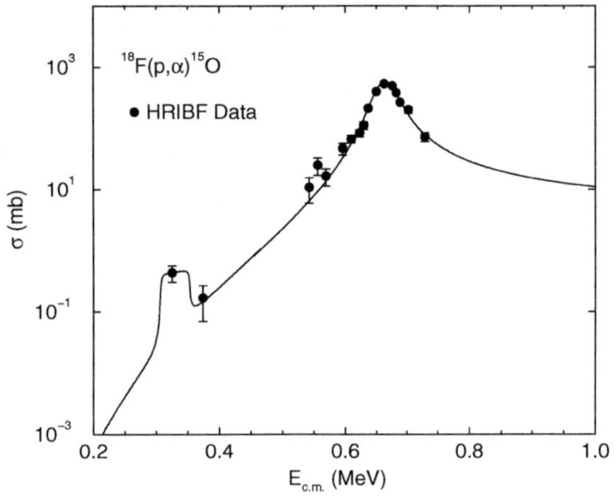

Fig. 5. The measured $^1H(^{18}F,\alpha)^{15}O$ cross-section is shown along with a fit to the data. Since the width of the 330 keV resonance is much less than the target energy loss, the curve appears "flat-topped" at these energies.

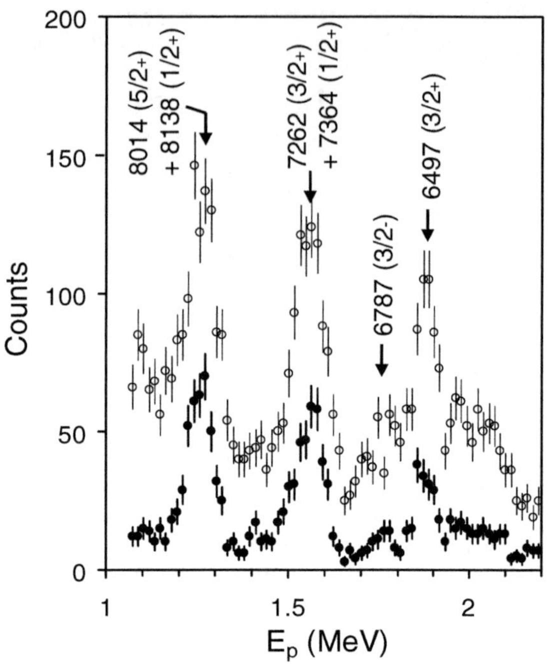

Fig. 6. $^2H(^{18}F,p)^{19}F$ spectra from the six SIDAR strips corresponding to a 147° laboratory angle as a function of proton energy, displayed in 15 keV bins. Open points are singles data and solid points are events in coincidence with the annular strip detector. Excitation energies are given in keV.

tner [16] of the $^{18}F(p,\alpha)^{15}O$ reaction, it has generally been believed that the reaction rate is dominated over a wide range of novae temperatures by a resonance at $E_{c.m.} = 330 \pm 6$ keV that arises from a ^{19}Ne level at $E_x = 6.741$ MeV. This analysis and all subsequent calculations had to be based upon uncertain estimates of the strength of this resonance. Such estimates may be incorrect by an order of magnitude or more [7]. We have made the first significant measurement of the strength of this important resonance by measuring the yield of the $^1H(^{18}F,\alpha)^{15}O$ reaction on and off resonance. Because the resonance energy was known well from previous studies and because the resonance is rather narrow ($\Gamma \sim 3$ keV), we chose to measure the thick target yield by covering the energy range $\Delta E_{c.m.} = 305$–350 keV within the target energy loss. In such a study, a measurement of the step height of the yield on resonance is directly related to the resonance strength of the state. The same technique as above was used to identify the events of interest, and the results are shown in fig. 5 along with the cross-sections from the 665 keV study. The best fit to the data was obtained for a resonance strength of $\omega\gamma = 1.48 \pm 0.46$ eV and a resonance energy of 332 ± 17 keV [15].

Additional constraints on the $^{18}F(p,\alpha)^{15}O$ reaction rate come from studies of the $^2H(^{18}F,p)^{19}F$ reaction. Levels in ^{19}Ne containing strong single-proton components should have mirror levels in ^{19}F with strong single-neutron components. The isospin mirrors to astrophysically important ^{19}Ne levels should thus be strongly populated in a study of the $^2H(^{18}F,p)^{19}F$ reaction. Furthermore from the observed magnitudes of the cross-sections to populate the various states, one can use isospin-symmetry arguments to constrain the proton-spectroscopic factors (and thus the proton widths) of important ^{19}Ne levels.

The $^2H(^{18}F,p)^{19}F$ reaction was studied at the HRIBF [17] by bombarding a 160(10) $\mu g/cm^2$ $(CD_2)_n$ target of 98% enrichment for $\sim$ 3 days with an isotopically pure, 108.49 MeV $^{18}F^{+9}$ beam at an intensity of $\sim 5\times10^5$/s. Using the SIDAR array of $\sim$ 500 μm thickness, light charged particles were detected in the laboratory angular range of 118°–157°, corresponding to "forward" center-of-mass angles in the range 8°–27°. The beam energy was selected to be high enough for direct reaction models, yet low enough to allow all the protons to be stopped in the SIDAR. A silicon strip detector at the focal plane of the Daresbury Recoil Separator (DRS) [18] was used to detect particle-stable recoils having $A = 19$ in coincidence with the SIDAR. This coincidence efficiency was essentially 100% of the +9 charge state fraction for recoil angles $< 1.6°$ and $> 70\%$ overall for particle-stable final states. Other recoils, from higher, α-decaying states in ^{19}F, were detected in coincidence just downstream from the target with an annular strip detector. This detector was also used for data normalization. Beam current normalization was achieved by directly counting beam particles at low intensity with a retractable silicon surface barrier detector placed temporarily at 0°. The overall uncertainty in normalization, estimated to be $\sim 10\%$, is owing mostly to uncertainty in target thickness. Independent internal energy calibrations were obtained for each laboratory angle by using excitation energies of the well-known levels at 1.554038(9), 4.377700(42), and 5.1066(9) MeV in ^{19}F. This allowed excitation energies in the region of interest to be determined with uncertainties ~ 10 keV.

Singles and coincidence spectra in the (α-decaying) region of importance for novae are shown for a laboratory angle of 147° in fig. 6. The coincidence spectra contain three main groups at about 6.5, 7.3, and 8.1 MeV excitation energy. The coincidence efficiencies are roughly 50%,

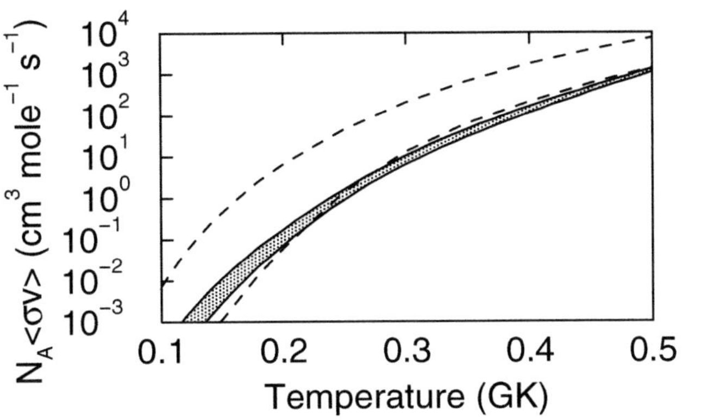

Fig. 7. The $^{18}F(p,\alpha)^{15}O$ reaction rate is plotted *vs.* temperature. The shaded band is from this work while the dashed band is from ref. [7].

60%, and 70%, respectively. Our internal energy calibrations allow identifications of these states which are consistent with known levels in ^{19}F, as indicated in fig. 6. Considerable strength was observed in the 6497 keV $3/2^+$ ^{19}F level, while we see no evidence for the 6528 keV state as suggested in ref. [19]. This indicates that the $\ell = 0$ strength near the proton threshold in ^{19}Ne may well be concentrated in a proton-bound state, or perhaps in the 8 keV resonance. Furthermore, no evidence was observed for a strong $3/2^+$ resonance reported previously at 7101 keV [20]. Most likely the mirror to the important ^{19}Ne resonance at 665 keV is thus contained in the doublet observed near ~ 7.3 MeV in fig. 6. This is consistent with the results of recent reanalysis of $^{15}N(\alpha,\alpha)^{15}N$ data [21].

Combining these various pieces of information on ^{19}Ne resonances, we have calculated improved values of the $^{18}F(p,\alpha)^{15}O$ reaction rate as a function of temperature [17]. In fig. 7, we plot our calculated rate as the shaded band along with the rate from ref. [7]. We find the rate to be dominated by resonances at $E_{c.m.} = 330$ and 665 keV for temperatures above 0.15 GK and by the resonance at 38 keV for temperatures below this. The rate is about two to five times smaller than that of Coc *et al.* [7] in the nova temperature range. We have used this updated $^{18}F(p,\alpha)^{15}O$ reaction rate in a nova nucleosynthesis calculation based on a post-processing approach with temperature and density histories of 28 zones of ejected material determined from hydrodynamics calculations of an explosion on the surface of a 1.25 solar mass white dwarf star [22]. The tools for this calculation are online in the Computational Infrastructure for Nuclear Astrophysics [23]. Using these new rates, we find roughly a factor of two more ^{18}F is produced than previously calculated, and thus ^{18}F observations are more likely than previously thought.

4 Studies with ^7Be beams

Interpretation of solar neutrino flux measurements requires an accurate knowledge of the $^7Be(p,\gamma)^8B$ cross-section. There have been over a dozen measurements of the reaction cross-section at a variety of energies, but all relatively precise measurements have used a radioactive (*i.e.*, decaying) ^7Be target and thus have similar systematic uncertainties. To achieve the desired precision, different approaches, which are not inhibited with the same systematic effects, are necessary.

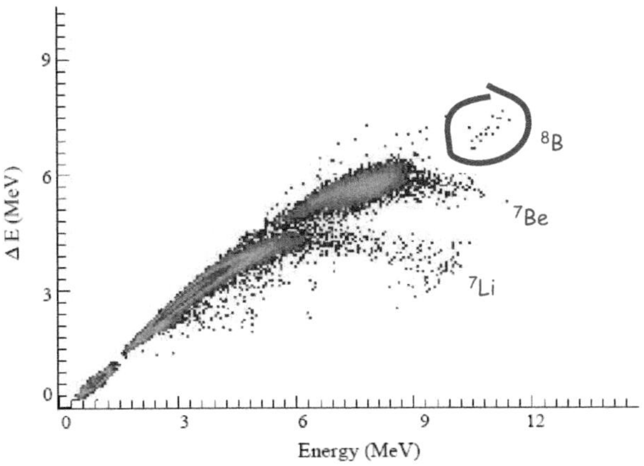

Fig. 8. ^8B recoils from the $^7Be(p,\gamma)^8B$ reaction are identified in an ionization counter at the focal plane of the DRS. Scattered low-energy projectiles of ^7Li and ^7Be are also observed. 22 ^8B events were identified in this demonstration experiment.

The present work [24] focuses on the development of a new approach for the measurement of $^7Be(p,\gamma)^8B$, using a ^7Be beam and a windowless hydrogen gas target at ORNL. The possible systematic errors for this measurement are different from those in normal kinematics using ^7Be targets. In this case, the target is known to be pure H_2 and to be of intransigent composition and thickness. Furthermore, the transmission target allows the recoils to be detected directly, and the systematic error in the number of fusions is decreased in the $^1H(^7Be,\gamma)^8B$ measurement, due to high recoil-detection efficiency and low background. However, low ^7Be beam current limits the statistical precision of the measurement. Due to the statistical limitations, our initial efforts will focus on the higher energy region of the cross-sections ($E_{c.m.} = 1$–2 MeV), where the cross-section is highest and where the largest discrepancy exists among the various models used to extrapolate the cross-section to stellar energies [25].

The ^7Be used to make the radioactive beam was produced at the Triangle Universities Nuclear Laboratory (TUNL) using the $^7Li(p,n)^7Be$ reaction. A chemical process was performed to purify the ^7BeO material, which was mixed with copper or silver in a powder matrix and pressed into a pellet for use in a sputter ion source at the HRIBF. The sputter source produced a low energy beam of $^7BeO^-$ which was injected into the tandem accelerator where it was broken up at the terminal by a gas stripper, and $^7Be^{1+}$ was mass analyzed and accelerated to the experiment.

Using this technique, a 12 MeV ^7Be beam was delivered to the HRIBF windowless hydrogen gas target [18] at an intensity of 2.5 ppA and mixed with ^7Li at a ratio of

$^7Be/^7Li \sim 1/7$. The windowless gas target is described in detail in ref. [18]. The ^8B recoils were tuned through the DRS, and counted and identified in the gas-filled ionization counter. The spectrum from the ion counter is shown in fig. 8. In this initial test measurement, 22 ^8B recoils were observed resulting in a rough cross-section measurement of 1.12 μb with 24% uncertainty at $E_{c.m.} = 1.5$ MeV. Further measurements with higher ^7Be beam intensities are planned to obtain a more statistically significant value.

Another stellar burning reaction studied at the HRIBF with ^7Be beams was the $^3He(^3He, 2p)^4He$ reaction. Approximately 86% of the hydrogen being processed to helium in the sun flows through the $^3He(^3He, 2p)^4He$ reaction, and thus its rate has a direct effect on our understanding of stellar burning and solar neutrino flux observations. The LUNA Collaboration measured the cross-section down to 16.5 keV, where the S-factor was observed to increase because of the effect of electron screening [26, 27]. Electron screening effectively reduces the height of the Coulomb barrier in the reaction resulting in an increase in the measured S-factor and is typically parameterized in terms of an electron screening potential, U_e. It was found in ref. [26] that the screening potential needed to reproduce the data was roughly twice the adiabatic (*i.e.*, upper) limit. While this discrepancy is not understood, it is clear that we must understand electron screening because the screening in the stellar plasma is quite different from that in laboratory studies. The cross-sections measured in the laboratory can only be related to the stellar reaction rate if the screening potential is understood. The observed discrepancy may, in part, be due to a broad ^6Be resonance near the $^3He + {}^3He$ threshold at 11.49 MeV. Understanding the level structure and determining the properties of missing ^6Be levels is thus important for understanding the low energy behavior of the $^3He(^3He, 2p)^4He$ reaction and the electron screening potential.

There is considerable evidence that unobserved levels do exist in this excitation energy region of ^6Be. First of all, several levels have been observed in the isospin mirror nucleus, ^6He, for which analogs have not been observed in ^6Be [28]. In particular, two ^6He levels at 14.6 and 15.5 MeV should have analog levels in ^6Be near or above the $^3He + {}^3He$ threshold at 11.49 MeV. Levels have been predicted at 12.8, 14.7, and 23.8 MeV [29]. Furthermore, a recent measurement at Notre Dame found tentative evidence for a ^6Be level at 9.6 MeV [30].

To search for these unobserved resonances, we have studied the $^2H(^7Be, t)^6Be$ reaction by bombarding a $\sim 1\,mg/cm^2$ CD_2 target with a 100 MeV $^7Be^{4+}$ beam. Typical beam intensities were 3×10^6 $^7Be/s$, and since the beam was fully stripped at the terminal, there was no ^7Li contamination in the beam. Tritons were detected and identified by the SIDAR array configured in telescope mode with 100 μm thick detectors backed by 500 μm thick detectors covering laboratory angles between $\sim 11°$ and 33°. With this range of laboratory angles and detector thicknesses, we should be sensitive to ^6Be levels between $E_x \sim 2$ and 13 MeV. A typical particle identification plot is shown in fig. 9. Unfortunately, it was found that a large number of high-energy ^3He ions are emitted, presumably from α transfer on to carbon in the target. These ^3He ions penetrate the stopping detector producing a back-bending locus in the particle-identification plot which interferes with the clean identification of tritons. To reduce this effect, a third layer (300 μm thick) of silicon-strip detectors was added to be used as veto detectors. To produce the plot in fig. 9, the absence of a veto detector hit is required. This condition was effective in reducing the punch-through ^3He ions around the triton group by about 95%. Nonetheless, some contamination of the triton group was still observed. Runs were also taken with similar thickness CH_2 targets from which it was observed that the number of tritons produced per beam particle was reduced by a factor of 2.4. Therefore, over 70% of the observed tritons are indeed due to ^7Be+d reactions. This data is currently in a preliminary stage of analysis so no conclusions have yet been drawn concerning the level structure of ^6Be.

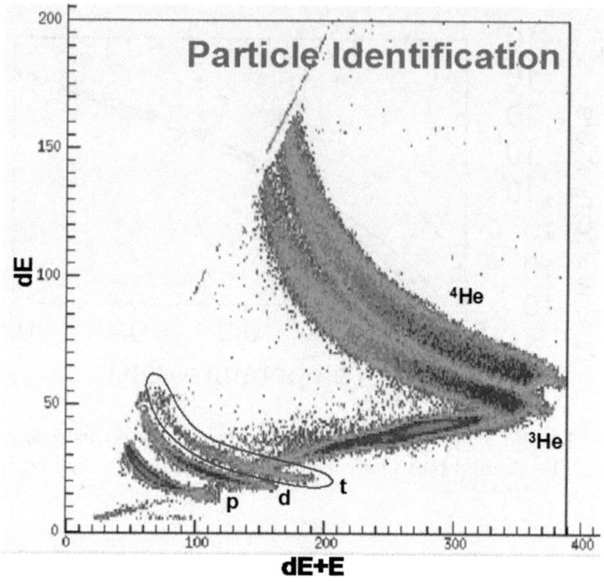

Fig. 9. A particle identification plot from the study of the $^2H(^7Be, t)^6Be$ reaction is shown. A veto condition has been applied which reduces the effect of "punch-through" ^3He ions near the triton locus by $\sim 95\%$.

5 Studies with neutron-rich beams

The r process produces roughly half of the elements heavier than iron via a series of neutron captures and β decays flowing through extremely neutron-rich nuclei. The reaction path and abundances produced are uncertain, however, because of the lack of information on the properties and structure of extremely neutron-rich nuclei. Recently, accelerated beams of many neutron-rich nuclei have become available at the HRIBF [3] making possible some of the first studies of reactions on nuclei along the r-process path. While no single measurement is likely to make a tremendous difference in r-process calculations, understanding the evolution of shell structure certainly will.

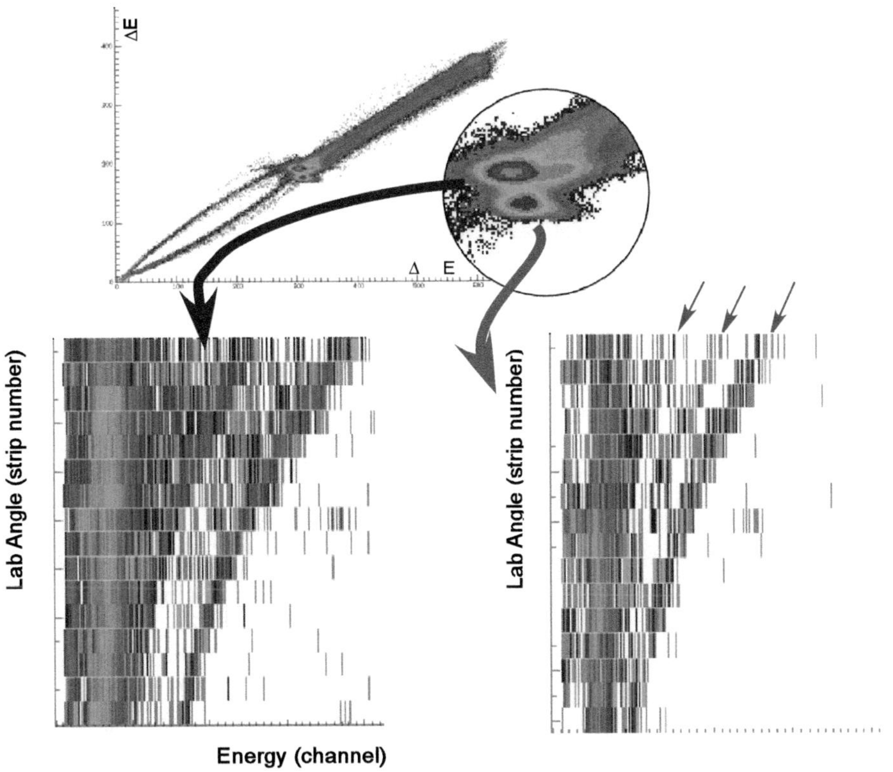

Fig. 10. The top plot shows clear separation of Ge from Se in the ionization counter. By requiring a coincidence with recoil Ge ions, protons detected by the SIDAR array from the ^{2}H(^{82}Ge, p)^{83}Ge reaction are identified.

The low-lying single-particle structure of neutron-rich nuclei near closed shells is especially important for understanding how the synthesis of elements in the r process may be modified by neutron-capture reactions following the fallout from nuclear statistical equilibrium [31]. The distribution of nuclei concentrates in tightly-bound isotopes near closed shells. As the material cools, neutron captures and β decays of these near-closed-shell nuclei alter the abundance pattern. Neutron capture reactions on neutron-rich, closed-shell nuclei are expected to be dominated by direct capture to bound states because of the small Q values for neutron capture and the low-level density in the compound nucleus. Direct capture rates on these nuclei depend sensitively on the structure of low-energy states —such as energy levels, spins, parities, electromagnetic transition probabilities, and single-particle spectroscopic factors— and typically cannot be accurately estimated in the absence of experimental data [32]. It is, therefore, critical that direct capture rate calculations be supplemented with experimental data near closed shells where the r-process abundances peak.

Low-lying single-neutron excitations in ^{83}Ge have been studied for the first time using the ^{2}H(^{82}Ge, p)^{83}Ge reaction [33]. Previously, only the half-life ($t_{1/2} = 1.85$ s) of the $N = 51$ nucleus had been measured. The $A = 83$ isotope of Ge has seven neutrons more than the last stable Ge nucleus, but it is only one neutron past the $N = 50$ closed shell. It is far enough from stability to lie on the path in some r-process models, but its low-lying spectrum is still expected to exhibit the simple characteristics of single-particle structure. With the (d,p) transfer reaction, neutron single-particle states are selectively populated, and proton angular distributions reveal orbital angular momenta and single-particle strengths of final states. Because the mass of ^{82}Ge has been measured, a measurement of the Q value of the reaction also determines the mass of ^{83}Ge.

A beam of ^{82}Ge ($t_{1/2} = 4.6$ s) was produced at the HRIBF by bombarding a UC target with a 42 MeV proton beam inducing fission of the uranium. Produced $A = 82$ ions (Se:Ge:As = 85:15:< 1) were accelerated to 330 MeV before delivery to the experiment. Beam currents of ^{82}Ge averaged 10^4/s. The isobarically mixed $A = 82$ beam was used to bombard a 430 μg/cm^2 CD_2 target. Protons from the (d,p) reaction were detected in the SIDAR array at backward laboratory angles ($\theta_{lab} = 105°$–$150°$) corresponding to forward angles in the center of mass ($\theta_{c.m.} = 36°$–$11°$). In coincidence with protons, beam-like recoils were detected and identified at forward angles ($\theta_{lab} < 1°$) in the gas filled ionization counter shown in fig. 1. The coincidence requirement was necessary to distinguish (d,p) protons induced by ^{82}Ge projectiles from those induced by ^{82}Se. This selection is shown in fig. 10. Events from ^{2}H(^{82}Ge, p)^{83}Ge and ^{2}H(^{82}Se, p)^{83}Se are clearly visible after the recoil selection.

The Q-value for populating the ground state of ^{83}Ge was found to be 1.47 ± 0.02 MeV resulting in a mass excess of $\Delta(^{83}\text{Ge}) = 61.25 \pm 0.26$ MeV, where the uncertainty is mostly the result of the uncertainty in the mass

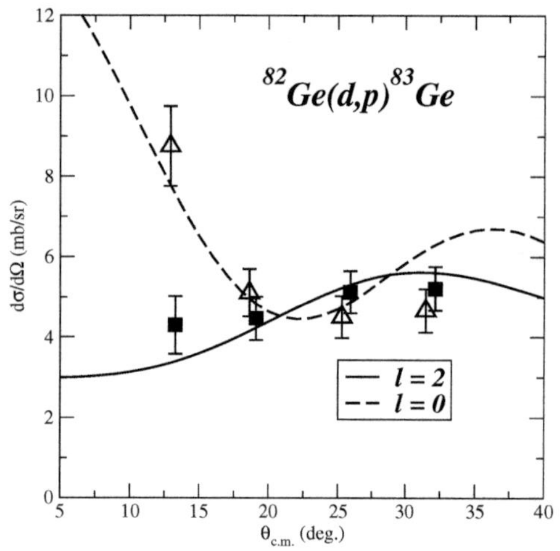

Fig. 11. Proton angular distributions as a function of c.m. angle for ^{83}Ge. Ground-state data (filled squares) fitted by $\ell = 2$ (solid curve); $E_x = 280\,\text{keV}$ data (open triangles) fitted by $\ell = 0$ (dashed curve).

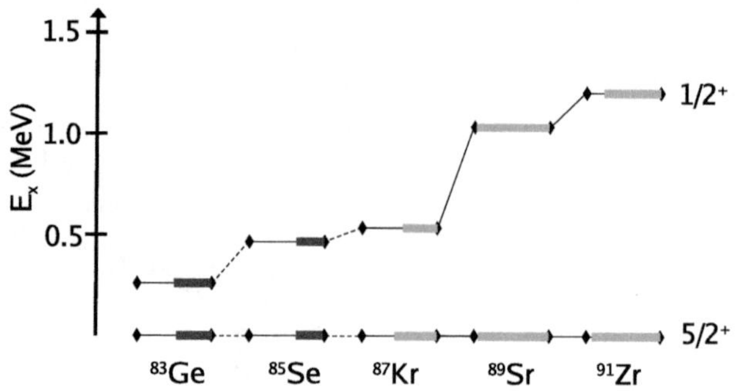

Fig. 12. Spectroscopic properties of the first two states of the even $Z < 40$, $N = 51$ isotones. The lengths of the thick lines represent measured spectroscopic factors. Data for ^{83}Ge and ^{85}Se are from the present work. Data for the other $N = 51$ isotones: ^{91}Zr [38], ^{89}Sr [39], ^{87}Kr [40].

of ^{82}Ge ($\pm 244\,\text{keV}$). The first excited state was found to be at $E_x = 280 \pm 20\,\text{keV}$. Angular distributions have been extracted for the ground and first excited state and are plotted in fig. 11.

Distorted-waves Born (DWBA) calculations have been performed to fit the angular distributions in fig. 11. In the present inverse kinematics study, with the silicon array covering backwards laboratory angles, elastic scattering could not be measured. Even so, the global optical model parameters as deduced by Lohr and Haeberli [34] (deuterons) and Varner *et al.* [35] (protons) are well suited for this mass and energy region. Using the DWBA code TWOFNR [36] and global parameter sets for the optical model potentials from refs. [34,35], the fits in fig. 11 were produced. The best fits supported a $J^\pi = 5/2^+$ assignment for the ground state and $1/2^+$ assignment for the first excited state with spectroscopic factors of 0.48 ± 0.14 and 0.50 ± 0.15, respectively.

Our second measurement with neutron-rich beams was a study of the ^{2}H(^{84}Se, p)^{85}Se reaction [37]. Since the reaction kinematics are nearly identical to the ^{82}Ge case, the detector setup was nearly the same as well. A 4.5 MeV/u $A = 84$ beam was used to bombard a $200\,\mu\text{g/cm}^2$ CD_2 target for nearly 10 days. The higher beam energy and thinner target used in this study resulted in a better energy resolution than in the ^{82}Ge study ($\Delta E_{c.m.} \sim 220\,\text{keV}$ compared with $\sim 300\,\text{keV}$ in the ^{82}Ge case). The beam consisted mainly of ^{84}Br (93%) and ^{84}Se (7%), and the total beam current was $\sim 9 \times 10^4$ pps. ^{85}Se levels at $E_x = 0, 0.462$, 1.114, and $\sim 1.44\,\text{MeV}$ were observed. DWBA fits to the ground and first excited states were again consistent with $5/2^+$ and $1/2^+$ spin/parity assignments with spectroscopic factors of 0.33 ± 0.08 and 0.30 ± 0.08, respectively. A summary of our results of (d, p) measurements on $N = 50$ nuclei is shown in fig. 12.

We are poised to extend these measurements to the next closed neutron shell at $N = 82$ by studying (d, p) reactions on ^{130,132}Sn. Transfer reactions using a beam of ^{132}Sn have been the subject of recent attention and are often singled out as prototypical experiments [41,42, 43] for next generation facilities such as the Rare Isotope Accelerator [44]. At the HRIBF, it is possible to accelerate beams of ^{130,132}Sn to energies around the Coulomb barrier with sufficient intensity to begin such studies.

The (d, p) cross-section to populate low lying, low angular momentum states is often greater at lower beam energies, due to kinematic matching conditions, and at the same time the reaction is cleaner as there are few reaction channels open. However, angular distributions below the Coulomb barrier become less distinctive, and it was not clear that ℓ values and spectroscopic factors could be extracted under such conditions. To examine these possible difficulties, we have performed a study of the ^{2}H(^{124}Sn, p)^{125}Sn reaction [45]. This reaction has been studied at several energies in normal kinematics and was therefore a good benchmark [46,47].

A beam of ^{124}Sn was accelerated to 562 MeV and focused onto a $100\,\mu\text{g/cm}^2$ target of deuterated polyethylene (CD_2). The target was turned 30° with respect to the beam, thus achieving an effective thickness of $200\,\mu\text{g/cm}^2$ and at the same time allowing emerging protons to be detected around $\theta_{lab} = 90°$ without being shadowed by the target frame. Protons were detected in two telescopes of silicon detectors. One telescope on the downstream side subtended $\theta_{lab} = 70°$–$102°$. The other telescope on the upstream side subtended $\theta_{lab} = 85°$–$110°$. The SIDAR array was mounted at more backward angles $\theta_{lab} = 130°$–$160°$. The telescopes consisted of a thin position-sensitive (ΔE) detector and a thicker stopping (E) detector ($1000\,\mu$m). The ΔE detectors were either 65 or $140\,\mu$m thick depending on the angles (and thus the proton energies) needed. Both ΔE detectors had 16 position-sensitive strips allowing the angle of proton emission to be determined to a precision of $\pm 0.5°$.

Data were collected for about 18 hours with a beam rate of 10^7 ^{124}Sn/s. The angles of particles measured in the downstream telescope are shown as a function of their energy in fig. 13. There are three clearly visible loci relating to elastically-scattered carbon atoms, deuterons, and

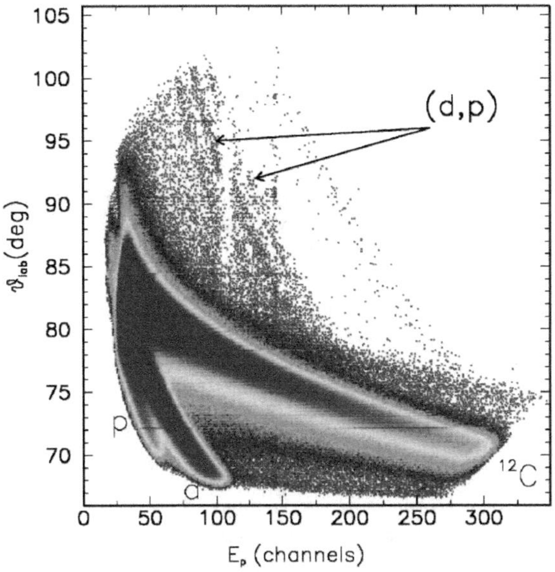

Fig. 13. Angle of particles measured in the $^{2}H(^{124}Sn, p)^{125}Sn$ study as a function of energy without any gates imposed. Elastic scattering of carbon atoms, deuterons, and protons from the target and reaction protons from the (d, p) reaction are indicated.

Table 1. Spectroscopic factors from this work and previous works for $^{124}Sn(d, p)^{125}Sn$. The quoted uncertainties include statistical, DWBA fitting effects, and systematic errors due to the normalization.

E_x (MeV)	J^{π}	This work	Ref. [47]	Ref. [46]
0.028	$3/2^+$	0.44(6)	0.53	0.44
0.215	$1/2^+$	0.33(4)	0.32	0.33
2.8	$7/2^-$	0.46(5)	0.52	0.54

protons from the target; additionally the distinctive loci resulting from the (d, p) reaction are discernible at larger angles. A break can be seen around channel 100 where protons following the (d, p) reaction have just enough energy to punch through the ΔE detector, leaving only a subthreshold signal in the E detector. Once an energy-loss gate is imposed on the protons, only protons from the (d, p) reaction remain in this plot [45].

^{125}Sn states at $E_x = 0.028$, 0.215, and 2.8 MeV were strongly populated in the study. The resolution in excitation energy was found to be approximately 200 keV, owing mainly to target thickness and beam spot size effects. It should be noted that the level density of low-lying states around ^{132}Sn is expected to be low, and thus such energy resolution should be more than sufficient for studies in this nuclide region. Angular distributions have been extracted for these levels [45] with the absolute normalization of the cross-sections determined from the observed angular distribution of elastically-scattered deuterons. The angular distributions from these states show distinctive structures which are characteristic of the angular momentum transferred in the reaction [45]. Spectroscopic factors have been extracted by comparing the measured angular distributions with calculations using TWOFNR. The results are shown in table 1. The values of the spectroscopic factors all agree well with those from the work of refs. [46,47] which were performed in normal kinematics. These results demonstrate the effectiveness of these techniques to determine spectroscopic properties of nuclei in this mass region at beam energies close to the Coulomb barrier.

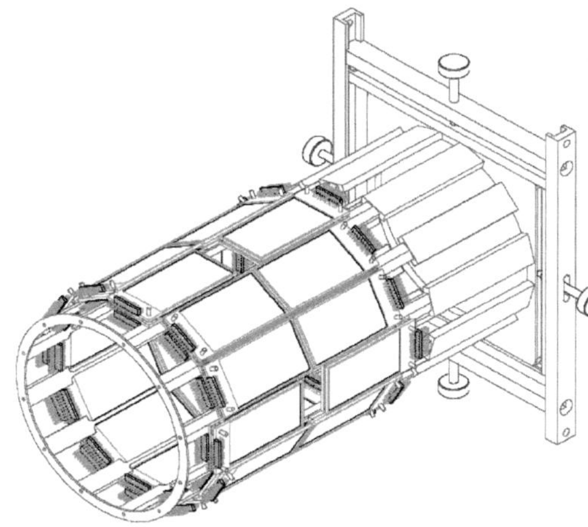

Fig. 14. The ORRUBA detector array showing the two rings of charged-particle telescopes. The gap between rings allows for insertion of the target.

After the encouraging results obtained in this demonstration measurement, we are developing a larger silicon-detector array specifically for the better detection and identification of charged particles emitted near 90° in the laboratory. The Oak Ridge Rutgers University Barrel Array (ORRUBA) [48] will be a large solid angle silicon detector array, capable of providing energy, angle, and particle identification information. ORRUBA will be comprised of two rings of 12 position-sensitive silicon detector telescopes, symmetrically covering angles forward and backward of 90° (see fig. 14). Each detector is ~ 8 cm long, and its width is divided into four 1 cm wide resistive strips, oriented parallel to the beam axis. Readouts from both ends of each strip allow measurement of the position of the interaction, allowing determination of the emission angle of the detected particles. Prototype detectors have arrived and are currently being tested.

6 Conclusions

Twenty years after Willy Fowler's historic call for data on the properties and interactions of radioactive nuclei [49], scientists have obtained a first glimpse of the possibilities that radioactive beams offer for nuclear astrophysics studies. The HRIBF is one of a few first-generation radioactive beam facilities that has provided some of these first glimpses. We have used accelerated beams of $^{17,18}F$ to better understand key reactions that power the nova outburst and produce observational tracers of the explosion such as ^{18}F. Our studies indicate that ^{18}F is destroyed in the novae environment by the $^{18}F(p, \alpha)^{15}O$ reaction at a lower rate than previously thought. We have used the tools available at `nucastrodata.org` to quantify the effect and calculate that roughly a factor of 2 more ^{18}F is ejected than previously thought.

We are using ^{7}Be beams to study reactions of importance to stellar burning. By bombarding a windowless hydrogen target with the ^{7}Be beam and detecting recoil ^{8}B nuclei at the focal plane of the Daresbury Recoil Separator, we are on the verge of making the first statistically-significant measurement of $^7\text{Be}(p,\gamma)^8\text{B}$ in inverse kinematics. Only after several measurements with different systematic uncertainties, can the community hope to obtain a truly precise measure of the $^7\text{Be}(p,\gamma)^8\text{B}$ reaction rate at stellar energies. We are also studying the unbound nucleus, ^{6}Be, via a measurement of the $^2\text{H}(^7\text{Be},t)^6\text{Be}$ reaction at 100 MeV. Levels in ^{6}Be are important for understanding the $^3\text{He}(^3\text{He},2p)^4\text{He}$ reaction and electron screening at stellar energies.

Finally, we are using radioactive beams of fission fragments to study neutron-rich nuclei near closed neutron closed shells. In particular, we are using the (d,p) reaction in inverse kinematics to probe single-particle excitations. We have made the first neutron-transfer measurement on an r process nucleus, ^{82}Ge, and are poised to continue our studies near doubly-magic ^{132}Sn. A new detector array, ORRUBA, is being developed to facilitate these studies.

Most of the experiments mentioned here were performed by the Radioactive Ion Beams for Explosive Nucleosynthesis Studies (RIBENS) Collaboration, which is a collaboration of over 30 scientists from 11 institutions including ORNL, Rutgers University, Tennessee Technological University, Colorado School of Mines, Ohio University, Yale University, and the University of North Carolina at Chapel Hill. This research was sponsored, in part, by the Laboratory Directed Research and Development Program of Oak Ridge National Laboratory, managed by UT-Battelle, LLC, for the U.S. Department of Energy under Contract No. DE-AC05-00OR22725.

References

1. M.S. Smith, K.E. Rehm, Annu. Rev. Nucl. Part. Sci. **51**, 91 (2000).
2. G.D. Alton, J.R. Beene, J. Phys. G **24**, 1347 (1998).
3. D.W. Stracener, G.D. Alton, R.L. Auble, J.R. Beene, P.E. Mueller, J.C. Bilheux, Nucl. Instrum. Methods Phys. Res. A **521**, 126 (2004) and references therein.
4. D.W. Bardayan *et al.*, Phys. Rev. Lett. **83**, 45 (1999).
5. C.J. Gross *et al.*, Nucl. Instrum. Methods Phys. Res. A **450**, 12 (2000).
6. M. Hernanz, J. José, A. Coc, J. Gómez-Gomar, J. Isern, Astrophys. J. Lett. **526**, L97 (1999).
7. A. Coc, M. Hernanz, J. José, J.-P. Thibaud, Astron. Astrophys. **357**, 561 (2000).
8. D.W. Bardayan *et al.*, Phys. Rev. C **62**, 042802(R) (2000).
9. R. Coszach *et al.*, Phys. Lett. B **353**, 184 (1995).
10. K.E. Rehm *et al.*, Phys. Rev. C **52**, R460 (1995); **53**, 1950 (1996).
11. D.W. Bardayan *et al.*, Phys. Rev. C **70**, 015804 (2004).
12. A. Galindo-Uribarri *et al.*, Nucl. Instrum. Methods Phys. Res. B **172**, 647 (2000).
13. N.-C. Shu, D.W. Bardayan, J.C. Blackmon, Y.-S. Chen, R.L. Kozub, P.D. Parker, M.S. Smith, Chin. Phys. Lett. **20**, 1470 (2003).
14. D.W. Bardayan *et al.*, Phys. Rev. C **63**, 065802 (2001).
15. D.W. Bardayan *et al.*, Phys. Rev. Lett. **89**, 262501 (2002).
16. M. Wiescher, K.-U. Kettner, Astrophys. J. **263**, 891 (1982).
17. R.L. Kozub *et al.*, Phys. Rev. C **71**, 032801(R) (2005).
18. R. Fitzgerald *et al.*, *Proceedings of the 6th International Conference on Radioactive Nuclear Beams*, Nucl. Phys. A **748**, 351 (2005).
19. N. de Séréville *et al.*, Phys. Rev. C **67**, 052801(R) (2003).
20. Y. Butt *et al.*, Phys. Rev. C **58**, 10(R) (1998).
21. D.W. Bardayan *et al.*, Phys. Rev. C **71**, 018801 (2005).
22. S. Parete-Koon *et al.*, Astrophys. J. **598**, 1239 (2003).
23. `http://www.nucastrodata.org`.
24. R. Fitzgerald, Ph. D. Thesis, University of North Carolina at Chapel Hill (2005).
25. R.H. Cyburt, B. Davids, B.K. Jennings, Phys. Rev. C **70**, 045801 (2004).
26. M. Junker *et al.*, Phys. Rev. C **57**, 2700 (1998).
27. R. Bonetti *et al.*, Phys. Rev. Lett. **82**, 5205 (1999).
28. D.R. Tilley *et al.*, Nucl. Phys. A **708**, 1 (2002).
29. J.J. Bevelacqua, Phys. Rev. C **33**, 699 (1986).
30. V. Guimaraes *et al.*, Nucl. Phys. A **722**, 341c (2003).
31. R. Surman, J. Engel, Phys. Rev. C **64**, 035801 (2001).
32. T. Rauscher *et al.*, Phys. Rev. C **57**, 2031 (1998).
33. J.S. Thomas *et al.*, Phys. Rev. C **71**, 021302(R) (2005).
34. J.M. Lohr, W. Haeberli, Nucl. Phys. A **232**, 381 (1974).
35. R.L. Varner, W.J. Thompson, T.L. McAbee, E.J. Ludwig, T.B. Clegg, Phys. Rep. **201**, 57 (1991).
36. University of Surrey modified version of the code TWOFNR of M. Igarashi, M. Toyama, N. Kishida (private communication).
37. J.S. Thomas, Ph. D. Thesis, Rutgers University (2005).
38. R.D. Rathmell, P.J. Bjorkholm, W. Haeberli, Nucl. Phys. A **206**, 459 (1973).
39. T.P. Cleary, Nucl. Phys. A **301**, 317 (1978).
40. K. Haravu, C.L. Hollas, P.J. Riley, W.R. Coker, Phys. Rev. C **1**, 938 (1970).
41. K.E. Rehm, in *Proceedings of the Second International Conference on Fission and Properties of Neutron-Rich Nuclei, St. Andrews, Scotland, 1999*, edited by J.H. Hamilton, W.R. Phillips, H.K. Carter (World Scientific, Singapore, 2000) p. 439.
42. *ISAC-II A Project for Higher Energies at ISAC (1999)*, `http://www.triumf.ca/ISAC-II/TRI-99-1.pdf`.
43. A.H. Wuosmaa, T. Al Tahtamouni, J.P. Schiffer, Nucl. Phys. A **746**, 267c (2004).
44. `http://www.orau.org/ria`.
45. K.L. Jones *et al.*, Phys. Rev. C **70**, 067602 (2004).
46. C.R. Bingham, D.L. Hillis, Phys. Rev. C **8**, 729 (1973).
47. A. Stromich, B. Steinmetz, R. Bangert, B. Gonsior, M. Roth, P. von Brentano, Phys. Rev. C **16**, 2193 (1977).
48. `http://www.orau.org/stewardship/`.
49. W.A. Fowler, in *Proceedings of the Accelerated Radioactive Beams Workshop*, edited by L. Buchmann, J.M. D'Auria (TRIUMF Report TRI-85-1, Vancouver, 1985) p. 1.

Eur. Phys. J. A **27**, s01, 107–115 (2006)
DOI: 10.1140/epja/i2006-08-015-8

EPJ A direct
electronic only

Beacons in the sky: Classical novae vs. X-ray bursts

J. José[1,a] and M. Hernanz[2]

[1] Institut d'Estudis Espacials de Catalunya (IEEC-UPC) and Departament de Física i Enginyeria Nuclear, Universitat Politècnica de Catalunya, Barcelona, Spain,
[2] Institut d'Estudis Espacials de Catalunya (IEEC-CSIC) and Institut de Ciències de l'Espai (CSIC), Bellaterra, Spain

Received: 22 June 2005 /
Published online: 24 February 2006 –

Abstract. Thermonuclear runaways are at the origin of some of the most energetic and frequent stellar cataclysmic events. In this review talk, we outline our understanding of the mechanisms leading to classical nova explosions and X-ray bursts, together with their associated nucleosynthesis. In particular, we focus on the interplay between nova outbursts and the Galactic chemical abundances (where ^{13}C, ^{15}N, and ^{17}O constitute the likely imprints of many nova outbursts during the overall 10 Gyr of Galactic history), the synthesis of radioactive nuclei of interest for gamma-ray astronomy (^{7}Be-^{7}Li, ^{22}Na, or ^{26}Al), the endpoint of nova nucleosynthesis, based on theoretical and observational grounds, and the recent discovery of presolar meteoritic grains, both in the Murchison and Acfer 094 meteorites, likely condensed in nova shells. Recent progress in the modeling of X-ray bursts as well as an insight into the input nuclear physics requests, for both novae and X-ray bursts, will also be presented.

PACS. 26.50.+x Nuclear physics aspects of novae, supernovae, and other explosive environments – 26.30.+k Nucleosynthesis in novae, supernovae and other explosive environments – 95.85.Pw γ-ray – 97.80.Gm Cataclysmic binaries (novae, dwarf novae, recurrent novae, and nova-like objects); symbiotic stars

1 Introduction

Classical novae and type Ia (or thermonuclear) supernovae have been extensively observed during the past two millennia. They constitute dramatic stellar explosions occurring in interacting binary systems, where a compact white dwarf (the stellar remnant of a Main Sequence star with a mass below ~ 10 M$_\odot$; white dwarfs have planetary dimensions and masses typically in the range 0.6–1.4 M$_\odot$) accretes material from a low-mass Main Sequence companion. In contrast, X-ray bursts have been discovered much recently, since a major fraction of their energy output is emitted in X-rays, and hence, detection requires the use of space observatories. About 65 Galactic low-mass X-ray binaries that exhibit such bursting behavior have been found since the independent discovery by Grindlay *et al.* (1976) [1] and Belian *et al.* (1976) [2]. In these systems, the explosion takes place in an even more compact stellar remnant: a neutron star (with a mass about 1–2 M$_\odot$, and a very small diameter, 20 to 30 km only; neutron stars formation accompanies some type II supernova explosions for stars more massive than ~ 10 M$_\odot$, but can be formed also in other processes such as the so-called "accretion induced collapse" of a white dwarf; see Swank *et al.* (1978) [3] for an account of the identification of a neutron star as the central emitting source in the bursting system 3U 0614 +09). In all these cataclysmic systems, mass transfer episodes, caused by Roche lobe overflows of the Main Sequence companion, proceed through the inner Lagrangian point of the system, forming an accretion disk around the compact star. A fraction of the material contained in the accretion disk ultimately falls on top of the white dwarf (as a result of angular momentum losses driven by dissipative forces in the disk), where it is gradually compressed up to degenerate conditions, thus leading to a thermonuclear runaway (TNR). The estimated number of nova events in our Galaxy is about $30 \pm 10\,\mathrm{yr}^{-1}$ (Shafter, 2002, [4]). Contrary to type Ia supernovae, for which the white dwarf is fully disrupted by the strength of the explosion, classical novae are expected to recur within a timescale of the order of 10^4–10^5 yr (Warner, 1995, [5]); in contrast, X-Ray bursts recur on timescales from hours to days (see Strohmayer and Bildsten, 2006, [6] for a recent review). These stellar explosions are characterized by a remarkable energy output, of the order of 10^{39} ergs (X-ray bursts), 10^{45} ergs (classical novae), and 10^{51} ergs (supernovae). A major difference between these cataclysmic events is the amount of mass ejected as well as the mean ejection velocities: whereas in a type Ia supernova, the whole star (cf., 1.4 M$_\odot$) is ejected at $> 10^4\,\mathrm{km\,s^{-1}}$, the

[a] e-mail: jjose@ieec.fcr.es

explosion in both classical novae and X-ray bursts is restricted to the accreted envelope. Hence, 10^{-4}–10^{-5} $M_\odot$ are typically ejected during nova outbursts, at mean ejection velocities ranging from 10^2–10^3 km s^{-1}. In contrast, it is likely that in a typical X-ray burst no mass ejection takes place, because of the extreme escape velocities from a neutron star (the energy required to escape from the strong gravitational field of a neutron star of mass M and radius R is $GMm_p/R \sim 200$ MeV/nucleon, whereas the nuclear energy released from fusion of solar-like matter into Fe-group elements is just ~ 5 MeV/nucleon).

2 Classical nova explosions

The canonical scenario of classical novae consists of an initially low-luminosity white dwarf accreting H-rich matter from a low-mass Main Sequence companion, at an accretion rate low enough to allow the building-up of an envelope where the TNR takes place. Accretion compresses matter to large densities, such that material becomes degenerate. Once H-ignition conditions are reached, a thermonuclear runaway takes place, since degenerate matter cannot expand when heated. The large energetic output produced by nuclear reactions cannot be only evacuated by radiation, and hence convection sets in and transports the β^+-unstable nuclei to the outer cooler regions where their decay provokes the nova outburst: the sudden release of energy from those β^+-decays increases the temperature and the entropy of the material; as a result, the envelope degeneracy is lifted and expansion sets in, paving the road to the ejection stage. Therefore, the TNR is halted by the envelope expansion rather than by fuel consumption. Another important effect of convection is that it transports fresh unburned material to the burning shell, and hence, non-equilibrium burning occurs and the resulting nucleosynthesis will be far from that of hydrostatic hydrogen burning. It is important to stress that significant amounts of CNO nuclei are required to let the ignition process really progress into a thermonuclear runaway and indeed to get an (energetic enough) explosion, and also to account for the inferred abundances from nova shells. Mixing episodes between the solar-like accreted envelope matter and the outer layers of the underlying white dwarf core have been invoked to account for the source of those CNO nuclei.

Nova explosions can occur in two types of white dwarfs: carbon-oxygen (CO) and oxygen-neon (ONe). The first one correspond to the end-product of a star with a mass below ~ 10 $M_\odot$, which undergoes two evolutionary stages, central hydrogen burning and central helium burning, leaving a carbon and oxygen-rich object (CO) supported by electron pressure (in contrast, neutron stars balance the strong gravitational force by means of neutron pressure). More massive progenitors undergo an additional stage, that is non-degenerate carbon ignition, leading to the formation of an oxygen and neon-rich core (ONe), with traces of magnesium and sodium. The cut-off mass, that somewhat distinguishes both white dwarf subclasses, is about 1.1 $M_\odot$. It is important to stress that historically, these cores were thought to be made of oxygen, neon and magnesium (the so-called ONeMg white dwarfs) on the basis of parametrized calculations of hydrostatic carbon burning (Arnett and Truran, 1969, [7]), but more recent, self-consistent models of sAGB (super-asymptotic giant branch) stars, following the thermally pulsing phases, have shown that the amount of ^{24}Mg in these cores is much smaller than expected (*i.e.*, smaller than ^{23}Na; see Domí nguez *et al.*, 1993, [8]; Ritossa *et al.*, 1996, [9]). It is also worth noting that these ONe white dwarfs have thick CO buffers on top of their very cores. Explosions on top of these objects (before a significant number of outbursts may erode the buffer) may be misclassified as non-neon (CO) novae, since their spectra will not show evidence for strong neon lines. These and other weird nucleosynthetic patterns have been predicted to accompany these explosions (see José *et al.*, 2003, [10] for details).

It is important to stress that the injection of "seed" nuclei in the Ne-Si mass region, resulting from mixing with an ONe white dwarf, withdraws the main argument in support of CNO-breakout in classical novae. Indeed, recent estimates of several CNO-breakout reactions (see, for instance, Davids *et al.*, 2003, [11] for ^{15}O(α,γ)), demonstrate that at the typical temperatures attained during nova outbursts the CNO cycle does not break out (contrary to what happens for X-ray bursts, for which peak temperatures are almost an order of magnitude higher than in classical nova outbursts).

In nova outbursts, the main nuclear activity is associated with (p,γ) and (p,α) reactions, running between the line of stable nuclei and the proton drip line. Some of these rates have been measured at the right energy range in the laboratory since at the typical temperatures achieved during nova outbursts, measurements of nuclear cross sections in the region of the Gamow peak are, except for a few exceptions, feasible. In this sense, classical nova are unique stellar explosions, and represent the only explosive site for which the nuclear physics input is (or will be in the near future) primarily based on experimental information (see José *et al.*, 2006, [12]).

2.1 The role of nova outbursts in the galactic alchemy

As discussed above, the moderately high peak temperatures achieved during nova explosions, $T_{\rm peak} \sim (2\text{–}3)\times 10^8$ K, suggest that abundance levels of the intermediate-mass elements in the ejecta must be significantly enhanced, as confirmed by spectroscopic determinations in well-observed nova shells. This raises the issue of the potential contribution of novae to the Galactic abundances, which can be roughly estimated as the product of the Galactic nova rate (~ 30 events yr^{-1}), the average ejected mass per nova outburst ($\sim 2\times10^{-5}$ $M_\odot$), and the Galaxy's lifetime ($\sim$ 10 Gyr). This order of magnitude estimate points out that novae scarcely contribute to the Galaxy's overall metallicity (as compared with other major sources, such as supernova explosions or AGB stars), but nevertheless they can substantially contribute to the synthesis of some largely overproduced species. Indeed, studies suggest that classical novae are likely sites for the synthesis of most of the Galactic ^{13}C, ^{15}N, and ^{17}O, whereas they can partially contribute to the Galactic abundances of other species with $A < 40$, such as ^7Li, ^{19}F, or ^{26}Al (Starrfield *et al.*, 1998, [13]; José and Hernanz, 1998, [14]). Because of the lower peak temperatures achieved in CO models (together with the lack of significant amounts of "seed" nuclei in the NeNa-MgAl region), the main nuclear activity in CO novae does not extend much beyond oxygen. In contrast, ONe models show a much larger nuclear activity, extending up to silicon (for 1.15 $M_\odot$ ONe) or argon (for 1.35 $M_\odot$ ONe). Hence, the presence of significantly large amounts of intermediate-mass nuclei in the spectra, such as phosphorus, sulfur, chlorine or argon, may reveal the presence of an underlying massive ONe white dwarf. Another trend derived from the analysis of the nucleosynthesis accompanying nova outbursts is the fact that the O/N and C/N ratios decrease as the mass of the white dwarf (and hence, the peak temperature attained during the explosion) increases.

2.2 The abundance pattern of nova shells

In order to constraint models, several works have focused on a direct comparison of the atomic abundances inferred from observations of the ejecta with the theoretical nucleosynthetic output (see José and Hernanz, 1998, [14]; Starrfield *et al.*, 1998, [13]; Kovetz and Prialnik, 1997, [15], and references therein; see also Yaron *et al.*, 2005, [16]). This comparison itself is not very restrictive since observations provide only information on elemental abundances (*i.e.*, for O we get the total contribution from ^{16}O, ^{17}O, and ^{18}O). Indeed, these atomic abundances are not absolute but relative to one element (often, hydrogen), and they cannot be directly inferred from observations: their determination involves modeling through photoionization codes or other numerical techniques. Nevertheless, and despite of the problems associated with the modeling of the explosion (Starrfield, 2002, [17]), there is, in general, good agreement between theory and observations as regards nucleosynthesis (*i.e.*, including atomic abundances —H, He, C, O, Ne, Na-Fe—, and a plausible endpoint for nova nucleosynthesis). In some cases, such as for PW Vul 1984, the agreement between observations and theoretical predictions (see José and Hernanz, 1998, [14], Table 5, for details) is quite remarkable. The reader is referred to Gehrz *et al.* (1998) [18] for an extended list of abundance determinations in the ejecta from novae.

Since the nuclear path is very sensitive to details of the evolution (chemical composition, extend of convective mixing, thermal history of the envelope...), the agreement between the inferred abundances and the theoretical yields suggests that the overall picture, in the framework of the thermonuclear runaway model, is reasonably well understood.

2.3 Presolar grains

Infrared (Evans, 1990, [19]; Gehrz *et al.*, 1998, [18]) and ultraviolet observations of the temporal evolution of nova light curves suggest that novae form grains in the expanding nova shells. Both CO and ONe novae behave similarly in the infrared right after the outburst. However, as the ejected envelope expands and becomes optically thin, such behavior dramatically changes: CO novae are typically followed by a phase of dust formation corresponding to a decline in visual light, together with a simultaneous rise in the infrared emission (Rawlings and Evans, 2002, [20]; Gehrz, 2002, [21]). In contrast, it has been argued that ONe novae (that involve more massive white dwarfs than CO novae) are not so prolific producers of dust as a result of the lower mass, high-velocity ejecta, where the typical densities can be low enough to enable the condensation of appreciable amounts of dust. Indeed, infrared measurements in a number of recent novae have revealed the presence of C-rich dust (Novae Aql 1995, V838 Her 1991, PW Vul 1984, ...), SiC (Novae Aql 1982, V842 Cen 1986, ...), hydrocarbons (Novae V842 Cen 1986, V705 Cas 1993, ...), or SiO2 (Novae V1370 Aql 1982, V705 Cas 1993). Remarkable examples, such as nova QV Vul 1987, exhibited simultaneous formation of all those types of dust (see Gehrz *et al.*, 1998, [18] for details on dust forming novae).

Since the first studies of dust formation in classical novae (Clayton and Hoyle, 1976, [22]), the identification of presolar nova grains in the laboratory (isolated from meteorites), presumably condensed in the shells ejected during the explosion, relied only on low $^{20}Ne/^{22}Ne$ ratios (with ^{22}Ne attributed to ^{22}Na *in situ* decay, rather than trapped ^{22}Ne, since noble gases do not condense into grains). But quite recently, five silicon carbide and two graphite grains (isolated from both the Murchison and Acfer 094 meteorites) that exhibit isotopic signatures characteristic of nova nucleosynthesis, have been identified (Amari *et al.*, 2001, [23]; Amari, 2002, [24]). These grains are characterized by low $^{12}C/^{13}C$ and $^{14}N/^{15}N$ ratios, ^{30}Si excesses and close-to or slightly lower-than-solar $^{29}Si/^{28}Si$ ratios, high $^{26}Al/^{27}Al$ ratios (determined only for two grains) and low $^{20}Ne/^{22}Ne$ ratios (only measured in the graphite grain KFB1a-161). This discovery provides valuable constraints for nova nucleosynthesis models (in contrast to observations of nova shells, for which only atomic abundances are inferred). Theoretical efforts to predict the expected imprints of nova outbursts on presolar grains have been conducted by several authors (Starrfield *et al.*, 1997, [25]; José *et al.*, 2003 [26]; 2004 [27]), including preliminary work on equilibrium condensation sequences in the ejected shells (José *et al.*, 2004, [27]). These studies suggest that classical novae may contribute to the known presolar corundum (Al_2O_3), spinel ($MgAl_2O_4$), enstatite ($MgSiO_3$), silicon carbide (SiC) and silicon nitride (Si_3N_4) grain populations. Aspects, such as the conditions that characterize grain formation in novae deserve more attention (see Shore and Gehrz, 2004, [28] for an insight into dynamic grain formation driven by photo-ionization).

2.4 Nova radioactivities

Among the isotopes synthesized during classical nova outbursts, several radioactive species have deserved special attention. Short-lived nuclei, such as ^{14,15}O and ^{17}F (and to some extent ^{13}N) have been identified as the key isotopes that power the expansion and further ejection of the envelope during nova outbursts through a sudden release of energy, a few minutes after peak temperature (Starrfield *et al.*, 1972, [29]). Other isotopes have been extensively investigated in connection with the predicted γ-ray output from novae (Clayton and Hoyle, 1974, [30]; Clayton, 1981, [31]; Leising and Clayton, 1987, [32]). Hence, ^{13}N and ^{18}F are responsible for the predicted prompt γ-ray emission (Hernanz *et al.*, 1999, [33]) at and below 511 keV, whereas ^{7}Be and ^{22}Na (Gómez-Gomar *et al.*, 1998, [34]), which decay much later (when the envelope is optically thin), are the sources that power line emission at 478 and 1275 keV, respectively. ^{26}Al is another important radioactive isotope that can be synthesized during nova outbursts, although only its cumulative emission can be observed because of its slow decay. We will briefly focus on the corresponding nuclear paths leading to the synthesis of the above-mentioned gamma-ray emitters, with special emphasis on the nuclear uncertainties associated with the relevant reaction rates.

2.4.1 ^{7}Be-^{7}Li

^{3}He$(\alpha,\gamma)^{7}$Be is the main reaction leading to ^{7}Be synthesis, which is transformed into ^{7}Li by means of an electron capture, emitting a γ-ray photon of 478 keV. The contribution of nova outbursts to the Galactic ^{7}Li content has been regarded as very controversial. Hints from the first pioneering estimate, in the framework of a simple parametric model (Arnould and Nørgaard, 1975, [35]), were confirmed by early hydrodynamic simulations (Starrfield *et al.*, 1978, [36]), assuming however an envelope in-place, thus neglecting the impact of the accretion phase on the evolution. These results were preliminary refuted later on, in terms of parametric one/two zone models (Boffin *et al.*, 1993, [37]), pointing out the critical role played by the ^{8}B(p,γ) reaction, not included in all previous works (*i.e.*, Arnould and Nørgaard, 1975, [35], Starrfield *et al.*, 1978, [36]), and claiming therefore for an unlikely synthesis of ^{7}Li in novae. The scenario was later revisited by Hernanz *et al.* (1996) [38] and José and Hernanz (1998) [14], who performed new hydrodynamic calculations, taking into account both the accretion and explosion stages, and a full reaction network (including ^{8}B(p,γ)). These studies confirmed that the *Be-transport mechanism* (Cameron, 1955, [39]) is capable of producing large amounts of ^{7}Li during nova explosions.

Among the key issues that may affect ^{7}Li synthesis in novae, one of the most critical ones is the final amount of ^{3}He that survives the early rise in temperature when the thermonuclear runaway ensues. In particular, the different timescales to reach T_{peak} achieved for CO and ONe novae, which deeply depend on the initial ^{12}C content in the envelope, lead to larger amounts of ^{7}Be in CO novae (which survives destruction through (p,γ) reactions because of the very efficient inverse photodisintegration reaction on ^{8}B). No relevant nuclear uncertainties in the domain of nova temperatures affect the main reaction rates for ^{7}Li synthesis. It is also worth noting that the potential contribution of classical novae to the Galactic ^{7}Li content is rather small (less than 15%, according to models). However, a nova contribution is required to match the ^{7}Li content in realistic calculations of Galactic chemical evolution (Romano *et al.*, 1999, [40]; Alibés *et al.*, 2002, [41]).

The observation of ^{7}Li in the ejecta accompanying nova outbursts is extraordinarily challenging. Recently, the presence of this elusive isotope in an ejected nova envelope has been claimed for the first time: an observed feature compatible with the doublet at 6708 Å of Li I has been reported in the spectra of V382 Vel (Nova Velorum 1999; see Della Valle *et al.*, 2002, [42]). However, it has been argued (Shore *et al.*, 2003, [43]) that the observed feature may correspond instead to another low-ionization emission centered at around 6705 Å, likely the doublet associated with N I. Attempts to detect the 478 keV line associated to ^{7}Be-^{7}Li in a classical nova outburst (with the GRS instrument on-board the SMM satellite (Harris *et al.*, 1991, [44]) or with the TGRS spectrometer on-board WIND (Harris *et al.*, 2001, [45])), have also been unsuccessful to date. Unfortunately, current prospects with ESA's INTEGRAL satellite are not much better, because of the small fluxes expected from novae and the limited sensitivity of the INTEGRAL detectors. The predicted detectability distance of the 478 keV line (a γ-ray signal that may last for about 2 months) is of the order of only ≤ 0.2 kpc for INTEGRAL/SPI (Hernanz and José, 2004, [46]). Future instruments, like the γ-ray lens MAX, will be better suited to attempt this challenging detection (see Hernanz and José, 2004, [47]).

2.4.2 ^{22}Na

The potential role of ^{22}Na for diagnosis of nova outbursts was first suggested by Clayton and Hoyle (1974) [30]. It decays to a short-lived excited state of ^{22}Ne (with a lifetime of $\tau = 3.75$ yr), which de-excites to its ground state by emitting a γ-ray photon of 1.275 MeV. Through this mechanism, nearby ONe novae may provide detectable γ-ray fluxes. Several experimental verifications of this γ-ray emission at 1.275 MeV from nearby novae have been attempted in the last twenty years, using balloon-borne experiments and detectors on-board satellites such as HEAO-3, SMM, or CGRO, from which upper limits on the ejected ^{22}Na have been derived. In particular, the observations performed with the COMPTEL experiment on-board CGRO of five recent Ne-type novae as well as observations of standard CO novae (Iyudin *et al.*, 1995, [48]), have led to an upper limit of 3.7×10^{8} $M_{\odot}$ for the ^{22}Na mass ejected by any nova in the Galactic disk. A limit that poses some constraints on pre-existing theoretical models of classical nova explosions.

Synthesis of ^{22}Na in novae proceeds through different reaction paths. In the ^{20}Ne-enriched envelopes of ONe novae (José *et al.*, 1999, [49]), it takes place through ^{20}Ne$(p,\gamma)^{21}$Na, followed either by another proton capture and a β^{+}-decay into ^{22}Na (cf., ^{21}Na$(p,\gamma)^{22}$Mg$(\beta^{+})^{22}$Na), or first decaying into ^{21}Ne before another proton capture ensues (*i.e.*, ^{21}Na(β^+)^{21}Ne(p, γ)^{22}Na). Other potential channels, such as proton captures on the seed nucleus ^{23}Na, play only a marginal role on ^{22}Na synthesis because of the much higher initial ^{20}Ne content in such ONe models. As for the main destruction channel at nova temperatures, ^{22}Na(p, γ)^{23}Mg competes favorably with ^{22}Na(β^+)^{22}Ne. Until very recently, important nuclear uncertainties affected the rates for ^{21}Na(p, γ)^{22}Mg and ^{22}Na(p, γ)^{23}Mg (José *et al.*, 1999, [49]), which translated into an uncertainty in the estimated ^{22}Na yields (and ultimately on the maximum detectability distance of the associated 1275 keV line). These uncertainties have been recently reduced by the first direct measurement of the ^{21}Na(p, γ)^{22}Mg rate with the DRAGON recoil separator facility at TRIUMF (Bishop *et al.*, 2003, [50]; D'Auria *et al.*, 2004, [51]; see also the indirect approach by Davids *et al.*, 2003, [52]), and by improving the ^{22}Na(p, γ)^{23}Mg rate through spectroscopies studies of the ^{23}Mg nucleus (populated by the reaction ^{12}C(^{12}C, n)^{23}Mg), carried out with Gammasphere at the Argonne National Laboratory (Jenkins *et al.*, 2004, [53]).

2.4.3 ^{26}Al

^{26}Al was discovered in the interstellar medium by the HEAO-3 satellite, through the detection of the 1.809 MeV γ-ray line (Mahoney *et al.*, 1982, [54]; 1984 [55]), a proof of active Galactic nucleosynthesis. This γ-ray feature is produced by the β^+-decay (τ = 1.04 Myr) of ^{26}Al^g (ground state) to the first excited state of ^{26}Mg, which de-excites to its ground state by emitting a 1.809 MeV photon. The synthesis of ^{26}Al requires moderate peak temperatures, of the order of 2×10^8 K, and a fast decline (Ward and Fowler, 1980, [56]), conditions that are achieved in typical nova outbursts. A detailed analysis of the way ^{26}Al synthesis proceeds, in the framework of 1D hydrodynamic calculations (José *et al.*, 1997, [57]; 1999 [49]) reveals the existence of several "seed" nuclei: in particular, ^{24,25}Mg and to some extent ^{23}Na and ^{20,22}Ne. The main nuclear reaction path leading to ^{26}Al synthesis is ^{24}Mg(p, α)^{25}Al(β^+)^{25}Mg(p, α)^{26}Al^g, whereas destruction is dominated by ^{26}Al^g(p, γ)^{27}Si.

A question raised during the *Nuclear Physics in Astrophysics II Workshop* concerns the importance of the uncertainties associated with ^{25}Mg(p, γ) as compared with those significantly affecting the ^{25}Al(p, γ) rate (Coc *et al.*, 1995, [58]; José *et al.*, 1999, [49]; see also Rowland *et al.*, 2004, [59] for a recent update of the ^{23}Na+p rates and their impact on ^{26}Al production in novae): according to a recent sensitivity study (Iliadis *et al.*, 2002, [60]) in nova nucleosynthesis, the uncertainty associated with the ^{25}Mg(p, γ) rates in the domain of nova nucleosynthesis (T = 0.1–0.4 GK) is rather limited (*i.e.*, a factor 0.6–1.8 deviation from the nominal rate). Hence, their impact on nova nucleosynthesis is negligible (cf., less than a factor of $\sim$ 2 change in the ^{25,26}Mg and ^{26}Al yields. In contrast, the ^{25}Al(p, γ) rate is affecting from a much larger uncertainty (a factor 0.01–100 deviation from the nominal rate). This huge uncertainty translates into an uncertainty in the contribution of novae to the Galactic ^{26}Al content (since this reaction determines the fraction of the nuclear path that proceeds through the isomeric ^{26}Al^m instead), an estimate that relies critically on two ingredients: the adopted surface composition for ONe white dwarfs (*i.e.*, the initial ^{24}Mg abundance), and the specific rates adopted for the reactions ^{26}Al^g(p, γ) and ^{27}Al(p, α); see José *et al.*, 1997, [57]). Whereas calculations performed by the Arizona group (Starrfield *et al.*, 1986, [61]; 1998 [13]; 2000 [62]) assume a composition based on hydrostatic models of C-burning nucleosynthesis (Arnett and Truran, 1969, [7]), highly enriched in ^{24}Mg (with a ratio ^{16}O:^{20}Ne:^{24}Mg of 1.5:2.5:1), the Barcelona group (José and Hernanz, 1998, [14]; José *et al.*, 1997, [57]; 1999 [49]; 2001 [63]) has adopted more recent estimates from stellar evolution calculations of intermediate-mass stars (Ritossa *et al.*, 1996, [9]), that result in a much smaller ^{24}Mg content (cf., ^{16}O:^{20}Ne:^{24}Mg is 10:6:1). Calculations based on this new ONe white dwarf composition, with an updated nuclear reaction network, suggest that the contribution of classical nova outbursts to the Galactic ^{26}Al abundance is small (less than $\sim$ 15%), in agreement with the results derived from the COMPTEL/CGRO map of the 1.809 MeV ^{26}Al emission in the Galaxy (see Diehl *et al.*, 1995, [64]), which points towards young progenitors (type II supernovae and Wolf-Rayet stars).

2.4.4 ^{13}N-^{18}F

The most intense emission predicted in γ-rays for classical novae is the 511 keV line and a continuum at lower energies with a cut-off due to photoelectric absorption at about 20–630 keV. This prompt emission, that lasts only for a few hours after T_{peak}, is caused by the sudden release of positrons from the β^+-decay of the short-lived species ^{13}N and ^{18}F, that annihilate with the surrounding electrons when the envelope is already transparent to γ-rays. The detection of this short duration emission is a challenge since this takes place before the nova optical discovery, which rules out any possibility of a pointed observation. Only instruments with a wide field-of-view (*e.g.*, BATSE and TGRS) have a chance to discover serendipitously this prompt emission, provided that they were pointing at the right place at the right time. Several attempts have been carried out to detect these γ-ray signatures from classical novae, from which only upper limits on the ^{18}F annihilation line were obtained. Detectability distance estimates of the 511 keV line with the INTEGRAL's spectrometer SPI and of the continuum with the INTEGRAL's imager IBIS are about 4 kpc (for 10 h of observation; see Hernanz *et al.*, 1999, [33]; Hernanz and José, 2004, [47]). However, the limited field-of-view of these instruments make this detection unlikely. Better chances can be expected with future instruments, such as EXIST (see Hernanz *et al.*, 2002, [65]).

The synthesis of ^{18}F (τ = 158 min) is mainly powered by ^{16}O(p, γ)^{17}F, which is followed either by

$^{17}F(p,\gamma)^{18}Ne(\beta^+)^{18}F$ or by $^{17}F(\beta^+)^{17}O(p,\gamma)^{18}F$. The dominant destruction channel is $^{18}F(p,\alpha)^{15}O$ plus a minor contribution from $^{18}F(p,\gamma)^{19}Ne$. The effect of the nuclear uncertainties associated with some of these rates (mainly $^{18}F(p,\alpha)^{15}O$, but also $^{18}F(p,\gamma)^{19}Ne$, $^{17}O(p,\alpha)^{14}N$, and $^{17}O(p,\gamma)^{18}F$. See discussion in Hernanz *et al.*, 1999, [33], and Coc *et al.*, 2000, [66]) translates into a significant uncertainty in the expected ^{18}F yields and, hence, in the corresponding γ-ray fluxes and maximum detectability distances. The situation has improved recently in which concerns the ^{17}O + p rates (see Fox *et al.*, 2004, [67], and Chafa *et al.*, 2005, [68]). Further advances to reduce the remaining uncertainties have been achieved through several nuclear physics experiments performed in Oak Ridge (see Bardayan, these proceedings, and references therein) and Louvain-la-Neuve (de Séréville *et al.* 2003 [69]).

2.5 The endpoint of nova nucleosynthesis

Theoretical estimates of the expected endpoint for novae nucleosynthesis suggest that nuclear activity stops around the mass region $A \sim 40$ (cf., calcium), in agreement with the abundance patterns inferred from observations of the ejecta. Indeed, spectroscopic abundance determinations of nova shells include silicon (Nova Aql 1982, QU Vul 1984), sulfur (Nova Aql 1982), chlorine (Nova GQ Mus 1983), argon and calcium (Nova GQ Mus 1983, Nova V2214 Oph 1988, Nova V977 Sco 1989 and Nova V443 Sct 1989), whereas no indication of a significant overproduction with respect to solar abundances has been ever reported for elements above Ca. It is worth mentioning that this conclusion relies on the fact that the temperatures attained in the envelope during the explosion are limited to $T_{\rm peak} < 4 \times 10^8$ K.

The nuclear activity in the Si-Ca region has been scarcely analyzed in detail in the context of classical nova outbursts (see Starrfield *et al.*, 1998, [13]; Iliadis *et al.*, 1999, [70]; Wanajo *et al.*, 1999, [71], and José *et al.*, 2001, [63]). It is powered by a leakage from the NeNa-MgAl region, where the activity is confined during the early stages of the explosion (see Rowland *et al.*, 2004, [59] for a recent paper on the lack of a closed NeNa-cycle in nova conditions). The main reaction that drives the nuclear activity towards heavier species (*i.e.*, beyond S) is mainly $^{30}P(p,\gamma)^{31}S$, either followed by $^{31}S(p,\gamma)^{32}Cl(\beta^+)^{32}S$, or by $^{31}S(\beta^+)^{31}P(p,\gamma)^{32}S$ (José *et al.*, 2001, [63]). The current $^{30}P(p,\gamma)^{31}S$ rate is based on Hauser-Feshbach estimates, which can be rather uncertain at the domain of nova temperatures. To test the effect of this uncertainty on the predicted yields, a series of hydrodynamic calculations has been performed (José *et al.*, 2001, [63]), modifying arbitrarily the nominal rate. Hence, for a high $^{30}P(p,\gamma)^{31}S$ rate (*i.e.*, 100 times the nominal one), the final ^{30}Si yields are dramatically reduced by a factor of 30, whereas for a low $^{30}P(p,\gamma)^{31}S$ rate (*i.e.*, 0.01 times the nominal one), the final ^{30}Si yields are slightly increased by a factor of 5, whereas isotopes above silicon are reduced by a factor of ~ 10, with a dramatic impact on estimates of the composition of the ejecta and of the chemical pattern of meteoritic grains to be condensed in the ejected shells. Attempts to reduce the uncertainty associated with this rate are currently in progress at several nuclear physics facilities, including ORNL (Oak Ridge), ANL (Argonne), and JYFL (University of Jyväskylä). In particular, a $^{12}C(^{20}Ne,n)^{31}S$ experiment to study proton-unbound levels in ^{31}S has been performed in ANL with gammasphere, with the goal to determine their corresponding spins and parities (see Jenkins *et al.*, 2005, [72]).

3 X-ray bursts

Accretion onto neutron stars in close binary systems can also give rise to some nova-like events known as X-ray bursts. Because of the stronger surface gravity in a neutron star (as compared with that of a white dwarf), temperatures and densities in the accreted envelope during X-ray bursts are typically an order of magnitude greater than in a typical nova outburst (see pioneering models by Woosley and Taam, 1976, [73]; Joss, 1977, [74], and Maraschi and Cavaliere, 1977, [75]). As a result, detailed nucleosynthesis studies involve several hundreds of isotopes (up to the so-called SnSbTe cycle. See Schatz *et al.*, 2001, [76]) and thousands of nuclear reactions. Indeed, the main reaction flow moves far away from the valley of stability, and even merges with the proton drip line beyond $A = 38$ (Schatz *et al.*, 1999, [77]). Until recently, because of computational limitations, many simulations of X-ray bursts were performed in the framework of reduced nuclear reaction networks. Examples include network endpoints around Ni (Woosley and Wallace, 1984, [78]; Taam, 1993, [79]; Taam, 1996, [80]), Ga (José and Moreno, 2003, [81]), Se (Hanawa *et al.*, 1983, [82]), Kr (Koike *et al.*, 1999, [83]), or Y (Wallace and Woosley, 1981, [84]). Indeed, Wallace and Woosley (1984) [85] reached ^{96}Cd, but in the context of a reduced 16-nuclei network. On the other hand, Schatz *et al.* (1999) [77], (2001) [76] have carried out very detailed nucleosynthesis calculations, with a complete reaction network up to the SnSbTe mass region, but using a simple one-zone approach. Similar criticism can be applied to other works, such as Moreno *et al.* (2001) [86], that reached Pd, and Koike *et al.* (2004) [87], whose network extended up to Bi. Recent hydrodynamic simulations with a reasonably extended network containing 298 isotopes (up to Te) have been performed by Fisker *et al.* (2004) [88]. An unprecedented attempt, coupling detailed hydrodynamic stellar models in one-dimension with a complete nuclear reaction network (up to 1300 isotopes, in the framework of an adaptive network) has been recently performed by Woosley *et al.* (2004) [89].

3.1 Nucleosynthesis in X-ray bursts

Contrary to classical novae, where the main nuclear activity is driven by proton-capture reactions in competition with β^+-decays, X-ray bursts are triggered by a combination of nuclear reactions, including H-burning (via rp-process) and He-burning (that initiates with the triple

α-reaction, and is followed both by the breakout of the CNO cycle by $^{14,15}O + \alpha$, plus a competition of proton captures and (α, p) reactions —the so-called αp-process). Initially, the CNO breakout is led by $^{15}O(\alpha,\gamma)^{19}Ne$ (see Fisker *et al.*, 2005, [90] for a recent study of the impact of the $^{15}O(\alpha,\gamma)$ rate on the bursting behavior of an accreting neutron star), which is followed by two consecutive proton-captures through ^{20}Na and ^{21}Mg, where the flow faces strong photodisintegration reactions on ^{22}Al. Following ^{21}Mg-decay, they flow shifts through $^{21}Na(p,\gamma)^{22}Mg$ (see D'Auria *et al.*, 2004, [51] for a recent update of this rate for X-ray bursts conditions). As the temperature rises, and enough ^{14}O is built up through the triple-α reaction, followed by $^{12}C(p,\gamma)^{13}N(p,\gamma)^{14}O$, an alternative path through $^{14}O(\alpha,p)^{17}F$ reaction dominates the flow (see Champagne and Wiescher, 1992, [91], and Woosley *et al.*, 2004, [89]), by-passing the $^{15}O(\alpha,\gamma)^{19}Ne$ link to ^{21}Na through $^{17}F(p,\gamma)^{18}Ne(\alpha,p)^{21}Na$, where $^{18}Ne(\alpha,p)^{21}Na$ represents the main path towards heavier species. The rates for these breakout reactions have huge associated uncertainties (see Görres *et al.*, 1995, [92], and Mao *et al.*, 1996, [93]) and are subject to intense experimental studies both at stable and radioactive beam facilities. Of special interest is the fact that H-burning continues after the peak of the explosion through the rp-process, which extends the main path much beyond ^{56}Ni, up to the SnSbTe region (see Fisker *et al.*, 2005, [90] for a detailed account of the main nuclear paths at different stages of the TNR).

A major drawback in the modeling of X-ray bursts comes from the lack of observational nucleosynthetic constraints (beyond obvious implications for the physics of the neutron star crusts). This results from the difficulty of ejection in the strong gravitational well of a neutron star. A recent attempt to overcome this limitation has been provided by high-resolution spectra obtained with XMM-Newton (Cottam *et al.*, 2002, [94]) that bring an insight into the chemical species present in the envelope at different epochs (determinations include C, N, O, Ne, or Fe, with different degrees of ionization; see also Chang *et al.*, 2005, [95] for a very recent study on formation of resonant atomic lines during X-ray bursts).

It is important to stress that, contrary to nova outbursts, convection is not that critical in the progress of the TNR for X-ray bursts. This can be tested by artificially turning-off convective transport in a series of numerical tests: whereas convection is the key ingredient to power the expansion and ejection stages in a nova outburst (through the transport of the short-lived species ^{13}N, $^{14,15}O$ and ^{17}F), an X-ray burst is not deeply affected by the lack of convection. Indeed, recent X-ray burst models (Woosley *et al.*, 2004, [89]) point out that convection sets in during a very short time (of the order of a few seconds).

3.2 Future challenges: superbursts

Whereas regular, type I X-ray bursts exhibit some common features in terms of duration, energetics, and recurrence times, a few extremely energetic events have been recently reported thanks to better performances in monitoring achieved with X-ray satellites (*i.e.*, BeppoSAX, Chandra, or XMM-Newton). The first observation of a *superburst* was reported by Cornelisse *et al.* (2000) [96] in the context of a "common" type I bursting source (cf., the BeppoSAX source 4U 1735 - 44). Several other sources have been identified since then, including 4U 1636 - 53 for which two superbursts have been reported already.

According to their distinctive characteristics, superbursts represent some sort of extreme X-ray bursts: they have long duration, with a typical (exponential) decay time ranging from 1 to 3 hours (including an extreme case, KS 1731 - 260, which lasted for more than 10 hours. See Kuulkers *et al.*, 2002, [97] for details), extremely energetic (about $\sim$ 1000 times more than a typical X-ray burst), and with much longer recurrence periods (4.7 yr for the system 4U 1636 - 53, the only one in which two superbursts have been observed to date. See Wijnands, 2001, [98]). Indeed, the durations and energetics of superbursts suggest that they result from thermonuclear flashes occurring in fuel layers at much greater depth than for typical X-ray bursts, more likely, in the C-rich ashes resulting from type I X-ray bursts (Cumming and Bildsten, 2001, [99]. See also Woosley and Taam, 1976, [73]; Taam and Picklum, 1978, [100], and Brown and Bildsten, 1998, [101]). Controversy remains as how much carbon is left after a type I burst: Schatz *et al.* (1999) [77], (2001) [76] have indeed shown than most of the C is burnt during the previous H/He burning. However, Cumming and Bildsten (2001) [99] concluded that even small amounts of carbon are enough to power a superburst (especially in neutron star oceans enriched from the heavy ashes driven by the rp-process). More theoretical efforts to test these ideas are required to shed light into this question.

4 Outlook

A final comment on nuclear physics input requirements as well as on reaction rates that need to be improved to overcome their huge associated uncertainties: the situation concerning nova modeling is quite well defined. Significant experimental information is already available and we are confident that, in the near future, novae will become the first explosive stellar site for which most (if not all) of the relevant nuclear physics input will be primarily based on experimental information. Main uncertainties are localized in only a handful of nuclear reaction rates (namely, $^{18}F(p,\alpha)$, $^{25}Al(p,\gamma)$ and $^{30}P(p,\gamma)$), for which several experiments have been already proposed in different facilities.

In contrast, the situation is much more complex as regards X-ray bursts: nuclear physics requirements include mass measurements along the rp-process path (see Clark *et al.*, 2004, [102] for a recent mass determination for ^{68}Se), and a much better knowledge of key reactions and effective lifetimes at the *waiting points* (in particular, $^{30}S(\alpha,p)$ and $^{34}Ar(\alpha,p)$; see Fisker *et al.*, 2004, [88]) under rp-process conditions. However, it is important to stress that, far from the waiting points, the situation is not well settled.

Indeed, recent hydrodynamic calculations (see details in D'Auria *et al.*, 2004, [51]) suggest that single nuclear reactions (in particular, ^{21}Na(p, γ)) are not so important, since the much higher temperatures achieved in X-ray bursts allow alternative paths for the main nuclear flows. The future looks bright for X-ray bursts physics, since future facilities like RIA in the U.S., or FAIR at GSI (Germany) have already identified the mass region of interest for these stellar explosions as part of their future research programs.

References

1. J. Grindlay *et al.*, Astrophys. J. **205**, L127 (1976).
2. R.D. Belian *et al.*, Astrophys. J. **206**, L135 (1976).
3. J.H. Swank *et al.*, Mon. Not. R. Astron. Soc. **182**, 349 (1978)
4. A.W. Shafter, in *Classical Nova Explosions*, edited by M. Hernanz, J. José, AIP Conf. Proc. **637**, 462 (2002).
5. B. Warner, in *Cataclysmic Variable Stars* (Cambridge University Press, Cambridge, 1995).
6. T. Strohmayer, L. Bildsten, in *Compact Stellar X-Ray Sources*, edited by W.H.G. Lewin, M. van der Klis (Cambridge University Press, 2006) in press.
7. W.D. Arnett, J.W. Truran, Astrophys. J. **157**, 339 (1969).
8. I. Domínguez *et al.*, Astrophys. J. **419**, 268 (1993).
9. C. Ritossa *et al.*, Astrophys. J. **460**, 489 (1996).
10. J. José *et al.*, Astrophys. J. **597**, L41 (2003).
11. B. Davids *et al.*, Phys. Rev. C **67**, 065808 (2003).
12. J. José *et al.*, Nucl. Phys. A, Special Volume, in press (2006).
13. S. Starrfield *et al.*, Mon. Not. R. Astron. Soc. **296**, 502 (1998).
14. J. José, M. Hernanz, Astrophys. J. **494**, 680 (1998).
15. A. Kovetz, D. Prialnik, Astrophys. J. **477**, 356 (1997).
16. O. Yaron *et al.*, Astrophys. J. **623**, 398 (2005).
17. S. Starrfield, in *Classical Nova Explosions*, edited by M. Hernanz, J. José, AIP Conf. Proc. **637**, 89 (2002).
18. R.D. Gehrz *et al.*, Publ. Astron. Soc. Pac. **110**, 3 (1998).
19. A. Evans, in *The Physics of Classical Novae*, edited by A. Cassatella, R. Viotti (Springer-Verlag, Berlin, 1990) p. 253.
20. J.M.C. Rawlings, A. Evans, in *Classical Nova Explosions*, edited by M. Hernanz, J. José, AIP Conf. Proc. **637**, 270 (2002).
21. R.D. Gehrz, in *Classical Nova Explosions*, edited by M. Hernanz, J. José, AIP Conf. Proc. **637**, 198 (2002).
22. D.D. Clayton, F. Hoyle, Astrophys. J. **203**, 490 (1976).
23. S. Amari *et al.*, Astrophys. J. **551**, 1065 (2001).
24. S. Amari, New Astron. Rev. **46**, 519 (2002).
25. S. Starrfield *et al.*, in *Astrophysical Implications of the Laboratory Study of Presolar Materials*, edited by T. Bernatowicz, E. Zinner, AIP Conf. Proc. **402**, 203 (1997).
26. J. José *et al.*, Publ. Astron. Soc. Aust. **20**, 351 (2003).
27. J. José *et al.*, Astrophys. J. **612**, 414 (2004).
28. S.N. Shore, R.D. Gehrz, Astron. Astrophys. **417**, 695 (2004).
29. S. Starrfield *et al.*, Astrophys. J. **176**, 169 (1972).
30. D.D. Clayton, F. Hoyle, Astrophys. J. **187**, L101 (1974).
31. D.D. Clayton, Astrophys. J. **244**, L97 (1981).
32. M.D. Leising, D.D. Clayton, Astrophys. J. **323**, 159 (1987).
33. M. Hernanz *et al.*, Astrophys. J. **526**, L97 (1999).
34. J. Gómez-Gomar *et al.*, Mon. Not. R. Astron. Soc. **296**, 913 (1998).
35. M. Arnould, H. Nørgaard, Astron. Astrophys. **42**, 55 (1975).
36. S. Starrfield *et al.*, Astrophys. J. **222**, 600 (1978).
37. H.M.J. Boffin *et al.*, Astron. Astrophys. **279**, 173 (1993).
38. M. Hernanz *et al.*, Astrophys. J. **465**, L27 (1996).
39. A.G.W. Cameron, Astrophys. J. **121**, 144 (1955).
40. D. Romano *et al.*, Astron. Astrophys. **352**, 117 (1999).
41. A. Alibés *et al.*, Astrophys. J. **571**, 336 (2002).
42. M. Della Valle *et al.*, Astron. Astrophys. **390**, 155 (2002).
43. S.N. Shore *et al.*, Astrophys. J. **125**, 1507 (2003).
44. M.J. Harris *et al.*, Astrophys. J. **375**, 216 (1991).
45. M.J. Harris *et al.*, Astrophys. J. **563**, 950 (2001).
46. M. Hernanz, J. José, in *Proceedings of the V INTEGRAL Workshop*, ESA SP-552, edited by U. Schönfelder, G. Lichti, C. Winkler (ESA Publ. Division: ESTEC, Noordwijk, 2004) p. 95.
47. M. Hernanz, J. José, New Astron. Rev. **48**, 35 (2004).
48. A.F. Iyudin *et al.*, Astron. Astrophys. **300**, 422 (1995).
49. J. José *et al.*, Astrophys. J. **520**, 347 (1999).
50. S. Bishop *et al.*, Phys. Rev. Lett. **90**, 162501 (2003).
51. J.M. D'Auria *et al.*, Phys. Rev. C **69**, 065803 (2004).
52. B. Davids *et al.*, Phys. Rev. C **68**, 055805 (2003).
53. D. Jenkins *et al.*, Phys. Rev. Lett. **92**, 031101 (2004).
54. W.A. Mahoney *et al.*, Astrophys. J. **262**, 742 (1982).
55. W.A. Mahoney *et al.*, Astrophys. J. **286**, 578 (1984).
56. R.A. Ward, W.A. Fowler, Astrophys. J. **238**, 266 (1980).
57. J. José *et al.*, Astrophys. J. **479**, L55 (1997).
58. A. Coc *et al.*, Astron. Astrophys. **299**, 479 (1995).
59. C. Rowland *et al.*, Astrophys. J. **615**, L37 (2004).
60. C. Iliadis *et al.*, ApJS **142**, 105 (2002).
61. S. Starrfield *et al.*, Astrophys. J. **303**, L5 (1986).
62. S. Starrfield *et al.*, ApJS **127**, 485 (2000).
63. J. José *et al.*, Astrophys. J. **560**, 897 (2001).
64. R. Diehl *et al.*, Astron. Astrophys. **298**, 445 (1995).
65. M. Hernanz *et al.*, New Astron. Rev. **46**, 559 (2002).
66. A. Coc *et al.*, Astron. Astrophys. **357**, 561 (2000).
67. C. Fox *et al.*, Phys. Rev. Lett. **93**, 081102 (2004).
68. A. Chafa *et al.*, Phys. Rev. Lett. **95**, 031101 (2005).
69. N. de Séréville *et al.*, Phys. Rev. C **67**, 052801 (2003).
70. C. Iliadis *et al.*, Astrophys. J. **524**, 434 (1999).
71. S. Wanajo *et al.*, Astrophys. J. **523**, 409 (1999).
72. D. Jenkins *et al.*, Phys. Rev. C **72**, 031303 (2005).
73. S.E. Woosley, R.E. Taam, Nature **263**, 101 (1976).
74. P.C. Joss, Nature **270**, 310 (1976).
75. L. Maraschi, A. Cavaliere, in *Highlights in Astronomy*, edited by E. Muller, vol. **4**, Part I (Reidel, Dordrecht) p. 127.
76. H. Schatz *et al.*, Phys. Rev. Lett. **86**, 3471 (2001).
77. H. Schatz *et al.*, Astrophys. J. **524**, 1014 (1999).
78. S.E. Woosley, R.K. Wallace, in *High Energy Transients in Astrophysics*, edited by S.E. Woosley (AIP Press, New York, 1984) p. 273.
79. R.E. Taam *et al.*, Astrophys. J. **413**, 324 (1993).
80. R.E. Taam *et al.*, Astrophys. J. **459**, 271 (1996).
81. J. José, F. Moreno, Nucl. Phys. A **718**, 602 (2003).
82. T. Hanawa *et al.*, Publ. Astron. Soc. Jpn. **35**, 491 (1983).
83. O. Koike *et al.*, Astron. Astrophys. **342**, 464 (1999).
84. R.K. Wallace, S.E. Woosley, Astrophys. J. Suppl. **45**, 389 (1981).

85. R.K. Wallace, S.E. Woosley, in *High Energy Transients in Astrophysics*, edited by S.E. Woosley (AIP Press, New York, 1984) p. 319.
86. F. Moreno *et al.*, Nucl. Phys. A **688**, 447 (2001).
87. O. Koike *et al.*, Astrophys. J. **603**, 242 (2004).
88. J.L. Fisker *et al.*, Astrophys. J. **608**, L61 (2004).
89. S.E. Woosley *et al.*, Astrophys. J. Suppl. **151**, 75 (2004)
90. J.L. Fisker *et al.*, Nucl. Phys. A **752**, 604 (2005).
91. A.E. Champagne, M. Wiescher, Annu. Rev. Nucl. Part. Sci. **42**, 39 (1992).
92. J. Görres *et al.*, Phys. Rev. C **51**, 392 (1995).
93. Z.Q. Mao *et al.*, Phys. Rev. C **53**, 1197 (1996).
94. J. Cottam *et al.*, Nature **420**, 51 (2002).
95. P. Chang *et al.*, Astrophys. J. **629**, 998 (2005).
96. R. Cornelisse *et al.*, Astron. Astrophys., **357**, L21 (2000).
97. E. Kuulkers *et al.*, Astron. Astrophys. **382**, 503 (2002).
98. R. Wijnands, Astrophys. J. **554**, L59 (2001).
99. A. Cumming, L. Bildsten, Astrophys. J., **559**, L127 (2001).
100. R.E. Taam, R. Picklum, Astrophys. J. **224**, 210 (1978).
101. E. Brown, L. Bildsten, Astrophys. J., **496**, 915 (1998).
102. J.A. Clark *et al.*, Phys. Rev. Lett. **92**, 192501 (2004).

Eur. Phys. J. A **27**, s01, 117–121 (2006)
DOI: 10.1140/epja/i2006-08-016-7

EPJ A direct
electronic only

Re-evaluating reaction rates relevant to nova nucleosynthesis from a nuclear structure perspective

D.G. Jenkins[1,a], C.J. Lister[2], R.V.F. Janssens[2], T.L. Khoo[2], E.F. Moore[2], K.E. Rehm[2], D. Seweryniak[2], A.H. Wuosmaa[2], T. Davinson[3], P.J. Woods[3], A. Jokinen[4], H. Penttila[4], G. Martínez-Pinedo[5,6], and J. Jose[6]

[1] Department of Physics, University of York, York YO10 5DD, UK
[2] Physics Division, Argonne National Laboratory, Argonne, IL 60439, USA
[3] School of Physics, University of Edinburgh, Edinburgh, UK
[4] Department of Physics, University of Jyväskylä, Jyväskylä, Finland
[5] ICREA, E-08010 Barcelona, Spain
[6] Institut d'Estudis Espacials de Catalunya (IEEC), E-08034, Barcelona, Spain

Received: 24 June 2005 /
Published online: 24 February 2006 –

Abstract. Conventionally, reaction rates relevant to nova nucleosynthesis are determined by performing the relevant proton capture reactions directly for stable species, or as has become possible more recently in inverse kinematics using short-lived accelerated radioactive beams with recoil separators. A secondary approach is to compile information on the properties of levels in the Gamow window using transfer reactions. We present a complementary technique where the states of interest are populated in a heavy-ion fusion reaction and their gamma decay studied with a state-of-the-art array of high-purity germanium detectors. The advantages of this approach, including the ability to determine resonance energies with high precision and the possibility of determining spins and parities from gamma-ray angular distributions are discussed. Two specific examples related to the ^{22}Na(p, γ) and ^{30}P(p, γ) reactions are presented.

PACS. 26.30.+k Nucleosynthesis in novae, supernovae, and other explosive environments – 21.10.Tg Lifetimes – 27.30.+t $20 \leq A \leq 38$

1 Introduction

Two key issues relating to nova nucleosynthesis have been recently identified. The former pertains to the probability of making direct observation of nova explosions through the detection of gamma rays following the beta decay of certain nuclear species produced in the explosion such as ^{22}Na and ^{26}Al [1]. In this respect, ^{22}Na which decays, with a 2.602 y half-life, into a short-lived excited state of ^{22}Ne, emitting a 1.275 MeV γ-ray, is seen as a particularly important diagnostic of nova explosions with the expectation that explosions within a few kiloparsecs of the Sun might provide detectable γ-ray fluxes associated with ^{22}Na decay. Efforts are ongoing to detect such cosmic γ-rays, in particular, with the recently launched ESA INTEGRAL mission [2]. In order to understand the likelihood of detecting cosmic gamma rays from ^{22}Na, it is important to quantify the processes responsible for the production and destruction of ^{22}Na in nova nucleosynthesis. Uncertainties in the rate of production of ^{22}Na have largely been lifted by a study of the ^{21}Na(p, γ) employing a radioactive ^{21}Na beam in inverse kinematics [3][1]. Prior to the work described here, there was rather more significant uncertainty in the destruction rate via the ^{22}Na(p, γ) reaction.

A second area of inquiry is the location of the endpoint for nova nucleosynthesis. Jose has identified the ^{30}P(p, γ) reaction as being the key determinant in this respect [5]. This reaction rate is determined on the basis of Hauser-Feshbach calculations as no relevant experimental measurements have been made. It is not known how reliable such a methodology might be in this case since the level density may not be high enough to make the necessary assumptions about the availability of suitable resonances in the Gamow window. Jose has shown that plausible variances in the reaction rate by a factor of 100 up or down have dramatic consequences for the endpoint of nova nucleosynthesis [5].

[a] e-mail: dj4@york.ac.uk

[1] We note that some of the remaining uncertainties in this reaction have very recently been removed by a gamma-ray spectroscopy study of ^{22}Mg using the Gammasphere array [4].

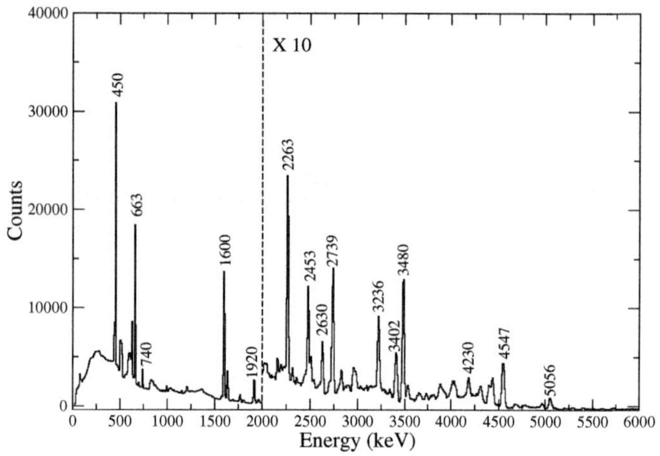

Fig. 1. Sum of double gates on the 450, 1600, 663 and 2739 keV transitions in the γ-γ-γ cube. Strong transitions in ^{23}Mg are labelled with their energy in keV.

2 Example 1: ^{22}Na(p, γ)

In the past, several methods have been employed in order to obtain the astrophysical reaction rate for the ^{22}Na(p, γ) reaction [6,7,8,9]. The key to such an analysis is a detailed knowledge of properties such as the excitation energy, spin and parity of levels in the unbound region. The conventional approach to this problem is to study the ^{22}Na(p, γ) directly by bombarding a specially prepared radioactive ^{22}Na target with protons and detecting the γ-rays following proton capture [6,7]. We have pursued a complementary approach in which particle-unbound states were populated in a heavy-ion fusion-evaporation reaction and their subsequent γ decay investigated with Gammasphere, a 4π high resolution γ-ray spectrometer consisting of 100 large-volume, high-purity germanium detectors comprising a total efficiency of around 9% for 1.33 MeV γ-rays [10]. This work is reported in more detail elsewhere [11].

A 10 pnA beam of ^{12}C was accelerated to 22 MeV by the ATLAS accelerator at Argonne National Laboratory and was incident on a 40 μg/cm^2 thick ^{12}C target. The resulting gamma decay was detected using the Gammasphere array. The fusion channels observed were single-proton, neutron or alpha emission leading to ^{23}Na, ^{23}Mg and ^{20}Na, respectively. A γ-γ matrix and a γ-γ-γ cube were produced and analysed to obtain information on the decay schemes. The construction of the decay schemes was straightforward given the small number of residual nuclei produced and their well-known decay schemes at low excitation energies [12]. An example of the quality of the data obtained is given in fig. 1.

It is worthwhile to reflect that large γ-ray spectrometers are most commonly designed for the study of excited states of heavy nuclei where high-multiplicity cascades ($\sim$ 20 photons) and relatively low energies ($\sim$ 1 MeV) are expected. By contrast, for the astrophysical application described here, the relevant cascades have both a relatively low multiplicity and γ-ray energies which may be above 10 MeV, meaning that particular attention needs to be paid to both energy and efficiency calibrations.

In obtaining accurate γ-ray energies, we have applied a correction for the non-linearity of the array as well as the finite recoil correction for large-energy γ-rays emitted from a light nucleus. In cases where two coincident transitions were crossed over by a third transition, the corrected energy sum was compared and found to be in agreement at the $\sim$ 0.5–1 keV level.

In order to assign a multipolarity to the observed transitions, a matrix was generated of γ-rays detected at all angles against those detected at 90° and a matrix of all γ-rays against those detected at 32° and 37°. The ratio (R_{DCO}) of the intensities of transitions in these two matrices when gating on the "all detector" axis was extracted. This ratio was around 0.9(1) for pure dipole transitions and around 1.7(2) for pure quadrupole transitions. Mixed $M1/E2$ dipole transitions may have various values depending on the value of the mixing ratio. As well as angular correlations, it was also possible to assign the spin/parity of states in ^{23}Mg, on the basis of their similarity in both energy and decay path to analogue states of well-established spin and parity in ^{23}Na [12], for which extensive additional spectroscopic information was obtained.

The high energy of many of the γ-rays observed implies very short (femtosecond) lifetimes which are readily extracted using the fractional Doppler-shift technique [13] since it may reasonably be assumed for high-lying, unbound states that the feeding is direct. Seven matrices were sorted containing un-Doppler–corrected gamma-ray energies observed at 32°, 50°, 80°, 90°, 100°, 130° and 148°, respectively, against Doppler-corrected energies observed at any angle. The peak centroids were obtained for a transition at all seven angles when gated on a transition on the Doppler-corrected "all detector" axis. This was used to calculate the observed Doppler shift and, hence, the fractional Doppler shift relative to the calculated maximum Doppler shift for the given beam and target. A model prescription was used to relate the fractional Doppler shift to the lifetime of the parent state. The gamma width of the state may be deduced from the lifetime.

2.1 Re-evaluation of the reaction rate

The Gammasphere array affords the possibility of determining the energy of the resonances to higher accuracy than that obtainable with a spectrometer. It also allows their decay path to be observed and through angular-distribution measurements, the spin/parity of these resonances may be inferred or at least restricted to some plausible range. The γ-ray energies, angular-correlation ratios, and lifetimes of proton-unbound states in ^{23}Mg are presented in table 1.

The resonance strengths were taken from the literature where known [14], or else were calculated from measured spectroscopic factors and calculated single-particle proton widths. The resonance strength for a state with spin, J, is given by

$$\omega\gamma = \omega \frac{\Gamma_p \Gamma_\gamma}{\Gamma_p + \Gamma_\gamma}, \tag{1}$$

Table 1. Spectroscopic information for proton-unbound states in ^{23}Mg relevant to the ^{22}Na(p, γ) reaction rate obtained in the present work, including spins/parities and the energy and angular distribution of de-exciting gamma rays. The resonance strengths are extracted in the manner described in the text.

E_x (keV) Endt 1998	E_x Present	$E_{\rm p}$(lab) (keV)	I_i^π	I_f^π	E_γ (keV)	τ (fs)	R_{DCO}	$\omega\gamma$ (meV)
7622(6)	7623.4(9)	45.8(16)	$9/2^+$	$5/2^+$	7172.5(9)	4(2)	1.57(24)	$1.7^{+2.5}_{-1.1} \times 10^{-13}$
7643(10)	7646.9(26)	73.5(30)	$3/2^+$	$5/2^+$	7196.0(26)		0.87(15)	$2.2^{+3.0}_{-1.4} \times 10^{-9}$
	7769.2(10)	198.2(19)	$(9/2^-)$	$9/2^+$	5054.8(6)	2(1)		$4.0_{-4.0}$
				$11/2^+$	2316.9(5)			
7783(3)	7779.9(9)	209.4(17)	$11/2^+$	$7/2^+$	5729.1(11)	< 1	1.42(11)	(5×10^{-2})
				$9/2^+$	5067.1(11)			
7783(3)	7784.6(11)	214.3(18)	$7/2^+$	$5/2^+$	7333.7(11)	10(3)	0.89(5)	$1.8^{+0.7}_{-0.7}$
7801(2)	unobs.		$5/2^+$					$2.2^{+1.0}_{-1.0}$
7857(2)	7851.5(14)	284.3(20)	$(7/2^+)$	$9/2^+$	5138.1(13)			$15.8^{+3.4}_{-3.4}$
8017(2)	8015.3(17)	455.5(23)	$(5/2^+$-$11/2^+)$	$9/2^+$	5300.2(9)			68^{+20}_{-20}
				$7/2^+$	5966.7(11)			
8166(2)	8159.7(20)	606.5(25)	$(5/2^+)$	$7/2^+$	6109.5(18)			235^{+33}_{-33}

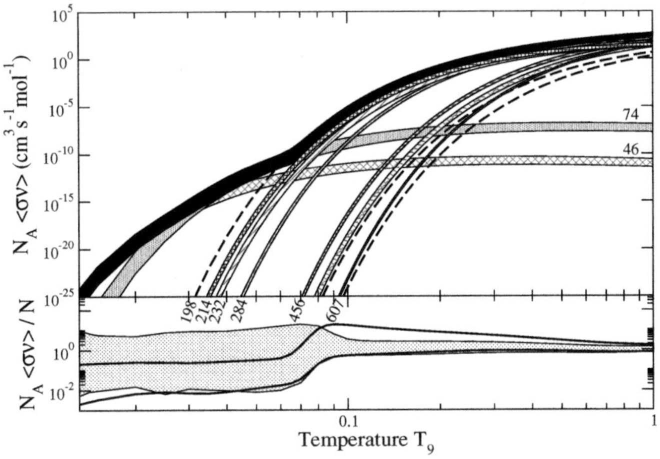

Fig. 2. The top panel presents individual contributions to the total reaction rate labelled by their corresponding proton energy, $E_{\rm p}$(lab). The bottom panel compares the bounds on the reaction rate determined in the present work (solid lines) with the uncertainties from the most recent compilation in the literature [14].

where

$$\omega = \frac{2J+1}{2(2J_T+1)}\,. \tag{2}$$

In the case where $\Gamma_p \ll \Gamma_\gamma$, this reduces to

$$\omega\gamma = \omega\Gamma_p\,. \tag{3}$$

Reaction rates were calculated from the individual resonance strengths and energies. The results of this analysis are presented in fig. 2. The changes to the reaction rate with respect to the compilation in the literature [14] have two principal origins. Firstly, we have fully constrained the spin and parity of the two near-threshold resonances showing that they are associated with $l = 2$ rather than $l = 0$ and hence, their contribution at lower temperatures will be lower than previously thought. Second, we have identified an additional state in the Gamow window at $E_{\rm p}(\rm lab) = 198\,keV$, which mirror symmetry considerations suggest may have $J^\pi = 9/2^-$. If this state did have this spin/parity, then it might, in principle, make a substantial contribution to the overall reaction rate. The contribution is, however, constrained by limited direct measurements made previously. We set an upper limit on the resonance strength for this new state of 4 meV. Clearly, it is a priority to determine the resonance strength of this state more precisely and further measurements are warranted.

2.2 Astrophysical implications

An analysis of the impact of the new rate for the ^{22}Na(p, γ) reaction on the amount of ^{22}Na ejected during nova outbursts has been performed. An evolutionary sequence for a nova outburst hosting an ONe white dwarf of 1.25 $M_\odot$, has been computed by means of a spherically symmetric, hydrodynamic, implicit code, in Lagrangian formulation, extensively used in the modeling of such explosions (see ref. [15] for details). If the resonance strength of the newly discovered state were to tend to its upper limit, then this would represent a reduction by roughly a factor of two in terms of the maximum detectability distance for the 1.275 MeV γ-ray from ^{22}Na. This would have important implications for ongoing satellite-based searches for gamma rays from novae such as the ESA INTEGRAL mission.

3 Example 2: ^{30}P(p, γ)

We have employed a similar methodology in our approach to improving the experimental situation for states in the Gamow window in ^{31}S relevant to the ^{30}P(p, γ) reaction. Excited states in ^{31}P and ^{31}S were produced at the same

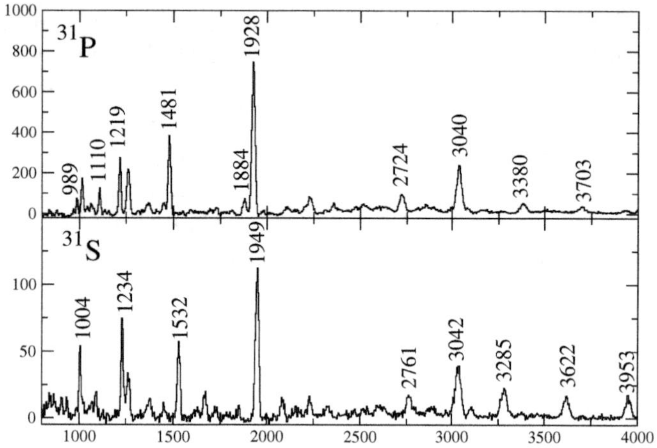

Fig. 3. The top spectrum shows transitions above the $7/2^+$ level in ^{31}P and is double-gated by the 1266 and 2148 keV transitions in the γ-γ-γ cube. The bottom spectrum contains transitions above the $7/2^+$ level in ^{31}S and is double-gated by the 1249 and 2102 keV transitions in the γ-γ-γ cube. Mirror transitions of interest are labelled with their energies in keV.

time through the ^{12}C(^{20}Ne, p) and ^{12}C(^{20}Ne, n) reactions using a 32 MeV beam from the ATLAS accelerator at Argonne National Laboratory. The resulting gamma decays were detected by Gammasphere [10]. Transitions in ^{31}S were rigorously identified by selecting ^{31}S recoils using the Fragment Mass Analyser (FMA) [16] and an ion chamber. This information was used in conjunction with a γ-γ matrix and a γ-γ-γ cube to develop level schemes for the two nuclei. An angular-correlation analysis was performed for the strongest gamma rays observed.

We were able to very cleanly identify the strongest transitions in ^{31}S when we demanded that the separated recoils had $A = 31$ and $Z = 16$. The statistics did not allow, however, the identification of weak or high-energy γ-rays which might be feeding the ground state, given the very low coincident efficiency. In order to develop the level scheme, we exploited the γ-ray coincidence data, which did not have a condition on the detection of ^{31}S residues. The analysis is, therefore, complicated by the presence of transitions with near identical energy in other nuclei which were more strongly populated in the experiment. For example, the first $5/2^+$ state in ^{31}S decays by a 2236 keV γ-ray, which is near degenerate with strong transitions in both ^{31}P and ^{30}Si, and so it is useless for a γ-γ analysis. We were able, however, to find clean gating conditions to observe transitions feeding the $3/2_1^+$, $5/2_2^+$, $7/2_1^+$ and $7/2_1^-$ levels in ^{31}S. An example of the quality of the data obtained, as well as an illustration of the close mirror symmetry of ^{31}S and ^{31}P is given in fig. 3.

A re-evaluation of the ^{30}P(p, γ) reaction rate is presently being prepared for publication. Essentially, we have identified many of the states in the Gamow window in ^{31}S, in particular, those with high spin. These are not directly relevant to the proton capture reaction. However, with these states eliminated from the analysis, it becomes possible to make a one-to-one matching of the remaining levels with their known mirror counterparts in ^{31}P. Taken together with calculations of proton spectroscopic factors for the *sd* shell model states, this allows many of the uncertainties in the reaction rate to be strongly reduced. A very important feature of this analysis has been the occurrence of striking mirror energy shifts between certain states in ^{31}S and ^{31}P, most notably for particular negative-parity states such as the first $9/2^-$ and $13/2^-$ levels. These mirror energy shifts are found, in some cases, to exceed 250 keV. It is believed that these shifts arise from the effects of the electromagnetic spin orbit interaction and their origin is discussed elsewhere [17]. These mirror energy shifts are by no means universal and appear to occur only for particularly pure stretched configurations —negligible energy shifts are observed between the first $7/2^-$ and $11/2^-$ levels, for example [17]. The understanding of nuclear structure effects such as these mirror energy shifts is therefore very important in determining the ^{30}P(p, γ) reaction rate.

3.1 Outlook for future measurements

We have considerably improved the experimental information for unbound states in ^{31}S. From the present measurements and careful comparisons with mirror states, there are no ambiguities on spin/parity assignments for the near-threshold resonances whose proton widths principally determine the reaction rate at lower temperatures. The uncertainties now rest principally in the excitation energy of the unmeasured resonances, which are 10–15 keV as determined in transfer reactions, and the unknown spectroscopic factors for low-spin negative-parity states in the Gamow window. The latter are very difficult to calculate since it would be necessary to have a full *sd*-*pf* shell model calculation. It would, therefore, be fundamentally difficult to improve on the measurement described here. The best approach would be to measure the ^{30}P(p, γ) reaction directly —in particular for low-spin negative-parity resonances. This would be a challenging measurement given the difficulty in producing a beam of ^{30}P using the ISOL technique, though "in-flight" production via the ^{30}Si(p, n) reaction may be possible. Given a modest radioactive beam of $\sim 10^6\,\mathrm{s}^{-1}$, a (^{3}He, d) reaction could be performed to measure spectroscopic factors for these states. The inaccurate spectroscopic factors are not, however, the only source of error since the error in the resonance energies must also be considered. The uncertainty in the resonance energies might, in principle, be reduced slightly, given a very careful transfer measurement. The inclusion of germanium detectors might be beneficial in this regard, but depending on the magnitude of the gamma branch, this might not be practical.

4 Conclusions

We have used a nuclear structure methodology to revisit the question of the ^{22}Na(p, γ) and ^{30}P(p, γ) reaction rates. The former work is published elsewhere [11], while the latter is being finalised for publication. We have shown that the complementary approach described here allows resonance energies to be obtained with high precision and ambiguities on spin/parity assignments to be lifted. Clearly, such a methodology is applicable to many other cases than the examples described here and, with modern 4π gamma-ray detector arrays, it is not difficult to perform very detailed spectroscopic studies on unbound states of *sd* and *sd-fp* shell nuclei.

References

1. J. Jose, A. Coc, M. Hernanz, Astrophys. J. **520**, 347 (1999).
2. M. Hernanz, in *Classical Nova Explosions*, edited by M. Hernanz, J. José (AIP, Melville, New York, 2002) p. 399.
3. S. Bishop, R.E. Azuma, L. Buchmann, A.A. Chen *et al.*, Phys. Rev. Lett. **90**, 162501 (2003).
4. D. Seweryniak *et al.*, Phys. Rev. Lett. **94**, 032501 (2005).
5. J. Jose, A. Coc, M. Hernanz, Astrophys. J. **560**, 897 (2001).
6. S. Seuthe *et al.*, Nucl. Phys. A **514**, 471 (1990).
7. F. Stegmüller, C. Rolfs, S. Schmidt, W.H. Schulte, H.P. Trautvetter, R.W. Kavanagh, Nucl. Phys. A **601**, 168 (1996).
8. S. Schmidt *et al.*, Nucl. Phys. A **591**, 227 (1995).
9. S. Kubono *et al.*, Z. Phys. A **348**, 59 (1994).
10. I.Y. Lee, Nucl. Phys. A **520**, 641c (1990).
11. D.G. Jenkins *et al.*, Phys. Rev. Lett. **92**, 031101 (2004).
12. R.B. Firestone, V.S. Shirley (Editors), *Table of Isotopes*, Vol. **1** (J. Wiley and Sons, 1996).
13. B. Cederwall *et al.*, Nucl. Instrum. Methods Phys. Res. A **353**, 591 (1994).
14. C. Angulo *et al.*, Nucl. Phys. A **656**, 3 (1999).
15. J. José, M. Hernanz, Astrophys. J. **494**, 680 (1998).
16. C.N. Davids, Nucl. Instrum. Methods Phys. Res. B **70**, 358 (1992).
17. D.G. Jenkins *et al.*, Phys. Rev. C **72**, 031303 (2005).

Eur. Phys. J. A **27**, s01, 123–128 (2006)
DOI: 10.1140/epja/i2006-08-017-6

EPJ A direct
electronic only

Evidence for p-process nucleosynthesis recorded at the Solar System abundances

T. Hayakawa[1,a], N. Iwamoto[2], T. Shizuma[1], T. Kajino[3], H. Umeda[4], and K. Nomoto[4]

[1] Advanced Photon Research Center, Japan Atomic Energy Research Institute, Kizu, Kyoto 619-0215, Japan
[2] Department of Nuclear Energy System, Japan Atomic Energy Research Institute, Tokai, Ibaraki 319-1195, Japan
[3] National Astronomical Observatory, Osawa, Mitaka, Tokyo 181-8588, Japan
[4] Department of Astronomy, School of Science, University of Tokyo, Tokyo 113-0033, Japan

Received: 17 June 2005 /
Published online: 13 March 2006 –

Abstract. Analyzing the Solar System abundance, we have found two universal scaling laws concerning the p- and s-nuclei with the same atomic number. They are evidence of the p-nucleus origin that has been discussed for long years. The first scaling is the correlation of the isotope abundances between a p-nucleus and an s-nucleus that is two neutron heavier than the p-nucleus. The abundance ratios of the s-nucleus to the p-nucleus are almost constant in a wide range of atomic numbers. This scaling indicates that the p-nucleus is dominantly produced from the s-nucleus by (γ, n) reactions in huge photon bath in supernova explosions. The second scaling indicates that the isotope abundance ratio between two p-nuclei with the same atomic number is almost unity. These two scalings are a new useful method to verify nucleosynthesis models for the p-process. We have calculated these ratios by a p-process nucleosynthesis based on a typical core-collapse supernova explosion model. The results are consistent with the scaling in the solar abundances. In addition, the scalings provide new concepts: a universality of the p-process, a rate meter for the s-process and a new nuclear cosmochronometer.

PACS. 26.30.+k Nucleosynthesis in novae, supernovae and other explosive environments – 98.80.Ft Origin, formation, and abundances of the elements – 91.65.Dt Isotopic composition/chemistry

1 Introduction

Light elements such as H, He were mainly produced by Big-Bang nucleosynthesis and heavy elements were dominantly synthesized by stellar nucleosynthesis after the formation of the Galaxy. The Solar System was formed from the interstellar medium (ISM), whose composition is provided from stellar nucleosynthesis episodes. The Solar System abundance is, therefore, an important record of stellar nucleosynthesis and the galactic chemical evolution (GCE). About 99% of elements heavier then the iron group were produced by two different neutron-capture reaction chains, *i.e.* the slow neutron-capture process (s-process) along the β-stability line, and the rapid neutron-capture process (r-process) in the neutron-rich side. The Solar System abundance shows specific indication that they have actually happened before the Solar System formation. The first evidence is the pairs of two abundance peaks near the neutron magic numbers $N = 50, 82, 126$. These two peaks correspond to the s- and r-processes [1]. The second evidence for the s-process is an empirical relation, $N \cdot \sigma \sim$ constant, where N and σ are, respectively, the solar abundance and the neutron capture cross-section [2, 3]. However, there are stable isotopes that cannot be synthesized by the neutron capture reactions because they are located in the neutron-deficient side of the β-stability line in the nuclear chart (see fig. 1). These isotopes are called the "p-nuclei". They have the feature that their isotope abundances are very rare (typically 0.1%–1%). The anti-correlation between the photodisintegration reaction rates and the solar abundances for the p-nuclei has been pointed out [4]. Nevertheless, their origin has long been discussed with many possible nuclear reactions, and their astrophysical sites have not been identified uniquely. The proposed nuclear processes are the rapid proton capture reactions in novae and type-I X-ray bursts in neutron stars (rp-process) [5,6], the proton-induced reactions by galactic cosmic rays [7], the photodisintegration reactions in supernova (SN) explosions (γ-process or p-process) [4,8, 9,10,11,12], and the neutrino-induced reactions in SN explosions (ν-process) [13,14]. The origin of the p-nuclei is crucial to our understanding of how the Solar System material formed and evolved. We here report two empirical scaling laws obtained from a careful analysis of the Solar

[a] e-mail: hayakawa.takehito@jaea.go.jp

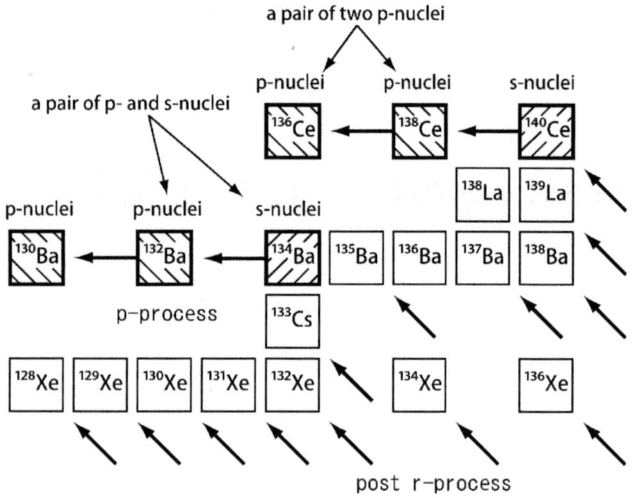

Fig. 1. Partial nuclear chart around the Xe and Ba isotopes. The nuclei located on the β-stability line are mainly synthesized by the s- and r-processes. In contrast, p-nuclei such as ^{130,132}Ba and ^{136,138}Ce are synthesized by the p-process. There are pairs of a p-nucleus and an s-nucleus that are two neutron heavier than the p-nucleus. We find a strong correlation between the solar abundances of these p- and s-nuclei.

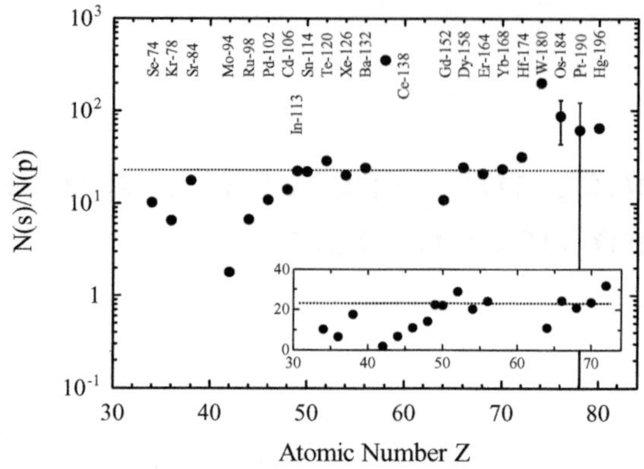

Fig. 2. Abundance ratios of pure s-nucleus to pure p-nucleus, $N(\mathrm{s})/N(\mathrm{p})$, in the Solar System. The p-nucleus is a two-neutron–deficient isotope from an s-nucleus with the same atomic number Z. The ratios are almost constant, $N(\mathrm{s})/N(\mathrm{p}) \approx 23$ in a wide region of atomic numbers. The inset displays the same quantities in the linear scale except for the ^{138}Ce and ^{180}W pairs which show large deviations from the scaling value ≈ 23. Deviations from the scaling and the uncertainty of the Os and Pt isotopes are discussed in the text.

System abundance [12], and discuss new concepts of "the universality of the p-process" and the rate meter for the s-process. We also calculate the p-process nucleosynthesis using a typical type-II SN model.

2 Analyzing the solar abundances and discovery of the first scaling

There are thirty-five p-nuclei in nature. Most p-nuclei are even-even nuclei consisting of even protons and even neutrons. There are 22 pairs of a p-nucleus and an almost pure s-nucleus that are two neutron heavier than the p-nucleus. The pure s-nuclei are dominantly synthesized by the s-process and shielded by stable isobars against the β^--decay after the freezeout of the r-process. Figure 1 shows a partial nuclear chart in the Xe-Ba region. A typical example is found in Ba isotopes: ^{132}Ba is a p-nucleus and ^{134}Ba is a pure s-nucleus shielded by an isobar ^{134}Xe against the β-decay.

We here discuss the isotope abundance ratios of these two isotopes. Taking the abundance ratio of the s-nucleus to the p-nucleus, $N(\mathrm{s})/N(\mathrm{p})$, where $N(\mathrm{s})$ and $N(\mathrm{p})$ means the solar abundances of the s- and p-nucleus, respectively, we find a clear correlation between them (see fig. 2). The ratios are almost constant and constrained at a constant value of $N(\mathrm{s})/N(\mathrm{p}) \approx 23$ in a wide region of atomic numbers except for the deviations. The reason for the deviations will be discussed later. This scaling has never been recognized in the literature quantitatively.

The first scaling shows a strong correlation between p- and s-nuclei with the same atomic number, which indicates that the origin of the p-nuclei is strongly correlated with the s-nuclei. This is consistent with the previous theoretical calculations that the p-nuclei are produced by the p-process (or γ-process) in SN explosions [15,16, 17]: namely, the pre-existing nuclei in massive stars are affected by the s-process during the pre-supernova evolutionary stage and the p-nuclei are subsequently produced from them by photodisintegration reactions such as (γ, n) reactions in a huge photon bath at extremely high temperatures in SN explosions. The previous calculations indicated that the p-nuclei are produced via two paths, direct (γ, n) reactions and the EC/β^+-decay from the neutron-deficient unstable nuclei which are first transmuted by successive photodisintegration reactions (γ, n), (γ, p), (γ, α) from heavier elements. The first scaling suggests that the former reactions are likely to play a more important role than the latter reactions. The charged particle reactions in the rp-process [5,6] and proton-induced reactions by cosmic rays [7] change the proton number of seed nuclei. In the ν-process, the charged current interaction that has a contribution larger than the neutral current interaction also changes the proton number [18,19]. Therefore, the scaling does not emerge from the dominant charged particle processes or the ν-process. The first scaling is, thus, a piece of evidence that the p-process is the most promising origin of the p-nuclei.

We finally discuss the reason for the deviations. The two large deviations for Ce and W can be explained by an exceptional contribution from the r-process because they are not shielded against β-decay. The heavy isotopes ^{140}Ce and ^{182}W are dominantly synthesized by the r-process. This fact indicates that the p-nuclei are not synthesized from the seed nuclei of the r-process mass distribution. In addition the effect of the neutron magic number $N = 82$ should contribute to the deviation for Ce. The small deviations for Cd, Sn and Gd may originate from a weak

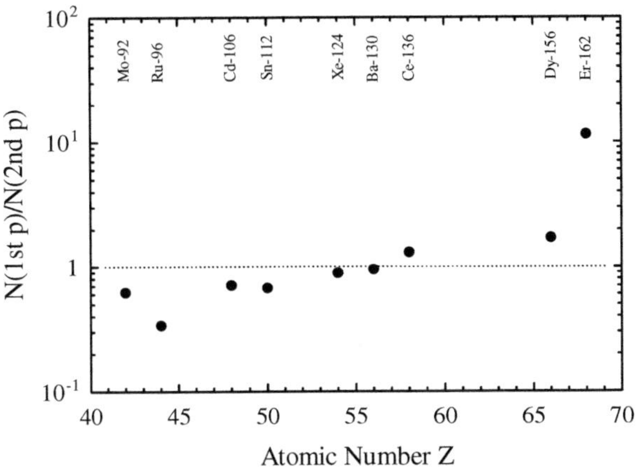

Fig. 3. Abundance ratios of two pure p-nuclei, N(1st p)/ N(2nd p). The first and second p-nuclei are, respectively, two- and four-neutron–deficient isotopes from an s-nucleus with the same atomic number.

branch of the s-process which contaminates the p-nuclei. The deviations of Mo and Ru are consistent with the underproduction of these isotopes in the previous calculations [15,16]. Their origin may be different with the other even-even p-nuclei. The deviations in the heavy-mass region may originate from the large uncertainties of the solar abundances. The uncertainties for ^{184}Os and ^{190}Pt are about 50% and 100%, respectively [20]. The measurement of their abundances with a high precision is desired.

3 Discovery of the second scaling

We find another scaling law between two p-nuclei with the same proton number. Nine elements have two p-nuclei. We define that the second p-nucleus is two neutron lighter than the first p-nucleus. For example, ^{132}Xe is the first p-nucleus and ^{130}Xe is the second p-nucleus (see fig. 1). Taking the abundance ratio, N(1st p)/N(2nd p), we also find a strong correlation between these isotope abundances (see fig. 3). The ratios are almost constant in a wide range of atomic numbers and they are almost unity, *i.e.* N(1st p)/N(2nd p) ≈ 1. This second scaling is useful to constrain the astrophysical condition of the p-process. There is a deviation for Er. It can be explained by a contamination from the β^--decay of ^{163}Dy under stellar s-process conditions [21,22].

4 Proposal of a novel concept of the universality of the p-process

The universality of nucleosynthesis processes is an important concept for understanding stellar nucleosynthesis. The Solar System was formed from the ISM originated from many nucleosynthesis episodes in stars in the Galaxy. The nucleosynthesis environments such as mass, metallicity and explosion energy are different and hence the produced abundance distribution may be different. The abundance distribution in the Solar System material is, thus, equal to the average of the stellar nucleosynthesis processes. However, astronomical observations for very metal-deficient stars have reported the "universal" abundance distribution for $Z > 56$ [23,24], which is in agreement with the abundance distribution of the r-process nuclides in the Solar System. These facts suggest a uniform site and/or uniform conditions for the synthesis of the r-process nuclei. Otsuki *et al.* could succeed in explaining the universality of the r-process by a neutrino-energized wind model [25].

The N(s)/N(p) ratios for the Solar System abundance are subject to galactic chemical evolution. The p-nuclei and s-nuclei were produced in different stellar environments. Thus, the mass distribution of synthesized nuclei may depend on astrophysical conditions. Nevertheless, the observed N(s)/N(p) ratios in the Solar System are almost constant in a wide region of atomic numbers. The observed ratios do not depend on the proton number. This leads to a novel concept, "the universality of the p-process", that the N(s)/N(p) ratios produced by individual p-process are constant in a wide range. We would like to stress that this universality indicates a uniform astrophysical condition or a nucleosynthesis process that does not depend on astrophysical conditions. The universality is, thus, essential for understanding the nucleosynthesis site of p-nuclei.

5 Proposal of a novel concept of rate meter for the s-process

The universality of the SN p-process is an important concept for understanding the chemical evolution of the Galaxy as well as the p-process nucleosynthesis. Figure 4 shows a schematic chart for the GCE. First, we would like to stress that the s-nuclei in the Solar System were mainly produced by the s-process in the low-mass AGB stars [3]. In contrast, p-nuclei are synthesized by the p-process in SNe. The astrophysical sites for the p- and s-processes are different. However, the isotope abundance ratios of the p- and s-nuclei have a strong correlation. This fact indicates that the average N(s)/N(p) ratios in the first scaling are proportional to the abundance synthesized by individual s-process and to the frequency of the formation of the AGB stars. Second, the s-process nucleosynthesis depends highly on the metallicity which increases along the evolution of the Galaxy [26,27]. These two facts concerning s-nuclei and the universality of p-nuclei indicate that the N(s)/N(p) ratios should depend on time.

Astronomical observations of the time variation of these ratios for various metallicity stars should constrain the galactic chemical evolution of s-nuclei and also provide new information on the metallicity dependence of the s-process nucleosynthesis [28]. The recent progress in spectroscopic studies of extremely metal-poor stars has enabled successfully isotope separation of several heavy elements [29,30]. It is of particular interest to observe the ancient metal-poor stars whose material had been affected by a single or a few SN p-processes. Since the primitive

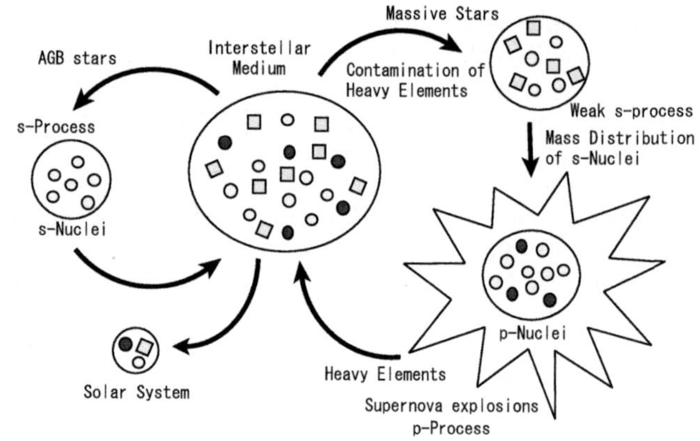

Fig. 4. A schematic chart for the galactic chemical evolution for p- and s-nuclei. The s-nuclei in the ISM are dominantly produced in the main s-process in the AGB stars. The massive stars have contamination of heavy elements from the ISM. The heavy elements are irradiated by neutrons in the weak s-process before SN explosions and the abundance distribution of the heavy elements is changed to that of the s-process. p-nuclei are synthesized in the SN p-process. The s-process depends strongly on metallicity. The $N(\mathrm{s})/N(\mathrm{p})$ ratio is proportional to the frequency of the s-process events and is time dependent.

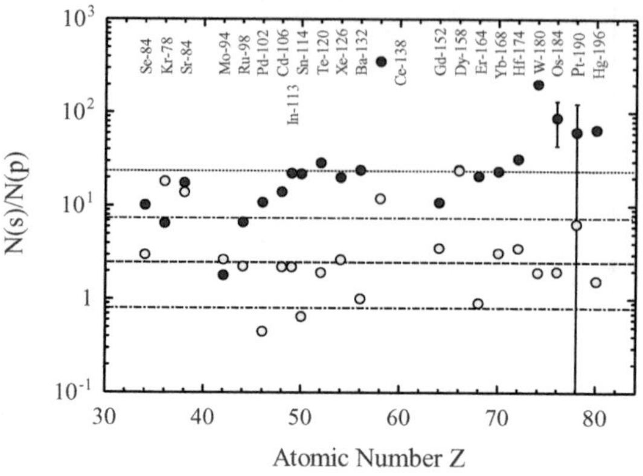

Fig. 5. Comparison of the calculated and observed abundance ratios, $N(\mathrm{s})/N(\mathrm{p})$. The filled and open circles mean the observed ratio in the Solar System and the calculated ratios. The uppermost dotted line is $N(\mathrm{s})/N(\mathrm{p}) = 23$. The dashed line displays the average value of the calculated $N(\mathrm{s})/N(\mathrm{p})$ ratios in the SN p-process model, and the two dot-dashed lines above and below this line are those multiplied by factor of 3 and 1/3, respectively.

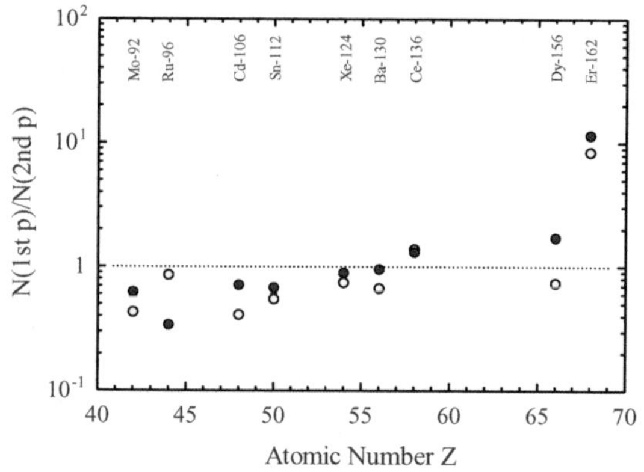

Fig. 6. Abundance ratios of two pure p-nuclei, $N(\text{1st p})/N(\text{2nd p})$. The first and second p-nuclei are, respectively, two- and four-neutron–deficient isotopes from an s-nucleus with the same atomic number Z. The filled circles stand for the observed ratios in the Solar System. The open circles stand for the calculated ratios.

gas is made of products of the Big-Bang nucleosynthesis or an explosive nucleosynthesis in the first generation population-III SNe, it does not contain any heavy s-nuclei. The abundance distribution of p-nuclei in metal-poor stars is, thus, expected to be very different from the solar abundance distribution, and the detection of p-nuclei by spectroscopically separating isotope abundances in these stars would be an urgent subject in future studies.

6 Supernova model calculations

For the photodisintegration-reaction p-process sites, some astrophysical sites have been proposed. They are 1) O/Ne layers in core collapse SNe [8,4], 2) He deflagration in C/O white dwarfs [9], 3) accretion disks around neutron stars or black holes [11,31]. Since one of the most probable site is the O/Ne layer in type-II SNe, we carry out nucleosynthesis calculations of the p-process in SNe [32]. The purpose of the calculations is to verify the robustness of the scalings in the Solar System abundance and to demonstrate the dependence of the calculated ratios, $N(\mathrm{s})/N(\mathrm{p})$ and $N(\text{1st p})/N(\text{2nd p})$, on astrophysical conditions. We use solar metallicity ($Z = Z_\odot$) progenitor models with 25 solar masses ($25M_\odot$) which exploded with an explosion energy of 10^{51} ergs. The s-processed abundances for the initial chemical composition are adopted. The calculated $N(\mathrm{s})/N(\mathrm{p})$ ratios are shown in fig. 5 by open circles. It is shown that they are almost constant in a wide region of atomic numbers, although the calculated ratios are lower than the observed values. This result is consistent with the observed scaling. The observed ratios in the light-mass region show a slight enhancement of p-nuclei, which may originate from progressively increasing roles of (γ,p) and (γ,α) reactions with decreasing atomic number and/or the production from heavier nuclei at high temperature. The calculated ratios are smaller than the observed ones by several factors because s-nuclei in the Solar System mainly originate from the AGB stars [3]. In contrast the relation $N(\text{1st p})/N(\text{2nd p}) \approx 1$ can be directly compared with the theoretical calculations of the SN p-process, and thus the second scaling can be used for strongly constraining the SN p-process models. We present the calculated $N(\text{1st p})/N(\text{2nd p})$ ratios in fig. 6. The calculated ratios (open circles) are consistent with the observed ratios (filled circles).

We further perform the p-process nucleosynthesis calculations for the 15 and 40 $M_\odot$ progenitors to study the progenitor mass dependence. The abundance patterns of the two ratios do not change drastically from those in the 25 $M_\odot$ models. This result indicates that the two ratios are almost independent of the progenitor mass of the massive stars. We calculate the p-process in the different metallicity ($Z = 0.05\ Z_\odot$) models with the same progenitor mass. The calculated result shows that the ratios are almost constant and independent of the metallicity. These results support the proposed universality of the p-process. The calculated results in the previous p-process studies were shown to compare directly with the Solar System abundance of p-nuclei [15,16,17], not in the form of $N(\mathrm{s})/N(\mathrm{p})$ or $N(\mathrm{1st\ p})/N(\mathrm{2nd\ p})$. The p-process calculations for different models constructed with different explosion energies or the $^{12}\mathrm{C}(\alpha,\gamma)^{16}\mathrm{O}$ reaction rate showed similar abundance distributions [15,16]. These results also support the universality of the p-process. Although these results indicate that p-nuclei in the Solar System are mainly produced by the p-process in type-II SNe, other astrophysical sites such as deflagrating white dwarf stars [9] and supernova-driven supercritical accretion disks [11,31] may also contribute to p-nuclei. Overproduction factors of p-process nuclei in realistic models of exploding stars were often a factor of a few below what is needed to explain the solar abundances. This may signal the p-process in some other environments as another producer of p-nuclei. We presume that such p-processes should also reproduce the two scalings.

7 Proposal of a new nuclear cosmochronometer

Long-lived radioactive nuclei are used as nuclear cosmochronometers, which are useful for an investigation of the nucleosynthesis process history along the GCE before the Solar System formation. The radioactive nuclei of cosmological significance are very rare and only six chronometers with half-lives in the range of the cosmic age 1–100 Gyr were known. They are ^{40}K [1] and ^{87}Rb [33] for the s-process or explosive nucleosynthesis in SNe, ^{176}Lu [34,35] for the s-process, and ^{187}Re [33,36,37], ^{232}Th and ^{238}U [1, 38] for the r-process. Historically, a new cosmochronometer with suitable half-life has not been proposed for the last thirty years. Recently, the two elements U and Th were detected for the first time [38] in a very metal-poor star. The actinide nuclei in very metal-poor stars were perhaps created in a single r-process of a SN explosion.

The universal scaling also plays a critical role in constructing a chronometer that can be applied to the analysis of pre-solar grains in primitive meteorites which had been affected strongly by a single or a few nucleosynthesis episodes. We here propose a new cosmochronometer ^{176}Lu (half-life 37.8 Gyr)-^{176}Hf-^{174}Hf of the p-process in the SN explosion. Although the ^{146}Sm and ^{92}Nb have already been proposed as possible chronometers of the p-process [39,40], their half-lives are shorter than the age

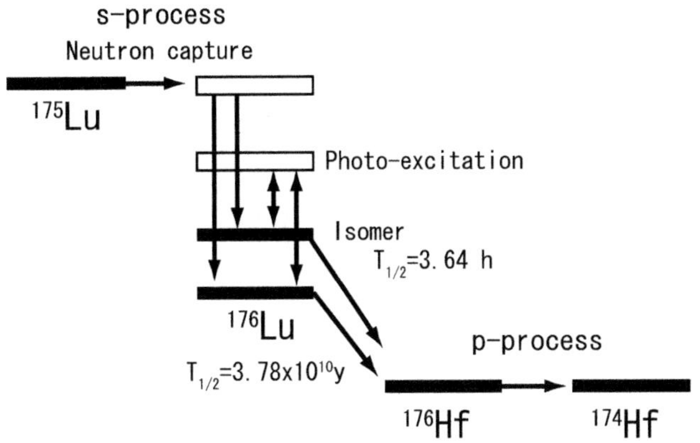

Fig. 7. ^{176}Lu-^{176}Hf-^{174}Hf chronometer. The ground state of ^{176}Lu decays to ^{176}Hf with a half-life of 3.78×10^{10} y. Although ^{176}Hf is produced by different nucleosynthesis paths, the first scaling indicates that the initial abundance of ^{176}Hf at the freezeout of the SN p-process can be calculated from the present abundance of ^{174}Hf.

of the Solar System. Therefore, our proposed chronometer becomes a unique p-process chronometer which has a suitable time scale of the order of the cosmic age ~ 10 Gyr.

A ^{176}Lu-^{176}Hf pair was previously proposed as an s-process chronometer [34], since ^{176}Lu is a pure s-nucleus. The daughter nucleus ^{176}Hf is also a pure s-nucleus and is located outside the main path of the s-process. An isomeric state in ^{176}Lu decays to ^{176}Hf with a short half-life of 3.64 hours. The experiments using stellar-energy neutrons have been carried out and the branching ratio between the ground state and the isomer in ^{176}Lu was measured. It was, however, pointed out that the solar abundances of ^{176}Lu cannot be explained by a classical s-process model [35]. The isomer is populated through intermediate states with high excitation energy by (γ,γ') reactions at high temperature, $T \sim 10^8$ K (see fig. 7). The decay rate of ^{176}Lu depends strongly on the temperature and the initial abundance at the end of a nucleosynthesis episode cannot be predicted by theoretical calculations. The system ^{176}Lu-^{176}Hf is considered to be useless as a chronometer, although it is a good thermometer [35].

The p-process chronometer of ^{176}Lu-^{176}Hf-^{174}Hf has the advantage that the initial abundance of ^{176}Hf can be calculated from the present abundance of ^{174}Hf by applying the first scaling if the pre-solar grain is affected strongly by a single SN event. The first scaling indicates that the abundance of ^{174}Hf is proportional to ^{176}Hf, although ^{176}Hf is produced through different nucleosynthesis paths. Our proposed chronometer of the p-process is, therefore, free from the uncertainty of the initial abundance. The time passing after the SN p-process can be calculated by

$$T = -\frac{T_{1/2}(^{176}\mathrm{Lu})}{\ln 2} \times \ln\left(\frac{N(^{176}\mathrm{Lu})}{N(^{176}\mathrm{Lu})+(N(^{176}\mathrm{Hf})-R_i(\mathrm{Hf})\times N(^{174}\mathrm{Hf}))}\right), \quad (1)$$

where, $N(^AZ)$ means the isotope abundance, and R stands for the $N(\mathrm{s})/N(\mathrm{p})$ ratio in the scaling in meteorites, which should be systematically measured or predicted by p-process calculations. Heavy elements such as Sr, Zr, Mo and Ba in a primitive material such as the pre-solar grains have already been successfully separated into isotopes including p-nuclei, whose origin is considered to be the ejecta of core collapse SN explosions [41,42]. Although the pre-solar grains would be likely to condense ^{176}Hf and ^{176}Lu from other regions of the star, the chemical composition of the grains enhanced by the products of the O/Ne layer may be found. The separation of the three isotopes, ^{174,176}Hf and ^{176}Lu, in the pre-solar grains is highly desirable.

8 Conclusion

In summary, we have presented two universal scaling laws concerning p- and s-nuclei in the Solar System abundance. They provide four novel concepts: a piece of evidence that the SN p-process is the most probable origin of p-nuclei, a universality that the abundance ratios $N(\mathrm{s})/N(\mathrm{p})$ of products by individual SN p-process are almost constant in a wide region of atomic numbers, a rate meter that the $N(\mathrm{s})/N(\mathrm{p})$ value is proportional to the frequency of the AGB s-process events, and a new nuclear cosmochronometer for the p-process. The scalings are useful for identifying the astrophysical sites of p-nuclei and limiting the contribution from other nuclear processes. We carry out typical type-II SN p-process calculations and the results support the universality of the p-process. Therefore our proposals provide new insights into the chemical evolution of the Galaxy as well as the SN p-process.

References

1. E.M. Burbidge, G.R. Burbidge, W.A. Fowler, F. Hoyle, Rev. Mod. Phys. **29**, 548 (1957).
2. P.A. Seeger, W.A. Fowler, D.D. Clayton, Astrophys. J. **11**, 121 (1965).
3. R. Gallino *et al.*, Astrophys. J. **497**, 388 (1998).
4. S.E. Woosley, W.M. Howard, Astrophys. J. Suppl. **36**, 285 (1978).
5. H. Schatz *et al.*, Phys. Rep. **294**, 167 (1998).
6. H. Schatz *et al.*, Phys. Rev. Lett. **86**, 3471 (2001).
7. J. Audouze, Astron. Astrophys. **8**, 436 (1970).
8. M. Arnould, Astron. Astrophys. **46**, 117 (1976).
9. W.M. Howard, B.S. Meyer, S.E. Woosley, Astrophys. J. **373**, L5 (1991).
10. M. Arnould, S. Goriely, Phys. Rep. **384**, 1 (2003).
11. S. Fujimoto *et al.*, Astrophys. J. **585**, 418 (2003).
12. T. Hayakawa *et al.*, Phys. Rev. Lett. **93**, 161102 (2004).
13. S.E. Woosley, D.H. Hartmann, R.D. Hoffman, W.C. Haxton, Astrophys. J. **356**, 272 (1990).
14. R.D. Hoffman, S.E. Woosley, G.M. Fuller, B.S. Meyer, Astrophys. J. **460**, 478 (1996).
15. N. Prantzos, M. Hashimoto, M. Rayet, M. Arnould, Astron. Astrophys. **238**, 455 (1990).
16. M. Rayet *et al.*, Astron. Astrophys. **298**, 517 (1995).
17. H. Utsunomiya *et al.*, Phys. Rev. C **67**, 015807 (2003).
18. S. Goriely, M. Arnould, I. Borzov, M. Rayet, Astron. Astrophys. **375**, L35 (2001).
19. A. Heger *et al.*, Phys. Lett. B **606**, 258 (2005).
20. P. De Bievre, P.D.P. Taylor, Int. J. Mass Spectrom. Ion Proc. **123**, 149 (1993).
21. K. Takahashi, K. Yokoi, Nucl. Phys. A **404**, 578 (1983).
22. M. Jung *et al.*, Phys. Rev. Lett. **69**, 2164 (1992).
23. C. Sneden *et al.*, Astrophys. J. **496**, 235 (1998).
24. C. Sneden *et al.*, Astrophys. J. **533**, L139 (2000).
25. K. Otsuki *et al.*, Astrophys. J. **533**, 424 (2000).
26. C.M. Raiteri, R. Gallino, M. Busso, Astrophys. J. **387**, 263 (1992).
27. M. Busso, R. Gallino, C.J. Wasserburg, Annu. Rev. Astron. Astrophys. **37**, 239 (1999).
28. W. Aoki *et al.*, Astrophys. J. **580**, 1149 (2002).
29. D.L. Lambert, C.A. Prieto, Mon. Not. R. Astron. Soc. **335**, 325 (2002).
30. W. Aoki *et al.*, Astrophys. J. **592**, L67 (2003).
31. S. Fujimoto *et al.*, Astrophys. J. **614**, 847 (2004).
32. N. Iwamoto, H. Umeda, K. Nomoto, *International Symposium on Origin of Matter and Evolution of Galaxies* (World Scientific, 2004) p. 493.
33. D.D. Clayton, Astrophys. J. **139**, 637 (1964).
34. J. Audouze, W.A. Fowler, D.N. Schramm, Nature Phys. Sci. **238**, 8 (1972).
35. H. Beer, F. Käppeler, K. Wisshak, R.A. Ward, Astrophys. J. Suppl. **46**, 295 (1981).
36. M. Arnould, K. Takahashi, K. Yokoi, Astron. Astrophys. **137**, 51 (1984).
37. T. Hayakawa *et al.*, Astrophys. J. **628**, 533 (2005).
38. R. Cayrel *et al.*, Nature **409**, 91 (2001).
39. J. Audouze, D.N. Schramm, Nature **237**, 447 (1972).
40. C.L. jr. Harper, Astrophys. J. **466**, 437 (1996).
41. M.J. Pellin *et al.*, Lunar Planet. Sci. **31**, 1917 (2000).
42. Q. Yin, S.B. Jacobsen, K. Yamashita, Nature **415**, 881 (2002).

Eur. Phys. J. A **27**, s01, 129–134 (2006)
DOI: 10.1140/epja/i2006-08-018-5

EPJ A direct
electronic only

(n, γ) cross-sections of light p nuclei

Towards an updated database for the p process

I. Dillmann[1,2,a], M. Heil[1], F. Käppeler[1], R. Plag[1], T. Rauscher[2], and F.-K. Thielemann[2]

[1] Institut für Kernphysik, Forschungszentrum Karlsruhe, Postfach 3640, D-76021 Karlsruhe, Germany
[2] Departement Physik und Astronomie, Universität Basel, Klingelbergstrasse 82, CH-4056 Basel, Switzerland

Received: 7 July 2005 /
Published online: 10 March 2006 –

Abstract. The nucleosynthesis of elements beyond iron is dominated by the s and r processes. However, a small amount of stable isotopes on the proton-rich side cannot be made by neutron capture and is thought to be produced by photodisintegration reactions on existing seed nuclei in the so-called "p process". So far most of the p-process reactions are not yet accessible by experimental techniques and have to be inferred from statistical Hauser-Feshbach model calculations. The parametrization of these models has to be constrained by measurements on stable proton-rich nuclei. A series of (n,γ) activation measurements on p nuclei, related by detailed balance to the respective photodisintegrations, was carried out at the Karlsruhe Van de Graaff accelerator using the ^{7}Li$(p,n)^7$Be source for simulating a Maxwellian neutron distribution of $kT = 25$ keV. We present here preliminary results of our extended measuring program in the mass range between $A = 74$ and $A = 132$, including first experimental (n,γ) cross-sections of ^{74}Se, ^{84}Sr, ^{120}Te and ^{132}Ba, and an improved value for ^{130}Ba. In all cases we find perfect agreement with the recommended MACS predictions from the Bao *et al.* compilation.

PACS. 25.40.Lw Radiative capture – 26.30.+k Nucleosynthesis in novae, supernovae, and other explosive environments – 27.50.+e $59 \le A \le 89$ – 27.60.+j $90 \le A \le 149$

1 Introduction

Astrophysical models can explain the origin of most nuclei beyond the iron group in a combination of processes involving neutron captures on long (s process) or short (r process) time scales [1,2]. However, 32 stable, proton-rich isotopes between ^{74}Se and ^{196}Hg cannot be formed in that way. Those p nuclei are 10 to 100 times less abundant than the s and r nuclei in the same mass region. They are thought to be produced in the so-called γ or p process, where proton-rich nuclei are made by sequences of photodisintegrations and β^+ decays [3,4,5]. In this scenario, pre-existing seed nuclei from the s and r processes are destroyed by photodisintegration in a high-temperature environment, and proton-rich isotopes are produced by (γ,n) reactions. When (γ,p) and (γ,α) reactions become comparable or faster than neutron emission within an isotopic chain, the reaction path branches out and feeds nuclei with lower charge number Z. The decrease in temperature at later stages of the p process leads to a freeze-out via neutron captures and mainly β^+ decays, resulting in the typical p-process abundance pattern with maxima at ^{92}Mo ($N = 50$) and ^{144}Sm ($N = 82$).

[a] e-mail: iris.dillmann@ik.fzk.de

The currently most favored astrophysical site for the p process is explosive burning in type-II supernovae. The explosive shock front heats the outer O/Ne shell of the progenitor star to temperatures of 2–3 GK, sufficient for providing the required photodisintegrations. Following the nucleosynthesis in such astrophysical models, good agreement with the required p production is found, with the exception of the low ($A < 100$) and intermediate ($150 \le A \le 165$) mass range, which are underproduced by factors of 3–4 [6]. Currently, however, it is not yet clear whether the observed underproductions are due to a problem with astrophysical models or with the nuclear physics input, *i.e.* reaction rates. Thus, a necessary requirement towards a consistent understanding of the p process is the reduction of uncertainties in nuclear data. By far most of the several hundreds of required photodisintegration rates and their inverses need to be inferred from Hauser-Feshbach statistical model calculations [7,8]. Experimental data can improve the situation in two ways, either by directly replacing predictions with measured cross-sections in the relevant energy range, or by testing the reliability of predictions at other energies when the relevant energy range is not experimentally accessible.

The role of (n,γ) reactions in the p process was underestimated for a long time, although it is obvious that they have an influence on the final p-process abundances.

Neutron captures compete with (γ, n) reactions and thus hinder the photodisintegration flux towards light nuclei, especially at lower-Z isotopes and even-even isotopes in the vicinity of branching-points. Rayet *et al.* [9] have studied the influence of several components in their p-process network calculations. Their nuclear flow schemes show that branching points occur even at light p nuclei, and are shifted deeper into the proton-rich unstable region with increasing mass and temperature. In contradiction to Woosley and Howard [3], who claimed for their network calculations that (n, γ) can be neglected except for the lightest nuclei ($A \leq 90$), Rayet *et al.* also examined the influence of neutron reactions for temperatures between $T_9 = 2.2$ and 3.2 GK by comparing overabundance factors if (n, γ) reactions on $Z > 26$ nuclides are considered or completely suppressed. As a result, the overabundances were found to change by up to a factor 100 (for ^{84}Sr) if the (n, γ) channel was artificially suppressed. This rather high sensitivity indicates the need for reliable (n, γ) rates to be used in p-process networks.

The influence of a variation of reaction rates on the final p abundances has also been studied previously [10, 11]. It turned out that the p abundances are very sensitive to changes of the neutron-induced rates in the entire mass range, whereas the proton-induced and α-induced reaction rates are important at low and high mass numbers, respectively.

A third reason for the determination of neutron capture rates of p nuclei are those cases where experimental photodissociation rates are not accessible. The respective astrophysical photodisintegration rate can then be inferred from capture rates by detailed balance [12]. This is the case for most stable p nuclei, which are separated from stable isotopes by a radioactive nucleus. While the reaction rate $^{A}X(\gamma, n)^{A-1}X$ can be determined by bremsstrahlung [13], the reaction $^{A+1}X(\gamma, n)^{A}X$ has to be measured via its inverse (n, γ) rate.

The present work comprises the first measurement of (n, γ) cross-sections for the p-process isotopes ^{74}Se, ^{84}Sr, ^{120}Te and ^{132}Ba at $kT = 25$ keV, and a re-measurement of ^{130}Ba. The direct determination of stellar (n, γ) rates requires a "stellar" neutron source yielding neutrons with a Maxwell-Boltzmann energy distribution. We achieve this by making use of the ^{7}Li$(p, n)^{7}$Be reaction. In combination with the activation or time-of-flight technique, this offers a unique tool for comprehensive studies of (n, γ) rates and cross-sections for astrophysics.

2 Experimental procedure

All measurements were carried out at the Karlsruhe 3.7 MV Van de Graaff accelerator using the activation technique. Neutrons were produced with the ^{7}Li$(p, n)^{7}$Be source by bombarding 30 μm thick layers of metallic Li on a water-cooled Cu backing with protons of 1912 keV, 30 keV above the reaction threshold. The resulting quasi-stellar neutron spectrum approximates a Maxwellian distribution for $kT = 25.0 \pm 0.5$ keV [14]. Hence, the proper stellar capture cross-section can be directly deduced from our measurement.

For all activations natural samples of the respective element were used. The selenium and tellurium samples were prepared from metal granules, whereas for the barium and strontium measurement thin pellets were pressed from powders of $Sr(OH)_2$, SrF_2, $SrCO_3$ and $BaCO_3$. In order to verify the stoichiometry, the powder samples were dried at 300 °C and 800 °C, respectively. All samples were enclosed in a 15 μm thick aluminium foil and sandwiched between 10–30 μm thick gold foils of the same diameter. In this way the neutron flux can be determined relative to the well-known capture cross-section of ^{197}Au [14].

Throughout the irradiation the neutron flux was recorded in intervals of 1 min using a ^{6}Li-glass detector for later correction of the number of nuclei, which decayed during the activation. The activations were carried out with the Van de Graaff accelerator operated in DC mode with a current of $\approx 100\,\mu$A. The mean neutron flux over the period of the activations was $\approx 1.5 \times 10^{9}$ n/s at the position of the samples, which were placed in close geometry to the Li target. The duration of the single activations varied between 3 h (for the partial cross-section to ^{85}Sr^{m}, $t_{1/2} = 67.6$ m) and 130 h (for determination of the 10.52 y ground state of ^{133}Ba).

3 Data analysis

The induced γ-ray activities were counted after the irradiation in a well-defined geometry using a shielded Ge detector in a low background area. Energy and efficiency calibrations have been carried out with a set of reference γ-sources in the energy range between 60 keV and 2000 keV. For the counting of the long-lived ^{133}Ba^{g} activity ($t_{1/2} = 10.52$ y) two fourfold segmented Clover detectors in close geometry were used [15].

A detailed description of the analysis procedure is given in refs. [16,17]. The number of activated nuclei A can be written as

$$A(i) = \Phi_{tot}\ N_i\ \sigma_i\ f_b(i), \tag{1}$$

where $\Phi_{tot} = \int \phi(t)dt$ is the time-integrated neutron flux and N_i the number of atoms in the sample. The factor f_b accounts for the decay of activated nuclei during the irradiation time t_a as well as for variations in the neutron flux. As our measurements are carried out relative to ^{197}Au as

Table 1. Stellar enhancement factors for different temperatures [12].

T (GK)	kT (keV)	SEF ^{74}Se	SEF ^{84}Sr	SEF ^{120}Te	SEF ^{130}Ba	SEF ^{132}Ba
0.3	26	1.00	1.00	1.00	1.00	1.00
2.0	172	1.01	1.02	1.10	1.23	1.16
2.5	215	1.02	1.06	1.18	1.33	1.22
3.0	260	1.03	1.09	1.25	1.42	1.28

Table 2. Maxwellian averaged cross-sections $\langle\sigma\rangle_{30}$ of all 32 p-process nuclei at $kT = 30\,\mathrm{keV}$. Values taken from this work are in bold. (1) Relative to Si $\equiv 10^6$. (2) Rescaled NON-SMOKER cross-sections accounting for known systematic deficiencies in the nuclear inputs [18]. (3) Xe abundances taken from ref. [19]. (4) Modified values [20]. (5) Preliminary cross-section.

Isotope	Solar Abundance(1)		Hauser-Feshbach prediction (mb)		Recommended values (mb) [18]
	Anders [21]	Lodders [22]	MOST [23]	NON-SMOKER [24]	
^{74}Se	5.50×10^{-1}	5.80×10^{-1}	304	207	**271 ± 15**
^{78}Kr	1.53×10^{-1}	2.00×10^{-1}	344	351	312 ± 26
^{78}Kr$\rightarrow^m$					92.3 ± 6.2
^{84}Sr	1.32×10^{-1}	1.31×10^{-1}	296 (4)	393	**300 ± 17**
^{84}Sr$\rightarrow^m$					**190 ± 10**
^{92}Mo	3.78×10^{-1}	3.86×10^{-1}	44	128	70 ± 10
^{94}Mo	2.36×10^{-1}	2.41×10^{-1}	87	151	102 ± 20
^{96}Ru	1.03×10^{-1}	1.05×10^{-1}	291	281	207 ± 8
^{98}Ru	3.50×10^{-2}	3.55×10^{-2}	370	262	173 ± 36 (2)
^{102}Pd	1.42×10^{-2}	1.46×10^{-2}	1061	374	373 ± 118 (2)
^{106}Cd	2.01×10^{-2}	1.98×10^{-2}	434	451	302 ± 24
^{108}Cd	1.43×10^{-2}	1.41×10^{-2}	260	373	202 ± 9
^{113}In	7.90×10^{-3}	7.80×10^{-3}	413	1202	787 ± 70
^{113}In$\rightarrow^m$					480 ± 160
^{112}Sn	3.72×10^{-2}	3.63×10^{-2}	208	381	210 ± 12
^{114}Sn	2.52×10^{-2}	2.46×10^{-2}	106	270	134.4 ± 1.8
^{115}Sn	1.29×10^{-2}	1.27×10^{-2}	212	528	342.4 ± 8.7
^{120}Te	4.30×10^{-3}	4.60×10^{-3}	340 (4)	551	**451 ± 18**(5)
^{120}Te$\rightarrow^m$					**61 ± 2**(5)
^{124}Xe	5.71×10^{-3}	6.57×10^{-3} (3)	593	799	644 ± 83
^{124}Xe$\rightarrow^m$					131 ± 17
^{126}Xe	5.09×10^{-3}	5.85×10^{-3} (3)	472	534	359 ± 51
^{126}Xe$\rightarrow^m$					40±6
^{130}Ba	4.76×10^{-3}	4.60×10^{-3}	561	730	**694 ± 20**(5)
^{132}Ba	4.53×10^{-3}	4.40×10^{-3}	300	467	**368 ± 25**(5)
^{132}Ba$\rightarrow^m$					**33.6 ± 1.7**(5)
^{136}Ce	2.16×10^{-3}	2.17×10^{-3}	227	495	328 ± 21
^{136}Ce$\rightarrow^m$					28.2 ± 1.6
^{138}Ce	2.84×10^{-3}	2.93×10^{-3}	160	290	179 ± 5
^{138}La	4.09×10^{-3}	3.97×10^{-3}	194	767	
^{144}Sm	8.00×10^{-3}	7.81×10^{-3}	37	209	92 ± 6
^{156}Dy	2.21×10^{-3}	2.16×10^{-3}	2025	1190	1567 ± 145
^{158}Dy	3.78×10^{-3}	3.71×10^{-3}	2188	949	1060 ± 400 (2)
^{162}Er	3.51×10^{-3}	3.50×10^{-3}	1818	1042	1624 ± 124
^{168}Yb	3.22×10^{-3}	3.23×10^{-3}	917	886	1160 ± 400 (2)
^{174}Hf	2.49×10^{-3}	2.75×10^{-3}	709	786	956 ± 283 (2)
^{180}W	1.73×10^{-3}	1.53×10^{-3}	722	707	536 ± 60
^{184}Os	1.22×10^{-3}	1.33×10^{-3}	697	789	657 ± 202 (2)
^{190}Pt	1.70×10^{-3}	1.85×10^{-3}	659	760	677 ± 82
^{196}Hg	4.80×10^{-3}	6.30×10^{-3}	493	372	650 ± 82 (2)

a standard, the neutron flux Φ_{tot} cancels out:

$$\frac{A(i)}{A(Au)} = \frac{\sigma_i\ N_i\ f_b(i)}{\sigma_{Au}\ N_{Au}\ f_b(Au)} \Longleftrightarrow$$
$$\sigma_i = \frac{A(i)\ \sigma_{Au}\ N_{Au}\ f_b(Au)}{A(Au)\ N_i\ f_b(i)}. \qquad (2)$$

The reference value for the experimental ^{197}Au cross-section in the quasi-stellar spectrum of the ^{7}Li$(p,n)^7$Be source is $586 \pm 8\,\mathrm{mb}$ [14]. By averaging the induced activities of the gold foils, one can determine the neutron flux Φ_{tot} at the position of the sample and deduce the experimental cross-section σ_i of the investigated sample as shown in eq. (2).

4 Results and discussion

4.1 General

In an astrophysical environment with temperature T, the neutron spectrum corresponds to a Maxwell-Boltzmann distribution

$$\Phi \sim E_n\ e^{-E_n/kT}. \qquad (3)$$

Table 3. Decay properties of the product nuclei [25]. Isotopic abundances are from ref. [26].

Reaction	Isot. abund. (%)	Final state	$t_{1/2}$	E_γ (keV)	I_γ (%)
^{74}Se(n, γ)^{75}Se	0.89 (0.04)	Ground state	119.79 ± 0.04 d	136.0	58.3 ± 0.7
				264.7	58.9 ± 0.3
^{84}Sr(n, γ)^{85}Sr	0.56 (0.01)	Ground state	64.84 ± 0.02 d	514.0	96 ± 4
		Isomer	67.63 ± 0.04 m	151.2 (EC)	12.9 ± 0.7
				231.9 (IT)	84.4 ± 2.2
^{120}Te(n, γ)^{121}Te	0.096 (0.001)	Ground state	19.16 ± 0.05 d	573.1	80.3 ± 2.5
		Isomer	154 ± 7 d	212.2 (IT)	81.4 ± 1.0
^{130}Ba(n, γ)^{131}Ba	0.106 (0.001)	Ground state	11.50 ± 0.06 d	123.8	29.0 ± 0.3
				216.1	19.7 ± 0.3
				373.2	14.0 ± 0.2
				496.3	46.8 ± 0.2
^{132}Ba(n, γ)^{133}Ba	0.101 (0.001)	Ground state	10.52 ± 0.13 y	356.0	62.1 ± 0.2
		Isomer	38.9 ± 0.1 h	275.9 (IT)	17.8 ± 0.6

The experimental neutron spectrum of the ^{7}Li$(p,n)^7$Be reaction approximates a Maxwellian distribution with $kT = 25$ keV almost perfectly [14]. But to obtain the exact Maxwellian averaged cross-section $\langle\sigma\rangle_{kT} = \frac{\langle\sigma v\rangle}{v_T}$ for the temperature T, the energy-dependent cross-section $\sigma(E)$ has to be folded with the experimental neutron distribution to derive a normalization factor $NF = \frac{\sigma}{\sigma_{exp}}$. The normalized cross-section in the energy range $0.01 \leq E_n \leq 4000$ keV was used for deriving the proper MACS as a function of thermal energy kT:

$$\frac{\langle\sigma v\rangle}{v_T} = \langle\sigma\rangle_{kT} = \frac{2}{\sqrt{\pi}} \frac{\int_0^\infty \frac{\sigma(E_n)}{NF} E_n \, e^{-E_n/kT} \, \mathrm{d}E_n}{\int_0^\infty E_n \, e^{-E_n/kT} \, \mathrm{d}E_n}. \quad (4)$$

In this equation, $\frac{\sigma(E_n)}{NF}$ is the normalized energy-dependent capture cross-section and E_n the neutron energy. The factor $v_T = \sqrt{2kT/m}$ denotes the most probable velocity with the reduced mass m.

Maxwellian averaged cross-sections have to be corrected by a temperature-dependent stellar enhancement factor

$$SEF(T) = \frac{\sigma^{star}}{\sigma^{lab}}. \quad (5)$$

The stellar cross-section $\sigma^{star} = \sum_\mu \sum_\nu \sigma^{\mu\nu}$ accounts for all transitions from excited target states μ to final states ν in thermally equilibrated nuclei, whereas the laboratory cross-section $\sigma^{lab} = \sum_0 \sum_\nu \sigma^{0\nu}$ includes only captures from the target ground state. In the investigated cases the thermal population effects in the stellar plasma at p-process temperatures are small for Se and Sr, but increase up to 1.42 for Ba (table 1).

4.2 Experimental results

For sample characteristics, activation features, and a detailed discussion of the Se and Sr results see ref. [17]. The results of the Te and Ba measurements in this paper are yet preliminary and correspond only to the cross-sections derived with the experimental neutron distribution at $kT = 25$ keV. Nevertheless, this value approximates the Maxwellian averaged cross-section at $kT = 30$ keV very well and can be used for comparison with other stellar cross-sections. The resulting MACS at 30 keV (for Se and Sr) and the experimental values for Te and Ba are shown in table 2. The extrapolation to higher (p-process) temperatures will not be discussed here and can also be found in ref. [17].

4.2.1 ^{74}Se(n,γ)^{75}Se

The ^{74}Se$(n,\gamma)^{75}$Se reaction was analyzed via the two strongest transitions in ^{75}As at 136.0 keV and 264.7 keV (table 3). The capture cross-section derived with the experimental neutron distribution is 281 ± 15 mb. The result for the stellar cross-section is $\langle\sigma\rangle_{30} = 271$ mb, in perfect agreement with the previously estimated value of 267 ± 25 mb from ref. [18].

4.2.2 ^{84}Sr(n,γ)^{85}Sr^{g,m}

In case of ^{84}Sr, neutron captures populate both the ground and isomeric state of ^{85}Sr. While ^{85}Sr^g decays can be identified via the 514 keV transition in ^{85}Rb, the decay of the isomer proceeds mainly via transitions of 232 keV and 151 keV. The isomeric state is 239 keV above the ground state and decays either via a 7 keV–232 keV cascade (internal transition, 86.6%) or directly by electron capture (13.4%) into the 151 keV level of the daughter nucleus.

The partial cross-section to the isomeric state can be easily deduced from the above-mentioned transitions at 151 keV and 232 keV and yields 189 ± 10 mb. The cross-section to the ground state was measured to 112 ± 8 mb, which leads to a total capture cross-section of 301 ± 17 mb.

The result for the total stellar cross-section of ^{84}Sr is $\langle\sigma\rangle_{30} = 300$ mb, 17% lower than the 368 ± 125 mb from ref. [18]. The partial cross-section to the isomer yields $\langle\sigma\rangle_{30}$(part) $= 190$ mb.

4.2.3 ^{120}Te(n,γ)^{121}Te^{g,m}

The Te samples were analyzed via the 576 keV γ line from the β^+ decay of ^{121}Te^g into ^{121}Sb. The partial cross-section of the isomeric state cannot be measured directly after the irradiation due to a huge Compton background around 210 keV, but after a waiting time of 80 d the expected 212 keV line from the IT to the ground state (88.6%) could be determined.

The preliminary result for the neutron capture to the ground state is 390 ± 16 mb, and 61 ± 2 mb for the partial cross-section to the isomeric state. This leads to a (preliminary) total (n,γ) cross-section of 451 ± 18 mb, which again is in good agreement with the estimated 420 ± 103 mb from ref. [18].

4.2.4 ^{130}Ba(n,γ)^{131}Ba

The ^{130}Ba cross-section can be determined via the transitions at 124 keV, 216 keV, 373 keV and 496 keV from the β^+ decay into ^{131}Cs. The resulting experimental cross-section is 694 ± 20 mb, which exhibits a much smaller uncertainty than the 760 ± 110 mb ones from ref. [27], which were derived at a filtered neutron beam.

4.2.5 ^{132}Ba(n,γ)^{133}Ba^{g,m}

The partial cross-section to the 38.9 h isomer in ^{133}Ba was measured via the 276 keV line (IT) to be 33.6 ± 1.7 mb. The total capture cross-section was determined with a Clover detector via the strongest EC decay transition into ^{133}Cs at 356.0 keV. The preliminary result is 368 ± 25 mb, in perfect agreement with the estimated 379 ± 137 mb from ref. [18].

4.3 Comparison with theory

Figure 1 shows a comparison of our experimental total capture cross-sections with $\langle\sigma\rangle_{30}$ values derived with various theoretical models [28,29,30,31,32,24,23,20]. For ^{74}Se and ^{84}Sr the experimental value shown is the MACS derived with the energy dependence of JEFF 3.0 [17,33], whereas the preliminary values for ^{120}Te, ^{130}Ba and ^{132}Ba shown here are only cross-sections derived with the experimental neutron distribution.

In the case of ^{130}Ba our experimental value agrees with the measurement of ref. [27]. In all other cases we find good agreement with the semi-theoretical values of Bao *et al.* [18], which are normalized NON-SMOKER cross-sections accounting for known systematic deficiencies in the nuclear inputs of the calculation.

5 Summary

We have presented the results of an ongoing experimental program to determine more precise p-process reaction

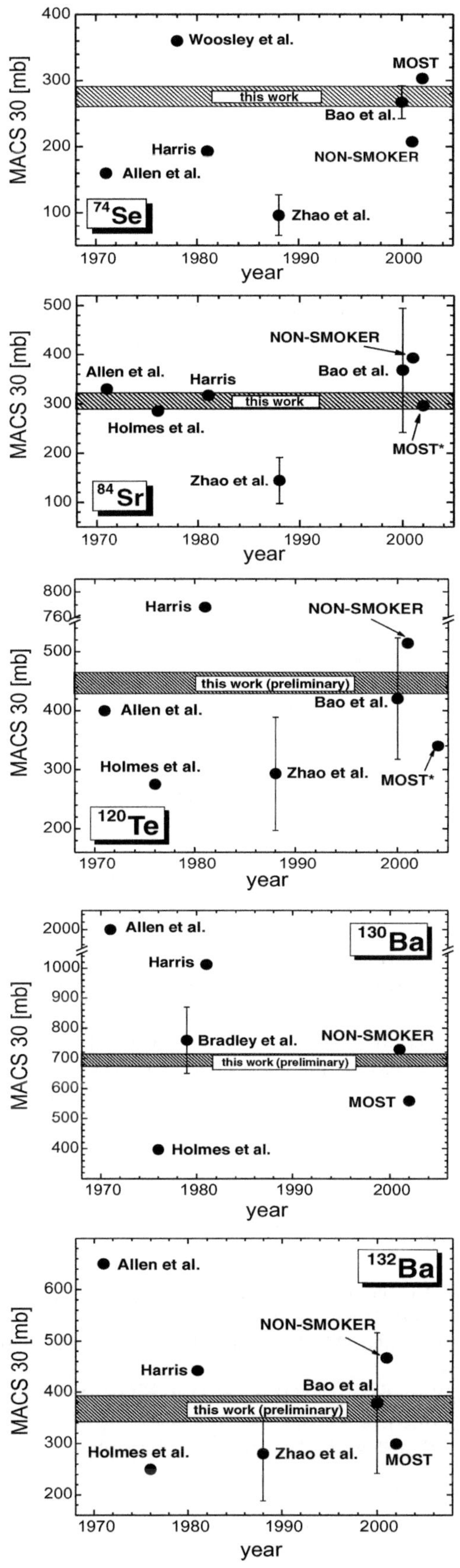

Fig. 1. Comparison of MACS30 predictions. MOST* marks modified values [20].

rates in the mass range $A = 70$–140. The (n,γ) cross-sections of the p nuclei ^{74}Se, ^{84}Sr, ^{120}Te, ^{132}Ba have been measured for the first time, including the partial cross-sections to the isomeric states in ^{85}Sr, ^{121}Te, and ^{133}Ba. A re-measurement of ^{130}Ba yielded a more precise total cross-section compared to the previous value [27].

As can be seen in table 2, experimental Maxwellian averaged cross-sections for ^{98}Ru, ^{102}Pd, ^{138}La, ^{158}Dy, ^{168}Yb, ^{174}Hf, ^{184}Os and ^{196}Hg are still missing. Thus, future efforts should be focussed on these measurements, as well as on an improvement of the accuracy of important isotopes like ^{92}Mo and ^{94}Mo.

6 KADoNiS —The Karlsruhe Astrophysical Database of Nucleosynthesis in Stars

The KADoNiS project is an online database for cross-sections in the s process and p process (`http://nuclear-astrophysics.fzk.de/kadonis/`). Its first part consists of an updated version of the Bao *et al.* compilation [18] for cross-sections relevant to the s process. A test launch of the KADoNiS webpage started in May 2005 with an online version of the original Bao *et al.* paper. By the end of June 2005 the first updated version was online. For the six isotopes $^{128-130}$Xe, ^{147}Pm, ^{151}Sm and ^{180}Ta^m the previously recommended semi-theoretical MACS were replaced by first experimental results. More than 40 isotopes (a list is available online) exhibit new measurements, which were included to re-evaluate the recommended MACS.

The KADoNiS data sheets include all necessary information for the respective (n,γ) reaction (recommended total and partial cross-sections, all available published values with references, energy dependence of the MACS for $5 < kT < 100$ keV, and the respective stellar enhancement factors).

The second part of the KADoNiS is planned to be a collection of experimental p-process reaction rates, including (n,γ), (p,γ), (α,γ) and their respective photodissociation rates. This part of the database has been launched in December 2005.

References

1. E.M. Burbidge, G.R. Burbidge, W.A. Fowler, F. Hoyle, Rev. Mod. Phys. **29**, 547 (1957).
2. K. Langanke, M. Wiescher, Rep. Prog. Phys. **64**, 1657 (2001).
3. S.E. Woosley, W.M. Howard, Astrophys. J. Suppl. Ser. **36**, 285 (1978).
4. S.E. Woosley, W.M. Howard, Astrophys. J. **354**, L21 (1990).
5. M. Rayet, M. Arnould, M. Hashimoto, N. Prantzos, K. Nomoto, Astron. Astrophys. **298**, 517 (1995).
6. T. Rauscher, A. Heger, R.D. Hoffman, S.E. Woosley, Astrophys. J. **576**, 323 (2002).
7. W. Hauser, H. Feshbach, Phys. Rev. **87**, 366 (1952).
8. T. Rauscher, F.-K. Thielemann, H. Oberhummer, Astrophys. J. **451**, L37 (1995).
9. M. Rayet, N. Prantzos, M. Arnould, Astron. Astrophys. **227**, 271 (1990).
10. T. Rauscher, Nucl. Phys. A **758**, 549c (2005).
11. W. Rapp, Report FZKA **6956**, Forschungszentrum Karlsruhe (2004).
12. T. Rauscher, F.-K. Thielemann, At. Data Nucl. Data Tables **75**, 1 (2000).
13. K. Vogt, P. Mohr, M. Babilon, J. Enders, T. Hartmann, C. Hutter, T. Rauscher, S. Volz, A. Zilges, Phys. Rev. C **63**, 055802 (2001).
14. W. Ratynski, F. Käppeler, Phys. Rev. C **37**, 595 (1988).
15. S. Dababneh, N. Patronis, P.A. Assimakopoulos, J. Görres, M. Heil, F. Käppeler, D. Karamanis, S. O'Brien, R. Reifarth, Nucl. Instrum. Methods A **517**, 230 (2004).
16. H. Beer, F. Käppeler, Phys. Rev. C **21**, 534 (1980).
17. I. Dillmann, M. Heil, F. Käppeler, T. Rauscher, F.-K. Thielemann, Phys. Rev. C **73**, 015803 (2006).
18. Z.Y. Bao, H. Beer, F. Käppeler, F. Voss, K. Wissshak, T. Rauscher, At. Data Nucl. Data Tables **76**, 70 (2000).
19. R. Reifarth, M. Heil, F. Käppeler, F. Voss, K. Wisshak, Phys. Rev. C **66**, 064603 (2002).
20. S. Goriely, private communication (2005).
21. E. Anders, N. Grevesse, Geochim. Cosmochim. Acta **53**, 1997 (1989).
22. K. Lodders, Astrophys. J. **591**, 1220 (2003).
23. S. Goriely, Hauser-Feshbach rates for neutron capture reactions (version 09/12/02), `http://www-astro.ulb.ac.be/Html/hfr.html`.
24. T. Rauscher, F.-K. Thielemann, At. Data Nucl. Data Tables **79**, 47 (2001).
25. National Nuclear Data Center, `www.nndc.bnl.gov/nudat2` (2004).
26. K.J.R. Rosman, P.D.P. Taylor, Pure Appl. Chem. **70**, 217 (1998).
27. T. Bradley, Z. Parsa, *Nuclear Cross Sections for Technology*, edited by J.L. Fowler, C.H. Johnson, C.D. Bowman (National Bureau of Standards, Washington D.C., 1979) p. 344.
28. B. Allen, J. Gibbons, R. Macklin, Adv. Nucl. Phys. **4**, 205 (1971).
29. J. Holmes, S.E. Woosley, W. Fowler, B. Zimmerman, At. Data Nucl. Data Tables **18**, 305 (1976).
30. S.E. Woosley, W. Fowler, J. Holmes, B. Zimmerman, At. Data Nucl. Data Tables **22**, 371 (1978).
31. M. Harris, Astrophys. Space Sci. **77**, 357 (1981).
32. Z. Zhao, D. Zhou, D. Cai, *Nuclear Data for Science and Technology*, edited by S. Igarasi (Saikon, Tokyo, 1988) p. 513.
33. NEA, Joint Evaluated Fission and Fusion General Purpose File: JEFF 3.0; online: `www.nea.fr/html/dbdata/eva/evaret.cgi` (2004).

Eur. Phys. J. A **27**, s01, 135–140 (2006)
DOI: 10.1140/epja/i2006-08-019-4

EPJ A direct
electronic only

Photodissociation of p-process nuclei studied by bremsstrahlung-induced activation

M. Erhard[1], A.R. Junghans[1,a], R. Beyer[1], E. Grosse[1,2], J. Klug[1], K. Kosev[1], C. Nair[1], N. Nankov[1], G. Rusev[1], K.D. Schilling[1], R. Schwengner[1], and A. Wagner[1]

[1] Forschungszentrum Rossendorf, Institut für Kern- und Hadronenphysik, Postfach 51 01 19, 01314 Dresden, Germany
[2] TU Dresden, Institut für Kern- und Teilchenphysik, 01062 Dresden, Germany

Received: 14 July 2005 /
Published online: 13 March 2006 –

Abstract. A research program has been started to study experimentally the near-threshold photodissociation of nuclides in the chain of cosmic heavy-element production with bremsstrahlung from the ELBE accelerator. An important prerequisite for such studies is the good knowledge of the bremsstrahlung distribution which was determined by measuring the photodissociation of the deuteron and by comparison with model calculations. First data were obtained for the astrophysically important target nucleus ^{92}Mo by observing the radioactive decay of the nuclides produced by bremsstrahlung irradiation at end-point energies between 11.8 MeV and 14.0 MeV. The results are compared to recent statistical model calculations.

PACS. 25.20.-x Photonuclear reactions – 25.20.Dc Photon absorption and scattering – 26.30.+k Nucleosynthesis in novae, supernovae, and other explosive environments

1 Introduction

The 35 neutron-deficient stable isotopes between Se and Hg that are shielded from the rapid neutron capture process by stable isobars, and that are bypassed by slow neutron captures of the s-process, are called p-process nuclei. They are thought to be produced during supernova explosions through chains of photodissociation reactions on heavy seed nuclei like (γ,n), (γ,p) and (γ,α). In proton-rich scenarios also (p,γ) reactions can occur. The temperatures are in the region of $T = (1\text{–}3)\cdot 10^9$ K. These temperatures need to occur on a short time scale to avoid the nuclei to be eroded by photodissociation reactions into light nuclei in the iron region. For a current review of the p-process see ref. [1]. In many network calculations of the p-process nucleosynthesis, Mo and Ru isotopes are produced with abundances lower than determined experimentally. ^{92}Mo is the second most abundant p-nucleus with a solar-system abundance of 0.378 relative to 10^6 Si atoms. Therefore, it is adequate to test if the photodissociation rates in the region of ^{92}Mo that are part of the nuclear physics input to the network calculations, are correct. We have set up an activation experiment with bremsstrahlung from the new ELBE accelerator at the Forschungszentrum Rossendorf, Dresden, to investigate the photodissociation of ^{92}Mo.

[a] e-mail: a.junghans@fz-rossendorf.de

2 Experimental setup

At the Forschungszentrum Rossendorf, Dresden, a new superconducting electron accelerator named ELBE (for Electron Linear accelerator of high Brilliance and low Emittance) has been built, which combines a high average beam current with a high duty cycle. The accelerator delivers electron beams of energies up to 40 MeV with average currents up to 1 mA for experiments studying photon-induced reactions. The micro-pulse repetition rate of the accelerator can be set between 1.6 MHz and 260 MHz. In addition, a macro-pulse of 0.1 ms to 35 ms with periods of 40 ms to 1 s, respectively, can be applied. The bremsstrahlung facility and the experimental area were designed such that the production of neutrons and the scattering of photons from surrounding materials are minimized [2]. A floorplan of the bremsstrahlung facility at ELBE is shown in fig. 1. The primary electron beam is focussed onto a thin foil made from niobium with various areal densities between 1.7 mg/cm^2 and 10 mg/cm^2 corresponding to $1.6\cdot 10^{-4}$ and $1\cdot 10^{-3}$ radiation lengths, respectively. After the radiator, the electron beam is separated from the photons by a dipole magnet and dumped into a graphite cylinder of 600 mm length and 200 mm diameter. A photo-activation site is located behind the beam dump using available photon fluxes of up to $10^{10}\,\mathrm{cm^{-2}\,s^{-1}\,MeV^{-1}}$. For in-beam studies with bremsstrahlung, a photon beam is formed by a collimator made from high-purity aluminum placed inside the concrete shielding of the accelerator hall. Photons

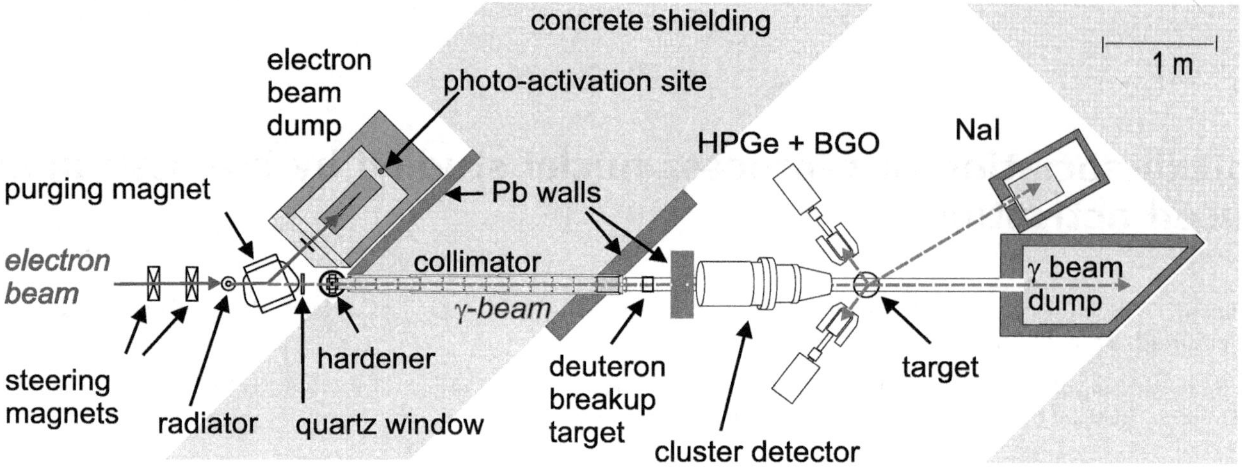

Fig. 1. Bremsstrahlung facility and experimental area for photon-scattering and photodissociation experiments at the ELBE accelerator. ^{197}Au and $H_3{}^{11}BO_3$ samples were irradiated at the target position in the bremsstrahlung cave. ^{nat}Mo and ^{197}Au samples were irradiated together at the photo-activation site.

scattered from a target are observed by means of high-purity germanium (HPGe) detectors. The photon flux at the target position amounts up to $10^8\,\mathrm{cm}^{-2}\,\mathrm{s}^{-1}\,\mathrm{MeV}^{-1}$ in the bremsstrahlung cave. All HPGe detectors are surrounded by escape-suppression shields consisting of bismuth-germanate (BGO) scintillation detectors. In order to determine the intensity, the spectral distribution and the degree of polarisation of the bremsstrahlung, the photodissociation of the deuteron $^2\mathrm{H}(\gamma,\mathrm{p})\mathrm{n}$ can be used. The count rate, the energy distribution and the azimuthal asymmetry of emitted protons is detected by four identical silicon semiconductor detectors placed symmetrically around the beam line at 90° with respect to the beam.

3 Photo-activation measurements

The number of radioactive nuclei $N_{act}(E_0)$ produced in a photo-activation measurement is proportional to the integral of the absolute photon flux $\Phi_\gamma(E,E_0)$ and the photodissociation cross-section $\sigma_{\gamma,x}(E)$ from the reaction threshold energy E_{thr} up to the bremsstrahlung spectrum end-point energy E_0. The symbol $x = n, p, \alpha$ stands for the emitted particle. N_{tar} is the number of the target atoms in the sample:

$$N_{act}(E_0) = N_{tar} \cdot \int_{E_{thr}}^{E_0} \sigma_{\gamma,x}(E) \cdot \Phi_\gamma(E,E_0)\,\mathrm{d}E\,. \quad (1)$$

The number of radioactive nuclei $N_{act}(E_0)$ is determined experimentally after irradiation with bremsstrahlung in a low-level gamma-counting setup using a 100% HPGe detector:

$$N_{act}(E_0) = Y(E_\gamma) \cdot \kappa_{corr} / \big(\varepsilon(E_\gamma) \cdot p(E_\gamma)\big). \quad (2)$$

$Y(E_\gamma), \varepsilon(E_\gamma), p(E_\gamma)$ are the dead-time and pile-up corrected full-energy peak counts of the observed transition, the absolute efficiency of the detector at the energy E_γ and the emission probability of the photon with energy E_γ, respectively. The factor κ_{corr} contains the relation of the detected decays in the measurement time t_{meas} to the number of radioactive nuclei present. Decay losses in the time t_{loss} in-between the bremsstrahlung irradiation and the beginning of the measurement as well as the decay during the irradiation time t_{irr} are taken into account. The symbol τ denotes the mean lifetime of the radioactive nucleus produced during the photo-activation:

$$\kappa_{corr} = \frac{\exp(t_{loss}/\tau)}{1-\exp(-t_{meas}/\tau)} \cdot \frac{t_{irr}/\tau}{1-\exp(-t_{irr}/\tau)}\,. \quad (3)$$

The constancy of the electron beam current and thus of the photo-activation rate was checked by monitoring the electron current both in the injector and in the beam dump. During a typical irradiation time of 8–16 hours there were no electron beam outages.

The setup can be used for photo-activation measurements in the following way: The sample (Mo) is activated in the high photon flux behind the beam dump together with a Au sample to measure an activation standard reaction, *e.g.* $^{197}\mathrm{Au}(\gamma,\mathrm{n})$ (photo-activation site in fig. 1). During the same experiment another Au sample is irradiated at the target position inside the bremsstrahlung cave. There, the absolute photon flux can be determined from the (γ,γ') yield of a sample containing ^{11}B which is irradiated in the same place. In ^{11}B the ground-state transition width of 4 levels is known with sufficient accuracy [3]. This measurement is done during the entire activation period with HPGe detectors. The cross-section of $^{197}\mathrm{Au}(\gamma,\mathrm{n})$ is then renormalized to give the measured activation yield with the absolute photon flux determined experimentally. With the renormalized $^{197}\mathrm{Au}(\gamma,\mathrm{n})$ cross-section and a simulated thick-target bremsstrahlung spectrum, the absolute photon flux at the photo-activation site behind the beam dump can be determined. From the absolute photon flux and the measured activation yield the cross-section normalization for photodissociation of p-process nuclei like ^{92}Mo can be determined.

3.1 Photon spectrum and end-point energy determination

The bremsstrahlung spectrum at the target position can be well described based on the bremsstrahlung cross-section of a thin target. Figure 2 shows theoretical bremsstrahlung cross-sections compared to the evaluation of Seltzer and Berger [4] for a niobium radiator. The calculations were made using formulae given by Schiff [5], Heitler [6], and Roche [7]; the last one being corrected and programmed by Haug [8], who also included an updated screening correction due to Salvat *et al.* [9]. Al-Beteri *et al.* performed a theory-motivated parametrization of experimental data as known in 1988 [10].

All curves shown give results that agree to within 5 percent of each other at the low-energy side of the spectrum, when atomic screening is taken into account. At the high-energy side at about $\simeq$ 1 MeV below the end point the descriptions differ to typically 20 percent. This will influence the activation yield around the reaction threshold, where the yield integral depends strongly on the overlap with the high-energy tail of the photon spectrum. For ^{197}Au(γ, n), *e.g.*, (cross-section calculated with ref. [11]) the yield integral calculated with the Al-Beteri cross-section [10] is about 25 percent higher at 100 keV above E_{thr}, compared to using the formula given by Schiff [5]. At $E_{thr}+1.5$ MeV the uncertainty due to the different photon spectra is below 5 percent.

The electron beam energy and the electron beam energy width determine the high-energy part of the bremsstrahlung spectrum. They need to be known with high precision. An uncertainty of the electron beam energy of $\pm$100 keV at energies around 10 MeV has significant influence on the yield integral. For ^{197}Au(γ, n) this error would change the yield integral by 20(10) percent at end-point energy $E_0 = E_{thr}+1.0(2.0)$ MeV. The data shown in this work were measured at $E_0 > 11.8$ MeV which is several MeV above the respective ^{197}Au(γ, n) and ^{92}Mo(γ, p) reaction thresholds. The electron beam energy was determined through the ion optical setting of the accelerator and the beam line and also online by measuring the deuteron breakup.

To a lesser extent the absolute photon flux in the experiment also depends on the electron beam energy width, which needs to be taken into account in the photon flux determination for reaction yields measured close to the reaction threshold. At ELBE, the energy width of the beam has been measured ion optically to be 60 keV (FWHM) during the measurements discussed here.

We have determined experimentally the end-point energy and also the spectral distribution of the bremsstrahlung by measuring proton spectra from deuteron breakup, shown in fig. 3. For details, see ref. [2]. The target is a thin deuterated polyethylene foil (areal density 4 mg/cm^2).

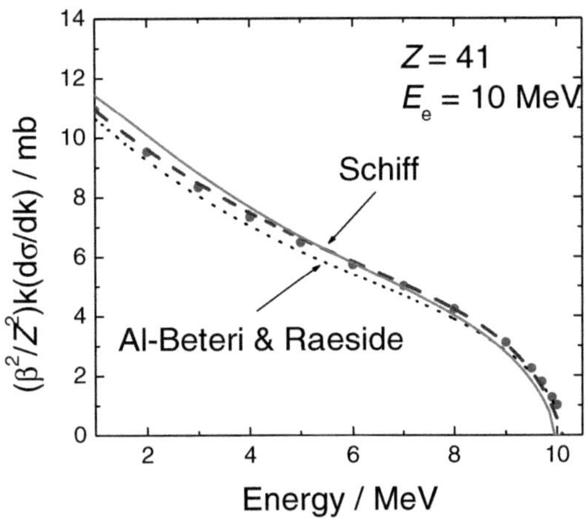

Fig. 2. Theoretical bremsstrahlung cross-sections in comparison with the evaluation of Seltzer and Berger (circles) [4] for a Nb radiator and electron kinetic energy of 10 MeV. The full line is from ref. [7], the dashed line is from ref. [6]. They were calculated using a program from E. Haug [8] that also includes a screening correction according to ref. [9]. The dotted line is from the parametrization [10].

Caused by the extended target size, the Si detectors register protons emitted at different angles causing a kinematic spreading of the proton energy distribution of 150–200 keV. This spread dominates the resolution of the simulated proton spectrum shown in fig. 3. The simulated proton spectra are based on different bremsstrahlung cross-section formulae. They include the energy resolution of the ELBE beam and the passage of the bremsstrahlung through the Al hardener in front of the collimator. Deviations below 2 MeV are due to photon and electron background in the silicon detectors. The uncertainty in the end-point energy determination is $\pm$100 keV. To improve

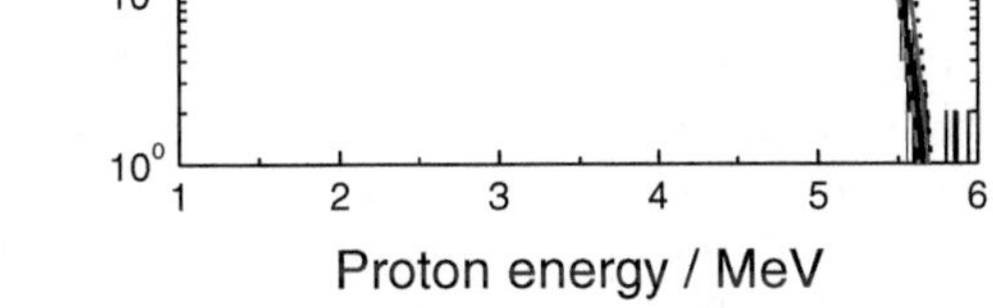

Fig. 3. Proton kinetic-energy spectrum from the photodissociation of a deuterated PE target (histogram) compared with simulated spectra based on the Al-Beteri (dotted), Roche (full), Bethe-Heitler (dashed) formulae; for references, see fig. 2.

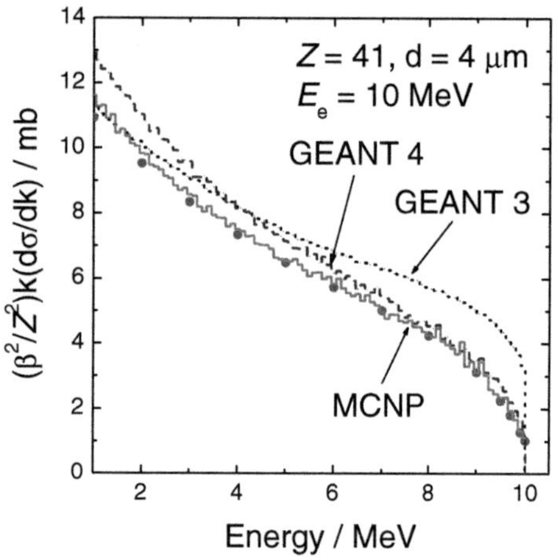

Fig. 4. Monte Carlo simulations [12] of bremsstrahlung spectra in comparison with the evaluation of Seltzer and Berger [4] (circles) for a Nb radiator and electron end-point energy of 10 MeV. The dotted histogram is calculated with GEANT3, the dashed histogram with GEANT4. The full histogram is calculated with MCNP4C2.

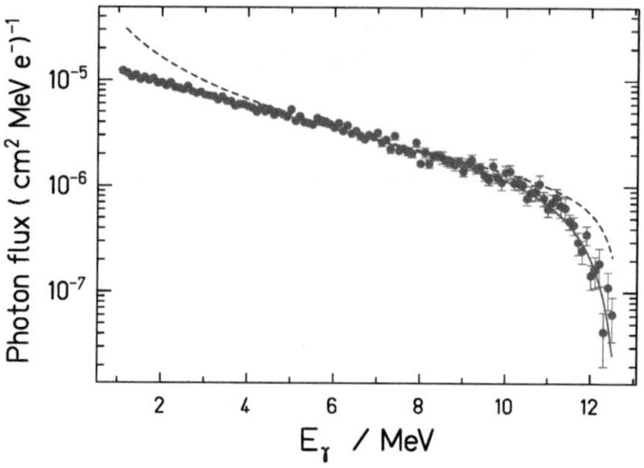

Fig. 5. Bremsstrahlung spectrum at the main photo-activation site behind the graphite beam dump as calculated with MCNP4C2 (circles) for an end-point energy of 12.6 MeV. The dashed line denotes a theoretical thin-target bremsstrahlung spectrum calculated according to Schiff [5]. It is normalized to the MCNP simulation at 6 MeV. The full line is a parametrization of the MCNP simulation.

the accuracy of the measurements we have moved the detectors further away from the target foil to reduce the effect of the reaction kinematics. Figure 3 shows that the spectrum can be well described with the Roche or Al-Beteri cross-sections, whereas the Bethe-Heitler formula is lower, as Coulomb correction terms are not included there.

We have also compared different Monte-Carlo Simulations to the tables of Seltzer and Berger, by making simple simulations of the absolute photon spectrum created in a thin Nb radiator. Figure 4 shows that both versions of GEANT [12] show appreciable differences from the Seltzer and Berger evaluation, while the MCNP4C2 (Monte Carlo N-Particle Transport Code) [12] simulation corresponds very closely to the data. MCNP uses the Seltzer and Berger evaluation [4]. For GEANT3 the CERN Program library long write-up states that Seltzer and Berger is used as well. GEANT4 takes bremsstrahlung cross-section input from the Evaluated Electron Data Library [13]. The simulations from GEANT4 and MCNP agree only at energies above 7 MeV when using 10 MeV electrons. From our study of the deuteron breakup we strongly favour the Roche code [7,8,9] and the concurrent tables of Seltzer and Berger [4], respectively the MCNP4C2 code.

With MCNP4C2 we have calculated the absolute photon flux at the photo-activation site behind the beam dump, see fig. 5. From the comparison with a thin target Schiff spectrum one can see how the shape of a thick target spectrum is changed due to creation of photon-electron cascades and multiple scattering. The thick target spectral shape is required to determine the absolute photon flux for the Mo samples that were irradiated at the photo-activation site with the bremsstrahlung produced in a thick graphite block. Based on a parametrization of the MCNP results the thick-target photon flux, that was used in the analysis of our Mo photo-activation data, was calculated.

3.2 ^{197}Au(γ, n) as activation standard

^{197}Au samples of approximately 200 mg have been irradiated at the target position together with an $H_3{}^{11}BO_3$ sample enriched in ^{11}B to 99.27 percent and mass 2.93 g. Activation measurements were performed at end-point energies of 11.8 MeV up to 14.0 MeV. The absolute photon flux was determined using known transition strengths in ^{11}B at $E_\gamma(\Theta_{lab} = 90°) = 4444$, 5019, 7283 and 8916 keV, respectively. Detectors were positioned at $\Theta_{lab} = 90°$ and $\Theta_{lab} = 127°$. Angular-correlation effects are important for the ^{11}B photon scattering yields as observed especially at 90°. Feeding corrections are estimated to be small for the high-energetic transitions (no feeding for the highest transition) used here, but will be included in the final data analysis.

The number of ^{196}Au nuclei produced during the activation was determined from decay measurements in a low-level counting setup with a 100% HPGe detector. The total efficiency was measured with the help of several calibration sources from Amersham and PTB [14] to 3 percent uncertainty in the energy range 150–2000 keV. A Cd absorber was used to minimize coincidence summing effects with X-rays emitted from the Au samples and some of the calibration sources. The efficiency was checked by GEANT3 simulations that were adjusted to the experimental data to give the efficiency as a function of photon energy. Coincidence summing corrections for the 333 keV and 356 keV decay lines of ^{196}Au were taken into account. The weaker transition at 426 keV into Hg that does not have coincidence summing was included in the analysis. The number of nuclei produced during the activation at the target position was normalized to the number of ^{197}Au target nuclei and to the photon flux at $E_\gamma = 8916$ keV.

The data for ^{197}Au(γ, n) are shown in fig. 6 in comparison with results that were calculated using the absolute photon flux determined experimentally and the theoretical ^{197}Au(γ, n) cross-section from the TALYS program [11]

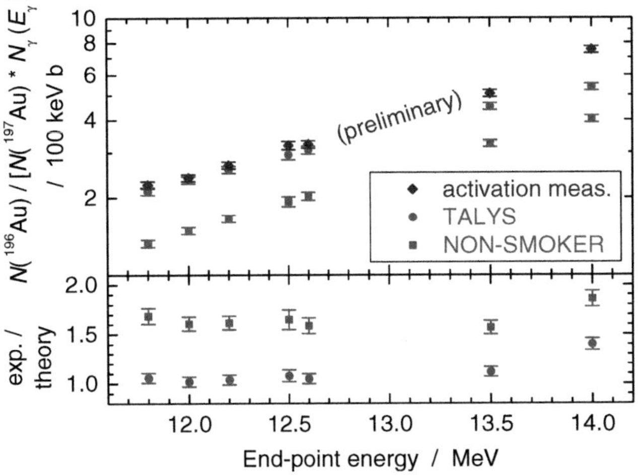

Fig. 6. Preliminary activation yield of $^{197}Au(\gamma, n)$ measured at the target position, see fig. 1. The experimental yield is normalized to the number of ^{197}Au atoms and to the absolute photon flux at the energy $E_\gamma = 8916\,keV$. The data are compared to yield integrals computed with the cross-sections from TALYS and NON-SMOKER using the absolute photon flux determined from known transitions in a sample containing ^{11}B.

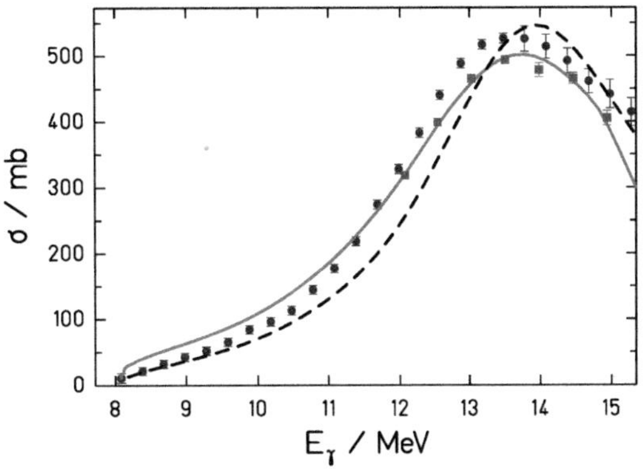

Fig. 7. Measured $^{197}Au(\gamma, n)$ from positron annihilation in flight compared with two model calculations. The squares denote the data from Berman *et al.*, the circles data from Veyssiere *et al.* The dashed line is the prediction from Rauscher and Thielemann [15], whereas the full line is calculated using the TALYS code from Koning *et al.* [11]. For references, see [16].

and from NON-SMOKER [15]. In the range below 13 MeV the theoretical result from TALYS is about 10 percent lower than the experimental values. The combined effect of the systematic uncertainties in the absolute photon flux related to the end-point energy, spectral shape of bremsstrahlung and electron beam energy resolution, as discussed above have not been finally determined yet. About 20 percent uncertainty does seem to be realistic, however. From the data shown in fig. 6 we conclude that the results from the NON-SMOKER code are considerably lower than observed experimentally. The yield integrals calculated with NON-SMOKER to end-point energies from 11.8 MeV to 14 MeV are only 60 to 80 percent of the yield integrals calculated with TALYS.

To investigate the discrepancy of the models, we compare in fig. 7 $^{197}Au(\gamma, n)$ cross-section data obtained with quasi-monoenergetic photons from positron annihilation in flight in comparison with the model calculations. Up to $E_\gamma = 13\,MeV$ the predictions from the NON-SMOKER code are systematically lower. TALYS closely matches the experimental data around the peak region of the GDR. The tails of the GDR are not described well by either model, therefore it is not straightforward to use $^{197}Au(\gamma, n)$ as an activation standard close to the reaction threshold around 8 MeV, as was also realized previously [17].

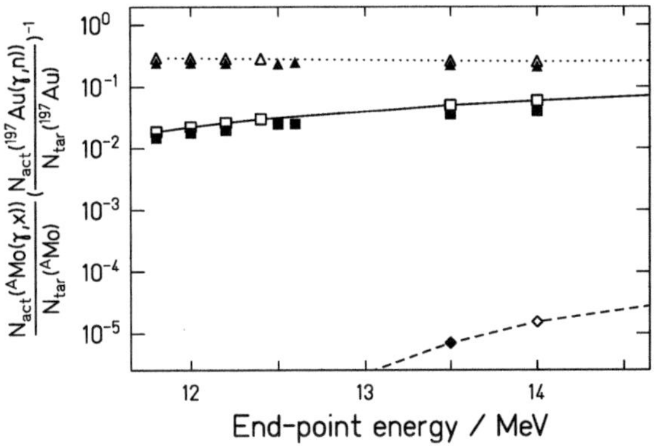

Fig. 8. Measured activation yields for different Mo isotopes at the photo-activation site as a function of the bremsstrahlung end-point energy. The data are normalized to the activation yield from $^{197}Au(\gamma, n)$ irradiated simultaneously. The full symbols denote the experimental yields of $^{100}Mo(\gamma, n)$ (triangles), $^{92}Mo(\gamma, p) + (\gamma, n)$ (squares), and $^{92}Mo(\gamma, \alpha)$ (diamond). The effect of different target masses is taken into account. The open symbols connected with lines to guide the eye represent yield integrals calculated with the photodissociation cross-sections from Rauscher and Thielemann [15]. The yield integrals of the reactions involving the Mo isotopes are divided by the $^{197}Au(\gamma, n)$ yield integral.

3.3 Photo-activation measurements of Mo isotopes

Natural samples of molybdenum (mass 2–4 g, disc diameter 20 mm) have been irradiated together with the Au samples as discussed above. We also did measurements with enriched Mo samples to study the dipole strength around the particle threshold, see ref. [18]. With an enriched ^{92}Mo sample we observed the $^{92}Mo(\gamma, \alpha)$ reaction at the rather low end-point energy of 13.5 MeV. As the absolute normalization of the photon flux at the photo-activation site is still in progress, fig. 8 shows the measured reaction yields relative to the experimental Au reaction yield as calculated in eq. (2). The data are normalized to the different number of target atoms in the samples. The experimental data points are compared with the yield integrals calculated with the simulated thick-target bremsstrahlung spectrum shown in fig. 5 and the NON-SMOKER photodissociation cross-sections. The yield integrals are calculated relative to the ^{197}Au(γ, n) yield integral. The ^{92}Mo(γ, α) data point is taken relative to the ^{92}Mo(γ, p) yield integral measured with the same target.

The data agree on a scale relative to ^{197}Au(γ, n) to typically 20–30 percent with the simulation. One can see from these measurements, that the (γ, p) reaction cross-section for the neutron-deficient isotope ^{92}Mo has about the same size as the (γ, n) and extends to lower energies. The ^{92}Mo(γ, p) reaction cross-section is dominant at energies below 12.6 MeV, as the ^{92}Mo(γ, n) channel is not open yet. The ^{91}Nb nuclei produced were identified by the 1205 keV transition following the β decay into ^{91}Zr. The population of the long-lived ($t_{1/2} = 680\,\mathrm{y}$) ground state in ^{91}Nb that cannot be easily detected in an activation measurement has to be taken into account, *e.g.* by statistical model calculations. With the TALYS code we have calculated that this effect is around 5–10 percent in the energy range measured. The contribution from ^{92}Mo(γ, n) above 12.6 MeV produces shorter-lived activity that can be identified with a pneumatic delivery system that is under development. The ^{100}Mo(γ, n) results will be used for comparison with Coulomb dissociation experiments done at GSI, Darmstadt [19]. The similarity of the relative data shown in fig. 8 as compared to the NON-SMOKER data [15] suggests already that the predicted underproduction of Mo/Ru isotopes might not be due to wrong photodissociation rates. Absolute data currently under analysis will allow to draw a firm conclusion.

4 Conclusion

First photodissociation measurements of the p-process nucleus ^{92}Mo have been performed at the new bremsstrahlung experiment at FZ Rossendorf, Dresden. The activation technique has been used to identify the different reaction products. The photodissociation reactions (γ, n), (γ, p), and (γ, α) have been observed. The bremsstrahlung spectrum has been studied using the photodissociation of the deuteron. This allowed to measure the end-point energy of the electron beam and to compare different formulae for thin-target bremsstrahlung. The absolute photon flux has been measured online by known transitions in ^{11}B and by using the reaction ^{197}Au(γ, n) as an activation standard. Preliminary results indicate that the absolute cross-section of ^{197}Au(γ, n) as calculated by the NON-SMOKER code is lower than the experimental data. This was also found in the energy region below 10 MeV in ref. [17]. The reaction yields for ^{92}Mo(γ, p) + (γ, n) relative to ^{197}Au(γ, n) agree to within 20–30 percent with the model calculations [15]. These reactions contribute to the possible destruction of the p-process nucleus ^{92}Mo.

We thank P. Michel and the ELBE crew for providing stable, high-intensity electron beams for the activation measurements and A. Hartmann and W. Schulze for continuous valuable technical support. We gratefully acknowledge theory discussions and help by H.W. Barz and E. Haug.

References

1. M. Arnould, S. Goriely, Phys. Rep. **384**, 1 (2003).
2. R. Schwengner *et al.*, Nucl. Instrum. Methods A **555**, 211 (2005); K.D. Schilling *et al.*, Institut für Kern- und Hadronenphysik, Forschungszentrum Rossendorf, Annual Report 2004.
3. F. Ajzenberg-Selove, Nucl. Phys. A **506**, 1 (1990).
4. S.M. Seltzer, M.J. Berger, At. Data Nucl. Data Tables **35**, 345 (1986).
5. L.I. Schiff, Phys. Rev. **83**, 252 (1951).
6. W. Heitler, *The Quantum Theory of Radiation* (Dover publications, New York, 1984) p. 242.
7. G. Roche, C. Ducos, J. Proriol, Phys. Rev. A **5**, 2403 (1972).
8. E. Haug, private communication.
9. F. Salvat, J.D. Martinez, R. Mayol, J. Parellada, Phys. Rev. A **36**, 467 (1987).
10. A.A. Al-Beteri, D.E. Raeside, Nucl. Instrum. Methods B **44**, 149 (1989).
11. A.J. Koning, S. Hilaire, M.C. Duijvestin, *Proceedings of the International Conferences on Nuclear Data for Science and Technology, ND2004, Santa Fe, USA, 2004*, AIP Conf. Proc. **769**, 177 (2005).
12. GEANT3: CERN Program Library Long Write-up W5013, GEANT4: `http://cern.ch/geant4`, MCNP4C2: `http://www.nea.fr/abs/html/ccc-0701.html`.
13. D.E. Cullen, S.T. Perkins, S.M. Seltzer, *Tables and Graphs of Electron Interaction Cross Section 10 eV to 100 GeV Derived from the LLNL Evaluated Electron Data Library (EEDL), Z = 1–100*, Lawrence Livermore National Laboratory, UCRL-50400, Vol. **31** (1991).
14. Physikalisch Technische Bundesanstalt, Fachbereich 6.1, Bundesallee 100, Braunschweig, Germany; Amersham: ISOTRAK AEA Technology QSA, Gieselweg 1, Braunschweig, Germany.
15. T. Rauscher, F.-K. Thielemann, At. Data Nucl. Data Tables **88**, 1 (2004).
16. B.L. Berman *et al.*, Phys. Rev. C **36** 1286 (1987); A. Veyssiere *et al.*, Nucl. Phys. A **159**, 561 (1970).
17. K. Vogt *et al.*, Nucl. Phys. A **707**, 241 (2002).
18. E. Grosse, G. Rusev *et al.*, these proceedings.
19. K. Sonnabend *et al.*, these proceedings.

Eur. Phys. J. A **27**, s01, 141–144 (2006)
DOI: 10.1140/epja/i2006-08-020-y

EPJ A direct
electronic only

^{106,108}Cd(p, γ)^{107,109}In cross-sections for the astrophysical p-process

Gy. Gyürky[1,a], G.G. Kiss[1,2], Z. Elekes[1], Zs. Fülöp[1], and E. Somorjai[1]

[1] Institute of Nuclear Research (ATOMKI), P.O. Box 51, H-4001 Debrecen, Hungary
[2] University of Debrecen, Hungary

Received: 26 July 2005 /
Published online: 10 March 2006 –

Abstract. The proton capture cross-sections of the two most proton-rich, stable isotopes of cadmium have been measured for the first time in the energy range relevant to the astrophysical p-process. The ^{106}Cd(p, γ)^{107}In and ^{108}Cd(p, γ)^{109}In cross-sections have been determined using the activation technique. Highly enriched as well as natural Cd targets have been irradiated with proton beams from both the Van de Graaff and Cyclotron accelerators of the ATOMKI. The cross-sections have been derived by measuring the γ-radiation following the β-decay of the ^{107}In and ^{109}In reaction products. The measurements were carried out in the energy range between 2.4 and 4.8 MeV which is the relevant energy region (Gamow window) for the astrophysical p-process. Preliminary results are presented here and are compared with the predictions of the Hauser-Feshbach statistical model calculations using the NON-SMOKER code.

PACS. 25.40.Lw Radiative capture – 26.30.+k Nucleosynthesis in novae, supernovae, and other explosive environments – 26.50.+x Nuclear physics aspects of novae, supernovae, and other explosive environments – 27.60.+j $90 \leq A \leq 149$

1 Introduction

Despite the tremendous experimental and theoretical efforts of recent years, the synthesis of the so called p-nuclei (the heavy, proton-rich isotopes which cannot be synthesized by neutron capture reactions in the s- or r-process) is still one of the least known processes of nucleosynthesis. It is generally accepted that the synthesis of the p-nuclei, the astrophysical p-process, involves mainly γ-induced reactions on abundant seed nuclei produced at earlier stages of nucleosynthesis by the s- (or to a less extent the r-) process [1]. During the p-process flow, material from the bottom of the valley of stability is driven to the proton-rich side by subsequent (γ, n) reactions. As the neutron separation energy increases along this path, charged-particle–emitting (γ, α) and (γ, p) reactions start to play a role contributing significantly to the final abundance distribution of p-nuclei.

The high-energy γ-photons necessary for the γ-induced reactions are available only in explosive nucleosynthetic scenarios. The generally accepted models locate the p-process in the deep O-Ne-rich layers of massive stars either in their pre-supernova or supernova phases where temperatures of a few times 10^9 K are reached.

The comprehensive modeling of the p-process requires, on the one hand, detailed information about the stellar environment (temperature, original seed abundances, burning time scale, etc.). On the other hand, nuclear physics plays also an important role. In the p-process modeling, the reaction rates of the thousands of nuclear reactions involved in nuclear reaction networks must be known. The reaction rates of the dominant γ-induced reactions are generally calculated with Hauser-Feshbach–type statistical models. The rates of γ-induced reactions can be calculated from the inverse capture reactions using the detailed balance theorem if the cross-sections of the capture reactions are known experimentally. While there are compilations for the (n,γ) cross-sections, very few charged-particle–induced reactions above the iron region have been investigated experimentally leaving the statistical model calculations largely untested.

Realizing the need for testing experimentally the statistical model calculations, several (p,γ) and a few (α,γ) reaction cross-sections have been measured in recent years, and the results have been compared with model predictions (see, *e.g.*, [2] and references therein). In general, the models are able to reproduce the experimental results within about a factor of two, however, some larger deviations are also found. The existing experimental database is still not enough to check the reliability of

[a] e-mail: gyurky@atomki.hu

model calculations globally, therefore further experimental data are highly needed.

Recently, the (α,γ) cross-section of ^{106}Cd has been measured in the energy range relevant to the p-process [3]. The results are compared with statistical model predictions using different input parameters, *e.g.* different optical potentials. The $\alpha+$nucleus optical potential is one of the key input parameters for the model calculations. This potential can directly be determined from (α,α) elastic-scattering experiments. Such an experiment is also in progress on ^{106}Cd, preliminary results are already available [4]. To give a complete description of ^{106}Cd, the measurement of its proton capture cross-section is also necessary.

In the present work the proton capture cross-sections of ^{106}Cd and ^{108}Cd (the other p-isotope of cadmium) have been measured. The ^{106,108}Cd$(p,\gamma)^{107,109}$In cross-sections have been measured using the activation technique and the results are compared with the predictions of the statistical model calculations using the NON-SMOKER code [5].

The relevant energy region (the Gamow window) for the investigated reactions is between 2.4 and 4.7 MeV (for a temperature of 3×10^9 K, typical for the p-process). The aim of the present work is to measure the capture cross-sections within the Gamow window. Consequently, the results can be compared with model predictions right at the astrophysically relevant energies.

2 Investigated reactions

Cadmium has 8 stable isotopes with mass numbers 106, 108, 110, 111, 112, 113, 114 and 116. The two lightest isotopes ^{106}Cd and ^{108}Cd are p-isotopes with low natural abundances of 1.25% and 0.89%, respectively. The proton capture of these two isotopes leads to unstable In isotopes (^{107}In and ^{109}In) decaying by β^+-decay or electron capture to ^{107}Cd and ^{109}Cd, respectively. For both isotopes the β-decay is followed by γ-radiation which makes it possible to determine the proton capture cross-section using the activation method. In this method Cd targets are irradiated by a proton beam and the capture cross-section is derived from the off-line measurement of the decay of reaction products.

Table 1 shows the decay parameters of the two In isotopes. Note, that only the strongest γ-radiations following the β-decay of the reaction products are listed.

Owing to the different decay patterns of the two reaction products, it is possible to measure both cross-sections in a single activation experiment if the target contains both ^{106}Cd and ^{108}Cd isotopes. Cd targets of natural isotopic abundance could in principle be appropriate for the cross-section determination. Proton-induced reactions on the heavier Cd isotopes, however, can be disturbing if they also lead to off-line γ-radiation. In the astrophysically relevant low-energy region the cross-sections of the two investigated capture reactions are very low, thus the elimination of any disturbing γ-radiation from the spectra is highly needed. Such disturbing γ-radiation can come, *e.g.*, from ^{110}Cd$(p,\gamma)^{111}$In or ^{113}Cd$(p,n)^{113}$In. Therefore, in order to avoid the disturbing activity produced on heavier Cd isotopes, enriched targets were used in the lower-energy region where the cross-section, and consequently the yield of the studied reactions, is very low.

Table 1. Decay parameters of ^{107}In and ^{109}In isotopes. Only the strongest γ-radiations following the β-decay of the reaction products which were used for the analysis are shown. The data for ^{107}In are taken from [6] and for ^{109}In from [7] with the exception of the ^{109}In half-life which is taken from a more recent work [8].

Product nucleus	Half life	Gamma energy (keV)	Relative intensity per decay (%)
^{107}In	32.4 ± 0.3 min	204.96	47.2 ± 0.3
^{109}In	4.168 ± 0.018 h	203.5	73.5 ± 0.5

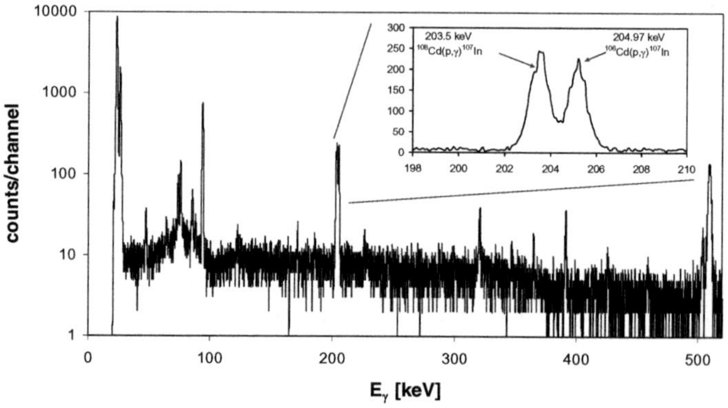

Fig. 1. Typical activation γ-spectrum taken after the irradiation of a natural Cd target with a 3.8 MeV proton beam. This spectrum was taken for 3 hours starting 8 minutes after the end of the irradiation. The inset shows the two resolved γ-peaks from the two reactions studied. The higher energy peaks visible in the spectrum are coming from beam-induced activities on heavier Cd isotopes.

3 Experimental procedure

3.1 Target preparation

The targets were prepared by evaporating natural or highly enriched Cd onto thin ($d=3\,\mu$m) Al foil. The enriched Cd consisted of 96.47% ^{106}Cd and 2.05% ^{108}Cd. Note, that this target material is enriched primarily in ^{106}Cd; however, it also contains more ^{108}Cd than natural Cd (0.89%), and the ratio of heavier Cd isotopes is reduced from 97.86% to 1.48%. The Cd powder was evaporated from a Mo crucible heated by electron bombardment. The Al foil was placed 5 cm above the crucible in a holder defining a circular spot with a diameter of 12 mm on the foil for Cd deposition. This procedure made it possible to determine the target thickness by weighting. The weight of the Al foil was measured before and after evaporation with a precision better than 5 μg and from the

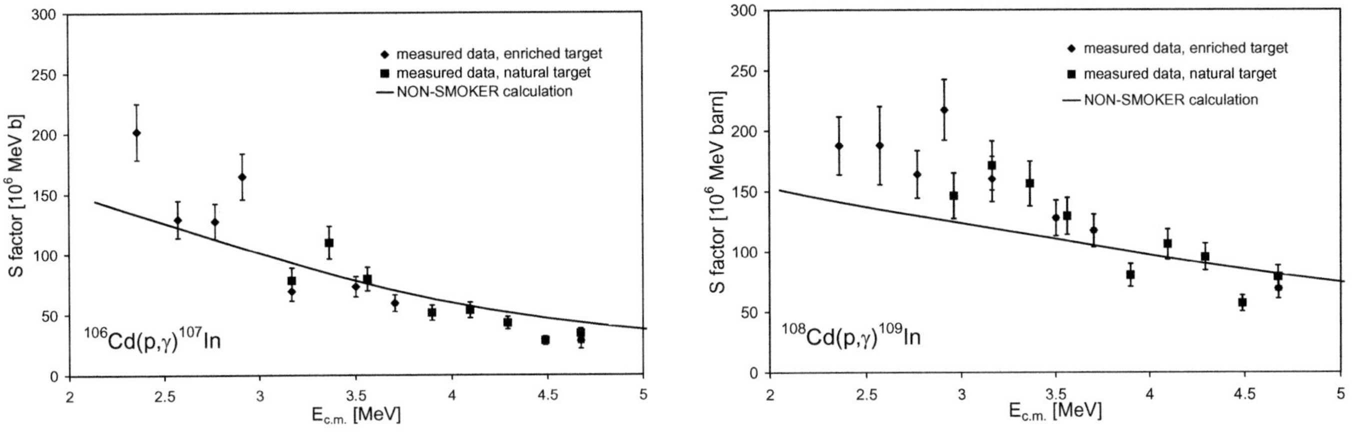

Fig. 2. Astrophysical S-factor of the ^{106}Cd(p, γ)^{107}In (left) and ^{108}Cd(p, γ)^{109}In (right) reactions as a function of the c.m. energy measured in the present work. Measurements carried out using natural and enriched Cd targets are represented by different symbols. The solid line is the statistical model prediction from the NON-SMOKER code [5].

difference the ^{106}Cd number density could be determined. Several natural and enriched targets were prepared with thicknesses varying between 100 and 600 μg/cm^2.

3.2 Irradiations

The energy range from $E_p = 2.4$ to 4.8 MeV was covered in 200 keV steps. The activations have been performed at the Van de Graaff accelerator of ATOMKI (low-energy part $E_p = 2.4$ to 3.6 MeV) and at the MGC cyclotron of ATOMKI (high-energy part $E_p = 3.6$ to 4.8 MeV). Each irradiation lasted about 10 hours and the beam current was restricted to 500 nA in order to avoid target deterioration. The current was kept as stable as possible but to follow the changes the current integrator counts were recorded in multichannel scaling mode in time steps of 10 s. This recorded current integrator spectrum was then used for the analysis solving the differential equation of the population and decay of the reaction products numerically.

A surface barrier detector was built into the chamber at $\Theta = 150°$ relative to the beam direction to detect the backscattered protons and to monitor the target stability. The RBS spectra were taken continuously and the number of counts in the Cd peak was checked regularly during the irradiation. Having the beam current restricted to 500 nA, no target deterioration was found within the 1% precision of the RBS measurement. This is also supported by the weight measurement of the target foils after irradiation.

The beam stop was placed 10 cm behind the target from where no backscattered particles could reach the surface barrier detector. The beam stop was directly water cooled.

In order to find any possible systematic difference between the results obtained at the two accelerators, the $E_p = 3.6$ MeV point has been measured with both accelerators, and perfect agreement was found (see sect. 4). Similarly, the energy points of $E_p = 3.2$ and 4.8 MeV have been measured twice using natural as well as enriched Cd targets. The results with enriched and natural targets are again in good agreement.

3.3 Detection of the induced γ-radiation

The γ-radiation following the β-decay of the produced In isotopes was measured with a HPGe detector of 40% relative efficiency. The target was mounted in a holder at a distance of 3.5 cm from the end of the detector cap. The whole system was shielded by 10 cm thick lead against laboratory background.

The γ-spectra were taken for at least 10 hours and stored regularly in order to follow the decay of the different reaction products.

The absolute efficiency of the detector was measured with calibrated ^{133}Ba, ^{60}Co and ^{152}Eu sources. Monte Carlo simulations have also been carried out to determine the detector efficiency. From the error of the efficiency determination with radioactive sources and from the deviation between the measured and simulated efficiencies, a final error of 9% has been assigned to the detector efficiency.

Figure 1 shows an off-line γ-spectrum taken after irradiation with 3.8 MeV protons in the first 3 h counting interval. The γ lines used for the analysis are indicated by arrows.

The strongest γ-radiations from the decay of the two In isotopes are very close to each other in energy (203.5 and 204.97 keV). However, the energy resolution of the HPGe detector at this low energy is about 0.8 keV (FWHM), hence the two peaks could be resolved (see the inset of fig. 1). Moreover, the different half-lives of the two reaction products make the separation even easier. In the case of enriched targets *e.g.*, the γ-radiation from ^{107}In (the reaction product of the highly enriched ^{106}Cd) is dominant at the beginning of the counting, but owing to its much shorter half-life, it decays out fast making the detection of the weaker ^{109}In decay easier.

Taking into account the detector efficiency and the relative intensity of the emitted gamma rays, coincidence summing effects were for both reactions well below 1% and were neglected.

4 Results and conclusion

The cross-sections of both investigated reactions have been determined in the astrophysically relevant energy range. The final analysis of the experimental data is still in progress, preliminary results are presented here. The measured cross-sections cover more than 3 orders of magnitude from about 3 to 5000 μb. Figure 2 shows the experimental results in terms of the astrophysical S-factor. The data points measured using natural or enriched targets are plotted with different symbols. The good agreement for the points measured both with natural and enriched targets at energies $E_p = 3.2$ and 4.8 MeV is clearly visible. Owing to the different energy calibration of the two accelerators (which was done after the measurement) the 3.6 MeV points which were measured with both accelerators do not coincide perfectly in energy. However, it is clearly visible that there is no systematic difference between the two points. The predictions of the Hauser-Feshbach statistical model code NON-SMOKER are also plotted for comparison. For both isotopes the model is able to reproduce fairly well the experimental result; however at low energies the model seems to underpredict the experimental results in the case of ^{106}Cd(p, γ)^{107}In and to a smaller extent also in the case of ^{108}Cd(p, γ)^{109}In. This might be the consequence of the slope of the theoretical curve which seems to be higher than what the experimental points show. It is also instructive to examine the dependence of the model predictions on the choice of input parameters such as nuclear level densities and optical model potentials. This work is still in progress and is beyond the scope of the present paper.

This work was supported by OTKA (T42733, T49245, F43408, D48283). Zs. Fülöp is a Bolyai fellow.

References

1. M. Arnould, S. Goriely, Phys. Rep. **384**, 1 (2003).
2. Gy. Gyürky *et al.*, Phys. Rev. C **68**, 055803 (2003).
3. Gy. Gyürky *et al.*, Nucl. Phys. A **758**, 517 (2005) (preliminary results).
4. G.G. Kiss *et al.*, these proceedings.
5. T. Rauscher, F.K. Thielemann, At. Data Nucl. Data Tables **79**, 47 (2001).
6. J. Blachot, Nucl. Data Sheets **89**, 213 (2000).
7. D. De Frenne, E. Jacobs, Nucl. Data Sheets **89**, 481 (2000).
8. Gy. Gyürky *et al.*, Phys. Rev. C **71**, 057302 (2005).

Eur. Phys. J. A **27**, s01, 145–148 (2006)
DOI: 10.1140/epja/i2006-08-021-x

EPJ A direct
electronic only

A study of alpha capture cross-sections of ^{112}Sn

N. Özkan[1,a], G. Efe[1], R.T. Güray[1], A. Palumbo[2], M. Wiescher[2], J. Görres[2], H.-Y. Lee[2], Gy. Gyürky[3], E. Somorjai[3], and Zs. Fülöp[3]

[1] Kocaeli University, Department of Physics, 41380 Umuttepe, Kocaeli, Turkey
[2] University of Notre Dame, Notre Dame, IN, 46556 USA
[3] Institute of Nuclear Research (ATOMKI) P.O. Box 51, H-4001 Debrecen, Hungary

Received: 1 August 2005 /
Published online: 10 March 2006 –

Abstract. The ^{112}Sn$(\alpha,\gamma)^{116}$Te reaction cross-section has been measured to test the applicability of statistical models, especially NON-SMOKER in the energy range of importance for the astrophysical p-process nucleosynthesis. The measurements were carried out at the Notre Dame FN Tandem Van de Graaff accelerator by means of the activation method. Enriched self-supporting foils were irradiated with alpha beams over the alpha bombarding energy range of 8 MeV to 12 MeV in 0.5 MeV steps. The induced activity was measured with a pair of large volume Ge Clover detectors in close geometry to maximize the detection efficiency. The preliminary results are compared with recent statistical model predictions using the code NON-SMOKER.

PACS. 25.40.Lw Radiative capture – 26.30.+k Nucleosynthesis in novae, supernovae, and other explosive environments – 82.20.Pm Rate constants, reaction cross sections, and activation energies

1 Introduction

The elements heavier than iron are mainly synthesized by mechanisms referred to as the slow neutron capture (s-process) and the rapid neutron capture (r-process). An additional mechanism, the p-process, is responsible for the production of the observed rare abundances of the proton-rich stable nuclides in the mass range $A \geq 74$. These so-called p-nuclei are shielded by stable nuclei from the production via the r- and s-process.

Early theories exist for the production of the p-nuclei. One of the earliest was proposed by B^{2}FH [1] who suggested that the p-process was a series of (p,γ) reactions on r- and s-process seeds. It was also suggested by Ito that the light p-nuclei could be synthesized at a high-temperature condition ($T_9 \sim 3$) from seeds by (p,γ) reactions while the heavy p-nuclei could be produced at a lower temperature ($T_9 \sim 2.5$) by (γ,n) reactions [2,3].

Current theories include synthesis from the destruction of pre-existing s- or r-nuclides by different combinations of (p,γ) captures, (γ,n), (γ,p) or (γ,α) reactions. Generally, p-nuclei are thought to be synthesized from an unstable progenitor through a chain of β-decays [4].

In the thermonuclear model, the p-process proceeds via a series of photodisintegration reactions, (γ,n), (γ,p) and (γ,α) on an existing heavy s- and r-seed in the temperature range 2–3 $\times 10^9$ K. The nuclides are driven by subsequent (γ,n) reactions to the neutron-deficient side where (γ,n) and (n,γ) reactions are in equilibrium. The reaction path is then dominated by the (γ,p) and/or (γ,α) reactions [5,6]. It is for this reason that radiative alpha capture reactions are of particular importance to the p-process.

Possible astrophysical scenarios for the p-process under consideration are massive stars in their pre-supernova or Type-II supernova phases, as well as their Type-Ib/Ic stages, and white dwarfs exploding as Type-Ia supernova in binary systems. Detailed p-process scenarios have been reviewed by Arnould and Goriely [7].

In comparison to the s- and r-, not much study is devoted to the p-process even though many discrepancies exist. In particular, the theoretical predictions of p-nuclei are frequently inconsistent with their observed abundances [7]. These abundance predictions are based on complex network calculations performed on thousands of nuclear reactions involving stable as well as unstable proton-rich nuclei within a given astrophysical environment. However, only very few of the associated reaction rates in the region of interest have been measured. The relevant reactions that have been investigated in the mass region beyond $A = 100$ include (α,γ) reactions on targets of ^{106}Cd [8], ^{112}Sn [9], ^{139}La [10], and ^{144}Sm [11]. Due to the lack of experimental data, p-process studies are based mostly on Hauser-Feshbach statistical models to predict the reaction rates. Therefore, it is crucial to determine

[a] e-mail: nozkan@kou.edu.tr

the cross-sections experimentally in order to test the reliability of the input parameters such as optical model potentials and nuclear level densities.

The primary aim of this paper is to extend the experimental database for simulation of the p-process by measuring the (α, γ) cross-sections on the p-nucleus ^{112}Sn. The photon-induced reaction cross-sections $^{116}\mathrm{Te}(\gamma, \alpha)^{112}\mathrm{Sn}$ ($Q = 0.928$ MeV) can also be calculated by using the detailed balance theorem. In order to test the applicability of statistical models, the deduced cross-section measurements are compared with one of the statistical models, NON-SMOKER [12]. A discussion of experimental procedure and a presentation of preliminary results follows.

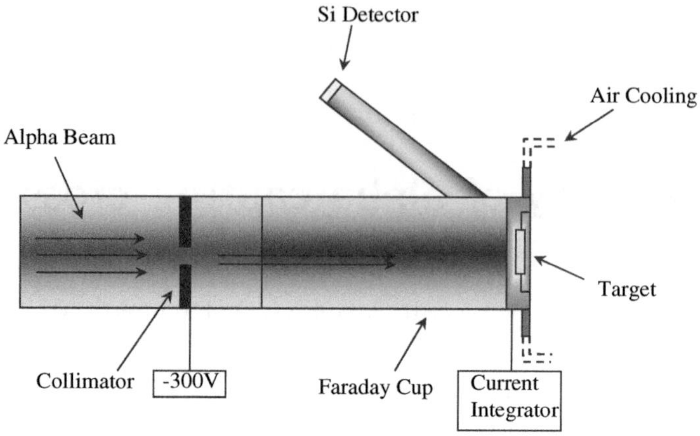

Fig. 1. A drawing of the components used in the beam line during the irradiation. The target was placed at the end of the beam line.

2 Experimental technique

The cross-section of $^{112}\mathrm{Sn}(\alpha, \gamma)^{116}\mathrm{Te}$ has been measured via the activation technique at the FN Tandem Van de Graaff accelerator at the University of Notre Dame, USA in the energy range 8–12 MeV. These energies are particularly interesting since they are relevant to the Gamow window for the high-temperature environments.

The activation method involves bombarding a stable isotope with projectiles to produce a radioactive species and the measurement of the residual radioactivity of the produced isotopes after the irradiation is stopped. This method provides a total cross-section for primary γ-ray transitions to particle bound states (which lead to the formation of the radioactive species in its ground state).

2.1 Targets

Isotopically enriched ^{112}Sn targets in the form of thin self-supporting foils of 2.2 mg/cm^2 were used for the measurements. The highly enriched (99.60 %) ^{112}Sn targets were prepared at Argonne National Laboratory via mechanical rolling. Metal ingots of enriched Sn are placed in a stainless steel pack in a rolling mill and initially rolled very slowly to prevent cracks. They are subsequently rotated by 90° each time while slowly tightening the plate gap until a uniform thickness is achieved [13].

2.2 Irradiation of the target

^{112}Sn targets were irradiated with an alpha beam in the energy range between 8 and 12 MeV in energy steps of 0.5 MeV.

A diagram of the experimental setup for the target irradiation is shown in fig. 1. In order to get an accurate measurement of the total number of charged particles hitting the target during the irradiation runs, the entire target chamber was designed as a Faraday cup (which was isolated from the rest of the beam line). The beam current was recorded in real time with a current integrator in time steps of 32 s, allowing fluctuations in the beam to be monitored. Throughout the irradiations with different α-beam energies, the typical current recorded was between 100 and 250 nA.

Electric suppression was employed to suppress the secondary electron emission from the target by a bias voltage of -300 V. Due to the relatively low melting point of ^{112}Sn and the thick targets, the target holder was air cooled during the irradiation to prevent target degradation. In addition, the target stability was also monitored online by detecting the backscattered alphas from the target using a collimated Si surface barrier detector at 135 degrees. The Si detector was calibrated using a mixed alpha source.

The length of irradiation was chosen based on the half-life of the (α, γ) activation product. The typical irradiation time was $\sim$ 8 h ($3 \times$ half-life) for low-energy measurements due to the steeply decreasing cross-sections at low beam energies.

2.3 Measurement of the residual radioactivity

After each irradiation, the target was taken to a remote low-background counting area. The counting setup was constructed in an area isolated from the target room so the detectors would not be saturated by the inevitable large flux of radiation produced during irradiation. Another advantage of having a separate counting area is that we were able to obtain a larger solid angle by utilizing a close geometry for the target and the detectors.

The detection system was composed of two Clover Ge detectors (Clover 1 and Clover 2), each detector having four HPGe crystals. Figure 2 shows the arrangement of detectors and the irradiated sample in the measurement. The two Clovers were utilized as a single detection unit, which is said to be operated in "direct" mode as described by ref. [14]. This implies that the photopeak detection efficiency of the system was the sum of each photopeak efficiency of one crystal (8 crystals in all). The nearly 4π detection geometry offers relatively high efficiency enabling the detection of low-energy gamma peaks at low

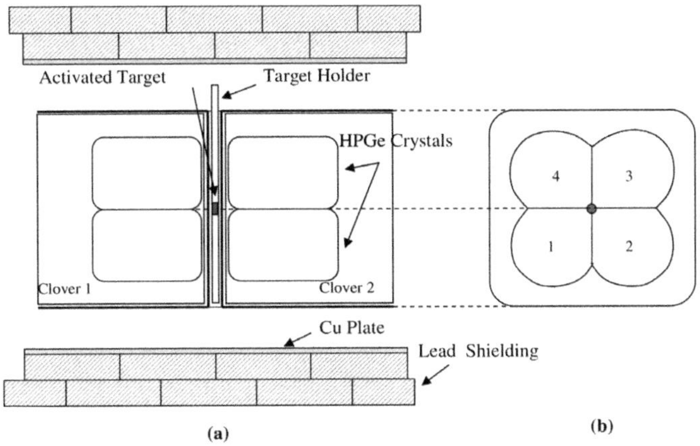

Fig. 2. The scheme of experimental setup used to measure the induced γ-ray activity. (a) Low-background counting area ensured by 4π lead bricks and Cu plates. (b) Target position viewed from the front window of the Clover 2.

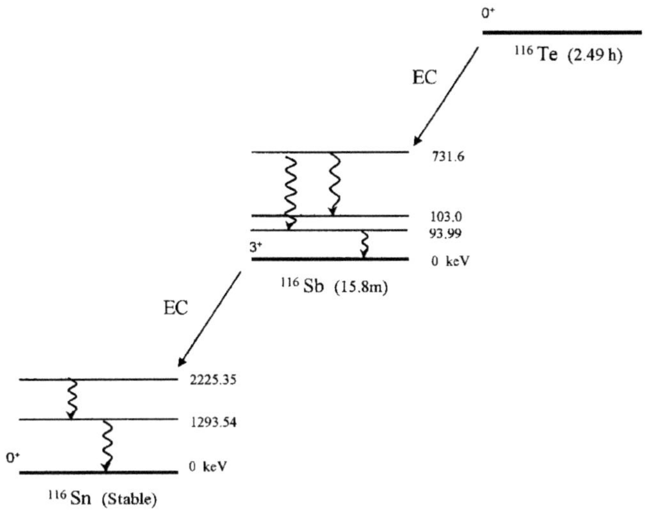

Fig. 3. A simple decay scheme of the residual nucleus, ^{116}Te; energies given in keV [15].

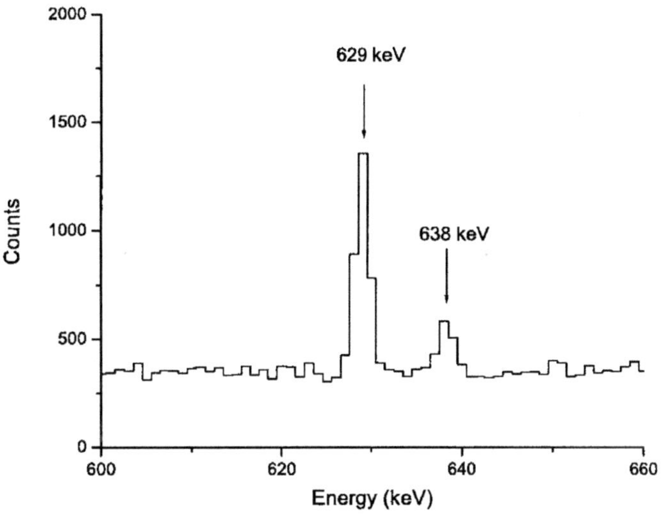

Fig. 4. The γ-ray spectrum in the relevant energy region obtained after activation with α beams of 12 MeV. The γ lines from the β^+-decay of the daughter nucleus ^{116}Te ($t_{1/2}$ = 2.49 h) used for cross-section measurements are indicated by arrows.

irradiation energies. Due to this geometry, corrections for angular-distribution effects were negligible.

In order to measure the induced γ-ray activity, the Clover Ge detectors were placed face to face in close geometry. The distance between the ends of the detectors was fixed at 4.9 mm for reproducibility of the counting geometry. To reduce the room background, the detectors were shielded with 5 cm of Pb and an inner Cu lining of 3 mm. After each irradiation, the activated target was placed at the center of the detectors and positioned at their common axis as shown in fig. 2. Depending on the count rate of the targets, the decays were observed in a time interval from 20 m to 8 h ($t_{1/2}$ = 2.49 h). For the dead-time correction, the output signal of a fixed-frequency (100 Hz) pulse generator was also fed into the electronics.

3 Experimental results

We were able to measure cross-sections from daughter and granddaughter decays since both the daughter nucleus ^{116}Te ($t_{1/2}$ = 2.49 h) and the granddaughter nucleus ^{116}Sb ($t_{1/2}$ = 15.8 m) are radioactive as seen in fig. 3. The γ-ray spectrum for the ^{112}Sn$(\alpha,\gamma)^{116}$Te reaction obtained at 12 MeV α-beam irradiation is presented in figs. 4 and 5, as an example of the γ-ray activities accumulated during and after termination of the activation measurements. The characteristic γ-transitions of 629 keV and 638 keV in ^{116}Sb populated through the β^+-decay of the reaction product ^{116}Te were used to measure the cross-section in the alpha-beam energy range 9.5–12.0 MeV, (fig. 4). For cross-section measurements at lower alpha-beam energies, we used two additional γ-decay transitions, 932 keV and 1294 keV, in ^{116}Sn populated by the ^{116}Sb secondary β^+-decay (fig. 5).

The total error of the measured cross-section values includes two components: statistical error based on counting statistics ($\leq$ 20%) and systematical error based on the errors in the procedural techniques such as the error in the efficiency measurement (10%), the error in the beam current integration (2%), and target thickness (5%). These components were added in quadrature.

Good agreement was obtained for the cross-sections resulting from the analysis of these four γ-decay transitions. These cross-sections are shown as a function of center-of-mass energy in fig. 6 in comparison with the predictions of NON-SMOKER calculations [12] (solid line). While good agreement can be observed at higher energies, the experimental data deviate considerably in the lower-energy range from the theoretical predictions.

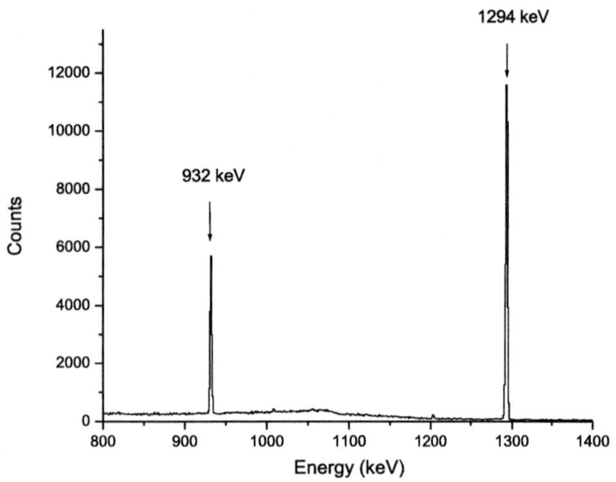

Fig. 5. The γ-ray spectrum in the relevant energy region obtained after activation with α beams of 12 MeV. The γ lines from the β^+-decay of the granddaughter nucleus ^{116}Sb ($t_{1/2} = 15.8$ m) used for cross-section measurements are also indicated by arrows.

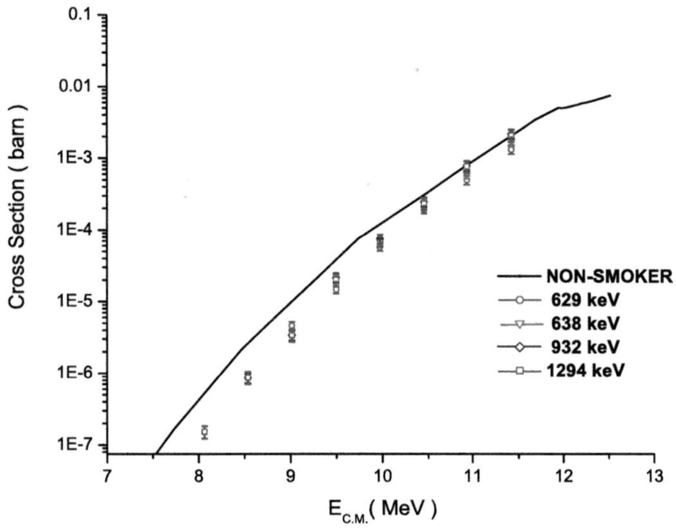

Fig. 6. The cross-sections of α-induced reactions on ^{112}Sn as a function of center-of-mass energy. The experimental results (individual points) as well as the predictions of NON-SMOKER calculations are shown [12].

In this work, the cross-sections for $^{112}\mathrm{Sn}(\alpha,\gamma)^{116}\mathrm{Te}$ have been measured in an energy range directly relevant to nuclear astrophysics. This energy range spans the Gamow window predicted for this reaction in the high-temperature environments; the Gamow window ranges between 6.8 MeV and 10.2 MeV for $T = 3 \times 10^9$ K. Further analysis of these data is in progress, but the present preliminary results are promising. These results will be compared with the statistical model calculations using different input parameters.

It is a pleasure to thank John P. Greene from Argonne National Laboratory for preparing the ^{112}Sn targets for us. This work was supported by The Scientific and Technical Research Council of Turkey TUBITAK - Grant TBAG-U/111 (104T2467) [16], the National Science Foundation NSF - Grant 0434844 and the Joint Institute for Nuclear Astrophysics JINA, PHY02-16783 [17], and The Hungarian Scientific Research Fund Programs OTKA - Grant T 42733, T 49245, F 43408, D 48283. Zs.F. is a Bolyai fellow.

References

1. E.M. Burbidge, G.R. Burbidge, W.A. Fowler, F. Hoyle, Rev. Mod. Phys. **29**, 547 (1957).
2. K. Ito, Prog. Theor. Phys. **26**, 990 (1961).
3. D.L. Lambert, Astron. Astrophys. Rev. **3**, 201 (1992).
4. R.N. Boyd, *Heavy Elements and Related New Phenomena*, edited by W. Greiner, R.K. Gupta (World Scientific, 1999) p. 893.
5. S.E. Woosley, W.M. Howard, Astrophys. J. Suppl. **36**, 285 (1978).
6. M. Rayet, Astron. Astrophys. **227**, 271 (1990).
7. M. Arnould, S. Goriely, Phys. Rep. **384**, 1 (2003).
8. Gy. Gyürky *et al.*, Nucl. Phys. A **758**, 517c (2005) (preliminary results).
9. N. Özkan, A.St.J. Murphy, R.N. Boyd, A.L. Cole, M. Famiano, R.T. Güray, M. Howard, L. Sahin, J.J. Zack, R. deHaan, J. Görres, M.C. Wiescher, M.S. Islam, T. Rauscher, Nucl. Phys. A **710**, 469 (2002).
10. E.V. Verdieck, J.M. Miller, Phys. Rev. **152**, 1253 (1967).
11. E. Somorjai *et al.*, Astron. Astrophys. **333**, 1112 (1998).
12. T. Rauscher, F.K. Thielemann, At. Data Nucl. Data Tables **79**, 427 (2001).
13. J.P. Greene, private communication.
14. S. Dababneh, N. Patronis, P.A. Assimakopoulos, J. Görres, M. Heil, F. Käppeler, D. Karamanis, S. O'Brien, R. Reifarth, Nucl. Instrum. Methods A **517**, 230 (2004).
15. R. Firestone, in *Table of Isotopes*, edited by V. Shirley (Wiley, New York, 1996).
16. `http://www.tubitak.gov.tr`.
17. `http://www.JINAweb.org`.

Eur. Phys. J. A **27**, s01, 149–152 (2006)
DOI: 10.1140/epja/i2006-08-022-9

EPJ A direct
electronic only

Photodissociation of neutron deficient nuclei

K. Sonnabend[a], M. Babilon, J. Hasper, S. Müller, M. Zarza, and A. Zilges

Institut für Kernphysik, TU Darmstadt, Schlossgartenstr. 9, D-64289 Darmstadt, Germany

Received: 20 June 2005 /
Published online: 7 March 2006 –

Abstract. The knowledge of the cross sections for photodissociation reactions like *e.g.* (γ, n) of neutron deficient nuclei is of crucial interest for network calculations predicting the abundances of the so-called p nuclei. However, only single cross sections have been measured up to now, *i.e.*, one has to rely nearly fully on theoretical predictions. While the cross sections of stable isotopes are accessible by experiments using real photons, the bulk of the involved reactions starts from unstable nuclei. Coulomb dissociation (CD) experiments in inverse kinematics might be a key to expand the experimental database for p-process network calculations. The approach to test the accuracy of the CD method is explained.

PACS. 25.20.-x Photonuclear reactions – 26.30.+k Nucleosynthesis in novae, supernovae and other explosive environments

1 Introduction

Most of the elements heavier than iron are produced by a sequence of neutron capture processes and β decays. The s-process takes place during stellar burning phases and is characterized by low neutron densities ($n_n \approx 2$–$4 \cdot 10^8\,\mathrm{cm}^{-3}$) and temperatures ($T \approx 1$–$3 \cdot 10^8\,\mathrm{K}$) [1]. Thus, the so-called s-process path is close to the valley of stability. In contrast, the r-process deals with high neutron densities ($n_n > 10^{20}\,\mathrm{cm}^{-3}$) and temperatures ($T \approx 2$–$3 \cdot 10^9\,\mathrm{K}$) and is thought to occur in explosive scenarios like, *e.g.*, supernovae [2,3].

However, several proton-rich isotopes between Se and Hg remain that cannot be produced during either of the processes. A complete list of these so-called p nuclei can be found in [4,5]. These nuclides are also produced during explosive events lasting a few seconds at temperatures of about 2–$3 \cdot 10^9\,\mathrm{K}$. The lighter p nuclei are thought to be produced by proton capture reactions during the so-called rp-process [6] while the heavier ones are created from s- and r-process seed nuclei by photodissociation processes like (γ, n), (γ, p), and (γ, α) reactions in a process sometimes referred to as γ-process [5].

During the γ-process the (γ, n) reactions compete with the (γ, p) and (γ, α) processes if the p and α separation energies of the produced proton-rich isotopes are low enough.

The special features corresponding to p-process nucleosynthesis in the $A \approx 100$ mass region are discussed in sect. 2. In sect. 3 the basics of Coulomb Dissociation experiments and the SIS/FRS/LAND setup at GSI Darmstadt are explained.

[a] e-mail: kerstin@ikp.tu-darmstadt.de

2 The A $\approx$ 100 mass region

The three most abundant p isotopes (^{92}Mo: 14.84%, ^{94}Mo: 9.25%, and ^{96}Ru: 5.52% natural elemental abundance) are found in the mass region $A \approx 100$. This region is the borderline between the rp- and γ-process, *i.e.*, it is not sure how the p nuclei in this region are produced: in one of the two processes or in both simultaneously. Network calculations including only the γ-process fail to reproduce the observed abundances by a factor of 20 [5], thus, leading to the assumption that both processes are responsible for the production of the Mo and Ru p nuclei.

However, the difficulties in reproducing the observed abundances might also stem from the nuclear physics part: most of the reaction rates being involved in the network calculations (compare fig. 1) are calculated by Hauser-Feshbach statistical model calculations (*e.g.*, [5]) due to the lack of experimental data in the astrophysically relevant energy region. The typical uncertainties of Hauser-Feshbach based calculations are about 20–30% (*e.g.*, [8]). However, if proton-rich nuclei in the vicinity of closed shells are concerned, it is not clear whether the involved level densities are high enough to legitimate this statistical approach. Thus, an experimental examination of the predicted reaction rates is highly desirable.

Different approaches are available and necessary to improve the experimental data base for the γ-process. While the (γ, n) cross sections in the energy regime of the Giant Dipole Resonance around 15 MeV have already been measured extensively several decades ago (see, *e.g.*, [9]), the knowledge about the astrophysically relevant energy region close above the n separation energy is rather scarce.

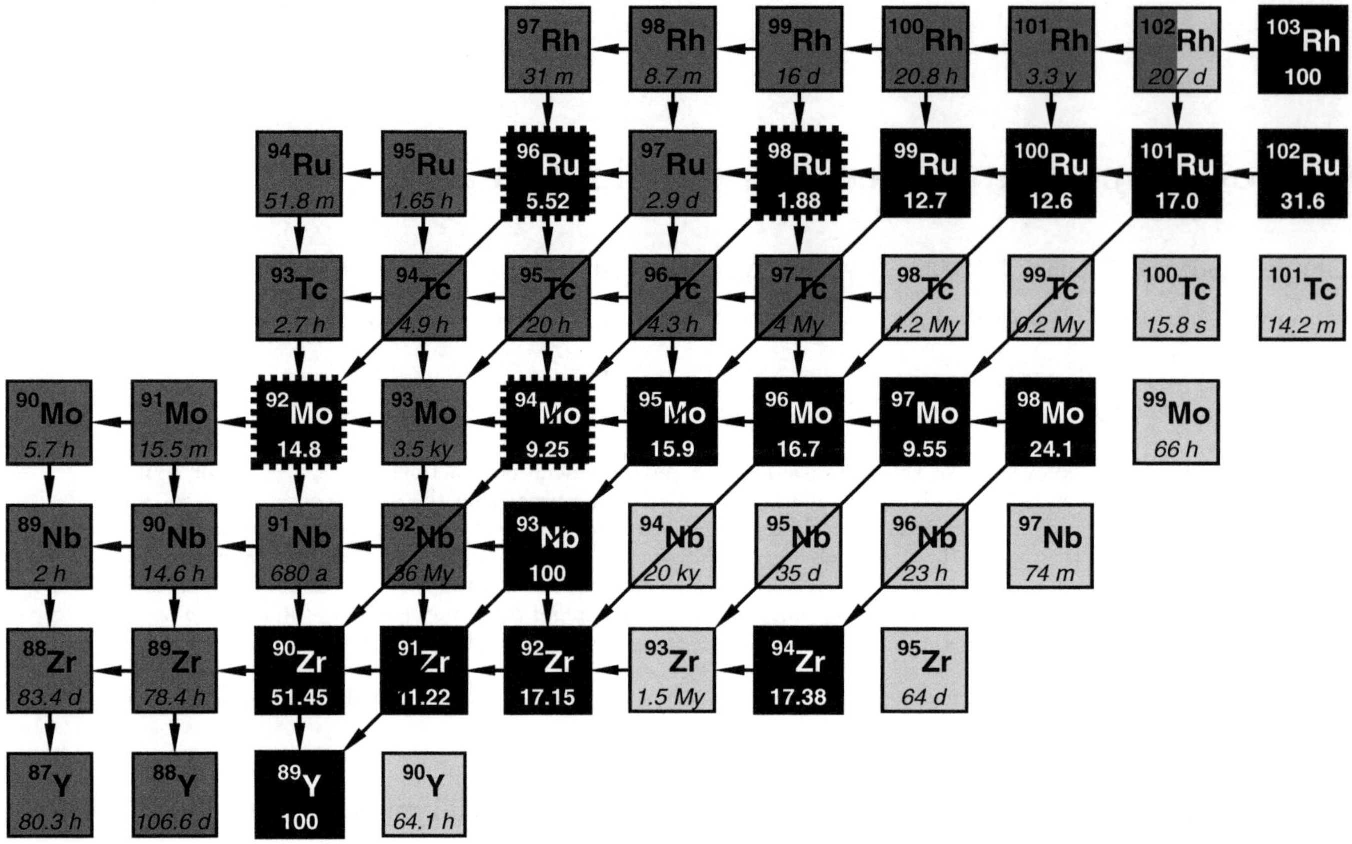

Fig. 1. Overview of the reactions involved in *p*-process nucleosynthesis in the mass region $A \approx 100$. The stable isotopes are printed in black boxes with their natural isotopic abundance. The *p* nuclei ^{92}Mo, ^{94}Mo, ^{96}Ru, and ^{98}Ru are marked with an indented margin. β^--unstable isotopes are printed in light grey, the dark grey boxes stand for β^+, ϵ-unstable isotopes both with their half-life under lab conditions. The relevance of the indicated (γ, n), (γ, p), and (γ, α) reactions in a *p*-process network was calculated using ref. [7].

Some efforts using continuous bremsstrahlung spectra have been made at the S-DALINAC at Darmstadt [10,11] and the ELBE setup at Rossendorf [12,13] to determine the reaction rates without any assumptions on the shape of the cross section's energy dependence. A determination of the reaction rates by an absolute cross section measurement is possible using monoenergetic photon beams produced by Laser Compton Backscattering [14].

However, both methods are limited to stable target nuclei. Here, Coulomb dissociation (CD) of fast radioactive beams in the Coulomb field of a high-Z target nucleus using virtual photons is a viable approach to measure the (γ, n) cross sections indirectly. Such experiments can be performed with the SIS/FRS/LAND facility at GSI [15] as described in the following section.

The experimental knowledge about the (γ, p) and (γ, α) reactions in the corresponding Gamow window is even worse. In fact, the experimental data is based on the observation of the time reversal (p, γ) and (α, γ) cross sections, respectively [16,17,18,19,20] for the proton-rich nuclei with mass numbers around 100. Due to the difficulties concerning the experimental accessibility of the (γ, α) reaction rates, a method using elastic α scattering has been established [21,22].

Likewise in the case of these charged particle reactions, CD provides an alternative possibility for experimental studies. The validity of the CD approach has been recently demonstrated for the case of the ^{7}Be(p, γ) reaction, where very good agreement has been found between direct and indirect methods [23,24].

3 Coulomb dissociation experiments

In Coulomb dissociation (CD) experiments the Coulomb field of a high-Z nucleus is used to excite the nuclei one is interested in (see fig. 2). Thus, CD experiments are always done in inverse kinematics with the high-Z nuclei as targets and the nuclei of interest as projectiles. This forced procedure yields the advantage that unstable isotopes can be observed if they are available as radioactive beams.

If, *e.g.*, $^{A}Z(\gamma, n)^{A-1}Z$ is the reaction of interest one has to study the reaction ^{208}Pb$(^{A}Z, ^{A-1}Z + n)^{208}$Pb using CD. The energy of the projectiles is choosen as high as possible due to several reasons. First of all, the higher the energy of the projectile the more the Coulomb field of the high-Z nucleus is distorted, thus, leading to a higher fraction of $E1$ excitations. Secondly, the used detection

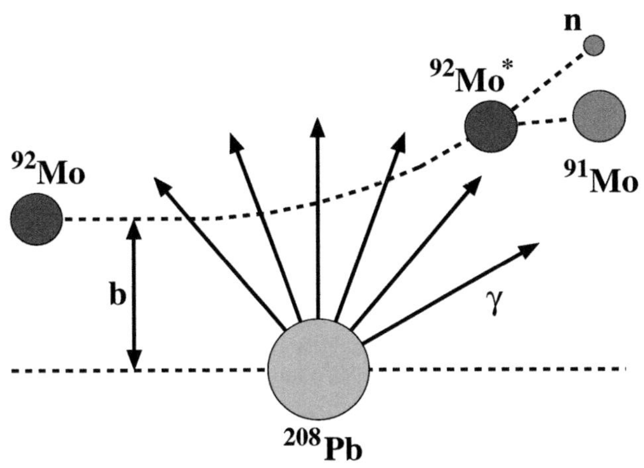

Fig. 2. Scheme of a Coulomb dissociation experiment in inverse kinematics observing the reaction ^{92}Mo(γ, n)^{91}Mo. The high-Z target composed of ^{208}Pb is shown with its surrounding field of virtual photons (marked with γ). The incoming ^{92}Mo projectile has an impact parameter b. After passing the Coulomb field of ^{208}Pb the now excited nucleus ^{92}Mo* de-excites by emitting a neutron and remains as ^{91}Mo.

systems cover 4π solid angles due to the fact that all reaction products are predominantly focussed in forward directions. At last, CD dominates the nuclear background under small scattering angles, hence, yielding a possibility to distinguish between these two contributions to the cross section.

The experimental method used at GSI Darmstadt, is to produce a high-energy stable or radioactive beam and to measure the breakup products in secondary targets with full kinematics, thus, allowing the reconstruction of the excitation energy by utilizing the invariant-mass method. All projectile-like decay products are detected, *i.e.*, all reaction products that have velocities close to the beam velocity. In this sense, the measurement is kinematically complete. Additionally, the γ-rays emitted by the excited projectile near the target position are measured.

The beam is delivered by the heavy-ion synchrotron SIS with energies up to 1 GeV/nucleon and intensities of about 10^{10} ions/s depending on the accelerated nuclei. Radioactive beams are produced by the in-flight method using a Be production target with a thickness of 4 g/cm^2. The fragment separator FRS [25] is used to select the fragments of interest according to their magnetic rigidity. Furthermore, scintillation detectors are placed in the FRS beam-line to determine the masses of the fragments by time-of-flight measurements.

In front and behind the target the position of the beam is determined by Si pin-diodes. The dimension of the beam spot and its emittance is defined by a system of active slits called ROLU [26]. The Pb target is placed at the beginning of the CsI detector [27,28] that is used to measure the γ-rays being emitted by the excited projectiles. Each of the single crystals covers a solid angle that is defined by the aim to realize the Doppler correction by suitable amplifications of the single signals. Thus, the energy of the emitted photon in the rest frame of the emitting source can be measured directly.

A large gap dipole magnet (ALADIN, see ref. [29]) separates the charged reaction products and the emitted neutrons. To determine the trajectories of the charged fragments, a position-sensitive Si pin-diode before the magnet and two large-area fiber detectors [30] behind the magnet are used. By defining the deflection angle in the magnetic dipole field, the magnetic rigidity of the particle is fixed. The velocity of the particles is measured with the time-of-flight (ToF) wall in combination with a thin organic plastic scintillator placed close to the target.

The LAND neutron detector [31] provides with its 2×2 m^2 active area a 100% acceptance for the emitted neutrons with kinetic energies up to 5.6 MeV. Using the two far-end sides of the one meter thick detector array time-of-flight and position information is available (description of the setup after [15]).

In the current experiment S295 the (γ, n) cross sections of the isotopes ^{92,93,94,100}Mo have been observed by CD. To study the stable isotopes ^{94}Mo and ^{100}Mo the corresponding beams were delivered by the synchrotron SIS. The beam of ^{93}Mo and ^{92}Mo nuclei was produced by a primary ^{94}Mo beam via one and two neutron removal, respectively. However, the cross sections of these two isotopes have been measured one after the other due to an easier analysis.

Additionally, a measurement with a C and a Sn target instead of ^{208}Pb was performed to subtract the nuclear contributions to the CD cross sections. To get full knowledge about the background conditions an empty-target run was also carried out. The whole beam-time lasted for nine days and was realized by a collaboration of Forschungszentrum Rossendorf, Forschungszentrum Karlsruhe, GSI Darmstadt, and TU Darmstadt. The evaluation of the data has just started, hence, it is not yet possible to show preliminary results.

4 Summary and outlook

The (γ, n) cross sections of the isotopes ^{92,93,94,100}Mo have been measured using the Coulomb dissociation technique at the SIS/FRS/LAND setup at GSI Darmstadt. To establish the accuracy of this method, the cross sections of the stable isotopes ^{92}Mo and ^{100}Mo are also determined using real photons provided by the bremsstrahlung setups at ELBE and S-DALINAC, respectively.

The expected agreement of the results should establish the accuracy of the CD method. Once this goal is achieved, CD measurements on many critical but unstable nuclei for the p-process can be envisaged. A first example is the isotope ^{93}Mo, that cannot be prepared as a target and was measured during the current experiment.

We thank the collaborators of experiment S295 from Forschungszentrum Rossendorf, Forschungszentrum Karlsruhe, and GSI Darmstadt for fruitful discussions and their support during beam-time. This work is supported by the DFG (contract SFB 634) and BMBF.

References

1. F. Käppeler, Prog. Part. Nucl. Phys. **43**, 419 (1999).
2. K.-L. Kratz, J.-P. Bitouzet, F.-K. Thielemann, P. Möller, B. Pfeiffer, Astrophys. J. **403**, 216 (1993).
3. G. Wallerstein, I. Iben, P. Parker, A.M. Boesgaard, G.M. Hale, A.E. Champagne, C.A. Barnes, F. Käppeler, V.V. Smith, R.D. Hoffman, F.X. Timmes, C. Sneden, R.N. Boyd, B.S. Meyer, D.L. Lambert, Rev. Mod. Phys. **69**, 995 (1997).
4. D.L. Lambert, V.V. Smith, M. Busso, R. Gallino, O. Straniero, Astrophys. J. **450**, 302 (1995).
5. M. Arnould, S. Goriely, Phys. Rep. **384**, 1 (2003).
6. H. Schatz, A. Aprahamian, J. Görres, M. Wiescher, T. Rauscher, J. Rembges, F.-K. Thielemann, B. Pfeiffer, P. Möller, K.-L. Kratz, H. Herndl, B. Brown, H. Rebel, Phys. Rev. **294**, 167 (1998).
7. S. Goriely, *Nuclear astrophysics data base* (2003) `http://www.astro.ulb.ac.be/Nucdata/`.
8. T. Rauscher, F.-K. Thielemann, At. Data Nucl. Data Tables **75**, 1 (2000).
9. B.L. Berman, S.C. Fultz, Rev. Mod. Phys. **47**, 713 (1975).
10. K. Vogt, P. Mohr, M. Babilon, J. Enders, T. Hartmann, C. Hutter, T. Rauscher, S. Volz, A. Zilges, Phys. Rev. C **63**, 055802 (2001).
11. K. Sonnabend, K. Vogt, D. Galaviz, S. Müller, A. Zilges, Phys. Rev. C **70**, 035802 (2004).
12. M. Erhard *et al.*, these proceedings.
13. G. Rusev *et al.*, these proceedings.
14. H. Utsunomiya, H. Akimune, S. Goko, M. Ohta, H. Ueda, T. Yamagata, K. Yamasaki, H. Ohgaki, H. Toyokawa, Y.-W. Lui, T. Hayakawa, T. Shizuma, S. Goriely, Phys. Rev. C **67**, 015807 (2003).
15. R. Palit, P. Adrich, T. Aumann, K. Bortzky, B. Carlson, D. Cortina, U. Datta Pramanik, T. Elze, H. Emling, H. Geissel, M. Hellström, K. Jones, J. Kratz, R. Kulessa, Y. Leifels, A. Leistenschneider, G. Münzenberg, C. Nociforo, P. Reiter, H. Simon, K. Sümmerer, W. Walus, Phys. Rev. C **68**, 034318 (2003).
16. T. Sauter, F. Käppeler, Phys. Rev. C **55**, 3127 (1997).
17. J. Bork, H. Schatz, F. Käppeler, T. Rauscher, Phys. Rev. C **58**, 524 (1998).
18. N. Özkan, A.S.J. Murphy, R.N. Boyd, A.L. Cole, M. Famiano, R.T. Güray, M. Howard, L. Şahin, J.J. Zach, R. deHaan, J. Görres, M.C. Wiescher, M.S. Islam, T. Rauscher, Nucl. Phys. A **710**, 469 (2002).
19. W. Rapp, M. Heil, D. Hentschel, F. Käppeler, R. Reifarth, H.J. Brede, H. Klein, T. Rauscher, Phys. Rev. C **66**, 015803 (2002).
20. G. Gyürky, Z. Fülöp, G. Kiss, Z. Máté, E. Somorjai, J. Görres, A. Palumbo, M. Wiescher, D. Galaviz, A. Kretschmer, K. Sonnabend, A. Zilges, T. Rauscher, Nucl. Phys. A **758**, 517 (2005).
21. Z. Fülöp, G. Gyürky, Z. Máté, E. Somorjai, L. Zolnai, D. Galaviz, M. Babilon, P. Mohr, A. Zilges, T. Rauscher, H. Oberhummer, G. Staudt, Phys. Rev. C **64**, 065805 (2001).
22. D. Galaviz, Z. Fülöp, G. Gyürky, Z. Máté, P. Mohr, T. Rauscher, E. Somorjai, A. Zilges, Phys. Rev. C **71**, 065802 (2005).
23. F. Schümann, F. Hammache, S. Typel, F. Uhlig, K. Sümmerer, I. Böttcher, D. Cortina, A. Förster, M. Gai, H. Geissel, U. Greife, N. Iwasa, P. Koczoń, B. Kohlmeyer, R. Kulessa, H. Kumagai, N. Kurz, M. Menzel, T. Motobayashi, H. Oeschler, A. Ozawa, M. Ploskoń, W. Prokopowicz, W. Schwab, P. Senger, F. Strieder, C. Sturm, Z.-Y. Sun, G. Surówka, A. Wagner, W. Waluś, Phys. Rev. Lett. **90**, 232501 (2003).
24. K. Sümmerer, these proceedings.
25. H. Geissel, P. Armbruster, K. Behr, A. Brünle, K. Burkard, M. Chen, H. Folger, B. Franczak, H. Keller, O. Klepper, B. Langenbeck, F. Nickel, E. Pfeng, M. Pfützner, E. Roeckl, K. Rykaczewski, I. Schall, D. Schardt, C. Scheidenberger, K.-H. Schmidt, A. Schröter, T. Schwab, K. Sümmerer, M. Weber, G. Münzenberg, T. Brohm, H.-G. Clerc, M. Fauerbach, J.-J. Gaimard, A. Grewe, E. Hanelt, B. Knödler, M. Steiner, B. Voss, J. Weckenmann, C. Ziegler, A. Magel, H. Wollnik, J. Dufour, Y. Fujita, D. Vieira, B. Sherrill, Nucl. Instrum. Methods B **70**, 286 (1992).
26. G. Stengel, *Entwicklung großflächiger Szintillatorfaserdetektoren und aktiver Blendensysteme*, Diploma Thesis, Institut für Kernphysik, Universität Frankfurt, unpublished, 1996.
27. I. Kraus, *Entwicklung eines CsI-Gammadetektors für Experimente mit radioaktiven Strahlen*, Diploma Thesis, Institut für Kernphysik, Universität Frankfurt, unpublished, 1999.
28. T. Lange, *Erprobung und Eichung eines CsI(Na)-Gamma-Detektor-Systems für Experimente mit radioaktiven Strahlen*, Diploma Thesis, Institut für Kernphysik, Universität Frankfurt, unpublished, 2001.
29. The ALADIN Collaboration, *The forward spectrometer ALADIN at the 4π detector*, GSI Nachrichten 02-89, 1989.
30. J. Cub, G. Stengel, A. Grünschloß, K. Boretzky, T. Aumann, W. Dostal, B. Eberlein, T. Elze, H. Emling, J. Holeczek, R. Holzmann, G. Ickert, J. Kratz, R. Kulessa, Y. Leifels, H. Simon, K. Stelzer, J. Stroth, A. Surowiec, E. Wajda, Nucl. Instrum. Methods A **402**, 67 (1998).
31. T. Blaich, T. Elze, H. Emling, H. Freiesleben, K. Grimm, W. Henning, R. Holzmann, G. Ickert, J. Keller, H. Klingler, W. Kneissl, R. Knig, R. Kulessa, J. Kratz, D. Lambrecht, J. Lange, Y. Leifels, E. Lubkiewicz, M. Proft, W. Prokopowicz, C. Schütter, R. Schmidt, H. Spies, K. Stelzer, J. Stroth, W. Walús, E. Wajda, H. Wollersheim, M. Zinser, E. Zude, Nucl. Instrum. Methods A **314**, 136 (1992).

Eur. Phys. J. A **27**, s01, 153–158 (2006)
DOI: 10.1140/epja/i2006-08-023-8

EPJ A direct
electronic only

Photonuclear reaction data and γ-ray sources for astrophysics

H. Utsunomiya[1,a], S. Goko[1], H. Toyokawa[2], H. Ohgaki[3], K. Soutome[4], H. Yonehara[4], S. Goriely[5], P. Mohr[6], and Zs. Fülöp[7]

[1] Department of Physics, Konan University, Japan
[2] Photonics Research Institute, AIST, Japan
[3] Institute of Advanced Study, Kyoto University, Japan
[4] JASRI/SPring-8, Japan
[5] Institut d'Astronomie et d'Astrophysique, Université Libre de Bruxelles, Belgium
[6] Strahlentherapie, Diakoniekrankenhaus, Schwäbisch Hall, Germany
[7] ATOMKI, Debrecen, Hungary

Received: 31 July 2005 /
Published online: 24 February 2006 –

Abstract. Direct determination of photoneutron cross sections of astrophysical importance has recently become possible with use of quasi-monochromatic γ beams produced in laser Compton backscattering (LCS) from relativistic electrons at AIST. The astrophysics to be discussed with the photodisintegration cross section are both stellar and big bang nucleosyntheses regarding the production of p-process and s-process nuclei as well as light elements. Synchrotron radiations from a 10 tesla superconducting wiggler (SCW) at SPring-8 serve as an ideal photon source to determine photoreaction rates. This paper covers the latest cross section measurements with the LCS photon beams and a feasibility study of determining (γ, x) $(\mathrm{x} = \mathrm{n}, \mathrm{p}, \alpha)$ reaction rates with the SCW radiation.

PACS. 25.20.-x Photonuclear reactions – 25.40.Lw Radiative capture – 26.30.+k Nucleosynthesis in novae, supernovae, and other explosive environments – 41.60.Ap Synchrotron radiation

1 Introduction

Recently real photon-induced reactions have attracted a revived interest in the context of nuclear astrophysics [1]. Currently efforts are made for determining photoneutron cross sections in the low-energy tail of the giant electric-dipole resonance, laboratory reaction rates by photoactivation, and $E1$ γ strength function below the neutron threshold. Measurements of (γ, α) reactions may not be impossible though an intense photon source is required. At present all experimental efforts have, as a matter of course, been limited to stable nuclei; real photon-induced reactions on unstable nuclei are far beyond our scope.

The direct determination of photodisintegration cross sections utilizes quasi-monochromatic γ-ray beams from laser Compton backscattering (LCS) that have recently become available at AIST [2]. Photodisintegration measurements are characterized by simplification and accuracy by employing the LCS γ beams, a bulk of target material, and a 4π-type neutron detector consisting of $BF_3/^3He$ proportional counters embedded in a poly- ethylene moderator. It is interesting to note that electric giant dipole resonance (GDR) was systematically studied in 1960s through 1980s with quasi-monochromatic photons from positron annihilation in flight in nuclear physics [3], whereas the GDR study is currently implemented by the LCS photons in nuclear astrophysics with emphasis on cross sections in the Gamow energy window for photonuclear reactions. The Gamow window lies immediately above the threshold energy in the neutron channel [4], while in the p or α channels it is shifted from particle threshold toward high energies for the Coulomb potential effect [5]. Some of the early measurements of GDR exhibit non-vanishing cross sections below neutron threshold, necessitating accurate measurements for nuclear astrophysics.

The astrophysics to be discussed with photodisintegration cross sections are both stellar and big bang nucleosyntheses: (a) p-process nuclei; (b) s-process nuclei; and (c) light nuclei. The group (a) is directly related to photonuclear reactions. In the group (b) for heavy nuclei with high level densities, photoneutron reactions are translated to neutron capture by the statistical model, whereas they are converted to radiative capture by the reciprocity theorem in the group (c) for light nuclei with low level densities.

In the p-process study, (γ, n) cross sections provide strong constraints on the $E1$ γ strength function from

[a] e-mail: `hiro@konan-u.ac.jp`

which stellar reaction rates are calculated within the framework of the Hauser-Feshbach model [6]. Both photoproduction and photodestruction cross sections are needed to discuss the p-process origin; photodestruction is by far difficult to measure because of the low natural abundance of p-nuclei. The $E1$ and $M1$ γ strength functions below neutron threshold in the tail of GDR are of unique astrophysical significance because they contribute to photodisintegration of nuclei in low-lying excited states that are thermally populated in stellar conditions. The stellar photodisintegration rate is larger by two to three orders of magnitude than the laboratory photodisintegration rate for a nucleus in the ground state. The reliability of the statistical model calculation is further enhanced if neutron optical potentials and nuclear level densities are constrained in different experiments like neutron capture, etc.

We limit the s-process study by means of photodisintegration to radioactive nuclei for which neutron capture leads to stable nuclei. In particular, photodisintegration is the most efficient way to constrain the neutron capture cross section for short-lived nuclei like ^{185}W ($t_{1/2}$ = 75 d) [7,8]. There are many such short-lived nuclei to study. Recently, neutron capture cross sections are measured for ^{151}Sm with a half-life of 90 years at n-TOF of CERN [9]. It was proposed [10] to measure photodisintegration cross sections for ^{152}Sm by scanning across low-lying excited states of the residual ^{151}Sm nucleus in order to get insight into the stellar enhancement factor.

Photoneutron cross sections so far measured with the LCS γ beams are categorized into the three groups as follows:

- (a) p-process nuclei: ^{93}Nb, ^{139}La, ^{181}Ta [6];
- (b) s-process nuclei: ^{80}Se, ^{108}Pd, ^{141}Pr, ^{152}Sm, ^{186}W [8,11], ^{187}Re [11], and ^{188}Os [11];
- (c) light nuclei: D [12], ^{9}Be [13].

A new photon source that is ideally suited to a systematic p-process study is under development at SPring-8 [14]. A beam of 8 GeV electrons passing through a 10 tesla superconducting wiggler (SCW) produces high-energy synchrotron radiation that mimics blackbody spectra at temperatures of billions of kelvin. It is shown that this SCW radiation can be used to determine laboratory photoreaction rates not only for (γ,n) but for (γ,α) and (γ,p) reactions.

2 Direct determination of (γ,n) cross sections

2.1 p-process

Among 35 nuclei classically referred to as p-nuclei, there are two rare odd-odd nuclei: ^{138}La (Z = 57, N = 81) and ^{180}Ta^m (Z = 73, N = 107). The p-process origin of ^{180}Ta, the only naturally occurring isomer and the rarest nucleus in the solar system, was investigated with focus on its production cross-sections of ^{181}Ta$(\gamma,\mathrm{n})^{180}$Ta [6]. Using a stellar rate constrained by the experimental cross sections, a 25 $M_\odot$ type-II supernova (SN-II) model with solar metalicity has confirmed that ^{180}Ta^m is a natural product of the p-process origin like the bulk of p-nuclides. Uncertainties remained both in experiment and theory: destruction cross sections of ^{180}Ta$^m(\gamma,\mathrm{n})^{179}$Ta on one hand and contributions from ν_e captures on ^{180}Hf [15] as well as the controversial s-process origins [16,17] on the other. The possibility of removing the experimental uncertainty is discussed in sect. 3.1.

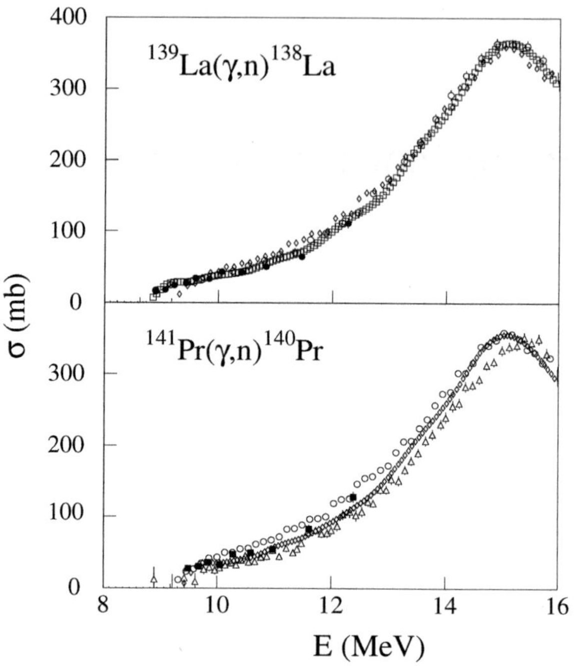

Fig. 1. Photodisintegration cross sections for ^{139}La and ^{141}Pr with the neutron magic number 82 (closed squares). Also shown are the previous data: open circles [19], open diamonds [20], and open squares [21] for ^{139}La; open triangles [22], open circles [20], and open diamonds [21] for ^{141}Pr.

Recently, cross sections for ^{139}La$(\gamma,\mathrm{n})^{138}$La were measured to address the underproduction problem of ^{138}La of thermonuclear origin in type-II supernovae (SNe-II) [18]. Figure 1 shows photoneutron cross sections measured for ^{139}La and ^{141}Pr with the magic neutron number 82 in comparison with the data taken previously [19,20,21,22]. The Hauser-Feshbach code MOST [23] with three different prescriptions of the $E1$ γ strength functions, namely the Lorentzian-type model [24], the Hybrid model [25] and the HFBCS + QRPA model [26] predicts a stellar rate $\lambda^*_{(\gamma\mathrm{n})} = 27 \pm 15\,\mathrm{s}^{-1}$ for ^{139}La at a typical p-process temperature $T = 2.5 \times 10^9$ K. The upper limit of the rate ($\lambda^*_{(\gamma\mathrm{n})} = 42\,\mathrm{s}^{-1}$) is given by the Lorentzian-type model, while the lower limit ($\lambda^*_{(\gamma\mathrm{n})} = 12\,\mathrm{s}^{-1}$), which is the previous standard value [18], by the HFBCS + QRPA model. The large uncertainty arises from different energy dependences of the three different models of the γ strength function below the neutron threshold.

To cure the underproduction problem completely within the p-process scenario of SNe-II, it is required that an increase in the production of ^{138}La needs to be complemented by a decrease in its destruction, resulting in a total enhancement factor of 20–25 [18]. The upper limit of the present rate may increase the stellar photoproduction rate of ^{138}La by a factor of 3.5. The complementary decrease is however unlikely from the viewpoint of a consistency in calculating the production and the destruction with the same model; for example, the Lorentzian model of the $E1$ γ strength function, which could increase the production rate, increases the destruction rate simultaneously. Therefore, the p-process in SNe-II is not favored to explain a significant fraction of the solar ^{138}La abundance. However, no definite conclusion can be drawn until the destruction rate of ^{138}La is constrained experimentally. A future measurement of ^{138}La$(\gamma, \mathrm{n})^{137}$La cross sections would be a challenge to experimentalists.

2.2 s-process

Photodisintegration cross sections were measured for ^{186}W, ^{187}Re and ^{188}Os, of which those for ^{188}Os were used to address a major uncertainty involved in the Re-Os cosmochronology [11]: namely, the correction factor F_σ for the contribution of the 9.5 keV state in ^{187}Os to the stellar neutron capture rate [27,28,29,30]. Figure 2 shows the present data in comparison with the data taken previously [31,32,33]. The result of the Hauser-Feshbach (HF) model calculations with different models of the $E1$ γ strength function, the level density, and the neutron optical potential are also shown in the figure. The Hartree-Fock + BCS (HFBCS) [34] and the back-shifted Fermi gas (BSFG) [35] models were used for the level density; The HFBCS + quasi-particle random phase approximation [26] (HFBCS + QRPA) and the Hybrid [25] models were used for the γ strength function; The JLMB [36] and Woods-Saxon [37] optical potentials were used.

Stellar neutron capture cross sections for ^{187}Os were calculated with the HF model parameters constrained by the present photodisintegration data and the laboratory neutron capture data available for ^{187}Os. Note that in this case, the present data sensitive to the $E1$ γ strength function combined with the capture data sensitive to the neutron optical potential and the level density can greatly improve the reliability of the HF model calculations. Except for the one with the Woods-Saxon potential labeled INP-3 in fig. 2, which turned out to significantly underestimate the experimental capture cross sections at low energies, all the HF model calculations provided stellar capture cross sections with small deviations. Thus, it was concluded that F_σ values at typical s-process temperatures (12–30 keV) are in the range of 0.86–0.94.

We investigated the uncertainty in the age of the Galaxy (T_G) associated with the constrained F_σ values in the Re-Os chronology. Using the simplest assumption of r-process nucleosynthesis yields of ^{187}Re varying exponentially in time, the probable range of the differential coefficient $\mathrm{d}T_G/\mathrm{d}F_\sigma$ was found to be $-(5.0$–$12.8)$ Gyr. Consequently, it was found that the associated uncertainty in T_G values is less than 1 Gyr; when the temperature dependences of F_σ and the Maxwellian-averaged cross sections

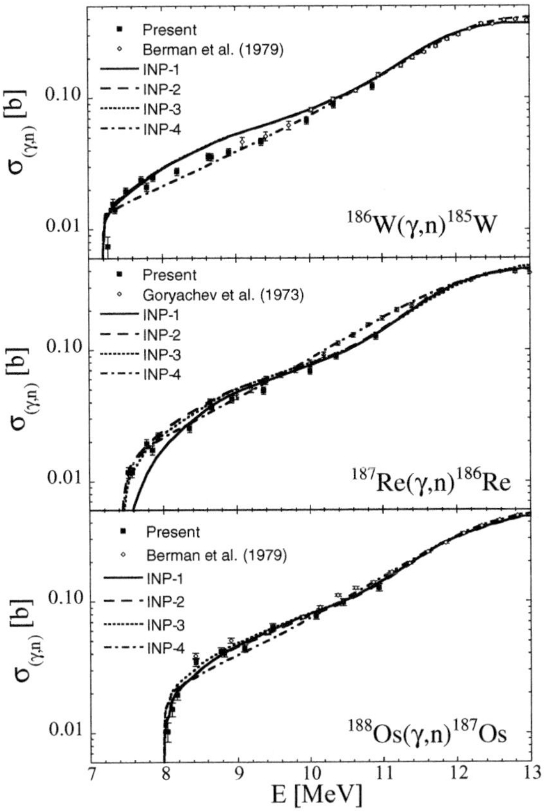

Fig. 2. Photodisintegration cross sections for ^{186}W, ^{187}Re, and ^{188}Os in comparison with the Hauser-Feshbach model calculations with different nuclear inputs. See text for details.

for ^{186}Os and ^{187}Os are considered, the uncertainties are less than 0.5 Gyr.

Nuclei at the s-process branching give a first approximation of the neutron density and temperatures of relevance in the s-process site [38]. Photodisintegration cross sections were measured for ^{186}W to evaluate neutron capture cross sections for an s-process branching nucleus ^{185}W ($t_{1/2} = 75$ d) [8]. The HF photodisintegration cross sections for ^{186}W with two very different statistical ingredients [7] were scaled to the present experimental cross sections with scaling factors being 1.0 and 0.77, respectively. By employing the same scaling factors for the HF neutron capture cross sections, a Maxwellian-averaged neutron capture cross section for ^{185}W was evaluated; $\sigma = 553 \pm 60$ mb at $kT = 30$ keV.

A discussion parallel to the previous classical-model analysis of the s-process flow at ^{185}W gave a higher neutron density $N_\mathrm{n} = 4.7^{+1.4}_{-1.1} \times 10^8\,\mathrm{cm}^{-3}$. This neutron density, which is still compatible with the branching at Nd-Pm-Sm ($A = 147$–149) [39], highlights incompatibility with the branching at Er-Tm-Yb ($A = 169$–171) [39] and Os-Ir-Pt ($A = 191$–193) [40]. On the other hand, in a realistic s-process model the present neutron capture cross section for ^{185}W, which is smaller than the previous value ≈ 700 mb [7], may enhance the overproduction of the s-only nucleus ^{186}Os.

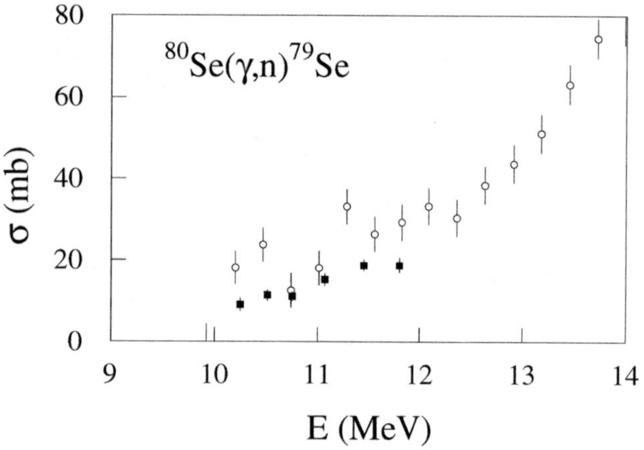

Fig. 3. Photodisintegration cross sections for ^{80}Se (closed squares). For comparison, the previous data (open circles) [42] taken with the positron annihilation γ-rays is also shown.

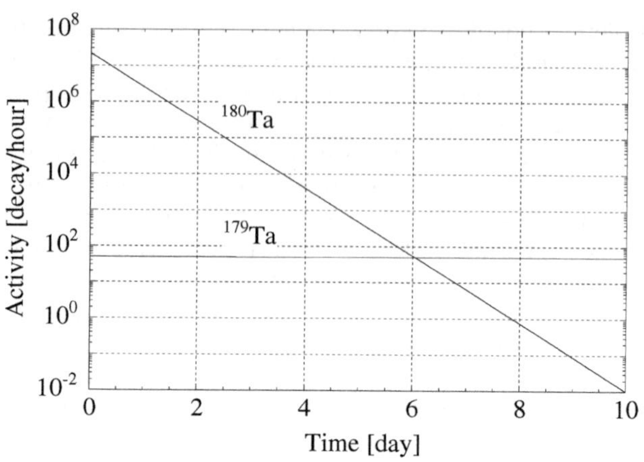

Fig. 4. The radioactivity of $^{180}Ta^{gs}$ and ^{179}Ta expected to be produced by irradiating 10 natural tantalum foils of 100 μm thickness each with the SCW radiation for 100 hours.

Long-lived radionuclides produced in the s-process are also of interest in the field of nuclear waste transmutation. The idea of nuclear transmutation is to transform them to either stable or short-lived nuclei by means of neutron capture. Photodisintegration can be used to evaluate neutron capture cross sections for nuclei in the former cases; they are ^{36}Cl [3×10^6 y], ^{79}Se [6.5×10^4 y], ^{93}Zr [1.53×10^6 y], ^{107}Pd [6.5×10^6 y], ^{151}Sm [90 y]. ^{151}Sm and ^{79}Se are nuclei at which s-process branching takes place in the main and weak s-process flows, respectively. Note that the terrestrial decay rate of ^{79}Se is shortened by four orders of magnitudes in the stellar condition at temperatures higher than 1×10^8 K due to decay from excited states [41].

Photodisintegration cross sections were measured for ^{80}Se, ^{108}Pd, and ^{152}Sm from the viewpoint of astrophysics and nuclear transmutation. Figure 3 shows results of the measurement for ^{80}Se. The present data significantly differ from the data taken with the positron annihilation γ-rays [42].

2.3 Light nuclei

Photodisintegration of light nuclei is of particular interest because it is directly connected to inverse radiative capture by the reciprocity theorem. ^{9}Be and D are two good examples [13,12]; the inverse $\alpha\alpha(n,\gamma)^9$Be reaction followed by $^9Be(\alpha, n)$ is considered to be most efficient in producing ^{12}C in the so-called neutrino-driven wind of SNe-II, dominating over the triple α reaction [43], while $p(n,\gamma)D$ belongs to big bang nucleosynthesis, producing the simplest two-nucleon system. Photodisintegration is an exclusive way to study the *three-body* reaction, where real photons strongly excited the $E1$ *resonance* state $(1/2^+)$ in ^{9}Be immediately above the n + ^{8}Be threshold. In contrast, virtual photons in the electron backward-scattering favored $M1$ excitation [44]. On the other hand, it is a unique way to study the $p(n,\gamma)D$ reaction, where the measurement successfully reduced uncertainties of $p(n,\gamma)D$ cross sections at energies relevant to big bang nucleosynthesis.

New measurements are planned for photodisintegration of D to determine 1) total $p(n,\gamma)D$ cross sections at neutron energies below 100 keV and 2) partial cross sections for s-wave and p-wave neutron capture. The former measurement is characterized by a long D_2O target extending across the polyethylene moderator, intense LCS γ beams with energies varied in a small step near the n + p threshold, and a photon difference method for data reduction. Previously, a measurement of the photon analyzing power was carried out in photodisintegration of deuterons using linearly polarized γ beams at the Duke facility [45]. The analyzing power data was used to determine relative $E1$ and $M1$ strengths of the photodisintegration cross section. In the new measurement, we plan to utilize linearly polarized LCS photons and fast (liquid scintillator) and slow (^{3}He + polyethylene) neutron detectors to determine absolute strengths of $M1$ and $E1$ cross sections from neutron angular distributions as well as the analyzing power.

3 Perspectives

3.1 Determination of photodisintegration rates

A 10 tesla superconducting wiggler is under development as an insertion light source at SPring-8. This light source produces high-energy synchrotron radiations with equivalent blackbody spectra at temperatures of billions of kelvin [14]. Recently, a study was made of using the SCW radiation to determine (γ, n) reaction rates. Photodestruction of the nature's rarest isotope $^{180}Ta^m$ with the natural abundance of 0.012% will be the most challenging experiment. Irradiation of natural tantalum foils with the intense beam of the SCW *blackbody* radiation produces ^{179}Ta [$t_{1/2} = 1.82$ y] less by a factor of the order of 10,000 than $^{180}Ta^{gs}$ [8.15 h] because of the unbalanced natural abundances. Figure 4 shows the expected radioactivity of $^{180}Ta^{gs}$ and ^{179}Ta produced under a proper irradiation condition [14]. Because of the very different half-lives, the radioactivity of $^{180}Ta^{gs}$ decreases at the same level as ^{179}Ta in the first 6 days and is below 1/1,000 in the next 4 days, while the activity of ^{179}Ta remains the same ($\sim$ 50 decays per hour) during this period. After a proper cooling time of about 10 days, specific hafnium KX-rays ($K_{\alpha 1}$: 59.32 keV, $K_{\alpha 2}$: 57.98 keV) emitted in the EC of ^{179}Ta can be measured under a low-radiation background.

Very recently a feasibility study was extended to (γ, α) and (γ, p) reactions. The event rate $N(t)$ is expressed by

$$N(t) = a_t \int \sigma(\varepsilon) n_\gamma(\varepsilon, T)\, \mathrm{d}\varepsilon. \qquad (1)$$

Here a_t is the areal number density of target nuclei [cm^{-2}], $\sigma(\varepsilon)$ is the photoreaction cross section for the emission of alpha particles or protons [cm^2], $n_\gamma(\varepsilon, T)$ is a flux of the black-body radiation at temperature T which is equivalent to the SCW synchrotron radiation [s^{-1}]. Note that $T = 4.4 \times 10^9$ K at the magnetic field of 10 tesla. The Gamow peak appears as a maximum in the integrand of eq. (1). The range of α particles and protons with the most probable energy defined at the Gamow peak was used to calculate a_t. Of the 233 reactions with cross sections compiled in [46] in the feasibility study, we found 35 (γ, α) and 36 (γ, p) reactions with the event rate larger than 10 counts per hour. For example, in the $^{96}\mathrm{Ru}(\gamma, \alpha)^{92}\mathrm{Mo}$ reaction the followings are obtained:

- The integral: 1.24×10^{-21} [cm^2 s^{-1}],
- The most probable α energy: 7.8 MeV,
- Range in Ru: 12.4 [μm],
- a_t: 1.22×10^{20} [cm^{-2}],
- Event rate: 5.43×10^2 [h^{-1}].

There are several interesting cases (^{74}Se, ^{96}Ru, ^{144}Nd, ^{152}Gd), where a theoretical relation between (γ, α) and (α, γ) measurements can be investigated. Most of the (γ, α) reactions produce stable nuclei so that a direct counting of emitted α-particles is necessary. The direct counting is possible by mounting target foils sufficiently thinner than the particle range inside a vacuum chamber and by surrounding each foil with particle detectors.

3.2 Call for international collaborations

Since photonuclear reactions have a variety of important facets in astrophysics, international collaborations in the following research categories can be called.

1) p-process: photodisintegration measurements, that are currently limited to (γ, n) reactions, need to be extended to an unexplored field of (γ, α) and (γ, p) reactions. Measurements of $E1$ and $M1$ γ strength functions in nuclear fluorescence experiments should be addressed in the context of nuclear astrophysics: that is, photonuclear reactions on excited states under stellar conditions.
2) s-process: Photodisintegration plays a complementary role to neutron capture in the s-process study for radioactive nuclei. For example, a complementary role is seen in (γ, n) for ^{152}Sm [stable] and (n, γ) for ^{151}Sm [$t_{1/2}$ = 90 y] as discussed in sect. 2.2. In particular, photodisintegration provides a unique opportunity to evaluate neutron-capture cross sections for short-lived nuclei provided that neutron capture leads to stable nuclei. One can find many nuclei in the chart of nuclides for a systematic study, *e.g.*, ^{64}Cu [12.7 h], ^{65}Zn [244 d], ^{70}Ga [21.2 min], ^{71}Ge [11.4 d], ^{74}As [17.8 d] etc. Although the $E1$ γ strength function is best probed by photodisintegration, remaining sources of uncertainties in the statistical model calculation are the neutron optical potential and the nuclear level density. If these quantities could be probed in separate experiments, the predictive power of the statistical model would be greatly improved.
3) light nuclei: Photodisintegration of light nuclei is a straightforward way to study inverse radiative capture based on the reciprocity theorem. Deuterium is a good research objective, where $\sim$ 100% linearly-polarized photons can be used to separate the $E1$ and $M1$ components of photodisintegration and thus the corresponding s-wave and p-wave neutron captures by proton.
4) photon sources and experimental technique: There are four fundamental factors for photon sources: intensity, monochromaticity, polarization, and energy variableness. An attempt of producing monochromatic γ-rays at the Institute Laue-Langevin utilizes neutron-capture γ-rays produced at the reactor and a bent Si 220 crystal as a monochromator [47]. The 10 tesla superconducting wiggler at SPring-8 with another unique feature of being equivalent to *blackbody* radiation would allow one to determine photoreaction rates not only for (γ, n) reactions [14] but also for (γ, α) and (γ, p) reactions.
5) theory and astrophysical modeling: Although important effort has been devoted in the last decades to measure reaction cross sections, theoretical predictions play a crucial role in the estimate of the reaction rates on stable as well as unstable nuclei for astrophysics applications. The nuclear ingredients to the reaction models, *i.e.* nuclear structure properties, optical model potentials, nuclear level densities, γ-ray strengths, should preferentially be estimated from microscopic global predictions based on sound and reliable nuclear models which, in turn, can compete with more phenomenological highly-parametrized models in the reproduction of experimental data. Well-targeted experiments are of prime importance to properly constrain theoretical models and consequently decrease the uncertainties in the predictions of astrophysics interest. A special care should be paid to the definition of the priorities and the sensitivity of the astrophysics observables to the nuclear ingredients. The relevance of the nuclear inputs should be dictated by astrophysics simulations which needs to be based on the state-of-the-art models for each problem considered. A simultaneous effort to improve astrophysics models is required.

This work was supported by the Japan Society of the Promotion of Science (Grant-in-Aid for Scientific Research (B), (C)) and The Ministry of Education, Culture, Sports, Science, and Technology (Grant-in-Aid for Exploratory Research), and the Japan Private School Promotion Foundation. S.G. is FNRS Research Associate.

References

1. H. Utsunomiya, P. Mohr, A. Zilges, M. Rayet, to be published in Nucl. Phys. A, nucl-ex/0502011.
2. H. Ohgaki, S. Sugiyama, T. Yamazaki, T. Mikado, M. Chiwaki, K. Yamada, R. Suzuki, T. Noguchi, T. Tomimasu, IEEE Trans. Nucl. Sci. **38**, 386 (1991).
3. S.S. Dietrich, B.L. Berman, At. Data Nucl. Data Tables **38**, 199 (1988).
4. P. Mohr, K. Vogt, M. Babilon, J. Enders, T. Hartmann, C. Hutter, T. Rauscher, S. Volz, A. Zilges, Phys. Lett. B **488**, 127 (2000).
5. P. Mohr, M. Babilon, D. Galaviz, K. Sonnabend, K. Vogt, A. Zilges, Nucl. Phys. A **719**, 90c (2003).
6. H. Utsunomiya, H. Akimune, S. Goko, M. Ohta, H. Ueda, T. Yamagata, K. Yamasaki, H. Ohgaki, H. Toyokawa, Y.-W. Lui, T. Hayakawa, T. Shizuma, E. Khan, S. Goriely, Phys. Rev. C **67**, 015807 (2003).
7. K. Sonnabend, P. Mohr, K. Vogt, A. Zilges, A. Mengoni, T. Rauscher, H. Beer, F. Käppeler, R. Gallino, Astrophys. J. **583**, 506 (2003).
8. P. Mohr, T. Shizuma, H. Ueda, S. Goko, Y. Makinaga, K.Y. Hara, H. Hayakawa, Y.-W. Lui, H. Ohgaki, H. Utsunomiya, Phys. Rev. **69**, 032801 (2004).
9. U. Abbondanno *et al.*, Phys. Rev. Lett. **93**, 161103 (2004).
10. A. Mengoni, *Proceedings of the International Conference on Nuclear Data for Science and Technology, Santa Fe, New Mexico (USA), 26 September - 1 October 2004*, AIP Conf. Proc. **769**, 1209 (2005).
11. T. Shizuma, H. Utsunomiya, P. Mohr, T. Hayakawa, S. Goko, A. Makinaga, H. Akimune, T. Yamagata, M. Ohta, H. Ohgaki, Y.-W. Lui, H. Toyokawa, A. Uritani, S. Goriely, Phys. Rev. C **72**, 025808 (2005), nucl-ex/0506027.
12. K.Y. Hara, H. Utsunomiya, S. Goko, H. Akimune, T. Yamagata, M. Ohta, H. Toyokawa, K. Kudo, A. Uritani, Y. Shibata, Y.-W. Lui, H. Ohgaki, Phys. Rev. D **68**, 072001 (2003).
13. H. Utsunomiya, Y. Yonezawa, H. Akimune, T. Yamagata, M. Ohta, M. Fujishiro, H. Toyokawa, H. Ohgaki, Phys. Rev. C **63**, 018801 (2001).
14. H. Utsunomiya, S. Goko, K. Soutome, N. Kumagai, H. Yonehara, Nucl. Instrum. Methods A **538**, 225 (2005).
15. S.E. Woosley, D.H. Hartmann, R.D. Hoffman, W.C. Haxton, Astrophys. J. **356**, 272 (1990).
16. R. Gallino, C. Arlandini, M. Busso, Astrophys. J. **497**, 388 (1998).
17. S. Goriely, N. Mowlavi, Astron. Astrophys. **362**, 599 (2000).
18. S. Goriely, M. Arnould, I. Bolzov, M. Rayet, Astron. Astrophys. **375**, L35 (2001).
19. R. Bergère, H. Beil, A. Veyssière, Nucl. Phys. A **121**, 463 (1968).
20. H. Beil, R. Bergère, P. Carlos, A. Lepretre, A. Veyssière, Nucl. Phys. A **172**, 426 (1971).
21. S.N. Beljaev, V.A. Semenov, Izv. Akad. Nauk SSSR **55**, 953 (1991).
22. R.E. Sund, V.V. Verbinski, H. Weber, L.A. Kull, Phys. Rev. C **2**, 1129 (1970).
23. S. Goriely, in *Nuclei in the Cosmos*, edited by N. Prantzos, S. Harissopulos (Editions Frontières, Gif-sur-Yvette, 1998) p. 314.
24. C.M. McCullagh, M.L. Stelts, R.E. Chrien, Phys. Rev. C **23**, 1394 (1981).
25. S. Goriely, Phys. Lett. B **436**, 10 (1998).
26. S. Goriely, E. Khan, Nucl. Phys. A **706**, 217 (2002).
27. S.E. Woosley, W.A. Fowler, Astrophys. J. **233**, 411 (1979).
28. K. Yokoi, K. Takahashi, M. Arnould, Astron. Astrophys. **117**, 65 (1983).
29. M. Arnould, K. Takahashi, K. Yokoi, Astron. Astrophys. **137**, 51 (1984).
30. K. Takahashi, Nucl. Phys. A **718**, 325c (2003).
31. B.L. Berman, M.A. Kelly, R.L. Bramblett, J.T. Caldwell, H.S. Davis, S.C. Fultz, Phys. Rev. **185**, 1576 (1969).
32. A.M. Goryachev, G.N. Zalesnyi, S.F. Semenko, B.A. Tulupov, Yad. Fiz. **17**, 463 (1973).
33. B.L. Berman, D.D. Faul, R.A. Alvarez, P. Meyer, D.L. Olson, Phys. Rev. C **19**, 1205 (1979).
34. P. Demetriou, S. Goriely, Nucl. Phys. A **695**, 95 (2001).
35. S. Goriely, J. Nucl. Sci. Tech. Suppl. **2**, 536 (2002).
36. E. Bauge, J.P. Delaroche, M. Girod, Phys. Rev. C **63**, 024607 (2001).
37. A.J. Koning, J.P. Delaroche, Nucl. Phys. A **713**, 231 (2003).
38. F. Käppeler, H. Beer, K. Wisshak, Rep. Prog. Phys. **52**, 945 (1989).
39. F. Käppeler, R. Gallino, M. Busso, G. Picchio, C.M. raiteri, Astrophys. J. **354**, 630 (1990).
40. P.E. Koehler, J.A. Harvey, K.H. Guber, R.R. Winters, S. Raman, J. Nucl. Sci. Tech. Suppl. **2**, 546 (2002).
41. K. Takahashi, K. Yokoi, At. Data Nucl. Data Tables, **36**, 375 (1987).
42. P. Carlos, H. Beil, R. Bergère, J. Fagot, A. Lepretre, A. Veyssière, G.V. Solodukhov, Nucl. Phys. A **258**, 365 (1976).
43. S.E. Woosley, R.D. Hoffman, Astrophys. J. **395**, 202 (1992).
44. H.-G. Clerc, K.J. Wetzel, E. Spamer, Nucl. Phys. A **120**, 441 (1968).
45. W. Tornow, N.G. Czakon, C.R. Howell, A. Hutcheson, J.H. Kelley, V.N. Litvinenko, S.F. Mikhailov, I.V. Pinayev, G.J. Weisel, H. Witala, Phys. Lett. B **574**, 8 (2003).
46. T. Rauscher, F.-K. Thielemann, At. Data Nucl. Data Tables **88**, 1 (2004).
47. P. Mutti, private communication.

5 Cross-Section Measurements and Nuclear Data for Astrophysics

Eur. Phys. J. A **27**, s01, 161–170 (2006)
DOI: 10.1140/epja/i2006-08-024-7

EPJ A direct
electronic only

CNO hydrogen burning studied deep underground

The LUNA Collaboration

D. Bemmerer[1,a], F. Confortola[2], A. Lemut[2], R. Bonetti[3], C. Broggini[1], P. Corvisiero[2], H. Costantini[2], J. Cruz[4], A. Formicola[5], Zs. Fülöp[6], G. Gervino[7], A. Guglielmetti[3], C. Gustavino[5], Gy. Gyürky[6], G. Imbriani[8], A.P. Jesus[4], M. Junker[5], B. Limata[8], R. Menegazzo[1], P. Prati[2], V. Roca[8], D. Rogalla[9], C. Rolfs[10], M. Romano[8], C. Rossi Alvarez[1], F. Schümann[10], E. Somorjai[6], O. Straniero[11], F. Strieder[10], F. Terrasi[9], and H.P. Trautvetter[10]

[1] Istituto Nazionale di Fisica Nucleare (INFN), Sezione di Padova, via Marzolo 8, 35131 Padova, Italy
[2] Dipartimento di Fisica, Università di Genova, and INFN, Genova, Italy
[3] Istituto di Fisica, Università di Milano, and INFN, Milano, Italy
[4] Centro de Fisica Nuclear da Universidade de Lisboa, Lisboa, Portugal
[5] INFN, Laboratori Nazionali del Gran Sasso, Assergi, Italy
[6] ATOMKI, Debrecen, Hungary
[7] Dipartimento di Fisica Sperimentale, Università di Torino, and INFN, Torino, Italy
[8] Dipartimento di Scienze Fisiche, Università di Napoli "Federico II", and INFN, Sezione di Napoli, Napoli, Italy
[9] Seconda Università di Napoli, Caserta, and INFN, Sezione di Napoli, Napoli, Italy
[10] Institut für Experimentalphysik III, Ruhr-Universität Bochum, Bochum, Germany
[11] Osservatorio Astronomico di Collurania, Teramo, and INFN, Sezione di Napoli, Napoli, Italy

Received: 3 July 2005 /
Published online: 8 March 2006 –

Abstract. In stars, four hydrogen nuclei are converted into a helium nucleus in two competing nuclear fusion processes, namely the proton-proton chain (p-p chain) and the carbon-nitrogen-oxygen (CNO) cycle. For temperatures above 20 million kelvin, the CNO cycle dominates energy production, and its rate is determined by the slowest process, the ^{14}N(p, γ)^{15}O radiative capture reaction. This reaction proceeds through direct and resonant capture into the ground state and several excited states in ^{15}O. High energy data for capture into each of these states can be extrapolated to stellar energies using an R-matrix fit. The results from several recent extrapolation studies are discussed. A new experiment at the LUNA (Laboratory for Underground Nuclear Astrophysics) 400 kV accelerator in Italy's Gran Sasso laboratory measures the total cross section of the ^{14}N(p, γ)^{15}O reaction with a windowless gas target and a 4π BGO summing detector, down to center of mass energies as low as 70 keV. After reviewing the characteristics of the LUNA facility, the main features of this experiment are discussed, as well as astrophysical scenarios where cross section data in the energy range covered have a direct impact, without any extrapolation.

PACS. 25.40.Lw Radiative capture – 26.20.+f Hydrostatic stellar nucleosynthesis – 29.17.+w Electrostatic, collective, and linear accelerators – 29.30.Kv X- and γ-ray spectroscopy

1 Introduction

Stars generate energy and synthesize chemical elements in thermonuclear reactions [1]. Initially, hydrogen is burned to helium, and then, depending on the mass and chemical composition of the star, also heavier elements can be synthesized.

Hydrogen burning in stars can proceed through several different mechanisms, namely the proton-proton chain (p-p chain), several catalytic cycles called the CNO (carbon–nitrogen–oxygen) cycles [2] I, II, III, and the Hot-CNO cycle, the neon-sodium and the magnesium-aluminium cycle [1]. The p-p I chain (in the following text simply called p-p chain) converts four protons into one ^{4}He nucleus; it is formed by the following nuclear reactions:

$$^1\mathrm{H}(\mathrm{p,e}^+\nu)^2\mathrm{H}(\mathrm{p},\gamma)^3\mathrm{He}(^3\mathrm{He,2p})^4\mathrm{He}\,.$$

The p-p II and III chains also convert four protons into one ^{4}He nucleus, but are much less likely than the p-p I chain, at solar temperature [3] but also at higher temperatures.

The CNO cycles I and II are given by the following chains of reactions, respectively:

$$^{12}\mathrm{C}(\mathrm{p},\gamma)^{13}\mathrm{N}(\beta^+)^{13}\mathrm{C}(\mathrm{p},\gamma)^{14}\mathrm{N}(\mathrm{p},\gamma)^{15}\mathrm{O}(\beta^+)^{15}\mathrm{N}(\mathrm{p},\alpha)^{12}\mathrm{C},$$
$$^{15}\mathrm{N}(\mathrm{p},\gamma)^{16}\mathrm{O}(\mathrm{p},\gamma)^{17}\mathrm{F}(\beta^+)^{17}\mathrm{O}(\mathrm{p},\alpha)^{14}\mathrm{N}(\mathrm{p},\gamma)^{15}\mathrm{O}(\beta^+)^{15}\mathrm{N}.$$

[a] e-mail: bemmerer@pd.infn.it

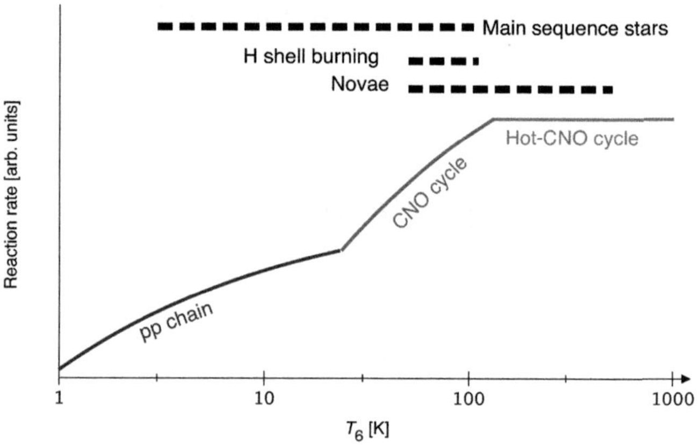

Fig. 1. The rate of energy generation of the three most important mechanisms of stellar hydrogen burning. The temperature of the transitions between the three regimes shown depends on the density and chemical composition of the star. Some relevant stellar burning scenarios [5] are indicated in the figure.

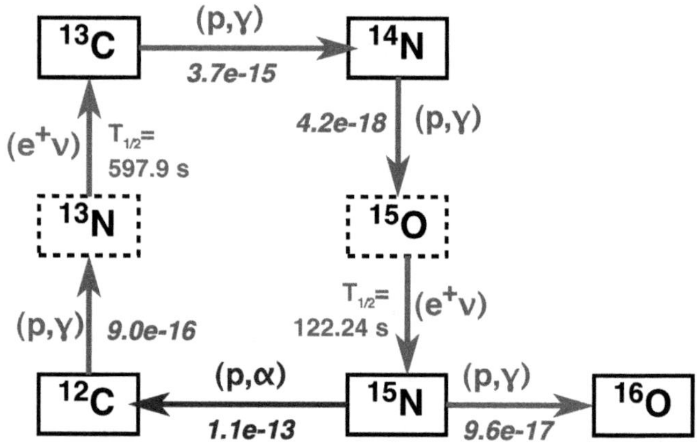

Fig. 2. The reactions of the CNO cycle [1]. Given are the NACRE [4] thermonuclear reaction rates $N_A\langle\sigma v\rangle$ in $\frac{\mathrm{cm}^3}{\mathrm{mol\cdot s}}$ at the temperature at the center of our sun ($T_6 = 16$).

These two cycles, as well as the less likely CNO cycles III and IV [1], also burn four protons into one ^{4}He nucleus. At higher stellar temperatures, the CNO cycles are supplanted by the so-called Hot-CNO cycles. The onset of the Hot-CNO cycles takes place when radiative capture on an unstable nuclide in the regular CNO cycles proceeds more rapidly than the β^+ decay of the same nuclide.

To give some approximate numbers, at low temperatures, $T_6 < 20$ (T_6 indicates the temperature of the burning site in the star in 10^6 K), energy production is dominated by the p-p chain (fig. 1). For $20 < T_6 < 130$, the CNO cycle I (for simplicity just called the CNO cycle) dominates, for a chemical composition like that of our sun. At $T_6 \approx 130$ (for a typical density of $100\,\frac{\mathrm{g}}{\mathrm{cm}^3}$), the rate of radiative proton capture on the unstable nuclide ^{13}N becomes faster than its β^+ decay, and the β-limited Hot-CNO cycle then dominates energy production.

Over the entire energy region where the CNO cycle dominates, the $^{14}\mathrm{N}(\mathrm{p},\gamma)^{15}\mathrm{O}$ reaction is its bottleneck (fig. 2). Therefore, the rate of this particular nuclear reaction determines the rate of the entire cycle.

The present work first proposes a nuclear energy range of interest for understanding stellar CNO burning. Extrapolations by different authors giving the rate of the CNO cycle at stellar energies are then reviewed. The most important features of the Laboratory for Underground Nuclear Astrophysics (LUNA) are given. A new experiment measuring the total cross section of the $^{14}\mathrm{N}(\mathrm{p},\gamma)^{15}\mathrm{O}$ reaction at energies $E = 70$–$230\,\mathrm{keV}$[1] is presented. The astrophysical impact of directly measured cross sections at such low energies is discussed.

Details of the $^{14}\mathrm{N}(\mathrm{p},\gamma)^{15}\mathrm{O}$ cross sections [6,7] obtained in the experiment described here will be published separately.

2 Which nuclear energy range is of astrophysical interest?

The rate of energy production in thermonuclear burning is obtained from the energy produced per reaction and the number of reactions taking place per second, called the rate. This Maxwellian averaged thermonuclear reaction rate is called $\langle\sigma v\rangle$ and is obtained by folding the Maxwell-Boltzmann velocity distribution, calculated for the temperature of the star, with the energy-dependent nuclear reaction cross section. More precisely, $\langle\sigma v\rangle$ is given by the relation [1]

$$\langle\sigma\rangle = \int_0^\infty \varphi(v)\cdot v\cdot\sigma(v)\;\mathrm{d}v, \tag{1}$$

where v is the relative velocity of the two reaction partners, $\varphi(v)$ the velocity distribution (given by the Maxwell-Boltzmann distribution) and $\sigma(v)$ the nuclear reaction cross section. In the following discussion, the center of mass energy E will be used instead of the relative velocity v.

For energies E far below the Coulomb energy, the cross section $\sigma(E)$ of a charged particle induced reaction drops steeply with decreasing energy due to the Coulomb barrier in the entrance channel:

$$\sigma(E) = \frac{S(E)}{E}e^{-2\pi\eta}, \tag{2}$$

where $S(E)$ is the astrophysical S-factor [1], and η is the Sommerfeld parameter with $2\pi\eta = 31.29\ Z_1Z_2\sqrt{\frac{\mu}{E}}$. Here Z_1 and Z_2 are the atomic numbers of projectile and target nucleus, respectively, μ is the reduced mass (in amu), and E is the center of mass energy (in keV).

The derivative $\frac{\mathrm{d}\langle\sigma v\rangle}{\mathrm{d}E}$ forms the so-called Gamow peak, and its maximum is found at the Gamow energy E_G. Because of the energy dependence of the cross section, the

[1] In the present work, E denotes the energy in the center of mass system, and E_p is the projectile energy in the laboratory system.

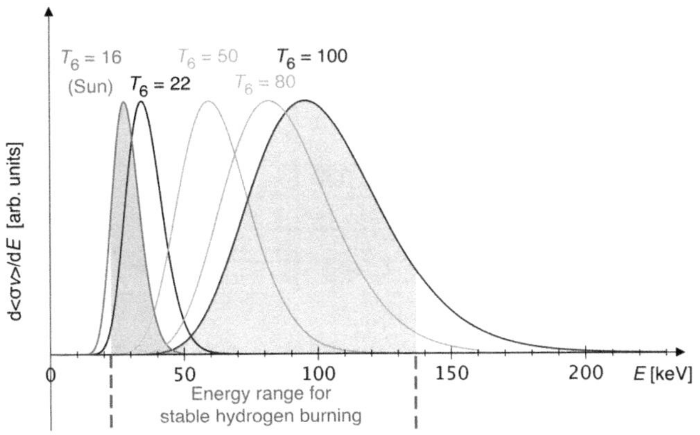

Fig. 3. Gamow peaks for the $^{14}N(p,\gamma)^{15}O$ reaction in stable hydrogen burning scenarios. The peaks have been normalized to equal height. The shaded areas cover 90% of the integral under the respective Gamow peak.

Gamow energy is generally much higher than the temperature k_BT (k_B: Boltzmann's constant) for the star. For example at solar temperature, $k_BT = 1.4\,keV$ and $E_G = 27\,keV$ for the $^{14}N(p,\gamma)^{15}O$ reaction (fig. 3).

Hydrogen burning in stars on the main sequence of the Hertzsprung-Russell diagram [1] takes place at temperatures of the order of $T_6 = 3$–100, the latter value for very heavy primordial stars [8]. Temperatures of $T_6 = 50$–80 are typical for the hydrogen burning shell of an asymptotic giant branch (AGB) star of mass $M = 2M_\odot$ ($M_\odot$: mass of our Sun) [9]. Higher temperatures are typical for explosive scenarios like novae [10] and X-ray bursts, which are not discussed here.

In the most recent solar model BS05 [11], the CNO cycle contributes only 0.8% of the solar luminosity, but a precise knowledge of its rate at $T_6 \approx 16$, the temperature at the center of our Sun, can help test stellar evolution theory [3]. Low mass stars leave the main sequence in the Hertzsprung-Russell diagram towards the end of their life. The luminosity at this turnoff point depends on the CNO rate and can be used to determine the age of the star [12]; the larger the rate, the fainter the turnoff luminosity. This can be used to give an independent lower limit on the age of the universe [13,14]. The stellar temperature at this turnoff point is of the order of $T_6 \approx 20$, depending on the star to be studied.

Using the temperatures indicated, one can propose an energy range of interest for understanding CNO hydrogen burning for the most important non-explosive stellar scenarios (fig. 3).

The cross section $\sigma(E)$ has a very low value at the resultant energies $E = 20$–140 keV, $\sigma(E) = 10^{-22}$–10^{-10} barn (eq. (2)). This prevents a direct cross section measurement in a laboratory at the earth's surface, where the signal to background ratio is too small because of cosmic ray interactions in detector, target, and shield. Hence, cross sections are measured at high energies and expressed as the astrophysical S-factor from eq. (2). The S-factor is then used to extrapolate the data to the relevant Gamow peak. Although $S(E)$ varies only slowly with energy for the direct nuclear reaction process, resonances and resonance tails may hinder an extrapolation, resulting in large uncertainties [1].

Therefore, the primary goal of experimental nuclear astrophysics remains to measure the cross section at energies inside the Gamow peak, or at least to approach it as closely as possible. The Laboratory for Underground Nuclear Astrophysics (LUNA) has been created for this purpose.

3 The $^{14}N(p,\gamma)^{15}O$ reaction

3.1 Situation up to the year 2000

Up to the year 2000, there have been many experimental studies of the $^{14}N(p,\gamma)^{15}O$ reaction at low energy (see, *e.g.*, [15,16,17,18,19,20]). The energy levels in the ^{15}O nucleus are known ([21], fig. 4), and it is also known that only capture into the ground state and three excited states in ^{15}O, at 5.181, 6.172, and 6.791 MeV, contributes significantly to the cross section at astrophysical energies [20].

Only one of the above named studies [16] obtained data that were at the edge of the astrophysically relevant energy region, with 50% statistical uncertainty for the cross section values. The other studies offer data only at energies above the astrophysical range, and generally, the

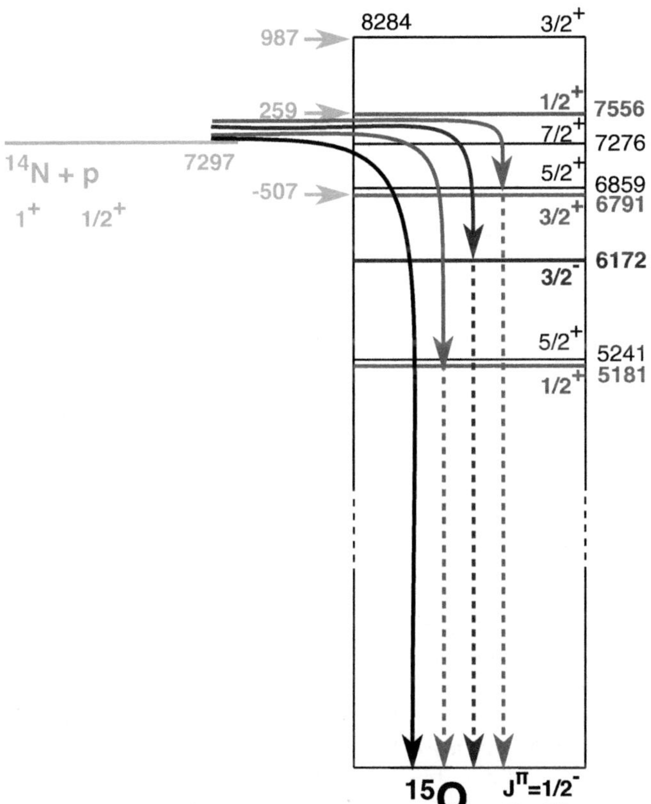

Fig. 4. Level scheme of ^{15}O up to 1 MeV above the $^{14}N + p$ threshold according to [21]. For levels shown bold, the level energies are taken from the LUNA solid target experiment [22].

results are extrapolated in the framework of the R-matrix model down to stellar energies. The standard cross section value used in reaction rate compilations [23,4] is mainly based on the data of the comprehensive study by Schröder *et al.* [20] and on the low energy data from ref. [16].

After the year 2000, the R-matrix results of Schröder *et al.* [20] have been revised by several works, on theoretical [24], indirect [25,26,27,28] and direct experimental [22, 29] grounds. The most dramatic revision was for capture to the ground state in ^{15}O; the following section focuses on this particular transition.

3.2 Recent R-matrix fits for radiative capture to the ground state in ^{15}O

The capture cross section into the ground state in ^{15}O is determined by destructive interference of direct capture amplitudes with resonant capture through the 6.79 MeV state.

The direct capture can be parameterized with an asymptotic normalization coefficient (ANC) C [30]. The total ANC for capture into the ground state $C = \sqrt{C^2_{p_{1/2}} + C^2_{p_{3/2}}}$ (with two proton orbitals contributing) has been experimentally determined through the ^{14}N(^{3}He, d)^{15}O reaction in two independent recent studies from Triangle Universities (TUNL) in 2002 [26] and from Texas A&M University (TAMU) in 2003 [27], with consistent results (table 1).

The most important parameter for resonant capture to the ground state via the 6.79 MeV state (acting as a subthreshold resonance in this case) is the width $\Gamma_{\gamma,6.79}$ of that state. This width has been measured with the Doppler shift attenuation method at TUNL in 2001 [25], and the obtained value could be confirmed in a Coulomb excitation study at RIKEN in 2004 [28] (table 2).

The LUNA 2004 study [22] measured the cross section for capture into the ground state in ^{15}O down to energies as low as E = 119 keV, much lower than any previous study for this transition, and directly confirmed the revised extrapolation for the ground state at those energies, inside the Gamow peak for some scenarios of stable hydrogen burning. Before, this revision had been based solely on theoretical and indirect considerations. In addition, the new low-energy data as well as previous data at higher energy, up to 2.5 MeV, from ref. [20][2] were used for a new R-matrix fit (fig. 5).

The TUNL 2005 study [29] gave experimental data that are consistent with ref. [22], albeit with larger error bars. This study used its own experimental data (E = 187–482 keV for the ground state) also for an R-matrix fit (fig. 5), without including higher energy data in their fit. For comparison, also the 2003 R-matrix fit by the TAMU group [27] that is based on their ANC measurement and normalized to the direct data from ref. [20] is included in the figure.

Figure 5 reveals interesting differences between the four extrapolations shown. The high $S(0)$ value from Schröder 1987 [20] is clearly dominated by the state at 6.79 MeV, here acting as a resonance 507 keV below the reaction threshold. All other extrapolations shown use a much smaller $\Gamma_{\gamma,6.79}$ value than Schröder 1987. Surprisingly, the fit by TUNL 2002 [26], not shown in the figure,

Table 1. Asymptotic normalization coefficient C for direct capture into the ground state in ^{15}O from different works.

Group	C [fm$^{-\frac{1}{2}}$]	Method	Data from
Angulo 2001 [24]	5.6	fit	[20]
TUNL 2002 [26]	7.9 ± 0.9	exp	[26,20]
TAMU 2003 [27]	7.3 ± 0.4	exp	[27,20]
LUNA 2004 [22]	7.3	fit	[22,20]
TUNL 2005 [29]	4.5 – 4.8	fit	[29]

Table 2. Gamma width of the state at 6.79 MeV in ^{15}O from different works.

Group	$\Gamma_{\gamma,6.79}$ [eV]	Method	Data from
Schröder 1987 [20]	6.3 ± 1.9	fit	[20]
Angulo 2001 [24]	1.75 ± 0.60	fit	[20]
TUNL 2001 [25]	$0.41^{+0.34}_{-0.13}$	exp	[25]
TAMU 2003 [27]	0.35	fit	[27,20]
RIKEN 2004 [28]	$0.95^{+0.60}_{-0.95}$	exp	[28]
LUNA 2004 [22]	0.8 ± 0.4	fit	[22,20]
TUNL 2005 [29]	1.7 – 3.2	fit	[29]

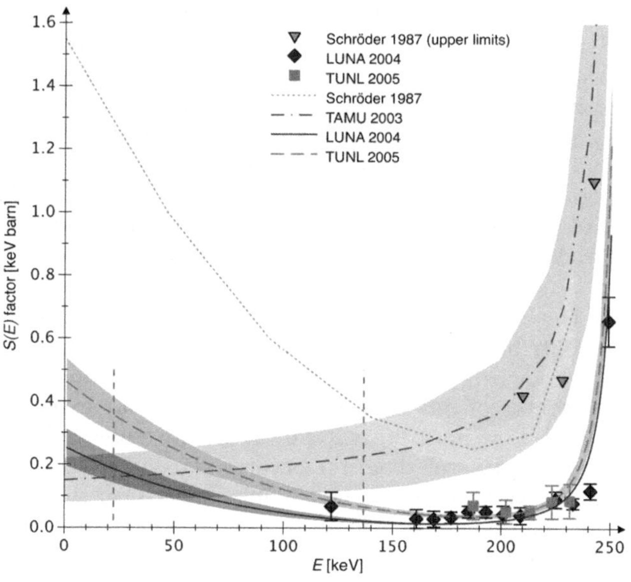

Fig. 5. Direct experimental data (inverted triangles: Schröder 1987 [20], upper limits; diamonds: LUNA 2004 [22]; squares: TUNL 2005 [29]) and R-matrix fits (lines) for capture to the ground state in ^{15}O. The shaded areas around the lines correspond to the relative error for the extrapolated $S(0)$ value quoted by each of the studies. The vertical lines correspond to the energy range for stable hydrogen burning defined in fig. 3.

[2] In addition to presenting new, low energy data, the LUNA 2004 work [22] corrected the Schröder ground state data [20] for the summing-in effect and included this corrected data in the R-matrix fit.

Table 3. Extrapolated $S(0)$-factor for radiative proton capture into three states in ^{15}O from different works.

Capture into ^{15}O state with E_x =	6.791	6.172	GS
Schröder 1987 [20]	1.41 ± 0.02	0.14 ± 0.05	1.55 ± 0.34
Angulo 2001 [24]	1.63 ± 0.17	$0.06^{+0.01}_{-0.02}$	$0.08^{+0.13}_{-0.06}$
TUNL 2002 [26]	1.17 ± 0.28	0.14 ± 0.03	1.67 ± 0.40
TAMU 2003 [27]	1.40 ± 0.20	0.13 ± 0.02	0.15 ± 0.07
Nelson 2003 [31]	1.50	0.16 ± 0.06	
LUNA 2004 [22]	1.35 ± 0.05		0.25 ± 0.06
TUNL 2005 [29]	1.15 ± 0.05	0.04 ± 0.01	0.49 ± 0.08

yields a similar rise of the S-factor to low energies, up to $S(0) = 1.67$ keV barn, even though that study took solely direct capture into account.

The main difference between the input parameters used by the LUNA 2004 [22] and the TUNL 2005 [29] studies is that LUNA obtained a $\Gamma_{\gamma,6.79}$ value that is much lower than the TUNL number (table 2). Both studies had left $\Gamma_{\gamma,6.79}$ as a free parameter to fit their experimental excitation functions. An analogous approach was used by the same two studies regarding the ANC of the ground state, where LUNA obtains a 50% higher value than TUNL. The 2003 R-matrix fit by the TAMU group [27] used a value for $\Gamma_{\gamma,6.79}$ that was very close to experiment, and the ANC used for the fit was obtained experimentally in the same work.

In summary, the results of different extrapolations (table 3) differ by more than the standard deviations quoted in the individual works, especially for capture into the ground state in ^{15}O, but also for capture into the other two states contributing significantly, those at 6.172 and 6.791 MeV. It is therefore worthwhile to attempt a direct measurement of the cross section at energies of astrophysical interest.

4 Laboratory for Underground Nuclear Astrophysics (LUNA)

The Laboratory for Underground Nuclear Astrophysics (LUNA) has been designed for cross section measurements at energies in or near the Gamow peak. It is located in the Gran Sasso underground laboratory (Laboratori Nazionali del Gran Sasso, LNGS[3]) in Italy. LUNA uses high current accelerators with small energy spread in combination with high efficiency detection systems, one of which is described below.

The Gran Sasso facility consists of three experimental halls and several connecting tunnels. Its site is protected from cosmic rays by a rock cover equivalent to 3800 m water. This shield suppresses the flux of cosmic ray induced muons by six orders of magnitude [32], resulting in a flux of muon induced neutrons of the order of $\Phi_{n_\mu} \approx 10^{-8} \frac{\text{n}}{\text{cm}^2 \cdot \text{s}}$ [33]. Because of neutrons from (α, n) reactions and spontaneous fission of ^{238}U taking place in the surrounding rock and concrete [34], the measured total neutron flux is higher, $\Phi_\text{n} \approx 4 \cdot 10^{-6} \frac{\text{n}}{\text{cm}^2\,\text{s}}$ [35]. This flux is three orders of magnitude below typical values for a laboratory at the surface of the earth.

This unique low background environment reduces the counting rate at 6.8 MeV in a germanium detector by at least a factor 2000, and at 6.5–8.0 MeV in a BGO detector by a factor 1600 [36]. For comparison, an active muon shield in a laboratory at the surface of the earth can reduce the background counting rate by about a factor 10–50 for E_γ = 7–11 MeV [37]. The shield provided by the Gran Sasso rock cover therefore offers a clear advantage, in particular at high γ energies, but also at low γ energies and for particle spectroscopy.

Taking advantage of the low laboratory background, at the 50 kV LUNA1 accelerator [38], the ^{3}He(^{3}He, 2p)^{4}He cross section was measured for the first time within its solar Gamow peak [39,40]. Subsequently, a windowless gas target setup and a 4π bismuth germanate (BGO) summing detector [41] have been used to study the radiative capture reaction ^{2}H(p, γ)^{3}He, also within its solar Gamow peak [42].

The 400 kV LUNA2 accelerator [43] has been used to study the radiative capture reaction ^{14}N(p, γ)^{15}O using titanium nitride (TiN) solid targets and a high purity germanium detector. The cross sections for the transitions to several states in ^{15}O, including the ground state, were measured down to $E = 119$ keV [44,22,45].

In order to extend the ^{14}N(p, γ)^{15}O cross section data to even lower energies, a gas target setup similar to the one used for the ^{2}H(p, γ)^{3}He study and an annular BGO detector have been installed at the LUNA2 400 kV accelerator [46].

5 LUNA ^{14}N(p, γ)^{15}O gas target experiment

A new measurement of the total cross section of the ^{14}N(p, γ)^{15}O reaction [6,7] has been performed in the Gran Sasso underground laboratory, at the LUNA2 400 kV accelerator [43]. The main features of the experiment are described in this section.

5.1 Setup

A schematic view of the setup is displayed in fig. 6. A three stage, differentially pumped, windowless gas target system (figs. 6 and 7) has been used. It is a modified version of the LUNA ^{2}H(p, γ)^{3}He setup [41], with a 120 mm long target cell. In the experiment, a proton beam of energy E_p = 80–250 keV and current up to 0.5 mA is provided by the LUNA2 400 kV accelerator and enters the target chamber through a sequence of long, narrow, water cooled apertures; the final aperture has a diameter of 7 mm, is 40 mm long and made from brass, with a copper cover on the side facing the ion beam. The target cell is fitted into the 60 mm wide bore hole at the center of an annular BGO detector having 70 mm radial thickness

[3] Web page: `http://www.lngs.infn.it`

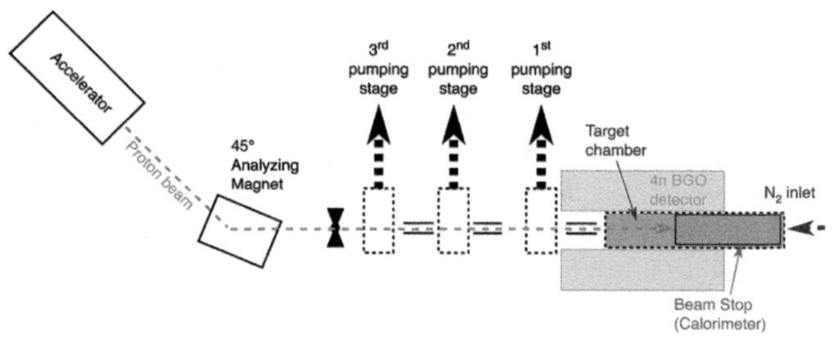

Fig. 6. Schematic view of the LUNA $^{14}N(p,\gamma)^{15}O$ gas target setup.

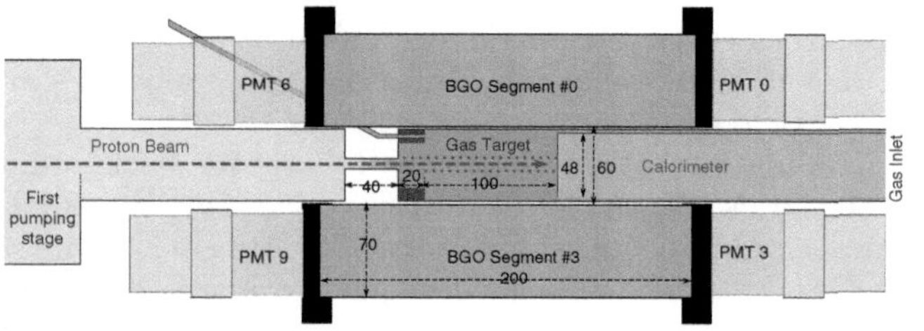

Fig. 7. Exploded view of the target chamber and 4π BGO detector. Dimensions are given in mm.

and 280 mm length. Also inside the BGO bore hole is a calorimeter (heated to 70 °C) for the measurement of the beam intensity, with a 41 mm thick block of oxygen free copper serving as the beam stop.

The target gas was 1.0 mbar nitrogen of chemical purity 99.9995% and natural isotopic composition, with 1.0 mbar helium gas of chemical purity 99.9999% used for monitor runs for ion beam induced background from the $^{13}C(p,\gamma)^{14}N$ reaction [36,6]. The first pumping stage is evacuated by a WS 2000 roots blower, leading to a pressure ratio between target and first pumping stage that is better than a factor 100. The second and third pumping stages are at 10^{-5} and 10^{-6} mbar pressure, respectively.

5.2 Target density

The target density without and with ion beam has been investigated in a dedicated study [47,6,7]. The target pressure was monitored with a capacitance pressure gauge with precision 0.1% and kept constant with a feedback system. The pressure profile within the target has been measured with similar precision and is flat to 4%. The temperature profile without incident ion beam has been measured to better than 1 K. To study the target density with an ion beam incident on the target, a collimated NaI detector was placed at an angle of 90° to the beam direction directly next to the target chamber, and the energy loss ΔE_{exp} of the ion beam inside the target chamber was measured with the resonance scan technique [48] using the $E_{\mathrm{p}} = 278$ keV resonance in the $^{14}N(p,\gamma)^{15}O$ reaction. The experiment was repeated for different pressures and beam currents I_{target} (fig. 8).

For high target pressure, therefore high power deposition per unit length in the target, there is a large relative effect on ΔE_{exp}. As is evident from fig. 8, the relative

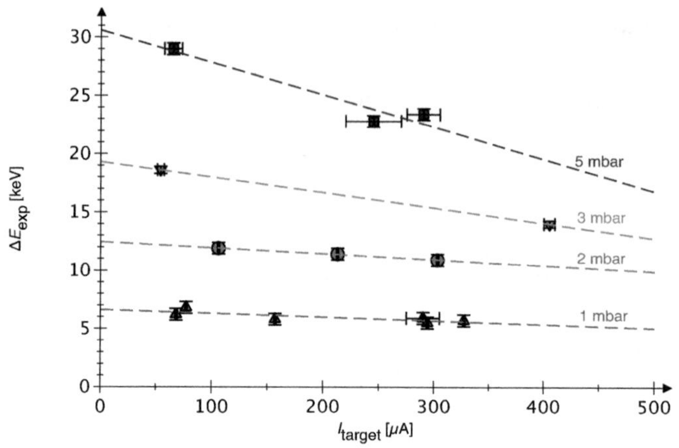

Fig. 8. Measured energy loss ΔE_{exp} in the nitrogen target gas as a function of I_{target} for different gas pressures. Triangles: 1 mbar; circles: 2 mbar; inverted triangles: 3 mbar; squares: 5 mbar.

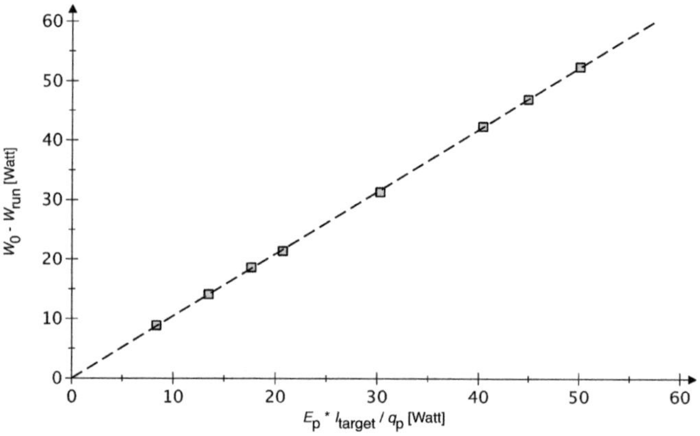

Fig. 9. Calorimetric power $W_0 - W_{\mathrm{run}}$ as a function of electrical power $\frac{E_{\mathrm{p}} \cdot I_{\mathrm{target}}}{q_{\mathrm{p}}}$, with q_{p} the charge of the proton. The dotted line is a fit to the data points.

change in ΔE_{exp} is also proportional to the beam current. Comparing ΔE_{exp} to the energy loss taken from the SRIM program [49], one obtains the particle density per unit volume. Consistent with the conclusions of ref. [48], the relative change in density was found to be proportional to the power deposited per unit length, which in the present case of small lateral straggling of the ion beam corresponds to the power deposited per unit volume.

5.3 Beam intensity

The intensity of the ion beam was measured with a calorimeter with constant temperature gradient [41]. The 41 mm thick copper beam stop forms the hot side of the calorimeter, that was kept at 70 °C with thermoresistors (power consumption typically 135 W). For the calibration of the calorimeter (fig. 9), the target chamber was used as a Faraday cup, a negative voltage was applied to the final collimator in order to repel secondary electrons, and the electrical target current I_{target} was measured with a standard current integrator. Electrical and calorimetric cur-

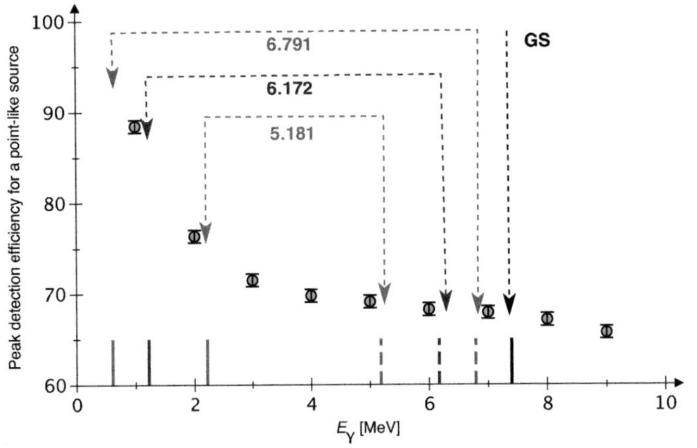

Fig. 10. Peak detection efficiency as a function of γ-ray energy [41]. The energies of the most important primary and secondary γ lines are indicated (primary: solid line, secondary: dashed line), for center of mass energy $E = 100$ keV.

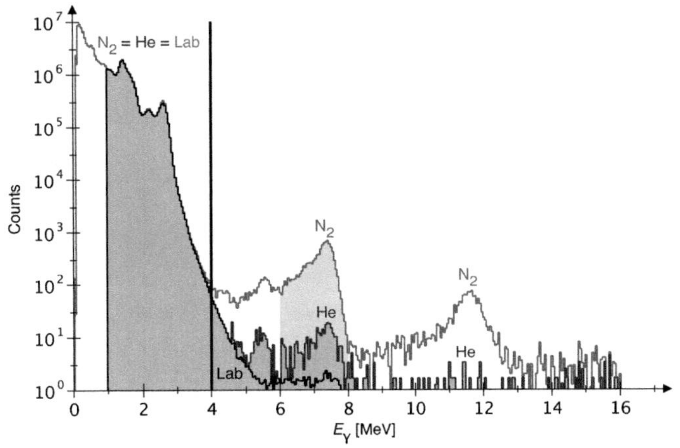

Fig. 11. N_2: Gamma-ray spectrum for $E_p = 140$ keV ($E = 127$ keV) with 1 mbar nitrogen gas, lifetime 47 hours, accumulated charge 45 coulomb. He: Same beam energy, 1 mbar helium in the target. For $E_\gamma < 4$ MeV, renormalized to equal lifetime with the N_2 run. For $E_\gamma > 4$ MeV, renormalized to equal charge and proton energy at the beam stop. Lab: Laboratory background without beam, renormalized to equal lifetime with the N_2 run.

rent were found to agree with a slope 5% different from unity, and no offset within errors.

5.4 Detection efficiency

The peak detection efficiency of the 4π BGO summing crystal as a function of γ-ray energy for a point-like source (fig. 10) has been given elsewhere [41].

For the analysis, all γ-rays detected in a region of interest (ROI) from 6 to 8 MeV (figs. 11, 12) are summed. Therefore, the peak from true coincidence summing of a primary and its associated secondary γ-ray is fully within the ROI, as well as the primary γ-ray at $E_\gamma = Q + E$ for capture into the ground state in ^{15}O (Q value $Q = 7.297$ MeV for ^{14}N(p, γ)^{15}O), the secondary γ-ray at 6.791 MeV and 80% of the peak area of the 6.172 MeV secondary γ-ray. This selection of the ROI renders the detection efficiency independent of the branching ratios for capture to the ground state and to the state at 6.791 MeV, and only weakly dependent on the branching ratio for capture to the state at 6.172 MeV. The efficiency depends more strongly on the branching ratio for capture to the state at 5.181 MeV, but the impact is small because of the low value of the branching to this state: 3.6% branching at the lowest measured point [45], and 0.8% extrapolated at zero energy [20]. Overall, the assumptions on these four branching ratios contribute 0.5% to the uncertainty in the detection efficiency.

The γ-ray detection efficiency for radiative capture to the states at 7.276 and 6.859 MeV in ^{15}O (fig. 4, extrapolated branching at zero energy 1 and 2%, respectively [20]) is 20% lower than for capture to the state at 6.791 MeV, because those states decay to the ground state via the 5.241 MeV state. The efficiency for capture into the 5.241 MeV state (extrapolated branching at zero energy 1% [20]) shows the same behavior. Capture into the three states at 7.276, 6.859 and 5.241 MeV in ^{15}O has been neglected in the present experiment. If one assumes three times higher branching ratios at low energy for these

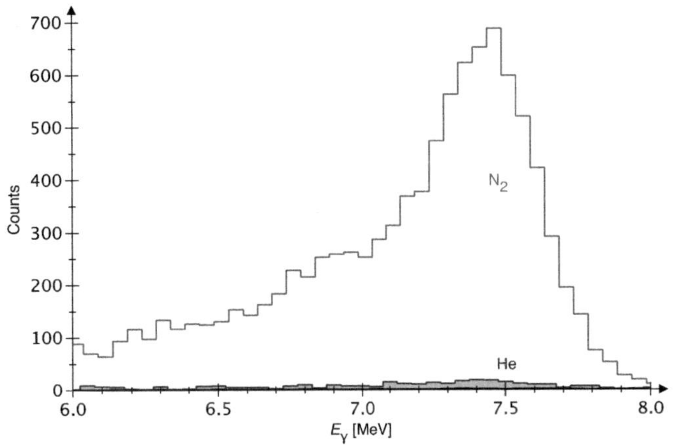

Fig. 12. Same spectrum as fig. 11, enlarged to the ROI. The Compton background to be subtracted for this spectrum corresponds to 5 counts per channel (not shown in the figure).

states than given by extrapolation [20], the total cross section obtained increases by 3%.

In summary, while the calculated detection efficiency does depend on the branching ratios for capture into the different states as taken from the LUNA solid target experiment [45] and from R-matrix extrapolations for low energy [22,24,20], this dependence is diluted by the particularities of the ^{15}O level scheme, the essentially flat peak detection efficiency curve for 5 MeV $< E_\gamma <$ 8 MeV, and the choice of a wide ROI, so that the resultant systematic uncertainty is 1% for reasonable and 3% for worst case assumptions on the uncertainties of the branching ratios.

The angular distribution $W(\vartheta)$ of the emitted γ-rays has been studied previously in the LUNA solid target experiment [50], above and below the $E_p = 278$ keV reso-

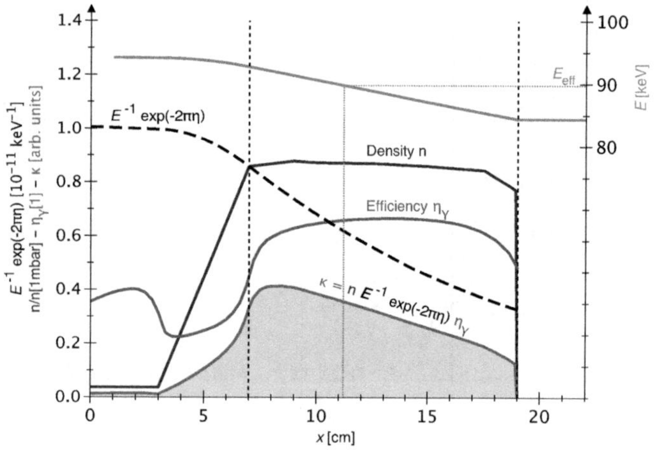

Fig. 13. Parameters used for the analysis in an example run at $E = 90\,\text{keV}$. Right axis: energy $E(x)$ [keV]. Left axis: $E(x)^{-1}\exp(-2\pi\eta)$ $[10^{-11}\,\text{keV}^{-1}]$. Effective target density $n(x)/n[1\,\text{mbar, nobeam}]$; absolute γ-ray detection efficiency η_γ; weighting factor κ for each piece of the target [arbitrary units].

nance; it can be parameterized as

$$W(\vartheta) \approx 1 + a_1 \cdot P_1(\cos(\vartheta)) + a_2 \cdot P_2(\cos(\vartheta)), \quad (3)$$

where $P_{1,2}$ are the first and second order Legendre coefficients and ϑ is the angle (in the center of mass system) between the ion beam and the direction of emission of the γ-ray.

All secondary γ-rays shown in fig. 10 were observed to be isotropic within errors, in agreement with theoretical expectation. For the primary γ-rays shown in fig. 10, theory predicts Legendre coefficients $a_2 < 0$ or $a_2 = 0$ for incident s- and p-waves. The data show all primary γ-rays to be isotropic within errors, with the exception of that from capture to the state at 6.791 MeV, where $a_2 \approx -0.8$ below the resonance. For all primary and secondary transitions, below the resonance the a_1 coefficient was found to be zero within errors [50].

An anisotropy with $a_2 < 0$ enhances emission perpendicular to the beam direction and therefore the detection efficiency for the low energy primary γ-ray, increasing the probability of it being detected in coincidence with the corresponding secondary γ-ray. For capture into the states at 6.791 and 6.172 MeV, the angular distribution of the primary γ-rays, while changing the shape of the spectrum, does not affect the detection efficiency, because the selection of the ROI ensures detection of both the secondary and the sum peak. For capture into the ground state (where theory predicts isotropy) and into the state at 5.182 MeV, there is an effect, but it is diluted because of the relatively small branching of those two states (combined less than 20%). The overall impact of the angular distributions on the detection efficiency is smaller than 3% without theoretical input and negligible when taking theory into account.

Using these inputs, the γ-ray detection efficiency η_γ can then be calculated for each point in the target, taking the solid angle and the effective detector thickness into account [6], with corrections for the attenuation of γ-rays in the vacuum vessel, the massive brass collimator and the massive copper beam stop. For the example shown in fig. 13, the detection efficiency is $\eta_\gamma = 0.592 \pm 0.020$, with the uncertainty given by the radioactive source used for the calibration (1.5%), the detector modeling (1%, [41]) and the branching ratios discussed above (1%).

5.5 Gamma-ray spectra and background

Using a dedicated setup, the γ-ray background has been studied previously to the actual experiment, identifying and localizing the major background sources [36,6]. Typical γ-ray spectra from the 4π BGO summing detector are shown in fig. 11, with the region of interest (ROI) for the $^{14}\text{N}(\text{p},\gamma)^{15}\text{O}$ study shaded in the (N_2) spectrum.

For $E_\gamma < 4\,\text{MeV}$, the spectrum is dominated by the laboratory background, whose counting rate in the ROI is constant and well known [36]. At higher γ-energies, the background induced by the ion beam is for most runs more important than the laboratory background. Background induced by the $^{13}\text{C}(\text{p},\gamma)^{14}\text{N}$ reaction ($Q = 7.551\,\text{MeV}$) leads to 7.7 MeV γ-rays, superimposed with the sum peak from the reaction to be studied. In order to evaluate the contribution from this reaction, monitor runs with helium gas in the target were performed at the same beam energy. The resulting monitor spectrum is then renormalized for equal charge with the nitrogen spectrum and for equal energy of the proton beam when arriving at the beam stop, where the ^{13}C background originates ((He) spectrum in fig. 11).

In the nitrogen (N_2) spectrum, the dominating peak in the ROI (fig. 12) is the sum peak at $E_\gamma - Q + E$. To the left of it are unresolved lines at 6.172 and 6.791 MeV, the energies of the secondary γ-rays. Outside the ROI, the peak at 5 MeV (fig. 11) is mostly from the secondary γ-ray at 5.181 MeV from the reaction to be studied, but partly also from the $^2\text{H}(\text{p},\gamma)^3\text{He}$ beam induced background reaction, as is revealed by the helium monitor run.

The broad structure at 12 MeV in the N_2 spectrum (fig. 11) results mainly from the $^{15}\text{N}(\text{p},\gamma)^{16}\text{O}$ reaction (the target gas has natural isotopic composition, 0.4% ^{15}N), but also from the $^{11}\text{B}(\text{p},\gamma)^{12}\text{C}$ beam induced background reaction. This last reaction also gives γ-rays at 16 MeV. All reactions leading to γ-rays of $E_\gamma > 8\,\text{MeV}$ [36] cause a small Compton continuum at lower energies. Its contribution is evaluated from a global fit to the helium monitor runs (after ^{13}C correction), and a correction factor is deduced, so that the high energy counts in each spectrum are used to calculate the Compton background for that same spectrum [6]. Finally, single lines from resonant background reactions producing γ-rays in the ROI [36] were fitted and subtracted for runs close to the resonance energy.

5.6 Data analysis

With the γ-ray detection efficiency η_γ, the effective target density n and therefore also the energy loss of the ion beam in the target (in the present case, typically 10 keV) known, a weighting factor $\kappa(x)$ is calculated for each point x in the target:

$$\kappa(x) \stackrel{\text{Def}}{=} n \cdot \frac{1}{E(x)} e^{-2\pi\eta} \cdot \eta_\gamma \qquad (4)$$

with $2\pi\eta$ the Sommerfeld parameter from eq. (2). The parameter $\kappa(x)$ (fig. 13) can then be used to calculate the effective energy E_{eff} and, using the measured yield Y, the astrophysical S-factor $S(E_{\text{eff}})$ [1,6]:

$$Y = \int_{x=0\,\text{cm}}^{28\,\text{cm}} \sigma(E(x)) \cdot n(x) \cdot \eta_\gamma(x)\,\mathrm{d}x = $$

$$= S(E_{\text{eff}}) \cdot \int_{x=0\,\text{cm}}^{28\,\text{cm}} \kappa(x)\,\mathrm{d}x\,. \qquad (5)$$

This analysis method requires an assumption on the energy dependence of the S-factor. In the present experiment, as a first step the analysis has been performed under the assumption of an S-factor that is constant over the energy interval given by the energy loss in the target. In a second step, the obtained energy dependence of the S-factor has been used as input for the renewed analysis.

Using this method, total cross section data with statistical uncertainties better than 10% has been obtained in the energy range $E = 70$–230 keV, energies lower than any previous study.

6 Astrophysical scenarios that can be better understood using data from the present experiment

The data obtained in the present experiment [6,7] can be used to directly evaluate the reaction rate for several important stellar scenarios, with negligible impact from the extrapolation applied for lower energies.

The derivative $\frac{\mathrm{d}\langle\sigma v\rangle}{\mathrm{d}E}$ of the reaction rate from eq. (1) has been calculated from the LUNA gas target experimental S-factor data [6], assuming a flat S-factor equal to the S-factor at $E = 70$ keV for $E < 70$ keV, where there is no data (fig. 14). For temperatures $T_6 \geq 60$, the data from the present experiment cover more than 50% of the Gamow peak, for $90 \leq T_6 \leq 300$, more than 90% of the Gamow peak, when one includes the strength of the $E = 259$ keV resonance that was also measured in the LUNA gas target experiment [7].

Low mass stars burn first hydrogen and then helium in their center. After the end of the helium burning phase, the star consists of a degenerate core of oxygen and carbon and two shells burning hydrogen and helium, respectively. This phase of stellar evolution is called the asymptotic giant branch (AGB) [51]. It is characterized by flashes of the helium burning shell that spawn convective mixing in a process called dredge-up. Such a dredge-up transports the products of nuclear burning from inner regions of the

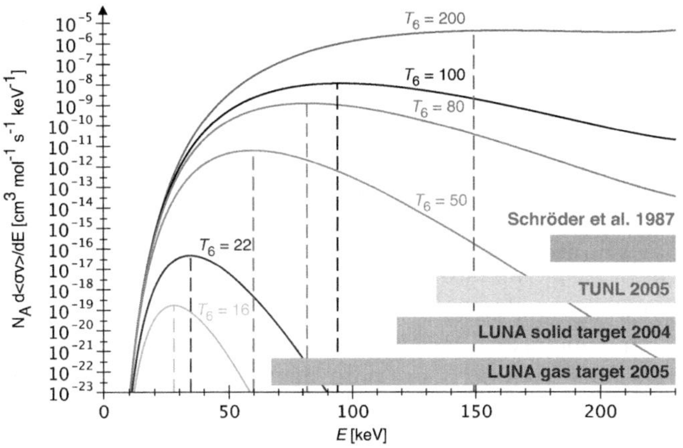

Fig. 14. Gamow peaks for several stellar temperatures discussed in the text. The horizontal bars correspond to the energy range where direct experimental data has been obtained in the study by Schröder *et al.* 1987 [20], the LUNA solid target experiment 2004 [22], the TUNL 2005 study [29], and the LUNA gas target experiment (present work).

star to its surface, where they are in principle accessible to astronomical observations.

The temperature in the hydrogen burning shell of an AGB star is of the order of $T_6 = 50$–80 for the example of a $2\,M_\odot$ star with metallicity $Z = 0.01$. It has been shown [9] that an arbitrary 25% reduction of the ^{14}N(p, γ)^{15}O rate with respect to the NACRE [4] rate leads to twice as efficient dredge-up of carbon to the surface of the star, because the rate of energy generation in the hydrogen burning shell becomes even lower than before, enhancing the disequilibrium between hydrogen and helium burning shell. The CNO rate suggested by the present study [6,7] is more than 25% below the NACRE [4] rate. Recent experimental data on the carbon producing triple-α reaction [52] result in a 10–20% decrease of its rate at temperatures relevant for helium shell burning, leading to a slightly lower production of carbon, reducing in a commensurate decrease of the amount of carbon transported to the stellar surface [9]. Still, the change in the ^{14}N(p, γ)^{15}O rate might lift a disagreement between model and observation for so-called carbon stars [53]: For low (*i.e.* $2\,M_\odot$) mass stars, models do not reproduce a sufficiently high dredge-up efficiency.

Recently, a simulation for a $5\,M_\odot$, $Z = 0.02$ AGB star [54] found stronger thermal flashes for a reduced CNO rate, consistent with the finding of ref. [9] for a $2\,M_\odot$, $Z = 0.01$ AGB star.

For a zero metallicity (population III) star of $1\,M_\odot$, after a sufficient amount of carbon has been created in the triple-α reaction, the CNO cycle is ignited in the so called CN flash. This CN flash takes place at $T_6 \approx 65$ and leads to a brief loop of the trajectory of the star in the Hertzsprung-Russell diagram [55]. With a CNO rate that is 40% lower than the NACRE [4] rate, this loop disappears [54]. Also, the first core helium flash in such a star was found to be less luminous than in the reference case, albeit with a higher core mass, as a result of a lower CNO rate [54].

Temperatures of $T_6 \approx 100$ correspond to CNO burning in heavy ($20\,M_\odot$) population III stars [8]. Explosive burning in novae [10] takes place at even higher temperatures, typically $T_6 \approx 200$. The $\frac{^{15}\mathrm{N}}{^{14}\mathrm{N}}$ isotopic ratio in nova ashes depends sensitively on the $^{14}\mathrm{N}(\mathrm{p},\gamma)^{15}\mathrm{O}$ rate [56]; the more precise rate that can be calculated from the cross sections obtained in the present study will reduce the uncertainty of the isotopic ratio.

In conclusion, data from the present study allow for the first time to directly evaluate the reaction rate for several scenarios of stable stellar hydrogen burning, as well as for explosive hydrogen burning.

During the experiment, D. Bemmerer was Wissenschaftlicher Mitarbeiter at the Institut für Atomare Physik und Fachdidaktik, Technische Universität Berlin, Germany. This work was supported in part by: INFN, TARI HPRI-CT-2001-00149, OTKA T 42733, BMBF (05CL1PC1-1), FEDER-POCTI/FNU/41097/2001, and EU RII3-CT-2004-506222.

References

1. C. Rolfs, W. Rodney, *Cauldrons in the Cosmos* (University of Chicago Press, Chicago, 1988).
2. H. Bethe, Phys. Rev. **55**, 434 (1938).
3. J.N. Bahcall, M.H. Pinsonneault, Phys. Rev. Lett. **92**, 121301 (2004).
4. C. Angulo *et al.*, Nucl. Phys. A **656**, 3 (1999).
5. M. Wiescher, J. Görres, H. Schatz, J. Phys. G **25**, R133 (1999).
6. D. Bemmerer, *Experimental study of the* $^{14}\mathrm{N}(\mathrm{p},\gamma)^{15}\mathrm{O}$ *reaction at energies far below the Coulomb barrier*, PhD Thesis, Technische Universität Berlin (2004).
7. A. Lemut, *Misura della sezione d'urto della reazione* $^{14}\mathrm{N}(\mathrm{p},\gamma)^{15}\mathrm{O}$ *ad energie di interesse astrofisico*, PhD Thesis, Università degli Studi di Genova (2005).
8. L. Siess, M. Livio, J. Lattanzio, Astrophys. J. **570**, 329 (2002).
9. F. Herwig, S. M. Austin, Astrophys. J. **613**, L73 (2004).
10. J. José, M. Hernanz, Astrophys. J. **494**, 680 (1998).
11. J.N. Bahcall, A.M. Serenelli, S. Basu, Astrophys. J. **621**, L85 (2005).
12. L. Krauss, B. Chaboyer, Science **299**, 65 (2003).
13. G. Imbriani *et al.*, Astron. Astrophys. **420**, 625 (2004).
14. E. Degl'Innocenti *et al.*, Phys. Lett. B **590**, 13 (2004).
15. D.B. Duncan, J.E. Perry, Phys. Rev. **82**, 809 (1951).
16. W. Lamb, R. Hester, Phys. Rev. **108**, 1304 (1957).
17. R.E. Pixley, *The reaction cross section of nitrogen 14 for protons between* 220 keV *and* 600 keV, PhD Thesis, California Institute of Technology (1957).
18. B. Povh, D.F. Hebbard, Phys. Rev. **115**, 608 (1959).
19. D.F. Hebbard, G.M. Bailey, Nucl. Phys. **49**, 666 (1963).
20. U. Schröder *et al.*, Nucl. Phys. A **467**, 240 (1987).
21. F. Ajzenberg-Selove, Nucl. Phys. A **523**, 1 (1991).
22. A. Formicola *et al.*, Phys. Lett. B **591**, 61 (2004).
23. E. Adelberger *et al.*, Rev. Mod. Phys. **70**, 1265 (1998).
24. C. Angulo, P. Descouvemont, Nucl. Phys. A **690**, 755 (2001).
25. P. Bertone *et al.*, Phys. Rev. Lett. **87**, 152501 (2003).
26. P.F. Bertone *et al.*, Phys. Rev. C **66**, 055804 (2002).
27. A. Mukhamedzhanov *et al.*, Phys. Rev. C **67**, 065804 (2003).
28. K. Yamada *et al.*, Phys. Lett. B **579**, 265 (2004).
29. R.C. Runkle *et al.*, Phys. Rev. Lett. **94**, 082503 (2005).
30. A.M. Mukhamedzhanov, C.A. Gagliardi, R.E. Tribble, Phys. Rev. C **63**, 024612 (2001).
31. S.O. Nelson *et al.*, Phys. Rev. C **68**, 065804 (2003).
32. S.P. Ahlen *et al.*, Phys. Lett. B **249**, 149 (1990).
33. H. Wulandari *et al.*, (2004), hep-ex/0401032.
34. H. Wulandari, J. Jochum, W. Rau, F. von Feilitzsch, Astropart. Phys. **22**, 313 (2004).
35. P. Belli *et al.*, Nuovo Cimento A **101**, 959 (1989).
36. D. Bemmerer *et al.*, Eur. Phys. J. A **24**, 313 (2005).
37. G. Müller *et al.*, Nucl. Instrum. Methods A **295**, 133 (1990).
38. U. Greife *et al.*, Nucl. Instrum. Methods A **350**, 327 (1994).
39. M. Junker *et al.*, Phys. Rev. C **57**, 2700 (1998).
40. R. Bonetti *et al.*, Phys. Rev. Lett. **82**, 5205 (1999).
41. C. Casella *et al.*, Nucl. Instrum. Methods A **489**, 160 (2002).
42. C. Casella *et al.*, Nucl. Phys. A **706**, 203 (2002).
43. A. Formicola *et al.*, Nucl. Instrum. Methods A **507**, 609 (2003).
44. A. Formicola *et al.*, Nucl. Phys. A **719**, 94c (2003).
45. A. Formicola, *A new study of* $^{14}\mathrm{N}(\mathrm{p},\gamma)^{15}\mathrm{O}$ *at low energy*, PhD Thesis, Ruhr-Universität Bochum (2004).
46. LUNA Collaboration, LNGS Annual Report, 159 (2003).
47. F. Confortola, *Studio della reazione* $^{14}\mathrm{N}(\mathrm{p},\gamma)^{15}\mathrm{O}$ *ad energie di interesse astrofisico*, Master's Thesis, Università degli Studi di Genova (2003).
48. J. Görres *et al.*, Nucl. Instrum. Methods **177**, 295 (1980).
49. J. Ziegler, `SRIM` version 2003.26, `http://www.srim.org` (2004).
50. H. Costantini, *Direct measurements of radiative capture reactions at astrophysical energies*, PhD Thesis, Università degli Studi di Genova (2003).
51. I. Iben, A. Renzini, Annu. Rev. Astron. Astrophys. **21**, 271 (1983).
52. H. Fynbo *et al.*, Nature **433**, 136 (2005).
53. I. Iben, Astrophys. J. **246**, 278 (1981).
54. A. Weiss, A. Serenelli, A. Kitsikis, H. Schlattl, J. Christensen-Dalsgaard, astro-ph/0503408 (2005).
55. A. Weiss, S. Cassisi, H. Schlattl, M. Salaris, Astrophys. J. **533**, 413 (2000).
56. C. Iliadis, A. Champagne, J. José, S. Starrfield, P. Tupper, Astrophys. J. Suppl. Ser. **142**, 105 (2002).

Eur. Phys. J. A **27**, s01, 171–176 (2006)
DOI: 10.1140/epja/i2006-08-025-6

EPJ A direct
electronic only

Pygmy dipole strength close to particle-separation energies —The case of the Mo isotopes

G. Rusev[1], E. Grosse[1,2,a], M. Erhard[1], A. Junghans[1], K. Kosev[1], K.-D. Schilling[1], R. Schwengner[1], and A. Wagner[1]

[1] Forschungszentrum Rossendorf, Institut für Kern- und Hadronenphysik, Postfach 510119, 01314 Dresden, Germany
[2] Technische Universität Dresden, Institut für Kern- und Teilchenphysik, 01062 Dresden, Germany

Received: 4 July 2005 /
Published online: 13 March 2006 –

Abstract. The distribution of electromagnetic dipole strength in ^{92,98,100}Mo has been investigated by photon scattering using bremsstrahlung from the new ELBE facility. The experimental data for well-separated nuclear resonances indicate a transition from a regular to a chaotic behaviour above 4 MeV of excitation energy. As the strength distributions follow a Porter-Thomas distribution much of the dipole strength is found in weak and in unresolved resonances appearing as fluctuating cross section. An analysis of this quasi-continuum —here applied to nuclear resonance fluorescence in a novel way— delivers dipole strength functions, which are combining smoothly to those obtained from (γ, n) data. Enhancements at 6.5 MeV and at ~ 9 MeV are linked to the pygmy dipole resonances postulated to occur in heavy nuclei.

PACS. 21.10.Pc Single-particle levels and strength functions – 24.10.Lx Monte Carlo simulations (including hadron and parton cascades and string breaking models) – 25.20.Dc Photon absorption and scattering – 26.50.+x Nuclear physics aspects of novae, supernovae, and other explosive environments

1 Dipole strength in heavy nuclei

The response of nuclei to dipole radiation is of special importance for the synthesis of the chemical elements in the cosmos: particle thresholds may be crossed in hot or explosive scenarios leading to the production of new nuclides from previously formed heavier ones by dissociation in the thermal photon bath. This is likely to be the main path for the generation of the approximately 30–40 neutron-deficient nuclides which cannot be produced in neutron capture reactions [1]. For the understanding and modelling of this so-called p-process the dipole strength function up to and near the particle thresholds has to be known accurately [2]. As shown previously [3], details of the dipole strength (now in n-rich nuclei) may as well have large consequences for the r-process path and also s-process branchings are influenced by nuclear excitations [4] induced by thermal photons.

The experimental knowledge [5] on dipole strength is reasonably well established for many heavy and medium mass nuclei in the region of the giant dipole resonance (GDR) by (γ, xn) studies, which often also cover the region directly above the neutron threshold S_n. At lower energies three features have been discussed to be of importance for processes in high-temperature cosmic environments:
a) the fall-off [6,7,8,9] of the $E1$-strength on the low-energy slope of the GDR;
b) the $E1$-strength between the ground-state (gs) and low energy excitations and its proper extension [10,11,12] into the regime a);
c) the occurrence of additional pygmy-resonances, [3,13,14,15], which are assumed to be not as broad as the GDR, but wider as compared to the average level distance D —thus forming an intermediate structure enclosing many levels. Their low energy may well enhance their contribution to photo-dissociation processes in spite of their relatively low strength as compared to the GDR.

In principle, also $M1$-transitions contribute to the dipole strength, but the average $M1$ strength is typically 1-2 orders of magnitude smaller as compared to $E1$, and they are frequently [1,2,3] not taken into account.

A further approximation has to be introduced to estimate dipole strengths for transitions not connected to the ground state; there are two possibilities proposed in the literature:
a) Strictly following a hypothesis set up by Brink and Axel [16] the strength of a transition connecting in a given nucleus two levels separated by an energy difference E_γ only depends on E_γ, on the transition type and on a statistical factor determined by the two spins.
b) As proposed by Kadmenskii [10], a temperature dependence of the strength function is introduced, which smears out the GDR-strength into the region below and effectively connects a certain excitation region above the ground state to the GDR domain.

[a] e-mail: e.grosse@fz-rossendorf.de

A combination of the two prescriptions has been tried [12], but this suffers from an inconsistency which may be of principal nature: At low energy the nuclear excitation is quantized, whereas with increasing energy statistical concepts from thermodynamics are more effective in describing the increasing complexity. Microscopic calculations [8,11] may allow to develop a consistent transition from the low to the high excitation region near the GDR and to properly account for other intermediate strength.

2 Nuclear resonance fluorescence

Information about energy-dependent dipole strength functions can be obtained from data on multi-step gamma-decays following n-capture [17] or direct reactions [12], from inelastic electron scattering [18] or elastic photon scattering [14,15,19]. For the region above the n-threshold the electromagnetic strength in a very large number of stable nuclei has been experimentally determined by observing the neutrons emitted after the excitation by quasi-monochromatic photons [5,20]. Data on this (γ, xn) process, taken as averages over a certain energy bin, allow to determine an (averaged) dipole strength function f_1 in this region by making use of the relation:

$$f_1(E) = \left(3\pi^2\hbar^2c^2E\right)^{-1} \cdot \sigma_\gamma(E), \tag{1}$$

where σ_γ describes the dipole dissociation of a spin 0 target by a photon of energy E. The strength function data obtained from the other methods suitable for the lower excitation energy should connect smoothly to (γ, xn) data to yield information about the dipole strength over a wide energy range from the ground state up to far above the particle emission thresholds. Of special importance here is the much discussed [7,10,11,12] question, if the extrapolation of the Lorentzian fit to the GDR fits dipole strength data also at and below the neutron threshold.

One experimental method has delivered interesting information about nuclear dipole strength; results from it have become more and more detailed with the improving measurements. Using a bremsstrahlung beam the scattering of photons with energies up to the GDR is observed by large volume Ge-detectors. The good resolution (3–5 keV) of such detectors in combination with Compton suppression shields limit the detector response matrix such that it becomes nearly completely diagonal. Thus the signal from elastically scattered photons identifies the energy of the incoming photon out of the bremsstrahlung continuum, as all nuclear levels with sufficient transition strength to the ground state are observed as narrow elastic scattering resonances up to the respective neutron emission thresholds; inelastic scattering to levels above the target ground states is also observed [21] and has to be considered in the data analysis.

From the knowledge of inelastic scattering via higher lying levels the feeding to a certain level can be accounted for in the analysis of the elastic cross section $\sigma_{\gamma\gamma}$:

$$I(E_R) = \int_R \sigma_{\gamma\gamma}(E)\mathrm{d}E - I_{feed} \tag{2}$$

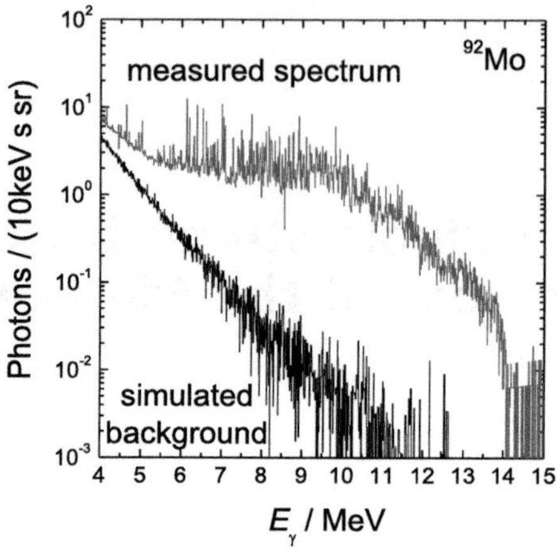

Fig. 1. Spectrum of bremsstrahlung photons scattered by ^{92}Mo into 127°. Above a background caused by atomic processes —whose height was determined from a Monte Carlo simulation— an accumulation of sharp lines near 7 MeV is observed as well as a strong quasi-continuum extending up to the endpoint energy of $\sim$ 14 MeV. Note that lines from identified background sources have been subtracted.

with the integral taken over the (narrow) resonance R centered at $E_\gamma = E_R$. The ground state (spin 0) width Γ_0 of such a resonance (spin 1) and I are related by

$$I(E_R) = 3\left(\frac{\pi\hbar c}{E_R}\right)^2 \cdot \frac{\Gamma_0^2}{\Gamma}, \tag{3}$$

where $\Gamma = \Gamma_0 + \Gamma_c$ is its total width and Γ_c the summed width of all decay processes competing to the decay back into the ground state. The contribution of the N_Δ resonances in an energy interval Δ to the dipole strength function can be calculated by

$$f_{1,\Delta} = \frac{1}{\Delta}\sum_{i=1}^{N_\Delta} \frac{\Gamma_{0,i}}{E_{R,i}^3}. \tag{4}$$

The present study on the Mo isotopes uses eq. (4) only for the single line spectra [21] observed below $E_R \simeq 4$ MeV with low endpoint bremsstrahlung, the other data require a more sophisticated analysis.

A non-negligible contribution to the scattering results from non-nuclear processes as Compton scattering with the subsequent bremsstrahlung reaching the detector; pair production leads to photon background in a similar way. Such processes can be simulated [19] to high accuracy and subtracted from the data as shown in figs. 1 and 2, to yield the "true" nuclear resonance fluorescence (nrf) cross section $\sigma_{\gamma\gamma}$ to be used in eq. (2). Only the resulting difference spectra contain information about the nuclear dipole strength. As the direct contributions from Thomson and Delbrück scattering are weaker by orders of magnitude, the scattering of MeV photons is primarily nrf. Only when Γ_c completely dominates Γ —*e.g.*, above the neutron threshold S_n— can the contribution of these direct scattering processes no longer be neglected. Contributions to the spectra from higher multipole radiation are identified by a different angular distribution. $M1$-transitions

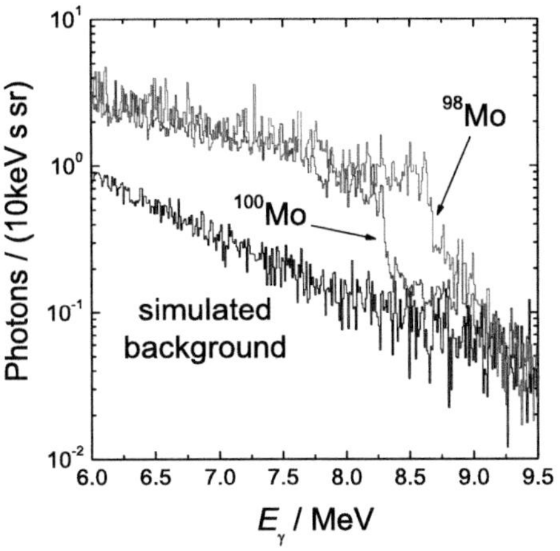

Fig. 2. Same as fig. 1 for ^{98}Mo and ^{100}Mo. The background simulation is the same for both isotopes. This allows a test of the determination of the continuum in the region between the two (different) neutron thresholds.

can be identified through the use of linearly polarized photons [22, 23, 24, 25].

3 Photon scattering experiments on Mo isotopes

The present paper reports on photon scattering experiments for the Mo isotopes with A = 92, 98 and 100. This rather wide range in neutron number N may allow a reasonable extrapolation to unstable isotopes; for the case of the pygmy resonance the $E1$-strength has been predicted [13] to vary strongly with N whereas its energy should weakly depend on N. The low-N stable Mo isotopes are p-process nuclides with a surprisingly high cosmic abundance thus making accurate information on the response of Mo isotopes to dipole radiation especially desirable. The photon scattering experiment at the new radiation source ELBE with its superconducting electron linac [26] was set up similar to previous nrf-studies [13, 14, 15, 27]. One of its special features is, that the bremsstrahlung emerges from a thin Nb-foil (approx. 5 mg/cm^2) bombarded with a beam of approx. 650 μA = $4 \cdot 10^{15}$ e/s. The electron beam is deflected into a well-shielded beam dump after passing the radiator. Highly enriched targets of ^{92}Mo, ^{98}Mo and ^{100}Mo have been used with masses of 2-3 g each. Electrons with an average momentum of 14 MeV/c and an rms momentum spread of 0.07 MeV/c were used in these experiments. For each target a second run with a lower electron momentum was performed such that the endpoint of the bremsstrahlung continuum stayed below the neutron and proton emission thresholds. The photon beam was limited transversely by a 2.5 m long conical Al-collimator, such that an approximate photon flux of about $10^7\gamma/(\text{s cm}^2\text{ MeV})$ hits the experimental targets. Four large high-purity germanium semiconductor detectors (enclosed by anti-Compton shields from BGO) set up at 90° and 127° registered the photons scattered by the target.

The detectors were shielded from unwanted background radiation by lead bricks positioned around the photon beam dump, near the collimator exit and around the beam tube at the target. A conical opening between target and Ge-detector, which was filled by only 2 cm of Pb to absorb the very intense low energy photons, determined the angle of observation and the solid angle. Details of the set-up and of the Monte Carlo simulations performed with the aim to optimize it are described elsewhere [19]. According to these simulations the main background contribution to the spectra of scattered photons (as shown in fig. 1) is due to bremsstrahlung produced from pair production and Compton scattering in the target. The simulation of this background (cf. figs. 1 and 2) could be made sufficiently accurate to allow the generation of pure nrf-spectra by subtracting the simulated non-resonant contribution from the experimental data.

Due to the 300 keV difference of the neutron binding energies of ^{98}Mo and ^{100}Mo a subtraction of the ^{100}Mo data from the ^{98}Mo data results in a pure nrf-spectrum in the range $S_n(^{100}\text{Mo})$ to $S_n(^{98}\text{Mo})$ —under the suggestive assumption that the non-resonant background is the same for both isotopes (see fig. 2). From the fact that this procedure leads within errors to the same nrf-strength in this energy bin as the Monte Carlo based subtraction, a test of the accuracy of the latter procedure is obtained. It should be noted here, that the subtraction explained above was performed with the data after their complete correction for detector response. Thanks to the high full energy efficiency of the Ge-detectors used and as result of the good Compton suppression by the BGO shields such a response correction could be performed without introducing large statistical uncertainties.

4 Dipole strength in isolated narrow resonances

In the nrf-spectra obtained from the raw data as shown in figs. 1 and 2 many isolated resonances could be analysed in the range from 4 MeV up to the endpoint energy of 13.2 MeV; the lower part of the spectra is discussed elsewhere [21]. For the three isotopes 299, 310 and 296 resonances, respectively, could be distinguished above 4 MeV; no strong lines can be identified above the respective neutron threshold. The ratios of the intensities

$$\frac{\mathrm{d}I}{\mathrm{d}\Omega} = \int_R \frac{\mathrm{d}\sigma(E)}{\mathrm{d}\Omega}\,\mathrm{d}E, \tag{5}$$

where the integral includes the full (narrow) resonance R, as observed at scattering angles 90° and 127°, have been compared to the values expected for a spin sequence 0-1-0 or 0-2-0 for excitation and de-excitation. Apparently nearly all of the transitions are due to spin 1 resonances. They are assumed to be $E1$, as from a previous experimental nrf-study on ^{92}Mo only 1 resonance is identified [22] to have positive parity; its $M1$-strength to the gs corresponds to $0.23\mu_N^2$: As in the neighbouring nucleus ^{90}Zr a

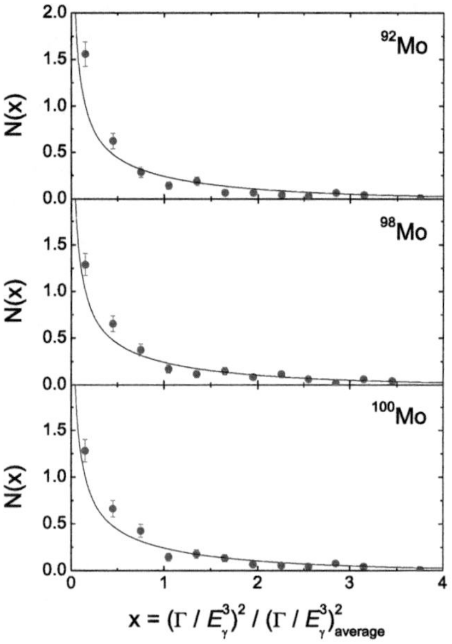

Fig. 3. Distributions of the transition widths as determined from each 50 transitions starting at 4 MeV, all reduced by the phase space factor E_γ^3. The drawn lines depict Porter-Thomas distributions.

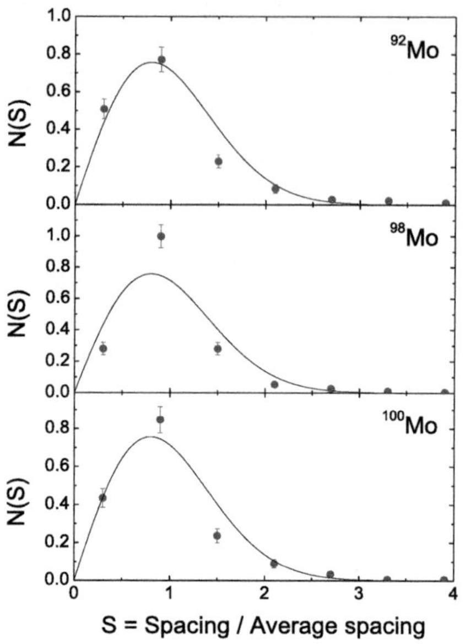

Fig. 4. Next neighbour distance distributions of the isolated narrow resonances above 4 MeV. Averages were formed over each 50 transition energies and the ratios of distance to average distance were collected in 7 bins. The drawn lines represent a Wigner distribution.

total $M1$ strength of $6.7\mu_N^2$ was found [23] to lie between 8 and 11 MeV, it is likely that a few more $M1$ resonances are also present in ^{92}Mo, but in any case most of the nrf-strength clearly is of $E1$ character. A similarly low $M1$ strength has been observed for ^{116}Sn and ^{124}Sn [25] and proposed for $N = 82$ nuclei [15].

For the well-identified resonances the reduced width distributions are displayed for the three isotopes in fig. 3 after a normalization to the respective average width taken over bins containing 50 dipole resonances each, observed in the range from 4 MeV up to the neutron binding energies. The distributions are in full agreement to Porter-Thomas distributions indicating chaotic statistics in the ground state transition strengths.

To study as well the statistical properties of the nearest neighbour spacings we also treat the resonances above 4 MeV in groups of 50 and determine the average spacing in each group. The actual spacings divided by this average are shown as dots in fig. 4 in comparison to Wigner distributions.

From comparison to the data taken at lower end-point energy, it is obvious that above 5 MeV nearly all of the identified transitions connect to the ground state. To obtain an estimate of the possible corrections necessary to account for the incorrect interpretation of non-gs transition energies as level energies we have performed respective Monte Carlo simulations of level sequences describing a Porter-Thomas or a Poisson case. Only a small distortion is caused, when the transition energies resulting from these simulations are (eventually erroneously) treated as level energies. In any case, our Mo data do closely resemble Wigner distributions, again pointing to chaotic statistics.

5 Level densities and fluctuating cross sections

The high level density in combination with Porter-Thomas fluctuations cause a large portion of the strength to appear in many weak transitions which are likely to be missed experimentally. This is why an average level density in its dependence on the excitation energy can only be determined from a fluctuation analysis on the basis of Porter-Thomas statistics [18]. Similarly, the dipole strength in a certain energy interval has to be obtained by integrating the complete nrf-spectra —*i.e.* all counts in discrete resonances and in the quasi-continuum in between, after subtraction of the non-resonant background. The accuracy of the determination of this background can be judged from what is presented in fig. 2 to be sufficiently high. As obvious from the analysis of the data shown in figs. 1 and 2, the average strength in the last MeV below S_n is approximately increased by at least a factor 3 when the continuum is included.

Another important correction has to account for inelastic scattering, *i.e.* transitions branching to excited states. Its effect can be identified from data taken at different endpoint energies of the bremsstrahlung spectrum. When the ^{100}Mo data taken with an endpoint energy of 8.3 MeV are compared to the data reaching up to the threshold of 7.8 MeV, the intensity distribution originating from these extra 500 keV of bremsstrahlung can be identified: More than 50 % of this intensity is observed as cascades with photons in the range of 3 of 5 MeV. Obviously the remaining intensity observed as gs-transitions has to be multiplied by a factor of 2-3 to obtain the full excitation strength. To obtain an approximation for this correction factor the assumption [27, 28] is adopted, that below S_n and S_p inelastic processes (*i.e.* branching) can on the average be accounted for by setting $\Gamma_c = 0.2\,\text{eV}$ (for the Zr region) in combination with the level density [29]. A more accurate correction is to be gained from the experiments at lower energy and by Monte Carlo simulations; obviously the approximation applied as described above cannot introduce extra structures.

This is the first time that high resolution nrf-spectra are analyzed such that not only the isolated resonances are included, but also the fluctuating quasi-continuum. This is accomplished by calculating the dipole strength function for the energy range up to S_n (as covered in this experiment) from the elastic component $\sigma_{\gamma\gamma}$ of nrf:

$$f_1(E) = \left(3\pi^2\hbar^2c^2E\right)^{-1} \cdot \frac{\Gamma}{\Gamma_0} \cdot \frac{1}{\Delta} \cdot \int_\Delta \sigma_{\gamma\gamma}(E_\gamma) \cdot \mathrm{d}E_\gamma\,, \quad (6)$$

where Δ is the interval selected around E for averaging the widths Γ_0 and Γ and the photon energy E_γ. The photon absorption cross section is thus derived from the observed elastic photon scattering cross section $\sigma_{\gamma\gamma}$ by correcting bin-wise for inelastic scattering (*i.e.* branching). This determination of f_1 then allows a quantitative comparison of nrf —and (γ, xn)— data (cf. eq. (1)), and both can be directly combined to extract a continuous dipole strength function.

6 The distribution of photon strength and pygmy resonances

The good compatibility of the nrf-strength (corrected for branching) directly below S_n and the (γ, xn) data directly above encourages a search for structure in $f_1(E)$ derived from the two data sets (see fig. 5). The data indicate an enhancement of the dipole strength at $\sim 9\,\text{MeV}$. In ^{92}Mo this possible pygmy resonance appears below and in ^{100}Mo it is above S_n; in ^{98}Mo the region directly above $S_n = 8.6\,\text{MeV}$ shows some irregularities. In an old tagged photon scattering experiment on natural Zr cross section enhancements at 9.1 and 11.6 MeV were found [27]; from the isotope enrichment and the n-threshold values it is argued, that they should originate from ^{90}Zr. Most of that strength was shown not to be $M1$ [23, 28]. The Mo data from ELBE as well as these results have to be compared to broad resonance-like structures seen [25] in Sn isotopes at 6.7 and between 8.0 and 8.7 MeV. The strength as extracted from an experiment [25] on ^{116}Sn an ^{124}Sn clearly stays below the extrapolation of the GDR-Lorentzian as only narrow isolated resonances had been analyzed. Although the broad structures seen in Sn by the tagged photon study [27] and in the new Ge-detector experiment [25] appear at nearly the same energies, the strength observed in the region of the broad pygmy structure differs by a factor of two between the two types of experiment. Obviously, an analysis of the well-isolated peak on the basis of eq. (4)

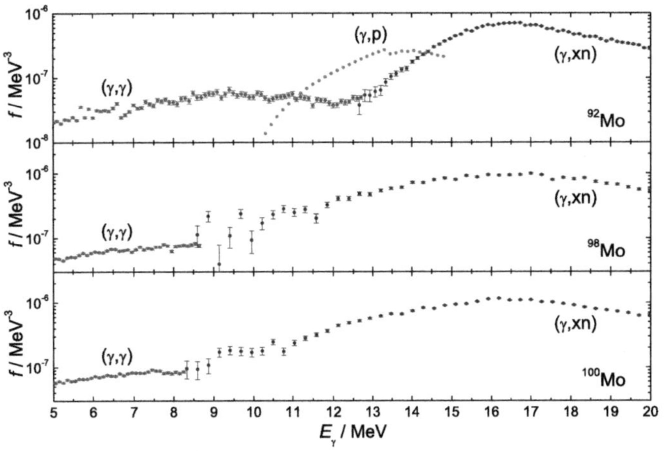

Fig. 5. Dipole strength functions determined from the photon scattering and the (γ, xn) data as described in the text. The nrf data are shown in bins of 100 keV; this makes the enhancement of strength in several single resonances near 6 MeV less obvious. In the case of ^{92}Mo also the (γ, p)-process has to be accounted for; this is indicated by including a respective cross section calculation [30] for the corresponding contribution to be added to obtain the total strength. Due to the weakness of quadrupole excitations the plotted f is effectively the dipole strength function f_1.

misses much of the strength which is accounted for by following a procedure characterized by eq. (6), which does not ignore the fluctuating quasi-background. The only other isotope chain studied in this range of A, the even Ge isotopes, show [24] no clear pygmy resonance appearing in the discrete spectra below S_n. The average strength function obtained from these resonances by using formula (4) amounts to $\sim 10^{-8}\,\text{MeV}^{-3}$ and clearly stays below the extrapolated Lorentzian extracted from the (γ, xn) data.

Recent HFB-QRPA calculations [8] give good fits to GDR data when the force Bsk-7 of Skyrme-type is used to describe the effective nucleon-nucleon interaction. Nevertheless it should be noted here, that $\sigma(\gamma, \text{n})$ directly above threshold is well described only for ^{94}Mo, whereas the calculation is below the experimental value [20] for ^{100}Mo by a factor 4. An extension of such calculations to lower E —including the pygmy region— seems interesting.

7 Conclusion

The response of nuclei to dipole radiation can well be studied by photon scattering investigated at a bremsstrahlung facility like ELBE. Using a sufficiently high endpoint energy and correcting the data for inelastic processes allows to directly combine the dipole strength functions f_1 obtained from the (γ, γ) (*i.e.* nrf) and the (γ, xn) data; together they span the full range from the ground state to the GDR. A comparison of $f_1(E)$ to a Lorentzian extrapolated from the GDR needs a more thorough discussion of the spreading of the GDR than is accomplished by just fitting near its maximum [5, 20]. Apparently, the large apparent width of the GDR in ^{100}Mo may be caused by a deformation of that nucleus; accounting for that by a two-resonance fit would reduce the low energy tail considerably. Calculations of the type presented recently [2,8,11] may help to clarify this point, especially when the nuclear deformation is included with sufficient accuracy.

Above $\sim 4\,\mathrm{MeV}$ the predicted level densities [29] increasingly surpass the number of identifiable resonances and apparently the levels of ^{92}Mo, ^{98}Mo and ^{100}Mo show signs of a chaotic structure: The next neighbour distance distributions of the clearly identified peaks are Wigner distributed and their ground-state transition widths follow Porter-Thomas distributions. From these facts one expects the photon scattering excitation functions —which in a bremsstrahlung beam are observed simultaneously over a wide range— to show Porter-Thomas fluctuations in case the detector resolution surpasses the average peak distance. As this is the case for most of the Mo data discussed here, a reasonable extraction of strength information should not ignore the fluctuating quasi-continuous part of the cross section.

Thus $f_1(E)$ was determined for the three Mo isotopes by using all scattering strength with the exception of the nonresonantly scattered photons, whose contribution to the spectra was calculated and subtracted. The f_1 data for the three Mo isotopes show a clear maximum at $\sim 9\,\mathrm{MeV}$ indicating the presence of a pygmy resonance, as was observed at this energy also in ^{90}Zr and slightly lower in energy in Sn isotopes. The 9 MeV structure is below S_n in ^{92}Mo and clearly above S_n in ^{100}Mo; in ^{98}Mo a cross section irregularity shows up at the neutron threshold. Intermediate structures are weakly showing up also between 6 and 7 MeV (as in ^{118}Sn and ^{124}Sn) and eventually also at 11 MeV (as in ^{90}Zr). This may be considered an indication for a sequence of pygmy resonances —not just one.

Dr P. Michel and the ELBE-Crew made these experiments possible with their strong commitment to deliver optimum beams. A. Hartmann and W. Schulze provided very valuable support during the difficult experiments. Intensive discussions with Dr F. Becvar, Dr F. Dönau and Dr R. Wünsch are gratefully acknowledged. The DFG has supported one of us (G.R.) under Do466/1-2 during the course of the studies presented here.

References

1. E.M. Burbridge *et al.*, Rev. Mod. Phys. **29**, 547 (1957); M. Arnould, S. Goriely, Phys. Rep. **384**, 1 (2003); T. Hayakawa *et al.*, Phys. Rev. Lett. **93**, 161102 (2004).
2. S. Goriely, Khan, Nucl. Phys. A **706**, 217 (2002).
3. S. Goriely, Phys. Lett. B **436**, 10 (1998).
4. F. Käppeler *et al.*, Rep. Progr. Phys. **52**, 945 (1989).
5. S.S. Dietrich, B.L. Berman, At. Data Nucl. Data Tables **52**, 199 (1989); B.L. Bermann *et al.*, Phys. Rev. C **36**, 1286 (1987).
6. C.M. McCullagh *et al.*, Phys. Rev. C **23**, 1394 (1981).
7. I. Kopecky, R.E. Chrien, Nucl. Phys. A **468**, 285 (1987); I. Kopecky, M. Uhl, Phys. Rev. C **41**, 1941 (1990).
8. S. Goriely *et al.*, Nucl. Phys. A **739**, 331 (2004).
9. H. Utsunomiya *et al.*, Phys. Rev. C **67**, 015807 (2003).
10. S.G. Kadmenskii *et al.*, Sov. J. Nucl. Phys. **37**, 165 (1983).
11. M. Arnould, S. Goriely, to be published in Nucl. Phys. A, doi:10.1016/j.nuclphysa.2005.02.116.
12. M. Guttormsen *et al.*, Phys. Rev. C **71**, 044307 (2005) and references therein.
13. P. van Isacker *et al.*, Phys. Rev. C **45**, R13 (1992).
14. N. Ryezayeva *et al.*, Phys. Rev. Lett. **89**, 272501 (2002).
15. R.D. Herzberg *et al.*, Phys. Rev. C **60**, 051307 (1999); A. Zilges *et al.*, Progr. Part. Nucl. Phys. **55**, 408 (2005).
16. D.M. Brink, Ph.D. Thesis, Oxford University (1955); P. Axel, Phys. Rev. **126**, 671 (1962).
17. L. Zanini *et al.*, Phys. Rev. C **68**, 014320 (2003); M. Krticka *et al.*, Phys. Rev. Lett. **92**, 172501 (2004).
18. A. Richter, Phys. Scr. **T5**, 63 (1983); G. Kilgus *et al.*, Z. Phys. A **326**, 326 (1987); P.G. Hansen *et al.*, Nucl. Phys. A **518**, 13 (1990).
19. R. Schwengner *et al.*, Nucl. Instrum. Methods A **555**, 211 (2005).
20. H. Beil *et al.*, Nucl. Phys. A **227**, 427 (1974), corrected according to ref. [5].
21. G. Rusev *et al.*, Phys. Rev. Lett. **95**, 062501 (2005); G. Rusev *et al.*, to be published.
22. F. Bauwens, Dissertation University of Gent (2000).
23. R.M. Laszewski *et al.*, Phys. Rev. Lett. **59**, 431 (1987).
24. A. Jung *et al.*, Nucl. Phys. A **584**, 103 (1995).
25. K. Govaert *et al.*, Phys. Rev. C **57**, 2229 (1998).
26. J. Teichert *et al.*, Nucl. Instrum. Methods A **507**, 354 (2003).
27. P. Axel *et al.*, Phys. Rev. C **2**, 689 (1970).
28. R. Alarcon *et al.*, Phys. Rev. C **36**, 954 (1987).
29. D. Bucurescu, T. von Egidy, to be published in J. Phys. G.
30. T. Rauscher, F.-K. Thielemann, At. Data Nucl. Data Tables **88**, 1 (2004).

Eur. Phys. J. A **27**, s01, 177–180 (2006)
DOI: 10.1140/epja/i2006-08-026-5

EPJ A direct
electronic only

Towards a high-precision measurement of the ^{3}He$(\alpha,\gamma)^7$Be cross section at LUNA

H. Costantini[1,a], D. Bemmerer[2], P. Bezzon[3], R. Bonetti[4], C. Broggini[2], M.L. Casanova[1], F. Confortola[1], P. Corvisiero[1], J. Cruz[5], Z. Elekes[6], A. Formicola[7], Zs. Fülop[6], G. Gervino[8], C. Gustavino[7], A. Guglielmetti[4], Gy. Gyürky[6], G. Imbriani[9], A.P. Jesus[5], M. Junker[7], A. Lemut[1], M. Marta[4], R. Menegazzo[2], P. Prati[1], E. Roca[10], C. Rolfs[11], M. Romano[10], C. Rossi Alvarez[2], F. Schümann[11], E. Somorjai[6], O. Straniero[9], F. Strieder[11], F. Terrasi[12], and H.P. Trautvetter[11]

1 Università degli Studi di Genova and INFN Genova, Dipartimento di Fisica, Via dodecaneso 33 16146, Genova, Italy
2 INFN Padova, Italy
3 INFN Laboratori Nazionali di Legnaro, Legnaro, Italy
4 Università di Milano, Istituto di Fisica and INFN Milano, Italy
5 Centro de Fisica Nuclear da Universidade de Lisboa, Lisboa, Portugal
6 ATOMKI, Debrecen, Hungary
7 INFN Laboratori Nazionali del Gran Sasso, Assergi, Italy
8 Università di Torino, Dipartimento di Fisica Sperimentale and INFN Torino, Italy
9 Osservatorio Astronomico di Collurania, Teramo and INFN Napoli, Italy
10 Università di Napoli and INFN Napoli, Italy
11 Institut für Experimentalphysik III, Ruhr-Universität Bochum, Bochum, Germany
12 Seconda Università di Napoli, Dipartimento di Scienze Ambientali, Caserta, Italy

Received: 24 October 2005 /
Published online: 7 March 2006 –

Abstract. The ^{3}He$(^4$He$,\gamma)^7$Be reaction is the key process for the production of ^{7}Be and ^{8}B neutrinos in the Sun. We have designed a new experimental setup to study this reaction with high accuracy at low energies using two different experimental techniques. The first method consists in measuring the prompt capture gamma-ray transitions with an ultra-low background germanium detector heavily shielded and placed at close distance from a ^{3}He windowless gas target. With another fully shielded large-volume germanium detector we will also measure the β-decay of the ^{7}Be residual nuclei. The aim of the experiment is to reduce the error on the astrophysical factor $S_{3,4}$ to 4%.

PACS. 25.40.Lw Radiative capture – 26.20.+f Hydrostatic stellar nucleosynthesis

1 Introduction

The solar neutrino flux resulting from the ^{7}Be$(p,\gamma)^8$B, depends on nuclear physics and astrophysics inputs [1]:

$$\Phi(B) = \Phi(B)(SSM) \cdot S_{3,3}^{-0.43} S_{3,4}^{0.84} S_{1,7}^{1} S_{e7}^{-1} S_{pp}^{-2.7} \cdot com^{1.4} opa^{2.6} dif^{0.34} lum^{7.2}. \quad (1)$$

The ^{3}He$(\alpha,\gamma)^7$Be reaction is one of the major source of uncertainty in determining the B solar neutrino flux and dominates over the present observational accuracy of 7% [2]. The foreseeable accuracy of the new generation solar neutrino experiments is 3%. This could illuminate about solar physics if the uncertainty on $S_{3,4}$ is reduced to a corresponding level. Moreover, the ^{3}He$(\alpha,\gamma)^7$Be reaction is important for understanding the primordial ^{7}Li abundance [3].

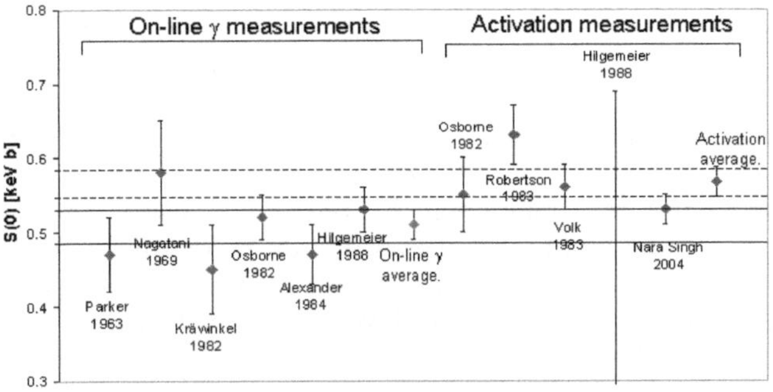

Fig. 1. Summary of past measurements performed with the on-line γ detection technique and the activation method. With the new activation measurement by Nara Singh *et al.* [4] the discrepancy between the two methods is reduced from 15% to 11%.

a e-mail: costant@ge.infn.it

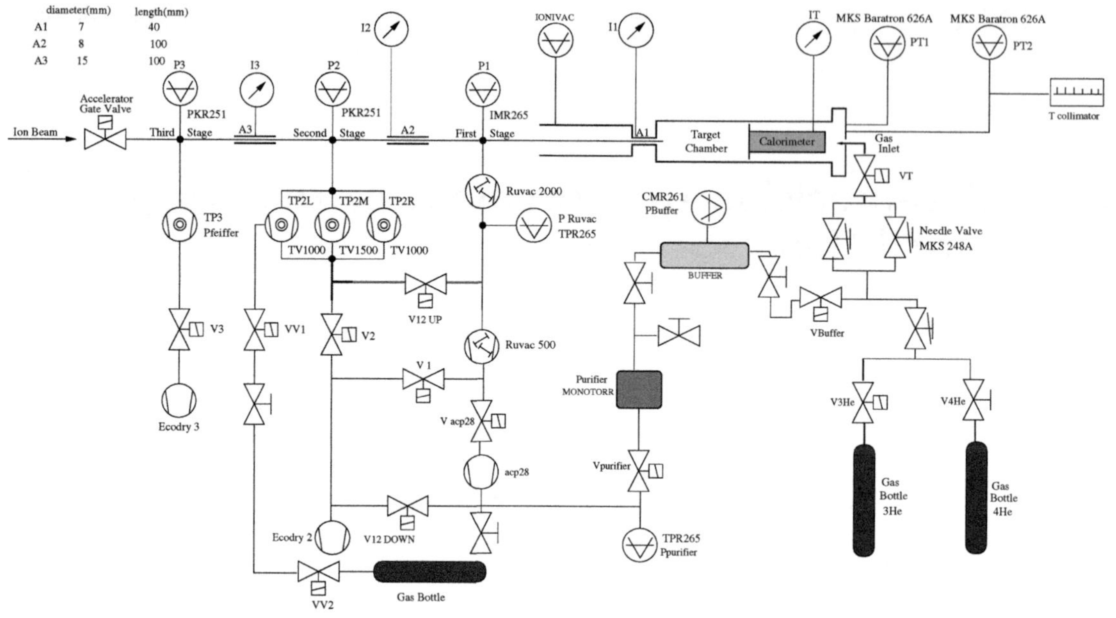

Fig. 2. Schematic diagram of the windowless gas target system.

In the last twenty years the $^3He(\alpha,\gamma)^7Be$ reaction have been measured using two techniques. In the first method direct α-capture γ rays were detected, while in the second the delayed ^7Be β decay γ rays were counted.

The average $S(0)$ S-factors, obtained with the two techniques, show a discrepancy of the order of 11% (see fig. 1).

Possible explanations for this discrepancy can be found in the systematic errors of the two methods: for example in the on-line γ measurement the low-energy angular distribution knowledge and the beam heating effect; in the activation measurement the beam-induced reactions on beam-stopper impurities leading to the production of ^7Be. Recently, Nara Singh *et al.* [4] studied the $^3He(\alpha,\gamma)^7Be$ reaction with high precision but with the activation technique only (see fig. 1). Therefore a new high-precision measurement of the $^3He(\alpha,\gamma)^7Be$ reaction, using both techniques at the same time, is highly desirable.

Here we report on a new measurement undergoing deep underground at the INFN (Istituto Nazionale di Fisica Nucleare) Laboratori Nazionali del Gran Sasso (LNGS)[1], 1500 m below the Gran Sasso Mountain (L'Aquila, Italy), where cosmic ray background is highly reduced by the natural shielding of the rock.

The measurement will be performed using both the direct γ detection and the activation techniques with the same experimental setup at the LUNA II 400 kV facility. The aim of the present work is to reduce the S-factor uncertainty down to 4%.

2 Experimental setup

The experiment will be performed using the LUNAII 400 kV accelerator. Details on the machine can be found in ref. [5]. Briefly, it consists of a radio frequency (RF) ion source, a singletron electrostatic extraction-accelerator system (embedded in a tank, which is filled with a gas mixture of N_2-CO_2 at 20 bar), a 45° magnet (30 cm radius) and a vertical steerer. The accelerator provides ion beam of approximately 500 μA protons and 250 μA He. The absolute beam energy is known with an accuracy of 0.3 keV and the energy spread and long-term energy stability were observed to be 100 eV and 5 eV/h.

We use an α beam in conjunction with a recirculating ^3He windowless gas target (a schematic diagram is shown in fig. 2). According to refs. [6,7,8,9] the deuterium and proton beam contamination have been found less in a ^4He beam than in an ^3He beam.

The beam enters the target chamber through three apertures of high gas flow impedance and is stopped on a beam calorimeter placed at the downstream part of the chamber. During the experiment the ^3He gas coming from the target is continuously recirculated. The gas is recovered from the first two pumping stages, cleaned through a heated getter gas purifier (Monotorr from SAES GETTER) and fed back into the target chamber. The purity of the gas inside the target is checked using a silicon detector that measures the scattered α-particles on the target atoms.

The beam current is measured through a calorimeter with constant temperature gradient. The power delivered by the beam is calculated as the difference between heating power without and with ion beam with an accuracy of 1%.

The target pressure is measured by a Baratron capacitance manometer (MKS model 127) in two different positions inside the gas target with an uncertainty of 0.25%. Typical gas pressure inside the target chamber is 1 mbar.

Due to the high beam current the local target density along the beam path can be lower than the density

[1] Web resource: `http://www.lngs.infn.it`

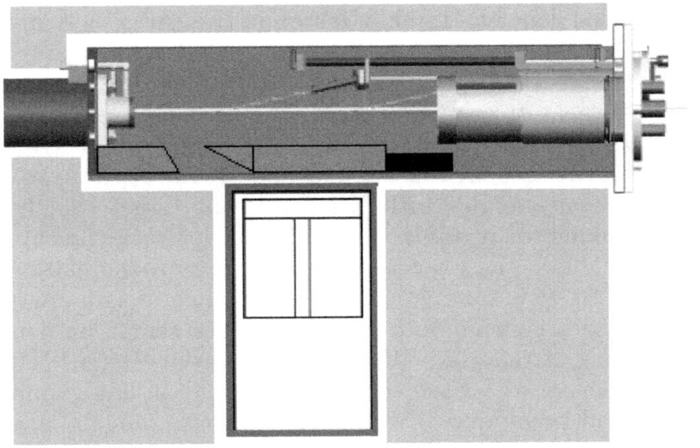

Fig. 3. Schematic view of the inner part of the target chamber. The internal lead collimator, the beam power calorimeter, the movable setup for the silicon detector and the long collimator before the carbon foil (at 20° with respect to the beam direction) are shown. The HpGe detector is positioned below the target chamber.

measured at the side of the through-the-pressure gauge. This effect, known as beam heating effect, depends on the power for unit length delivered by the beam on the gas target [10].

To avoid systematic uncertainties the target density, together with the beam current, is measured through α-Rutherford scattering cross section with a silicon detector positioned inside the target chamber (see fig. 3) with an uncertainty of 0.1%.

To optimize the detector time of life and the target chamber geometry we have decided to use a double scattering setup. The α-particles are first scattered by the gas atoms in the target chamber and subsequently by a carbon foil of $15\,\mu g/cm^2$, put at 20° in respect to the beam direction (see fig. 3).

The effective density profile as a function of the position along the beam direction, the beam current and target pressure will be obtained with a series of dedicated measurements with silicon detector with an estimated accuracy better than 1%.

3 The on-line γ detection technique

The $^3He(\alpha,\gamma)^7Be$ reaction is an α-capture reaction that can occur through electromagnetic decay to the ^7Be ground state (DC→0) or to the first excited state (DC → 429 keV) with the emission of γ's of energy $E_\gamma = 1586\,keV + E_{cm}$ or $E_\gamma = 1157\,keV + E_{cm}$, respectively (in the latter with the subsequent emission of the 429 keV γ).

To measure the cross section we will detect the two primary transitions (1.2 and 1.6 MeV) using a 135% ultra low background Canberra HpGe detector positioned under the target chamber in very close geometry. Since the energy of the primary γ-transitions is in the energy region of the natural radioactive isotopes we will build a copper and lead shielding around the detector of $0.3\,m^3$. Passive shielding is particularly effective underground since the muons flux, coming from cosmic rays that, at surface, produces secondary γ-rays in the lead shielding, is reduced by six orders of magnitude thanks to the natural shielding of the Gran Sasso mountain. The expected attenuation factor for the ^{40}K 1.46 MeV γ is 10^{-5}–10^{-6}, according to a GEANT4 simulation [11] where the complete geometry of the target and of the shielding has been considered. In order to reduce the background on the detector, low activity materials have been used in the construction of the target chamber, silicon detector support and calorimeter. In particular the target chamber is made by OFC copper and no welding materials have been used in the chamber assembly.

According to the DC model calculations [12] the DC→0 and DC→429 keV angular distributions are dominated by $E1$ transitions that can occur through s- or d-waves. Most recent angular distributions measurements [13] showed a small anisotropy of these transitions, manifesting interference effects of both partial wave contributions also at the lowest measured energy ($E_{cm} = 148\,keV$). It should be noted, however, that the DC model predictions depend sensitively on the s- and d-wave phase shifts in the $^3He + {^4He}$ elastic scattering channel, which are known experimentally only at $E_{cm} \geq 1.4\,MeV$. A detector placed at 55° with respect to the beam direction, would become almost independent of angular distribution anisotropy.

Therefore we have put an internal lead collimator, inside the target chamber (see fig. 3), designed in such a way that the detector collects mostly the γ's emitted at 55°. At the same time, this design reduces the effective target length seen by the detector to approximately 15 cm corresponding to a beam energy loss ΔE of 8 keV for 1 mbar of ^3He gas. The internal lead collimator has been also designed to shield the detector from beam-induced γ's coming from reactions occurring on the entrance collimator and on the calorimeter cap (beam stopper).

Taking into account the angular distribution measured, and considering the extreme cases of isotropy and full anisotropy, we have estimated that with our setup the angular distribution systematic effect is reduced to less than 4%.

With direct method we will explore, in the first phase of the experiment, the energy region $\Delta E_{cm} = 100$–$170\,keV$. With the detection efficiency of the apparatus previously described and a typical α-beam current of $200\,\mu A$ the expected counting rate for the 1.6 MeV γ will be about 100 counts/day at $E_{cm} = 100\,keV$ and 2000 counts/day at $E_{cm} = 170\,keV$.

4 The activation technique

An alternative way to measure directly the γ rays produced by the $^3He(\alpha,\gamma)^7Be$ reaction is to detect the γ's from the ^7Be electron-capture decay to the 478 keV state in ^7Li. Since the ^7Be residual nuclei are produced inside the gas target are moving in the beam direction, they are implanted into a removable copper calorimeter cap

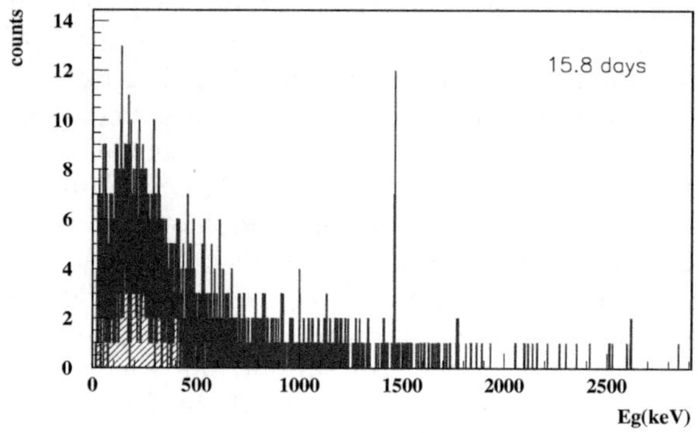

Fig. 4. Background γ-ray spectrum obtained with the ultra-low background 125% HpGe detector at the low activity laboratory at LNGS. The measuring life-time is 15.8 days.

Table 1. Expected counting rate for the 480 keV ^{7}Be γ after an irradiation time of one day, 200 μA α-current and 1 mbar of gas target pressure.

E_{lab}	E_{cm}	σ(nb)	no. ev/week
280	120	3.1	60
327	140	7.2	150
373	160	16.8	320

(diameter = 6 cm, see fig. 3). To prevent ^{7}Be nuclei eventually escaping from the cap, a covering foil will be inside the chamber wall. The same setup described in sect. 3 will be used for the activation method.

After α-beam irradiation, the cap and the foil ^{7}Be activity will be measured with 125% HpGe detector. This detector is installed at the Low Activity Laboratory of the LNGS and is completely shielded by 15 cm of lead and 10 cm of copper. The background in the region of interest for the ^{7}Be decay γ is about 2 counts/day (see fig. 4). Preliminary measurements aimed to investigate the calorimeter copper purity to search for possible parasitic reactions, as ^{6}Li(p, γ)^{7}Be, ^{6}Li(d, n)^{7}Be and ^{10}B(p, α)^{7}Be, have been done at Atomki (Debrecen, Hungary). A sample of a calorimeter cap has been irradiated with p and d beams. No ^{7}Be nuclei have been detected at the detection limit of the test setup (0.3 ppm). Further measurements to investigate the beam purity, will be performed directly at the LUNAII accelerator bombarding the cap with α-beam at different energies with ^{4}He gas in the target chamber and looking for possible ^{7}Be nuclei produced in the cap.

In table 1 the expected counting rate of the 480 keV decay ^{7}Be γ, assuming an irradiation time of 1 day with an α-beam current of 200 μA and a gas target pressure of 1 mbar, is reported. We will measure the ^{3}He(α,γ)^{7}Be cross section with both the activation and on-line gamma detection techniques at selected energies, ranging from E_{cm} = 100–170 keV, comparing directly the two methods.

This work was supported by INFN and partially by LUNA TARI project (RII-CT-2004-506222) and by OTKA T42733 and T049245.

References

1. V. Castellani *et al.*, Phys. Rep. **281**, 309 (1997).
2. J.N. Bahcall, M.H. Pinsonneault, Phys. Rev. Lett. **92**, 121301 (2004).
3. A. Coc *et al.*, Astrophys. J. **600**, 544 (2004).
4. B.S. Nara Singh *et al.*, Phys. Rev. Lett. **93**, 262503 (2004).
5. A. Formicola *et al.*, Nucl. Instrum. Methods A **507**, 609 (2003).
6. H. Volk *et al.*, Z. Phys. A **310**, 91 (1983).
7. R.G.H. Robertson *et al.*, Phys. Rev. C **27**, 11 (1983).
8. J.L. Osborne *et al.*, Nucl. Phys. A **419**, 115 (1984).
9. M. Hilgemeier *et al.*, Z. Phys. A **329**, 243 (1988).
10. J. Goerres *et al.*, Nucl. Instrum. Methods **177**, 295 (1980).
11. The RD44 Collaboration, *GEANT4: An Objected-Oriented Toolkit for Simulation in HEP*, CERN/LHCC 95-70 (1995).
12. T.A. Tombrello, P.D. Parker, Phys. Rev. **131**, 2582 (1980).
13. H. Krawinkel *et al.*, Z. Phys. A **304**, 307 (1982).

Eur. Phys. J. A **27**, s01, 181–185 (2006)
DOI: 10.1140/epja/i2006-08-027-4

EPJ A direct
electronic only

Radiative and non-radiative electron capture from carbon atoms by relativistic helium ions

A. Gójska[1,a], D. Chmielewska[1], P. Rymuza[1], J. Rzadkiewicz[1], Z. Sujkowski[1], T. Adachi[2], H. Fujita[2], Y. Fujita[2], Y. Shimbara[2], K. Hara[3], Y. Shimizu[3], H.P. Yoshida[3], Y. Haruyama[4], J. Kamiya[5], H. Ogawa[6], M. Saito[7], and M. Tanaka[8]

[1] The Andrzej Soltan Institute for Nuclear Studies, 05-400 Otwock - Świerk, Poland
[2] Department of Physics, Osaka University, Toyonaka, Osaka 560-0043, Japan
[3] Research Center for Nuclear Physics, Osaka University, Ibaraki, Osaka, 567-0047, Japan
[4] Laboratory of Applied Physics, Kyoto Prefectural University, 1 Hangicho, Shimogamo, Sakyo-ku, Kyoto, 606-8522, Japan
[5] IPNS (Institute of Particle and Nuclear Studies), KEK (High Energy Accelerator Research Organization), Oho 1, Tsukuba, Ibaraki 305-0801, Japan
[6] Department of Physics, Nara Women's University, Kitauoya-nishimachi, Nara 630-8506, Japan
[7] Kyoto Prefectural University, Kyoto 606-8522, Japan
[8] Kobe Tokiwa Jr. College, Nagata, Kobe, 653-0824, Japan

Received: 11 July 2005 /
Published online: 29 March 2006 –

Abstract. Interaction of radiation with helium atoms and/or ions is of interest in various astrophysical applications. The reverse reactions of fast 150 MeV/amu $^3He^{++}$ ions with solid C targets have been studied at the isochronous cyclotron of the RCNP in Osaka. The singly ionized helium ions resulting from capture of the target electrons to the projectile were observed with the use of large magnetic spectrograph, Grand Raiden, set at $\theta = 0°$ with respect to the beam. The yield ratios of singly-to-doubly ionized helium ions emerging from thin carbon foils, He^+/He^{++}, have been measured as a function of the foil thickness. Extrapolating the results to zero target thickness permits to determine the cross section values separately for electron stripping from $^3He^+$ ions and for electron capture to $^3He^{++}$ ions. The radiative and non-radiative contributions to the capture cross section were determined in a separate experiment in which the 84 keV (C.M.) photons were observed in coincidence with the He^+ ejectiles. The results are compared with theoretical predictions. Need for improved calculations is noted for all the three processes involved, *i.e.* for the radiative and non radiative electron capture as well as for the electron stripping.

PACS. 25.40.Lw Radiative capture – 34.70.+e Charge transfer

1 Introduction

The main processes occurring during the passage of ions through matter are the electron capture from the target to the vacant states of the projectile and the ionization (stripping) of the bound electrons from the passing ion (see [1] for a general review). Cross sections for these processes depend sharply on the velocity of the projectile as well as on the atomic number of the projectile and of the target. There seems to be no satisfactory theoretical description of the ionization process for fast projectiles. Rather crude approximations are used, applicable in limited energy and Z_T, Z_P ranges, where Z_T, Z_P are the atomic numbers of the target atom and of the projectile, respectively. The classical Bohr theory [2] for low Z_T and not very fast ions predicts:

$$\sigma_{STRIP} \propto \frac{Z_T^2}{Z_P^2 v^2}, \tag{1}$$

where σ_{STRIP} is the stripping cross section, v is the projectile velocity.

The ionization cross section for fast ions interacting with multi electron target atoms requires more precise theory. The quantum description of the stripping process was given by Gillespie [3,4]:

$$\sigma_{STRIP} \propto \frac{I}{v^2}, \tag{2}$$

where I is the collision strength. A phenomenological expression for $I(Z_T)$ can be found in [5].

Two very different mechanisms contribute to the electron capture: the radiative one, in which the excess energy

[a] e-mail: gojska@ipj.gov.pl

is carried away by a photon, and the non-radiative one, NREC. Because of the conservation laws the latter process can occur only for the capture of electrons bound in the target atom. The competing radiative transfer of bound atomic electrons is referred to as the Radiative Recombination, RR, while the capture of free electrons is called the Radiative Electron Capture, REC. The REC process can be considered as the time-reversed photoelectric effect on the partly ionized projectile atom. The REC cross section, σ_{REC}, grows quickly with Z_P and decreases rapidly with v. For high but non-relativistic projectile energy:

$$\sigma_{REC} \propto \frac{Z_T Z_P^{4.5}}{v^5} . \tag{3}$$

In contrast, the NREC process occurs mainly at the velocity matching condition $v \approx v_e$, where v_e is the velocity of the captured electron, bound in the target atom. For $v \gg v_e$:

$$\sigma_{NREC} \propto \frac{Z_T^5 Z_P^5}{v^{11}} . \tag{4}$$

A simple approach to describe the non-radiative electron capture cross section is the Oppenheimer-Brinkmann-Kramer (OBK) theory [1]. The screening corrections are included in [6,7], while the relativistic effects are described in [1]. Because of the sharp dependence of electron capture cross section on velocity as well as on Z_P the measurements for fast light ions are difficult and require very refined techniques. Recent experimental information on interaction of fast helium ions with various gaseous and solid targets can be found in [8,9]. The authors have measured the stripping and the capture cross section for ^{3}He ions with energies up to 43.4 MeV/amu. The $\sigma_{CAP}/\sigma_{STRIP}$ ratios were measured for a number of thick targets in [5]. The present work extends this information to much higher ^{3}He energy, 150 MeV/amu, for $Z_T = 6$. This is the first measurement for such high-energy light projectile. Preliminary results were published in [10,11]. Similar measurements for $Z_T = 79$ are described in [12]. The present results are compared with theoretical capture and stripping cross sections.

2 Experiment

2.1 Total capture and stripping cross sections

There is a huge difference in the magnitude of stripping and capture cross sections for a fast light ion traversing a solid. As a result, the electron captured by the ion in one collision is very quickly lost in the subsequent one. In effect, the charge state of the ion reaches the equilibrium conditions in very thin layers of the solid. In order to determine the interaction cross sections the target thickness used has to be significantly smaller than the equilibrium thickness. This is of the order of 100 μg/cm^2 for solid carbon. A method to determine the total capture as well as the stripping cross section is to carry out an extrapolation experiment in which the yield ratio, $Y(^3\text{He}^+)/Y(^3\text{He}^{++})$, of the singly-to-doubly ionized He ions emerging from the

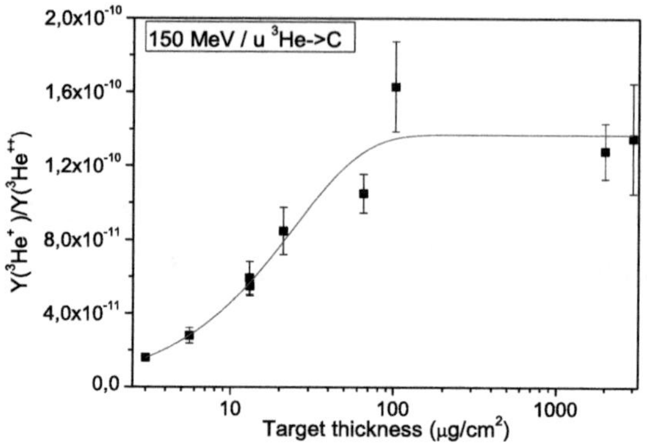

Fig. 1. The measured $Y(^3\text{He}^+)/Y(^3\text{He}^{++})$ yield ratio as a function of target thickness for C target. The errors are statistical only. The curve is a fit of eq. (6) to the data.

target is measured as a function of the target thickness. For thicknesses larger than the saturation value, x_{sat}, this ratio is equal to the cross section ratio [5]:

$$\frac{Y(^3\text{He}^+)}{Y(^3\text{He}^{++})}(x > x_{sat}) = \frac{\sigma_{CAP}}{\sigma_{STRIP}} . \tag{5}$$

Extrapolating the yield ratio to zero target thickness permits to separately determine the respective cross section values.

The $^3\text{He}^{++}$ beam has been accelerated to 150 MeV/amu in the AVP cyclotron at the Research Center for Nuclear Physics (RCNP) in Osaka. The experimental set-up was identical to that used to study the (^{3}He, t) nuclear charge exchange reactions [13]. The singly ionized $^3\text{He}^+$ ions were detected together with tritons in the focal plane of the magnetic spectrometer Grand Raiden, set at 0° with respect to the beam. The $^3\text{He}^{++}$ beam was fully intercepted by a Faraday cup placed in the first dipole magnet of the spectrometer. The intensity ratio of singly-to-doubly ionized helium ions leaving the carbon target was measured as a function of the target thickness (fig. 1). Carbon targets: 3, 5.6, 13, 13.1, 21, 65, 102, 2000, 3000 μg/cm^2 thick were used. The errors in thicknesses were estimated to be between 5 and 10%. Two sets of data were obtained under largely changed spectrograph settings used in the singles and coincidence experiments. A very satisfactory matching of these sets is observed. This builds up confidence in the proper reduction of possible systematic errors. The capture and stripping cross sections can be determined by fitting the simple function to the measured yield ratio *versus* target thickness:

$$\frac{Y(^3\text{He}^+)}{Y(^3\text{He}^{++})} = a\left[1 - \exp(-bx)\right], \tag{6}$$

where $a = \frac{\sigma_{CAP}}{\sigma_{STRIP}+\sigma_{CAP}} \sim \frac{\sigma_{CAP}}{\sigma_{STRIP}}$ cross section ratio, $b = \sigma_{STRIP} + \sigma_{CAP} \sim \sigma_{STRIP}$, and x denotes the target thickness in μg/cm^2 multiplied by the number density of the target atoms. The implicit assumption for eqs. (5) and (6) is that the change in the $^3\text{He}^{++}$ intensity after penetrating the target foil is negligible.

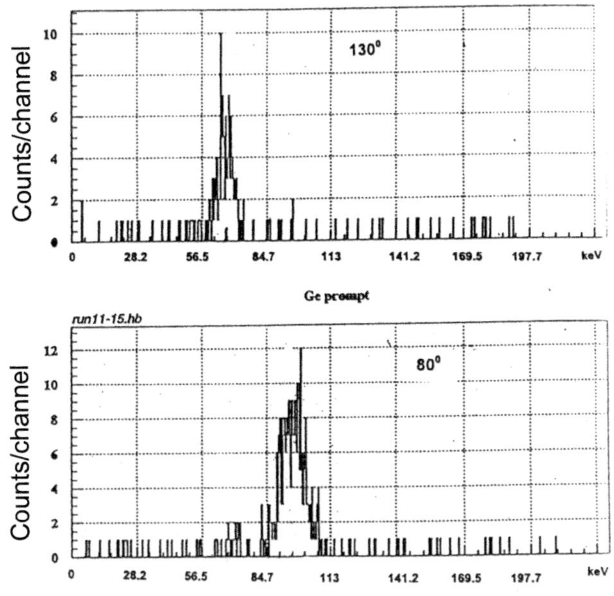

Fig. 2. The coincidence REC photon spectra.

2.2 Coincidence experiment

The extrapolation experiment measures the total electron capture cross section which is the sum of the non-radiative and radiative components

$$\sigma_{CAP} = \sigma_{REC} + \sigma_{NREC}. \tag{7}$$

The relative contributions of these components can be determined in a coincidence experiment in which the He^{+} ejectiles are registered in coincidence with the REC photons. Neglecting the electron binding energy in helium ion the energy of these photons corresponds to the energy of electrons in the centre-of-mass system of the projectile. The REC photons were measured in coincidence with $^{3}He^{+}$ ions by the Ge detectors set at $\theta = 80°$ and $\theta = 130°$ with respect to the beam. The centre-of-mass energy is $E_{REC} = 84$ keV. Figure 2 shows the measured spectra. The REC photons are strongly anisotropic [14,15]. In order to correct for this effect the coincidence yields were analysed according to the formula:

$$\frac{\sigma_{REC}}{\sigma_{tot}} = \frac{N(REC)}{(3/2)\varepsilon \sin^2\theta N(^3\mathrm{He}^{++})} \tag{8}$$

where *Y(REC)* is the coincidence photon yield, $Y(^3\mathrm{He}^{++})$ is the beam intensity, θ is the angle between the direction of observation and the direction of the beam and ε is the efficiency of the photon detection.

3 Results and discussion

The measured values for the three cross sections are:
$\sigma_{STRIP} = (821 \pm 60)$ kb,
$\sigma_{REC} = (64 \pm 12)\,\mu$b,
$\sigma_{NREC} = (47 \pm 8)\,\mu$b.
These values are compared below with the respective theoretical predictions.

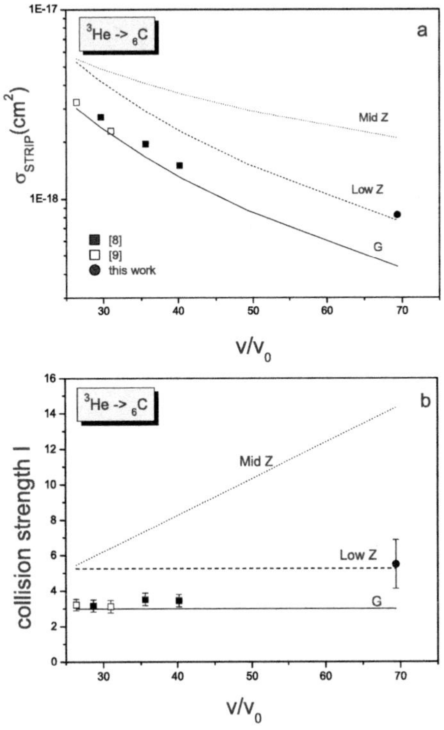

Fig. 3. The experimental (points) and calculated electron stripping cross section (a) and collision strength (b) as a function of ion velocity, in units of $v_0 = 2.19 \cdot 10^8$ cm/s. The solid line is Gillespie, the dotted and dashed lines are Bohr, approximations for medium- and low-Z targets, respectively.

3.1 Electron stripping from He^{+} ions

The existing data on the stripping cross section in the energy range 17.3–43.3 MeV/amu together with the present value at 150 MeV/amu are collected in fig. 3a. The theoretical values calculated according to Bohr and Gillespie approximations are included for comparison. Figure 3b shows the same data in the collision strength representation. None of the theories used reproduces the data in a satisfactory way; the deviation of the Gillespie approximation, though suggestive, is not very significant for this light target.

3.2 Non-radiative electron capture

The earlier experiments determining the total capture cross sections for He^{++} ions have been done [8,9] in the energy range 17.3–43.3 MeV/amu. At these low energies the capture is strongly dominated by the non-radiative process and the radiative electron capture can be neglected. Figure 4 presents these data together with the present value for σ_{NREC} at 150 MeV/amu as a function of the projectile velocity. The data are compared with theoretical calculations. The predictions of Nikolaev underestimate the effect dramatically. The discrepancy increases with velocity. The second order Oppenheimer-Brinkmann-Krammer (B2) approximation [1] yields values close to

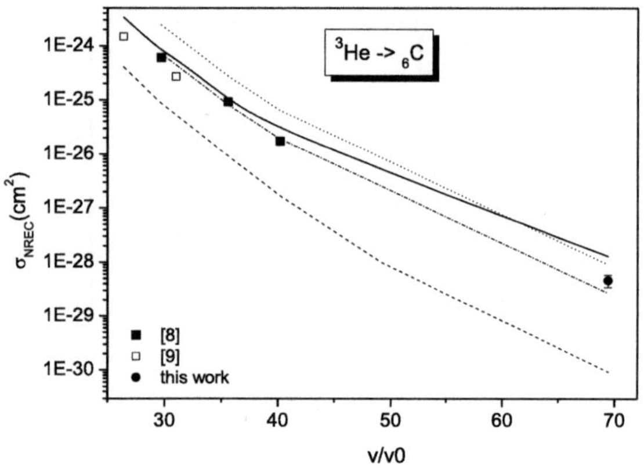

Fig. 4. The experimental (points) and calculated electron capture cross section as a function of projectile velocity. The solid line is from eikonal theory, dotted line is from OBK and dash-dot line from B2 approximations, dashed line refers to Nikolaev formula.

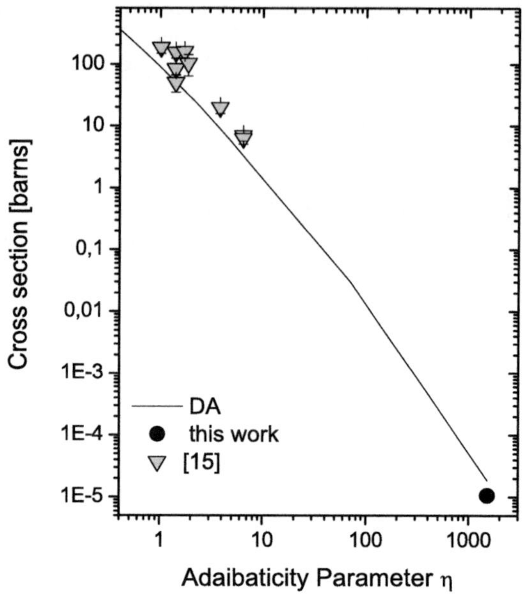

Fig. 5. The REC electron capture cross section per target electron measured for bare ions in collisions with light target atoms. The results are plotted as a function of the adiabaticity parameter and compared with the result of the Dipole Approximation (DA) theory.

the experiment though it slightly underestimates the effect at high velocity, while the refined eikonal theory deviates from the data in opposite direction with the velocity increasing.

3.3 Radiative electron capture

The radiative electron capture has never been measured for as light a projectile as helium. The present value can be compared with theoretical one calculated relativistically for all the shells: $\sigma_{REC} = 130\,\mu$b.

The non-relativistic analytical expression of Eichler [1] gives a lower value of 111 μb. The implicit assumption is that thc electron in the carbon target can be considered as free and uncorrelated. The assumption seems to be plausible at the high helium velocity. It remains to be seen whether the rather large difference between the calculated and the experimental value can be related to the difference between the radiative recombination effect for He^{++} - C and the REC effect for free electrons.

4 Radiative electron capture by fast projectiles vs. the adiabaticity parameter

The available systematics [15] of REC cross section for projectiles ranging from light ones up to as heavy as the hydrogen-like uranium shown in fig. 5 can be presented in the form of a universal curve as a function of the adiabaticity parameter:

$$\eta = \left(\frac{v}{v_e}\right)^2, \tag{9}$$

where v_e denotes the captured electron velocity in the projectile. The earlier data, obtained mainly for very heavy projectiles, correspond to $\eta < 10$ and to $\sigma_{REC} > 1$ b. The present experiment extends this picture till $\eta = 1500$ and to the cross section several orders of magnitude lower, $\sigma_{REC} = (64 \pm 12)\,\mu$b. The theoretical value is about twice larger.

5 Summary and conclusions

Cross sections for the three main processes occurring for He ions traversing solid carbon have been determined at 150 MeV/amu bombarding energy. The processes in question are the stripping of electron from singly charged He ions, He^+, and the radiative and non-radiative capture of electrons from the target to the vacant states in the He^{++} ions. This measurement, done at the highest bombarding energy ever used for such light ions, was feasible thanks to the use of advanced techniques of the modern nuclear physics. The results significantly extend the systematics of cross section *versus* energy data for all the three processes and permit to test the theoretical predictions in a sensitive region. Analysis of this systematics results in the following statements:

– There is no theory available to satisfactorily describe the energy dependence of the stripping cross section.

– The non-radiative electron capture from carbon to He^{++} ions is more closely described by the relatively simple second order Oppenheimer-Brinkmann-Kramer approximation than by the more sophisticated eikonal theory. The latter seems to fail at high projectile velocities.

– The measured cross section for radiative transfer of the bound electron from the carbon target to the helium projectile is factor of two smaller than that calculated under the assumption of the REC process *i.e.* of the free electrons being captured in the time-reversed photoelectric effect. This is the first and the only measurement for radiative process, be it the Radiative Recombination or the Radiative Electron Capture, for such a light projectile at high adiabaticity parameter. Naively, the two processes should be practically identical at high energies. There might, however, be some constraints due to, *e.g.*, the angular momentum coupling, which change the radiative recombination probability in comparison with the probability of free electron capture. The similarity of the binding energies in carbon atoms and in He^+ ions should also be noted. Clearly, there is a need for more studies of this difficult to measure, yet highly interesting effect.

The interaction of helium ions with free electrons is of primary astrophysical interest [10,16]. The REC photons due to radiative electron capture by bare He^{++} ions may provide an observable for detecting these ions in the intergalactic space [17]. The fully ionized matter is otherwise unobservable by the usual optical methods. The present experiment yields the cross section value for this process in the observationally most interesting region of $E(\text{electron}) \leq 100\,\text{keV}$.

References

1. J. Eichler, Phys. Rep. **193**, 165 (1990).
2. N. Bohr, K. Dan, Vidensk. Selsk. Mat.-Fys. Medd. **18**, No. 8 (1948).
3. G.H. Gillespie, Phys. Rev. A **18**, 1967 (1978).
4. G.H. Gillespie, Phys. Rev. A **26**, 2421 (1982).
5. K. Dennis *et al.*, Phys. Rev. A **50**, 3992 (1994).
6. V.S. Nikolaev, Zh. Eksp. Theor. Fiz. **51**, 1263 (1966).
7. G. Lapicki *et al.*, Phys. Rev. A **22**, 1896 (1980).
8. I. Katayama *et al.*, Phys. Lett. A **92**, 385 (1982).
9. I. Katayama *et al.*, Phys. Rev. A **53**, 242 (1996).
10. Z. Sujkowski, *Proceedings of the International Symposium on Advances in Nuclear Physics, Bucharest, Romania, 1999*, edited by D. Poenaru, S. Stoica (World Scientific, 2000) p. 91.
11. D. Chmielewska, *Proceedings of the International Winter Meeting on Nuclear Physics, Bormio, Italy*, Suppl. No. 116 (2000) p. 90.
12. A. Gojska *et al.*, Nucl. Instrum. Methods B **235**, 368 (2005).
13. H. Akimune *et al.*, Phys. Rev. C **52**, 604 (1995).
14. E. Spindler *et al.*, Phys. Rev. Lett. **42**, 832 (1979).
15. Th. Stoehlker, Phys. Rev. A **51**, 2098 (1995).
16. Z. Sujkowski, Nucl. Phys. A **719**, 266c (2003).
17. D. Chmielewska, Z. Sujkowski, these proceedings.

Eur. Phys. J. A **27**, s01, 187–192 (2006)
DOI: 10.1140/epja/i2006-08-028-3

EPJ A direct
electronic only

Evidence for a host-material dependence of the n/p branching ratio of low-energy d+d reactions within metallic environments

A. Huke[1,a], K. Czerski[1,2], T. Dorsch[1], A. Biller[1], P. Heide[1], and G. Ruprecht[1,3]

[1] Institut für Atomare Physik und Fachdidaktik, Technische Universität Berlin, Hardenbergstr. 36, 10623 Berlin, Germany
[2] Institute of Physics, University of Szczecin, Szczecin, Poland
[3] TRIUMF, Vancouver, Canada

Received: 6 July 2005 /
Published online: 16 March 2006 –

Abstract. Angular distributions and the neutron-proton branching ratio of the mirror reactions ^{2}H(d, p)^{3}H and ^{2}H(d, n)^{3}He have been investigated using different self-implanted deuterized metallic targets at projectile energies between 5 and 60 keV. Whereas the experimental results obtained for the transition metals Zr, Pd, Ta and also Al do not differ from those known from gas-target experiments, an enhancement of the angular anisotropy in the neutron channel and an attenuation of the neutron-proton branching ratio have been observed for the (earth)alkaline metals Li, Sr and Na at deuteron energies below 20 keV. Experimental results are discussed with consideration of the special problems arising from the properties of these chemically very reactive target materials. A first theoretical effort explaining simultaneously both n/p asymmetry effects based on an induced polarization of the reacting deuterons within the crystal lattice is presented.

PACS. 24.70.+s Polarization phenomena in reactions – 25.45.Hi Transfer reactions – 26.20.+f Hydrostatic stellar nucleosynthesis – 89.30.Jj Nuclear fusion power

1 Introduction

The d+d reactions have been investigated for decades because of their simplicity, fundamental importance and possible application in energy generation technology (*e.g.*, see the compilation [1]). Two of the three exit channels ^{2}H(d, p)^{3}H and ^{2}H(d, n)^{3}He generating high energetic particles are mediated by the strong interaction with a branching ratio of about 1 below 50 keV while the third one ^{2}H(d, γ)^{4}He is an electromagnetic transition suppressed by $> 10^7$. Close to the reaction threshold there are two 1^- resonances in the compound nucleus ^{4}He. They can be excited by deuterons with an orbital angular momentum of 1. For this reason the extraordinary strong anisotropy of the angular distribution of the ejectiles even at the lowest energies can be observed. These facts are well known mostly from experiments on gas and polyethylene targets.

There is multiple evidence that the physical environment where the nucleus is embedded can influence nuclear interactions, *e.g.* this is employed in nuclear condensed matter physics. Also in astrophysics the prolongation of the life period of ^{7}Be in the stellar plasma plays an important role in the solar model, *e.g.* [2] and references therein.

[a] e-mail: huke@physik.tu-berlin.de

In order to investigate the environmental impact on nuclear reactions, we experiment with the d+d reactions in metallic environments. We have already found a strongly enhanced electron screening effect leading to a gross increase of the effective cross section by abatement of the Coulomb barrier due to the metal electrons [3,4] which was later reconfirmed [5]. Angular distributions and relative intensities of the proton and neutron channels investigated for d+d reactions taking place in Al, Zr, Pd and Ta targets were, however, in agreement with the results of gas-target experiments. Here we present new results obtained for Sr, Li and Na targets giving a first evidence for an alteration of the neutron-proton branching ratio and the angular distributions. The experiment was published in [6,7] and now a first theoretical explanation for this surprising observation can be presented [8,9].

2 Experimental results

The experiment has been carried out at an electrostatic cascade accelerator at beam energies below 60 keV maintained by a highly stabilized power supply. The deuterium beam ions were generated by an RF ion source with final currents at the target $< 200\,\mu$A depending on the energy. The principal set-up of the detection facility is outlined

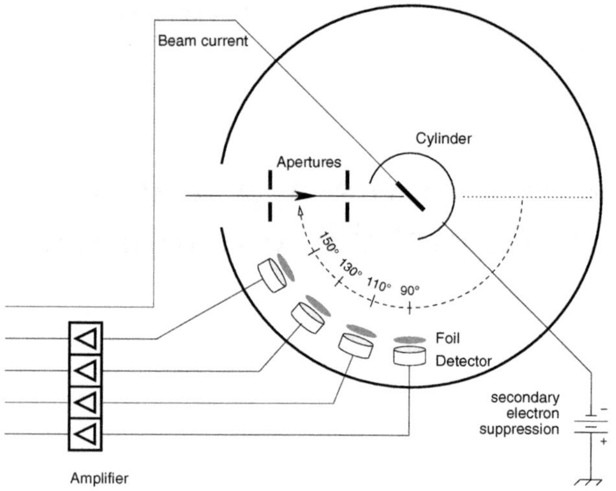

Fig. 1. The set-up in the target chamber.

in fig. 1. After having traversed electrostatic quadrupoles and magnetic dipoles the beam is finally formed by two apertures with a diameter of 1 cm. The targets were pure metal disks becoming self-implanted deuterium targets under the deuteron irradiation. Four Si-detectors with an active surface diameter of 1 cm at a distance of 10 cm and the laboratory angles of 90°, 110°, 130° and 150° were used for the detection of all charged particles, p, t, ^{3}He, of the reactions ^{2}H(d, p)t and ^{2}H(d, n)^{3}He [4]. The detectors needed to be shielded from the backscattered deuterons in order to prevent a congestion of them and the data acquisition system. Therefore grounded Al foils of thicknesses from 120–150 μg/cm^2 were placed in front of the detectors insulated from them. The thickness is sufficient in order to stop backscattered deuterons up to 60 keV. The solid angles surveyed by the detectors were determined with a radioactive α source.

The low-energy part of some representative spectra from the 90° detector is depicted in fig. 2 magnifying the two lines of the recoil nuclei ^{3}He and t. The integral counting number of the spectra are normalized to unity in order to make them comparable. The energies above the peaks are the kinetic energies of the ejectiles in the laboratory system. They drop for increasing projectile energies, which is especially significant for the back angle positions. The gray filled spectra are from Ta targets, while the black and gray step lines are for Sr and Li and Na, respectively. The two plots compare the form of the spectral lines at a low projectile energy of 8 keV to a high energy of 30 keV. At 8 keV the t-line of Ta is well separated while the ^{3}He-line sits on an exponential background. The background is subtracted by fitting an exponential function to the lowest energy part and then by extrapolating it to the high energies. The spectral lines for Sr are already broader with an enhanced low-energy tail leading to an overlap of both lines. This effect becomes even stronger for Li and Na. At 30 keV the Ta lines are broader but the tails of the Sr, Li and Na lines are much more distinctive. The overlap of the two lines is even higher. For Li and Na the ^{3}He-line is hardly more than an edge. The p-line at 3 MeV has

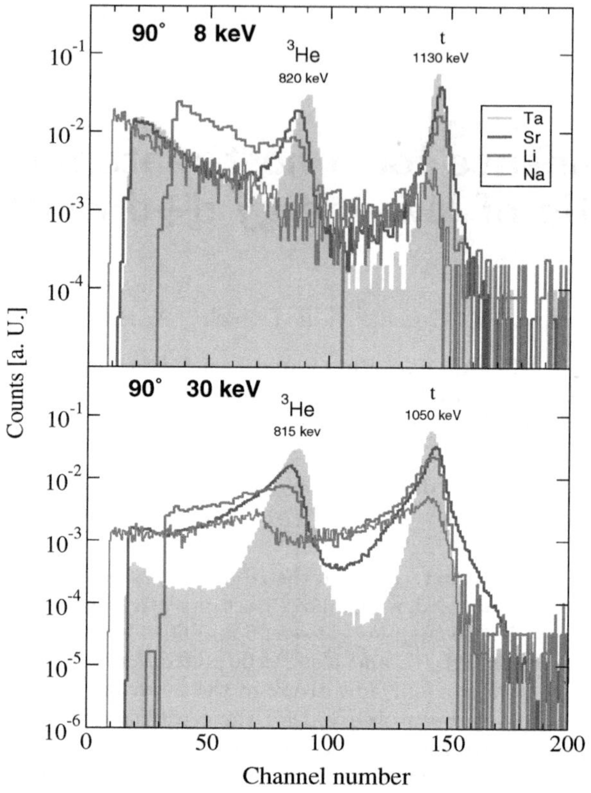

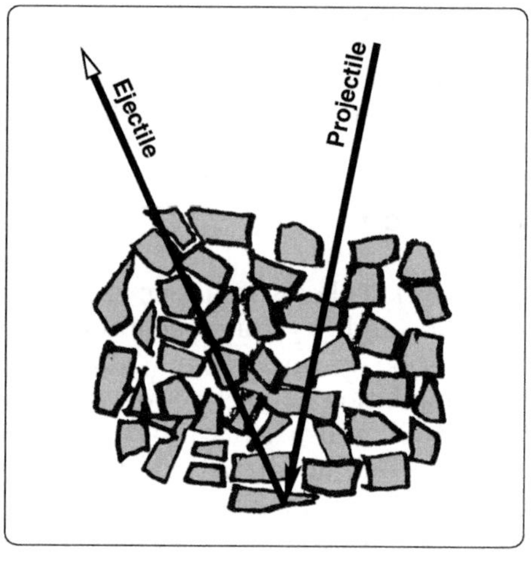

Fig. 2. Normalized spectra from the 90° detector obtained at deuteron energies 8 and 30 keV. The low-energy tail complicates the discrimination and is caused by embrittlement which becomes stronger for more reactive metals. The sketch shows how different paths through the target explain the tails.

also a long low-energy tail but it vanishes before the t-line. The appearance and the properties of these tails can be explained by a phenomenon known from the physical chemistry of the metal hydrides, called *embrittlement* [10] which means that the crystal structure of the metal is bursted by the recrystallization process that accompanies the formation of the metal hydride crystal. Reactive metals change their crystal structure while forming the metal hydride. If the hydration precedes not in a thermal equilibrium and relatively slowly, the material cannot compensate the tension of the recrystallization process and bursts. Since deuteron implantation is far off the thermal equilibrium, embrittlement is a hardly avoidable concomitant phenomenon for reactive metals. How embrittlement effectuates the tails is elucidated with the sketch in fig. 2. Assuming the projectile travels through the target along a path covering many empty regions, the energy loss becomes smaller and consequently the nuclear reactions occur deeper below the target surface than in the case of compact materials. Therefore the ejectiles that in turn can travel through more compact target regions loose more energy additionally contributing to the low-energy tail of the particle spectrum. The increase of the tail with the projectile energy arises from the simultaneous increase in the range of the projectiles. The material dependence can be explained, too. Ta is almost a noble metal with low reactivity but nonetheless able to chemically bind hydrogen to high amounts. It just stretches its lattice but does not recrystallize like the highly reactive metals of the groups I and II of the periodic system. So there is no embrittlement and hardly a tail visible. On the other hand, the effects of embrittlement and the tail increase from Sr over Li to Na with decreasing electron negativity. The symptoms were even visible, *e.g.* dust particles crumbled from a strontium target, the thickness of a sodium target grew considerably. The low-energy tail formation complicates the integration of the spectral lines till infeasibility in the case of Na. The ^{3}He-line sits on the tail of the t-line. All efforts to describe and extrapolate the tail of the t-line to the lower energies analytically failed, since the form of the lines is dependent on the nucleus species, ejectile and projectile energy. Uncertainties in the integral of the spectral lines are taken into account in the errors additionally to the counting statistic. Consequently they are the dominating error source. If in doubt, counts were attributed to the ^{3}He-line only, gaining a conservative estimate of the n/p branching ratio at least. Fortunately, the tails are small at the low projectile energies where the asymmetry in the branching ratios becomes observable. The problems of integrating the overlapping spectral lines cannot be circumvented by the use of detector telescopes for particle identification. The ΔE-detector of the usual semiconductor detector telescopes would already absorb the recoil nuclei.

Another problem affecting the results comes from a property of RF ion sources inherent to their design. The extraction of the ions from the plasma within the source is done with an electric field which is formed by an extraction channel made from pure aluminium inside a ceramic cylinder. This extraction channel presets the direction, focus and flux profile of the beam prior to the lenses in front of the acceleration line. Because of its contact to the plasma and its small dimensions in order to throttle the gas loss it is a wear out commodity. The burn down proceeds not necessarily symmetrical relative to the geometrical axis. Thus, the beam direction and flux profile can gradually change during the wear out causing a shift of the flux distribution within the beam spot on the target.

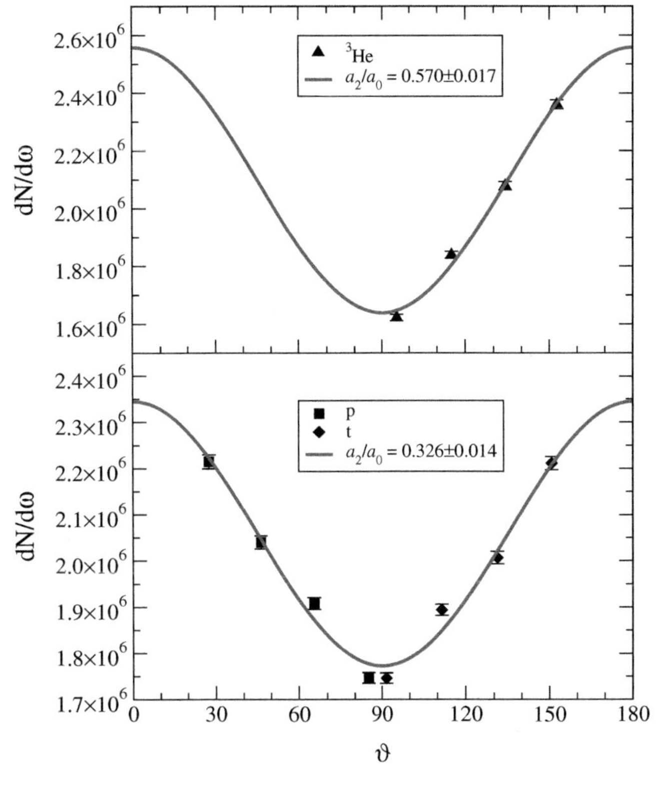

Fig. 3. Angular distribution of the ^{2}H(d, p)^{3}H and ^{2}H(d, n)^{3}He reactions obtained at the deuteron energy $E_d = 20$ keV for Sr.

If this shift occurs within the plain stretched by the detectors and the target, the relative counting number among the detectors is altered leading to a likewise distortion of the angular distribution. Here a close detector geometry was set up facing a large beam spot, thus aggravating the problem. On the other hand a set-up in far geometry with a small beam spot would suffer from low counting rates.

Neglecting $l \geq 2$ contributions, the angular distribution can be described as follows:

$$\frac{\mathrm{d}\sigma}{\mathrm{d}\omega}(\vartheta) = A_0 + A_2 \cos^2 \vartheta\,. \tag{1}$$

Because of the identical bosons in the entrance channel the angular distribution is symmetric around 90°. ϑ and ω are the polar angle and the solid angle in the CM system, respectively. Since the experimentally determined thick target yield for reactions far below the Coulomb barrier is dominated by the high energy contributions a similar expression is valid for the differential counting number

$$\frac{\mathrm{d}N}{\mathrm{d}\omega}(\vartheta) = a_0 + a_2 \cos^2 \vartheta\,. \tag{2}$$

The expansion coefficients a_0 and a_2 now include constant factors describing detector and target properties and the number of incident projectiles. Consequently, the anisotropy coefficient can be defined as $\frac{a_2}{a_0} \approx \frac{A_2}{A_0}$. A measurement at 20 keV for Sr exemplary shows the results

Table 1. Anisotropy and branching ratios tabulated over the deuteron energy in keV in the laboratory system.

$E_{\mathrm{d}}^{\mathrm{lab}}$	$\frac{a_2}{a_0}\,(^3\mathrm{He})$	$\frac{a_2}{a_0}\,(\mathrm{p})$	$\frac{\sigma_{\mathrm{n}}}{\sigma_{\mathrm{p}}}$
Ta			
8.01	0.33±0.04	0.22±0.04	1.007±0.015
10.01	0.42±0.02	0.225±0.019	0.995±0.007
12.02	0.430±0.017	0.210±0.015	0.990±0.005
14.02	0.459±0.019	0.227±0.016	0.998±0.006
17.08	0.492±0.016	0.226±0.014	0.996±0.005
25.05	0.504±0.005	0.251±0.004	1.0030±0.0014
30.02	0.538±0.007	0.281±0.006	1.013±0.002
35.02	0.553±0.007	0.288±0.006	1.017±0.002
40.03	0.582±0.006	0.300±0.005	1.0187±0.0019
45.02	0.633±0.006	0.357±0.005	1.0209±0.0018
50.02	0.692±0.006	0.384±0.005	1.0230±0.0017
Sr			
7.01	0.49±0.05	0.20±0.04	0.837±0.013
8.01	0.54±0.02	0.208±0.016	0.909±0.006
10.01	0.458±0.018	0.245±0.015	0.948±0.006
12.02	0.455±0.012	0.194±0.010	0.923±0.004
14.02	0.372±0.015	0.182±0.013	0.991±0.006
17.02	0.440±0.012	0.214±0.010	0.971±0.004
20.02	0.473±0.014	0.283±0.012	0.991±0.005
25.04	0.506±0.013	0.247±0.011	0.992±0.004
30.04	0.505±0.015	0.286±0.005	1.025±0.003
35.19	0.587±0.007	0.274±0.005	1.021±0.002
40.02	0.669±0.007	0.369±0.004	1.0356±0.0019
45.03	0.550±0.004	0.269±0.003	1.0361±0.0014
50.03	0.640±0.005	0.334±0.004	1.0295±0.0015
55.02	0.684±0.005	0.327±0.004	1.0478±0.0014
Li			
8.02	1.9±1.9	0.25±0.06	0.83±0.19
10.02	0.9±0.3	0.30±0.03	0.79±0.05
12.02	0.9±0.3	0.41±0.03	0.77±0.04
14.03	1.3±0.3	0.66±0.03	0.85±0.05
17.02	0.59±0.14	0.304±0.011	1.02±0.04
25.01	0.96±0.19	0.39±0.04	0.98±0.04

for the differential counting number and the corresponding fitting function in fig. 3. The fit is computed with a non-iterative generalized linear fitting algorithm employing singular value decomposition thus allowing for more accurate values and better error handling. The data points obtained for the protons are included. As can be seen protons and tritons follow the same angular distribution. One observes a significantly stronger angular anisotropy for the neutron channel. Due to the strong energy dependence of the reaction cross sections, the branching ratio of the two mirror reactions can be obtained by dividing the thick target yields Y and correspondingly the total counting numbers N:

$$\frac{\sigma_{\mathrm{d(d,n)^3He}}}{\sigma_{\mathrm{d(d,p)t}}} \approx \frac{Y\,(^3\mathrm{He})}{Y(\mathrm{p})} = \frac{N\,(^3\mathrm{He})}{N(\mathrm{p})} = \frac{a_0\,(^3\mathrm{He}) + \frac{1}{3}a_2\,(^3\mathrm{He})}{a_0(\mathrm{p}) + \frac{1}{3}a_2(\mathrm{p})}. \tag{3}$$

When calculating with the fitting coefficients one must consider that they are not independent variables. Then the Gaussian error propagation formula needs to be completed by a term containing the off-diagonal element of the covariance matrix from the fit.

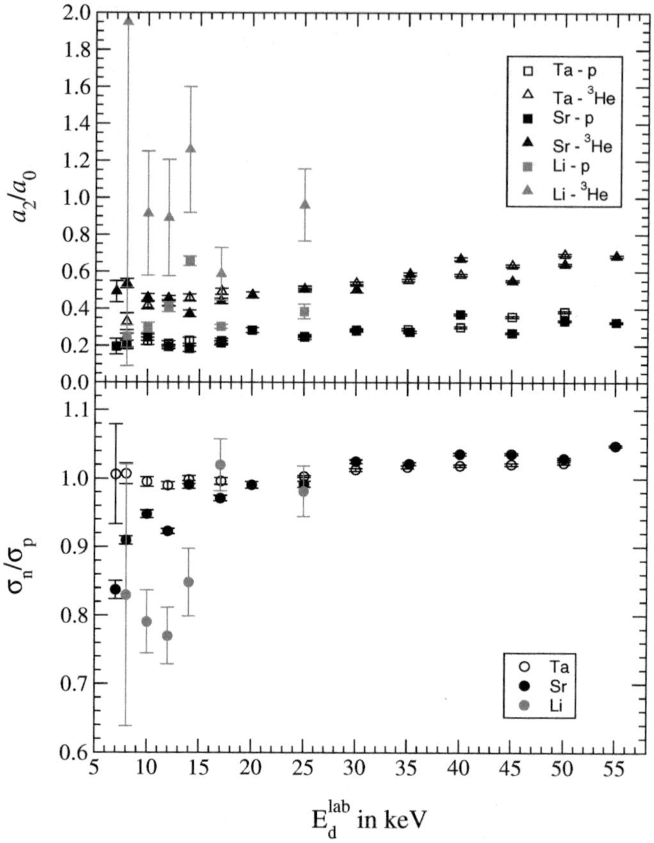

Fig. 4. The upper part displays the anisotropy from the detection of the p and ^{3}He ejectiles. The lower part shows the branching ratio of the two mirror reactions.

The results are listed in table 1 and plotted in fig. 4. The branching ratios and angular distributions for Ta agree with the results of the gas target experiment [11] and due to the much higher target nuclei density the data set has a notedly higher precision and extends to lower energies ([4], fig. 1). This agreement applies likewise to Al, Zr and Pd. Not so for Sr and Li. While for p there are no peculiarities, for ^{3}He the anisotropy raises at lower energies. Simultaneously, the n branch is suppressed. This is better cognizable in the rescaled fig. 5 for Sr. The results for protons and tritons are concordant. The low quality of the Li points results from the ambiguity of the integration of the spectral lines with large low-energy tails. For the same reason, the spectra obtained for Na could not be analyzed quantitatively, though the spectra indicate a strong suppression of the neutron-proton ratio at low energies, too. Because of the relative alteration in the counting numbers of the detectors from the burn down of the extraction channel the angular anisotropies obtain a gradually increasing positive offset corresponding to the sequence of the measurement campaign. Starting with Ta the value of the offset increased for Sr, Li, Zr and finally culminated at Al. Therefore the anisotropy data for Sr in table 1 and figs. 4, 5 was renormalized to the ones of Ta. Note: the anisotropy values in fig. 3 are the original ones. These corrections, however, hardly affect the branching

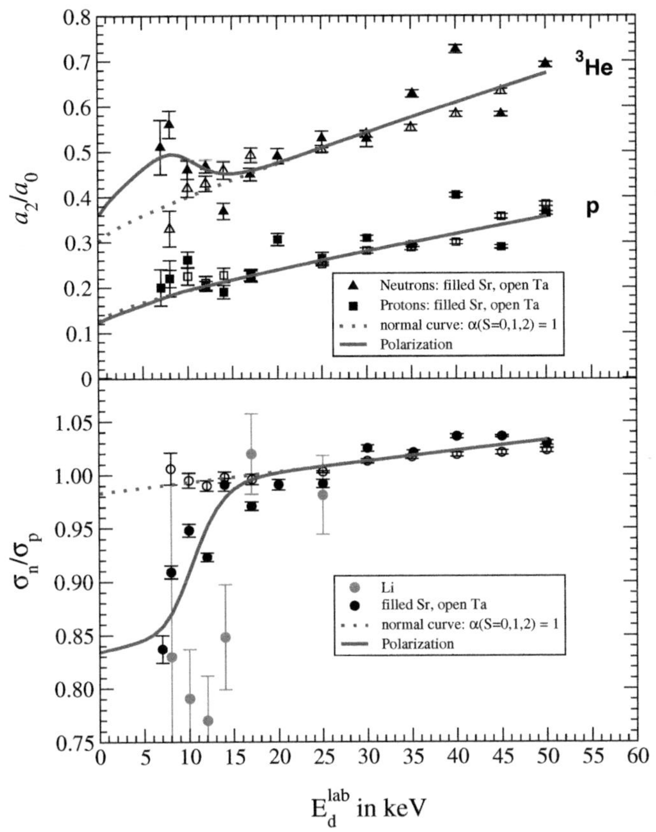

Fig. 5. The dashed line represents the normal curve. The solid lines result from a deuteron polarization corresponding to a suppression of the $S = 0$ channel at lower energies.

ratios (3). From the anisotropy presented in fig. 5 it can be seen that there are significant deviations from the smooth progression of the curve at 40 and 45 keV. They originate from suboptimal beam focus adjustment with accordingly beam flux shifts because of necessary beam current reductions at higher deuteron energies. The deviations occur likewise for protons and ^{3}He. Altogether an additional systematic error is introduced. For details refer to [6].

3 Theoretical considerations and discussion

Multiple scattering of the ejectiles in the thick target could possibly redirect leaving particles depending on the nucleus species and thereby change the detection rate. In order to test this, a Monte Carlo simulation was performed tracing the path of the reaction ejectiles [12]. The target bulk is assumed to be amorphous where the reaction occurs at a randomly selected depth weighted with the stopping power function and the cross section. The ejectile on his part is emitted in a direction determined by a random function regarding the angle dependent cross section. On its way it suffers hits on scattering centers and loses meanwhile energy to electrons. The ejectiles hitting onto the detectors were counted and set in relation to the same sort of ejectiles which would have reached the detector if there were no subsequent scattering. The differences in the target materials are manifested merely in the stopping power coefficients. The calculations were performed for Li, Sr and Ta. Only for high energies above 20 keV at the backward angle 130° a significant alteration can be noticed causing a slight additional anisotropy of the ^{3}He ejectiles. This can be understood from the reaction kinematics predicting an enlarged energy drop for massive ejectiles at backward angles which is further amplified by the higher energy loss and stronger scattering of the double charged ^{3}He nuclei altogether preferring the 130°-detector across from the 45° tilted target. This behaviour is a direct consequence of the concrete detector target geometry and strongly dependent on it. This also means that there is no modification at the low energies. So multiple scattering can be excluded from the possible trivial reasons for the material dependent n/p branching ratio. Moreover, the long low energy tail of the spectral lines was not obtained by the simulation which is not unexpected because of its origin from the embrittlement. Instead, the simulation disclosed that the tiny tail at the high energy side of the spectral lines, only visible with high spectral resolution, is from multiple scattering.

The embrittlement cannot be responsible for the observed anomalous asymmetry in the branching ratios, since the effects of embrittlement like tail formation rise with the projectile energy in contradiction to the deviations in the branching ratio. This is also valid for conceivable weird surface textures. Such was reconfirmed on a Ta target with an artificial blemished surface showing no differences to the usual results for Ta. Furthermore, anisotropic symmetries in crystal structures cause effects like optical activity and piezo and pyro electricity [13]. So this could be a conceivable reason for the experimental observations. Enantiomorphy is a necessary condition for such effects. However, the point groups belonging to LiD, SrD_2 and NaD do not allow for this. Transient magnetic fields can cause an alignment of the projectile nucleus relative to the interior magnetic field of ferromagnetic target materials [14,15]. Despite unfulfilled requisites, this still not fully understood phenomenon could be considered. An experiment on a ferromagnetic Fe target showed no effect.

From the theoretical point of view the cross sections for the mirror reactions ^{2}H(d, p)^{3}H and ^{2}H(d, n)^{3}He at deuteron energies below 100 keV can be described with 16 collision matrix elements, corresponding to S,P,D-waves in the entrance channel. The matrix elements for incoming D-waves cannot be omitted as frequently asserted since they are mandatory to describe the angular anisotropy down to the lowest energies. The values of the matrix elements are relatively well known and were obtained by fitting experimental cross sections, vector and tensor analyzing powers measured in gas target experiments [16, 17]. The differential cross section for both reactions can be presented by a coherent superposition of all sixteen matrix elements [8] (dashed line in fig. 5) and agrees with our results obtained for Al, Zr, Pd and Ta. In the case of Sr (also for Li) a polarization of the deuterons in the crystal lattice had to be assumed. A suppression of the channel spin $S =$ 0 (spins of the deuterons are anti-parallel) and allowing the other channels with spins $S = 1, 2$ to be undisturbed permits to describe simultaneously the enhancement of the angular anisotropy of the $^2H(d,n)^3He$ reaction and the decrease of the n/p branching ratio at very low energies down to 0.83. The results of corresponding calculations are presented in fig. 5 as full lines. Here we have assumed that the deuteron polarization takes place gradually below the Fermi energy (for Sr about 25 keV), reaching its maximum value already below 10 keV. A strong quenching of the neutron channel might also be explained by different screening energies for relative angular momentum $L = 0, 1$. Such was also tested and proved to be in contradiction to the experimental results. The full calculations will be subject of a forthcoming publication, also refer to [8].

4 Conclusion

We presented a first experimental evidence for an alteration of the branching ratios in the d+d fusion reactions obtained in an accelerator experiment which can be theoretically explained by polarization of the reacting deuterons in the crystal lattice. Several other conceivable but rather trivial causes could be excluded. Albeit the deeper reason for the deuteron polarization on its part is still unknown. A distinctiveness of the (earth)alkaline metals is the formation of an ionic bond to hydrogen, which might be a starting point for a possible explanation based on the spin-spin interaction mediated by electrons. The conditions in high vacuum make up for metal oxides and hydroxides in the chemical reactive targets [18] which could also be responsible for the observed effects by means of their special electron configuration and would be as such non-trivial and interesting, too.

So with the spin polarization is here another new way how the environmental electron configuration can influence immediate nuclear processes. The dense bound and free electrons in the metal can abate the Coulomb barrier in a dynamic process prior to the reaction generating a gross enhancement of the cross section which can still not be described by theory to this extent [19]. Both are dynamic processes. Static processes with respect to the electrons like alterations in the β decay of light nuclei are easier comprehensible. The screening from the electrons in the molecule CH_3T causes a shift in the spectral distribution of the β particles [20]. The removal of the two 1s electrons of ^7Be in the hot stellar plasma significantly increases the half-time for the decay by electron capture. Since the 1s electrons contribute by far the most to the electron density in the nucleus the modification of the orbitals of outer shell valence electrons would have only little influence on the decay parameters.

Our findings also provide a first independent support for the claim in cold fusion that requires a heavily alteration of the d+d reaction channels in contradiction to the results obtained for gas targets. Thus making it, together with the enhanced electron screening in metals [19], more credible although further efforts are necessary. An experiment with more sophisticated particle detection techniques is in progress in order to refine the data.

References

1. D.R. Tilley, H.R. Weller, G.M. Hale, Nucl. Phys. A **541**, 1 (1992).
2. Zs. Fülöp, Gy. Gyürky, E. Somorjai, D. Schürmann *et al.*, Nucl. Phys. A **758**, 697c (2005).
3. K. Czerski, A. Huke, P. Heide, M. Hoeft, G. Ruprecht, in *Nuclei in the Cosmos V, Proceedings of the International Symposium on Nuclear Astrophysics, Volos, Greece, July 6-11, 1998*, edited by N. Prantzos, S. Harissopulos (Editions Frontières, Paris, 1998) p. 152.
4. K. Czerski, A. Huke, A. Biller, P. Heide, M. Hoeft, G. Ruprecht, Europhys. Lett. **54**, 449 (2001).
5. F. Raiola *et al.*, Phys. Lett. B **547**, 193 (2002).
6. A. Huke, PhD Thesis, Technische Universität Berlin (2002); `http://edocs.tu-berlin.de/diss/2002/huke_armin.htm`.
7. A. Biller, K. Czerski, P. Heide, M. Hoeft, A. Huke, G. Ruprecht, in *Verhandlungen der DPG*, Vol. **1** (DPG-Frühjahrstagung, Göttingen, 1997) p. 28.
8. T. Dorsch, Diploma Thesis, Institut für Atomare Physik und Fachdidaktik der Technischen Universität Berlin (2004).
9. A. Huke, K. Czerski, T. Dorsch, P. Heide, *Proceedings of the International Conference on Condensed Matter Nuclear Science, Marseille* (2004).
10. W.M. Mueller, J.P. Blackledge, G.G. Libowitz (Editors), *Metal Hydrides* (Academic Press, New York, London, 1968).
11. R.E. Brown, N. Jarmie, Phys. Rev. C **41**, 1391 (1990).
12. A. Biller, Diploma Thesis, Institut für Atomare und Analytische Physik der Technischen Universität Berlin (1998).
13. M. Wagner, *Gruppentheoretische Methoden in der Physik* (Vieweg, Braunschweig, Wiesbaden, 1998).
14. K. Dybdal, J.S. Forster, N. Rud, Phys. Rev. Lett. **43**, 1711 (1979).
15. K.-H. Speidel, Phys. Lett. B **324**, 130 (1994).
16. H. Paetz gen. Schieck, S. Lemaitre, Ann. Phys. (Leipzig) **2**, 503 (1993).
17. O. Geiger, S. Lemaître, H. Paetz gen. Schieck, Nucl. Phys. A **586**, 140 (1995).
18. A. Huke, K. Czerski, P. Heide, Nucl. Phys. A **719**, 279c (2003).
19. K. Czerski, A. Huke, P. Heide, G. Ruprecht, Europhys. Lett. **68**, 363 (2004).
20. C.K. Hargrove, D.J. Paterson, J.S. Batkin, Phys. Rev. C **60**, 034608 (1999).

Eur. Phys. J. A **27**, s01, 193–196 (2006)
DOI: 10.1140/epja/i2006-08-029-2

EPJ A direct
electronic only

New measurement of ^{7}Be half-life in different metallic environments

B.N. Limata[1,a], Zs. Fülöp[2], D. Schürmann[3], N. De Cesare[4], A. D'Onofrio[4], A. Esposito[4], L. Gialanella[1], Gy. Gyürky[2], G. Imbriani[1], F. Raiola[3], V. Roca[1], D. Rogalla[4], C. Rolfs[3], M. Romano[1], E. Somorjai[2], F. Strieder[3], and F. Terrasi[4]

[1] Dipartimento di Scienze Fisiche, Università Federico II and INFN Sezione di Napoli, Italy
[2] ATOMKI, H-4001 Debrecen, POB 51, Hungary
[3] Institut für Experimentalphysik III, Ruhr-Universität Bochum, Germany
[4] Facoltà di Scienze Ambientali della Seconda Università di Napoli, Caserta Italy

Received: 29 July 2005 /
Published online: 7 March 2006 –

Abstract. ^{7}Be decays via electron capture and therefore its half-life is expected to depend on the electron density at the nucleus. We measured the ^{7}Be half-life in palladium, tungsten, zirconium and tantalum metals, in order to investigate the influence of the quasi-free electrons in metals on the probability of electron capture. The ^{7}Be samples were obtained implanting a 4.85 MeV pure radioactive ^{7}Be ion beam. The ^{7}Be half-life was determined measuring the 478 keV gamma decay following the electron capture by means of HPGe detectors. In order to reduce systematic errors, we planned to perform independent measurements in three different laboratories: Naples (Italy), Bochum (Germany) and Debrecen (Hungary). On the basis of the first results, we do not find a ^{7}Be half-life change within the experimental errors of 0.4%.

PACS. 21.10.Tg Lifetimes – 23.40.-s β decay; double β decay; electron and muon capture – 27.20.+n $6 \leq A \leq 19$

1 Introduction

In the stellar environment, ^{7}Be is produced by ^{3}He$(\alpha,\gamma)^7$Be and destroyed by ^{7}Be$(\mathrm{p},\gamma)^8$B, which is responsible for the high-energy solar neutrino flux, and, in competition, by ^{7}Be decay. ^{7}Be decays to ^{7}Li via electron capture, populating either the ^{7}Li ground state or, in 10% of the cases, the first excited state, which decays to the ground state emitting a 478 keV gamma ray.

The half-life of the nuclei such as ^{7}Be which decays by electron capture depends on the electron density at the nucleus. Therefore, the decay probability of these nuclei in stellar plasma environment differs from their terrestrial values. Starting from 1949 [1], many experiments were done in order to measure variations of the ^{7}Be half-life depending on the host chemical environment. Although the half-life changes are in most cases lower than 0.2% [2], some authors found larger variations [3,4,5].

Recently, it was discovered that the electron screening in d(d, p)t reaction for deuterated metals is much higher than the screening measured in d(d, p)t using deuterium gas target [6]. This feature has been explained by the Debye plasma model applied to the quasi-free electrons in metals: these electrons form an electron cloud around the nucleus with a Debye radius which is about a factor 10 smaller than the atomic radius [7]. According to this picture, ^{7}Be nuclei implanted in metallic host materials may probe a higher electron density, that would provide a test case for the calculations of the ^{7}Be decay in the solar plasma.

We implanted ^{7}Be in different metallic targets and we measured the half-life of an electron capture in order to highlight a detectable half-life variation depending on the different quasi-free electronic density. The choice of the host materials was based on the measurements of the electron screening potential for d(d, p)t in deuterated metals [6]: palladium, tungsten and tantalum metals show a large screening potential, whereas for zirconium the screening effect is smaller.

In previous experiments, with the exception of [8], secondary ^{7}Be beams have been used and therefore both the projectile and the ^{7}Be recoil have been implanted at roughly the same depth. As a consequence, the electron distribution around ^{7}Be nuclei may be altered due to a) the chemical contamination of the beam; b) the lattice damage induced by the high implantation currents. In this experiment, we use a different approach, *i.e.* a pure ^{7}Be beam is implanted in host materials in order to virtually reduce to zero the implantation damage and avoid changes of the sample stoichiometry. Table 1 shows the experimen-

[a] e-mail: `limata@na-infn.it`

Table 1. Implantation procedure of earlier ^{7}Be half-life measurements.

Author	Primary beam	Host material, ^{7}Be range	^{7}Be half-life (d) or ratio (%)
F. Lagoutine [8]	^{7}Be (Energy not reported)	metallic Al, –	53.17 ± 0.07
E.B. Norman [9]	40 MeV ^{3}He	grafite, –	53.107 ± 0.022
	10 MeV H	boron nitride, –	53.174 ± 0.037
	45 MeV ^{7}Li	Ta, –	53.195 ± 0.052
		Au, –	53.311 ± 0.042
T. Ohtsuki [3]	16 MeV H	C_{60} cages, –	52.68 ± 0.05
	γ irradiation	metallic ^{9}Be, –	53.12 ± 0.05
A. Ray [4]	7 MeV H	Au, 1.8 μm	$(T_{1/2}(\mathrm{Au}) - T_{1/2}(\mathrm{Al_2O_3}))/T_{1/2}(\mathrm{Al_2O_3})$
		Al_2O_3, 2 μm	$= (0.72 \pm 0.07)\%$
Zhou [5]	^{7}Li/^{7}Be mixed beam	Pd, 23.6 μm	$(T_{1/2}(\mathrm{Au}) - T_{1/2}(\mathrm{Pd}))/T_{1/2}(\mathrm{Pd})$
		Au, 20.9 μm	$= (0.8 \pm 0.2)\%$
Liu [10]	3.2 MeV H	natural beryllium, 2 μm	$(T_{1/2}(\mathrm{Au}) - T_{1/2}(\mathrm{Be}))/T_{1/2}(\mathrm{Be})$
		natural gold, 1 μm	$\leq 0.12\%$
Present work, [11]	^{7}Be	Pd, 2.9 μm	see table 3
		W, 2.7 μm	
		Zr, 4.3 μm	
		Ta, 3.1 μm	

tal procedure of the ^{7}Be implantation of the earlier and present ^{7}Be half-life measurements.

Moreover, the ^{7}Be half-life has been independently measured in 3 different laboratories in order to reduce the effects of possible systematic errors.

2 Experimental procedure

The production of a pure ^{7}Be beam at the 3MV Tandem accelerator in Naples is described elsewhere [12]. Shortly, ^{7}Be is produced via $^7\mathrm{Li(p,n)}^7\mathrm{Be}$ using a 11.7 MeV proton beam with a current intensity of 20 μA, delivered by the ATOMKI Cyclotron of Debrecen (Hungary). ^{7}Be is then extracted from the metallic Li target by means of radiochemical methods [12] at the isotopic laboratory of the Ruhr-Universität Bochum (Germany). The activity is collected into a cathode of the sputtering ion source which is used to produce a $^7\mathrm{BeO}^-$ beam. This molecular ion beam is injected in the accelerator and the emerging $^7\mathrm{Be}^{2+}$ beam is selected as a $^7\mathrm{Be}^{4+}$ beam, after passing a thin C foil. This procedure suppresses the Li contamination in the beam and allows to obtain a high-purity ^{7}Be beam. The beam intensity was about 7 ppA. A pair of scanning magnets provided a uniform implantation. The scheme of the experimental setup is shown in fig. 1.

The implanted metallic materials are Pd, W, Zr, Ta with activities ranging from 3.6 to 36 kBq. The kinetic beam energy $E_{lab} = 4.85$ MeV provided about 3–4 μm ion stopping range (table 1), which is deeper than the layers where one expects surface contaminations and oxidation of the metal. The ^{7}Be half-life of the implanted samples was independently measured in the laboratories of Bochum, Debrecen and Naples. The 478 keV gamma emission was detected with HPGe detectors. A ^{60}Co source was included in all measuring setups (except for the W Naples measurement). In this work we show the first experimental results of Debrecen and Naples groups. The details of these two setups are summarized in table 2.

Table 2. Experimental details of the Debrecen and Naples ^{7}Be half-life measurements.

	Debrecen	Naples
HPGe Detector efficiency		
(3″ × 3″ NaI scale %)	20	28
Pb shield thickness (cm)	5	5
Sample to HPGe distance (cm)	3	0
^{60}Co to HPGe distance (cm)	3	10
Analyzed samples	Pd-W	Pd-W
	Zr-Ta	
γ spectra time		
integration (h)	12	1–6
Measuring time (days)	37–40	50–100
	74–37	

The γ spectra were automatically stored at every 1, 6 or 12 hours, depending on the activity of the sample. As an example, fig. 2 shows a typical spectrum containing the ^{7}Be decay peak, 511 keV annihilation peak and the two γ lines of ^{60}Co reference source.

In principle, possible sources of systematic errors are: a) efficiency variation during these long measurements; b) pileup effect which may increase the half-life value; c) acquisition dead time; d) absolute time shifts. The a), b), c) errors can be avoided if the half-life value is evaluated by normalising the ^{7}Be counts to another independent source. To this purpose, we used two ^{60}Co peaks and the 1460 keV natural background peak of ^{40}K. As regards d), the error was kept below 0.01%. For example, the internal clock of the acquisition PC of Naples setup was compared at the beginning of each run to the Greenwich

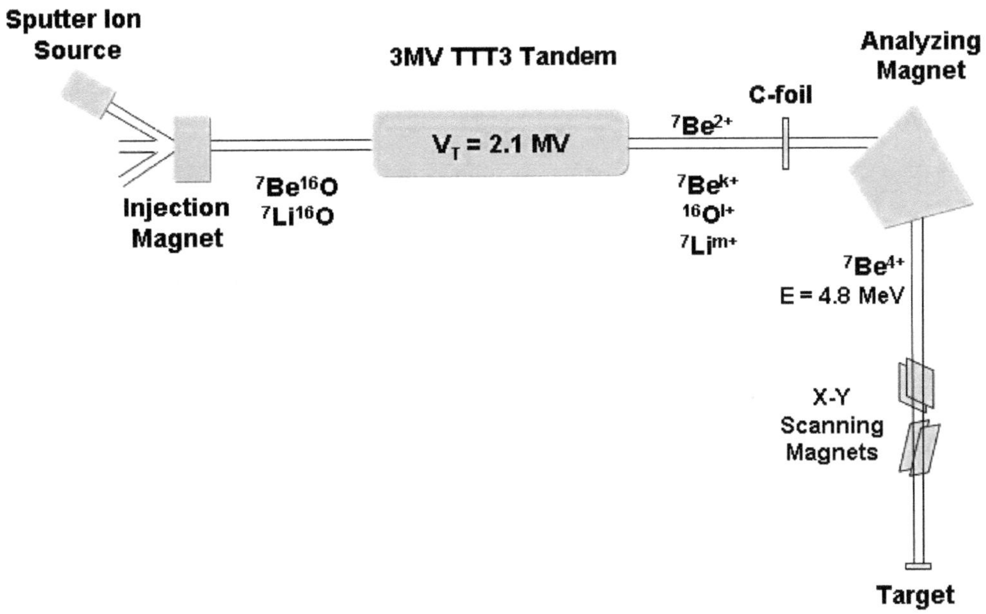

Fig. 1. Scheme of the ^{7}Be implantation setup at the TTT3 Tandem accelerator of the Federico II University of Naples.

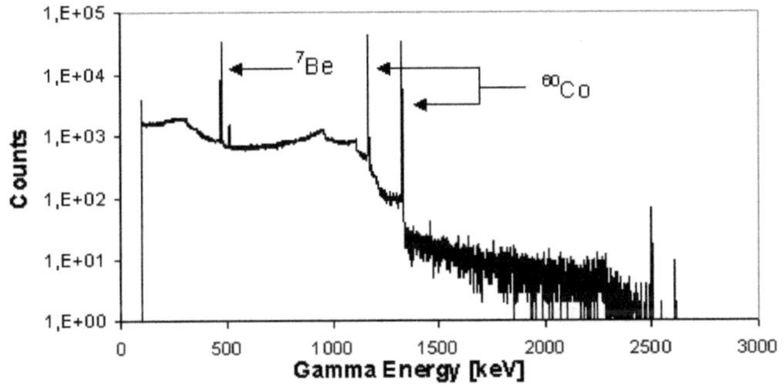

Fig. 2. Gamma spectrum of Pd sample measured in Naples (Δt =1 h).

reference time (National Institute of Standards and Technology web site: `www.tf.nist.gov/service/its.htm`).

Finally, it should be noted that the choice of the experimental setup, the data acquisition and the analysis procedure were completely left to the different groups which independently draw out their results.

3 Results

The samples measured in Naples were Pd, W, Zr. The data analysis was completed for the Pd and the W targets, whereas the analysis of the Zr sample is still in progress. The complete data analysis of Debrecen workgroup is available for all the implanted targets W, Pd, Zr and Ta. All the results are shown in table 3. As an example, the details of data analysis at Naples are presented below.

Every spectrum has been analysed fitting a linear background below the ^{7}Be and the reference peaks. We obtain a set of values $\{A_n^{exp} \pm \sigma_n\}$ which are the time integrals of the peaks after a linear background subtraction. This analysis has been done for both ^{7}Be and the reference peaks. A least-square function Q^2,

$$Q^2 = \sum_n \left[\frac{A_n^{exp} - A_n^{th}}{\sigma_n}\right]^2, \tag{1}$$

was minimized using the experimental values A_n^{exp} and the fit function A_n^{th}:

$$\begin{aligned} A_n^{th} &= \int_{start}^{stop} A_0\epsilon e^{-t/\tau}\,\mathrm{d}t - \int_{(\Delta t)_{dead}} A_0\epsilon e^{-t/\tau}\,\mathrm{d}t \\ &= \int_{start}^{start+\Delta t_{live}} A_0\epsilon e^{-t/\tau}\,\mathrm{d}t \\ &\overset{\epsilon=const}{=} A_0\epsilon\tau e^{-t_{start}/\tau}(1 - e^{-\Delta t_{live}/\tau}), \end{aligned} \tag{2}$$

where A_0 is the initial activity, ϵ is the detection efficiency and Δt_{live} and Δt_{dead} are the acquisition live and dead time, respectively. The free parameters are $(A_0\epsilon)$ and τ. The uncertainty on τ is determined by means of a χ^2 analysis.

Similar expressions were used for the ratio with the reference sources.

For each implanted sample, we evaluated the ^{7}Be half-life by fitting both the ^{7}Be counts and the ratio between

Table 3. First experimental results of Debrecen and Naples groups. The ^{7}Be half-life values evaluated with and without normalization are reported.

	Workgroup	^{7}Be half-life [d]			
Normalization method		Pd	W	Zr	Ta
Without normalization	Naples	53.14 ± 0.05	53.65 ± 0.06		
	Debrecen	53.28 ± 0.14	52.95 ± 0.15	53.01 ± 0.05	53.21 ± 0.34
	Average	$\mathbf{53.16 \pm 0.05}$	$\mathbf{53.55 \pm 0.06}$		
Norm. to 1173 keV ^{60}Co line	Naples	53.13 ± 0.14	–		
	Debrecen	53.18 ± 0.22	53.32 ± 0.21	53.04 ± 0.15	52.67 ± 0.36
	Average	$\mathbf{53.14 \pm 0.12}$			
Norm. to 1332 keV ^{60}Co line	Naples	53.31 ± 0.14	–		
	Debrecen	53.30 ± 0.23	53.07 ± 0.21	53.02 ± 0.15	52.70 ± 0.36
	Average	$\mathbf{53.31 \pm 0.12}$			
Norm. to 1460 keV ^{40}K line	Naples	–	54.72 ± 0.94		
	Debrecen	53.41 ± 0.48	53.39 ± 0.43	53.12 ± 0.20	53.46 ± 0.51
	Average		$\mathbf{53.62 \pm 0.39}$		

^{7}Be and the reference source peaks. The half-life values obtained without normalization are in agreement with the normalized ones. The values and the errors evaluated using the normalized data take into account some systematic errors like the pileup effect, the efficiency variation and the acquisition dead time. A direct comparison between Naples and Debrecen data is possible for Pd and W metals. Unfortunately, for the W metal, the precision of the Naples measurement is not sufficient, since no additional reference source was available in the experimental setup.

The results have to be compared to the ^{7}Be half-life adopted value [2] of 53.22 ± 0.06 days. Although the data analysis has to be finalized for all the implanted metals, on the basis of the first Debrecen and Naples data above, we do not find ^{7}Be half-life change within the experimental errors (0.3-0.4%).

This result is compatible at a 2σ level with the predictions of the Debye model applied to the free electrons in metals, provided the following additional assumptions are made: a) the contribution of the free electrons capture is negligible; b) the Debye screening does not affect the electron steady wave functions of ^{7}Be atoms. In this case, in fact, the ^{7}Be screening potential U_D due to the free electrons is $U_D = -1 \times 4 \times U_{d+d}$, where 4 is Beryllium atomic number and U_{d+d} is the electron screening value measured in d(d, p)t reaction in metallic environments. Between the metals considered, the highest U_{d+d} value is 800 eV for Palladium [6], which gives $U_D = -3.2$ keV. Since the electron capture process scales with the energy squared, the reduction factor of the decay probability is $[(862 - 3.2)/862]^2 = 0.992$, *i.e.* 0.8%, which correspond to a longer half-life([1]).

Finally, our results seem not to confirm the findings of [5] where a correlation between the screening and the ^{7}Be half-life was observed having opposite sign respect to the predictions of the Debye model. However, a realistic model of the electron capture decay of ^{7}Be in metallic environment should be developed to compare these experimental results with the data of the electron screening in metals.

This work was supported by OTKA T042733, F043408, D048283, T049245. Zs. F. is a Bolyai fellow.

References

1. E. Segré, C.E. Wiegand, Phys. Rev. **75**, 39 (1949).
2. M.M. Bé *et al.*, *Table of Radionuclide (comments on evaluation)*, Monographie BIPM-5 (2004) Bureau International des poids et des mesures.
3. T. Ohtsuki *et al.*, Phys. Rev. Lett. **93**, 112501 (2004).
4. A. Ray *et al.*, Phys. Lett. B **455**, 69 (1999).
5. S.H. Zhou *et al.*, Chin. Phys. Lett. **22** No. 3, 565 (2005).
6. F. Raiola *et al.*, Phys. Lett. B **547**, 193 (2002).
7. F. Raiola *et al.*, Phys. Lett. A **719**, 37c (2003).
8. F. Lagoutine *et al.*, Int. J. Appl. Radiat. Isot. **26**, 131 (1975).
9. E.B. Norman *et al.*, Phys. Lett. B **519**, 15 (2001).
10. Z.Y. Liu *et al.*, Chin. Phys. Lett. **20** No. 6, 829 (2003).
11. Zs. Fülöp *et al.*, Nucl. Phys. A **758**, 697c (2005).
12. L. Gialanella *et al.*, Nucl. Instrum. Methods B **197**, 150 (2002).

[1] The energy release of the ^{7}Be electron capture decay is 862 keV.

Eur. Phys. J. A **27**, s01, 197–200 (2006)
DOI: 10.1140/epja/i2006-08-030-9

EPJ A direct
electronic only

Study of the ^{106}Cd$(\alpha,\alpha)^{106}$Cd scattering at energies relevant to the p-process

G.G. Kiss[1,2,a], Zs. Fülöp[1], Gy. Gyürky[1], Z. Máté[1], E. Somorjai[1], D. Galaviz[3,b], A. Kretschmer[3], K. Sonnabend[3], and A. Zilges[3]

[1] Institute of Nuclear Research (ATOMKI), P.O. Box 51, H-4001 Debrecen, Hungary
[2] University of Debrecen, Debrecen, Hungary
[3] Technische Universität Darmstadt, D-64289 Darmstadt, Germany

Received: 26 July 2005 /
Published online: 8 March 2006 –

Abstract. The elastic scattering cross section of ^{106}Cd$(\alpha,\alpha)^{106}$Cd has been measured with high accuracy at energies of $E_{c.m.} \approx 15.5$, 17, and 19 MeV. The optical potential for the system ^{106}Cd $\otimes\alpha$ has been derived at energies above and below the Coulomb barrier. Predictions for the ^{106}Cd$(\alpha,\gamma)^{110}$Sn capture cross section at astrophysically relevant energies are presented and compared to the experimental data measured recently.

PACS. 24.10.Ht Optical and diffraction models – 25.55.-e ^{3}H-, ^{3}He-, and ^{4}He-induced reactions – 25.55.Ci Elastic and inelastic scattering – 26.30.+k Nucleosynthesis in novae, supernovae and other explosive environments

1 Introduction

The nucleosynthesis of nuclei above the iron peak proceeds mainly by neutron capture in the s- and r-process. However, there are 35 other, stable, proton-rich, the so-called p-nuclei, which cannot be produced via neutron capture reactions [1]. The production of the p-nuclei proceeds mainly via photon-induced reactions in the O/Ne layers of type-II supernovae. The s and r seed nuclei are disintegrated by (γ,n), (γ,p) and (γ,α) reactions in the high photon flux of the explosion.

Calculations for the p-process involve more than 1000 nuclei in a network that requires more than 10000 reaction rates [2]. Almost none of these reaction rates has been measured and the calculations rely completely on the statistical model. One of the input parameters in statistical model calculations to determine (γ,α) reaction rates is the alpha-nucleus optical potentials. However, the uncertainties shown by the alpha-nucleus potentials at astrophysically relevant energies are large [3,4]. Experimental informations are therefore required to reduce the uncertainties in the calculation of (γ,α) reaction rates.

In principle, the alpha-nucleus potentials can be determined from alpha elastic scattering experiments. The feasibility of such a measurement is, however, limited in general because the experimentally determined cross section at energies below the Coulomb barrier shows only a small deviation from the Rutherford cross section and the results have ambiguities. In recent years, however, alpha-nucleus potential parameters of ^{144}Sm, ^{92}Mo, ^{112,114}Sn have been successfully derived at ATOMKI [4,5,6]. A new experiment on ^{106}Cd, the most proton-rich stable isotope of Cd, helps to better understand the behavior of the alpha-nucleus optical potential as a function of the mass number and energy.

The choice of the measured energies at about 15.5, 17 and 19 MeV has the following reason. The Gamow-window for (γ,α) reactions at $T_9 \approx 2$–3 is in the range of $E_\gamma \approx 5$–10 MeV corresponding to 4–9 MeV for the inverse ^{106}Cd$(\alpha,\gamma)^{110}$Sn reaction. Recently the (α,γ) capture cross section on ^{106}Cd has been measured in the upper part and above the Gamow-window [7,8]. The experimental determination of the nuclear part of the optical potential at this astrophysical energy, however, is impossible, because of the dominating Coulomb interaction. The height of the Coulomb barrier is about 18.2 MeV. The aim of the present work is to determine the optical potential at the lowest possible energies, moreover, at several energies above and below the Coulomb barrier to be able to extrapolate the optical potential parameters to the astrophysically relevant energy region. We finally compare the measured (α,γ) cross section of ^{106}Cd with the pre-

[a] e-mail: ggkiss@atomki.hu
[b] *Present address*: NSCL Michigan State University, 1 Cyclotron Lab East Lansing MI 48824-1321 USA.

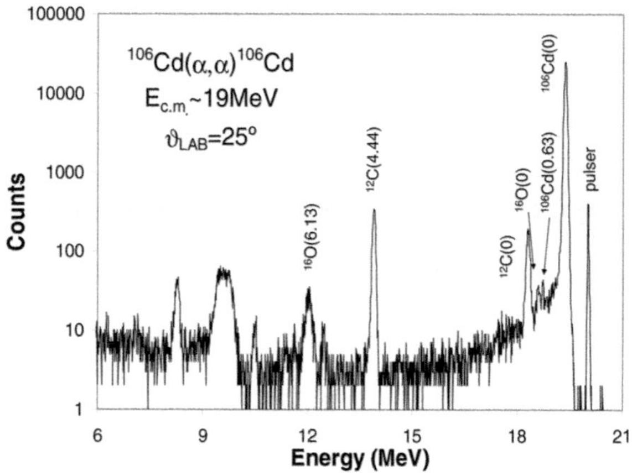

Fig. 1. Typical spectrum of $^{106}\mathrm{Cd}(\alpha,\alpha)^{106}\mathrm{Cd}$ at $\vartheta = 25°$. Elastic scattering on target contaminations (mainly ^{12}C from the carbon backing) and inelastically scattered particles are clearly separated from the elastic peak. The pulser peak used for the dead time correction is also shown.

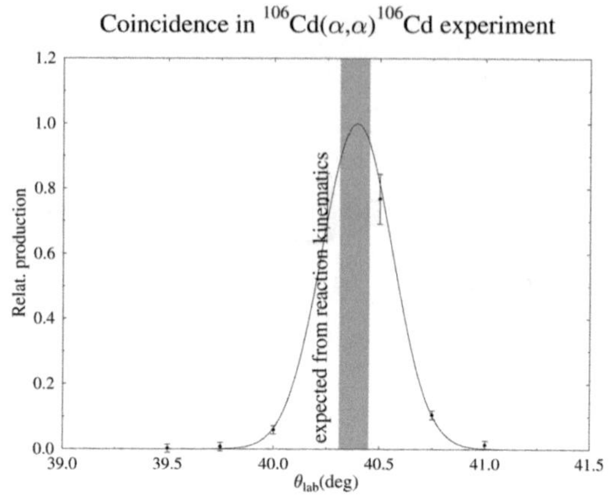

Fig. 2. Relative yield of ^{12}C recoil nuclei in coincidence with elastically scattered alpha-particles. The gray area presents the angle and the uncertainties expected from the reaction kinematics. A Gaussian fit to the experimental data (solid line) is shown to guide the eye.

dictions of statistical model calculations using the optical potential parameters derived in this work.

2 Experimental setup and procedure

The scattering experiment was performed at the cyclotron laboratory at ATOMKI, Debrecen. Complete angular distributions between 20° and 170° were measured in steps of 1° ($20° \le \vartheta \le 100°$), 1.5° ($100° \le \vartheta \le 140°$) and 2° ($140° \le \vartheta \le 170°$) at alpha energies of $E_{Lab} =$ 16.13 MeV, 17.65 MeV and 19.61 MeV. The beam intensity was 150 pnA. A typical spectrum of $^{106}\mathrm{Cd}(\alpha,\alpha)^{106}\mathrm{Cd}$ reaction is shown in fig. 1.

The highly enriched ($\approx 97\%$) cadmium targets were produced by evaporation at the target laboratory at ATOMKI. A thin carbon foil ($\approx 20\,\mu\mathrm{g/cm}^2$) was used as backing. The thickness of the target was roughly $250\,\mu\mathrm{g/cm}^2$. The target was mounted on a remotely controlled target ladder in the centre of the scattering chamber. The stability of the target was monitored during the whole experiment to avoid systematic uncertainties from changes in the target.

An aperture of 2×6 mm was mounted on the target holder to check the beam position and size of the beam spot before and after every change of beam energy or current. We optimized the beam until not more than 1% of the total beam current could be measured on this aperture. As a result, the horizontal size of the beam spot was smaller than 2 mm during the whole experiment which is very important for the precise determination of the scattering angle.

Taking into account the Q values of the open reaction channels, particle ID was not necessary. For the measurement of the angular distribution we used four surface barrier detectors with an active area of 50 mm^2. The detectors were mounted on upper and lower turntables, the angular distance between two detectors on the same turntable was 10°. The solid angles of the detector pairs were $\Delta\Omega = 1.63 \times 10^{-4}$ and $\Delta\Omega = 1.55 \times 10^{-4}$. The ratios of the solid angles of the different detectors were checked by measurements at overlapping angles with an accuracy of better than 1%.

Additionally, two detectors were mounted at the wall of the scattering chamber at a fixed angle of $\vartheta = \pm 15°$ with respect to the beam direction. These detectors were used as monitor detectors during the experiment to normalize the measured angular distribution and to determine the precise position of the beam spot. The solid angle of these detectors was $\Delta\Omega = 8.1 \times 10^{-6}$.

The signals from the detectors were amplified using charge-sensitive preamplifiers, which were mounted directly at the scattering chamber. The output signal was led to the main amplifier and fed into an analog-digital converter. For the coincidence measurements (see below) the bipolar signals of the main amplifiers were fed into timing single-channel analyzers, and the unipolar outputs were gated using linear gate stretchers.

The angular calibration of the setup is of crucial importance for the precision of the scattering experiments at energies close to the Coulomb barrier because the Rutherford cross section depends very sensitively on the angle. Small uncertainties of 0.1° in the determination of the scattering angle lead to uncertainties of 2% in the cross section at forward angles. To determine the scattering angle precisely, we measured kinematic coincidences between elastically scattered alpha-particles and the corresponding ^{12}C recoil nuclei using a pure Carbon foil as target. One detector was placed at $\vartheta = 80°$ and the signals from the elastically scattered alpha particles on ^{12}C were selected as gates for signals from another detector which moved around the expected ^{12}C recoil angle $\vartheta = 40.2°$, fig. 2. shows the relative yield of ^{12}C recoil nuclei in coincidence with elastically scattered alpha particles as a function of the ^{12}C recoil angle. In this way the final angular uncertainties of our setup was determined to be 0.07°.

Table 1. Parameters of the real and imaginary part of the alpha-nucleus optical potential of ^{106}Cd.

a^* (MeV fm^3)	b^* (fm^3)	$J_{R,0}$	ω	W_V (MeV)	R_V (fm)	a_V (fm)	W_S (MeV)	R_S (fm)	a_S (fm)
377.99	−0.6519	266.91	0.987	−2.879	1.744	0.347	339.01	1.262	0.206

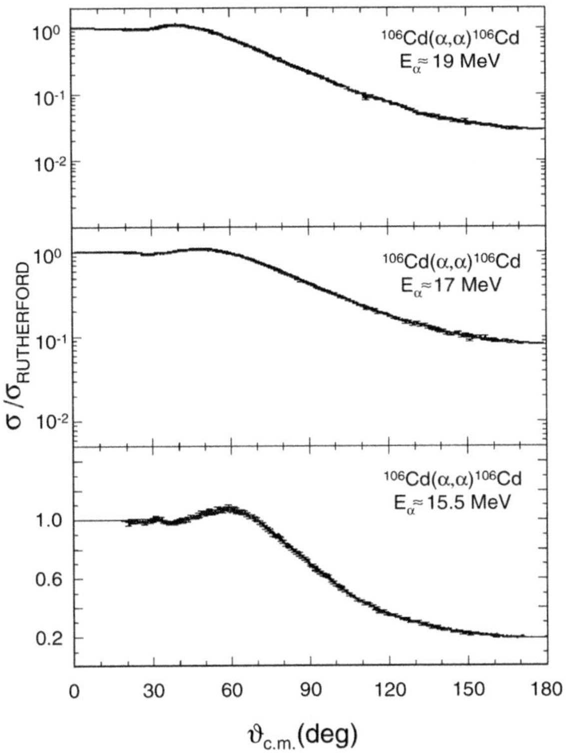

Fig. 3. Experimental cross section of ^{106}Cd$(\alpha,\alpha)^{106}$Cd at $E_{c.m.} \approx$ 19, 17 and 15.5 MeV normalized to the Rutherford cross section.

The count rates $N(\vartheta)$ in the four detectors have been normalized to the number of counts in the monitor detectors $N_{MON.}(\vartheta = 15°)$:

$$\left(\frac{d\sigma}{d\Omega}\right)(\vartheta) = \left(\frac{d\sigma}{d\Omega}\right)_{MON.} \frac{N(\vartheta)}{N_{MON.}} \frac{\Delta\Omega_{MON.}}{\Delta\Omega}, \qquad (1)$$

where $\Delta\Omega$ is the solid angle of the detector. The cross section at the monitor detectors is given by the Rutherford cross section owing to the low scattering angle. The beam was stopped in a Faraday cup and the beam current was measured by a current integrator.

The absolute cross sections cover five orders of magnitude in the measured angular range. However the statistical uncertainties of each data point changes only from $\leq 0.3\%$ (forward angles) to about 1%–2% (backward angles). The experimental cross section normalized to the Rutherford cross section is shown in fig. 3.

3 Optical potential parameters

In order to determine the alpha nucleus potential of ^{106}Cd, we have performed our analysis in the framework of the Optical Model (OM). The optical potential takes the form

$$U(r) = V_C(r) + V(r) + iW(r), \qquad (2)$$

where $V_C(r)$ is the Coulomb potential, $V(r)$ and $W(r)$ are the real and imaginary parts of the nuclear potential, respectively. The description of $V(r)$ is done using the double-folding procedure, in which both nuclei interact via an effective nucleon-nucleon interaction in the well-established DDM3Y parametrization [9,10]. The real part of the nuclear potential is based on this double-folding potential $V_f(r)$, in which two small corrections in strength (λ) and width ($\omega \approx 1.0$) have been applied:

$$V(r) = \lambda V_f(r/\omega). \qquad (3)$$

The parameter ω is introduced to modify the width of the potential. Through this rearrangement, it is possible to correct the deviations between the proton and neutron density distributions within the nucleus. For stable light nuclei with $Z = N$ there is no need for such a parameter. In case of medium or heavy nuclei with a neutron-to-proton ratio of $N/Z \geq 1.2$ it is necessary to take this correction into consideration.

The strength parameter λ has been described by a linear form:

$$\lambda = \frac{a^* + b^* E_{c.m.}}{J_{R,0}}. \qquad (4)$$

The coefficients a^* and b^* are listed in table 1. The volume integral of the potential $J_{R,0}$ for $\lambda = 1.0$ and the corresponding ω are also listed. The weak energy dependence of the volume integral through b^* reduces the uncertainties of the extrapolation to the astrophysically relevant energy region.

For a comparison of different potentials we use the integral parameters such as the volume integral per interacting nucleon pair J_R and the root-mean-square (rms) radius $r_{rms,R}$, which are given by

$$J_R = \frac{1}{A_p A_T} \int V(r)\, d^3r, \qquad (5)$$

$$r_{rms,R} = \sqrt{\frac{\int V(r) r^2\, d^3r}{\int V(r)\, d^3r}}, \qquad (6)$$

for the real part of the potential $V(r)$ and the corresponding equations hold for $W(r)$. The Coulomb potential is taken in the usual form of a homogeneously charged sphere. In the imaginary part of the nuclear potential we have tested different parameterizations. It turned out that the best fit to our experimental data is given by the combination of volume (V) and surface (S) Wood-Saxons potentials. The relative weight between the volume and surface

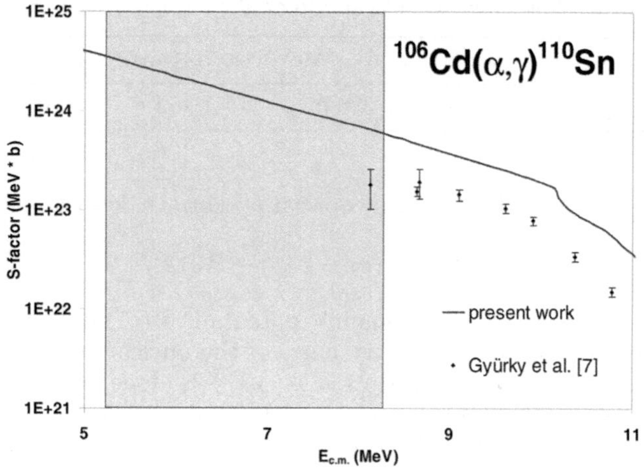

Fig. 4. Astrophysical S-factor of $^{106}\mathrm{Cd}(\alpha,\gamma)^{110}\mathrm{Sn}$ capture reaction. The experimental data from [7,8] are compared to the optical potential obtained from the analysis of the scattering data. The gray area shows the energy region relevant to the p-process.

terms of the imaginary part of the nuclear potential is $J_{I,V} = 0.22 J_{I,S}$, as found in a study of the elastic scattering data in the $A \approx 100$ mass region [11]. This dominance of the surface Woods-Saxon term at energies close to the Coulomb barrier provides a better description of the alpha capture data at the astrophysically interesting energy window. The calculations were performed using the A0 code [12]. The resulting best fit parameters are shown in table reftab:1. For details of the fitting procedure see [13].

Due to the astrophysical interest, the laboratory (α,γ) reaction cross section on ^{106}Cd nucleus has been measured close to the Gamow-window [7,8]. The preliminary astrophysical S-factor of the reaction $^{106}\mathrm{Cd}(\alpha,\gamma)^{110}\mathrm{Sn}$ is shown in fig. 4. In addition, the predictions from statistical model calculations using alpha-nucleus optical potential derived in this work are as input parameters for the NON-SMOKER code [14] are plotted as well.

It is also instructive to compare the results of the present experiment with the calculated scattering cross sections using different global alpha-nucleus potentials. This work is still in progress and it is beyond the scope of the present paper.

This work was supported by OTKA (T042733, F043408, D048283, T049245, T038404) and DFG (FOR 272/2-2 and SFB632) Zs. F. is a Bolyai fellow.

References

1. M. Arnould, S. Goriely, Phys. Rep. **384**, 1 (2003).
2. T. Rauscher, A. Heger, R.D. Hoffman, S.E. Woosley, Astrophys. J. **576**, 323 (2002).
3. E. Somorjai, Zs. Fülöp, A.Z. Kiss, C.E. Rolfs, H.-P. Trautvetter, U. Greife, M. Junker, S. Goriely, M. Arnould, M. Rayet, T. Rauscher, H. Oberhummer, Astron. Astrophys. **333**, 1112 (1998).
4. P. Mohr, T. Rauscher, H. Oberhummer, Z. Máté, Zs. Fülöp, E. Somorjai, M. Jaeger, G. Staudt, Phys. Rev. C **55**, 1523 (1997).
5. Zs. Fülöp, Gy. Gyürky, E. Somorjai, L. Zolnai, D. Galaviz, M. Babilon, P. Mohr, A. Zilges, T. Rauscher, H. Oberhummer, Phys. Rev. C **64**, 065805 (2001).
6. D. Galaviz, Zs. Fülöp, Gy. Gyürky, Z. Máté, P. Mohr, T. Rauscher, E. Somorjai, A. Zilges, Phys. Rev. C **71**, 065802 (2005).
7. Gy. Gyürky, Zs. Fülöp, G.G. Kiss, Z. Máté, E. Somorjai, J. Görres, A. Palumbo, M. Wiescher, D. Galaviz, A. Kretschmer, K. Sonnabend, A. Zilges, T. Rauscher Nucl. Phys. A **758**, 517c (2005).
8. Gy. Gyürky, Z. Elekes, Zs. Fülöp, G.G. Kiss, E. Somorjai, J. Görres, A. Palumbo, M. Wiescher, W. Rapp, N. Özkan, R.T. Gürray, T. Rauscher, in preparation.
9. G.R. Satchler, W.G. Love, Phys. Rep. **55**, 183 (1979).
10. A.M. Kobos, B.A. Brown, R. Lindsay, G.R. Satchler, Nucl. Phys. A **425**, 205 (1984).
11. T. Rauscher, in *Proceedings of the IX Workshop on Nuclear Astrophysics* (1998).
12. H. Abele, University of Tübingen, computer code A0, unpublished.
13. D. Galaviz, PhD Thesis, TU Darmstadt (2004).
14. T. Rauscher, F.K. Thielemann, At. Data Nucl. Data Tables **79**, 47 (2001).

Eur. Phys. J. A **27**, s01, 201–204 (2006)
DOI: 10.1140/epja/i2006-08-031-8

EPJ A direct
electronic only

Study of fission fragments produced by $^{14}N + ^{235}U$ reaction

M. Yalçınkaya[1,a], E. Ganioglu[1], M.N. Erduran[1], B. Akkus[1], M. Bostan[1], G. Gürdal[2], S. Ertürk[3], D. Balabanski[4], G. Rainovski[4], M. Danchev[4], R. Dragomirova[4], A. Minkova[4], K. Vyvey[5], R. Beetge[6], R.W. Fearick[6], G.K. Mabala[6], D.G. Roux[6], W. Whittaker[6], B.R.S. Babu[7], J.J. Lawrie[7], S. Naguleswaran[7], R.T. Newman[7], C. Rigolet[7], J.V. Pilcher[7], F.D. Smith[7], and J.F. Sharpey-Shafer[7]

[1] Istanbul University, Sciences Faculty, Department of Physics, 34459 Vezneciler, Istanbul, Turkey
[2] WNSL, Yale University, New Haven, CT 06520-8124, USA
[3] Nigde University, Science and Art Faculty, Department of Physics, Nigde, Turkey
[4] Faculty of Physics, Sofya St. Kliment Ohridsky University of Sofia, BG-1164 Sofia, Bulgaria
[5] Institut voor Kern-en Stralingsfysica, University of Leuven B-3001 Leuven, Belgium
[6] Department of Physics, University of Cape-Town, 7701 Cape Town, South Africa
[7] National Accelerator Centre, 7131 Faure (Cape Town), South Africa

Received: 8 July 2005 /
Published online: 8 March 2006 –

Abstract. This work was performed to understand the structure of neutron-rich fission fragments around the 130 mass region. A thin ^{235}U target was bombarded by a ^{14}N beam with 10 MeV/A from the Separated Sector Cyclotron at the iThemba Laboratory for Accelerator Based Sciences, Cape Town, South Africa. The main goal was to detect and identify fission fragments and to obtain their mass distribution by using solar cell detectors in the AFRODITE (African Omnipurpose Detector for Innovative Techniques and Experiments) spectrometer. The X-rays emitted from fission fragments were detected by LEP (Low Energy Photon) detectors and γ-rays emitted from excited states of the fission fragments were detected by CLOVER detectors in the spectrometer.

PACS. 25.70.Jj Fusion and fusion-fission reactions – 21.10.Gv Mass and neutron distributions

1 Introduction

The studies for heavy-ion–induced fusion-fission reaction have attracted a great deal of attention in recent years due to the expansion of the knowledge and understanding of the structure of neutron-rich fission fragments. There are two mechanisms of the fusion-fission reaction when bombarding a heavy target with a high-energy beam well above the Coulomb barrier: low-energetic fission and deep-inelastic processes. Asymmetric fission as a result of the former mechanism is a process dominated by shell effects and the heavy fragment is a neutron-excessive nucleus having approximately 50 protons and 82 neutrons. Since the beam energy is well above the Coulomb barrier, a few nucleons are evaporated and the mass of the fragments is shifted towards the line of stability. Thus, the interplay between the target-projectile combination and the variation of the projectile energy provides a possibility of moving the centroid of the fragment mass distribution throughout the (N, Z)-plane and accessing nuclei which cannot be produced in spontaneous fission or through symmetric fission [1,2]. Therefore, other reaction mechanisms needed to be utilized in these cases.

In a previous study, Yu *et al.* [3] investigated the reaction ^{12}C + ^{238}U at 20 MeV/A and demonstrated that with the increase of the beam energy deep-inelastic processes begin to compete with fusion-fission reactions for which, the asymmetric fission channel is open when using actinide targets.

Recently, in the low-energy proton-induced fission of actinides, it has also been demonstrated that there exist at least two independent deformation paths for fission process; one leads to a symmetrically elongated scission configuration, and the other leads to a compact scission configuration with reflection asymmetry [4].

In this work, we present an experiment to investigate nuclei around the 130 mass region utilizing the ^{14}N + ^{235}U reaction at 10 MeV/A. In order to obtain the fragment mass distribution directly from the reaction given above, a thin target and solar cell array have to be used. This arrangement will also give opportunity to correct Doppler shift due to the fission fragments decaying in flight as well as direct fragment identification.

[a] e-mail: yalcinm@istanbul.edu.tr

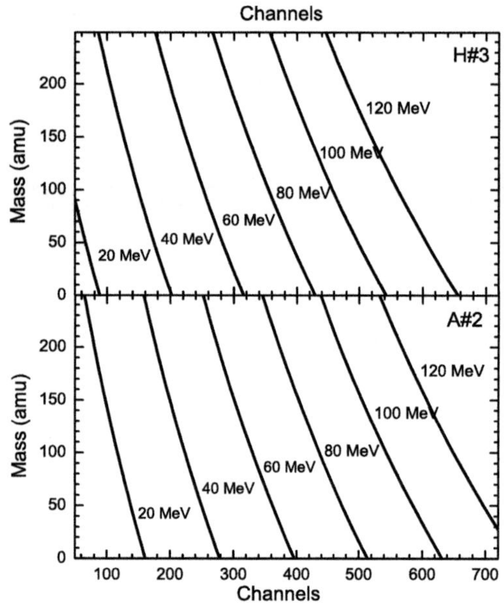

Fig. 1. The Channel-Energy-Mass dependency obtained for the H3 and A2 solar cell detectors.

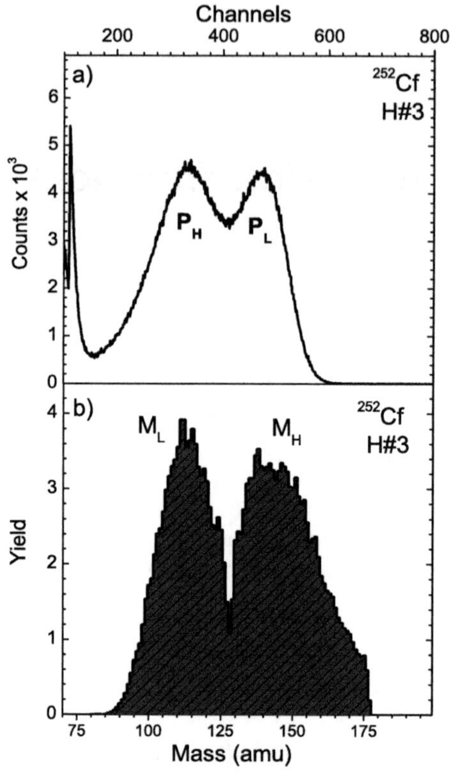

Fig. 2. a) The pulse height spectrum of ^{252}Cf obtained with the H3 solar cell detectors. b) The unfolded mass distribution of ^{252}Cf normalized to 200.

2 Experiment

We used ^{14}N + ^{235}U reactions in order to produce fission fragments. The 140 MeV ^{14}N beam for these experiments was delivered by the Separated Sector Cyclotron at the iThemba Laboratory for Accelerator Based Sciences, Cape Town, South Africa. The ^{14}N projectile was chosen in order to somewhat enhance the production of odd-Z isotopes. The γ decay and the X-rays after the reaction were detected by the AFRODITE spectrometer, which consisted of seven Compton-suppressed CLOVER detectors and eight large-area LEPs detectors. This spectrometer is described in detail in ref. [5].

The reaction chamber had Mylar windows. All this allowed to have a detection limit for the X-rays as low as 8–10 keV. A thin 500 μg/cm^2 target of ^{235}U was used in the experiment. Solar cell detectors, positioned at forward angles, were operated together with the AFRODITE spectrometer, which allowed measurement of fragment–γ–X-ray coincidences.

3 Results

The response of solar cell detectors have to be calculated since the energy of incoming fragments from the fission is not exactly proportional to the deposited energy due to the Pulse Height Defects (PHD). Due to PHD, the energy of fission fragment, E, can be related to the signal observed by surface barrier semiconductor/solar cell detector used for the detection of heavy ions:

$$E = (a + a^{'}M)x + (b + b^{'}M), \tag{1}$$

where, M is the mass of fragment, a, $a^{'}$, b and $b^{'}$ are the coefficients of the charged-particle detector used. These coefficients were determined by using Schmitt's Calibration Method described in [6]. In this calibration procedure, pulse height spectra of the fragments from ^{252}Cf spontaneous-fission source were obtained by using the solar cells. Then using the iterative processing algorithm suggested by Houry [7], the mass distributions of ^{252}Cf were obtained. The Energy-Channel-Mass dependency obtained for H3 and A2 solar cells are shown in fig. 1. In figs. 2a and 3a, the pulse height spectra and in figs. 2b and 3b, the unfolded mass distributions of ^{252}Cf are shown. In figs. 2b and 3b, the mass distributions were normalized to 200 and smoothed with using 3 channel averaged method. The calibration and unfolding results plotted in figs. 2 and 3 are based on average neutron multiplicity for each fragment [8] and the well-known ^{252}Cf mass distribution [9,10,11].

As can be easily seen, there are two prominent humps determined as

$\langle A_L \rangle$ = 108–109 amu and $\langle A_H \rangle$ = 144 amu in the mass distribution corresponding to the light and heavy fission fragments, respectively, are in good agreement with a literature values $\langle A_L \rangle = 108.9 \pm 0.5$ amu and $\langle A_H \rangle = 143.1 \pm 0.5$ within the experimental accuracy [9].

Solar cell detectors calibrated using the method given above were then used for detecting fission fragments from the ^{14}N + ^{235}U fusion-fission reaction. As has been already indicated above, all solar cell detectors are positioned in forward angles allowing only one of the two fis-

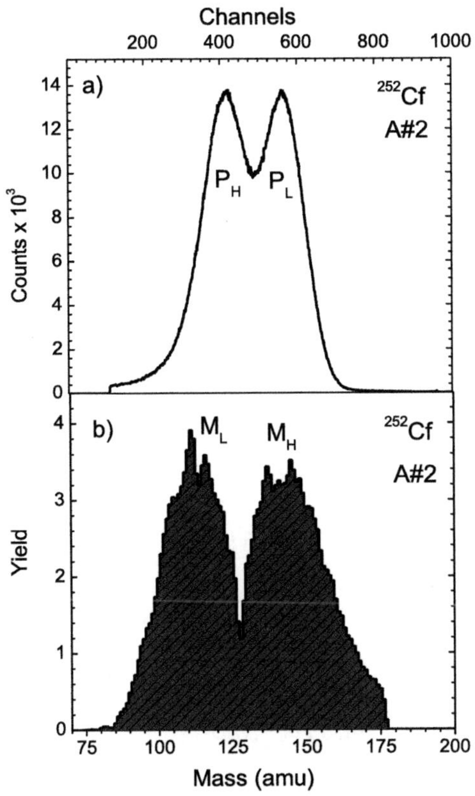

Fig. 3. a) The pulse height spectrum of ^{252}Cf obtained with the A2 solar cell detectors. b) The unfolded mass distribution of ^{252}Cf normalized to 200.

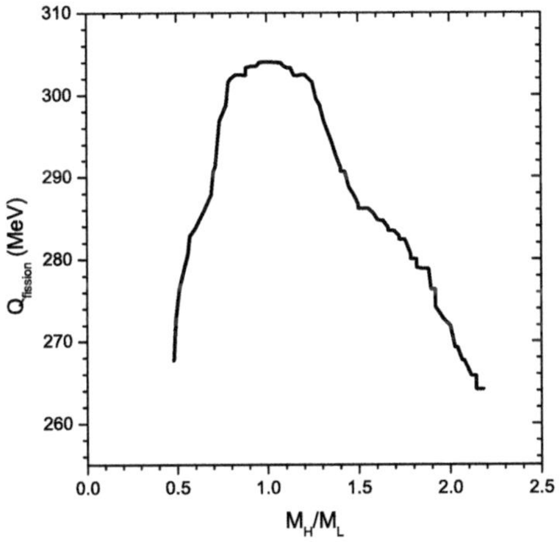

Fig. 4. The M_H/M_L dependency of $Q_{fission}$ calculated for the ^{245}Es fissioning nucleus.

sion fragments to be detectable. In order to determine the energy of the undetected complementary fragment, the total kinetic energy release from the fission and total excitation energy of the fragments have to be known. The total kinetic energy of fission fragments was taken from the Viola systematics [12] with mass asymmetry dependency [13], while fragment masses were taken from the Möller Mass Table [14] and then they were used to deter-

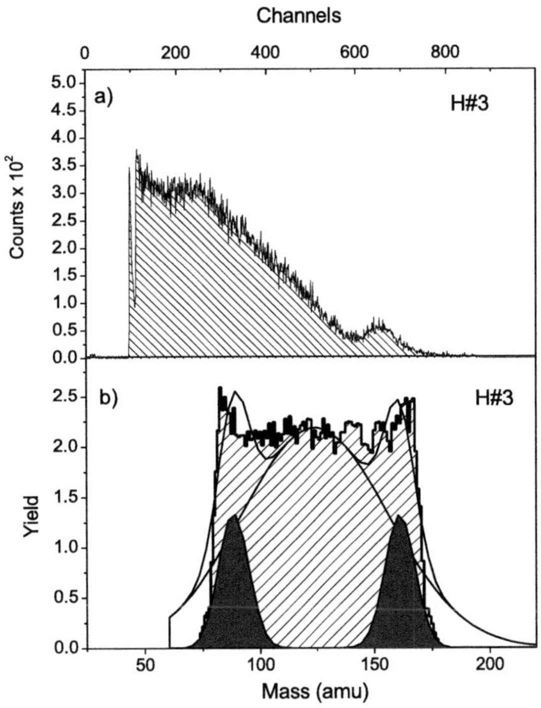

Fig. 5. a) The pulse height spectrum obtained with the H3 solar cell detector for the $^{14}N+^{235}U$ reaction. b) The fitted unfolded mass distribution of $^{14}N+^{235}U$ normalized to 200.

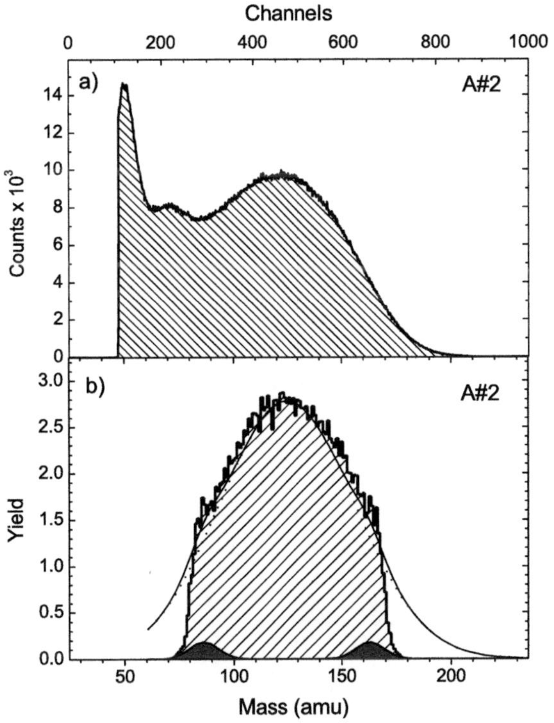

Fig. 6. a) The pulse height spectrum obtained with the A2 solar cell detector for the $^{14}N+^{235}U$ reaction. b) The fitted unfolded mass distribution of $^{14}N+^{235}U$ normalized to 200.

mine the maximum reaction energy as a function of the mass ratio M_H/M_L, where M_H and M_L are the mass of the heavy and the light fragment, respectively. Figure 4 shows the fission Q-value dependency on the M_H/M_L ratio for ^{245}Es assuming that the number of pre-fission neutron evaporation $\nu_{pre} = 3.69$. The average number of neutrons evaporated ν_{pre} and ν_{post} were taken from ref. [15].

The pulse height spectra of the fission fragments are shown in figs. 5a, 6a and the unfolded mass distributions of the fission products are shown in figs. 5b, 6b. The unfolded mass distributions were obtained with the same method mentioned above using the iterative processing algorithm of Houry [7].

4 Conclusion

The mass distributions obtained from solar cell detectors for ^{14}N + ^{235}U reactions in figs. 5b and 6b were subjected to further investigation by fitting the whole mass distributions to sum of three Gaussian functions. The masses of fragments formed in the area are in the range of 75–180 amu. The thick solid curve in figs. 5b and 6b is the results of the fit assuming that there are one dominant symmetric and two asymmetric components. These Gaussians are centered around 124 amu with $\sigma = 57$ (for the A2 solar cell) and $\sigma = 64$ (for the H3 solar cell) for the symmetric and around 87 amu and 162 amu with $\sigma = 13$ for both the asymmetric components. In figs. 5b and 6b, the dark grey Gaussian-like distributions can be attributed to the shell closures, which have been observed in ref. [16] and ref. [17].

It can be concluded that the asymmetric channel could be opened, hence the neutron-rich nuclei heavier than those produced in the spontaneous fission could be obtained [18,19,20]. It could also be concluded that by employing solar cells in the array improves the mass selectivity by direct fragment identification.

This work was supported by the Research Fund of the University of Istanbul, Project numbers UP-12/040199 and UP-8/270598.

References

1. M-G Porquet *et al.*, Acta Phys. Pol. B **27**, 179 (1996).
2. D.L. Balabanski *et al.*, *The Nucleus New Physics for the New Milennium* (Kluwer Academic, Plenum Publishers, New York, 2000) p. 63, ISBN 0-306-46302.
3. W. Yu *et al.*, Phys. Rev. C **36**, 2396 (1987).
4. S. Goto *et al.*, J. Nucl. Radiochem. Sci. **3**, No. 1, 63 (2002).
5. J.F. Sharpey-Schafer, in the *Structure of the Vacuum and Elementary Matter, Widerness, South Africa*, edited by H. Stöcker, A. Gallmann, J.H. Hamilton (World Scientific Singapore, 1997) p. 656.
6. H.W. Schmitt, Phys. Rev. B **137**, 837 (1965).
7. M. Houry, PhD Thesis, University of Paris, No d'ordre: 6033 (1998).
8. S.L. Whetstone, Phys. Rev. **114**, 581 (1959).
9. J. van Aarle *et al.*, Nucl. Phys. A **578**, 77 (1994).
10. J.L. Durell, *Proceedings of the International Conference on the Spectroscopy of Heavy Nuclei, Crete, Greece*, Inst. Phys. Conf. Ser. **105**, 307 (1989).
11. F. Goennenwein, *The Nuclear Fission Process* (CRC Press, 1993) p. 287.
12. V.E. Viola *et al.*, Phys. Rev. C **31**, 1550 (1985).
13. D.J. Hinde, Phys. Rev. C **45**, 1229 (1992).
14. P. Möller, At. Data Nucl. Data Tables **59**, 185 (1995).
15. W.U. Schroder, J.R. Huizenga, Nucl. Phys. A **502**, 473c (1989).
16. S.I. Mulgin *et al.*, Phys. Lett. B **462**, 29 (1999).
17. H. Baba *et al.*, Eur. Phys. J. A, **462**, 281 (1998).
18. P.J. Nolan *et al.*, Annu. Rev. Nucl. Part. Sci. **45**, 561 (1994).
19. I. Ahmed, W.R. Phillips, Rep. Prog. Phys. **58**, 1415 (1995).
20. J.H. Hamilton *et al.*, Prog. Part. Nucl. Phys. **35**, 365 (1995).

Eur. Phys. J. A **27**, s01, 205–215 (2006)
DOI: 10.1140/epja/i2006-08-032-7

EPJ A direct
electronic only

Indirect techniques in nuclear astrophysics

Asymptotic Normalization Coefficient and Trojan Horse

A.M. Mukhamedzhanov[1,a], L.D. Blokhintsev[2], B.A. Brown[3], V. Burjan[4], S. Cherubini[5], C.A. Gagliardi[1], B.F. Irgaziev[6], V. Kroha[4], F.M. Nunes[3], F. Pirlepesov[1], R.G. Pizzone[5], S. Romano[5], C. Spitaleri[5], X.D. Tang[7], L. Trache[1], R.E. Tribble[1], and A. Tumino[5]

[1] Cyclotron Institute, Texas A&M University, College Station, TX 77843, USA
[2] Institute of Nuclear Physics, Moscow State University, Moscow, Russia
[3] N.S.C.L. and Department of Physics and Astronomy, Michigan State University, East Lansing, MI, USA
[4] Nuclear Physics Institute of Czech Academy of Sciences, Prague-Řež, Czech Republic
[5] DMFCI, Università di Catania, Catania, Italy and INFN - Laboratori Nazionali del Sud, Catania, Italy
[6] Physics Department, National University, Tashkent, Uzbekistan
[7] Physics Division, Argonne National Laboratory, Argonne, IL 60439, USA

Received: 21 June 2005 /
Published online: 15 March 2006 –

Abstract. Owing to the presence of the Coulomb barrier at astrophysically relevant kinetic energies it is very difficult, or sometimes impossible, to measure astrophysical reaction rates in the laboratory. That is why different indirect techniques are being used along with direct measurements. Here we address two important indirect techniques, the asymptotic normalization coefficient (ANC) and the Trojan Horse (TH) methods. We discuss the application of the ANC technique for calculation of the astrophysical processes in the presence of subthreshold bound states, in particular, two different mechanisms are discussed: direct capture to the subthreshold state and capture to the low-lying bound states through the subthreshold state, which plays the role of the subthreshold resonance. The ANC technique can also be used to determine the interference sign of the resonant and nonresonant (direct) terms of the reaction amplitude. The TH method is unique indirect technique allowing one to measure astrophysical rearrangement reactions down to astrophysically relevant energies. We explain why there is no Coulomb barrier in the sub-process amplitudes extracted from the TH reaction. The expressions for the TH amplitude for direct and resonant cases are presented.

PACS. 26.20.+f Hydrostatic stellar nucleosynthesis – 21.10.Jx Spectroscopic factors and asymptotic normalization coefficients – 25.55.Hp Transfer reactions – 27.20.+n $6 \leq A \leq 19$

1 Introduction

For better understanding stellar evolution, cross sections of astrophysically relevant nuclear reactions should be known at the Gamow energy with an accuracy better than 10% [1]. The presence of the Coulomb barrier for colliding charged nuclei makes nuclear reaction cross sections at astrophysical energies so small that their direct measurements in laboratories is very difficult, or even impossible. That is why direct measurements are being done at higher energies and then extrapolated down to the Gamow energy. Such an extrapolation procedure can cause an additional uncertainty. Also for nuclear reactions studied in laboratory, the electron clouds surrounding the interacting nuclei lead to a screened cross section which is larger than the "bare" nucleus one (see [2,3,4,5] and references therein). The enhancement factor is determined by the electron screening potential which is a model dependent quantity and its value in the laboratory is different from the one present in the stellar environment. There are four often used indirect techniques: the asymptotic normalization coefficient (ANC) method [6], Coulomb breakup processes [7,8], Trojan Horse (TH) [9,5] and the surrogate reactions method (see [10] and references therein). In this work we address only two indirect techniques, the ANC and TH methods.

2 ANC method

The ANC method has been suggested in [11,12] and can be used to determine the astrophysical factors for peripheral radiative capture processes. The method can be applied for analysis of direct radiative capture processes

[a] e-mail: akram@comp.tamu.edu

leading to final loosely bound states. Due to small binding energies and strong Coulomb barrier, the direct capture reactions are peripheral. In previous papers [11,12,13] it has been pointed out that the overall normalization of the cross section for a direct radiative capture reaction at low binding energy is entirely defined by the ANC of the final bound state wave function into the two-body channel corresponding to the colliding particles. The ANC technique turns out to be very productive for analysis of the astrophysical processes in the presence of the subthreshold state [14]. Here we address some applications of the ANC method in the presence of the subthreshold state. We also demonstrate how ANC technique can be used to determine the interference sign of the direct and resonant amplitudes for some important astrophysical radiative capture reactions.

2.1 Definition of the ANC

We present first some useful equations for the ANC. Let us consider a virtual decay of nucleus c into two nuclei a and b. First we introduce the overlap function I of the bound state wave functions of particles c, a, and b [15]:

$$I^c_{ab}(\mathbf{r}) = \langle \varphi_a(\zeta_a)\,\varphi_b(\zeta_b)|\varphi_c(\zeta_a,\zeta_b;\mathbf{r})\rangle = \sum_{l_c m_{l_c} j_c m_{j_c}} i^{l_c}\langle J_a M_a j_c m_{j_c}|J_c M_c\rangle \times \langle J_b M_b\, l_c m_{l_c}|j_c m_{j_c}\rangle Y_{l_c m_{l_c}}(\hat{\mathbf{r}})\, I^c_{abl_cj_c}(r). \quad (1)$$

Here φ_i, ζ_i, J_i and M_i are the bound state wave function, a set of internal coordinates including spin-isospin variables, spin and spin projection for nucleus i. Also r is the relative coordinate of the centers of mass of nuclei a and b, $\hat{\mathbf{r}} = \mathbf{r}/r$, j_c, m_{j_c} are the total angular momentum of particle b and its projection in the nucleus $c = (ab)$, l_c, m_{l_c} are the orbital angular momentum of the relative motion of particles a and b in the bound state $c = (ab)$ and its projection, $\langle j_1 m_1 j_2 m_2|j_3 m_3\rangle$ is a Clebsch-Gordan coefficient, $Y_{l_c m_c}(\hat{\mathbf{r}})$ is a spherical harmonic, and $I^c_{abl_cj_c}(r)$ is the radial overlap function which includes the antisymmetrization factor due to identical nucleons. The summation over l_c and j_c is carried out over the values allowed by angular momentum and parity conservation in the virtual process $c \to a + b$. The asymptotic normalization coefficient $C^c_{abl_cj_c}$ defining the amplitude of the tail of the radial overlap function $I^c_{abl_cj_c}(r)$ is given by [15]

$$I^c_{abl_cj_c}(r) \overset{r>R_N}{\longrightarrow} C^c_{abl_cj_c}\frac{W_{-\eta_c,l_c+1/2}(2\kappa_{ab}r)}{r}, \quad (2)$$

where R_N is the nuclear interaction radius between a and b, $W_{-\eta_c,l_c+1/2}(2\kappa_{ab}r)$ is the Whittaker function describing the asymptotic behavior of the bound state wave function of two charged particles, $\kappa = \sqrt{2\,\mu_{ab}\,\varepsilon_c}$ is the wave number of the bound state $c = (ab)$, μ_{ab} is the reduced mass of particles a and b, ε_c is the binding energy of the bound state (ab) and $\eta_c = Z_a\, Z_b\, e^2\, \mu_{ab}/\kappa$ is the Coulomb parameter of the bound state (ab), $Z_i\, e$ is the charge of particle i. We use the system of units such that $\hbar = c = 1$. There is another definition of the ANC, the most model independent one. The elastic $a+b$ scattering amplitude in the channel (l_c, j_c) has a pole in the momentum plane [14]

$$M_{l_cj_c}(k) = \frac{S_{l_cj_c} - 1}{2\,i\,k} \overset{k\to k_p}{\longrightarrow} \frac{1}{2\,i\,k_p}\frac{W_{l_cj_c}}{k - k_p}, \quad (3)$$

corresponding to the bound state $c = (ab)$ for $k_p = i\,\kappa$ and to the resonance for $k_p = k_R$, where $k_R = k_0 - i\,k_I$ is the resonance location in the momentum plane. Here, $S_{l_cj_c}$ is the elastic matrix element of the S-matrix. The residue in the pole $W_{l_cj_c}$ is

$$W_{l_cj_c} = -(-1)^{l_c}\, i e^{i\pi\eta_c}\,(C^c_{abl_cj_c})^2, \quad k_p = i\,\kappa, \quad (4)$$

$$W_{l_cj_c} = -(-1)^{l_c}\, i\,(C^c_{abl_cj_c(R)})^2, \quad k_p = k_R. \quad (5)$$

For narrow resonances, $k_I \ll k_0$,

$$(C^c_{abl_cj_c(R)})^2 = (-1)^{l_c}\frac{\mu_{ab}}{k_1}e^{\pi\,\eta_0}\, e^{2i\,\delta_{l_cj_c}(k_0)}\,\Gamma_{l_cj_c}. \quad (6)$$

Here η_0 is the Coulomb parameter for the resonance at momentum k_0, $\delta_{l_cj_c}(k_0)$ is the potential (non-resonant) scattering phase shift taken at the momentum k_0. Thus the residue in the bound state or resonance pole is expressed in terms of the ANC and for the resonance the ANC can be expressed in terms of the partial resonance width [14]. Note that eq. (3) holds only for k in the closest vicinity of the pole. For elastic scattering at positive energies in the presence of the Coulomb barrier, the elastic scattering amplitude with the bound state pole behaves (in the R-matrix approach) as

$$M_{l_cj_c}(k) \overset{k\to 0}{\longrightarrow} = -\frac{1}{2\,k}\, e^{-2i(\phi_{l_c} - \sigma_{l_c})}\,\frac{\Gamma_c}{E + \varepsilon_c + i\,\Gamma_c/2}, \quad (7)$$

where

$$\Gamma_c = 2\,P_{l_c}(E)\,\gamma^2_c. \quad (8)$$

Here $P_{l_c}(E)$ is the penetrability through the Coulomb-centrifugal barrier, ϕ_{l_c} is the solid sphere scattering phase shift in the partial wave l_c and $\sigma_{l_c} = \sum_{n=1}^{l_c} \tan^{-1}(\frac{\eta_c}{n})$, r_0 is the channel radius, γ^2_c is the effective (observable) reduced width:

$$\gamma^2_c = \frac{1}{2\mu_{ab}}\frac{W_{-\eta_c,l_c+1/2}(2\kappa r_0)}{r_0}\,(C^c_{abl_cj_c(r)})^2. \quad (9)$$

Thus at positive energies, $E \to +0$ due to the presence of the Coulomb-centrifugal barrier the elastic scattering amplitude behaves as the resonant scattering amplitude with the resonance width expressed in terms of the ANC. At positive energies the elastic scattering cross section in the presence of the bound state and the barrier behaves as the high-energy tail of the resonance located at energy $E = -\varepsilon_c$. That what is called the "subthreshold" resonance. However, it is not a resonance because the real resonance is located at complex energies on the second energy sheet, while the subthreshold resonance is just the bound state located on the first energy sheet at negative energy, corresponding to the bound state. At negative energies (positive imaginary momenta) eq. (9) reduces to eq. (3). Definitions of the ANC dictate the experimental methods of its determination. The ANC can be determined from peripheral transfer reactions which are dominated by the tail of the overlap function. Equation (3) offers another possibility to determine the ANC, namely, by extrapolating the elastic scattering amplitude (or equivalently the phase shift) to the bound state pole [16].

2.2 ANC and astrophysical processes

i) For peripheral direct radiative capture reaction $a+b \to c+\gamma$ to the final state $l_c j_c$ proceeding through the EL transition, the cross section is

$$\sigma \sim |\langle I^{c}_{abl_cj_c}(r)|r^L|\psi_{k_il_i}(r)\rangle|^2 \approx |C^{c}_{abl_cj_c}|^2|\langle \frac{W_{-\eta_c,l_c+1/2}(2\kappa_{ab}r)}{r}|r^L|\psi_{k_il_i}(r)\rangle|^2. \quad (10)$$

Here L is the multipolarity of transition, $\psi_{k_i\,l_i}(r)$ is the initial $a{+}b$ scattering wave function with the relative momentum k_i in the partial wave l_i. Thus the ANC determines the overall normalization of the direct radiative capture cross sections.

ii) The elastic scattering amplitude (7) describes the elastic scattering through the intermediate bound state $c=(ab)$. Assume that it is an excited state. Then, when the excited bound state is formed it can decay into the ground state by emitting a photon. In this case we have the radiative capture process which is called the capture to the ground state through the subthreshold resonance. The amplitude of this process is given by

$$M_{l_cj_c}(k) \stackrel{k\to 0}{\longrightarrow}= -\frac{1}{2\,k}\,e^{-2i(\phi_{l_c}-\sigma_{l_c})}\,\frac{\Gamma_c^{1/2}\,\Gamma_\gamma^{1/2}}{E+\varepsilon_c+i\,\Gamma_c/2}. \quad (11)$$

Here $|\Gamma_\gamma^{1/2}|^2$ gives the radiative width for the transition from the excited bound state $\to$ ground state. Thus in the presence of an excited bound state close to threshold, two different radiative capture processes can occur: direct capture to this excited bound state or capture to the low-lying bound states through this subthreshold bound state (capture through the subthreshold resonance). In what follows we present some astrophysical reactions in the presence of the subthreshold state.

2.3 ANC for $^{14}N + p \to {}^{15}O$ and the astrophysical S-factor for $^{14}N(p,\gamma)^{15}O$

The $^{14}\mathrm{N}+p \to {}^{15}\mathrm{O}+\gamma$ reaction is a notorious example of an important astrophysical reaction where the subthreshold state plays a dominant role. This reaction is one of the most important processes in the CNO cycle. As the slowest reaction in the cycle, it defines the rate of energy production [1] and, hence, the lifetime of stars that

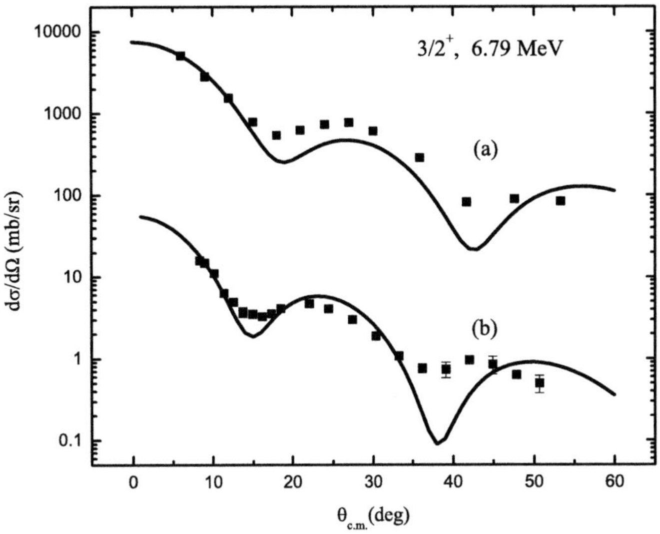

Fig. 1. The $^{14}\mathrm{N}(^3\mathrm{He},d)^{15}\mathrm{O}$ differential cross sections. The squares are data points and the solid lines are the DWBA calculations normalized to the experimental measurements in the main peaks; (a) our data, (b) our fit of the angular distribution measured in ref. [17].

are governed by hydrogen burning via CNO processing. The $^{14}\mathrm{N}(p,\gamma)^{15}\mathrm{O}$ reaction proceeds through direct capture to the subthreshold state $3/2^+$, 6.79 MeV (binding energy 504 keV) and, possibly, via direct capture to the ground state and resonant capture through the first resonance and subthreshold resonance at $E_s = -504$ keV. The overall normalization of the direct capture is defined by the corresponding ANC. The ANC for the subthreshold state $E_s = -504$ keV also determines the partial proton width of the subthreshold resonance. In order to determine the ANCs for $^{14}\mathrm{N}+p \to {}^{15}\mathrm{O}$, the $^{14}\mathrm{N}(^3\mathrm{He},d)^{15}\mathrm{O}$ proton transfer reaction has been measured at an incident energy of 26.3 MeV. Angular distributions for proton transfer to the ground and five excited states were obtained. Angular distributions of deuterons from the $^{14}\mathrm{N}(^3\mathrm{He},d)^{15}\mathrm{O}$ reaction leading to the most important transition to the fourth excited state $3/2^+$, 6.79 MeV in $^{15}\mathrm{O}$ measured by us at an incident energy of 26.3 MeV and in [17] measured at an incident energy of 20 MeV, together with our DWBA fits are shown in fig. 1. The proton ANC that we obtain for the $^{14}\mathrm{N}+p \to {}^{15}\mathrm{O}(3/2^+, 6.79\,\mathrm{MeV})$ is $C^2 = 27.1 \pm 6.8\,\mathrm{fm}^{-1}$. Using our ANCs, we calculated the astrophysical factor and reaction rates for the $^{14}\mathrm{N}(p,\gamma)^{15}\mathrm{O}$ process. The capture to the $3/2^+$, 6.79 MeV state dominates all others and the calculated astrophysical factor is $S(0) = 1.40 \pm 0.20$ keV b. The calculated and experimental $S(E)$-factors for the transition to this subthreshold state are presented in fig. 2. The uncertainty in $S(0)$ is entirely determined by the ANC of this state and the 13% systematic uncertainty in the experimental $S(E)$-factor [18]. We find that the astrophysical factor for the capture to the ground state is $S(0) = 0.15 \pm 0.07$ keV b. The total calculated astrophysical factor at zero energy is $S(0) = 1.70 \pm 0.22$ keV b, which is in excellent agreement with the S-factor $S(0) = 1.70 \pm 0.22$ keV b obtained

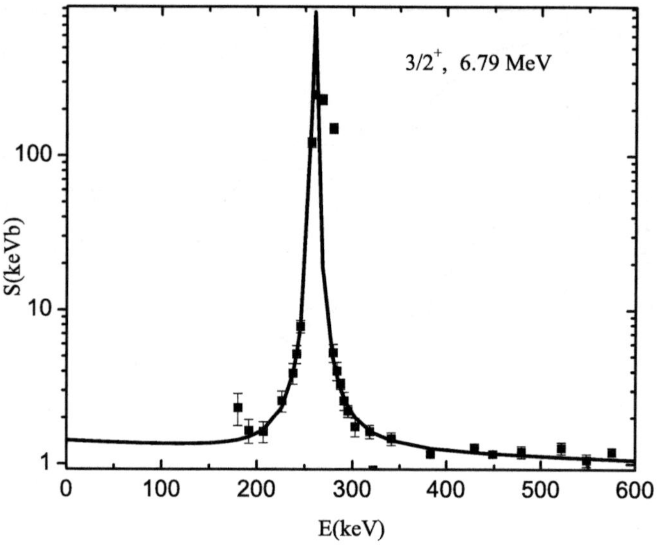

Fig. 2. The $^{14}N(p,\gamma)^{15}O$ astrophysical S-factor for capture to the fourth excited state ((c): $3/2^+$, 6.79 MeV), which includes the incoherent sum of the resonant and nonresonant terms. The squares are data points [18]; the solid lines represent the calculated S-factor using our measured ANC.

from recent direct measurements performed at LUNA [19]. The lower astrophysical factor of the $^{14}N(p,\gamma)^{15}O$ reaction leads to an increase in the age of the main-sequence turnoff in globular clusters [20].

2.4 ANC and interference of direct and resonant amplitudes

To demonstrate how the information about the ANC can be used to determine the interference sign of the resonant and direct amplitudes of the radiative capture process we use the R-matrix approach. Let us consider the radiative capture reaction $a + b \to c + \gamma$.

The R-matrix radiative capture amplitude to a state of nucleus c with a given spin J_f and relative orbital angular momentum of the bound state l_f is given by the sum of resonant and nonresonant (direct capture) amplitudes [21]:

$$U_{Il_f J_f J_i} = U^R_{Il_f J_f J_i} + U^{NR}_{Il_f J_f J_i}. \qquad (12)$$

Interference effects only occur in eq. (12) if the resonant and nonresonant amplitudes have the same channel spin I and orbital angular momentum l_i. In the one-level, one-channel approximation, the resonant amplitude for the capture into the resonance with energy E_{rn} and spin J_i, and subsequent decay into the bound state with the spin J_f, is given by

$$U^R_{Il_i J_f J_i} = -i\, e^{i(\phi_{l_i} - \sigma_{l_i})}\, \frac{[\Gamma^{J_i}_{Il_i}(E)]^{1/2}\, [\Gamma^{J_i}_{\gamma J_f}(E)]^{1/2}}{E - E_{rn} + i\,\frac{\Gamma_{J_i}}{2}}. \qquad (13)$$

Here J_i is the total angular momentum of the colliding nuclei a and b in the initial state, J_a and J_b are the spins of nuclei a and b, and I, k, and l_i are their channel spin, relative momentum and orbital angular momentum in the initial state. $U_{Il_i J_f J_i}$ is the transition amplitude from the initial continuum state (J_i, I, l_i) to the final bound state (J_f, I). Also $[\Gamma^{J_i}_{Il_i}(E)]^{1/2}$ is real and its square, $\Gamma^{J_i}_{Il_i}(E)$, is the observable partial width of the resonance in the channel $a + b$ with the given set of quantum numbers, $[\Gamma^{J_i}_{\gamma J_f}(E)]^{1/2}$ is complex and its modulus square is the observable radiative width:

$$\Gamma^{J_i}_{\gamma J_f}(E) = \left|[\Gamma^{J_i}_{\gamma J_f}(E)]^{1/2}\right|^2. \qquad (14)$$

The energy dependence of the partial and radiative widths is given by

$$\Gamma^{J_i}_{Il_i}(E) = \frac{P_{l_i}(E)}{P_{l_i}(E_{Rn})}\, \Gamma^{J_i}_{Il_i}(E_{Rn}), \qquad (15)$$

and

$$\Gamma^{J_i}_{\gamma J_f}(E) = \left(\frac{E + \varepsilon_f}{E_{Rn} + \varepsilon_f}\right)^{2L+1} \Gamma^{J_i}_{\gamma J_f}(E_{Rn}), \qquad (16)$$

respectively. Here, $\Gamma^{J_i}_{Il_i}(E_{Rn})$ and $\Gamma^{J_i}_{\gamma J_f}(E_{Rn})$ are the experimental partial and radiative resonance widths, ε_f is the proton binding energy of the bound state in nucleus A, L is the multipolarity of the gamma quanta emitted during the transition, and $\Gamma_{J_i} \approx \sum_I \Gamma^{J_i}_{Il_i}$. In a strict R-matrix approach

$$[\Gamma^{J_i}_{\gamma J_f}(E)]^{1/2} = 2\,[P_{l_i}(E)]^{1/2}\, \gamma^{J_i}_{\gamma J_f}. \qquad (17)$$

Here the radiative reduced-width amplitude $\gamma^{J_i}_{\gamma J_f}$ is given by the sum of the internal and external (or channel) reduced-width amplitudes:

$$\gamma^{J_i}_{\gamma J_f} = \gamma^{J_i}_{\gamma J_f}(int) + \gamma^{J_i}_{\gamma J_f}(ch). \qquad (18)$$

Hence the total radiative width is

$$\left|[\Gamma^{J_i}_{\gamma J_f}(E)]\right| = \left|[\Gamma^{J_i}_{\gamma J_f}(E)]^{1/2}_{int} + [\Gamma^{J_i}_{\gamma J_f}(E)]^{1/2}_{ch}\right|^2, \qquad (19)$$

$$[\Gamma^{J_i}_{\gamma J_f}(E)]^{1/2}_{int,ch} = 2\,[P_{l_i}(E)]^{1/2}\, \gamma^{J_i}_{\gamma J_f}(int, ch). \qquad (20)$$

While the internal reduced-width amplitude is real, the channel reduced-width amplitude is complex [21] and is defined as

$$\begin{aligned} \gamma^{J_i}_{\gamma J_f}(ch) &= i^{l_i + L - l_f + 1}\, e^{i(\omega_{l_i} - \phi_{l_i})}\, \frac{1}{k}\, \mu_{ab}{}^{L+1/2} \\ &\times \left(\frac{Z_a e}{m_a^L} + (-1)^L\, \frac{Z_b e}{m_b^L}\right) \times \sqrt{\frac{(L+1)(2L+1)}{L}} \\ &\times \frac{1}{(2L+1)!!}\, (k_\gamma a)^{L+1/2}\, C_{J_f I l_f}\, \sqrt{\Gamma^{J_i}_{bIl_i}(E_R)} \\ &\times ([F_{l_i}(k,a)]^2 + [G_{l_i}(k,a)]^2) \times W_{l_f}(2\kappa a)\, (l_i 0\, L0 | l_f 0) \\ &\times U(L l_f\, J_i I;\, l_i J_f)\, J_L(l_i\, l_f). \end{aligned} \qquad (21)$$

The nonresonant capture amplitude is given by

$$U^{NR}_{Il_i J_f J_i} = -(2)^{3/2}\, i^{l_i+L-l_f+1}\, e^{i(\omega_{l_i}-\phi_{l_i})}\, \frac{1}{k}\, {\mu_{ab}}^{L+1/2} \times \left(\frac{Z_a\, e}{m_a^L} + (-1)^L\, \frac{Z_b\, e}{m_b^L}\right) \sqrt{\frac{(L+1)(2\,L+1)}{L}} \times \frac{1}{(2\,L+1)!!}(k_\gamma\, a)^{L+1/2}\, C_{J_f I l_f}\, F_{l_i}(k, r_0) \times G_{l_i}(k, r_0)\, W_{-\eta_f, l_f+1/2}(2\,\kappa\, r_0) \times \sqrt{P_{l_i}}(l_i 0\, L0|l_f 0)\, U(L\, l_f\, J_i\, I;\, l_i\, J_f) \times J^{'}_L(l_i\, l_f), \tag{22}$$

$$P_{l_i}(E) = \frac{k\, r_0}{F^2_{l_i}(k, r_0) + G^2_{l_i}(k, r_0)}, \tag{23}$$

where F_{l_i} and G_{l_i} are the regular and singular (at the origin) solutions of the radial Schrödinger equation, $\kappa = \sqrt{2\mu_{ab}\,\varepsilon_f}$ is the wave number, and $k_\gamma = E + \varepsilon_f$ is the momentum of the emitted photon. Integrals $J_L(l_i\, l_f)$ and $J^{'}_L(l_i\, l_f)$ are expressed in terms of F_{l_i}, G_{l_i} and Whittaker function $W_{-\eta_f, l_f+1/2}$ and are given in [21,22]. Both the channel radiative width and nonresonant amplitude are normalized in terms of the ANC, $C_{J_f I l_f}$, which defines the amplitude of the tail of the bound state wave function of nucleus c projected onto the two-body channel $a+b$ with the quantum numbers J_f, I, l_f. Such a normalization is physically transparent: both quantities describe peripheral processes and, hence, contain the tail of the overlap function of the bound wave functions of c, a and b, whose normalization is given by the corresponding ANC. Note that in the R-matrix method the internal nonresonant amplitude is included into the resonance term. Also, in the conventional R-matrix approach the channel radiative width and nonresonant amplitude are normalized in terms of the reduced width amplitude, which is not directly observable and depends on the channel radius. However, it is more convenient to express the normalization of the nonresonant amplitude in terms of the ANC that can be measured independently [14]. Then only the radial matrix element depends on the channel radius. As we can see from eqs. (21) and (22) the relative phase of the channel radiative width and the nonresonant amplitude is fixed because only the ANC has unknown phase factor. Thus by measuring the ANC for the bound state we are able to fix the absolute normalization of the channel radiative width and nonresonant amplitude simultaneously.

2.5 Interference of the resonant and nonresonant amplitudes for the $^{11}C(p,\gamma)^{12}N$ astrophysical radiative capture reaction

The evolution of very low-metallicity, massive stars depends critically on the amount of CNO nuclei that they produce. Alternative paths from the slow $3\,\alpha$ process to produce CNO seed nuclei could change their fate. The $^{11}C(p,\gamma)^{12}N$ reaction is an important branch point in one such alternative path. At energies appropriate to stellar evolution of very low-metallicity, massive stars, nonresonant capture to the ground state and interference of the second resonance and the nonresonant terms determine the reaction rate. The ANC for $^{12}N \to {}^{11}C + p$ has been determined from peripheral transfer reaction $^{14}N(^{11}C, {}^{12}N)^{13}C$ at 10 MeV/nucleon [22]. The contributions from the second resonance and interference effects were estimated using the R-matrix approach with the measured asymptotic normalization coefficient and the latest value for the radiative width of the second resonance [23]. The ANC gives useful information not only about the overall normalization of the direct capture amplitude, but also about the radiative width of the resonances. According to eqs. (20), the channel part of the radiative width may be determined from the ANC. Since the channel part is complex, $[\Gamma_{\gamma J_f\, J_i}(E)]^{1/2}_{ch} = \lambda + i\tau$, while the internal part of the radiative width amplitude is real, *i.e.* $[\Gamma_{\gamma J_f\, J_i}(E)]^{1/2}_{int} = \nu$, the total radiative width is given by

$$\Gamma_{\gamma J_f\, J_i}(E) = (\lambda + \nu)^2 + \tau^2. \tag{24}$$

The relative phase of λ and ν is, *a priori*, unknown, so these real parts may interfere either constructively or destructively. Hence, τ^2 always provides a lower limit for the radiative width and additional stronger limits may be obtained if assumptions are made about the interference between the two real contributions. For constructive interference of the real parts, the channel contribution gives a stronger lower limit. In the case of destructive interference, if $|\lambda| > |\nu|$, the channel contribution gives an upper limit for the radiative width. These limits depend on only one model parameter, the channel radius.

Recently, a measurement at RIKEN [23] found the gamma width to be 13.0 ± 0.5 meV. Using the measured ANC we find that for a channel radius of $r_0 = 5.0$ fm, $\Gamma_{\gamma J_f\, J_i}(E_R)_{ch} = 54$ meV. Taking into account the experimental value of the total radiative width, one can find the internal contribution from

$$\Gamma_{\gamma J_f\, J_i}(E_R) = \left|\Gamma_{\gamma J_f\, J_i}(E_R)^{1/2}_{ch} + \Gamma_{\gamma J_f\, J_i}(E_R)^{1/2}_{int}\right|^2. \tag{25}$$

There are two solutions, 15 and 112 meV. Assuming that the second value is too high [24], we conclude that the internal part of the radiative width is 15 meV, and destructive interference between the real parts of the channel and internal contributions gives the experimental value, 13 meV. In this case, the channel contribution alone represents an upper limit for the radiative width, while the square of the imaginary part of the channel contribution, 1.8 meV, gives a lower limit. The relative phase between the direct capture amplitude and the channel contribution to the radiative width of the second resonance is fixed in the R-matrix approach. Therefore, when the channel contribution to the radiative width dominates, the sign of the interference effects may be determined unambiguously. For $^{11}C(p,\gamma)^{12}N$, we find that the nonresonant and resonant capture amplitudes interfere constructively below the resonance and destructively above it. It has important consequences on the reaction rates for ^{12}N production. In particular, the reaction sequence $^7Be(\alpha,\gamma)^{11}C(p,\gamma)^{12}N$

will provide a means to produce CNO nuclei, while bypassing the 3 α reaction, in lower-density environments than previously anticipated [25].

2.6 Interference of the resonant and nonresonant amplitudes for the ^{13}N(p, γ)^{14}O astrophysical radiative capture

$^{13}\mathrm{N}(p,\gamma)^{14}\mathrm{O}$ is one of the key reactions which trigger the onset of the hot CNO cycle. This transition occurs when the proton capture rate on ^{13}N is faster, due to increasing stellar temperature ($\geq 10^8$ K), than the ^{13}N β-decay rate. The rate of this reaction is dominated by the resonant capture to the ground state of ^{14}O through the first excited state of ($E_R = 0.528$ MeV). However, through constructive interference, direct capture below the resonance makes a non-negligible contribution to the reaction rate. We have determined this direct contribution by measuring the asymptotic normalization coefficient for $^{13}\mathrm{N}+p \rightarrow {}^{14}\mathrm{O}$ (0.0 MeV). This ANC has been determined from the peripheral reaction $^{14}\mathrm{N}(^{13}\mathrm{N},{}^{14}\mathrm{O})^{13}\mathrm{C}$ [26]. The radiative capture cross section was estimated using an R-matrix approach with the measured asymptotic normalization coefficient and the latest resonance parameters. What is not known is the sign of the interference term between the resonant and nonresonant components of the radiative capture amplitudes. As we have mentioned it is possible to sometimes infer the sign of the interference to be used in R-matrix calculations of the radiative capture cross section if the ANC is known even in the absence of direct experimental data. Such is the case for the reaction being considered here. At energies below the resonance, the channel part, which depends on the ANC, has the same sign as the nonresonant amplitude leading to the constructive interference of these two terms. From eqs. (21) and (20) we find $[\Gamma^{J_i}_{\gamma\, J_f}(E_R)]^{1/2}_{ch} = 0.90 + i\,0.02\,\mathrm{eV}^{1/2}$ and the channel radiative width $|[\Gamma^{J_i}_{\gamma\, J_f}(E_R)]_{ch}| = 0.81$ eV at the resonance energy and the channel radius $r_0 = 5$ fm. The total resonance radiative width is $|[\Gamma^{J_i}_{\gamma\, J_f}(E)]|| = |[\Gamma^{J_i}_{\gamma\, J_f}(E)]^{1/2}_{int} + [\Gamma^{J_i}_{\gamma\, J_f}(E)]^{1/2}_{ch}|^2$. Thus there are two possible solutions for the internal part, a large negative value $[\Gamma^{J_i}_{\gamma\, J_f}(E)]^{1/2}_{int(1)} = -2.73\,\mathrm{eV}^{1/2}$ and a small positive value $[\Gamma^{J_i}_{\gamma\, J_f}(E)]^{1/2}_{int(2)} = 0.93\,\mathrm{eV}^{1/2}$. The first solution leads to the destructive interference with the non-resonant component at energies below the resonance, but it yields a high internal radiative width, $|\Gamma^{J_i}_{\gamma\, J_f}(E)]_{int}| = 7.48$ eV. The second solution leads to the constructive interference with the non-resonant component at energies below the resonance peak. We select this second solution because it is corroborated by the microscopic calculations [27], where it has been shown that the internal and external parts of the $E1$ matrix elements have the same sign and very close magnitudes. Our choice is also supported by the single-particle calculations [28,26]. Due to this constructive interference we find the S-factor for $^{13}\mathrm{N}(p,\gamma)^{14}\mathrm{O}$ to be larger than previous estimates. Consequently, the transition from the cold to hot CNO cycle for novae would be controlled by the slowest proton capture reaction $^{14}\mathrm{N}(p,\gamma)^{15}\mathrm{O}$.

3 Trojan Horse

The Trojan Horse method (THM) is a powerful indirect method which selects the quasi-free (QF) contribution of an appropriate three-body reaction performed at energies well above the Coulomb barrier to extract a charged particle two-body cross section at astrophysical energies free of Coulomb suppression. The THM has been suggested by Baur [9] and has been advanced and practically applied by a group from the Universitá di Catania working at the INFN-Laboratori Nazionali del Sud in Catania in collaboration with other institutions (see [5] and references therein). The THM has already been applied many times to reactions connected with fundamental astrophysical problems [29,30] such as $^7\mathrm{Li}(p,\alpha)^4\mathrm{He}$, $^6\mathrm{Li}(d,\alpha)^4\mathrm{He}$, $^6\mathrm{Li}(p,\alpha)^3\mathrm{He}$, and many others, see [5] and references therein.

Let us consider the TH reaction

$$a + A \rightarrow y + b + B, \tag{26}$$

where $a = (xy)$. The subreaction of interest is

$$x + A \rightarrow b + B. \tag{27}$$

In the TH method the incident particle a is accelerated to energies above the Coulomb barrier. After penetration through the barrier the projectile breaks into $x+y$ leaving the fragment x to interact with target A, while the second fragment-spectator y leaves carrying away the excess energy. By a proper choice of the final particle kinematics, the THM allows one to extract the cross section of the sub-process (27). However, the extracted amplitude of the reaction (27) in the THM is half-off-energy shell because the initial particle x in the sub-process (27) is off-the-energy shell. It has been suggested in the original paper [9] that the virtuality of particle x is compensated for by the higher momentum components in the Fermi motion of the fragments x and y inside the projectile a. However, high momentum components means that the distance between the fragments is so small that the interaction between the fragments is not negligible and the mechanism of the reaction is more complicated than the QF one. Instead, the virtuality of particle x in the extracted cross section is significantly compensated if we take into account the binding energy of the fragments x and y in the projectile a [31].

The THM allows one to determine both direct and resonant reactions (27). As an example of the result achieved using the THM, we present in fig. 3 the astrophysical factor for the $^3\mathrm{He}(d,p)^4\mathrm{He}$ process determined from the $^3\mathrm{He}(^6\mathrm{Li},\alpha\, p)^4\mathrm{He}$ TH reaction [32]. The TH resonant cross section (full dots) is normalized to the direct experimental data (open circles and open triangles) at energies near the resonance peak. The black on-line solid line is the result of a fit of the TH data (see ref. [32] for details), showing the trend of the bare nucleus $S(E)$-factor, while the blue on-line solid line is obtained by interpolating the screened direct data.

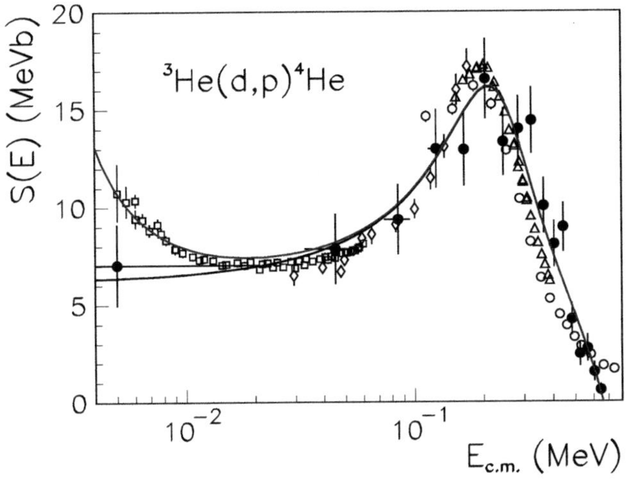

Fig. 3. (Colour on-line) The $^3\mathrm{He}(d,p)^4\mathrm{He}$ astrophysical S-factor determined from the TH reaction. The open circles and open triangles are direct experimental data; the full dots are the TH data. The black solid line represents the behavior of the bare nucleus $S(E)$-factor, resulting from a fit on the TH data, while the solid blue line is interpolation of the direct data.

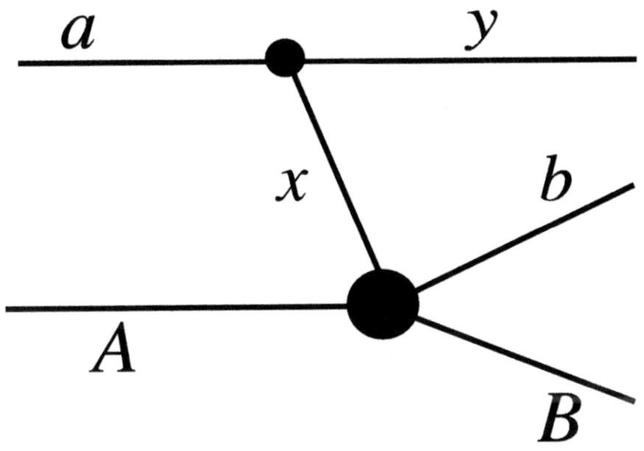

Fig. 4. Pole diagram describing the quasi-free mechanism.

3.1 TH reaction amplitude

A simple mechanism describing the TH process is the so-called QF process shown in fig. 4. In the quasi-free process it is assumed that the incident particle (assume incident particle is A) interacts with one of the fragments of $a = (xy)$, say with x which is considered to be "quasifree", while the second fragment is considered to be a "passive" spectator which is not involved in the process. Thus the interaction of the spectator y with x and A in the knockout process is neglected. The fact that the fragment x is not free is taken into account by folding the quasi-free reaction amplitude with the Fourier component of the (xy) bound-state wave function which takes into account the Fermi motion of x in the bound state $a = (xy)$.

In this section we present a derivation of the TH reaction amplitude from the general $2 \to 3$ reaction amplitude for the TH process (26). A general expression for the amplitude of the reaction is given by

$$M = \langle \chi_{bB}^{(-)} \chi_{yF}^{(-)} \varphi_y \varphi_b \varphi_B \big| \Delta V_f (1 + G^+ \Delta V_i) \big| \varphi_A \varphi_a \chi_i^{(+)} \rangle \tag{28}$$

$$= \langle \chi_{bB}^{(-)} \chi_{yF}^{(-)} \varphi_y \varphi_b \varphi_B \big| (\Delta V_f \, G^+ + 1) \Delta V_i \big| \varphi_A \varphi_a \chi_i^{(+)} \rangle. \tag{29}$$

The amplitudes (28) and (29) are the post and prior forms of the exact amplitude. Let us consider the post form. Here, G^+ is the total Green function of the system $a + A$, $\chi_i^{(+)}$ is the distorted wave describing the scattering wave function of $a + A$ in the initial state of the reaction, $\chi_{bB}^{(-)}$ is the distorted wave describing the scattering of particles $b + B$ in the final state: the distorted wave χ_{yF}^{-} describes the distorted wave of the spectator y and the center of mass of the system $F = b{+}B$ in the final state. For the moment we assume that Coulomb interactions are screened. Eventually we can take the limit of the screening radius to infinity. Also φ_i is the bound state wave function of nucleus i,

$$\Delta V_i = V_{aA} - U_{aA}, \tag{30}$$

$$\Delta V_f = V_{bB} - U_{bB} + V_{yF} - U_{yF}, \tag{31}$$

V_{ij} and U_{ij} are the interaction potential and optical potential between particles i and j. For example, $V_{aA} = V_{xA} + V_{yA}$. To extract the amplitude of the subprocess $x + A \to b + B$, which is the final goal of the TH method, we note that the Hamiltonian of the system $a + A$ is

$$H = H_{aA} + H_a + H_A = H_{xA} + H_{yF} + H_x + H_A + H_y, \tag{32}$$

where H_i is the internal Hamiltonian of nucleus i and $H_{ij} = T_{ij} + V_{ij}$ is the Hamiltonian of the relative motion of nuclei i and j, T_{ij} is their relative kinetic energy operator and V_{ij} is their interaction potential. The total Green's function operator can be written as

$$G^+ = \frac{1}{E - H_{aA} - H_a - H_A + i0} \tag{33}$$

$$= \frac{1}{E - H_{xA} - H_{yF} - H_{xyA}^0 + i0} \tag{34}$$

$$= \frac{1}{E - H_{xA} - T_{yF} - U_{yF} - \Delta V_{yF} - H_{xyA}^0 + i0} \tag{35}$$

$$= \tilde{G}^+ + G^+ \, \Delta V_{yF} \, \tilde{G}^+, \tag{36}$$

Here $\Delta V_f = V_{yF} - U_{yF}$, $V_{yF} = V_{yx} + V_{yA}$, $H_{xyA}^0 = H_x + H_y + H_A$ and

$$\tilde{G}^+ = \frac{1}{E - H_{xA} - T_{yF} - U_{yF} - H_{xyA}^0 + i0} \tag{37}$$

We substitute eq. (37) into (28) and drop the term $\Delta V_f \, G^+ \, \Delta V_{yF} \, \tilde{G}^+ \, \Delta V_i$ as the higher order term in the perturbation expansion over ΔV. Then we get from eq. (28)

$$M = \langle \chi_{bB}^{(-)} \chi_{yF}^{(-)} \varphi_y \varphi_b \varphi_B \big| \Delta V_f (1 + \tilde{G}^+ \, \Delta V_i) \big| \varphi_A \varphi_a \chi_i^{(+)} \rangle. \tag{38}$$

To single out the TH subprocess amplitude we replace $\Delta V_i = V_{xA} + V_{yA} - U_{aA}$ by V_{xA} and $\Delta V_f = V_{yF} - U_{yF} + V_{bB} - U_{bB}$ by $\Delta V_{bB} = V_{bB} - U_{bB}$. Then the amplitude (28) becomes

$$M = \langle \chi_{bB}^{(-)} \chi_{yF}^{(-)} \varphi_y \varphi_b \varphi_B | \Delta V_{bB} \,(1 + \tilde{G}^{+}\, V_{xA}) | \varphi_A \varphi_a \chi_i^{(+)} \rangle$$
$$= \langle \chi_{bB}^{(-)} \chi_{yF}^{(-)} \varphi_y \varphi_b \varphi_B | \Delta V_{bB} \,(1 + G_{xA}^{+}\, V_{xA}) | \varphi_A \varphi_a \chi_i^{(+)} \rangle. \quad (39)$$

Here

$$G_{xA}^{+} = \frac{1}{E_{xA} - H_{xA} + i0} \quad (40)$$

and E_{xA} is the relative kinetic energy of particles x and A. The appearance of G_{xA}^{+} in eq. (40) is due to

$$\langle \chi_{yF}^{(-)} \varphi_y \varphi_b \varphi_B | \tilde{G}^{+} = \langle \chi_{yF}^{(-)} \varphi_y \varphi_b \varphi_B | G_{xA}^{+}. \quad (41)$$

Equation (39) reveals a very important result. It contains a factor $1 + G_{xA}^{+} V_{xA}$. For the on-shell case, the relative momentum of particles x and A $\mathbf{p}_{xA} = \mathbf{k}_{xA}$, where $\mathbf{k}_{xA}$ is the $x - A$ on-shell relative momentum related with their relative kinetic energy as $E_{xA} = p_{xA}^2/(2\,\mu_{xA})$. Correspondingly,

$$(1 + G^{+} V_{xA}) | e^{i\mathbf{k}_{xA}\cdot\mathbf{r}_{xA}} \rangle = \chi_{\mathbf{k}_{xA}}^{+}(\mathbf{r}_{xA}) \quad (42)$$

is the scattering wave function of particles x and A interacting via the optical potential V_{xA}. We assume at the moment that all the Coulomb interactions are screened. However, in the TH method the entry particle x is not free because it is in the bound state $a = (xy)$, *i.e.* the momentum of x is not fixed. In other words, x is off-the-energy shell because $E_{xA} \neq p_{xA}^2/(2\,\mu_{xA})$. For the off-shell case

$$(1 + G^{+} V_{xA}) | e^{i\mathbf{p}_{xA}\cdot\mathbf{r}_{xA}} \rangle = \chi_{(os)\mathbf{k}_{xA},\mathbf{p}_{xA}}^{+}(\mathbf{r}_{xA}) \quad (43)$$

is the so-called off-shell scattering function.

3.2 TH method for direct reactions

We first consider the direct subreaction (27). We assume that this reaction proceeds through the transfer of particle c from A to x (it can be also considered as a particle transfer from x to A), *i.e.* $A = (Bc)$ and $b = (xc)$. The "pole" diagram corresponding to the on-shell reaction describing the particle c transfer mechanism with the $x - A$ rescattering in the initial and $b - B$ rescattering in the final state is shown in fig. 5. This diagram describes the DWBA amplitude. To simplify eq. (39) in the case of the direct transfer subprocess, we insert the projection operators $\sum |\varphi_x\rangle\langle\varphi_x|$, $\sum |\varphi_B\rangle\langle\varphi_B|$ and $\sum |\varphi_c\rangle\langle\varphi_c|$ into the bra and ket states. The sum is taken over discrete states and an integral is used for the continuum states of the corresponding nucleus. We leave in the projection operator only the ground state projections $|\varphi_x\rangle\langle\varphi_x|$, $|\varphi_B\rangle\langle\varphi_B|$ and $|\varphi_c\rangle\langle\varphi_c|$ assuming that only the ground states of x, B and c contribute to the reaction. If necessary the excited states

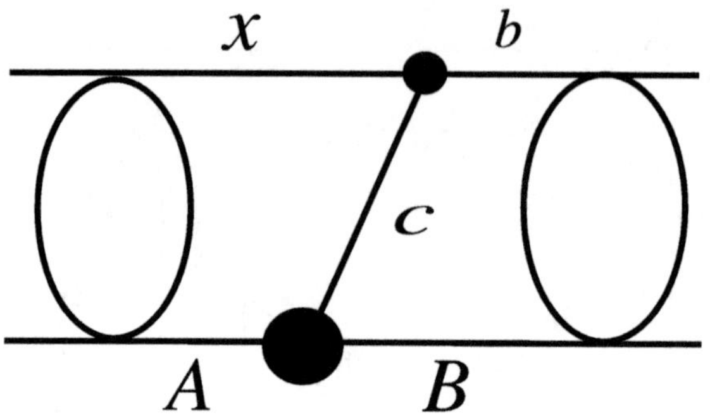

Fig. 5. Pole diagram describing the direct reaction $x + A \to b + B$ mechanism. Bubbles show the initial and final state interactions.

can also be taken into account. Then we get

$$\langle \varphi_y \varphi_b \varphi_B | \Delta V_{bB} \,(1 + G_{xA}^{+}\, V_{xA}) | \varphi_A \varphi_a \rangle$$
$$\approx \langle \varphi_b | \varphi_c \varphi_x \rangle \langle \varphi_B | \Delta V_{bB} | \varphi_B \rangle$$
$$\times (1 + \langle \varphi_x | G_{xA}^{+} | \varphi_x \rangle \langle \varphi_x | V_{xA} | \varphi_x \rangle)$$
$$\times \langle \varphi_c \varphi_B | \varphi_A \rangle \langle \varphi_x \varphi_y | \varphi_a \rangle \quad (44)$$

We introduce the overlap functions $I_{\beta\gamma}^{\alpha} = \langle \varphi_\beta\, \varphi_\gamma | \varphi_\alpha \rangle$ and use the approximation $\langle \varphi_x | V_{xA} | \varphi_x \rangle \approx U_{xA}$; also we use the approximation

$$\langle \varphi_x | G_{xA}^{+} | \varphi_x \rangle \approx G_{xA}^{(U)+} = (E_{xA} - T_{xA} - U_{xA} + i0)^{-1}. \quad (45)$$

Note that $\langle \varphi_B | \Delta V_{bB} | \varphi_B \rangle \approx V_{xB} + V_{cB} - U_{bB}$, where V_{jB} is the interaction potential between the point like nuclei $j = x, c$ and B. All the neglected terms are higher order in the perturbation theory over ΔV. Then we get in lowest order for the TH amplitude with the subprocess described by the direct transfer reaction (27):

$$M = \langle \chi_{yF}^{(-)} [\chi_{bB}^{(-)} I_{xc}^{b} | \Delta V_{bB} |\, I_{cB}^{A} \,(1 + G_{xA}^{+}\, V_{xA})]\, I_{xy}^{a} \chi_i^{(+)} \rangle. \quad (46)$$

The expression in the brackets is the amplitude of subreaction (27) which is the final goal of the TH. To see it we just rewrite (46) in momentum space:

$$M = \int \frac{\mathrm{d}\mathbf{p}_{yF}}{(2\,\pi)^3} \frac{\mathrm{d}\mathbf{p}_{xA}}{(2\,\pi)^3}\, \chi_{yF}^{*(-)}(\mathbf{p}_{yF}) M^{sub}(\mathbf{k}_{bB}, \mathbf{p}_{xA}) I_{xy}^{a}(\mathbf{p}_{xy})$$
$$\times \chi_i^{(+)}(\mathbf{p}_{xA}), \quad (47)$$

where

$$\mathbf{p}_{xy} = \frac{m_y \mathbf{p}_x - m_x \mathbf{p}_y}{m_x + m_y} = \frac{m_y}{m_x}\mathbf{p}_a - \mathbf{p}_y. \quad (48)$$

Also note that in the center of mass of TH reaction $a + A \to y + b + B$ the relative momentum is given by $\mathbf{p}_{aA} = \mathbf{p}_a$ and $\mathbf{p}_{yF} = \mathbf{p}_y$. We denote by $\mathbf{p}_i$ ($\mathbf{k}_i$) the momentum of the virtual (real) particle i and by $\mathbf{p}_{ij}$ ($\mathbf{k}_{ij}$) the relative momentum of virtual (real) particles i and j. Also $\chi_i^{(+)}(\mathbf{p}_{xA}) \equiv \chi_{\mathbf{k}_{aA}}^{(+)}(\mathbf{p}_{xA})$, *i.e.* it is the Fourier component of the $a - A$ scattering wave function with the incident momentum $\mathbf{k}_{aA}$ which in the center of mass of the TH reaction is just $\mathbf{k}_a$. Correspondingly $\chi_{yF}^{(-)}(\mathbf{p}_{yF}) \equiv \chi_{\mathbf{k}_{yF}}^{(-)}(\mathbf{p}_{yF})$.

The half-off-the-energy shell amplitude of the subprocess (27) is given by

$$M^{sub}(\mathbf{k}_{bB},\mathbf{p}_{xA}) = \langle \chi_{bB}^{(-)} I_{xc}^{b} | \Delta V_{bB} | I_{cB}^{A} \chi_{(os)\mathbf{k}_{xA},\mathbf{p}_{xA}}^{+} \rangle. \tag{49}$$

The virtuality of the entry particle x of this amplitude results in the fact that the relative momentum of particles x and A in the initial state of reaction (27) $p_{xA} \neq \sqrt{2\,\mu_{xA}\,E_{xA}}$. Due to the off-shell entry particles amplitude (49) does not have the Gamow penetration factor. We would like to underscore that from $E_{xA} + Q = E_{bB}$ for positive $Q > 0$ for reaction (27) at $E_{xA} \to 0$, $E_{bB} \approx$ const. Hence the off-shell scattering function $\chi_{(os)\mathbf{k}_{xA},\mathbf{p}_{xA}}^{+}$ is the only E_{xA} dependent factor in M^{sub} at $E_{xA} \to 0$. The off-shell scattering function is a universal factor which does not depend on the specifics of the direct reaction. Rewriting the matrix element in eq. (49) in the momentum representation gives

$$M^{sub}(\mathbf{k}_{bB},\mathbf{p}_{xA}) = \int \frac{\mathrm{d}\mathbf{p}_{bB}}{(2\,\pi)^3} \frac{\mathrm{d}\mathbf{p}_{xA}'}{(2\,\pi)^3} \chi_{\mathbf{k}_{bB}}^{*(-)}(\mathbf{p}_{bB}) \times I_{xt}^{*b}\Big(\mathbf{p}_x' - \frac{m_x}{m_b}\,\mathbf{p}_b\Big)\, \Delta V_{bB}\, I_{cB}^{A}\Big(\mathbf{p}_B - \frac{m_B}{m_A}\,\mathbf{p}_A'\Big) \times \chi_{(os)\mathbf{k}_{xA},\mathbf{p}_{xA}}^{+}(\mathbf{p}_{xA}'). \tag{50}$$

Approximation $\Delta V_{bB} \approx V_{cB}$, which works for $m_x > m_c$, is enough for us to investigate the dependence of $M^{sub}(\mathbf{k}_{bB},\mathbf{p}_{xA})$ on E_{xA} for arbitrary masses of x and c. Using this approximation we get from eq. (50)

$$M^{sub}(\mathbf{k}_{bB},\mathbf{p}_{xA}) = \int \frac{\mathrm{d}\mathbf{p}_{bB}}{(2\,\pi)^3} \frac{\mathrm{d}\mathbf{p}_{xA}'}{(2\,\pi)^3} \chi_{\mathbf{k}_{bB}}^{*(-)}(\mathbf{p}_{bB}) \times I_{xc}^{*b}\Big(\mathbf{p}_x' - \frac{m_x}{m_b}\,\mathbf{p}_b\Big)\, W_{cB}^{A}\Big(\mathbf{p}_B - \frac{m_B}{m_A}\,\mathbf{p}_A'\Big) \times \chi_{(os)\mathbf{k}_{xA},\mathbf{p}_{xA}}^{+}(\mathbf{p}_{xA}'). \tag{51}$$

Here $W_{cB}^{A}(\mathbf{p}_{cB})$ is the form factor determined by

$$W_{cB}^{A}(\mathbf{p}_{cB}) = \int \mathrm{d}\mathbf{r}_{cB} e^{-i\,\mathbf{p}_{cB}\cdot\mathbf{r}_{cB}}\, V_{cB}(r_{cB})\, I_{cB}^{A}(\mathbf{r}_{cB}). \tag{52}$$

The Fourier component of the off-shell scattering function $\chi_{(os)\mathbf{k}_{xA},\mathbf{p}_{xA}}^{+}(\mathbf{r}_{xA})$ is given by

$$\chi_{(os)\mathbf{k}_{xA},\mathbf{p}_{xA}}^{+}(\mathbf{p}_{xA}') = \delta(\mathbf{p}_{xA}' - \mathbf{p}_{xA}) + G_0^{+}(p_{xA}'; E_{xA}) \times T(\mathbf{p}_{xA}',\mathbf{p}_{xA}; E_{xA}), \tag{53}$$

$$G_0^{+}(p_{xA}'; E_{xA}) = \frac{1}{E_{xA} - p_{xA}'^2/2\,\mu_{xA} + i0}, \tag{54}$$

$T(\mathbf{p}_{xA}',\mathbf{p}_{xA}; E_{xA})$ is the off-shell $x - A$ scattering amplitude. Amplitude $M^{sub}(\mathbf{k}_{bB},\mathbf{p}_{xA})$ extracted from the THM should be compared with the corresponding on-shell reaction amplitude

$$M^{onsh}(\mathbf{k}_{bB},\mathbf{k}_{xA}) = \int \frac{\mathrm{d}\mathbf{p}_{bB}}{(2\,\pi)^3} \frac{\mathrm{d}\mathbf{p}_{xA}}{(2\,\pi)^3} \chi_{\mathbf{k}_{bB}}^{*(-)}(\mathbf{p}_{bB}) \times I_{xt}^{*b}\Big(\mathbf{p}_x - \frac{m_x}{m_b}\,\mathbf{p}_b\Big)\, \Delta V_{bB}\, I_{cB}^{A}\Big(\mathbf{p}_B - \frac{m_B}{m_A}\,\mathbf{p}_A\Big) \times \chi_{\mathbf{k}_{xA}}(\mathbf{p}_{xA}). \tag{55}$$

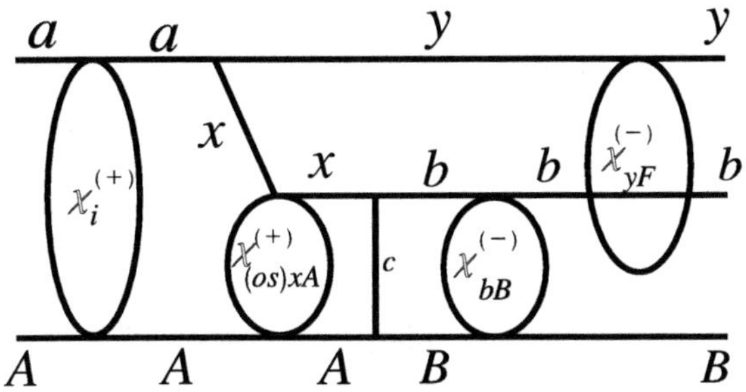

Fig. 6. Diagram describing the TH reaction $a + A \to y + b + B$ proceeding through the direct subprocess $x + A \to b + B$ mechanism. Bubbles show initial and final state interactions and the off-shell scattering function

Equations (47) and (49) is our final result. The diagram corresponding to this amplitude (47) is shown in fig. 6. Equation (47) is a general expression for the TH reaction amplitude which contains the half-off-shell direct subprocess amplitude and the initial and final state rescatterings. As we can see the subprocess amplitude is not factorized, but instead is folded with the initial and final state distorted waves and the overlap function for $a \to y + x$. Note that if the initial and distorted waves in the momentum space are replaced by delta-functions, eq. (47) just becomes a trivial plane wave impulse approximation described by the diagram of fig. 3.

3.3 TH for resonant reactions

In sect. 3.1 we derived a general expression, eq. (39), for the amplitude of the TH reaction (26) which is valid for both direct and resonant subprocesses (27). Here we consider the resonant TH reactions, *i.e.* we assume that the subprocess (27) proceeds through the intermediate resonance F^*. Our goal is to relate the half-off-shell and on-shell resonant amplitudes. Note that it is easier to relate the off-shell and on-shell resonant reactions than the direct ones. The resonant TH amplitude can be extracted from eq. (39) in a straightforward manner because it contains the Green's operator G_{xA}^{+}. Below we demonstrate how to do it. For simplicity here we neglect the initial and final state interactions.

The TH amplitude of the reaction (26), which proceeds through the resonance state F^* in the intermediate system $x + A$, is given by

$$M = M^{sub(R)}(\mathbf{k}_{bB},\mathbf{p}_{xA})\, I_{xy}^{a}(\mathbf{p}_{xy}). \tag{56}$$

Here $M^{sub(R)}$ is the amplitude of the resonant subprocess (27). Usually in practical calculations the overlap function I_{xy}^{a} is expressed in terms of the corresponding single-particle bound state wave function φ_{xy}:

$$I_{xy}^{a} = S_{xy}^{1/2}\, \varphi_{xy}. \tag{57}$$

Here, S_{xy} is the spectroscopic factor of the bound state (xy) in a with given quantum numbers. For simplicity we do not write down symbols corresponding to the quantum

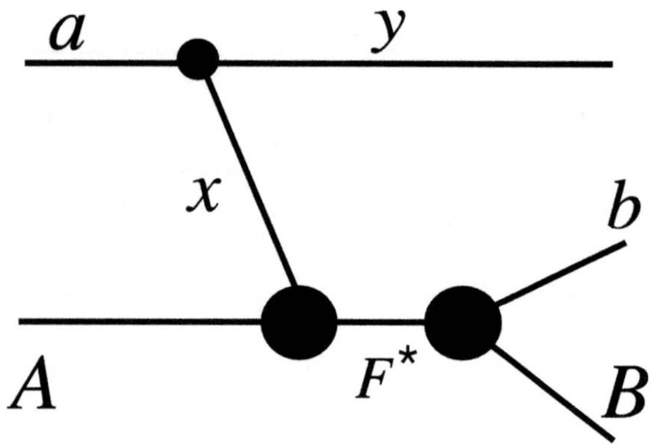

Fig. 7. Diagram describing the resonant reaction $a + A \to y + b + B$.

numbers and assume that $S_{xy} = 1$. In the momentum space the bound state wave function is given by

$$\varphi_{xy}(\mathbf{p}_{xy}) = -2\mu_{xy}\frac{W(\mathbf{p}_{xy})}{p_{xy}^2 + \kappa_{xy}^2}, \quad (58)$$

$$W(\mathbf{p}_{xy}) = \int \mathrm{d}\mathbf{r}\, e^{-i\mathbf{p}_{xy}\cdot\mathbf{r}}\, V_{xy}(r)\, \varphi_{xy}(\mathbf{r}) \quad (59)$$

$$= \left(-\varepsilon_a - \frac{p_{xy}^2}{2\mu_{xy}} \right) \varphi_{xy}(\mathbf{p}_{xy}). \quad (60)$$

Now we can find the virtuality factor

$$\sigma_x = E_x - \frac{p_x^2}{2m_x} \quad (61)$$

of the virtual particle x using the energy and momentum conservation laws in both vertices $a \to x + y$ and $x + A \to F^*$. After simple algebraic transformations we get

$$\sigma_x = E_{xA} - \frac{p_{xA}^2}{2\mu_{xA}} = -\frac{1}{2\mu_{xy}}\left[p_{xy}^2 + (\kappa_{xy}^a)^2\right] < 0. \quad (62)$$

Thus we derived a very important result for the relative kinetic energy of particles x and A E_{xA} in the TH method: $E_{xA} < p_{xA}^2/2\mu_{xA}$, *i.e.* always $k_{xA} < p_{xA}$, where $k_{xA} = \sqrt{2\,\mu_{xA}\,E_{xA}}$ is the $x - A$ relative on-shell momentum. The half-off-shell resonant reaction amplitude in the TH method is described by the diagram shown in fig. 7 and is given by

$$M^{sub(R)}(\mathbf{k}_{bB},\, \mathbf{p}_{xA}; E) = -\frac{1}{2}(4\pi)^2\sqrt{\frac{1}{\mu_{bB}k_{bB}}}$$
$$\times \sum_{m_0=-l_0}^{l_0} Y_{l_0m_0}(\hat{\mathbf{k}}_{bB})Y_{l_0m_0}^*(\hat{\mathbf{p}}_{xA})\, e^{i\delta_{fl_0}(k_{bB})}$$
$$\times \frac{\sqrt{\Gamma_{bB}(E_{bB}, r_0)}\, w_{l_0}(p_{xA}, k_{xA(R)})}{E_{xA} - E_{xA}^{(R)}}. \quad (63)$$

Here $k_{xA(R)} = \sqrt{2\,\mu_{xA}\,E_{xA}^{(R)}}$, $\mathbf{k}_{bB}$ is the on-shell relative momentum of particles b and B in the final state, $l_0(m_0)$ is the resonance orbital angular momentum (its projection), $Y_{l_0m_0}$ is the corresponding spherical harmonics, $\hat{\mathbf{r}} = \mathbf{r}/r$, δ_{fl_0} is the nonresonant (potential) scattering phase shift of particles b and B in the final state. The off-shell form factor

$$w_{l_0}(p_{xA}, k_{xA(R)}) = \int_0^\infty \mathrm{d}r\, r^2\psi_{nl_0}^{(R)}(r)\, V(r) j_{l_0}(p_{xA}\, r)$$
$$= \left(E_{xA}^{(R)} - E_{p_{xA}}\right)\int_0^\infty \mathrm{d}r\, r^2\psi_{nl_0}^{(R)}(r)\, j_{l_0}(p_{xA}\, r)$$
$$= \left(E_{xA}^{(R)} - E_{p_{xA}}\right)\psi_{nl_0}^{(R)}(p_{xA}). \quad (64)$$

Here $\psi_{nl_0}^{(R)}(r)$ is the resonant Gamow radial wave function, $\psi_{nl_0}^{(R)}(p_{xA})$ is its Fourier component, $j_{l_0}(p_{xA}\, r)$ is the spherical Bessel function, $E_{p_{xA}} = p_{xA}^2/2\,\mu_{xA}$, n is the principal quantum number.

Let us write down the well known expression for the on-shell Breit-Wigner resonance amplitude for the resonant process $x + A \to b + B$

$$M^{(R)}(\mathbf{k}_{bB},\, \mathbf{k}_{xA}; E) = -\frac{1}{4}(4\pi)^2\sqrt{\frac{1}{\mu_{bB}k_{bB}}}\sqrt{\frac{1}{\mu_{xA}k_{xA}}}$$
$$\times \sum_{m_0=l_0}^{l_0} Y_{l_0m_0}(\hat{\mathbf{k}}_{bB})Y_{l_0m_0}^*(\hat{\mathbf{k}}_{xA})\, e^{i\delta_{fl_0}(k_{bB})}e^{i\delta_{fl_0}(k_{xA})}$$
$$\times \frac{\sqrt{\Gamma_{bB}(E_{bB}, r_0)}\,\sqrt{\Gamma_{bB}(E_{xA}, r_0)}}{E_{xA} - E_{xA}^{(R)}}, \quad (65)$$

where $\mathbf{k}_{xA}$ is the on-shell relative momentum of the initial particles x and A and $\mathbf{k}_{bB}$ is the on-shell relative momentum of the final particles b and B. In the R-matrix method the resonance width contains the Coulomb-centrifugal barrier penetrability factor which exponentially decreases with energy. Hence for $E_{xA} \to 0$ the resonant amplitude $M^R \sim \sqrt{P_{l_0}(k_{xA})}\,\tilde{M}^R$. Just this factor makes it difficult or impossible to measure resonant reactions at astrophysically relevant energies. Now we compare the half-off-shell resonant amplitude, eq. (63), and the on-shell amplitude, eq. (65). The half-off-shell amplitude contains the form factor $w_{l_0}(p_{xA}, k_{(xA)R})$. The barrier factor should come from the integral representation in eq. (64), namely from $j_{l_0}(p_{xA}r)$. However, $j_{l_0}(p_{xA}r)$ does not contain the Coulomb penetration factor and does not depend on the on-shell momentum k_{xA}. Hence in limit $k_{xA} \to 0$ the off-shell form factor does not go to zero. We underscore that it is very important that always in the TH reaction $p_{xA} > k_{xA}$. Comparing eqs. (63) and (65), we get

$$M^{(R)}(\mathbf{k}_{bB},\, \mathbf{k}_{xA}; E_{xA}) = -\frac{1}{2}e^{i\delta_{(xA)l_0}(k_{xA})}\sqrt{\frac{1}{\mu_{xA}k_{xA}}}$$
$$\times \frac{\sqrt{\Gamma_{xA}(E_{xA}, r_0)}}{w_{l_0}(p_{xA}, k_{xA(R)})}M^{sub(R)}(\mathbf{k}_{bB},\, \mathbf{p}_{xA}; E_{xA}). \quad (66)$$

Note the only difference between the half-off-shell and the on-shell resonant amplitudes is the appearance of the form factor $w_{l_0}(p_{xA}, k_{xA(R)})$. Now we give the expression for the on-shell resonant cross section which can be derived from the TH half-off-shell resonant cross section

$$\sigma(E_{xA}) = \frac{\mu_{xA}k_{xA}\,\mu_{bB}k_{bB}}{(2\pi)^2}\frac{1}{k_{xA}^2}\frac{1}{4\pi}\int \mathrm{d}\Omega_{\mathbf{k}_{bB}} \times \int \mathrm{d}\Omega_{\mathbf{k}_{xA}}\left|M^{(R)}(\mathbf{k}_{bB},\, \mathbf{k}_{xA}; E_{xA})\right|^2 \quad (67)$$

$$= \frac{1}{4}\frac{\mu_{bB}k_{bB}}{(2\pi)^2}\frac{1}{k_{xA}^2}\frac{1}{4\pi}\frac{\Gamma_{xA}(E_{xA}, r_0)}{|w_{l_0}(p_{xA}, k_{xA(R)})|^2} \times \int \mathrm{d}\Omega_{\mathbf{k}_{bB}}\int \mathrm{d}\Omega_{\mathbf{p}_{xA}}\left|M^{sub(R)}(\mathbf{k}_{bB},\, \mathbf{p}_{xA}; E_{xA})\right|^2. \quad (68)$$

4 Summary

In this work we have addressed two important indirect techniques in nuclear astrophysics use, asymptotic normalization coefficient (ANC) and Trojan Horse (TH) methods. Both techniques allow one to determine the astrophysical factors at Gamow peak or even at zero energy avoiding the extrapolation procedure. The ANC method determines the overall normalization of the peripheral radiative capture processes. The ANC technique becomes especially powerful for astrophysical processes proceeding through a subthreshold state —a loosely bound state. In this case the ANC determines both the overall normalization of the direct radiative capture to the subthreshold state and the resonance partial width for captures through the subthreshold resonance. We demonstrated the application of the ANC technique for the key CNO cycle reaction $^{14}\mathrm{N}(p,\gamma)^{15}\mathrm{O}$. The ANC method turns out to be useful also for determination of the sign of the interference term of the resonant and nonresonant radiative capture amplitudes. We demonstrated it for two important CNO cycle reactions: $^{11}\mathrm{C}(p,\gamma)^{12}\mathrm{N}$ and $^{13}\mathrm{N}(p,\gamma)^{14}\mathrm{O}$.

The TH method allows one to determine the astrophysical factors for astrophysical reactions, both direct and resonant. In practical applications the astrophysical factor extracted from the TH reaction is available in a wide energy range from astrophysical energies to higher energies. Its absolute normalization is determined by normalization of the TH astrophysical factor to the one obtained from direct measurements at higher energies. Assuming that the energy dependence of the TH astrophysical factor is correct, one can determine the absolute astrophysical factor at astrophysical energies. In this work we have derived a general expression for the TH reaction amplitude which takes into account the off-shell effects and initial and final state interactions. The direct and resonant TH reactions are considered separately. We derived the TH amplitude for direct subreactions in terms of the off-shell scattering wave function. The energy dependence of this wave function determines the energy dependence of the TH astrophysical factor for an arbitrary direct reaction mechanism. We connect the TH resonant cross section with the on-shell resonant cross section. We intend to use the derived equations to calculate the absolute astrophysical factors.

This work was supported by the U.S. Department of Energy under Grant No. DE-FG03-93ER40773, the U.S. National Science Foundation under Grant No. INT-9909787 and Grant No. PHY-0140343, ME 385(2000) and ME 643(2003) projects NSF and MSMT, CR, project K1048102 and grant No. 202/05/0302 of the Grant Agency of the Czech Republic, and by the Robert A. Welch Foundation.

References

1. C. Rolfs, W.S. Rodney, *Cauldrons in the Cosmos* (The University of Chicago Press, 1988) p. 368.
2. H.J. Assenbaum, K. Langanke, C. Rolfs, Z. Phys. A **327**, 461 (1987).
3. F. Streider *et al.*, Naturwissenschaften **88**, 461 (2001).
4. LUNA Collaboration (C. Casella *et al.*), Nucl. Phys. A **706**, 203 (2002).
5. C. Spitaleri *et al.*, Phys. Rev. C **69**, 055806 (2004).
6. A.M. Mukhamedzhanov *et al.*, Phys. Rev. C **67**, 065804 (2003).
7. G. Baur *et al.*, Nucl. Phys. A **458**, 188 (1986).
8. T. Motobayashi *et al.*, Phys. Rev. Lett. **73**, 2680 (1994).
9. G. Baur, Phys. Lett. B **178**, 135 (1986).
10. W. Younes, H.C. Britt, Phys. Rev. C **67**, 024610 (2003).
11. A.M. Mukhamedzhanov, N.K. Timofeyuk, JETP. Lett. **51**, 282 (1990).
12. H.M. Xu *et al.*, Phys. Rev. Lett. **73**, 2027 (1994).
13. C.A. Gagliardi *et al.*, Phys. Rev. C **59**, 1149 (1999).
14. A.M. Mukhamedzhanov, R.E. Tribble, Phys. Rev. C **59**, 3418 (1999).
15. L.D. Blokhintsev, I. Borbely, E.I. Dolinskii, Fiz. Elem. Chastits At. Yadra **8**, 1189 (1977).
16. L.D. Blokhintsev *et al.*, Phys. Rev. C **48**, 2390 (1993).
17. P.F. Bertone *et al.*, Phys. Rev. C **66**, 055804 (2002).
18. U. Schröder *et al.*, Nucl. Phys. A **467**, 240 (1987).
19. A. Formicola *et al.*, Phys. Lett. B **591**, 61 (2004).
20. P.F. Bertone *et al.*, Phys. Rev. Lett. **87**, 152501 (2001).
21. F.C. Barker, T. Kajino, Aust. J. Phys. **44**, 369 (1991).
22. X.D. Tang *et al.*, Phys. Rev. C **67**, 015804 (2003).
23. T. Minemura *et al.*, RIKEN Accel. Prog. Rep. **35**, 58 (2002).
24. P. Descouvemont, Nucl. Phys. A **646**, 261 (1999).
25. M. Wiescher *et al.*, Astrophys. J. **343**, 352 (1989).
26. X.D. Tang *et al.*, Phys. Rev. C **69**, 055807 (2004).
27. P. Descouvemont, D. Baye, Nucl. Phys. A **500**, 155 (1989).
28. P. Descouvemont, Nucl. Phys. A **646**, 261 (1999).
29. C.J. Copi, D.N. Schramm, M.S. Turner, Science **267**, 192 (2005).
30. L. Piau, S. Turck-Chieze, Astrophys. J. **566**, 419 (2002).
31. A. Tumino *et al.*, Phys. Rev. C **67**, 065803 (2003).
32. M. La Cognata *et al.*, Phys. Rev. C **72**, 065802 (2005).

Eur. Phys. J. A **27**, s01, 217–220 (2006)
DOI: 10.1140/epja/i2006-08-033-6

EPJ A direct
electronic only

Can the neutron-capture cross sections be measured with Coulomb dissociation?

Á. Horváth[1,a], K. Ieki[2], Á. Kiss[1], A. Galonsky[3,4], M. Thoennessen[3,4], T. Baumann[3], D. Bazin[3], C.A. Bertulani[5], C. Bordeanu[6], N. Carlin[7], M. Csanád[1], F. Deák[1], P. DeYoung[3,8], N. Frank[3,4], T. Fukuchi[2], Zs. Fülöp[9], A. Gade[3], D. Galaviz[3], C. Hoffman[3,10], R. Izsák[1], W.A. Peters[3,4], H. Schelin[11], A. Schiller[3], R. Sugo[2], Z. Seres[12], and G.I. Veres[1]

[1] Department of Atomic Physics, Eötvös Loránd University, 1117 Budapest Pázmány Péter sétány 1/A, Hungary
[2] Department of Physics, Rikkyo University, 3 Nishi-Ikebukuro, Toshima, Tokyo 171, Japan
[3] National Superconducting Cyclotron Laboratory, Michigan State University, East Lansing, MI 48824, USA
[4] Department of Physics and Astronomy, Michigan State University, East Lansing, MI 48824, USA
[5] Department of Physics, University of Arizona, 1118E 4th Street, Tucson, AZ 85721, USA
[6] Department of Physics, University of Washington, Seattle, WA 98195, USA
[7] Instituto de Física, Universidade de São Paulo, Caixa Postal 66318, 05315-970, São Paulo, Brazil
[8] Department of Physics and Engineering, Hope College, Holland, MI 49423-9000, USA
[9] ATOMKI Institute of Nuclear Research, P.O.B. 51 H-4001, Debrecen, Hungary
[10] Department of Physics, Florida State University, Tallahassee, FL 32306, USA
[11] Centro Federal de Educação Tecnológica do Paraná, Avenue Sete de Setembro 3165, 80 230-901 Curitiba, Paraná, Brazil
[12] KFKI Research Institute for Particle and Nuclear Physics, P.O.B. 49, Budapest, 114, Hungary

Received: 25 August 2005 /
Published online: 28 March 2006 –

Abstract. In this paper we present first results from a ^{8}Li electromagnetic neutron-breakup experiment. Specific reactions studied were Pb(^{8}Li,^{7}Li+n)Pb and C(^{8}Li,^{7}Li+n)C at 41 MeV/nucleon beam energy. This is an effort to compare the results of a Coulomb dissociation experiment with the well determined (n,γ) reaction cross sections at astrophysical energies. The angular dependence of the cross section above 7 degree, which is the grazing angle of ^{8}Li-Pb system, is similar in shape for lead and carbon and approximately proportional to $A^{2/3}$ in magnitude indicating that the nuclear dissociation is the main component in this region. At very forward angles the angular distributions differ significantly and the electromagnetic dissociation dominates for the lead, although the nuclear contribution is not negligible.

PACS. 25.40.Lw Radiative capture – 25.70.De Coulomb excitation

1 Introduction

Many astrophysical nuclear processes, such as the r-process, involve neutron capture by short-lived nuclei. Direct measurement of these reactions are not available because one cannot make a target of these nuclei. However, there are other methods of obtaining information of the cross sections. One example is the asymptotic normalization coefficient (ANC) method [1]. Another approach is Coulomb breakup where the inverse reaction is investigated. This inverse method has been applied already in several experiments and generally good agreements with theoretical estimates are found for dissociation cross sections [2].

However, there have not been experimental investigations of the validity of the assumptions associated with extracting information from the inverse reaction. For example, it is known that the excited states of the resulting nucleus following neutron capture plays an important role [3]. The radioactive beam contains only ground state nuclei. Therefore, when the inverse method is applied, the contribution of capture to an excited state should be taken into account theoretically.

Another concern relates to the calculation of the neutron capture cross section from the experimental results of the dissociation. The neutron breakup can occur from either nuclear or electromagnetic interaction. The contribution of the nuclear breakup has to be measured and removed from the total yield. The proper way to do this needs verification.

This paper will present first experimental results to check the validity of the inverse method. The reaction investigated was the neutron capture of ^{7}Li. This nucleus

[a] e-mail: akos@ludens.elte.hu

is stable, and the neutron capture has been measured directly in several experiments [3,4] including bombardment by thermal and fast neutrons. These results showed that the capture obey a $1/v$ law below 254 keV where a resonance peak is located.

As the (n,γ) reaction for ^{7}Li is well known, the inverse process can be used to provide better understanding of the Coulomb dissociation process. This is the first detailed study of a Coulomb dissociation reaction (energy dependence and angular distributions) where the associated, well-established (n,γ) reaction is known.

2 The method of Coulomb dissociation and its experimental difficulties

The dissociation of nuclei passing through the strong electric field of a heavy nucleus can be described as an absorption of a virtual photon (γ) from the target field, followed by emission of a neutron [5]. The (γ,n) reaction can be then related to the inverse (n,γ) reaction of astrophysical interest via detailed balance. The (γ,n) cross section is much larger than the corresponding (n,γ) cross section at astrophysical energies because of the difference in wave numbers between γ and neutron. As the electromagnetic interaction is well known, this simple, powerful model is often used.

When applying the method, one will face several questions that can result in systematic errors. One must consider that the capture by a ground state nucleus can lead to a ground state or excited states [3]. Another question concerns the excited states of the capturing nuclei. This will give difficulties if it occurs in the dissociation process, since the dissociation into the ground state should be measured. But for those captures when the contributions from excited states are small, they can be theoretically corrected. A detailed experimental investigation can show the importance of this circumstance and would provide a way to correct for it.

The most important systematic ambiguity is the nuclear contribution in the breakup. Since the inverse of the neutron capture is only electromagnetic dissociation, the contribution of the nuclear breakup should be well understood. The general way to estimate and subtract the contribution relies on an empirical method. The most used depends on $A^{1/3}$ systematics. In this model, one can assume that the neutron removal can occur only at the circumference of the target viewed from the beam direction. The angular dependence of the two types of dissociation is well known. The nuclear dissociation can occur at larger scattering angles compared to the Coulomb dissociation that is limited to angles below the grazing angle. Therefore, the angular distribution of the breakup on several targets is one way to determine the correction factors for a particular target for the electromagnetic dissociation [6].

Another aspect is the calculation of the virtual photon numbers. This is theoretically well established but depends on the closest approach of the colliding nuclei. With the precise measurement of the scattering angle, it is possible to select different impact parameters of the reaction.

As the model of Coulomb dissociation is based on first-order perturbation theory, some questions still arise due to higher-order effects (*i.e.*, dynamical effects and relativistic effects) [7]. The neutron capture generally is dominated by an $E1$ transition, but in the breakup there are a number of $E2$ virtual photons available. Experimental results of longitudinal momentum distribution of the fragment showed that the $E1$-$E2$ process can interfere [8].

These questions have never been verified experimentally and left room for systematical errors. Although these points were discussed many times for the ^{8}B breakup reaction, which is a very important reaction in the solar neutrino problem, and several efforts have been made to clarify nuclear or $E2$ contributions [9], no attempt have been made to study Coulomb dissociation reaction systematics in detail.

3 The experimental procedure

The experiment was performed at the Coupled Cyclotron Facility of the National Superconducting Cyclotron Laboratory at Michigan State University. A ^{8}Li beam of 41.2 MeV/nucleon bombarded a lead (carbon) target of 56.7 (28.8) mg/cm^2 thickness, respectively. The experimental setup is shown in fig. 1. The direction of the incoming ^{8}Li was measured with a pair of Cathode Readout Drift Chamber (CRDC) detectors [10]. A thin plastic scintillator placed in front of the target provided the start pulse for the time of flight (ToF). The neutrons were detected with the Modular Neutron Array (MoNA) which consists of 9 layers of 16 scintillator bars, 10 cm $\times$ 10 cm $\times$ 2 m each [11,12]. The charged particles were deflected by the Sweeper Magnet [13] and their positions were measured after the magnet with two CRDCs separated by 1 m. Thin and thick plastic scintillators were used to identify the particles by their energy loss. The thin plastic detector also provides trigger signal for the measurement. Measurements without any target were performed to deduce contributions from reactions in the start detector.

In the Coulomb dissociation reaction, the relative energy between ^{7}Li and n is determined by their relative velocities event by event. The neutron velocity vector is deduced from the position of the scintillator bar hit in MoNA, and the ToF between the target and the particular bar. Horizontal position at MoNA is calculated from the time difference between the two photomultipliers attached at the left and right ends of each bar. The relation between time difference and position was carefully examined with cosmic-ray events. The vertical and longitudinal position of the neutron was determined from the position of the bar. The ToF was calibrated with prompt γ-rays from a thick target. The velocity was deduced from ToF between the pulse of the thin plastic scintillator and the mean time of the two photomultipliers.

To deduce the velocity of the ^{7}Li particle, more elaborate efforts were necessary. As the primary purpose of the Sweeper Magnet was to sweep the charged particles off and pass the neutrons through its large vertical gap, the magnetic field is not uniform over the whole region. To

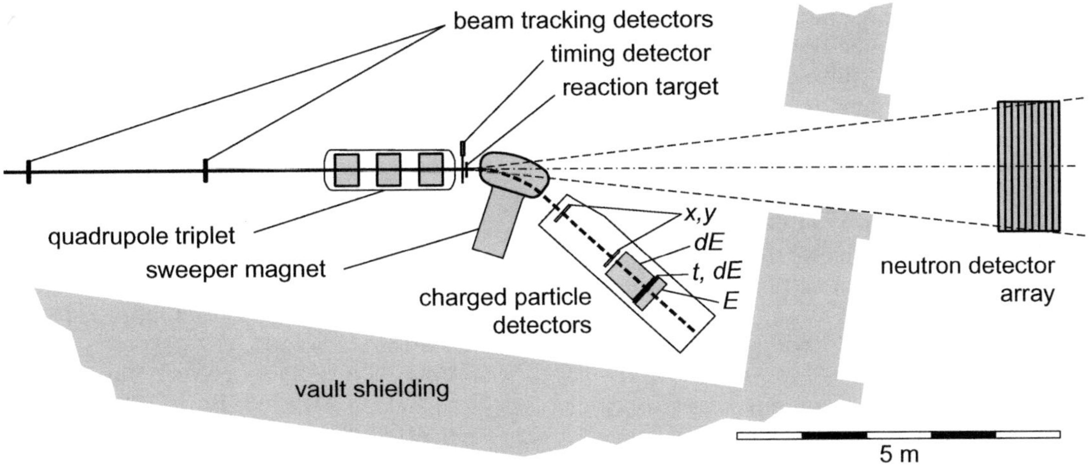

Fig. 1. Experimental setup; "x, y": CRDC detectors, "t, dE": thin plastic detector, "E": thick plastic detector.

understand the characteristics of the charged particle trajectory in the Sweeper Magnet, the program "*COSY infinity*" [14] was employed. From the measured magnetic field, COSY can produce a forward map, which relates position, direction and the rigidity of the particle at the target to the position and direction after the magnet. Because of the nonuniformity of the field, there are significant deviations between the calculated trajectory with the forward map and the actual trajectory for the particles with large angle or with large momentum difference relative to the particle at the central trajectory. This problem can be solved by using multiple maps for various incident parameters. However, to trace back the particle from the detector after the magnet to the target position, this procedure did not allow COSY to create an inverse map. We employed the novel technique of a neural network [15]. As the position data of the particle at the target position is also known from the beam tracking CRDCs event by event, we have enough information to deduce the rigidity and direction of the particle at the target. A Monte Carlo simulation using the forward map provides a set of events, which is used to train the neural network. Obtained parameters were further tested with another set of simulated events. These steps were repeated until a satisfactory conversion was obtained.

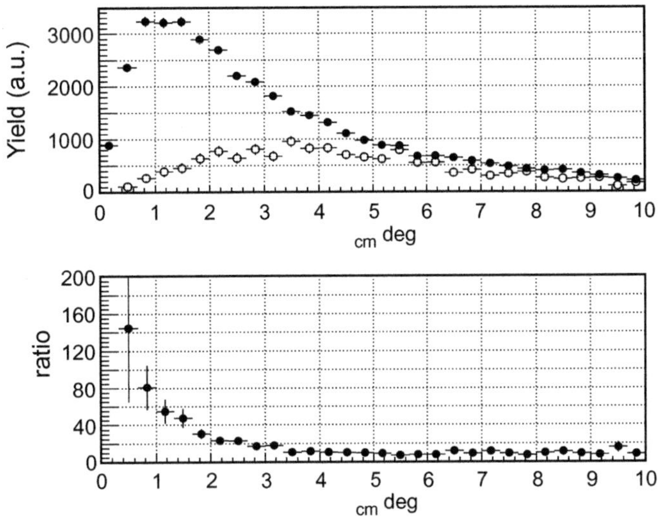

Fig. 2. Angular distribution of n-^{7}Li coincidence at 41 MeV/A. Upper panel: solid symbols are for the lead target and the open circles are for the carbon target. Carbon data is multiplied by factor 6.7 (see text). Lower panel: Pb/C ratio of the angular distributions.

4 Results and discussion

Using this technique, the momentum and the velocity vector of the ^{7}Li particles were successfully deduced. The results for angular distributions are displayed in fig. 2. The upper panel shows the net yields for the lead target and the carbon target, respectively, normalized by the number of incoming particles and the target thickness. The contribution from reactions in the thin scintillator in front of the target was subtracted using "blank target" runs. The carbon yield is multiplied by a factor of 6.7 for comparison purpose. The angle of the center of mass velocity of the n-^{7}Li system relative to the initial ^{8}Li particle is displayed in the center of mass system of ^{8}Li+target. Although the angular dependent acceptance effect is not corrected yet, a clear difference of the angular distributions for the two targets can be seen. For the lead target, there is a clear enhancement at forward angles less than 6 degrees. This shows the dominance of the Coulomb dissociation process at forward angles. Small yields below 1 degree may correspond to the cutoff of virtual photons *i.e.*, the number of virtual photons whose energy is high enough to dissociate ^{8}Li (2.033 MeV) decreases significantly at the large impact parameter. The carbon target data, on the other hand, shows a rather flat distribution with a maximum at around 3–4 degree. Above 7 degree, which is the grazing angle of the ^{8}Li+Pb system at 41 MeV/A, the two distributions have the same shape, indicating the nuclear dissociation dominates in these angular regions for both targets. Using the factor 6.7 mentioned above, the magnitude of the two angular distributions matches above 6 degrees.

The lower panel of the figure shows the ratio between two distributions. This ratio may give a better insight of the target dependence, as the finite acceptance effect is cancelled out in part. For large angles, the ratio is almost a constant. This constant 6.7 corresponds to $A^{2/3}$ scaling of lead over carbon. With the often used $A^{1/3}$ factor, the two distributions differ in magnitude. As the angle decreases, the ratio gradually becomes larger. At very forward angle, this ratio exceeds 140 which is close to the Z^2 ratio of the targets. Further analysis, such as angular correlation of the breakup particles (n and ^{7}Li) and incident energy dependence, which is now underway, will give more insight to $E2$ contribution.

In summary, we showed the angular distributions of the Coulomb dissociation for C and Pb target at 41 MeV/A ^{8}Li beam energy. Above the grazing angle of the reaction, the nuclear dissociation becomes dominant. The Coulomb dissociation process dominates over nuclear dissociation at forward angles, although nuclear contribution is not negligible below 4 degrees. The present data suggests that a more careful investigation of the angle and energy dependence of the Coulomb dissociation reaction method is clearly necessary to deduce the inverse (n, γ) reaction cross section, which is important for the astrophysical nuclear process.

The authors would like to thank to Karlheinz Langanke for emphasizing the importance of the question. Support of the National Science Foundation under grant No. PHY01-10253, PHY-0354920 and of the OTKA under grant Nos. T42733, T38404, T043585, T049837 are gratefully acknowledged.

References

1. A.M. Mukhamedzhanov *et al.*, Nucl. Phys. A **725**, 279 (2003).
2. G. Baur *et al.*, Prog. Part. Nucl. Phys. **51**, 487 (2003).
3. Y. Nagai *et al.*, Phys. Rev. C **71**, 055803 (2005).
4. Y. Nagai *et al.*, Astrophys. J. **381**, 444 (1991); M. Heil *et al.*, Astrophys. J. **507**, 997 (1998).
5. G. Baur, C.A. Bertulani, Nucl. Phys. A **458**, 188 (1986).
6. N. Fukuda *et al.*, Phys. Rev. C **70**, 054606 (2004).
7. S. Typel *et al.*, Nucl. Phys. A **613**, 147 (1997).
8. B. Davids *et al.*, Phys. Rev. Lett. **81**, 209 (1998).
9. T. Motobayashi *et al.*, Phys. Rev. Lett. **73**, 2680 (1994); F. Schümann *et al.*, Phys. Rev. Lett. **90**, 232501 (2003).
10. J. Yurkon *et al.*, Nucl. Instrum. Methods A **422**, 291 (1999).
11. T. Baumann *et al.*, Nucl. Instrum. Methods A **543**, 517 (2005).
12. B. Luther *et al.*, Nucl. Instrum. Methods A **505**, 33 (2003).
13. M.D. Bird *et al.*, IEEE Trans. Appl. Supercond. **15**, 1252 (2005).
14. K. Makino, M. Berz, Nucl. Instrum. Methods A **427**, 338 (1999).
15. *E.g.*, R. Brun *et al.*, *ROOT Users Guide version 4*, Chapt. 5 (2005).

Eur. Phys. J. A **27**, s01, 221–225 (2006)
DOI: 10.1140/epja/i2006-08-034-5

EPJ A direct
electronic only

Study of the ^{9}Be(p, α)^{6}Li reaction via the Trojan Horse Method

S. Romano[1,2,a], L. Lamia[1,2], C. Spitaleri[1,2], C. Li[1], S. Cherubini[1,2], M. Gulino[1,2], M. La Cognata[1,2], R.G. Pizzone[1], and A. Tumino[1,2]

[1] Laboratori Nazionali del Sud - INFN, Catania, Italy
[2] Dipartimento di Metodologie Fisiche e Chimiche per l'Ingegneria, Università di Catania, Catania, Italy

Received: 1 July 2005 /
Published online: 13 March 2006 –

Abstract. The Trojan Horse Method has been applied to the ^{2}H(^{9}Be, ^{6}Liα)n three-body reaction in order to investigate the ^{9}Be(p, α)^{6}Li two-body reaction, which is involved in the study of light element abundances (lithium, beryllium and boron). A coincidence measurement was performed in order to identify the presence of the quasi-free mechanism in the three-body reaction, needed for the application of the method. The astrophysical $S(E)$-factor was extracted and compared to direct data. No information about electron screening effects can be extracted due to the poor resolution of the indirect data.

PACS. 24.10.-i Nuclear reaction models and methods – 26.20.+f Hydrostatic stellar nucleosynthesis

1 Introduction

In recent years the abundances of light elements lithium, beryllium and boron (LiBeB) have been increasingly used as diagnostics between different scenario for primordial or stellar nucleosynthesis. As reported in [1], beryllium primordial abundances can provide a powerful test to discriminate between homogeneous and inhomogeneous primordial nucleosynthesis. Moreover, the study of beryllium abundances in young stars, together with lithium and boron, can provide a strong test for understanding stellar structure and discriminate between possible non-standard mixing processes in stellar interiors [2]. In both stellar and primordial environments, however, LiBeB are mainly destroyed by proton-capture reactions via the (p, α) channel with a Gamow energy E_G ranging from 10 keV (for stellar nucleosynthesis) to 100 keV (for primordial nucleosynthesis). These energies are low if compared with the Coulomb barrier E_C usually of the order of MeV, thus implying the reactions take place via tunnel effect with an exponential decrease of the cross section to nano or pico barn values. The behavior of the direct cross sections are usually extrapolated at the astrophysical interest region from higher energies by using the definition of the astrophysical factor

$$S(E) = E \cdot \sigma(E) \cdot \exp(2\pi\eta) \tag{1}$$

(where η is the Sommerfeld parameter) which varies smoothly with energy. Nevertheless this extrapolation procedure can introduce some uncertainties due, for example, to the presence of unexpected subthreshold resonances or electron screening effects [3].

However, in recent years, many indirect methods [4, 5,6,7,8,9,10] have been developed in order to extract the $S(E)$-factor without extrapolations. In particular the Trojan Horse Method (THM) [8,9,10] represents a powerful tool which select the quasi-free (QF) contribution of a suitable three-body reaction under appropriate kinematical conditions. The energy in the entrance channel of the three-body reaction is chosen well above the Coulomb barrier to extract the two-body cross section at astrophysical energies free of Coulomb suppression. In the present paper we present the results of the study of the two-body ^{9}Be(p, α)^{6}Li reaction, obtained through the application of THM to the three-body ^{2}H(^{9}Be, ^{6}Liα)n reaction.

2 The Trojan Horse Method

The basic idea of the THM [8,9,10] is to extract a two-body $a + b \rightarrow c + d$ reaction cross section from the QF contribution of a suitable three-body $a + x \rightarrow c + d + s$ reaction. Here the x nucleus shows a strong $b \oplus s$ cluster structure and, in the Impulse Approximation (IA) description, only b interacts with a, whereas s is considered to be spectator to the virtual two-body reaction.

The Plane Wave Impulse Approximation (PWIA) leads to a factorization of the three-body reaction cross section:

$$\frac{\mathrm{d}^3\sigma}{\mathrm{d}E_c \mathrm{d}\Omega_c \mathrm{d}\Omega_d} \propto KF \left(\frac{\mathrm{d}\sigma}{\mathrm{d}\Omega}\right)^{off} \cdot |G(\boldsymbol{p}_s)|^2, \tag{2}$$

[a] e-mail: romano@lns.infn.it

where KF is a kinematical factor, $(d\sigma/d\Omega)^{off}$ is the off-energy-shell differential cross section for the two-body $a(b,c)d$ reaction and $|G(\boldsymbol{p}_s)|^2$ is the s momentum distribution in x. Under these assumptions, if $|G(\boldsymbol{p}_s)|^2$ is known (KF is calculated), a quantity which is proportional to the two-body cross section can be extracted from a measurement of the three-body $d^3\sigma/dE_c d\Omega_c d\Omega_d$ cross section.

The above-mentioned assumptions and the validity tests carried out from the data analysis are fully discussed in [10] and references therein.

Moreover in the THM approach [9,10] the initial projectile velocity is compensated for by the binding energy of particle b inside x. Thus the two-body reaction can be induced at very low relative energies. If the incoming energy E_a is chosen high enough to overcome the Coulomb barrier in the entrance channel of the three-body reaction, both Coulomb barrier and electron screening effects are negligible in the two-body THM data. The cluster b is brought into the nuclear interaction region and the x nucleus is considered as a Trojan Horse.

We stress that in view of the various approximations involved in the THM and in particular of the assumption that off-energy-shell effects are negligible, one cannot extract the absolute value of the two-body cross section. However, the absolute value can be obtained through normalization to the direct data available at energies above the Coulomb barrier.

Moreover, as already mentioned, the THM data are not affected by electron screening. Therefore, once the behavior of the absolute bare $S_b(E)$-factor from the two-body cross-section is extracted, a model-independent estimate of the screening potential U_e can be obtained from comparison with the direct screened $S(E)$-factor.

The aim of the present experiment was to extract the cross section of the ^{9}Be+p → ^{6}Li+α reaction after selecting the QF contribution of ^{9}Be+d → ^{6}Li+α+n reaction. The deuteron was used like trojan horse nucleus, due to its p⊕n structure [11]; in this framework the proton acts like participant while the neutron is the spectator to the virtual two-body reaction.

3 Experimental procedure

The experiment was performed at the Laboratori Nazionali del Sud in Catania. The SMP Tandem Van de Graaf accelerator provided a 22 MeV ^{9}Be beam which was accurately collimated in order to have a spot diameter of about 2 mm and intensities up to 2–5 pnA. A deuterated polyethylene target (CD_2) of about 190 μg/cm^2 was placed at 90° with respect to the beam direction. A silicon ΔE-E telescope was placed at about 70 cm from the target with an angle of about 45° for a continuous monitoring of the target thickness during the experiment. The particle detection was performed by using two silicon ΔE-E telescopes, with a position-sensitive detector (PSD) as E. The telescopes were placed at opposite sides with respect to the beam direction at a distance from the target $d = 25$ cm. The detection angular ranges were 11.5°–25.5° (telescope devoted to ^{6}Li detection) and 14.5°–28.5° (telescope for ^{4}He detection). The displacement of the experimental setup was chosen by means of a Monte Carlo simulation in order to cover the whole QF angular range. The trigger for the event acquisition was given by the coincidence of two particles hitting the two PSDs respectively. Energy and position signals for the detected particles were processed by standard electronics together with the coincidence relative time and sent to the acquisition system for on-line monitoring of the experiment and data storage for the off-line analysis.

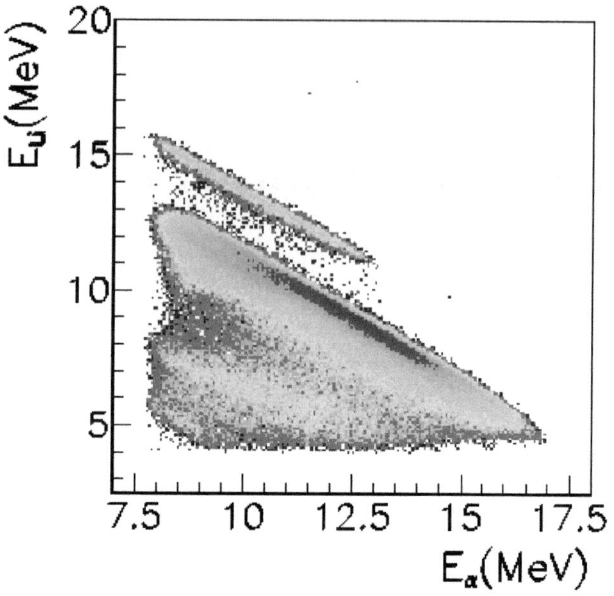

Fig. 1. Experimental kinematic locus E_{Li} *vs.* E_α for the coincidence events.

In order to perform position calibration, grids with a number of equally spaced slits were placed in front of each PSD. A correspondence between position signal from the PSDs and detection angle of the particle was then established. Energy calibration was performed by means of a standard three-peak α source and α and ^{6}Li particles from ^{12}C(^{6}Li, α)^{14}N and ^{12}C(^{6}Li, ^{6}Li)^{12}C reactions.

4 Data analysis

4.1 Three-body reaction identification

After the position and energy calibration, Li and α particles detected in coincidence were selected with the standard ΔE-E technique. The kinematic locus (E_{Li} *vs.* E_α) (fig. 1) was reconstructed in very good agreement with the simulation. Moreover the experimental Q-value spectrum for the three-body reaction was reconstructed under the assumption of mass number 1 for the undetected third particle. The result is shown in fig. 2, where it is evident a peak centered at about −0.1 MeV according to the expected theoretical value. The results of fig. 1 and fig. 2 make us confident on the identification of the three-body reaction exit channel. Events below the peak in the

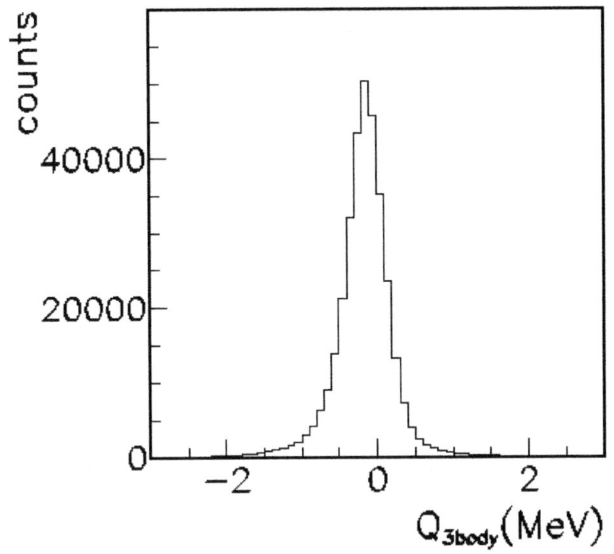

Fig. 2. Experimental Q-value spectrum for the three-body reaction ^{9}Be + d → ^{6}Li + α + n, with cuts in the kinematical locus in fig. 1.

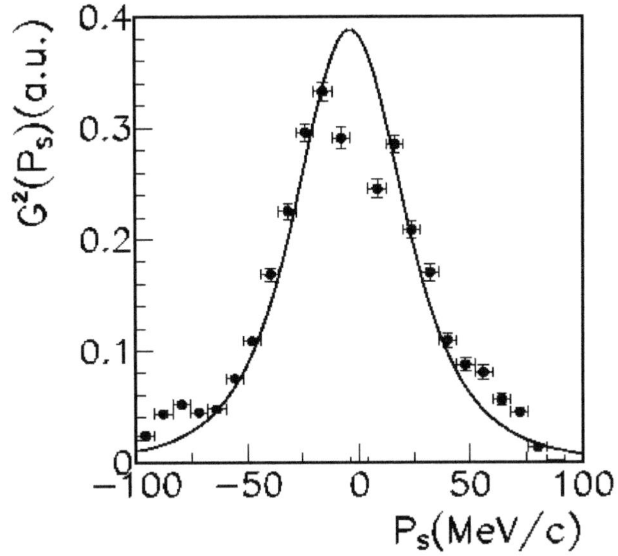

Fig. 3. Comparison between experimental (dots) and theoretical Húlthen function (solid line) for the neutron momentum distribution. Error bars are due to the statistical error.

Q-value spectrum (fig. 2) were selected for the further analysis.

4.2 QF mechanism identification: neutron momentum distribution

According to the theory of the THM [9,10] the energy of ^{9}Be was chosen to overcome the Coulomb barrier in the entrance channel of the three-body reaction. This means that both Coulomb and electron screening effects are negligible in the two-body reaction data. Thus, the term $(\mathrm{d}\sigma/\mathrm{d}\Omega)^{off}$ in eq. (2) represents the nuclear part of the differential cross section for the virtual two-body reaction ^{9}Be(p, α)^{6}Li that in post collision prescription occurs at an energy

$$E_{cm} = E_{^6\mathrm{Li}\text{-}\alpha} - Q_{2b}, \tag{3}$$

where $E_{^6\mathrm{Li}\text{-}\alpha}$ is the ^{6}Li-α relative energy and Q_{2b} is the two-body Q-value.

In order to reconstruct the neutron momentum distribution $|G(\boldsymbol{p}_s)|^2$, a small ^{6}Li-α relative energy window (about 100 keV) was selected. In such a small energy windows the $(\mathrm{d}\sigma/\mathrm{d}\Omega)^{off}$ can be considered constant. Thus the experimental $|G(\boldsymbol{p}_s)|^2$ distribution was extracted by dividing the three-body coincidence yield by the kinematic factor. The result is compared with the theoretical one in fig. 3. The agreement between experimental and theoretical momentum distribution represents a very strong check for the existence of the QF mechanism in the present data.

4.3 Validity tests for the THM and the astrophysical S(E)-factor

After the identification of the QF mechanism only events with spectator momentum $|P_s| < 30$ MeV/c were considered.

A first test of validity of the THM approach is represented by the comparison between the indirectly extracted angular distributions and the direct behavior. The relevant angle in order to get the indirect angular distributions, *i.e.* the emission angle for the alpha-particle in the ^{6}Li-α center of mass system, can be calculated according to the relation [12]

$$\theta_{cm} = \arccos \frac{(\boldsymbol{v}_p - \boldsymbol{v}_t) \cdot (\boldsymbol{v}_C - \boldsymbol{v}_\alpha)}{|\boldsymbol{v}_p - \boldsymbol{v}_t||\boldsymbol{v}_C - \boldsymbol{v}_\alpha|}, \tag{4}$$

where the vectors $\boldsymbol{v}_p, \boldsymbol{v}_t, \boldsymbol{v}_C, \boldsymbol{v}_\alpha$ are the velocities of projectile, transferred proton, and the outgoing ^{6}Li and α-particles, respectively. These quantities can be calculated from their corresponding momenta in the laboratory system, where the momentum of the transferred particle is equal and opposite to that of neutron spectator, due to the quasi-free assumption [12]. The angular distributions test was performed for different ^{6}Li-α relative energy intervals and normalized to the direct data [13,14]. An example of the result is shown in fig. 4. The error bars include both statistical and normalization errors. The two-body cross section is in arbitrary units and the solid lines show the behavior of direct angular distribution [13,14]. The quite fair agreement between the two trends makes us confident on the validity of the IA.

A second validity test consists in the comparison between the behavior of the indirect excitation function with the direct one. Therefore, by using the eq. (2), the quantity $(\mathrm{d}\sigma/\mathrm{d}\Omega)^{off}$ has been extracted. It has to be emphasized that in the present case the obtained cross section is the nuclear part, the Coulomb barrier being already overcome in entrance channel. In order to do the comparison the indirect two-body cross section was multiplied by Coulomb penetration function, given in terms of regular and irregular Coulomb functions (see [9,10,15] and references

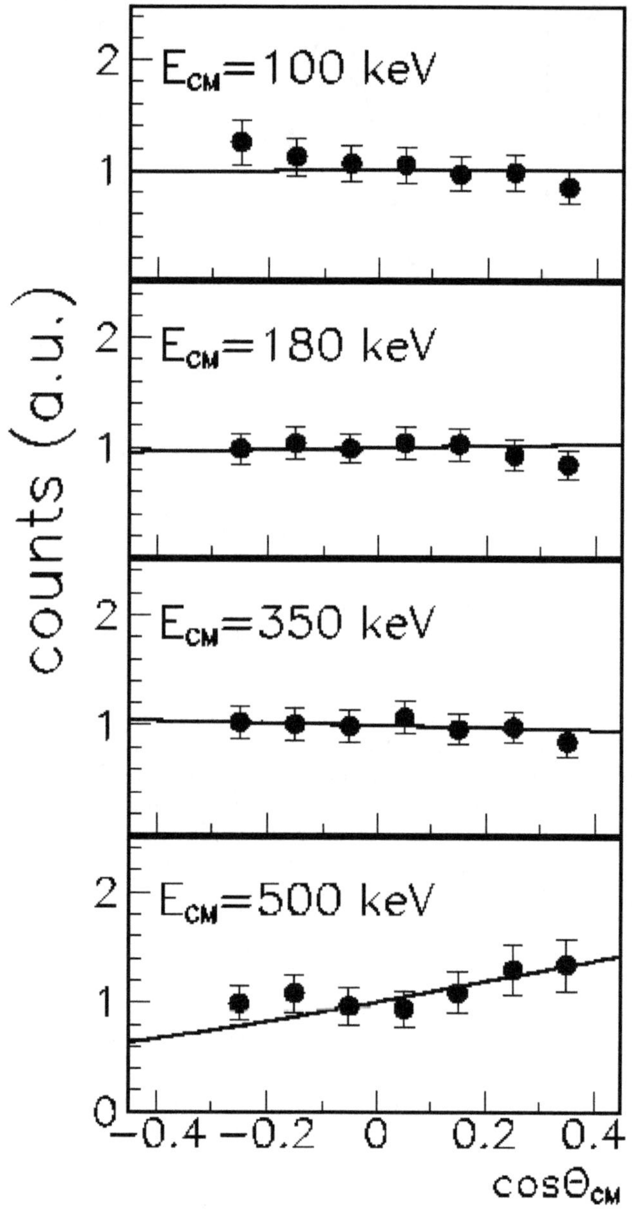

Fig. 4. Example of angular distribution extracted at different E_{cm} via the THM (dots) compared to the direct data (solid lines) [13,14].

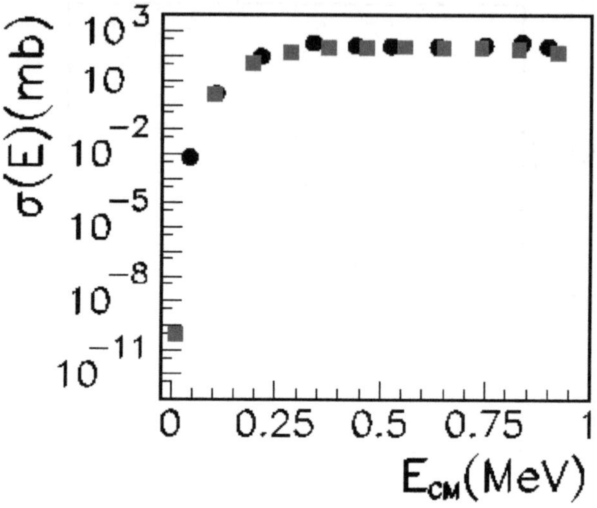

Fig. 5. Comparison between the THM indirect excitation function (square symbols) for the ^{9}Be(p, α)^{6}Li reaction and the direct behavior (dots) [13].

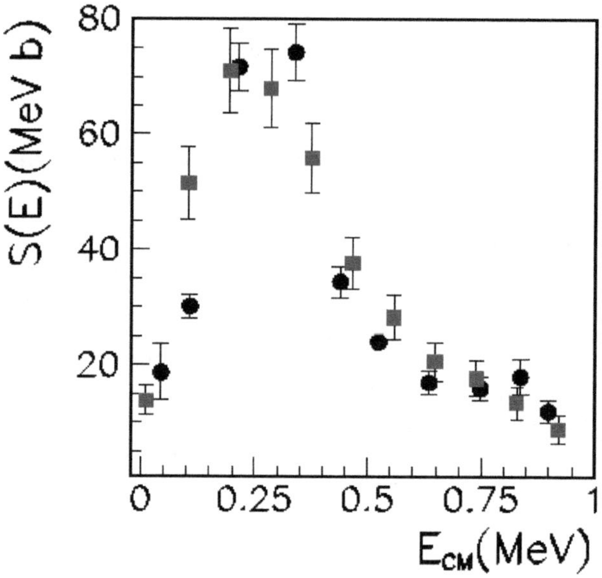

Fig. 6. The bare-nucleus astrophysical $S(E)$-factor extracted via the THM (square symbols) compared with the direct one (dots) [13].

therein). The resulting two-body cross-section $\sigma(E)$ is shown in fig. 5 (square symbols) where direct data [13] are also reported (dots). The normalization to direct behavior was performed in the region around $E_{cm} = 700$ keV. The good agreement between the two data sets is a necessary condition for the further extraction of the astrophysical $S(E)$-factor by means of THM.

At the end of this second test on the data we can conclude that the PWIA analysis is able to correctly describe the studied process.

According to eq. (1) the bare-nucleus $S_b(E)$-factor was extracted. The result is shown in fig. 6 and is compared with direct data. Both sets of data were averaged out at the same energy bin of 90 keV.

5 Conclusion

The indirect study of the ^{9}Be(p, α)^{6}Li reaction was performed in the astrophysical energy region by applying the THM to the ^{2}H(^{9}Be, ^{6}Liα)n three-body break-up process. The results obtained represent an additional validity test of the method at sub-Coulomb energies. In particular the behavior of the indirect $S(E)$-factor (fig. 6) show the presence of the expected low-energy resonance at $E_{cm} \sim 0.27\,\text{MeV}$, corresponding to the $6.87\,\text{MeV}$ $J = 1^-$ level of ^{10}B. It should be noticed that the resonance width in the indirect data is larger than in direct ones. This can be connected with the energy resolution (around 90 keV) of the present experiment which is poorer with respect to the direct one. This means that presently it is not possible to extract any information about screening effects. An upgraded experimental setup might improve the present results and give useful information for astrophysical applications.

References

1. R.N. Boyd, T. Kajino, Astrophys. J. **336**, L55 (1989).
2. A. Stephens *et al.*, Astrophys. J. **491**, 339 (1997).
3. C. Rolfs, W.S. Rodney, *Cauldrons in the Cosmos* (University of Chicago Press, Chicago, 1988).
4. G. Baur, H. Rebel, J. Phys. G **20**, 1 (1994) and references therein.
5. G. Baur, H. Rebel, Annu. Rev. Nucl. Part. Sci. **46**, 321 (1996).
6. A.M. Mukhamedzhanov *et al.*, Phys. Rev. C **56**, 1302 (1997).
7. A.M. Mukhamedzhanov *et al.*, Phys. Rev. C **63**, 024612 (2001).
8. G. Baur, Phys. Lett. B **178**, 135 (1986).
9. C. Spitaleri *et al.*, Phys. Rev. C **60**, 055802 (1999).
10. C. Spitaleri *et al.*, Phys. Rev. C **69**, 055806 (2004).
11. M. Zadro *et al.*, Phys. Rev. C **40**, 181 (1989).
12. M. Jain *et al.*, Nucl. Phys. A **153**, 49 (1970).
13. A.J. Sierk, T.A. Tombrello, Nucl. Phys. A **210**, 341 (1973).
14. D. Zahnow *et al.*, Z. Phys. A **359**, 211 (1997).
15. S. Typel, H. Wolter, Few-Body Syst. **29**, 7 (2000).

Eur. Phys. J. A **27**, s01, 227–232 (2006)
DOI: 10.1140/epja/i2006-08-035-4

EPJ A direct
electronic only

Re-evaluation of the low-energy Coulomb-dissociation cross section of ^{8}B and the astrophysical S_{17} factor

K. Sümmerer[a]

For the S223 Collaboration[b]
GSI Darmstadt, Germany

Received: 22 July 2005 /
Published online: 10 March 2006 –

Abstract. An exclusive measurement of the Coulomb breakup of ^{8}B into ^{7}Be+p at 254 A MeV was used to infer the low-energy ^{7}Be(p, γ)^{8}B cross section. Particular emphasis was placed on the angular correlations of the breakup particles which demonstrate clearly that $E1$ multipolarity dominates within the angular cuts selected for the analysis. The deduced astrophysical S_{17} factors exhibit good agreement with the most recent direct ^{7}Be(p, γ)^{8}B measurements.

PACS. 25.40.Lw Radiative capture – 25.60.-t Reactions induced by unstable nuclei – 25.70.De Coulomb excitation – 26.65.+t Solar neutrinos

1 Introduction

The flux measured in neutral-current interactions of high-energy solar neutrinos by the Sudbury Neutrino Observatory (SNO) [1,2] is in general agreement with the flux predicted by the standard solar model (SSM, refs. [3,4]). It is essential, however, to further reduce the uncertainty of nuclear inputs to the SSM in order to refine its predictions. One critical quantity is the ^{7}Be(p, γ)^{8}B cross section at solar energies since it is linearly related with the high-energy solar neutrino flux stemming from ^{8}B β-decay.

In recent years, many attempts have been undertaken to measure this cross section (or, equivalent, the astrophysical S_{17} factor) with high-precision in direct-proton-capture measurements using radioactive ^{7}Be targets [5,6,7,8]. A completely different approach with different systematic errors is Coulomb dissociation (CD) of ^{8}B in the electromagnetic field of a high-Z nucleus. Such measurements have been performed at intermediate [9,10] and high energies [11]. This contribution reports on a CD experiment similar to that of ref. [11], but with an improved experimental technique. Preliminary results of this study have been published earlier [12]. In the present contribution we present a re-evaluation of the published data and show that the efficiency to detect low-energy break-up events was slightly overestimated. As a consequence, the lowest data points are increased by about 6–10% which suggests a different theoretical model to extrapolate to zero energy than used in ref. [12]. We have recently published a full account of the present work [13].

2 Theoretical calculations

Realistic theoretical calculations of the CD of ^{8}B are essential for several reasons. From a practical point of view, the relatively bad energy resolution of the CD method requires to simulate, *e.g.*, the effect of cross talk between neighboring energy bins, of the finite size and resolution of the tracking detectors etc. These simulations require a CD event generator that is reasonably close to reality so that the remaining differences between the measured and simulated cross-section distributions can be attributed to the S_{17} factor. As input to the event generator we have to specify a nuclear model for ^{8}B and choose a method to calculate Coulomb dissociation.

The simplest model for ^{8}B is that of a p-wave proton coupled to an inert ^{7}Be core with $I^\pi = 3/2^-$ to form the ^{8}B ground state with $I^\pi = 2^+$. With this model we obtain astrophysical S-factors as a function of the proton-^{7}Be relative energy, E_{rel}, as shown in fig. 1. The non-resonant

[a] e-mail: `k.suemmerer@gsi.de`

[b] Members of the S223 Collaboration are F. Schümann and F. Strieder (Universität Bochum); S. Typel, K. Sümmerer, F. Uhlig, H. Geissel, P. Koczon, N. Kurz, E. Schwab, P. Senger, and Zhi-Yu Sun (GSI Darmstadt); F. Hammache (IPN Orsay); D. Cortina (Universidad Santiago de Compostela); A. Förster, H. Oeschler, and C. Sturm (TU Darmstadt); I. Boettcher, B. Kohlmeyer and M. Menzel (Universität Marburg); M. Gai (University of Connecticut); U. Greife (Colorado School of Mines); N. Iwasa (Tohoku University); R. Kulessa, M. Ploskon, W. Prokopowicz, G. Surowka, and W. Walus (Krakow University); H. Kumagai, T. Motobayashi, and A. Ozawa (RIKEN); A. Wagner and E. Grosse (FZ Rossendorf).

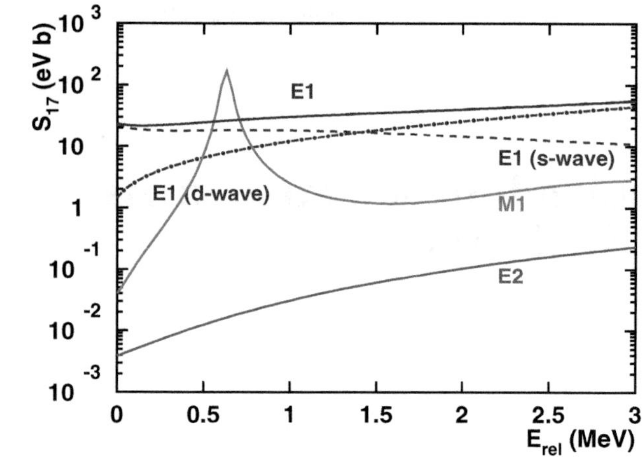

Fig. 1. Theoretical S_{17} factors from a simple potential model of ^{8}B and their decomposition into contributions from various partial waves.

direct capture into the ^{8}B ground state proceeds mainly via *s*- and *d*-wave captures and $E1$ γ-emission. Capture of *p*- and *f*-waves followed by $E2$ emission plays an insignificant role, in particular at solar energies. The resonant component proceeds through the 1^+ resonance at 633 keV above threshold which decays mainly by $M1$ emission and is limited essentially to a narrow region around the resonance energy.

Coulomb dissociation of ^{8}B on ^{208}Pb at 254 A MeV is calculated in the semi-classical model in first-order perturbation theory (PT), as described in more detail elsewhere [14,15,16]. This approach should be a good approximation at the high bombarding energy used. Due to the large number of $E2$ photons present in the virtual photon spectrum seen by the ^{8}B projectile, one can assume the $E2$ component to be significantly enhanced compared to the direct-capture case. This will be investigated experimentally in our experiment.

3 Experimental procedures

The ^{8}B secondary beam was produced at the SIS/FRS radioactive beam facility at GSI [17] by fragmenting a 350 A MeV ^{12}C beam in a 8 g/cm^2 Be target and separating it from contaminant ions in a 1.4 g/cm^2 wedge-shaped Al degrader placed in the FRS intermediate focal plane. Typical ^{8}B beam intensities in front of KaoS were 5×10^4 ions per 4 s spill; the only contaminant consisted of about 20% ^{7}Be ions which could be identified event by event with the help of a time-of-flight measurement. For this purpose a 3 mm thick plastic scintillator detector was installed in the transfer line between FRS and KaoS, about 85 m upstream from the breakup target, to serve as a time-of-flight (ToF) start detector. Positions and angles of the secondary beam incident on the Pb breakup target were measured with the help of two parallel-plate avalanche counters (PPAC) located at 308.5 cm and 71 cm upstream from the target, respectively. The detectors had areas of $10 \times 10\,\mathrm{cm}^2$ and allowed to track the incident ^{8}B beam with about 99% efficiency and with position and angular resolutions of 1.3 mm and 1 mrad, respectively. In addition, they provided a ToF stop signal with a resolution of 1.2 ns (FWHM). The ^{8}B energy at the target was 254 A MeV and was limited by the maximum bending power of the KaoS spectrometer.

A schematic view of the experimental setup used in the present experiment to detect the breakup of ^{8}B in semi-complete kinematics at the KaoS spectrometer at GSI is shown in fig. 2. Apart from the PPAC tracking detectors mentioned above, it consisted of i) the ^{208}Pb break up target; ii) two pairs of Si strip detectors; iii) the magnets of the KaoS spectrometer; iv) two large-area multi-wire proportional chambers (MWPC); v) a ToF wall serving as a trigger detector.

Downstream from the Pb target (52 mg/cm^2 of ^{208}Pb), the angles and positions as well as the energy losses of the outgoing particles were measured with two pairs of single-sided Si strip detectors (SSD, 300 μm thick, 100 μm pitch) located at distances of about 14 cm and 31 cm downstream from the target.

The KaoS magnetic spectrometer [18] consisted of a large-aperture quadrupole and a horizontally focussing dipole magnet. To avoid multiple scattering of the fragments in air, the chamber inside the quadrupole and dipole magnets was filled with He gas at 1 bar pressure.

Behind the magnets, two large-area MWPC were installed as close to the focal plane as possible. One chamber, with horizontal and vertical dimensions of 60 cm and 40 cm, respectively, detected the position of protons behind KaoS. The other one, 120 cm wide and 60 cm high, was set to detect the ^{8}B non-interacting beam and the ^{7}Be fragments.

Behind the focal plane and parallel to it, a plastic-scintillator wall with 30 elements (each 7 cm wide and 2 cm thick) was installed and used for trigger purposes. The wall was subdivided into two sections. Coincident signals in the left-hand (proton) part and in the right-hand (ion) part of the wall indicated a break-up event ("breakup" trigger). Singles hits in the right hand section were interpreted as "beam" triggers and recorded with a down-scale factor of 1000.

4 Data reduction and results

The experiment described in the present paper recorded events from three different sources:
i) break-up events originating in the Pb target;
ii) down-scaled beam particles;
iii) background from a variety of sources (*e.g.*, cosmic rays).
Though event classes i) and ii) are mainly correlated with a corresponding trigger type ("breakup" trigger for class i), "beam" trigger for class ii)) we have checked if by chance the trigger types and event classes were mixed in rare cases, and have corrected for that.

The coincident p and ^{7}Be signals resulting from breakup in the ^{208}Pb target were identified among the

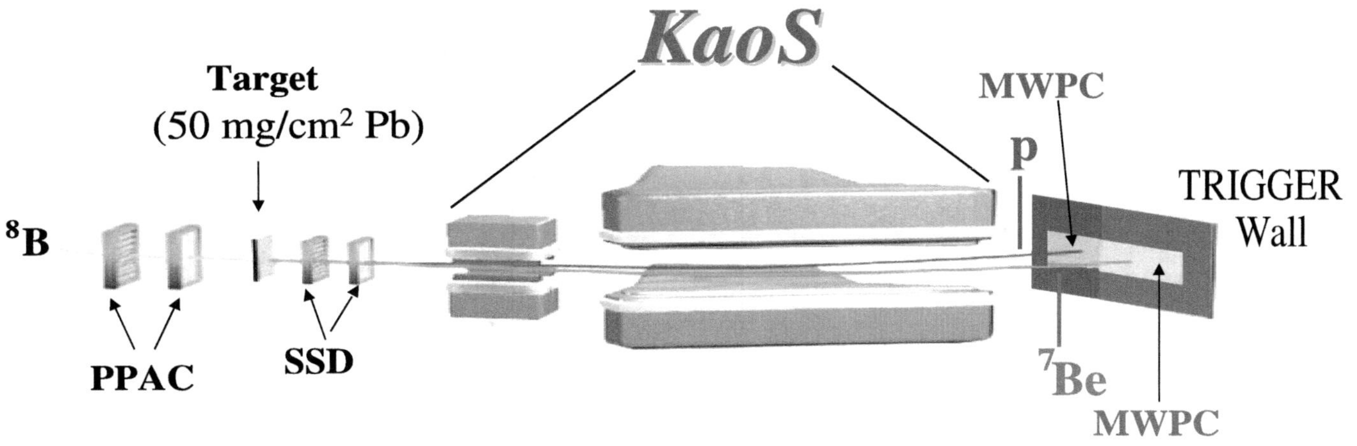

Fig. 2. Artist's view of the experimental setup. Shown schematically are the beam-tracking detectors (PPAC) in front of and the fragment-tracking Si strip detectors (SSD) behind the Coulomb-breakup target. Proton and ^{7}Be positions in the focal plane of the KaoS magnetic spectrometer are determined by large-area multi-wire chambers (MWPC) followed by a scintillator-paddle wall for trigger purposes.

class i) events ("breakup" trigger) in several successive steps:

1) The ΔE-ToF condition was applied to select only incident ^{8}B ions (see above).
2) A multiplicity of $m \geq 2$ in each SSD was required. This required that at least one empty strip was found between two respective hit clusters.
3) A 3σ-window around the ΔE peak corresponding to the energy loss of ^{7}Be in each SSD selected those events where ^{8}B was converted into ^{7}Be.
4) The coincident protons were found among all events with $\Delta E < 500$ keV in each SSD the trajectories of which converged towards the target. Moreover, their closest distance to the ^{7}Be trajectory was required to lie inside a volume given in x and y by the size of the target (± 18 mm in x- and ± 12 mm in y-direction) and having a z-value along the beam axis of ± 25 mm around the target (located at $z = 0$).

The inclusive ΔE spectra resulting from conditions 1 and 2 above are shown by the thin intermediate line in fig. 3, whereas conditions 3 and 4 lead to the full innermost histograms in fig. 3. This procedure removed all break-up events in layers of matter other than the target and led to a practically background-free measurement.

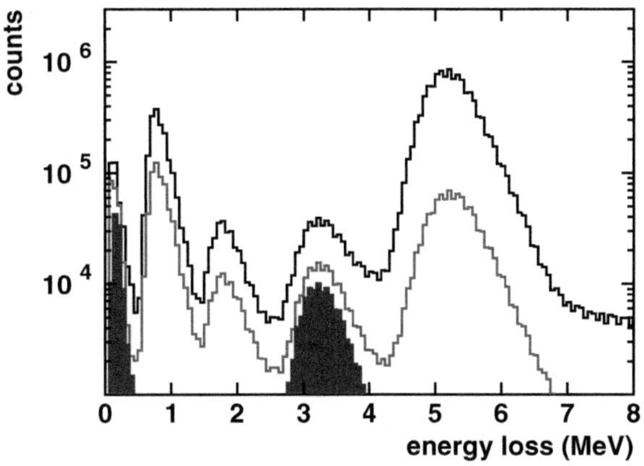

Fig. 3. Energy-deposition of the incident ^{8}B ions and of the break-up fragments in the third Si strip detector. The outermost contour corresponds to all events. The thin intermediate contour depicts events where an incident ^{8}B particle is correlated with multiplicity $m = 2$ in each SSD. The full innermost histograms are obtained by requiring a ΔE-cut on ^{7}Be in each SSD plus a p-^{7}Be vertex inside the target volume (see text).

4.1 Invariant-mass reconstruction

The p-^{7}Be relative energy, E_{rel}, is derived from the total energies, E_i, of the particles i (i = p, Be), their 3-momenta, p_i, and the p-Be opening angle, θ_{17}, according to

$$E_{rel} = \sqrt{(E_{\mathrm{Be}} + E_{\mathrm{p}})^2 - p_{\mathrm{Be}}^2 - p_{\mathrm{p}}^2 - 2p_{\mathrm{Be}}p_{\mathrm{p}}\cos(\theta_{17})}. \quad (1)$$

To reconstruct a break-up event, the p and ^{7}Be hits in each SSD have to be separated by at least one empty strip. Since this affects the efficiency for identifying a break-up event for low E_{rel}, we have to make sure that the GEANT simulation accurately reproduces this efficiency. This has been achieved by introducing a weighting function in GEANT that gradually increases the efficiency for detecting two separated hits from zero to one over the appropriate distance for each detector so that experimental and simulated distance distributions look alike. In fig. 4 we plot the inclusive horizontal distances between proton and ^{7}Be hits in the first SSD. One can observe that experiment and simulation yield very similar distributions. It should be emphasized that in our earlier data analysis a step function of this efficiency was assumed that jumped from zero to full efficiency at a fixed distance of 0.4 mm in each SSD. This is visualized by the dashed histogram in

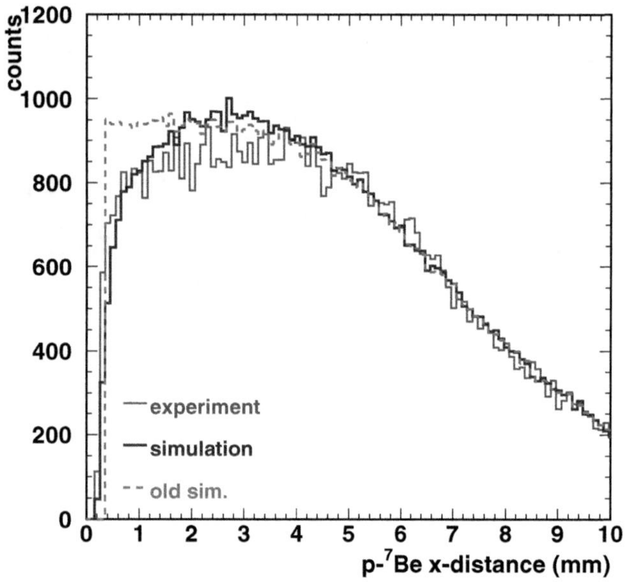

Fig. 4. Horizontal (x-) distances between proton and ^{7}Be hits in the first Si strip detector. The thin line shows the distribution of experimental distances, the thick one results from the present GEANT simulation. The dashed histogram shows the GEANT simulation that was used to evaluate our previous results [12].

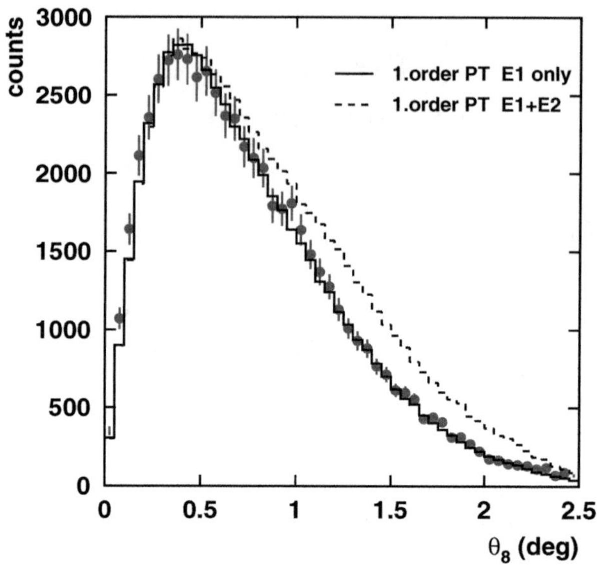

Fig. 5. Scattering angle θ_8 of the excited ^{8}B prior to breakup, as reconstructed from the proton and ^{7}Be vectors. The full histogram has been calculated in first-order perturbation theory assuming pure $E1$ multipolarity, the dashed one assuming $E1$+$E2$ multipolarity.

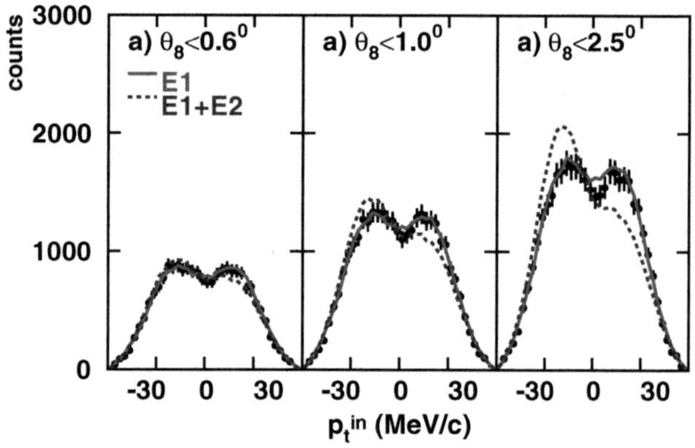

Fig. 6. In-plane transverse momenta, p_t^{in}, of the breakup protons for three different cuts in θ_8. The theoretical curves (full lines: $E1$ multipolarity, dashed lines: $E1$+$E2$ multipolarity) have been calculated in first-order perturbation theory. They were normalized individually to the data points in each frame.

fig. 4; it clearly shows that we overestimated the GEANT detection efficiency for small p-^{7}Be distances (small E_{rel}) in our previous paper [12].

4.2 Angular distributions

In the following we will present some angular distributions that can be shown to be sensitive to an $E2$ amplitude in CD. With ^{8}B* we denote the reference system of the (excited) ^{8}B prior to breakup. Its momentum vector is reconstructed from the measured proton and ^{7}Be momentum vectors. The angle θ_8 is the laboratory scattering angle of ^{8}B* relative to the incoming ^{8}B beam. The polar angles, θ_{cm}, and the azimuthal angles, ϕ_{cm}, of the break-up protons are measured in the rest frame of the ^{8}B* system. In the same way, one can calculate the transverse proton momentum vector in the reaction plane (p_t^{in}).

Figure 5 shows the θ_8 distribution, in comparison to two-model calculations using first-order PT. The full histogram denoting pure $E1$ multipolarity follows the data points very well, even to very large angles. The dashed histogram, where both $E1$ and $E2$ with their full theoretical strength were assumed, deviates from the data points markedly for θ_8 values above about 0.7 degrees. Note that the theoretical histograms were folded with the experimental response. We conclude that this distribution already indicates $E1$ dominance.

We present in fig. 6 the distribution of p_t^{in} for three different upper limits in θ_8, 0.6°, 1.0°, and 2.5°. In classical Rutherford scattering, this corresponds to impact parameters of 30 fm, 19 fm, and 7 fm, respectively. Relative energies between p and ^{7}Be up to 1.5 MeV were selected. The experimental data for all three θ_8-cuts can be reproduced well by a PT calculation that includes only $E1$ multipolarity (full histograms in fig. 6, the theoretical curves were normalized individually to the data points). If $E1$-plus-$E2$ multipolarity is used in the PT calculation, the different impact-parameter dependences of $E1$ and $E2$ multipolarity lead to markedly different shapes for the different θ_8-cuts (dashed histograms in fig. 6). In particular for large values of θ_8, the latter distributions show a strong anisotropy around $p_t^{in} = 0$ which is absent in our data. This is a very sensitive test of $E2$ admixture and again shows no substantial $E2$ contribution.

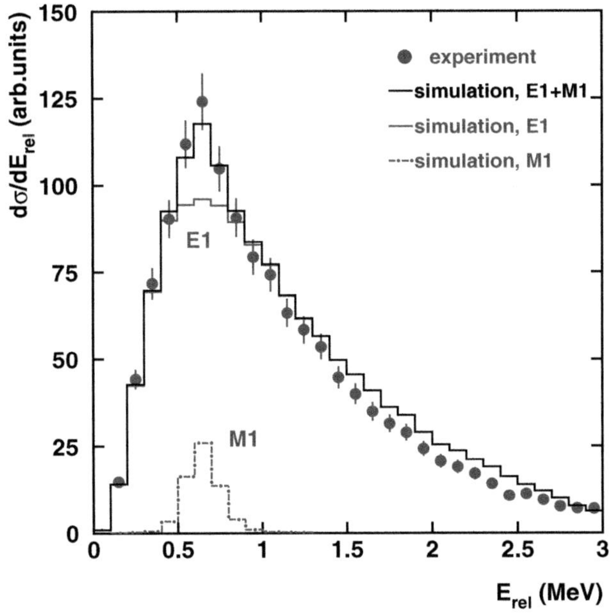

Fig. 7. Energy-differential Coulomb-dissociation yields for equal-sized E_{rel} bins of 100 keV each. The thick outermost histogram results from our GEANT simulation including $E1$ and $M1$ multipolarity, scaled by a factor of 0.82. The thin (dot-dashed) histograms show the separate contributions from $E1(M1)$ multipolarity.

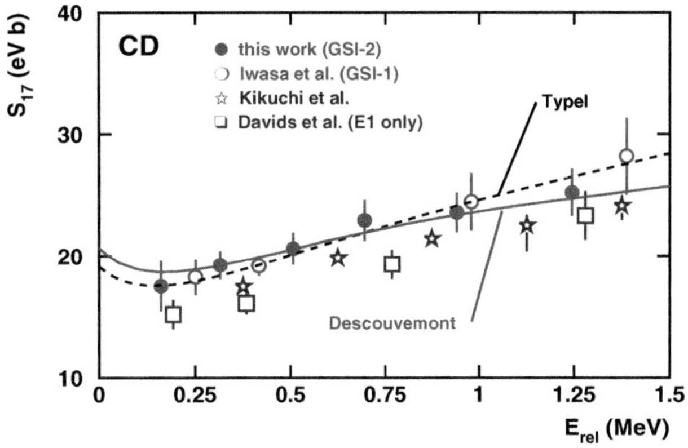

Fig. 8. Comparison between S_{17} values from Coulomb-dissociation experiments. The full (open) circles indicate the present (previous) GSI CD experiment labelled GSI-2 (GSI-1). Open stars depict ref. [9], open squares ref. [10] ($E2$ contribution subtracted). The theoretical curves are described in the text.

4.3 Energy-differential dissociation yields

The measured momentum vectors of the outgoing p and ^{7}Be particles allowed to calculate E_{rel} according to eq. (1), from which we have constructed the energy-differential dissociation yields of the excited ^{8}B* system prior to breakup (fig. 7). In line with our findings of a negligible $E2$ contribution discussed above, we compare this spectrum to a simulated one that contains contributions from $E1$ and $M1$ multipolarities only. The latter contribution was calculated using the $M1$ resonance parameters as determined by Filippone *et al.* [19].

In plotting fig. 7, we have restricted the Rutherford scattering angles θ_8 to values below 1.0° to ensure both dominance of CD and reduction of the effect of any possible $E2$ contribution. After determining an absolute $E1$ normalization factor of $f = 0.82$, the experimental and simulated distributions agree rather well. Small deviations between the data points and the black histogram indicate discrepancies between the assumed S_{17} factor from our potential model and the true one, as will be discussed in the next section.

5 The astrophysical S_{17} factor

The measured quantity in CD of ^{8}B is the distribution of energy-differential cross sections, fig. 7. This distribution is related to S_{17} via a theoretical model. We assume that at the high incident energy used in our experiment and for the low Q-value of the reaction, first-order perturbation theory is adequate to describe Coulomb dissociation. In analyzing our results, we also assume that the GEANT simulations describe all experimental effects quantitatively, in particular the feeding of neighboring bins due to the relatively bad E_{rel} resolution. Any remaining discrepancies between the two histograms in fig. 7 are attributed to a deviation of the true $E1$ S_{17} factor from the one used in our simulation. Thus, the true S_{17} factor for each bin was obtained by multiplying the theoretical one (averaged over the bin width) by the ratio of observed and simulated counts. The resulting S_{17} factors as a function of E_{rel} are visualized in fig. 8.

5.1 Comparison with other CD experiments

Figure 8 shows our astrophysical S_{17} factors in comparison with those from the other three CD experiments published so far [9,11,10] (the data of ref. [10] represent their $E1$-S_{17} factors after subtraction of the $E2$ contribution). The CD S_{17} factors are in reasonable agreement, though both the Kikuchi *et al.* [9] and the Davids *et al.* [10] data are systematically lower. We note also that our earlier CD experiment [11] and the present one are in good agreement in this energy range, marked discrepancies occur only at higher E_{rel} values. Compared to the results given in ref. [12], the lowest three data points have been modified to become larger by 6.7%, 10%, and 5.8%, respectively, due to a smaller efficiency to detect two separate break-up products in the SSD, as explained in subsect. 4.1 and visualized in fig. 4. The remaining data points remain largely unaffected. As a consequence, the slope of our S_{17} factors as a function of E_{rel} becomes smaller and fits much better than previously to the energy dependence of Descouvemont's cluster model [20,21].

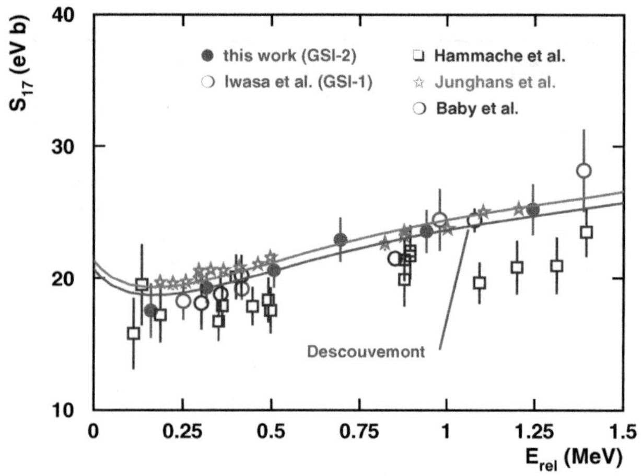

Fig. 9. S_{17} from this work in comparison with the (p, γ) experiments of ref. [5] (squares), ref. [8] (stars), and ref. [7] (open circles). The latter data were corrected for the contribution of the $M1$ resonance by the authors. The theoretical curves are from Descouvemont [21] and have been fitted to the Seattle data (upper curve) and the present data (lower curve), respectively. See text for more details.

5.2 Comparison with direct-capture experiments

Figure 9 compares our data to those of the recent $^{7}Be(p, \gamma)^{8}B$ measurements where the authors have subtracted the contribution from the $M1$ resonance (refs. [5, 7, 8]). With the modifications of the lowest-E_{rel} data points discussed above, our dataset follows now closely the (p, γ) data of Junghans *et al.* [8] over their *entire* energy range.

5.3 Extrapolation to zero relative energy

To extrapolate to zero energy, we have chosen the recent cluster-model calculation of Descouvemont [21] (we refer to this model below as D04). In this refined approach, the curve resulting from the Minnesota force (MN) is closer to the experimental data and has been used in fig. 9 to fit both the Seattle data and our present results over the energy range up to $E_{rel} = 1.5$ MeV. The fits yield practically identical results within their respective errors.

Our previous data set [12] was found to be best compatible with the potential-model calculation of Typel as discussed in subsect. 2 of the present paper or in Davids and Typel [14] (referred to below as DT03). It is obvious that with the modified low-energy data points of the present paper, the agreement with this model is less satisfactory. The black dashed curve in fig. 8 visualizes a fit of this theory to our data.

When we fit our lowest 8 data points, up to $E_{rel} = 2$ MeV, to the D04 model, we obtain $S_{17}(0) = 20.6 \pm 0.8$ eV b. A systematic error of 5.6% has to be added, yielding $S_{17}(0) = 20.6 \pm 0.8(stat) \pm 1.2(syst)$ eV b. Not included in these numbers is the theoretical uncertainty given by Descouvemont [21] as about 5–10% depending on the relative energy.

6 Conclusions

We conclude that at sufficiently high incident energy, a high-resolution exclusive CD experiment can provide a rather precise value for the low-energy $^{7}Be(p, \gamma)^{8}B$ cross section. By setting tight constraints to the scattering angle, θ_8, and analyzing proton-^{7}Be angular correlations, a significant contribution from $E2$ multipolarity could be excluded. Contrary to our earlier publication [12], our reanalyzed results for the astrophysical S_{17} factor follow closely the energy dependence as predicted by the refined cluster-model description of Descouvemont [21]. This finding is in line with the most recent measurements of the $^{7}Be(p, \gamma)^{8}B$ reaction. The combined statistical and systematic errors of our fit value for $S_{17}(0)$ amounts to 6.6%; a similar error contribution of about 5% comes from the model uncertainty [21].

The authors wish to thank K.-H. Behr, K. Burkard, and A. Brünle for technical assistance. Vivid discussions with B. Davids, P. Descouvemont, M. Hass, and A. Junghans are gratefully acknowledged.

References

1. Q.R. Ahmad *et al.*, Phys. Rev. Lett. **89**, 011301; 011302 (2002).
2. SNO Collaboration, arXiv:nucl-ex/0502021.
3. J.N. Bahcall, M.H. Pinsonneault, Phys. Rev. Lett. **92**, 121301 (2004).
4. S. Turck-Chieze *et al.*, Phys. Rev. Lett. **93**, 211102 (2004).
5. F. Hammache *et al.*, Phys. Rev. Lett. **86**, 3985 (2001).
6. F. Strieder *et al.*, Nucl. Phys. A **696**, 219 (2001).
7. L.T. Baby *et al.*, Phys. Rev. Lett. **90**, 022501 (2003).
8. A.R. Junghans *et al.*, Phys. Rev. C **68**, 065803 (2003).
9. T. Kikuchi *et al.*, Phys. Lett. B **391**, 261 (1997); T. Kikuchi *et al.*, Eur. Phys. J. A **3**, 213 (1998).
10. B. Davids *et al.*, Phys. Rev. C **63**, 065806 (2001).
11. N. Iwasa *et al.*, Phys. Rev. Lett. **83**, 2910 (1999).
12. F. Schümann *et al.*, Phys. Rev. Lett. **90**, 232501 (2003).
13. F. Schümann *et al.*, Phys. Rev. C **73**, 015806 (2006).
14. B. Davids, S. Typel, Phys. Rev. **68**, 845802 (2003).
15. C.A. Bertulani, Phys. Rev. C **49**, 2688 (1994); Z. Phys. A **356**, 293 (1996).
16. S. Typel *et al.*, Nucl. Phys. A **613**, 147 (1997), implemented as code CDXS by S. Typel (2002).
17. H. Geissel *et al.*, Nucl. Instrum. Methods Phys. Res. B **70**, 286 (1992).
18. P. Senger *et al.*, Nucl. Instrum. Methods Phys. Res. B **327**, 393 (1993).
19. B.W. Filippone *et al.*, Phys. Rev. C **28**, 2222 (1983).
20. P. Descouvemont *et al.*, Nucl. Phys. A **567**, 341 (1994).
21. P. Descouvemont, Phys. Rev. C **70**, 065802 (2004).

Eur. Phys. J. A **27**, s01, 233–236 (2006)
DOI: 10.1140/epja/i2006-08-036-3

EPJ A direct
electronic only

Study of the ^{26}Si(p, γ)^{27}P reaction through Coulomb dissociation of ^{27}P

Y. Togano[1,a], T. Gomi[1], T. Motobayashi[2], Y. Ando[1], N. Aoi[2], H. Baba[1], K. Demichi[1], Z. Elekes[2,3], N. Fukuda[2], Zs. Fülöp[3], U. Futakami[1], H. Hasegawa[1], Y. Higurashi[2], K. Ieki[1], N. Imai[2], M. Ishihara[2], K. Ishikawa[4], N. Iwasa[5], H. Iwasaki[6], S. Kanno[1], Y. Kondo[4], T. Kubo[2], S. Kubono[7], M. Kunibu[1], K. Kurita[1], Y.U. Matsuyama[1], S. Michimasa[7], T. Minemura[2], M. Miura[4], H. Murakami[1], T. Nakamura[4], M. Notani[7], S. Ota[8], A. Saito[1], H. Sakurai[6], M. Serata[1], S. Shimoura[7], T. Sugimoto[4], E. Takeshita[1], S. Takeuchi[2], K. Ue[6], K. Yamada[1], Y. Yanagisawa[2], K. Yoneda[2], and A. Yoshida[2]

[1] Department of Physics, Rikkyo University, Tokyo 171-8501, Japan
[2] RIKEN (The Institute of Physical and Chemical Research), Saitama 351-0198, Japan
[3] Institute of Nuclear Research (ATOMKI), 4001 Debrecen, Hungary
[4] Department of Physics, Tokyo Institute of Technology, Tokyo 152-8551, Japan
[5] Department of Physics, Tohoku University, Miyagi, 980-8578, Japan
[6] Department of Physics, University of Tokyo, Tokyo 113-0033, Japan
[7] Center for Nuclear Study (CNS), University of Tokyo, Saitama 351-0198, Japan
[8] Department of Physics, Kyoto University, Kyoto 606-8502, Japan

Received: 20 June 2005 /
Published online: 10 March 2006 –

Abstract. The Coulomb dissociation of ^{27}P was studied experimentally using ^{27}P beams at 57 MeV/nucleon with a lead target. The $E2$ gamma decay width of the first excited state in ^{27}P was determined to be $(2.8 \pm 0.5) \times 10^{-5}$ eV. The total $(M1 + E2)$ gamma decay width of the state, which is of astrophysical interest, is estimated by combining the experimental result and a shell model calculation. The reaction rate of ^{26}Si(p, γ)^{27}P is deduced from the estimated total gamma decay width. The astrophysical implication derived from estimated gamma decay width is discussed.

PACS. 25.60.-t Reaction induced by unstable nuclei – 25.70.De Coulomb excitation – 26.30.+k Nucleosynthesis in novae, supernovae and other explosive environments – 27.30.+t $20 \leq A \leq 38$

1 Introduction

The mapping of the isotope ^{26}Al in the interstellar medium through the detection of its 1.809 MeV γ line by satellites [1] has increased the interest in ^{26}Al nucleosynthesis. This is a trace of ongoing nucleosynthesis in the universe because lifetime of ^{26}Al (10^6 yr) is much shorter than the timescale of the cosmic evolution. Consequently, the nuclear physics of the isotopes around ^{26}Al is of importance.

The ^{26}Si(p, γ)^{27}P reaction is relevant to the synthesis of ^{26}Al. The nuclide ^{26}Si β decays to the isomeric state of ^{26}Al which subsequently β decays to the ground state of ^{26}Mg, thus no γ-ray is emitted. The characteristic 1.809 MeV γ line is only emitted by the β decay of ^{26}Al in the ground state. Therefore, the synthesis of ^{26}Si was not thought to contribute the γ-ray emission of ^{26}Al. Recently, it has been suggested that high temperature environment with $T_9 \geq 0.4$ is hot enough to establish an equilibrium between the isomeric state and the ground state in ^{26}Al [2]. Therefore, ^{26}Si destruction by proton capture is important to determine the amount of the ground state in ^{26}Al produced by the equilibrium.

Under this high temperature the capture reaction rate, $N_A\langle\sigma v\rangle$, is expected to be dominated by the resonant capture via the first excited state in ^{27}P [3]. Resonant capture rates for isolated, narrow resonances are given by

$$N_A\langle\sigma v\rangle = 1.54 \times 10^{11}(\mu T_9)^{-3/2}\omega\gamma \exp\left[-11.605\frac{E_R}{T_9}\right] \tag{1}$$

with N_A Avogadro's number, μ the reduced mass in amu, T_9 the temperature in unit of GK, $\langle\sigma v\rangle$ the thermally averaged nuclear cross section, and E_R the resonance energy in MeV [4]. The resonance strength $\omega\gamma$ is given by

$$\omega\gamma = \frac{2J+1}{(2J_p+1)(2J_T+1)}\frac{\Gamma_p\Gamma_\gamma}{\Gamma_{tot}}, \tag{2}$$

[a] e-mail: toga@ne.rikkyo.ac.jp

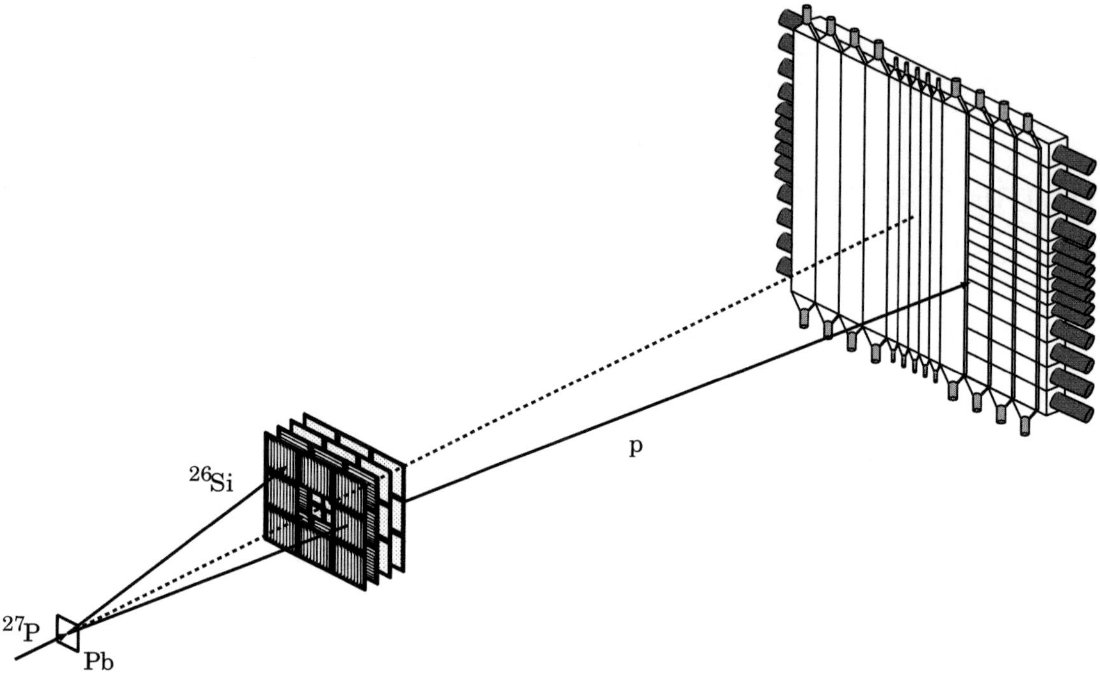

Fig. 1. A schematic view of the experimental setup. The entire system is in vacuum.

where J is the spin of the compound nucleus, and J_p and J_T are the spins of the proton and target, respectively. Γ_p and Γ_γ are the proton widths and the gamma decay widths of the resonance. The total width Γ_{tot} is the sum of the partial widths.

A direct measurement of the relevant capture cross section is the preferable method for determining the proton capture reaction rates. However, the cross section becomes quite small at stellar energies due to the Coulomb barrier and radioactive beams of sufficient intensities are not available. Therefore, indirect methods must be employed to extract the information about the reaction rate.

The Coulomb dissociation at intermediate energies is an alternative method to study the radioactive capture reactions of astrophysical interest at low energies [5,6,7]. The process can be regarded as photodisintegration by virtual photons, which is essentially the inverse of the radiative capture process [8]. The extracted electromagnetic transition probability is directly determines the Γ_γ. The Coulomb breakup process is characterized by a large cross section which helps to compensate for the small capture cross section.

The present paper reports on an experimental study of the dissociation of ^{27}P in the Coulomb field of a Pb nucleus. The Γ_γ values of the first excited state in ^{27}P, which have the largest contribution to the resonant capture rate of ^{26}Si(p, γ)^{27}P, has been extracted.

2 Experiment

The experiment was performed at the RIKEN Accelerator Research Facility. A ^{27}P beam was produced via the projectile fragmentation of a 115-A-MeV ^{36}Ar beam incident on a 460 mg/cm^2 thick Be target, and separated by the RIKEN Projectile-fragment Separator (RIPS) [9]. The typical ^{27}P intensity was about 2800 counts per second and the energy was 57 A MeV with an energy spread of about 1.5%. It bombarded a 125 mg/cm^2 thick lead target. The isotopic purity of ^{27}P in the secondary beam is about 1%. The major contributions are ^{26}Si, ^{25}Al, and ^{24}Mg. The particle identification for secondary beams was performed event by event by means of the time-of-flight (TOF)-ΔE method using a 0.5 mm thick plastic scintillator located at the second focal plane of the RIPS. Two sets of parallel plate avalanche counters were also placed at the final focal plane of the RIPS to extrapolate the position and angle of the beam at the target.

The isotope ^{27}P were excited by the lead target and disintegrated to ^{26}Si and proton. Figure 1 shows the detector system for the measurement of breakup products. The emission angles of these products were measured at a position-sensitive silicon telescope located at 48 cm downstream of the target. The silicon telescope consists of four layers of detectors with 0.5 mm thickness. Each layer was of eight silicon detectors with 50 $\times$ 50 mm^2 effective areas on 56 $\times$ 56 mm^2 frames. The eight detectors in a layer formed a 3×3 matrix with a hole in the center. The count rate in the telescope was suppressed to 3000 counts per second by introducing the hole whereas the total beam rate was about 3×10^5 count per second. The detectors in the first and second layers have 5 mm wide strip electrodes, which enables to measure the hit positions of the products. The energy of the ^{26}Si was also measured by the silicon telescope. ^{26}Si was stopped at the fourth layer, and identified using the ΔE-E method.

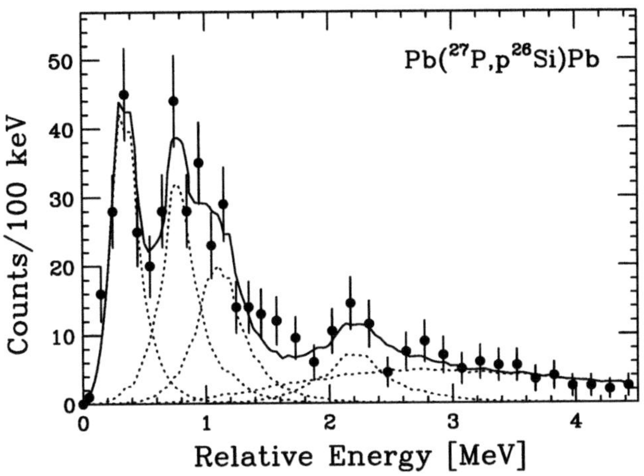

Fig. 2. Relative energy spectrum of the ^{27}P breakup on Pb (full circle). The data was fitted by the detector responses simulated using Geant4 code [10]. The dashed curves and solid curve represent the components and sum of the elements.

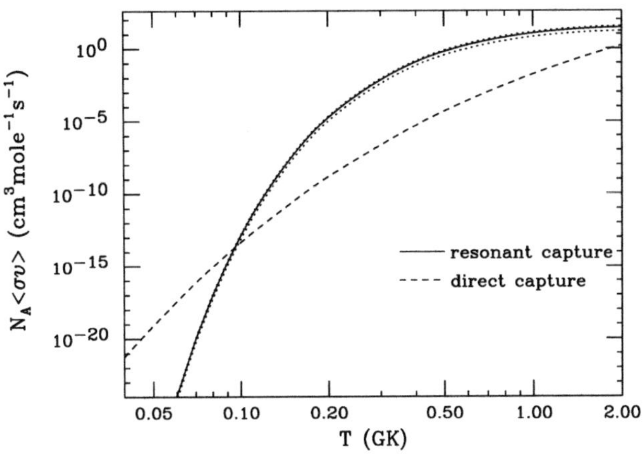

Fig. 3. The reaction rate of $^{26}Si(p,\gamma)^{27}P$ as a function of the temperature of stars. The solid and dotted curve represent our result and the range of its error, respectively. The dashed line shows the direct capture component estimated by J.A. Caggiano *et al.* [3].

The energy of the proton, which penetrated the silicon telescope, was determined with a plastic scintillator hodoscope placed at the 2.8 m downstream of the target by measuring the TOF. The hodoscope with an active area of $1 \times 1\,m^2$, consists of thirteen 5 mm thick ΔE- and sixteen 60 mm thick E-plastic scintillators. The outgoing proton was stopped in the E counters after passing through the ΔE counters. The proton was identified by the TOF-ΔE and TOF-E methods.

The relative energy E_{rel} between ^{26}Si and proton was extracted by combining the positions and energies of the products. The energy E_{rel} corresponds to the center-of-mass energy, and thus the excitation energy of ^{27}P E_x can be expressed as

$$E_x = E_{rel} + E_s, \quad (3)$$

where E_s represents the separation energy for the decay channel (0.859 MeV for ^{27}P). Therefore, by measuring the E_{rel} the excitation energy E_x can be determined.

3 Result and discussion

The relative energy spectrum is shown in fig. 2. The full circles represent the experimental data. The solid curve represents the best fit with five contributions shown by the dashed curves. The detector responses were simulated by the GEANT4 code [10]. The peak at 0.31 MeV corresponds to the known first excited state at 1.2 MeV in ^{27}P. The bump at around 1 MeV may be respectively due to the known second excited state at 1.6 MeV and an unknown one at 2.0 MeV. The peak at 2.2 MeV corresponds to the known state at 3.4 MeV. The direct breakup component, which distributes from 0.8 MeV to 4.5 MeV, corresponds to the non-resonant proton capture process. The cross section σ is related to the astrophysical S-factor

$$S = \sigma E \exp(2\pi\eta), \quad (4)$$

where η denotes the Sommerfeld parameter. We assume that the non-resonant capture is dominated by the $E1$ transition and the astrophysical S-factor is independent of the energies.

The Coulomb dissociation cross section for the first excited state in ^{27}P was determined to be 14.8 ± 2.7 mb from the yield of the peak. The error includes the statistical one and ambiguity of the detection efficiency. Supposing the spin and parity of the state is $3/2^+$ from the level scheme of the mirror nucleus ^{27}Mg [11], the transition between the first excited state and ground state ($1/2^+$) is induced by the $M1$ and $E2$ multipolarities. This means that the Γ_γ is a sum of a $E2$ component ($\Gamma_\gamma(E2)$) and an $M1$ component ($\Gamma_\gamma(M1)$). Since the Coulomb dissociation is highly sensitive to the $E2$ component compared with the $M1$ transition [12], the experimental cross section is expected to be exhausted by the $E2$ excitation. $\Gamma_\gamma(E2)$ is determined to be $(2.8 \pm 0.5) \times 10^{-5}$ eV form the measured cross section.

To deduce the $\Gamma_\gamma(M1)$, the $E2/M1$ mixing ratio δ of the first excited state in ^{27}P was estimated by the combination of the known mixing ratio [13] of the mirror nucleus ^{27}Mg and the double ratio R defined as $R = \delta_{T_Z+}/\delta_{T_Z-}$, where δ_{T_Z+} and δ_{T_Z-} represent the mixing ratios of the mirror pair. To estimate the ratio R, we performed shell model calculations using the USD effective interaction [14] with conventional values of the effective charges in the *sd*-shell region, $e_p = 1.3e$ and $e_n = 0.5e$ [15]. The reliability of the R calculated by the shell model was evaluated to be 60% (2σ) by comparing with the experimentally known values. From the calculated R and the known δ of ^{27}Mg, the mixing ratio for ^{27}P was estimated to be 0.020 ± 0.012. The $\Gamma_\gamma(M1)$ was derived from the mixing ratio and thus the total Γ_γ estimated to be $(3.7 \pm 2.2) \times 10^{-3}$ eV. The error includes the experimental error and ambiguity come from the estimation of $M1$ contribution. This value is consistent with the previous estimation [3].

The reaction rate of the $^{26}Si(p,\gamma)^{27}P$ was calculated using the extracted Γ_γ value. Figure 3 shows the temperature dependence of the reaction rate. The resonance

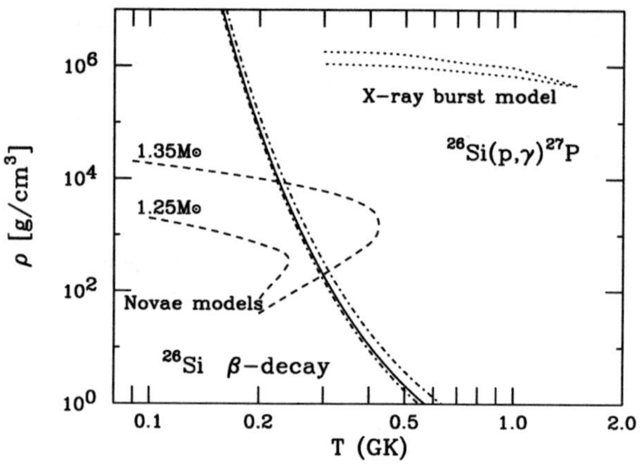

Fig. 4. Temperature-density boundary (solid line) at which the ^{26}Si(p, γ)^{27}P reaction and the competing β decay are of equal strength assuming a hydrogen mass fraction $X_H = 0.5$. Dot-dashed curves represent the range of error. The dashed curves and dotted line indicate T-ρ profiles of two novae sequences and an X-ray burst, respectively.

parameters are taken from ref. [3] except for Γ_γ. The solid and dotted curves represent the present result of the reaction rate and range of its error. The dashed line denotes the direct capture component of reaction rate calculated by J.A. Caggiano *et al.* [3] based on a shell model. This figure indicates that the resonant capture is the dominant process above 0.1 GK, the temperature region of novae and X-ray bursts temperature [16].

The competition between ^{26}Si(p, γ)^{27}P reaction and ^{26}Si β decay can be discussed by using the extracted reaction rate. The competition depends on the density, temperature, and mass fraction of proton in stars. The solid linc in fig. 4 represents the temperature and density condition for which (p, γ) reaction and the competing β decay are of equal strength. The dot-dashed lines show the error of the present estimate. The solid curve was calculated using our result and assuming a hydrogen mass fraction of $X_H = 0.5$. In the region above the solid curve the proton capture reaction dominates, while below the solid line the nuclei ^{26}Si are exhausted by β decay. The dashed lines denote temperature-density profiles for the two novae sequences, whose masses are 1.25 $M_\odot$ and 1.35 $M_\odot$ [17,18], and the dotted line shows one for an X-ray burst [19] of a 1.4 $M_\odot$ neutron star. The novae sequence evolves in time from larger to smaller densities, on the other hand the one for X-ray burst evolves from smaller to larger densities. It can be seen that the (p, γ) reaction dominates in the heavy novae when they are at around the peak temperature, whereas the reaction hardly occurs in the light novae. On the other hand, the X-ray burst is dominated by the (p, γ) reaction in their all phases.

4 Summary

The Coulomb dissociation of ^{27}P was experimentally studied to investigate the ^{26}Si(p, γ)^{27}P reaction in explosive hydrogen burning in novae and X-ray bursts. The first excites state in ^{27}P, which has the largest contribution among the resonances of ^{27}P to the reaction rate, was observed by measuring the relative energy between ^{26}Si and proton.

The $E2$ component of Γ_γ for first excited state was experimentally determined to be $(2.8\pm0.5)\times10^{-5}$ eV. We estimated the total Γ_γ to be $(1.3\pm0.8)\times10^{-3}$ eV by the measured $E2$ component together with the $E2/M1$ mixing ratio estimated by the one for the mirror nuclei ^{27}Mg with the help of a shell model calculation. This is consistent with the previously estimated value by J.A. Caggiano *et al.* [3]. The present result indicates non-negligible effects of the ^{26}Si(p, γ)^{27}P reaction to the nuclear burning in heavy novae and X-ray bursts.

Analysis for higher excited states, which may affect the reaction rate at high temperature, is in progress.

We thank the staff of RIKEN accelerator for their excellent operation of the beam delivery. This work is supported by Research Fellowships of the Japan Society for the Promotion of Science for Young Scientists.

References

1. R. Diehl *et al.*, Astron. Astrophys. Suppl. Ser. **97**, 181 (1993).
2. A. Coc *et al.*, Phys. Rev. C **61**, 015801 (1999).
3. J.A. Caggiano *et al.*, Phys Rev. C **64**, 025802 (2001).
4. W.A. Fowler *et al.*, Annu. Rev. Astron. Astrophys. **5**, 525 (1967).
5. T. Motobayashi *et al.*, Phys. Rev. Lett. **73**, 2680 (1994).
6. N. Iwasa *et al.*, Phys. Rev. Lett. **83**, 2910 (1999).
7. T. Gomi *et al.*, Nucl. Phys. A **734**, E77 (2004).
8. G. Baur *et al.*, Nucl. Phys. A **458**, 188 (1986).
9. T. Kubo *et al.*, Nucl. Instrum. Methods B **70**, 309 (1992).
10. S. Agostinelli *et al.*, Nucl. Instrum. Methods A **506**, 250 (2003).
11. P.M. Endt, Nucl. Phys. A **521**, 1 (1990).
12. K. Langanke *et al.*, Phys. Rev. C **49**, 1771 (1994).
13. M.J.A. de Voigt *et al.*, Nucl. Phys. A **186**, 365 (1972).
14. B.H. Wildenthal, Prog. Part. Nucl. Phys. **11**, 5 (1984).
15. B.A. Brown *et al.*, Annu. Rev. Nucl. Sci. **38**, 29 (1988).
16. C. Iliadis *et al.*, Astrophys. J. **524**, 434 (1999).
17. J. Jose *et al.*, Astrophys. J. **520**, 347 (1999).
18. C. Iliadis *et al.*, Astrophys. J. Suppl. **142**, 105 (2002).
19. O. Koike *et al.*, Astron. Astrophys. **342**, 464 (1999).

Eur. Phys. J. A **27**, s01, 237–242 (2006)
DOI: 10.1140/epja/i2006-08-037-2

EPJ A direct
electronic only

Breakup of loosely bound nuclei as indirect method in nuclear astrophysics: ^{8}B, ^{9}C, ^{23}Al

L. Trache[1,a], F. Carstoiu[2], C.A. Gagliardi[1], and R.E. Tribble[1]

[1] Cyclotron Institute, Texas A&M University, College Station, TX 77843-3366, USA
[2] National Institute of Physics and Nuclear Engineering H. Hulubei, Bucharest, Romania

Received: 25 August 2005 /
Published online: 10 March 2006 –

Abstract. We discuss the use of one-nucleon breakup reactions of loosely bound nuclei at intermediate energies as an indirect method in nuclear astrophysics. These are peripheral processes, therefore we can extract asymptotic normalization coefficients (ANC) from which reaction rates of astrophysical interest can be inferred. To show the usefulness of the method, three different cases are discussed. In the first, existing experimental data for the breakup of ^{8}B at energies from 30 to 1000 MeV/u and of ^{9}C at 285 MeV/u on light through heavy targets are analyzed. Glauber model calculations in the eikonal approximation and in the optical limit using different effective interactions give consistent, though slightly different results, showing the limits of the precision of the method. The results lead to the astrophysical factor $S_{17}(0) = 18.7 \pm 1.9$ eV · b for the key reaction for solar neutrino production ^{7}Be(p, γ)^{8}B. It is consistent with the values from other indirect methods and most direct measurements, but one. Breakup reactions can be measured with radioactive beams as weak as a few particles per second, and therefore can be used for cases where no direct measurements or other indirect methods for nuclear astrophysics can be applied. We discuss a proposed use of the breakup of the proton drip line nucleus ^{23}Al to obtain spectroscopic information and the stellar reaction rate for ^{22}Mg(p, γ)^{23}Al.

PACS. 25.60.-t Reactions induced by unstable nuclei – 25.60.Gc Breakup and momentum distributions – 26.65.+t Solar neutrinos – 26.30.+k Nucleosynthesis in novae, supernovae, and other explosive environments

1 Introduction

Radiative proton capture reactions are important in nuclear astrophysics, and a large number of reaction chains were found to be needed in nucleosynthesis calculations for static or explosive hydrogen burning scenarios (see *e.g.* [1,2]). This means that more data involving proton capture on unstable nuclei are necessary. In some cases direct experiments are possible, but in many more they are impossible with the present techniques and even with those of the foreseeable future. We have to rely on indirect methods instead. In this presentation we discuss such an indirect method, and we shall concentrate on three particular cases, first to demonstrate the feasibility, then to show the strengths of the method and its limits.

Part of the results discussed were published before, when we originally proposed to extract astrophysical S-factors from one-nucleon-removal (or breakup) reactions of loosely bound nuclei at intermediate energies or later [3,4,5]. In the present paper, first we use the well studied case of ^{8}B breakup as a benchmark to demonstrate the usefulness of the method and show the possibilities of the Glauber reaction model used. We show that existing experimental data at energies between 30 and 1000 MeV/nucleon [6,7,8,9,10] on a range of light and heavy targets translate into consistent values of the ANC, which is then used to determine the astrophysical factor S_{17} (which gives the reaction rate for the ^{7}Be(p, γ)^{8}B reaction of crucial importance for the solar neutrino question). We show that the precision of the method is limited to about 10% by our ability to compute absolute cross sections. Second, we use the same technique for ^{9}C breakup data at 285 MeV/nucleon [7] to determine S_{18} (which gives the rate for the ^{8}B(p, γ)^{9}C reaction of importance for explosive hydrogen burning) with reasonable accuracy. For a third case, a proposed experiment for the breakup of ^{23}Al is discussed to show that the method is particularly well adapted to rare isotope beams produced using fragmentation. Spectroscopic information is sought in this case. In particular we seek to determine the spin and parity of the ground state of the dripline nucleus ^{23}Al and the ANC, which will be then used to calculate the reaction rates for ^{22}Mg(p, γ)^{23}Al. The last part of the present paper will concentrate on this latter case, which has not been discussed before.

[a] e-mail: livius_trache@tamu.edu

2 The reaction model

The method is based on data showing that the structure of halo nuclei is dominated by one or two nucleons orbiting a core [11,12]. Consequently, we use the fact that the breakup of halo or loosely bound nuclei is essentially a peripheral process, and therefore, the breakup cross-sections can give information about the wave function of the last nucleon at large distances from the core. More precisely, asymptotic normalization coefficients (ANCs) can be determined. We show that there exists a favorable kinematical window in which breakup reactions are highly peripheral and are dominated by the external part of the wave function and, therefore, the ANC is the better quantity to be extracted. The approach offers an alternative and complementary technique to extracting ANCs from transfer reactions [13].

In the breakup of loosely bound nuclei at intermediate energies, a nucleus $B = (Ap)$, where B is a bound state of the core A and the nucleon p, is produced by fragmentation from a primary beam, separated and then used to bombard a secondary target. In measurements, the core A is detected, measuring its parallel and transverse momenta and eventually the gamma-rays emitted from its de-excitation. Spectroscopic information can be extracted from these experiments, such as the orbital momentum of the relative motion of the nucleon and the contribution of different core states, typically comparing the measured momentum distributions with those calculated with Glauber models. The integrated cross sections can be used to extract absolute spectroscopic factors [12] or the ANC [3]. The latter approach has the advantage that it is independent of the geometry of the proton binding potential. We note that the ANC C^B_{Ap} for the nuclear system $A + p \leftrightarrow B$ specifies the amplitude of the tail of the overlap function of the bound state B in the two-body channel (Ap) (see, for example [13] and references therein). Fortunately, this ANC is all we need to determine the astrophysical S-factor for the radiative proton capture reaction $A(p,\gamma)B$ which is a highly peripheral process. Details about the reaction model are published elsewhere [5,14].

3 Three particular cases

3.1 Breakup of ^{8}B to determine the S_{17} astrophysical factor

The calculations presented in [3] have been extended and refined. The Coulomb part of the dissociation cross section was refined by including the final state interaction into calculations and new data on the breakup of ^{8}B are analyzed [8,9,10]. Also a new set of calculations for the breakup of ^{8}B were made using five sets of different effective NN interactions. We describe the breakup of ^{8}B (and in the next subsection of ^{9}C) in terms of an extended Glauber model. The loosely bound ^{8}B (^{9}C) nucleus is moving on a straight line trajectory and the proton and the ^{7}Be (^{8}B) core making it, interact independently with the target. The breakup cross sections depend on the proton-target and core-target interactions and on the relative p-core motion. The wave function of the ground state of ^{8}B (^{9}C) is a mixture of $1p_{3/2}$ and $1p_{1/2}$ orbitals, around a ^{7}Be (^{8}B) core. The total ANC $C^2_{tot} = C^2_{p_{3/2}} + C^2_{p_{1/2}}$ can be extracted from the measured breakup cross sections.

The calculations reproduce well all the measured parallel and transverse momentum distributions measured so far, on light or heavy targets, giving us confidence in the Glauber model used. We show that the reaction is peripheral in various degrees, depending on the energy and target used. The ^{8}B ANC is extracted from existing breakup data at energies between 30-1000 MeV/nucleon and on different targets ranging from C to Pb [6,7,8,9,10]. Two approaches were used. The first is a potential approach. To obtain the folded potentials needed in the S-matrix calculations we used the JLM effective nucleon-nucleon interaction [15], using the procedure and the renormalizations of ref. [16]. We applied this technique for energies below 285 MeV/nucleon only and on all targets. In a second approach, the Glauber model in the optical limit was used. The breakup process is treated as multiple elementary interactions between partners' nucleons, and the cross sections and the complex scattering amplitudes are taken from the literature. Calculations were done using different ranges for the elementary interactions: zero range, 1.5 fm ("standard"), 2.5 fm and individual ranges for each NN component ("Ray") [17]. No new parameters were adjusted. The contribution of the ^{7}Be core excitation was calculated for each target and at each energy using the data from an experiment which disentangle it [10], and corrected for in all cases. For details on the procedure see [5]. In fig. 1 we show that from the widely varying breakup cross sections (panel a)) on all targets and at so different energies, we extract ANCs which are consistent with a constant value (panel b)). However, we see that a certain dependence on the NN interaction used exists, which points to the limitations of our present knowledge of the effective nucleon-nucleon interactions.

If we take the unweighted average of all 31 determinations we find an ANC $C^2_{tot}(JLM) = 0.483 \pm 0.050\,\text{fm}^{-1}$ (fig. 1). The value is in agreement with that determined using the (^{7}Be, ^{8}B) proton transfer reactions at 12 MeV/u [18,19]. The two values agree well, in spite of the differences in the energy ranges and in the reaction mechanisms involved. The ANC extracted leads to the astrophysical factor $S_{17}(0) = 18.7 \pm 1.9\,\text{eV} \cdot \text{b}$ for the key reaction for solar neutrino production ^{7}Be(p, γ)^{8}B. The uncertainties quoted are only the standard deviation of the individual values around the average, involving therefore the experimental and theoretical uncertainties. This 10% error bar is probably a good measure of the precision we can claim from the method at this point in time, due essentially to the uncertainties in the cross section calculations. The $S_{17}(0)$ value we extract is also in agreement with those extracted from indirect methods and with most of the direct determinations (see the discussions in [20,21,22]), but one which stands out in its claim of a larger value and very small error [23]. There are currently many

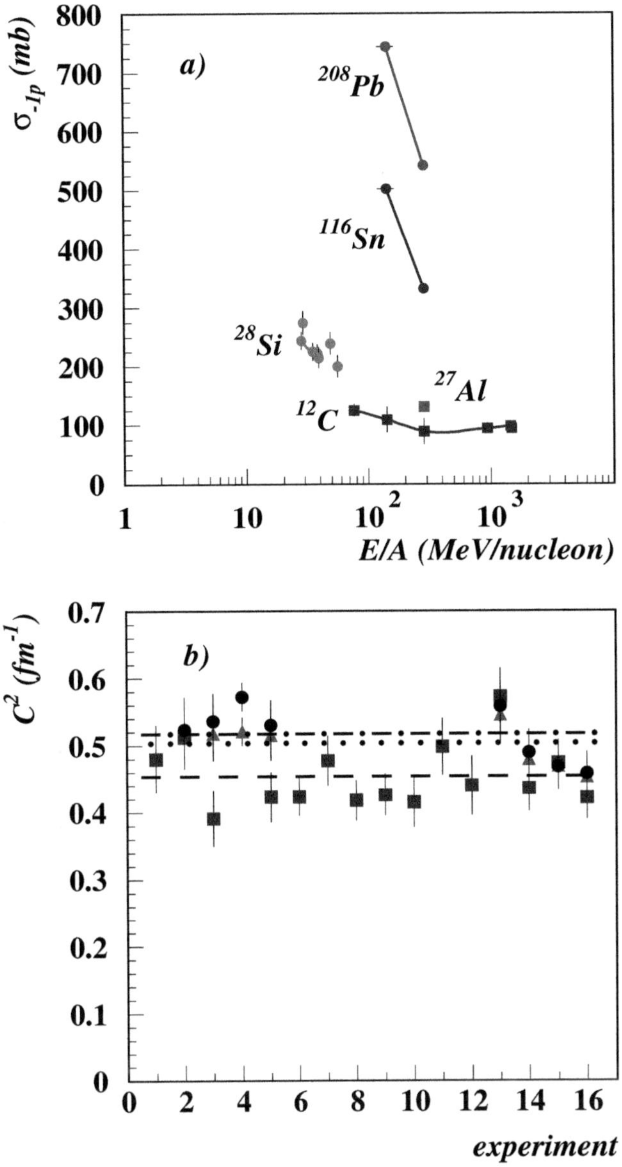

Fig. 1. a) The one-proton-removal cross sections on C, Al, Sn and Pb targets, depending on energy. b) The ANCs determined from the breakup of ^{8}B at 28-1000 MeV/nucleon using the data above and various effective interactions: JLM (squares), "standard" (circles) and "Ray" (triangles). The dashed, dotted and dash-dotted lines are the averages of the three interactions above, in that order. List of experiments in ref. [5].

evaluations of existing or new data and variations occur in the central values and uncertainties of the determinations. It is difficult to quote all of them and is not our intention to do so here. However, we notice that our average value of $S_{17}(0)$ is very close to the "low" values obtained from Coulomb dissociation data and some direct data $S_{17}(0) = 18.6 \pm 0.4(\text{exp})\pm 1.1(\text{syst})\,\text{eV}\cdot\text{b}$ [21]. It is also in reasonable agreement with the average value obtained by Cyburt *et al.* [24] $S_{17}(0) = 20.8 \pm 0.6(\text{stat})\pm 1.0(\text{syst})\,\text{eV}\cdot\text{b}$ using all radiative capture data in the assumption they are completely independent. The difference between our value and the value obtained from the direct measurement of Junghans *et al.* [23] $S_{17}(0) = 22.1 \pm 0.6(\text{stat})\pm 0.6(\text{theor})\,\text{eV}\cdot\text{b}$ still exists and is only relevant if the small uncertainty of the latter is true, given the fact that it involves extrapolation. It would, of course, be interesting to understand why the results differ. The difficulties encountered by the direct methods, both experimental (very small cross sections, difficult targets, etc...) and theoretical (extrapolations), are known (see, *e.g.*, ref. [25]). One important factor in any indirect determination of the astrophysical S-factor is that of the accuracy of the theoretical calculations involved. Much effort is done currently, *e.g.*, to investigate the accuracy of the absolute values of the calculations used in the analysis of the Coulomb dissociation experiments [26, 27]. We did our part above, using different NN interactions. Our central value is about 1σ lower than the average central value obtained by Cyburt *et al.* [24] in a recent analysis that uses all of the best available capture data, under the assumption that they are independent. Including the uncertainty quoted by Cyburt *et al.* our results are consistent at the 1σ level.

3.2 Breakup of ^{9}C to determine S_{18}

The same procedures have been applied for ^{9}C to determine the astrophysical S_{18} factor for the reaction ^{8}B$(p,\gamma)^9$C. The reaction is important in hot *pp*-chains as it can provide a starting point for an alternative path across the $A = 8$ mass gap [2]. The ANC for $^9\text{C} \rightarrow {}^8\text{B}+p$, has been determined using existing experimental data for the breakup of ^{9}C projectiles at 285 MeV/u on four different targets: C, Al, Sn and Pb [7]. No experimental data are available here for momentum distributions. The introduction of the final state interaction in the Coulomb dissociation part does not change the result by much, compared with our previous analysis [4]. We find now $C^2_{p_{3/2}} + C^2_{p_{1/2}} = 1.26 \pm 0.13\,\text{fm}^{-1}$. To calculate the astrophysical S-factor we use the potential model. We find $S_{18}(0) = 47 \pm 6\,\text{eV}\cdot\text{b}$. A very weak dependence on energy is observed: $S(E) = 47.3 - 15.1E + 7.34E^2$ (E in MeV). This result is in very good agreement with other determinations [28,8], but not with one from Coulomb dissociation [29], a fact that we do not understand. We underline that for this case the precision achieved from this determination is the best so far and is sufficient for astrophysical purposes.

3.3 Breakup of ^{23}Al and the consequences on the ^{22}Mg$(p,\gamma)^{23}$Al stellar reaction rate

Space-based gamma-ray telescopes have the ability to detect γ-rays of cosmic origin. They already provided strong and direct evidence that nucleosynthesis is an ongoing process through the detection of transitions in the decay of ^{26}Al, ^{56}Ni, ^{44}Ti, etc. Among the expected γ-ray emitters is ^{22}Na ($T_{1/2} = 2.6\,$y) produced in the thermonuclear runaway and the high-temperature phase in the so-called ONe novae (oxygen-neon novae) through the

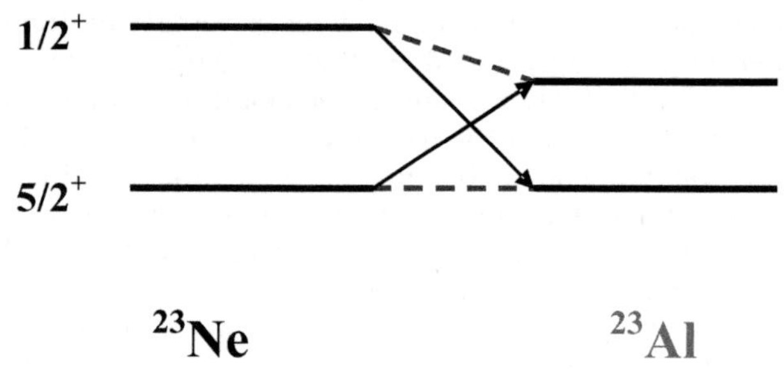

Fig. 2. The level inversion in ^{23}Al suggested in refs. [37,38].

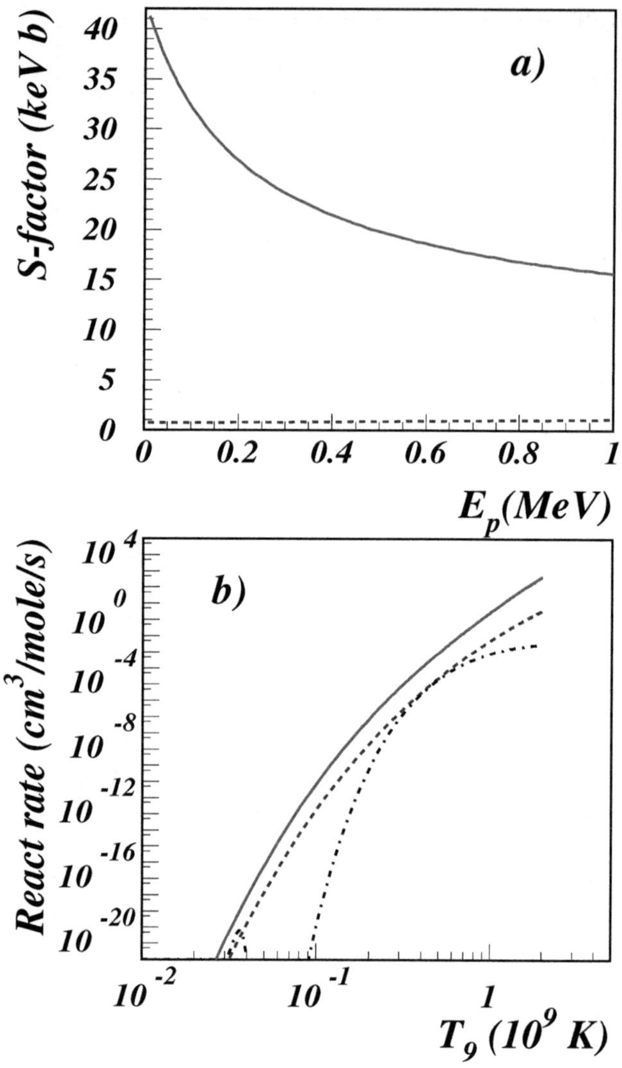

Fig. 3. a) The astrophysical S-factor for the ^{22}Mg(p, γ)^{23}Al reaction, calculated assuming $J^{\pi} = 5/2^+$ (dashed line), or $J^{\pi} = 1/2^+$ (orbital inversion, full line) for the g.s. of ^{23}Al. b) The corresponding reaction rate calculated for the case of non-inversion (dashed), or inversion (full line). The dash-dotted line shows the resonant contribution of the 1st excited state in ^{23}Al.

reaction chain ^{20}Ne(p, γ)^{21}Na(p, γ)^{22}Mg(β, γ)^{22}Na (NeNa cycle) [30,31,32]. Measurements, however, have not detected the 1.275 MeV gamma-ray following the decay of ^{22}Na and have only been able to set an upper limit on its production, a limit which is below the theoretical predictions (see, for example, [33,34] and references therein). This discrepancy may arise from a poor knowledge of the reaction cross sections employed in the network calculations for the rp-process. In particular, it was proposed that the precursor ^{22}Mg can be depleted by the radiative proton capture reaction ^{22}Mg(p, γ)^{23}Al [35], which can result in a serious reduction of the ^{22}Na abundance. The reaction is dominated by direct capture and resonant capture through the first excited state in ^{23}Al. There is no direct measurement of the cross section at stellar energies because it is impossible to make a ^{22}Mg ($T_{1/2} = 3.86$ s) target and difficult to obtain an intense ^{22}Mg beam. Therefore, currently the rate of this reaction is estimated based on the mass and resonance energy determined experimentally [36] and assuming that the spins and parities are as in the mirror system ^{23}Ne.

The nucleus ^{23}Al is a weakly bound proton-rich nucleus ($S_p = 0.123(19)$ MeV) close to the drip line. Recent measurements of the reaction cross sections for $N = 10$ isotones and $Z = 13$ isotopes around 30 MeV/nucleon on a ^{12}C target found a remarkable enhancement for ^{23}Al, which led the authors to the conclusion that it is one of the rare proton halo nuclei [37]. This is explained with a presumed level inversion between the $2s_{1/2}$ and $1d_{5/2}$ orbitals (fig. 2). The inversion was further supported by several microscopic nuclear structure calculations that find $J^{\pi} = 1/2^+$ for the ^{23}Al ground state [38]. If the above-mentioned inversion is correct, it will affect the radiative capture cross section much more strongly than any other uncertainties. Indeed, assuming such an inversion, we recalculate the astrophysical S-factor (fig. 3a) and the stellar reaction rate (fig. 3b) for the ^{22}Mg(p, γ)^{23}Al reaction and find an increase of 30 to 50 times over the current estimate of the rate for the temperature range $T_9 = 0.1 - 0.3$. Clearly then, it is important to determine the spin and parity of the low-lying levels in ^{23}Al. It is important for both nuclear structure and for its consequences for nuclear astrophysics. As a further complication, the NNDC data base gives $J^{\pi} = 3/2^+$ for the ground state of ^{23}Al [39].

We proposed the use of intermediate-energy one-proton removal reactions on a light target as a means to determine the structure of the ^{23}Al ground state. Such reactions have proven to be a reliable spectroscopic tool, with advantages in particular for the case of weakly bound isotopes, close to the drip lines [12,14]. We calculate that for ^{12}C(^{23}Al, ^{22}Mg) at 60 MeV/nucleon, the parallel momentum distribution is some 2 times narrower for a $2s_{1/2}$ orbital than for a $1d_{5/2}$ orbital (fig. 4) and the associated cross section is about a factor two larger. We intend to compare the calculated momentum distributions and cross sections with the experimental ones and determine the spin and parity of the ^{23}Al ground state. We shall derive the related ANCs and from them, the astrophysical S-factor. Calculations for the momentum distributions have

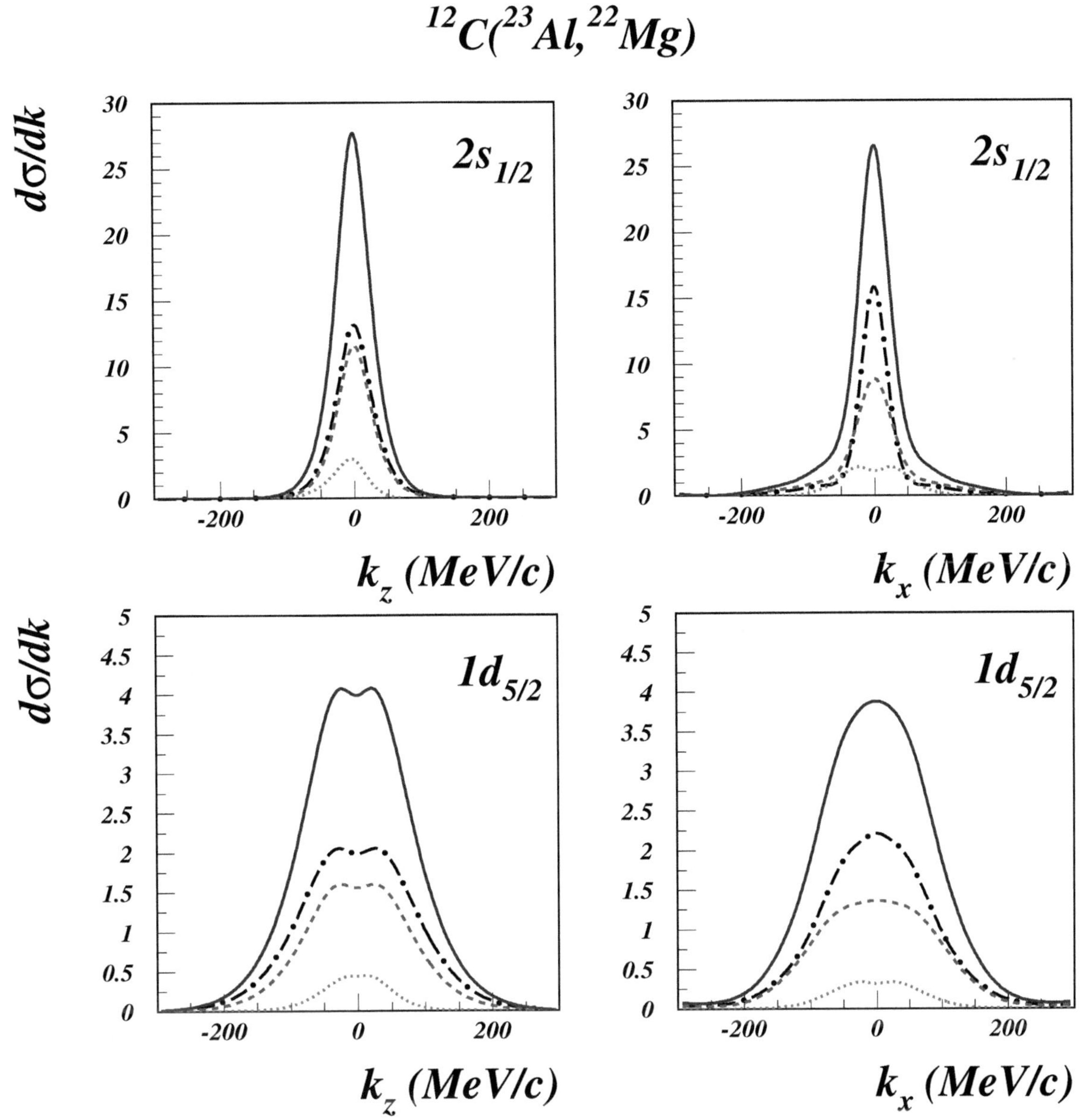

Fig. 4. The calculated parallel (left panels) and transverse (right panels) momentum distributions for the proton-removal (breakup) of 60 MeV/nucleon ^{23}Al on a light target in the two spin assumptions. The curves show the different contributions: stripping (dash-dotted), diffraction dissociation (dashed) and Coulomb (dotted). The full lines are the sum of all contributions. They were all calculated assuming pure $1d_{5/2}$ or $2s_{1/2}$ orbitals, respectively.

been performed with procedures similar to those used previously [3,14]. The first step is the calculation of the single-particle density in ^{22}Mg using a spherical HF+BCS calculation with the density energy functional of Beiner and Lombard [40]. The experimental proton separation energy in ^{23}Al, $S_p = 0.123$ MeV, was reproduced. There are two possibilities for the spin-parity of the ground state: $J^\pi = 5/2^+$ or $1/2^+$. Glauber model calculations have therefore been performed for each case, assuming pure $1d_{5/2}$, or $2s_{1/2}$ orbitals, in order to demonstrate the sensitivity of the technique. Scattering functions defining the stripping and diffraction transition operators were generated with double-folding potentials using the JLM effective interaction, renormalized as above. Calculations were done for breakup on a ^{12}C target to minimize the Coulomb effects.

We draw two conclusions from the calculations:

1) The two possible assignments may be resolved on the basis of the inclusive cross sections and momentum distributions. For example the cross section drops by a factor of two if a $1d_{5/2}$ state is assumed rather than $2s_{1/2}$. This is easy to understand because the low-binding energy and the lack of a centrifugal barrier in the case of a $2s_{1/2}$ orbital leads to a much longer tail of the radial wave function than for the case of the $1d_{5/2}$ orbital. Moreover, the very peripheral character of single-nucleon removal reactions means that it is the asymptotic part of the wave

function that dictates the cross section and momentum distribution. In the case of the latter, the widths of the momentum distributions differ by a factor of two for both the parallel and transverse momenta, reflecting the different behavior of the tails of the wave functions. Cross sections of 97 mb ($2s_{1/2}$), and of 42 mb ($1d_{5/2}$) were found for the two ground-state spin-parity assignments. The corresponding widths (FWHM) of the distributions are predicted to be 60 MeV/c, and 180 MeV/c, respectively.

2) The shape of momentum distributions is extremely selective —narrow for a $2s_{1/2}$ state and broad with a flat top and a small central dip for removal of a $1d_{5/2}$ (fig. 4).

A study of one-proton removal from ^{23}Al should, therefore, allow the spin-parity of the ground state of ^{23}Al to be deduced. Measurements of the cross sections and momentum distributions in coincidence with gamma-rays from the ^{22}Mg core will allow us to disentangle the detailed structure of the wave function, and in particular to deduce the spectroscopic factors for the various configurations. This spectroscopic information will also be valuable to determine if ^{23}Al is deformed or spherical. Currently experimental studies of Coulomb dissociation of ^{23}Al are carried out at RIKEN to determine the gamma-ray width of its first excited state. The results in ref. [41] $\Gamma_\gamma = 7.2 \pm 1.4 \cdot 10^{-7}$ eV agree with the value estimated by us using single-particle wave functions for the $E2$ transition: $\Gamma_\gamma = 6 \cdot 10^{-7}$ eV and used in the present calculation of the resonant term (shown in fig. 3b) for the capture reaction rate ^{22}Mg$(p,\gamma)^{23}$Al.

4 Conclusions

In conclusion, we have shown that one-proton-removal reactions at intermediate energies can be used to obtain astrophysical S-factors at stellar energies for radiative proton capture reactions. Difficult or impossible direct measurements for nuclear astrophysics at very low energies can be replaced by indirect measurements with radioactive beams at larger energies. We find that a kinematic window exists at 30–150 MeV/nucleon where the reactions are peripheral and the relevant ANC can be determined. The method is particularly useful because it can be used for rare isotopes, for poor-quality radioactive beams obtained from fragmentation, with cocktail beams and with low intensity beams. It was shown that breakup at intermediate energies can be studied with beams as low as a few particles/s [42]. Our results from the use of different NN interactions remind us of the fact that the precision of all indirect methods depends not only on the precision of the experiments but also on the accuracy of the calculations. Our findings may give a measure of the present status of accuracy.

This work was supported in part by the U.S. Department of Energy under Grant No. DE-FG03-93ER40773, by the Romanian Ministry for Research and Education under contract No. 555/2000, and by the Robert A. Welch Foundation.

References

1. J.N. Bahcall, M.H. Pinsoneault, S. Basu, Astrophys. J. **555**, 990 (2001).
2. M. Wiescher *et al.*, Astrophys. J. **343**, 352 (1989).
3. L. Trache, F. Carstoiu, C.A. Gagliardi, R.E. Tribble, Phys. Rev. Lett. **87**, 271102 (2001).
4. L. Trache *et al.*, Phys. Rev. C **66**, 035801 (2002).
5. L. Trache, F. Carstoiu, C.A. Gagliardi, R.E. Tribble, Phys. Rev. C **69**, 032802(R) (2004).
6. F. Negoita *et al.*, Phys. Rev. C **54**, 1787 (1996).
7. B. Blank *et al.*, Nucl. Phys. A **624**, 242 (1997).
8. J. Enders *et al.*, Phys. Rev. C **67**, 064301 (2003).
9. R. Warner *et al.*, to be published in Phys. Rev. C.
10. D. Cortina-Gil *et al.*, Nucl. Phys. A **720**,3 (2003).
11. I. Tanihata, J. Phys. G **22**, 157 (1996).
12. P.G. Hansen, B.M. Sherrill, Nucl. Phys. A **693**, 133 (2001).
13. A.M. Mukhamedzhanov *et al.*, Phys. Rev. C **63**, 024612 (2001).
14. E. Sauvan *et al.*, Phys. Lett. B **491**, 1 (2000); Phys. Rev. C **69**, 044603 (2004).
15. J.P. Jeukenne, A. Lejeune, C. Mahaux, Phys. Rev. C **16**, 80 (1977).
16. L. Trache *et al.*, Phys. Rev. C **61**, 024612 (2000).
17. L. Ray, Phys. Rev. C **20**, 1857 (1979).
18. A. Azhari *et al.*, Phys. Rev. Lett. **82**, 3960 (1999).
19. G. Tabacaru *et al.*, Phys. Rev. C **73**, 025808 (2006).
20. F. Schumann *et al.*, Phys. Rev. Lett. **90**, 232501 (2003) and references therein.
21. B. Davids, S. Typel, Phys. Rev. C **68**, 045802 (2003).
22. F. Hammache *et al.*, Nucl. Phys. A **746**, 370c (2004).
23. A.R. Junghans *et al.*, Phys. Rev. C **68**, 065803 (2003).
24. R.H. Cyburt, B. Davids, B.K. Jennings, Phys. Rev. C **70**, 045501 (2004).
25. F. Strieder *et al.*, Eur. Phys. J. A **3**, 1 (1998).
26. H. Esbensen, G.F. Bertsch, K.A. Snover, Phys. Rev. Lett. **94**, 042502 (2005).
27. C. Bertulani, Phys. Rev. Lett. **94**, 072701 (2005).
28. D. Beaumel *et al.*, Phys. Lett. B **514**, 226 (2001).
29. T. Motobayashi, Nucl. Phys. A **718**, 101c (2002); RIKEN Accelerator Progress Report 1999, **33**, 64 (2000) and 2002, **36**, 64 (2003).
30. S. Starrfield, J.W. Truran, M. Wiescher, W.M. Sparks, Mon. Not. R. Astron. Soc. **296**, 502 (1998).
31. J. Jose, A. Coc, M. Hernanz, Astrophys. J. **520**, 347 (1999).
32. S. Wanajo, M. Hashimoto, K. Homono, Astrophys. J. **523**, 409 (1999)
33. I. Iyundin *et al.*, Astron. Astrophys. **300**, 422 (1995).
34. R. Diehl, Nucl. Phys. A **718**, 52c (2003) and references therein.
35. M. Wiescher *et al.*, Nucl. Phys. A **484**, 90 (1988).
36. J. Caggiano *et al.*, Phys. Rev. C **64**, 025802 (2001).
37. X.Z. Cai *et al.*, Phys. Rev. C **65**, 024610 (2002).
38. H.-Y. Zhang *et al.*, Chin. Phys. Lett. **19**, 1599 (2002); **20**, 46; 1234 (2003).
39. *National Nuclear Data Center*, BNL. `http://www.nndc.bnl.gov`.
40. M. Beiner, R.J. Lombard, Ann. Phys. (N.Y.) **86**, 262 (1974).
41. T. Gomi *et al.*, Nucl. Phys. A **718**, 508c (2003); **734**, Suppl. 1, E77 (2004); **758**, 761c (2005).
42. V. Maddalena *et al.*, Phys. Rev. C **63**, 024613 (2001).

Eur. Phys. J. A **27**, s01, 243–248 (2006)
DOI: 10.1140/epja/i2006-08-038-1

EPJ A direct
electronic only

Validity test of the Trojan Horse Method applied to the ^{7}Li + p $\longrightarrow \alpha + \alpha$ reaction via the ^{3}He break-up

A. Tumino[1,2,a], C. Spitaleri[1,2], M.L. Sergi[1,2], V. Kroha[3], V. Burjan[3], S. Cherubini[1,2], Zs. Fülöp[4], M. La Cognata[1,2], L. Lamia[1,2], J. Novác[3], R.G. Pizzone[2], S. Romano[1,2], E. Somorjai[4], S. Tudisco[1,2], and J. Vincour[3]

[1] Dipartimento di Metodologie Fisiche e Chimiche per l'Ingegneria, Universitá di Catania, Catania, Italy
[2] INFN Laboratori Nazionali del Sud, Catania, Italy
[3] Nuclear Physics Institute of ASCR, Rez near Prague, Czech Republic
[4] Institute of Nuclear Research of Hungarian Academy of Sciences (ATOMKI), Debrecen, Hungary

Received: 20 June 2005 /
Published online: 10 March 2006 –

Abstract. The Trojan Horse Method (THM) was applied to the ^{3}He+^{7}Li interaction in order to investigate the quasi-free ^{7}Li(p, α)^{4}He reaction. The three-body experiment was performed at 33 MeV corresponding to a ^{7}Li-p relative energy ranging from 50 keV to 7 MeV. The extracted ^{7}Li(p, α)^{4}He quasi-free cross-section was compared with the behavior of direct data, as well as with the result of a previous THM investigation on the ^{7}Li(p, α)^{4}He reaction off the neutron in ^{2}H. A good agreement between data sets shows up throughout the energy range investigated, providing a very important validity test of the pole approximation for the THM.

PACS. 24.10.-i Nuclear-reaction models and methods – 25.40.-h Nucleon-induced reactions

1 Introduction

In the last years, the Trojan Horse Method (THM) [1, 2,3,4,5,6,7,8,9,10,11,12] has proven to be a successful indirect approach for studying charged-particle two-body reactions relevant for astrophysics. The Trojan Horse Method selects the quasi-free contribution of an appropriate three-body $A + a \rightarrow c + C + s$ reaction performed at energies well above the Coulomb barrier to extract the relevant two-body $A + x \rightarrow c + C$ reaction cross-section at astrophysical energies free of Coulomb suppression as well as electron screening effects. The method is based on previous studies on quasi-free (QF) scattering and reactions [13,14,15,16,17], where evidence for a QF mechanism at energies of few tens of MeV was observed and justified by the high momentum transferred to the outgoing particles, due to the high Q-value of the reactions involved. When applying the method, kinematical conditions are appropriately chosen in order that s be spectator to the process and the x-s binding energy "enough" to compensate for the A+a relative motion. The prescription to calculate the accessible astrophysical energy region is given by

$$\Delta E_{qf} = E_{Ax} - B_{xs} \pm E_{xs}, \qquad (1)$$

where E_{Ax} is the beam energy in the center of mass of the two-body A-x system, B_{xs} represents the binding energy for the x-b system and E_{xs} describes their inter-cluster motion within a chosen cutoff in momentum. A key role is thus played by the "Trojan Horse nucleus" a, whose wave function is required to have a large amplitude for a $x - s$ cluster configuration. However, a general view of the influence of its bound structure on the validity of the pole approximation for the THM is still missing. Recently this problem was addressed within an experimental program to indirectly study relevant reactions destroying Li isotopes in stellar environments. In particular the ^{7}Li(p, α)^{4}He reaction, key reaction in the so-called "Li problem", was first investigated via ^{2}H break-up and then by picking the proton up from ^{3}He. From the first investigation via the ^{2}H(^{7}Li, $\alpha\alpha$)n break-up process, very interesting results were obtained [3,6,7,18]. Two resonances, associated with the group of ^{8}Be states between 19.9 and 20.2 MeV and that around 22.2 MeV of excitation energy, contributing to the ^{7}Li-p direct cross section, could be observed also in the indirect ^{7}Li-p excitation function [18]. Moreover, the behavior of the THM $S(E)$-factor at low energy was found to be in very good agreement with that of standard measurements in the region where screening effects are negligible. Estimates of the bare nucleus S_0 factor and electron screening potential U_e were obtained in PWIA [6] as well as applying the more sophisticated MPWBA [5,7], confirming the values from direct measurement, but with smaller errors. Direct values are indeed affected by larger

[a] e-mail: tumino@lns.infn.it

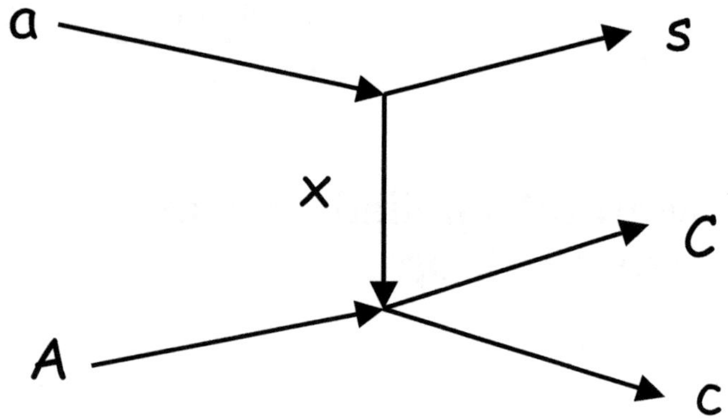

Fig. 1. Diagram representing the quasi-free $A+a \to c+C+s$ reaction; the nucleus A interacts only with the cluster x, leaving particle s as spectator to the process.

uncertainties due to the extrapolation procedure employed to get them [19]. In the present paper the THM results of the ^{7}Li(p, α)^{4}He reaction via the ^{7}Li(^{3}He, $\alpha\alpha$)^{2}H three-body experiment analysed in PWIA are presented. A ^{7}Li-p relative energy range (0.2–7 MeV) including the mentioned resonances, was populated. Results may be eventually different due to the presence of a different "Trojan Horse nucleus" (^{3}He instead of ^{2}H). Moreover a charged spectator, as in the present case, may introduce some distortions in the exit channel. A good agreement between direct and indirect two-body excitation functions also in this case, would provide a very important validity test of the pole approximation as well as of invariance of the reaction amplitude for the two-body channel.

2 Plane-Wave Impulse Approximation

In the phase space region where the QF mechanism is expected to be present, the Impulse Approximation [20, 21] can be applied, which describes the quasi free $A+a \to c+C+s$ reaction by a Pseudo Feynman diagram (see fig. 1).

A pole of the diagram refers to the break-up of the target nucleus a into the clusters x and s, and the other one contains the information on the virtual $A+x \to c+C$ two-body process. In order that this description be valid, the following conditions have to be fulfilled [20]:

- The momentum transfer sufficiently high or equivalently the associated wavelength sufficiently smaller (less than 1 fm). Consequently the $A - x$ interaction can be considered confined.
- The incident center-of-mass energy higher than the binding energy of clusters x-s.

In the Plane-Wave Impulse Approximation (PWIA) the cross section of the three-body reaction can be factorized into two terms corresponding to the two poles of fig. 1 [22, 23] and it is given by

$$\frac{\mathrm{d}^3\sigma}{\mathrm{d}E_c \mathrm{d}\Omega_c \mathrm{d}\Omega_C} \propto KF \left(\frac{\mathrm{d}\sigma}{\mathrm{d}\Omega}_{cm}\right)^{off} \cdot |\Phi(\boldsymbol{p}_s)|^2, \qquad (2)$$

where

- $[(\mathrm{d}\sigma/\mathrm{d}\Omega)_{cm}]^{off}$ is the off-energy-shell differential cross section for the two body $A(x,c)C$ reaction at the center-of-mass energy E_{cm} given in post collision prescription by
$$E_{cm} = E_{c\text{-}C} - Q_{2b}, \qquad (3)$$
where Q_{2b} is the two body Q-value of the $A+x \to c+C$ reaction and $E_{c\text{-}C}$ is the relative energy between the outgoing particles c and C.
- KF is a kinematical factor containing the final-state phase-space factor and it is a function of masses, momenta and angles of the outgoing particles:
$$KF = \frac{\mu_{Aa} m_c}{(2\pi)^5 \hbar^7} \frac{p_C p_c^3}{p_{Aa}} \left[\left(\frac{\boldsymbol{p}_{Bx}}{\mu_{Bx}} - \frac{\boldsymbol{p}_{Cc}}{m_c}\right) \cdot \frac{\boldsymbol{p}_c}{p_c}\right]^{-1}. \qquad (4)$$
- $\Phi(\boldsymbol{p}_s)$ is the momentum distribution of the spectator particle. In PWIA it is given by the Fourier transform of the radial wave function $\chi(\boldsymbol{r})$ for the x-s inter-cluster motion, usually described in terms of Hänkel, Eckart and Hulthén functions depending on the x-s system properties. Within momentum values of 40–50 MeV/c, its shape cannot be distinguished by those ones associated with more sophisticated approaches.

In the experimental work reported in the present paper, the validity conditions of the IA appear to be fulfilled. Indeed ^{3}He has a quite high energy of 33 MeV (610 MeV/c in momentum) corresponding to an associated de Broglie wavelength of 0.32 fm, much smaller than the ^{3}He mean square radius of 1.95 fm. Once measured the three-body cross-section, one can extract the cross-section for the $A+x$ interaction from eq. (2), under the assumption that off-energy-shell effects are negligible. However, this assumption has to be verified from comparison with direct data.

3 The experiment

The ^{7}Li(^{3}He, $\alpha\alpha$)^{2}H experiment was performed at the Nuclear Physics Institute of ASCR in Rez, near Prague. A 33 MeV ^{3}He cyclotron beam was delivered onto an isotopically enriched lithium fluoride target (^{7}Li $\approx$ 95%). Two silicon ΔE-E telescopes for α's identification, consisting of 20 μm ΔE- and 1000 μm position-sensitive E-detector, were placed on opposite sides with respect to the beam direction covering the laboratory angles 94.6° to 109.4° and 44.3° to 59.7°. The angular ranges were chosen in order to cover momentum values p_s of the undetected deuteron ranging from about -100 MeV/c to about 100 MeV/c. This assures that the bulk of the QF contributions for the break-up process of interest falls inside the investigated regions, allowing also to cross check the method outside the relevant phase-space regions. The trigger for the event acquisition was given by the coincidences between the two telescopes.

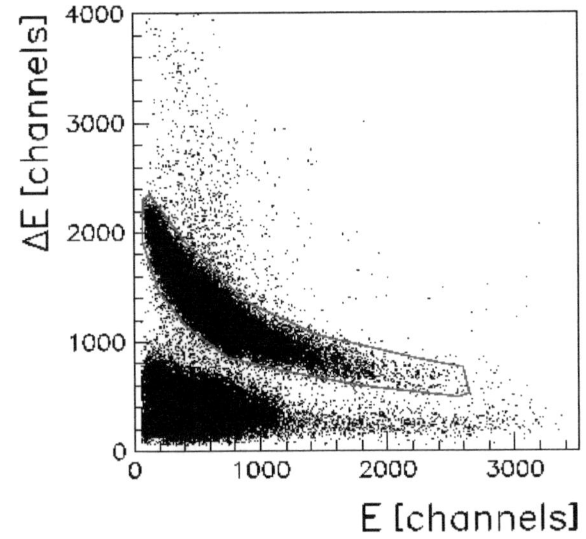

Fig. 2. Example of ΔE-E 2D-plot; the graphical cut to select the α-particles is shown as full line.

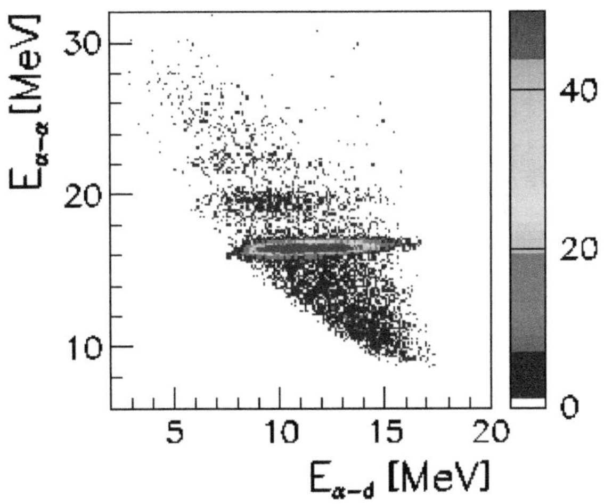

Fig. 4. Coincidence events in the α-α *vs.* α-d relative energy plane.

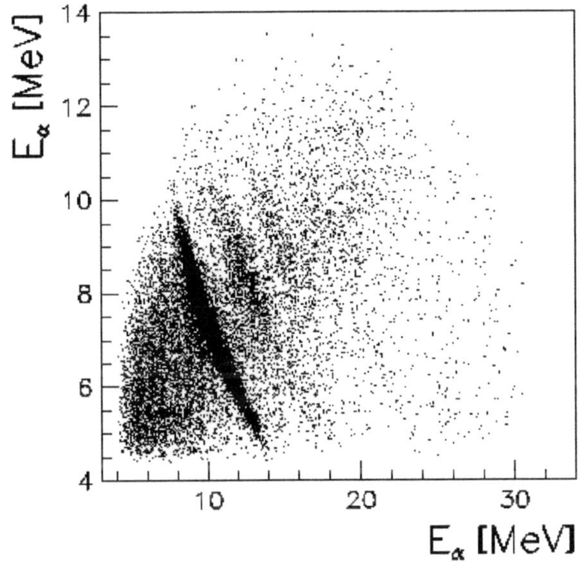

Fig. 3. Locus of events in the E_α *vs.* E_α plane. The correlated group of events corresponds to ^{8}Be excited states, feeding the two α's in the exit channel.

4 Data analysis and results

The energy and position calibration of the telescopes was performed using data acquired in preliminary runs of the ^{3}He + ^{197}Au elastic scattering. A standard three-peak α source was also used. In order to identify the channel of interest and to choose the kinematical conditions where the quasi-free process is dominant, α loci were selected in the ΔE-E two-dimensional plots. An example of the ΔE-E plot is shown in fig. 2, with the graphical cut to select the α-particles. The kinematics were reconstructed under the assumption of a deuteron as third particle, leading to the locus of events in the E_α *vs.* E_α plane shown in fig. 3. It reproduces very well the one calculated by using a three-body kinematical calculation. At least three correlated groups of events can be recognized in the figure, corresponding to excited states of ^{8}Be, which decay into the two α's in the exit channel. The first group from the left is fed by the 16.6 and 16.9 MeV states, the second one is due to the states between 19.9 and 20.2 MeV, while the third one corresponds to the 22.2 and 22.9 MeV levels. These states have all isospin $T = 0$ and natural parity. The Q-value spectrum for these events shows a peak at about 12 MeV, which refers to the α+α+d channel of interest whose calculated Q-value is 11.85 MeV. However, the Q-value for the ^{7}Li(p, α)^{4}He two-body reaction (17.35 MeV) is larger than the excitation energies of the ^{8}Be states contributing to the first group of events. Thus these states will not appear as resonances in the ^{7}Li-p excitation function. The resulting spectra make us confident on the quality of the performed calibration, and on the possibility to well identify the α+α+d channel. Another representation of the coincidence events is given by projecting them onto the 2D-plot of relative energies for any two of the three final particles. The 2D-plot for α-α *vs.* α-d relative energies is shown in fig. 4. In this representation very clear horizontal loci show up, corresponding to the groups of ^{8}Be states recognized in fig. 2. As expected, neither vertical nor diagonal loci associated with the excitation of ^{6}Li, seem to be present, because no α decaying states can contribute in this α-d relative energy region.

The ^{8}Be excited states can be populated via quasi-free mechanism or via sequential decay. A way to investigate the reaction mechanism involved and to disentangle quasi-free coincidence events from other, is to investigate the behavior of the coincidence yield depending on the deuteron momentum p_s for all coincidence events, in the whole angular range covered by the detectors. Relative

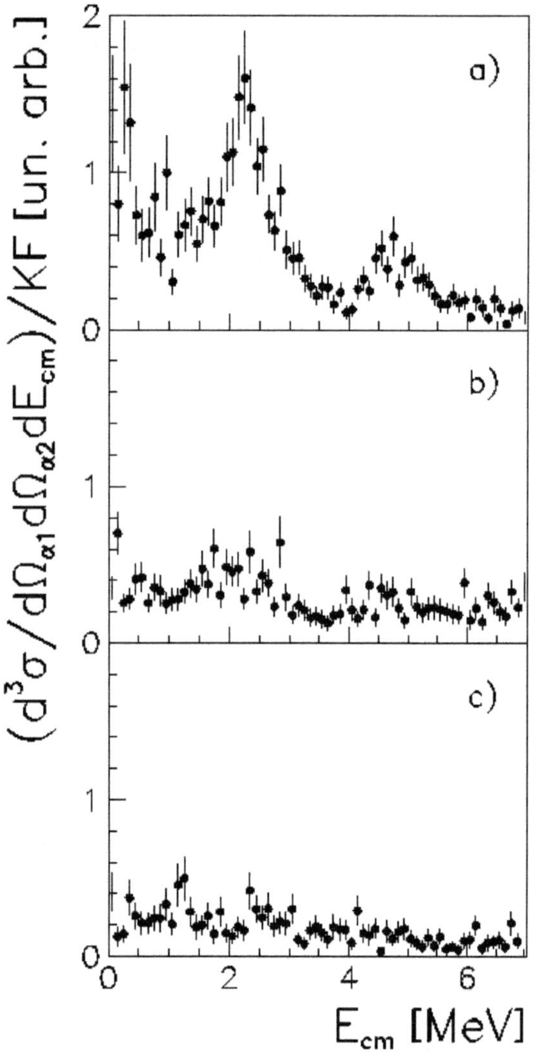

Fig. 5. Coincidence yield projected onto the ^{7}Li-p energy axis (E_{cm}) for different p_s ranges: $0\,\mathrm{MeV}/c \le |p_s| \le 30\,\mathrm{MeV}/c$ (a), $30\,\mathrm{MeV}/c \le |p_s| \le 60\,\mathrm{MeV}/c$ (b) and $|p_s| \ge 60\,\mathrm{MeV}/c$ (c).

energy $E_{^7\mathrm{Li}\text{-}p}$ spectra ($E_{^7\mathrm{Li}\text{-}p}$ corresponds to the E_{cm} variable of eq. (2)) divided by the phase-space contribution were reconstructed for different ranges of the deuteron momentum p_s. Within $0\,\mathrm{MeV}/c \le |p_s| \le 30\,\mathrm{MeV}/c$ (fig. 5a) and $30\,\mathrm{MeV}/c \le |p_s| \le 60\,\mathrm{MeV}/c$ (fig. 5b) momentum ranges, the coincidence yield appears to be quite high in particular close to the $E_{^7\mathrm{Li}\text{-}p}$ resonant window. Moving a bit far in momentum ($|p_s| > 60\,\mathrm{MeV}/c$), the coincidence yield drastically decreases as shown in fig. 5c.

A strong correlation between coincidence yield and deuteron momentum p_s shows up, a necessary condition for the dominance of the quasi-free mechanism in the region approaching zero deuteron momentum. But in case of resonances in the low p_s region, this result might be not a sufficient condition, since the correlation can be partially depending on the resonant behavior, regardless of its sequential decay- or quasi free-origin. An observable which turns out to be more sensitive to the reaction mechanism is the shape of the experimental momentum distribution for the deuteron. The experimental momentum distribution was reconstructed in Plane-Wave Impulse Approximation (PWIA) by applying the energy sharing method [24] to our coincidence data selecting ^{7}Li-p relative energy windows of 100 keV. The ^{7}Li-p relative energy was calculated in post collision prescription from eq. (3). Data were analyzed in PWIA, applying the factorization of eq. (2), and taking as $\Phi(\boldsymbol{p}_s)$ the Fourier transform of the radial wave function for the p-d intercluster motion inside ^{3}He, described in terms of a Hulthén function

$$\Phi(\boldsymbol{r}) = \sqrt{\frac{ab(a+b)}{2\pi(a-b)^2}}\,\frac{1}{r}\left(e^{-ar} - e^{-br}\right) \tag{5}$$

with parameters $a = 0.2317\,\mathrm{fm}^{-1}$ and $b = 1.202\,\mathrm{fm}^{-1}$ [18]. Dividing the three-body coincidence yield by the kinematical factor, we are left with a quantity proportional to the product of the momentum distribution with the differential ^{7}Li-p two-body cross-section. Since in the restricted relative energy ranges of 100 keV, the differential two-body cross-section can be considered constant, the quantity above reflects the behavior of the experimental momentum distribution in arbitrary units. The result is reported in fig. 6. The full line superimposed onto the data represents the shape of the theoretical Hulthén function, which is normalized to the experimental maximum. A quite good agreement shows up, making us confident that in the experimentally selected kinematical region, the quasi-free mechanism gives the main contribution to the ^{3}He + ^{7}Li reaction and it can be selected without

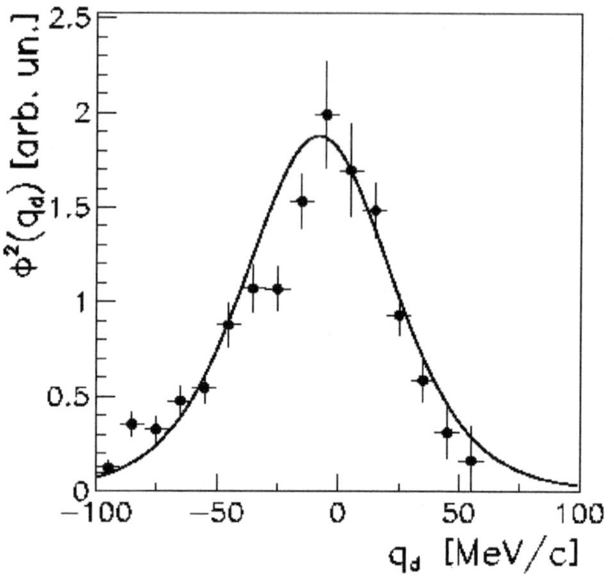

Fig. 6. Experimental deuteron momentum distribution. The full line represents the shape of the theoretical Hulthén function in momentum space.

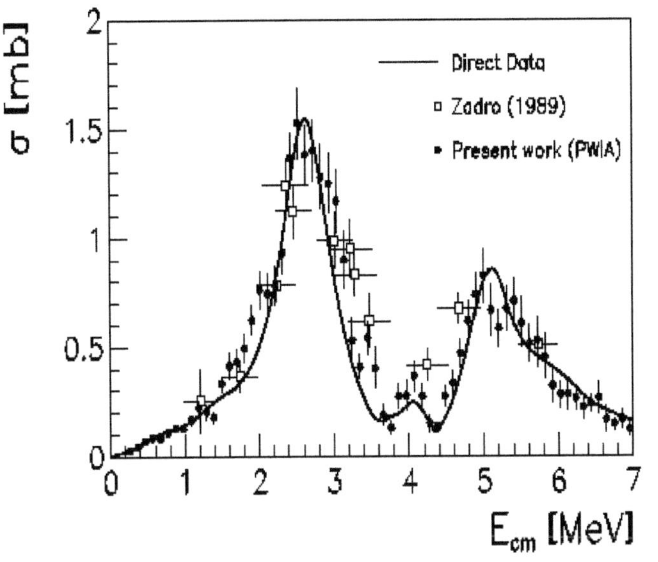

Fig. 7. Quasi-free data from the present experiment (full dots), direct cross-section (full line) from [19,26,27] and previous quasi-free data from ^{2}H break-up (open squares) [18].

significant interference with contaminant sequential decay processes. The further analysis was performed by considering coincidence events with a neutron momentum ranging between −30 and 30 MeV/c. Following the PWIA approach, the two-body cross-section was derived dividing the selected three-body coincidence yield by the result of a Monte Carlo calculation reproducing the behavior of the $KF\Phi(\boldsymbol{p}_s)$ product (see eq. (2)). The geometrical efficiency of the experimental setup as well as the detection thresholds were accounted for in the calculation. An error calculation for the ^{7}Li-p relative energy was also performed giving a value ranging from 80 to 120 keV, the minimum estimate corresponding to the phase space region where the lens effect is more efficient [25]. In the extracted off-energy shell ^{7}Li-p two-body cross-section, penetrability effects were included before the comparison with direct data [19,26,27]. Both s and p waves in the entrance channel had to be considered, the s-wave describing the non-resonant behavior, while the p-wave being responsible of the resonant contributions. Thus Coulomb as well as $l = 1$ centrifugal barrier were accounted for in the procedure. In order to perform the comparison, direct data were integrated over the same θ_{cm} angular region covered in the present experiment (50°–70°), θ_{cm}, being the emission angle for one outgoing α-particle in the α-α center-of-mass system [4,7,8,11]. The comparison is shown in fig. 7, where full dots represent our data, while the behavior of direct data from [19,26,27] is shown as full line, both sets averaged out at the same energy bin of 100 keV comparable with the uncertainty. The normalization to the direct data was performed in the ^{7}Li-p relative energy region between 2 and 3 MeV. Our former data via ^{2}H break-up [18] are also reported in the figure as open squares. Data sets agree quite well throughout the investigated range, including the resonant regions. It is important to underline that with a single normalization parameter it was possible to reproduce the direct cross section over a wide ^{7}Li-p relative energy range. The good agreement validates the pole approximation for this experiment together with the PWIA approach and, importantly, the present results with ^{3}He as "Trojan Horse nucleus" agree with the previous investigation of the ^{7}Li(p, α)^{4}He reaction off the neutron in ^{2}H. Therefore the invariance of the reaction amplitude for this two-body process is confirmed within the experimental errors.

5 Conclusions

The ^{7}Li(p, α)^{4}He reaction was investigated by selecting the quasi-free contribution to the ^{3}He(^{7}Li, αα)^{2}H three-body reaction performed at 33 MeV. The two-body cross-section was deduced in PWIA approach and compared with the direct behavior as well as with previous indirect data from the ^{7}Li(d, αα)n [18]. The good agreement between the sets of data suggests that ^{3}He is a good "Trojan Horse nucleus", in spite of its quite high binding energy (5.85 MeV) and that the validity of the pole approximation, at least for the ^{7}Li+p interaction, is not "Trojan Horse nucleus" dependent. Although the present analysis in PWIA does not include Coulomb distortions due to the charged spectator, these effects seem not so important. The possibility to employ the simple PWIA is a relevant result, because it allows to simplify the theoretical formulation. However, from recent THM investigations [7,11, 12] it comes out that more sophisticated approaches can be needed at sub-Coulomb energies in order to extract the cross-section of astrophysical interest. An important effect to be considered is the Coulomb distortion in the particle wave functions of the two-body exit channel. Further investigation of this experimental work with theoretical approaches based on the DWBA formalism is thus needed in order to better understand these important aspects.

This work was supported in part by OTKA T49245 and IN64269.

References

1. G. Baur, Phys. Lett. B **178**, 135 (1986).
2. S. Cherubini, V.N. Kondratyev, M. Lattuada, C. Spitaleri, Dj. Miljanić, M. Zadro, G. Baur, Astrophys. J. **457**, 855 (1996).
3. C. Spitaleri, M. Aliotta, S. Cherubini, M. Lattuada, Dj. Miljanić, S. Romano, N. Soić, M. Zadro, R.A. Zappalà, Phys. Rev. C **60**, 055802 (1999).
4. C. Spitaleri, M. Aliotta, P. Figuera, M. Lattuada, R.G. Pizzone, S. Romano, A. Tumino, C. Rolfs, L. Gialanella, F. Strieder, S. Cherubini, A. Musumarra, Dj. Miljanić, S. Typel, H.H. Wolter, Eur. Phys. J. A **7**, 181 (2000).
5. S. Typel, H. Wolter, Few-Body Syst. **29**, 7 (2000).

6. M. Aliotta, C. Spitaleri, M. Lattuada, A. Musumarra, R.G. Pizzone, A. Tumino, C. Rolfs, F. Strieder, Eur. Phys. J. A **9**, 435 (2000).
7. M. Lattuada, R.G. Pizzone, S. Typel, P. Figuera, Dj. Miljanić, A. Musumarra, M.G. Pellegriti, C. Rolfs, C. Spitaleri, H.H. Wolter, Astrophys. J. **562**, 1076 (2001).
8. C. Spitaleri, S. Typel, R.G. Pizzone, M. Aliotta, S. Blagus, M. Bogovac, S. Cherubini, P. Figuera, M. Lattuada, M. Milin, Dj. Miljanić, A. Musumarra, M.G. Pellegriti, D. Rendić, C. Rolfs, S. Romano, N. Soić, A. Tumino, H.H. Wolter, M. Zadro, Phys. Rev. C **63**, 005801 (2001).
9. A. Tumino, C. Spitaleri, S. Cherubini, A. Di Pietro, P. Figuera, M. Lattuada, A. Musumarra, M.G. Pellegriti, R.G. Pizzone, S. Romano, C. Rolfs, S. Tudisco, S. Typel, Nucl. Phys. A **718**, 499 (2003).
10. C. Spitaleri, S. Cherubini, A. Del Zoppo, A. Di Pietro, P. Figuera, M. Gulino, M. Lattuada, Dj. Miljanić, A. Musumarra, M.G. Pellegriti, R.G. Pizzone, C. Rolfs, S. Romano, S. Tudisco, A. Tumino, Nucl. Phys. A **719**, 99 (2003).
11. A. Tumino, C. Spitaleri, A. Di Pietro, P. Figuera, M. Lattuada, A. Musumarra, M.G. Pellegriti, R.G. Pizzone, S. Romano, C. Rolfs, S. Tudisco, S. Typel, Phys. Rev. C **67**, 065803 (2003).
12. C. Spitaleri, L. Lamia, A. Tumino, R.G. Pizzone, S. Cherubini, A. Del Zoppo, P. Figuera, M. La Cognata, A. Musumarra, M.G. Pellegriti, A. Rinollo, C. Rolfs, S. Romano, S. Tudisco, Phys. Rev. C **63**, 055806 (2004).
13. M. Jain, P.G. Roos, H.G. Pugh, H.D. Holgrem, Nucl. Phys. A **153**, 49 (1970).
14. J. Kasagi, T. Nakagawa, N. Serine, T. Tohei, H. Ueno, Nucl. Phys. A **239**, 233 (1975).
15. N. Arena, D. Vinciguerra, F. Riggi, C. Spitaleri, Lett. Nuovo Cimento **17**, 231 (1976).
16. M. Lattuada, F. Riggi, C. Spitaleri, D. Vinciguerra, Phys. Rev. C **26**, 1330 (1982).
17. S. Barbarino, M. Lattuada, F. Riggi, C. Spitaleri, D. Vinciguerra, Phys. Rev. C **21**, 1104 (1980).
18. M. Zadro, Dj. Miljanić, C. Spitaleri, G. Calvi, M. Lattuada, F. Riggi, Phys. Rev. C **40**, 181 (1989).
19. S. Engstler, G. Raimann, C. Angulo, U. Greife, C. Rolfs, U. Schröder, E. Somorjai, B. Kirch, K. Langanke, Z. Phys. A **342**, 471 (1992).
20. G.F. Chew, Phys. Rev. **80**, 196 (1950).
21. N.S. Chant, P.G. Roos, Phys. Rev. C **15**, 57 (1977).
22. U.G. Neudatchin, Y.F. Smirnov, At. Energy Rev., **3**, 157 (1965).
23. G. Jacob, Th.A. Maris, Rev. Mod. Phys. **38**, 121 (1966).
24. N. Arena, D. Vinciguerra, F. Riggi, C. Spitaleri, Nuovo Cimento **45**, 405 (1978).
25. G. Baur, H. Rebel, Annu. Rev. Nucl. Part. Sci. **46**, 321 (1996).
26. G.S. Mani *et al.*, Nucl. Phys. **60**, 588 (1964).
27. Y. Cassagnou *et al.*, Nucl. Phys. **33**, 449 (1962).

Eur. Phys. J. A **27**, s01, 249–254 (2006)
DOI: 10.1140/epja/i2006-08-039-0

EPJ A direct
electronic only

Indirect measurement of the ^{15}N(p, α)^{12}C reaction cross section through the Trojan-Horse Method

M. La Cognata[1,2], S. Romano[1,2], C. Spitaleri[1,2,a], R. Tribble[3], L. Trache[3], S. Cherubini[1,2], Changbo Fu[3], L. Lamia[1,2], A. Mukhamedzhanov[3], R.G. Pizzone[1,2], C. Rolfs[4], G. Tabacaru[3], and A. Tumino[1,2]

[1] Dipartimento di Metodologie Fisiche e Chimiche per l'Ingegneria - Università di Catania, Catania, Italy
[2] Laboratori Nazionali del Sud - INFN, Catania, Italy
[3] Cyclotron Institute, Texas A&M University, College Station, TX, USA
[4] Ruhr-Universität, Bochum, Germany

Received: 28 June 2005 /
Published online: 13 March 2006 –

Abstract. The low-energy, bare-nucleus cross section for ^{15}N(p, α)^{12}C is extracted by means of the Trojan-Horse Method applied to the ^{2}H(^{15}N, α^{12}C)n reaction at $E_{beam} = 60$ MeV. The astrophysical S-factor is compared to the direct data in the same energy region. A nice agreement shows up down to 100 keV, while the low-energy behaviour of the S-factor suggests a smaller rate than is reported in the literature.

PACS. 24.10.-i Nuclear Reaction Models and Methods – 26.20.+f Hydrostatic Stellar Nucleosynthesis

1 Introduction

The importance of ^{19}F in astrophysics is two-fold: in red giants fluorine is enhanced by large factors with respect to the solar abundance; on the other hand, the fluorine abundances observed in giants can constrain AGB star models [1]. Primary sources of uncertainty in predicting fluorine abundances in AGB stars come from the adopted reaction rates and the treatment of nucleosynthesis during the thermal pulses and the interpulse periods [2].

The reaction rate for the ^{15}N(p, α)^{12}C affects the production of fluorine since it removes both protons and ^{15}N nuclei from the ^{19}F production chain [1]. Calculations show that even the highest experimentally observed surface fluorine abundances are reproduced for some masses and metallicities, but this occurs for C/O values much larger than what experimental data suggest [1]. In fact ^{12}C and ^{19}F are correlated since they are both dredged up to the stellar surface in correspondence of the third dredge-up. The ^{15}N(p, α)^{12}C reaction rate introduces at least an 8% uncertainty in the fluorine surface abundance [1], since NACRE [3] and CF88 [4] rates show up a factor 2 difference.

As the ^{15}N(p, α)^{12}C experimental cross section is measured down to 80 keV [5], much larger than the Gamow peak energy (8 keV), only extrapolations are available in the low energy region.

[a] e-mail: spitaleri@lns.infn.it

1.1 The Trojan-Horse Method

The measurement of nuclear cross sections at ultra-low energies is a very difficult task because of the presence of the Coulomb barrier exponentially damping the cross section. Therefore extrapolation is necessary when data at astrophysical energies are unavailable, thus introducing additional uncertainties related to poorly known low-energy or subthreshold resonances and to the electron screening effect [6]. The extrapolation procedure is usually performed on the astrophysical factor

$$S(E) = E \cdot \exp(2\pi\eta) \cdot \sigma(E), \qquad (1)$$

where η is the Sommerfeld parameter, $S(E)$ being a smoothly varying function of the energy even in the sub-Coulomb region. A possible way to avoid extrapolation from data at energies much higher than the Gamow window is the application of the Trojan-Horse Method.

The Trojan-Horse Method (THM) [7,8] is an indirect method that allows for the extraction of the bare nucleus cross section of a two-body reaction of astrophysical interest,

$$A + x \to c + C, \qquad (2)$$

by selecting the Quasi-Free (QF) contribution of the cross section of a suitable three-body process

$$A + a \to c + C + s, \qquad (3)$$

where x acts as participant and s as a spectator to the two-body reaction. If the energy of the projectile A is higher than the Coulomb barrier in the $A+a$ channel, then the reaction (2) can be induced inside the nuclear interaction field. The nucleus a, having a high probability for the $x \oplus s$ configuration, constitutes the so-called *Trojan-Horse Nucleus* since it brings the cluster x inside the target A nuclear field. Therefore the Coulomb barrier and electron screening effect are negligible and a measurement of the cross section at ultra-low energies is possible.

In the simple Plane-Wave Impulse Approximation (PWIA) post-form approach [8] the two-body bare nucleus cross section is extracted from the measured three-body cross section using the equation

$$\frac{\mathrm{d}^3\sigma}{\mathrm{d}E_C \mathrm{d}\Omega_C \mathrm{d}\Omega_c} \propto KF\,|G(p_s)|^2 \frac{\mathrm{d}\sigma_l^N}{\mathrm{d}\Omega}, \qquad (4)$$

where KF is a kinematical factor, $|G(p_s)|^2$ the momentum distribution of spectator particle s inside a and $\mathrm{d}\sigma_l^N/\mathrm{d}\Omega$ the *nuclear* (off-energy-shell) cross section, under the hypothesis of a dominant l-contribution. In fact, since the reaction (2) is induced inside the short-range nuclear field the penetration probability of the Coulomb barrier has to be introduced in order to compare the THM cross section with the direct data from literature in the energy region below the Coulomb barrier [9]:

$$P_l(k_{Ax}R) = \frac{1}{F_l^2(k_{Ax}R) + G_l^2(k_{Ax}R)}, \qquad (5)$$

where F_l and G_l are the regular and irregular Coulomb wave functions, k_{Ax} and R the $A-x$ relative wave number and interaction radius, respectively.

In the present work, the measurement of the bare-nucleus cross section for the $^{15}\mathrm{N}(\mathrm{p},\alpha)^{12}\mathrm{C}$ reaction is presented in the astrophysically relevant energy region, from 600 keV down to 5 keV. The indirect two-body cross section is extracted in arbitrary units from the $^2\mathrm{H}(^{15}\mathrm{N},\alpha^{12}\mathrm{C})\mathrm{n}$ coincidence yield in the PWIA approach and is normalized to the direct data to get the absolute cross section. In the measured energy region, the $^{15}\mathrm{N}(\mathrm{p},\alpha)^{12}\mathrm{C}$ cross section is dominated by the 1^- resonance at 314 keV [3], therefore this study allows one to perform a complete validity test of the THM below the Coulomb barrier.

2 The experiment

The experiment was performed at the Texas A&M University Cyclotron Institute. The K500 superconducting cyclotron provided a 60 MeV $^{15}\mathrm{N}$ beam with a spot size on target of 1 mm and intensities up to 5 nA, impinging on a 200 μg/cm^2 thick CD_2 target. A simple experimental setup was used: a telescope (A) made up of an ionization chamber and a silicon position-sensitive detector (PSD) to discriminate carbon nuclei, and a couple of silicon PSD's (B, C) on the opposite side with respect to the beam direction. Two kinds of event were recorded: coincidences between A and B or between A and C. The angular condition were selected in order to maximize the quasi-free contribution: table 1 shows the laboratory angles for all the coincidence detectors.

Table 1. Laboratory central angles (θ_0), angular range spanned ($\Delta\theta$), solid angle subtended ($\Delta\Omega$) and thickness (d) for each detector.

Detector	θ_0 (deg)	$\Delta\theta$ (deg)	$\Delta\Omega$ (msr)	d (mm)
ΔE-A	15.0	–	–	50
PSD-A	15.0	11.1	8.7	0.492
PSD-B	12.2	13.1	11.2	0.984
PSD-C	32.2	16.7	20.0	0.984

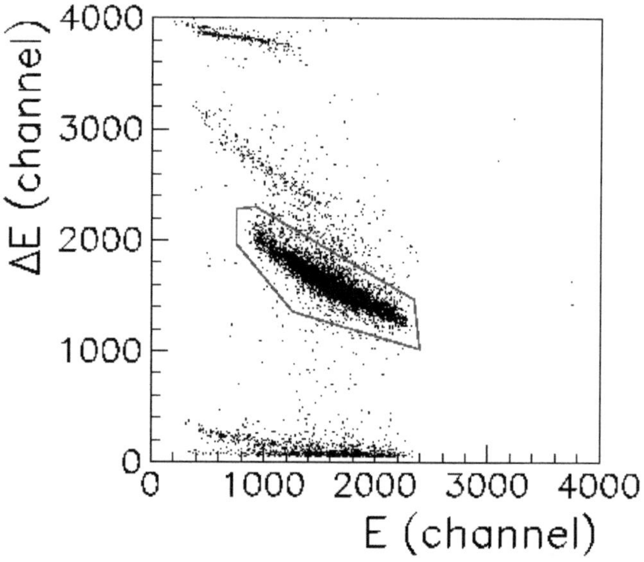

Fig. 1. Typical ΔE-E spectrum for the A telescope. The graphical selection shown in the picture marks the locus of carbon nuclei.

The ionization chamber was filled with 60 mbar butane and was closed on both side by 1.5 μm thick Mylar foil windows; overall energy resolution turned out to be about 15%, therefore one could not discriminate $^{12}\mathrm{C}$ nuclei from $^{13}\mathrm{C}$ and an off-line kinematical selection was required. A typical ΔE-E spectrum is shown in fig. 1. The carbon locus is selected by means of a graphical cut as shown in the figure. The alpha-particle discrimination is obtained by selecting the kinematical locus of the three-body reaction.

The spectator particle momentum range spanned in the experiment is such that a comparison of the coincidence yield for small p_s (less than 20 MeV/c) and larger p_s can be performed, in order to check that the resonances occurring in the 2-body reaction are fed through a quasi-free channel and not through sequential decay of intermediate nuclei.

2.1 Data analysis

After detector calibration, the first step of the analysis is the selection of the events corresponding to the three-body reaction $^2\mathrm{H}(^{15}\mathrm{N},\alpha^{12}\mathrm{C})\mathrm{n}$. This is accomplished through a

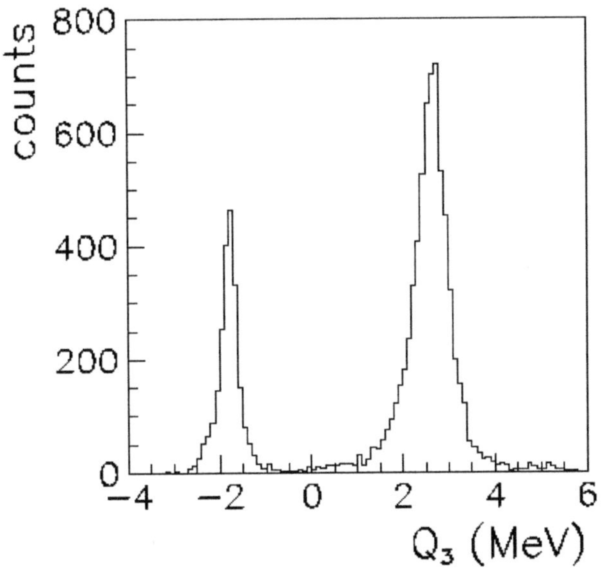

Fig. 2. Three-body Q-value spectrum. Two peaks corresponding to the ground state and first excited state (4.4389 MeV) of ^{12}C are clearly visible.

selection of coincidence events and of the reaction kinematical locus. For the selected events the Q-value is calculated by imposing energy balance for the three-body reaction. In fig. 2 the Q-value spectrum is reported, showing two peaks corresponding to the ground state and first excited state of ^{12}C (Q-values 2.74 MeV and −1.70 MeV respectively).

The good agreement between the experimental and the theoretical Q-values confirms the accuracy of the calibration. In the subsequent analysis only the events corresponding to the high-energy peak are retained in order to extract the ^{15}N(p, α_0)^{12}C cross section. In fact, the ^{15}N(p, α_1)^{12}C reaction is of negligible astrophysical relevance in the context outlined in the introduction, since its cross section is much lower than the ^{15}N(p, α_0)^{12}C one inside the Gamow peak [5]. For this reason in the following sections the 0 will be dropped.

The second step of the THM data analysis is the study of the reaction mechanisms feeding the exit channel. This is a necessary step to disentangle the QF from the sequential decay mechanisms. Such a selection is fulfilled by investigating the correlation between the relative energies ($E_{\alpha\text{-}n}$ *vs.* $E_{^{12}\text{C-}\alpha}$ and $E_{^{12}\text{C-}n}$ *vs.* $E_{^{12}\text{C-}\alpha}$, as shown in figs. 3a and 3b, respectively) of the detected particles and between $E_{^{12}\text{C-}\alpha}$ and the neutron momentum p_s (fig. 4).

In the α-n *vs.* ^{12}C-α (fig. 3a) and ^{12}C-n *vs.* ^{12}C-α (fig. 3b) relative energy two-dimensional plots very clear vertical loci appear, corresponding to excited states of ^{16}O at 12.44 MeV, 13.09 MeV and 13.26 MeV. An additional contribution to the three-body cross section due to the sequential decay of the excited states of ^{13}C at 6.68 MeV, 7.49 MeV, 7.55 MeV, 7.67 MeV (where the last three levels are not resolved) is also visible, corresponding to the horizontal loci in fig. 3b. Such background processes give a negligible contribution to the coincidence yield in the astrophysical energy region, since in the phase space explored there is no contribution coming from ^{13}C sequential decay at ^{12}C-α relative energies lower than 6 MeV (as it will be shown below zero energy in the ^{15}N(p, α)^{12}C channel corresponds to 4.966 MeV in the ^{12}C-α relative energy).

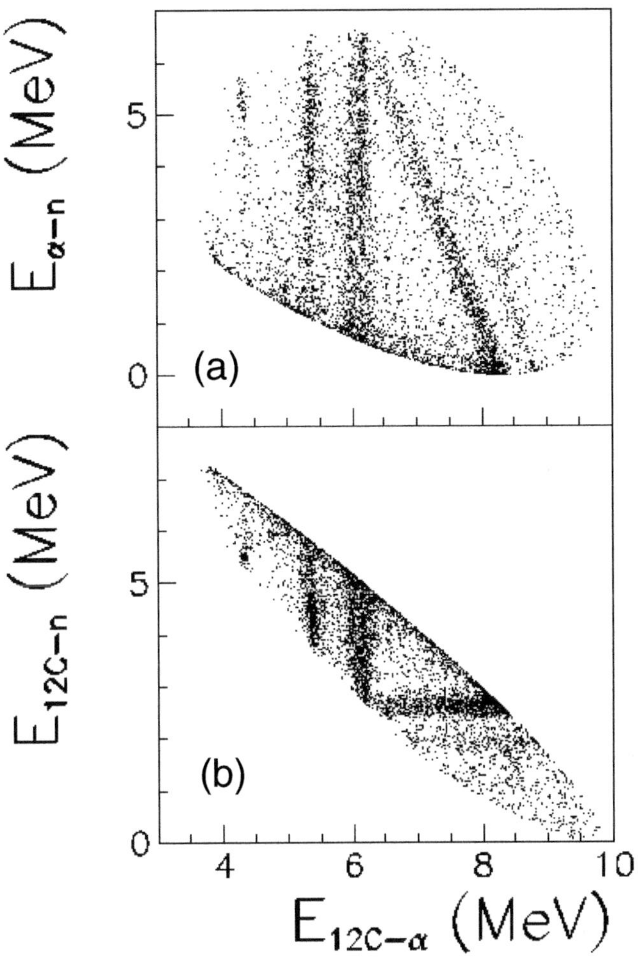

Fig. 3. (a) α-n *vs.* ^{12}C-α and (b) ^{12}C-n *vs.* ^{12}C-α relative energy two-dimensional plot. Very clear vertical loci appear, corresponding to excited states of ^{16}O at 12.44 MeV, 13.09 MeV and 13.26 MeV. The contribution to the three-body cross section due to the sequential decay of the excited states of ^{13}C at 6.68 MeV, 7.49 MeV, 7.55 MeV, 7.67 MeV is also apparent (the horizontal loci in (b)).

On the other hand, according to the previous discussion, the occurrence of sequential mechanisms in the ^{12}C-α channel cannot be ruled out by studying the relative energy two-dimensional plots since the same excited state of ^{12}C+α system can be formed through both a quasi-free reaction mechanism and a sequential one. Therefore an investigation of the $E_{^{12}\text{C-}\alpha}$ *vs.* p_s two-dimensional plot is necessary in order to establish the presence of the quasi-free contribution and disentangle the sequential mechanism.

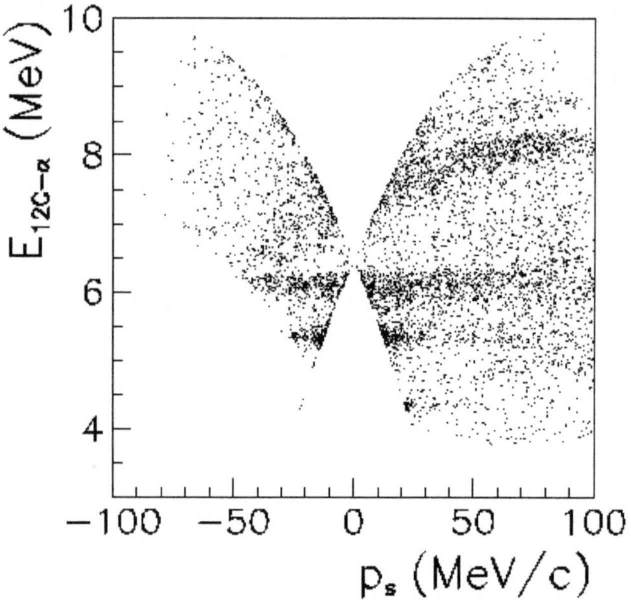

Fig. 4. ^{12}C-α relative energy *vs.* neutron momentum p_s two-dimensional plot. The sharp horizontal lines corresponds to the excited states of ^{16}O at 12.44 MeV, 13.09 MeV and 13.26 MeV. Negligible background coming from ^{13}C sequential decay shows up in the astrophysically relevant energy range.

Figure 4 shows the ^{12}C-α relative energy *vs.* neutron momentum p_s in a two-dimensional plot. The sharp horizontal lines correspond to the excited states of ^{16}O at 12.44 MeV, 13.09 MeV and 13.26 MeV. At higher $E_{^{12}\text{C-}\alpha}$ the occurrence of the ^{13}C sequential decay is clearly visible, but its contribution in the astrophysically relevant energy region is absolutely negligible. From the comparison of the loci corresponding to the different level decays, the quasi-free origin of the ^{16}O levels in fig. 4 is apparent. In fact, while the kinematical locus connected to the ^{13}C decay is still clearly visible at high neutron momentum, the contributions due to the ^{16}O decay is barely visible (in comparison with the background) showing a definite correlation of the coincidence yield for the cited ^{16}O levels with the spectator neutron momentum.

As a further check to determine if the resonances occurring in the ^{12}C-α channel are fed through a quasi free mechanism and not through sequential decay of intermediate nuclei, the $E_{c.m.}$ relative energy for the ^{15}N(p,α)^{12}C reaction, defined as [8]

$$E_{c.m.} = E_{^{12}\text{C-}\alpha} - Q_{2\text{-}body}, \quad (6)$$

is reported as a function of the momentum p_s of the spectator particle for all coincidence events. In order to explore the behaviour of the coincidence yield as a function of the momentum p_s, the relative energy $E_{c.m.}$ spectra divided by the phase-space contribution were reconstructed for different ranges of the neutron momentum p_s. Within the range 0 to 20 MeV/c (fig. 5, top panel) the coincidence yield appears to be dominated by the decay of the three levels of ^{16}O at 12.44 MeV, 13.09 MeV and 13.26 MeV. For larger p_s, the coincidence yield decreases as shown in fig. 5 (middle panel) and fig. 5 (bottom panel), and such resonances become barely visible with respect to the background. These experimental data provide clear evidence for a strong correlation between coincidence yield and neutron momentum p_s, a necessary condition for the dominance of the quasi-free mechanism in the region approaching zero spectator momentum. In fact this is expected

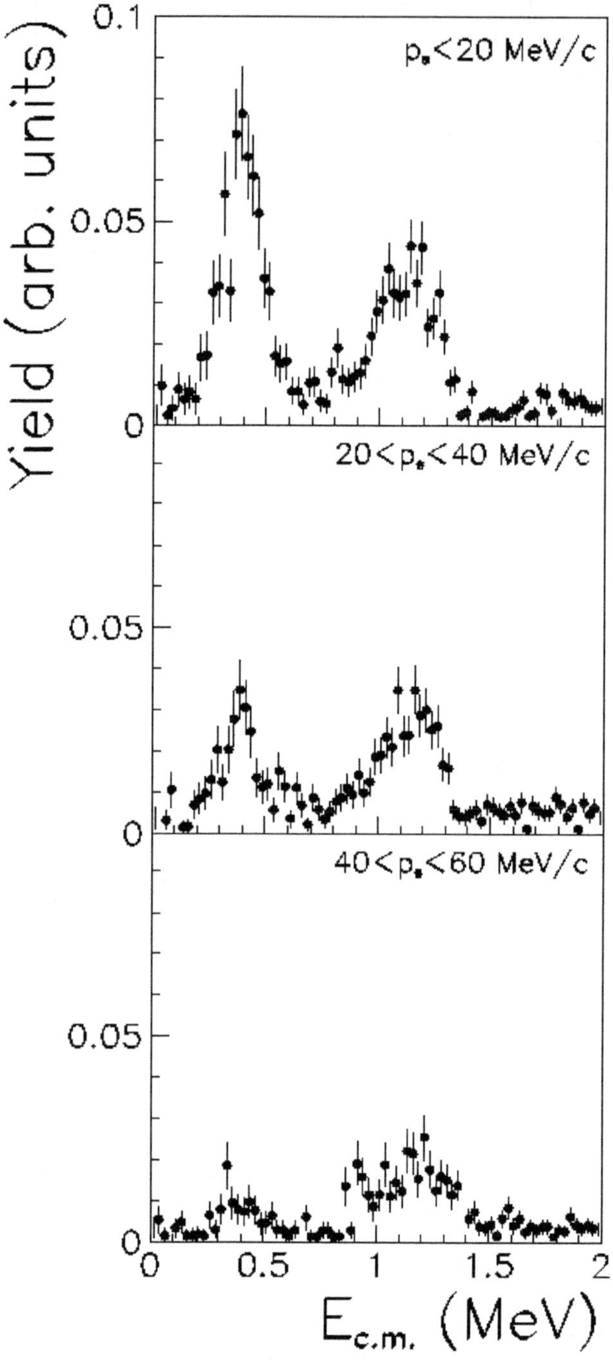

Fig. 5. Correlation spectra between $E_{c.m.}$ and neutron momentum p_s. The excited levels of ^{16}O at 12.44 MeV and 13.09 MeV, 13.26 MeV (not resolved) are clearly visible.

Table 2. Fitting parameters of the function used to parameterize the bare nucleus THM S-factor, the sum of a 2nd-order polynomial and a Breit Wigner, in which the centroid is fixed at the 1^- resonance energy of 319 keV. In the last column the χ^2 per degree of freedom is also reported.

p_0 (MeVb)	p_1 (MeVb/keV)	p_2 (MeVb/keV^2)	FWHM (keV)	W (MeVb)	χ^2/N
-256 ± 6	-0.82 ± 0.14	$(0.30 \pm 0.03) \cdot 10^{-2}$	190 ± 2	$(0.3193 \pm 0.0016) \cdot 10^8$	2.6

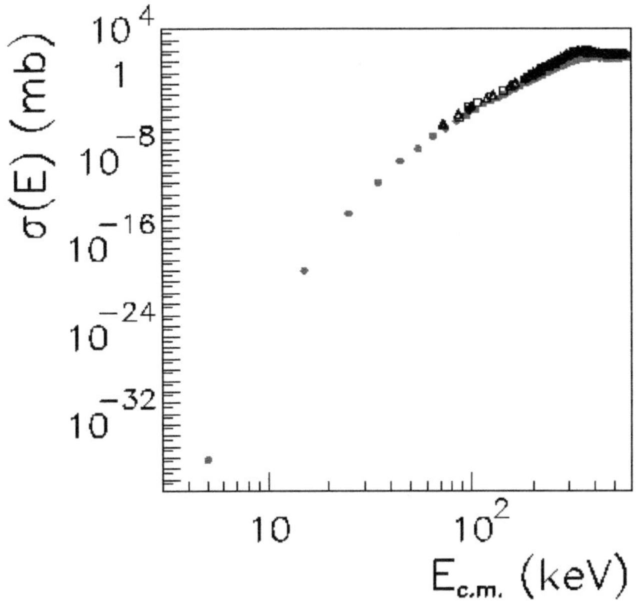

Fig. 6. THM cross section (full dots) together with direct data from [5, 10, 11] in the energy region between 0 and 600 keV. The lowest energy point correspond to a center-of-mass energy of 5 keV.

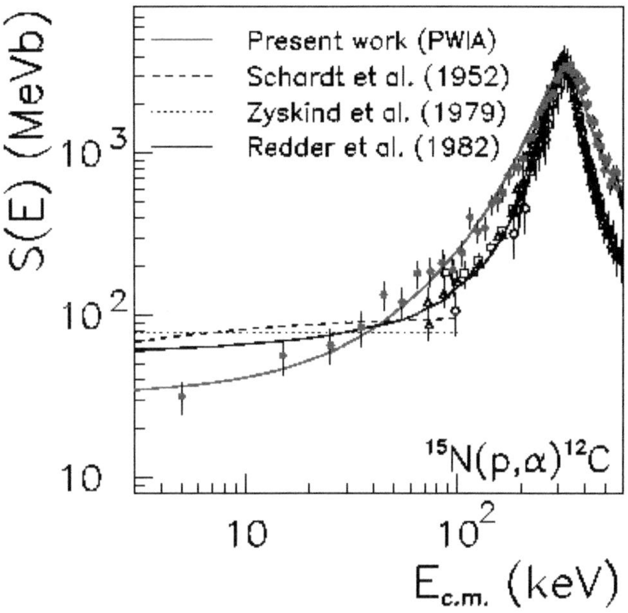

Fig. 7. THM astrophysical $S(E)$ (full red dots) factor together with direct data from [5, 10, 11] in the low energy region. A fit to the THM bare nucleus S-factor has been performed by means of the sum a Breit-Wigner and a 2nd-order polynomial (parameters in table 2). For comparison, the extrapolation from [5, 10, 11] are reported.

when the relative motion inside the Trojan-Horse nucleus is in s-wave, as is the case of the deuteron.

3 Results and discussion

Once the quasi-free contribution has been selected, the bare nuclear cross section, $d\sigma_l^N/d\Omega$, is extracted by means of eq. (4). A p_s window of 20 MeV/c around zero spectator momentum is adopted in accordance with the previous result.

After penetration probability correction, the THM cross section and astrophysical $S(E)$ factor are extracted and are reported together with direct data [5, 10, 11] in fig. 6 and in fig. 7: the error bars include only the statistical uncertainties. Here only a minor part of the data is shown, coming from a single pair of detectors (A-C coincidences) and only the simple PWIA approach has been reported: the MPWBA analysis [9, 12] is currently in progress.

The THM cross section is normalized to direct data (equating the peak values for the 319 keV resonance) for getting the cross section in absolute units. By fitting $S(E)$ from the indirect data with the sum of a 2nd-order polynomial and a Breit-Wigner (the fit parameters are reported in table 2), $S_{bare}(0)$ is determined to be $S_{bare}(0) = 32 \pm 6$ MeVb (see fig. 7). This is in clear disagreement with direct data extrapolation [5, 10, 11], giving a much higher value $S_{bare}(0) = 65 \pm 7$ MeVb [5, 3]. The present data set also suffers from a systematic error $\sim$ 20% arising from normalization procedure of the indirect data to the direct ones. The THM $S_{bare}(0)$ value is likely to be an overestimate of the *true* value because of the poorer resolution of THM data with respect to direct measurement, as shown by the larger FWHM of the 319 keV resonance (fig. 7). Such a result tends to reduce the rate for the ^{15}N(p, α)^{12}C in the astrophysically relevant energy region, confirming the value reported in CF88 in contrast with the one cited in NACRE compilation.

The origin of such a disagreement can be traced back to the electron screening effect [6]. In fact the direct data from [5, 10, 11] were all measured before the middle 80's, when the problem of electron screening in nuclear reactions studied in laboratory was first analyzed [13, 14]. The cross section determined by [5, 10, 11] and therefore the extrapolation reported in [3] and in [4] and the reaction rates are likely to be affected by the enhancement due to atomic effects of nearly 10% [13] at 80 keV (corresponding to the lowest measured energy).

Thus the extrapolation was performed on data having a much larger value and a different low-energy behaviour with respect to the bare nucleus cross section, which is the relevant nuclear input parameter in astrophysical codes. In addition, since a large number of low-energy cross-section measurements have shown that the electron screening potential is much higher than what the adiabatic limit provides (for example, for the ^{3}He(d, p)^{4}He the electron screening potential turned out to be a factor two higher than the adiabatic limit [15]), an even higher overestimate of the ^{15}N(p, α)^{12}C S-factor inside the Gamow window is quite likely. As a consequence a further analysis of the ^{15}N(p, α)^{12}C low-energy data is needed with through both direct and indirect methods.

References

1. M. Lugaro *et al.*, Ap. J. **615**, 934 (2004).
2. A. Renda *et al.*, Mon. Not. R. Astron. Soc. **354**, 575 (2004).
3. C. Angulo *et al.*, Nucl. Phys. A **656**, 3 (1999).
4. G.R. Coughlan, W.A. Fowler, At. Data Nucl. Data Tables **40**, 283 (1988).
5. A. Redder *et al.*, Z. Phys. A **305**, 325 (1982).
6. F. Strieder *et al.*, Naturwissenschaften **88**, 461 (2001).
7. G. Baur, Phys. Lett. B **178**, 135 (1986).
8. C. Spitaleri *et al.*, Phys. Rev. C **69**, 055806 (2004).
9. C. Spitaleri *et al.*, Phys. Rev. C **63**, 055801 (2001).
10. A. Schardt *et al.*, Phys. Rev. **86**, 527 (1952).
11. J.L. Zyskind *et al.*, Nucl. Phys. A **320**, 404 (1979).
12. S. Typel *et al.*, Ann. Phys. (N.Y.) **305**, 228 (2003).
13. H.J. Assenbaum *et al.*, Z. Phys. A **327**, 461 (1987).
14. G. Blüge *et al.*, Z. Phys. A **333**, 219 (1989).
15. M. Aliotta *et al.*, Nucl. Phys. A **690**, 790 (2001).

6 Nuclear Structure Far from Stability

Eur. Phys. J. A **27**, s01, 257–267 (2006)
DOI: 10.1140/epja/i2006-08-040-7

EPJ A direct
electronic only

Nuclear structure far off stability —Implications for nuclear astrophysics

H. Grawe[1,a], A. Blazhev[1,2], M. Górska[1], R. Grzywacz[3], H. Mach[4], and I. Mukha[1,5,6]

[1] GSI, Planckstr. 1, D-64291 Darmstadt, Germany
[2] University of Sofia, Sofia, Bulgaria
[3] Department of Physics and Astronomy, University of Tennessee, Knoxville, TN 37996, USA
[4] ISV, Uppsala University, Nyköping, Sweden
[5] RRC Kurchatov Institute, RU-123481 Moscow, Russia
[6] Katholieke Universiteit Leuven, Leuven, Belgium

Received: 20 June 2005 /
Published online: 15 March 2006 –

Abstract. The single-particle structure and shell gap of ^{100}Sn as inferred from previous in-beam γ-ray spectroscopy has been confirmed in recent studies of seniority and spin-gap isomers by $\gamma\gamma$, $\beta\gamma$, $\beta p\gamma$, $p\gamma$ and $2p\gamma$ spectroscopy. The results for ^{94,95}Ag, ^{98}Cd and its $N = 50$ isotones ^{96}Pd and ^{94}Ru stress the importance of large-scale shell model calculations employing realistic interactions for the isomerism, *np-nh* excitations, seniority mixing and $E2$ polarisation of the ^{100}Sn core. The strong monopole interaction of the $\Delta l = 0, 1$ spin/isospin-flip partners $\pi g_{9/2}$-$\nu g_{7/2}$ along the $N = 50$ isotones and the $\pi f_{5/2}$-$\nu g_{9/2}$ pair of nucleons along the $Z = 28$ Ni isotopes are decisive for the evolution of the shell structure towards ^{100}Sn and ^{78}Ni. It can be traced back to the tensor force in the effective nucleon-nucleon interaction and provides a straightforward explanation for new shells in neutron-rich light nuclei, implying qualitative predictions for new $N = 32, 34$ subshells in Ca isotopes, persistence of the ^{78}Ni proton and neutron shell gaps and non-equivalence of the $g_{9/2}$ valence mirror Ni isotopes and $N = 50$ isotones. This is corroborated by recent experimental data on ^{56,58}Cr and $^{70-76}$Ni. The implication of monopole driven shell evolution for apparent spin-orbit splitting towards $N \gg Z$ and structure along the astrophysical r-path between $N = 50$ and $N = 82$ is discussed.

PACS. 21.60.Cs Shell model – 23.20.-g Electromagnetic transitions – 27.30.+t $16 \leq A \leq 68$ – 27.60.+j $78 \leq A \leq 132$ – 26.50.+x Nuclear physics aspects of the r-process

1 Introduction

The evolution of shell structure towards exotic nuclei with extreme isospin has become, and will be for the future, the major topic of experimental studies and the main challenge for realistic in-medium nucleon-nucleon (NN) interactions as used in large-scale shell model and mean field theory. Especially on the neutron-rich side of the Segré chart, where the drip line for $A \geq 80$ is far beyond reach for experiments, the understanding of the underlying shell driving mechanism is of key importance for astrophysics applications as, *e.g.*, the r-process. Two scenarios with differing experimental signature have been proposed to describe the shell structure of nuclei on the pathway towards large N/Z ratios. The first is based on the larger radial extension and a softer neutron potential. This shifts large-l orbitals upward in energy and reduces the spin-orbit (SO) splitting, which is proportional to the potential gradient, for nucleon orbitals probing the nuclear surface [1,2]. Thus for medium-heavy and heavy nuclei the harmonic oscillator (HO) unique-parity shells are restored. The changing shell structure evolves smoothly with A and N/Z and only large variations of these parameters as expected towards the neutron dripline will have substantial effects. The second scenario originates from the strong monopole shifts of selected shell model orbits which have been ascribed to the $\sigma\tau$ central and the tensor force of the NN interaction [3,4,5,6]. This will be discussed in sect. 3 and recent experimental evidence for monopole driven shell structure and apparent SO splitting from the $^{100-132}$Sn and ^{48}Ca to ^{78}Ni regions is presented in sects. 2 and 4. The application to r-path nuclei is discussed in a qualitative way in sect. 5.

[a] Conference presenter e-mail: h.grawe@gsi.de

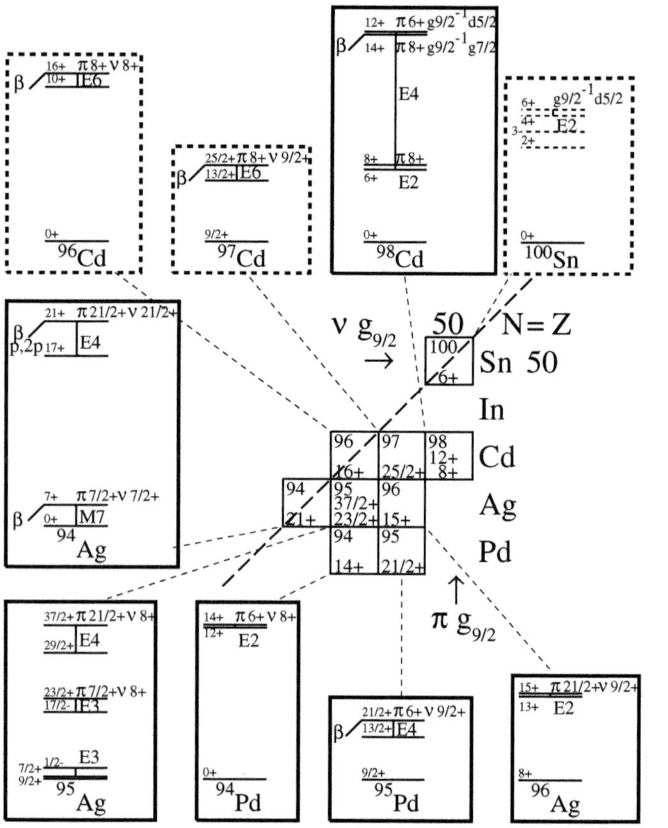

Fig. 1. Experimentally established (solid frames) and shell model predicted (dashed frames) spin-gap isomers close to ^{100}Sn.

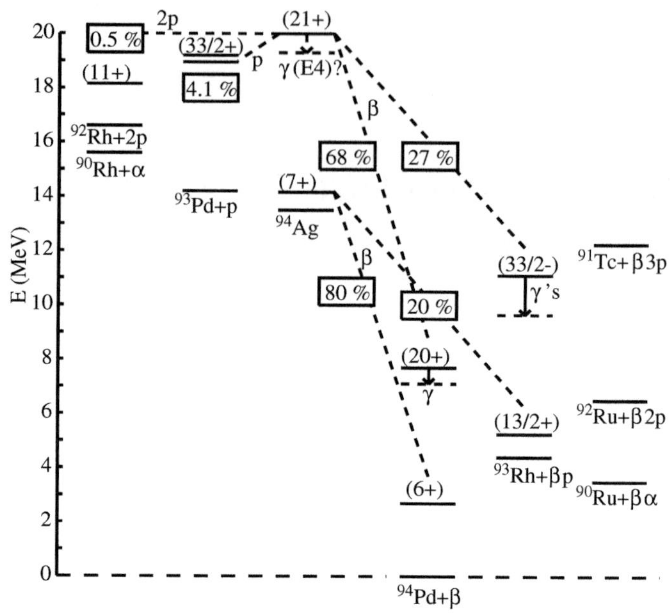

Fig. 2. Decay channels for the $I^\pi = (21^+)$ isomer in ^{94}Ag.

2 Spin-gap isomerism in the ^{100}Sn region

In spite of the missing experimental information on binding energies and excited states in ^{100}Sn and its one-particle (hole) neighbours the shell structure as inferred from more remote nuclei by shell model aided extrapolation is well established [7,8,9]. It shows a remarkable similarity to ^{56}Ni one major shell below at $N = Z$ and close to the proton dripline. This implies the existence of analogous structure features such as softness with respect to $L = 2$ core excitations, a pronounced Gamow-Teller resonance due to the $\pi g_{9/2}$-$\nu g_{7/2}$ respective $\pi f_{7/2}$-$\nu f_{5/2}$ proton-neutron orbitals, a substantial amount of *np-nh* configurations besides the doubly-magic configuration in the ground state of ^{100}Sn, high-spin isomers, and possibly even proton decay. The study of seniority and spin-gap isomers provides a sensitive probe of single particle energies, residual interaction, core excitation and shell gaps as demonstrated in recent experiments on ^{98}Cd [10] and ^{94}Ag [11]. The status of observed isomers and shell model predicted ones and their main structure features in terms of the leading proton-neutron $g^n_{9/2}$ configuration are summarised in fig. 1. Besides the well known isomers ^{96}Ag, $I^\pi = (15^+)$ [12], ^{94}Pd, $I^\pi = 14^+$ [13] and ^{95}Pd, $I^\pi = 21/2^+$ [14,15], recently the cases of ^{95}Ag, $I^\pi = (23/2^+)$ and $I^\pi = (37/2^+)$ [16] were reported between the $N = 50$ and the $N = Z$ lines below ^{100}Sn.

The spin-gap isomers with the highest spins, as, *e.g.*, the $I^\pi = (37/2)^+$ state in ^{95}Ag, which exhaust the $\pi\nu g^n_{9/2}$ configuration, are not predicted to be isomeric in the proton-neutron $(p_{1/2}, g_{9/2})$ valence space [17] and require excitation of the ^{100}Sn core. This has been studied recently in a dedicated prompt-delayed experiment at the EUROBALL IV array, where the $I^\pi = (12^+)$ core excited isomer in ^{98}Cd, the two-proton hole neighbour of ^{100}Sn, was identified [10]. The decay pattern exhibits a striking analogy to ^{54}Fe, two proton holes from ^{56}Ni [9,18]. The level scheme, the core excited isomer and the $E2$ transition rates are excellently reproduced by a large scale shell model (LSSM) calculation in the $(0g, 1d, 2s)$ model space allowing for up to $4p4h$ excitations of the ^{100}Sn core. Values of 6.46 (15) MeV for the ^{100}Sn shell gap were inferred from the excitation energy of the $I^\pi = (12^+)$ $E4$ isomer, and a small proton polarisation charge of $\delta e_\pi \leq 0.2e$ was extracted from the $\pi g^{-2}_{9/2}$; $I^\pi = (8^+)$ $E2$ isomeric decay [10,19]. It should be noted that the LSSM, which exhibits good predictive power for isomers and $E2$ strengths in the whole region between $N = 50$ and the $N = Z$ line (see below ^{94}Ag and sect. 4.2), also predicts an $I^\pi = 14^+$ $E6$ isomer in ^{98}Cd and an $I^\pi = 6^+$ $E2$ isomer in ^{100}Sn. Like the $I^\pi = 25/2^+$ and 16^+ $E6$ isomers in ^{97,96}Cd, which are expected [17] and searched for since long, they await future experimental verification (fig. 1).

A second example for the sensitivity of spin-gap isomers to subtle structure details such as single particle energies, shell gaps and proton-neutron $(\pi\nu)$ interaction even in LSSM calculations is provided by the $I^\pi = (21^+)$ isomer in ^{94}Ag. This state features exotic properties with respect to spin, excitation energy and decay modes (fig. 2) to a degree, which is unprecedented in the Segré chart [11, 20,21,22]. The isomerism is not predicted in the pure valence hole space below ^{100}Sn but requires inclusion of

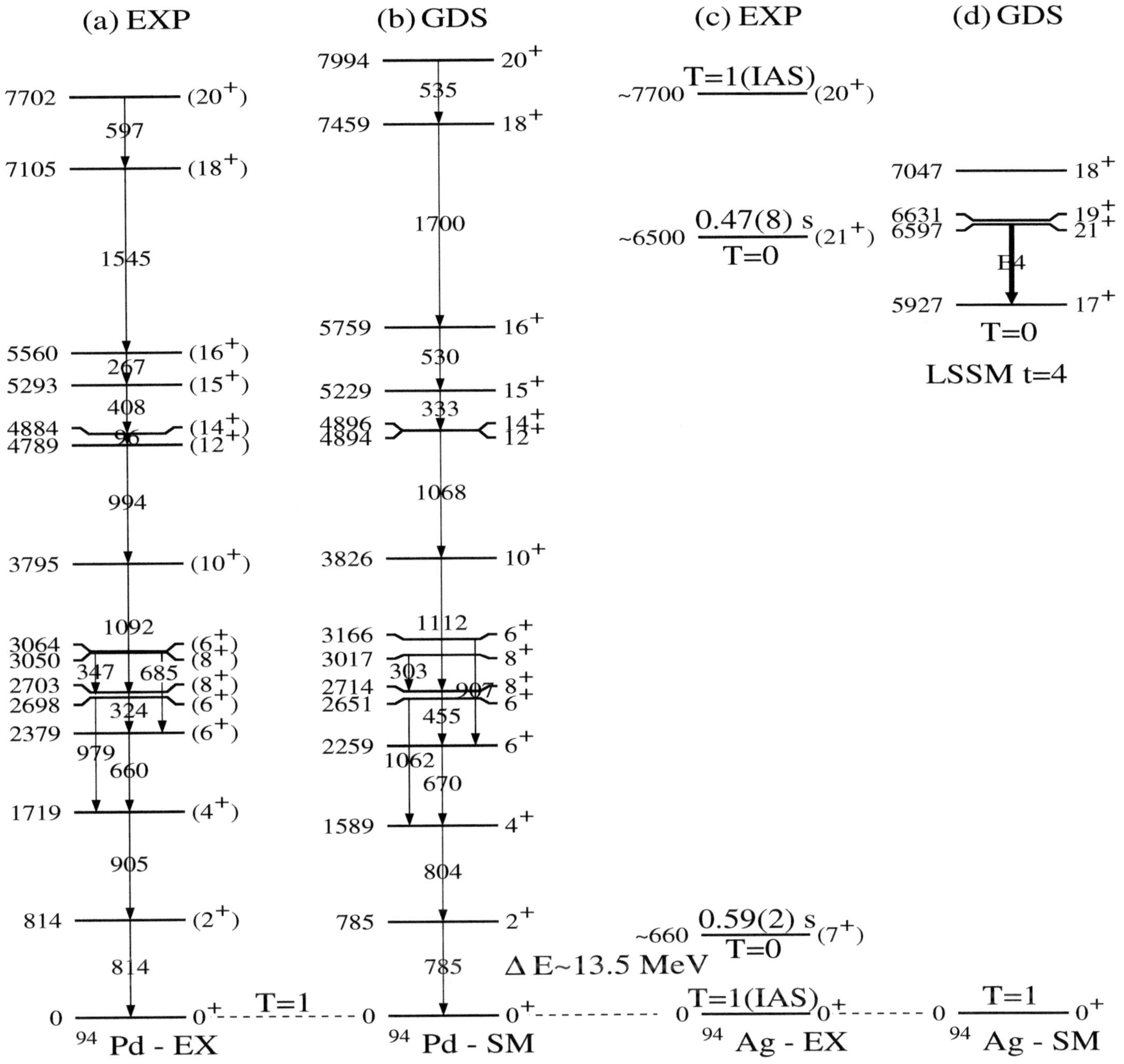

Fig. 3. Experimental and large-scale shell model level scheme for ^{94}Ag and its $\beta\gamma$ daughter ^{94}Pd. The $T = 0$ states in ^{94}Ag are normalised to the $T = 1$ isobaric analogue states (IAS).

core excitations in the $(0g, 1d, 2s)$ model space. Excellent agreement between LSSM calculations and experiment is observed without any specific modification of the LSSM input used in the $N = 50$ to $N = Z$ region. The quality of agreement is demonstrated in fig. 3 for the $T = 1$ daughter states in ^{94}Pd populated in the $\beta\gamma$ decay of the $I^{\pi} = (21^+)$ and 7^+ isomers [11]. The decay scheme of the high-spin state as shown in fig. 2 allows for a number of different and due to the high-excitation energy and spin quite exotic decay modes, which were observed in follow-up experiments as $\beta\gamma$ [11], $\beta p\gamma$ [20], $p\gamma$ [21], $2p\gamma$ [22] channels. It should be emphasised that detection of the exotic decay modes was only due to the high spin of the isomer which enabled application of high-spin γ-decay spectroscopy via $p\gamma\gamma$ and $2p\gamma\gamma$ coincidence techniques [21,22]. Properties and decay modes of the isomer can be summarised as follows:

- highest spin (21^+) and energy (6.5 MeV) of a β decaying isomer in the Segré chart;
- spin-gap isomer with maximum aligned proton and neutron spin and leading configuration $(\pi g_{9/2}^{-3})_{21/2}\ (\nu g_{9/2}^{-3})_{21/2}$;
- isomerism reproduced in LSSM including $4p4h$ excitations of the ^{100}Sn core;

- one-proton emitter ($l = 4, 6$) with extremely small spectroscopic factor ($< 10^{-5}$ for $l = 4$);
- correlated two-proton emission dominates as sequential decay is hindered by small spectroscopic factor and large γ-decay widths;
- two-proton width largely enhanced due to a highly deformed shape isomer.

It is obvious that the characteristics of the one-proton and correlated two-proton emission are beyond the scope of even the LSSM calculations. Nevertheless in the simplest valence hole space the spectroscopic factor for $l = 4$ proton emission is only 4% and the quadrupole moment of the parent state is larger by more than a factor of two relative to the daughter. It should be noted that the high-spin isomers $I^\pi = (37/2^+)$ in ^{95}Ag and $I^\pi = (12^+, 14^+)$ in ^{98}Cd are further candidates for direct proton emission.

The success of the shell model in the description of the high-spin isomerism is due to the fact that the $\pi\nu$ realistic interaction is well tuned for the model space and gives a correct description of the evolution of single-particle energies. In fig. 4 the extrapolation of single neutron energies from experimentally known values in ^{88}Sr respective ^{90}Zr to ^{100}Sn is shown for two different realistic interactions GG [23] and MHJ [24]. The agreement with experimentally identified single-neutron states in $N = 51$ nuclei and among the two interactions demonstrates the reliability of this shell model aided extrapolation. The deviations as, *e.g.*, the $d_{5/2}$-$g_{7/2}$ inversion in ^{101}Sn show the systematic uncertainty of the procedure, which in this specific case is due to an improper $\pi g_{9/2}\nu g_{7/2}$ monopole in the MHJ interaction, which will be further discussed in sect. 5.

It is the monopole part of the $\pi\nu$ interaction that determines the evolution of the neutron single particle (hole) energies and the $N = 50$ shell gap upon filling of the $\pi 0g_{9/2}$ orbit from the experimentally known $Z = 40$ region towards $Z = 50$ (fig. 4). The monopole for a specific multiplet (j, j') is defined by

$$V^m_{jj'} = \sum_J (2J+1)\langle jj'J|V|jj'J\rangle / \sum_J (2J+1), \qquad (1)$$

which gives rise to the single-particle energy evolution between two shell closures CS and CS' [9]

$$\epsilon_j^{CS} = \epsilon_j^{CS'} + \sum_{j'} (2j'+1-\delta_{jj'})V^m_{jj'}. \qquad (2)$$

The Kronecker symbol applies for $T = 1$ and identical orbitals to maintain the Pauli principle. This simple formula can be used to calculate the single particle energies for ^{100}Sn from the experimentally known ones in ^{90}Zr or ^{88}Sr as shown in fig. 4 for neutrons and a given residual interaction. It should be noted that eq. (2) holds only for closed j' shells, *i.e.* in the example of fig. 4 ($j' = \pi g_{9/2}$) for ^{90}Zr and ^{100}Sn, in between due to configuration mixing the trend deviates from a strict scaling with the $j' = \pi g_{9/2}$ occupation and the exact progression can be inferred from a full shell model calculation (see fig. 4). In fig. 4 the spin-flip pairs $\pi 0g_{9/2}\nu 0g_{7/2}$, which are also spin-orbit partners ($\Delta l = 0$), and $\pi 0g_{9/2}\nu 1d_{3/2}$ ($\Delta l = 2$) exhibit much

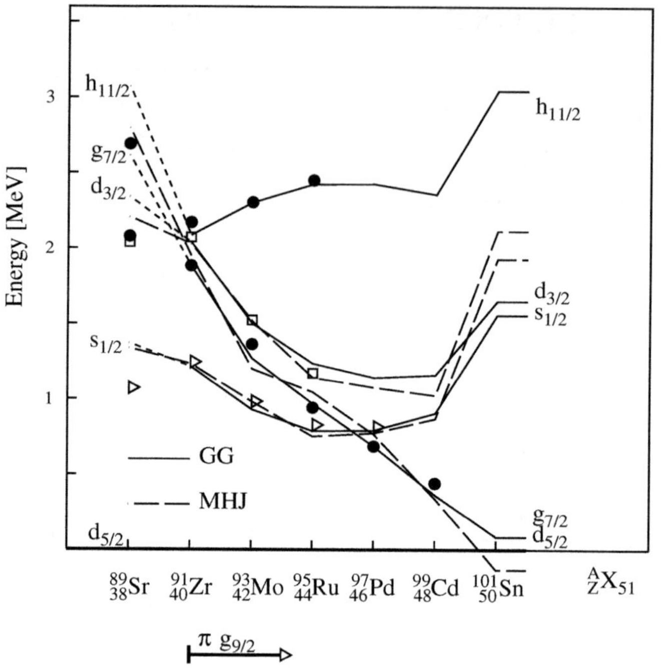

Fig. 4. Evolution of $N = 51$ single-particle energies relative to the $\nu d_{5/2}$ ground state for two different realistic interactions (GG) [23] and (MHJ) [24].

larger downward steps from ^{90}Zr to ^{100}Sn as compared to the $\nu 1d_{5/2}$ reference state, *i.e.* comparatively larger monopoles. This is a very general feature of the $\pi\nu$ interaction which is especially strong when the corresponding radial wave functions have good overlap [5,9] (see sect. 3). We also note that the evolution of neutron single-particle energies from ^{90}Zr ($Z = 40$) to ^{100}Sn ($Z = 50$) and single hole energies from ^{132}Sn to ^{122}Zr is described by the same interaction only slightly modified due to the different core mass (see sect. 5).

3 Monopole driven shell structure in neutron-rich nuclei

Strong monopole drifts are not restricted to the examples shown in fig. 4 but have been experimentally observed all over the Segré chart, the most prominent being the $\Delta l = 0$ spin-orbit $\pi\nu$ pairs $0p_{3/2}$-$0p_{1/2}$, $0d_{5/2}$-$0d_{3/2}$, $0f_{7/2}$-$0f_{5/2}$, $0g_{9/2}$-$0g_{7/2}$ and the $\Delta l = 1$ spin-flip pairs $0p_{1/2}$-$0d_{5/2}$, $0d_{3/2}$-$0f_{7/2}$, $0f_{5/2}$-$0g_{9/2}$, $0g_{7/2}$-$0h_{11/2}$. They are summarised in recent reviews [3,5,9,25] and can be traced back to the $\sigma\tau$ and tensor parts of the NN interaction [3, 4]. The empirical evidence translates into the following criteria for strong monopoles: i) the interacting nucleons are spin-flip partners with ii) $\Delta l = 0, 1, 2$ and iii) should have the same number of nodes in their radial wave functions to optimize the overlap. These features are also borne out in realistic interactions as derived from effective NN potentials fitted to scattering data via standard many-body techniques [24] as shown in fig. 17 of [9]. They suffer, however, from the fact that due to the neglect of three-body

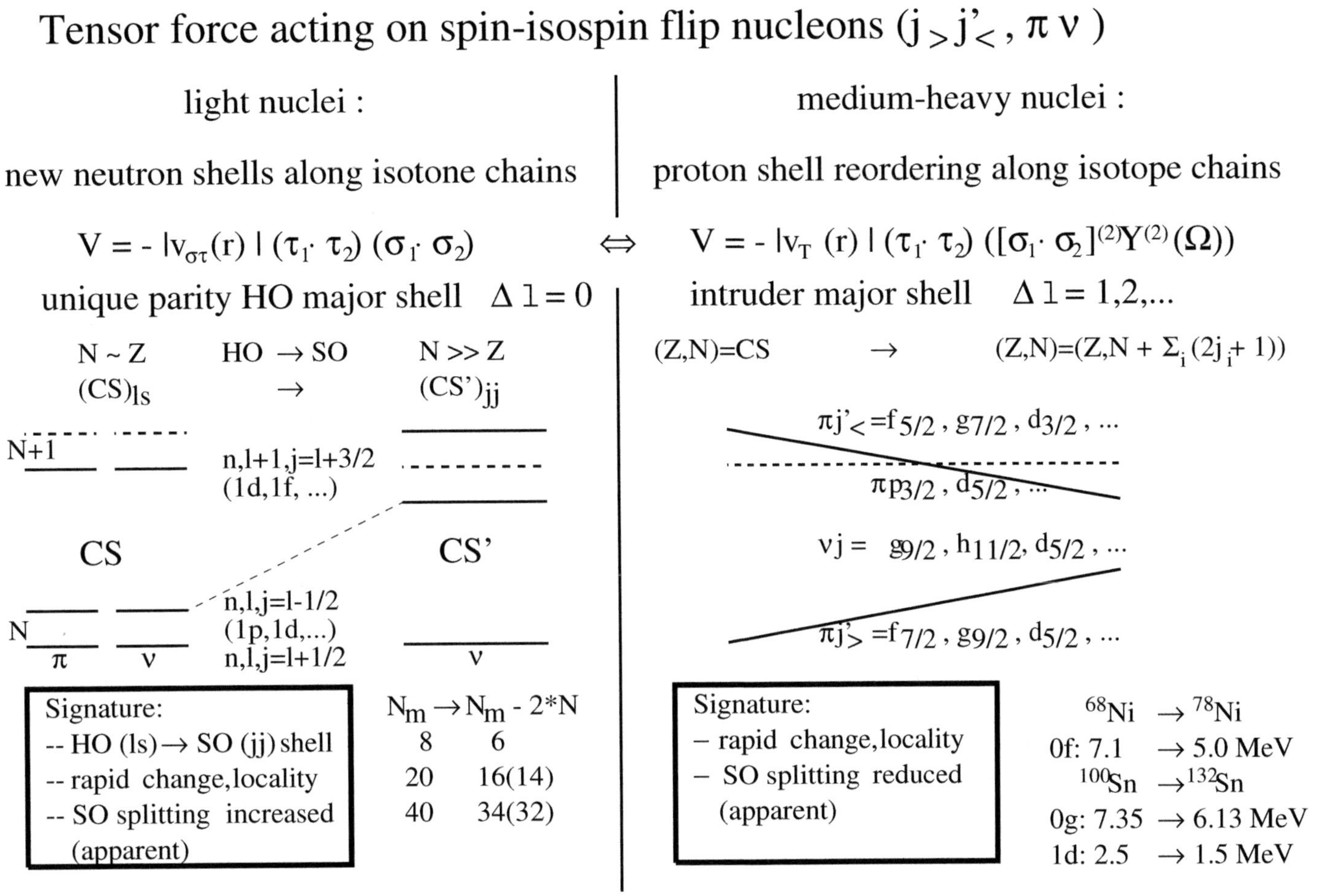

Fig. 5. Schematic illustration of signatures for tensor force driven shell evolution.

effects the monopole part is not determined well and has to be tuned to experimental shell evolution, which hampers their predictive power. The dramatic impact of monopole drifts and the sensitivity to subtle details of the interaction is due to the factor $(2j'+1)$ in eq. (2) which is large in filling (emptying) a high-spin orbital j' and translates monopole corrections of about 100 keV into MeV.

On the NN interaction level the strong monopoles are due to the tensor force

$$V = -|v_T(r)|(\tau_1 \cdot \tau_2)([\sigma_1 \cdot \sigma_2]^{(2)} Y^{(2)}(\Omega)), \tag{3}$$

which can be shown to reduce within a unique parity harmonic oscillator shell to [4] to the central $\sigma\tau$ force

$$V = -|v_{\sigma\tau}(r)|(\tau_1 \cdot \tau_2)(\sigma_1 \cdot \sigma_2) \tag{4}$$

3.1 New shells in light nuclei and shell reordering in medium-heavy nuclei

In fig. 5 the shell driving signatures of these interactions are compared schematically.

– In light nuclei within unique harmonic-oscillator shells starting with an ls-closed $N=Z$ doubly magic nucleus as ^{16}O or ^{40}Ca and proceeding towards $N \gg Z$ along an isotonic chain by proton removal the HO magic neutron number N_m is changed to $N_m - 2 \cdot N$, where N is the HO principal quantum number. This is due to the fact that upon emptying the $\pi j = l + 1/2$ orbit the $\nu j = l - 1/2$ becomes less bound due to the strong $\Delta l = 0$ (spin-orbit partners) $\sigma\tau$ monopole. Before the empty $\pi j = l - 1/2$ orbit releases the $\nu j' = l + 1 - 1/2$ orbit of the adjacent shell by action of the tensor $\Delta l = 1$ (spin-flip) monopole and stabilises the N_m neutron shell for $Z = N_m - 2 \cdot N$ as observed in ^{14}C, ^{36}S and ^{34}S, which demonstrates the isospin symmetry of the scenario. This converts an ls closed shell $N_m = 8$ (^{16}O), 20 (^{40}Ca), 40 (^{68}Ni) into jj closures $N'_m = 6, (14)16, (32)34$. The ambiguity is due to the presence of a $j = 1/2$ shell which exhibits a strong $T = 1$, $J = 0$ pairing matrix element (identical with the monopole for $j = 1/2$), and thus opens another gap upon filling or emptying this orbit. The scenario shown on the left side of fig. 5 gives rise to a new shell closure chart as shown in refs. [5,6,9]. Simultaneously the isotones in the HO oscillator semi-magic nuclei due the shell quenching and particle-hole excitations across the shell develop deformation as observed in ^{12}Be (fig. 20 in ref. [9]), ^{32}Mg [26] and ^{66}Fe [27]. Further abundant experimental evidence is presented in refs. [3,5,6,9].

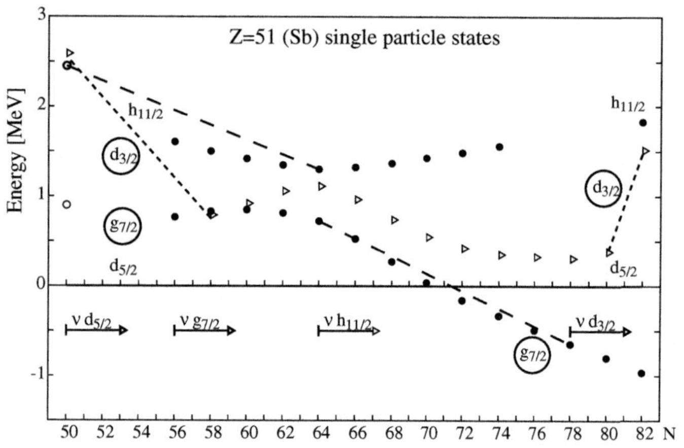

Fig. 6. Apparent spin-orbit splitting for $Z = 51$ single proton orbitals with reference to the $\pi d_{5/2}$ state. Experimental values are denoted by full circles, extrapolated ones by open circles. Dashed lines are drawn to guide the eye.

– In medium-heavy nuclei starting from a jj-closed doubly magic nucleus as ^{68}Ni or ^{100}Sn and going neutron-rich along an isotopic (Ni or Sn) chain by adding neutrons the apparent spin-orbit splitting of proton $\pi j = l \pm 1/2$ orbits is reduced when filling the $\nu j = l+1+1/2$ due to the $\Delta l = 1$ tensor monopole. It can be shown that the following identity holds [4]:

$$(2j_> + 1)V^m_{j'j_>} + (2j_< + 1)V^m_{j'j_<} = 0 \qquad (5)$$

for $T = 0, 1$, *i.e.* for $\pi\nu$, where in the present scenario j' is a neutron orbit and $(j_>, j_<)$ denote proton levels or vice versa as it is again symmetric in isospin. The shell evolution is schematically illustrated on the right side of fig. 5. Representative experimental numbers are given for the $Z = 28$ (Ni) and $Z = 50$ (Sn) proton levels in fig. 5. It should be noted that the values for ^{78}Ni and ^{100}Sn have been extrapolated by shell model aided extrapolation assuming realistic interactions. As for heavier nuclei the spin-orbit force increases so much that it does not determine the shell gap alone, in spite of the SO split reduction by the tensor force the shell gap is only reduced but not fully quenched. The effect is experimentally well established as shown in fig. 16 b,c of ref. [9] and in fig. 6 below for the $\nu g_{9/2}\pi f$ and the $\nu h_{11/2}\pi g$. This will be further discussed in sects. 4 and 5.

3.2 Apparent spin-orbit splitting driven by monopole shift

The signatures listed in fig. 5 and with respect to SO splitting may be compared to the scenario described in the introduction (sect. 1). They are not restricted to neutron-rich nuclei as they are symmetric in isospin, *i.e.* they hold for exchange of protons and neutrons though their importance to proton-rich nuclei is limited due to the close-lying dripline. The sign of the change in SO splitting reverses upon filling/emptying of an orbit. It is in contrast for the shell quenching scenario described in sect. 1 i) not determined by the gradient in N/Z ratio and ii) not restricted to neutron orbits. This is illustrated in fig. 6 for the $Z = 51$ (Sb) single particle states when filling the major neutron shell from ^{100}Sn to ^{132}Sn. In the following qualitative discussion it should be kept in mind that the neutron orbitals are not filled successively but due to configuration mixing partly in parallel, and that spectroscopic factors have been measured consistently only for the $\pi g_{7/2}$ and $\pi h_{11/2}$ states [28]. Striking monopole drifts can be observed for the $\pi\nu g_{7/2}$-$h_{11/2}$ and the $\pi\nu d_{3/2}$-$d_{5/2}$ pair of nucleons.

– It is known since long that the $\pi g_{7/2}$ is more strongly bound relative to the $\pi d_{5/2}$ reference state as soon as the $\nu h_{11/2}$ is filled above $N = 64$ with a net effect of ~ 1.9 MeV. The same monopole determines the downsloping of the $\pi h_{11/2}$ upon filling of the $\nu g_{7/2}$ between $N = 50$ and 64 by ~ 1.2 MeV. The ratio of the net effects according to eq. (2) is close to that of the multiplicities $(2j+1)$ of the filled neutron orbitals, namely 12/8. The exact trend is distorted in this case as the $\nu g_{7/2}$ is also acting on the $\pi d_{5/2}$ reference state. From eq. (5) one would expect that the $\pi g_{9/2}$ spin-orbit partner would be lifted up by filling of the $\nu h_{11/2}$ which should result in a reduced πg SO splitting. The latter effect is masked and compensated, however, by the strong (see fig. 4) $\pi g_{9/2}$-$\nu g_{7/2}$ monopole, as the $\nu g_{7/2}$ is filled before and/or in parallel.

– With respect to apparent spin-orbit splitting the $\pi d_{3/2}$-$d_{5/2}$ distance is a much better study object. The splitting reduces from $N = 50$ to 56 when the $\nu d_{5/2}$ orbit is filled, it increases from $N = 78$ to 82 when the filling of the $\nu d_{3/2}$ binds the reference level $\pi d_{5/2}$ more strongly than the $\pi d_{3/2}$. As it is the identical monopole which rules the shift, the ratio of down- and up-shift should be 4/6. In fact the ratio is smaller and this is due to two other neutron orbitals that are being filled in between. From $N = 56$ to 64 the splitting increases due to the $\nu g_{7/2}$ filling effect on the $\pi d_{5/2}$ reference while from $N = 64$ to 76 it reduces again due to the $\pi d_{3/2}$-$\nu h_{11/2}$ monopole. Again the multiplicity factor 12/8 results in a net decrease of the SO splitting, if the two different monopoles involved have a similar value.

As the sign of the SO splitting always places the $j_<$ level above the $j_>$ it can be concluded that a major part of the observed SO reduction for proton levels by adding neutrons from $N = Z$ towards larger N/Z values within a full major shell is related to the trivial (2j+1) weighting factor of eq. (2) that always favours $j_<$ over $j_>$ energetically as discussed for proton states along the Sn isotopes. The opposite holds for neutron levels when protons are removed towards larger N/Z along isotonic chains. This is not in contradiction to the recently observed SO reduction along $N = 82$ from ^{144}Sm to ^{132}Sn [28], as this covers only the lowest $\pi 1d_{5/2}$ and $\pi 0g_{7/2}$ part of the $Z = 50$–82 proton shell leaving aside the effect by the $\pi 1d_{3/2}$ and $\pi 0h_{11/2}$ orbitals.

4 Towards ^{78}Ni

4.1 From ^{48}Ca to ^{78}Ni

The heaviest nucleus with doubly-magic features and an ls-closed HO neutron shell is ^{68}Ni at $N = 40$. The neutron shell gap has been discussed in many publications and was found to be small and to disappear at more than two nucleons distance from ^{68}Ni [9,27,29,30]. This can be understood within the framework of the scenarios shown in sect. 3 and fig. 5. Towards the neutron-rich Ca isotopes the removal of $\pi f_{7/2}$ protons will release the $\nu f_{5/2}$ neutrons to close the $N = 40$ gap. Experimentally deformation was observed already in ^{66,67}Fe [27,31] and ^{64}Cr would correspond to ^{32}Mg one major shell lower. The $N = 40$ gap according to this scenario would shift to $N = 32, 34$ in the Ca isotopes. Recently relativistic Coulomb excitation experiments were performed on the $N = 32, 34$ Cr [32] and Ti [33] isotopes. In both cases clear evidence for a $N = 32$ subshell was observed in the $B(E2, 2^+ \rightarrow 0^+)$ transition strength, which corroborates an earlier conclusion from excited states in ^{54}Ti [34]. On the other hand evidence for the $N = 34$ closure was not seen and it may develop only in the Ca isotopes.

Beyond ^{68}Ni the doubly-magic $N = 50$ nucleus ^{78}Ni has been the subject of numerous experimental studies with respect to the persistence of the $N = 50$ shell and its relevance for the astrophysics r-path. Early β-decay results seem to indicate a substantial shell quenching [35], while in-beam experiments on $N \sim 50$ Ge-Se isotopes [36] and isomer studies following fragmentation [37,38,39,40] give evidence for the persistence of the $N = 50$ shell. Decisive for both the $Z = 28$ and $N = 50$ shell gaps in ^{78}Ni is according to the tensor force scenario sketched on the right hand side of fig. 5 the monopole part of the spin-flip $\Delta l = 1$ $\pi 0f_{5/2}\nu 0g_{9/2}$ pair of nucleons. In Ni isotopes ($Z = 28$) beyond $N = 40$ by filling of the $\nu 0g_{9/2}$ shell the $\pi 0f_{5/2}$ orbit is bound more strongly than the adjacent $\pi 1p_{3/2}$ and $\pi 0f_{7/2}$ and eventually crosses the $\pi 1p_{3/2}$ to enter the shell gap. This was experimentally observed up to $N = 44$ in the β decay of odd-A Ni isotopes [41]. Governed by the same monopole, along $N = 50$ the removal of $\pi 0f_{5/2}$ protons will release the $\nu 0g_{9/2}$ stronger than $\nu 1d_{5/2}$ which will reduce the gap. Recent shell model extrapolations of the $Z = 28$ and $N = 50$ shell gaps from ^{68}Ni and ^{90}Zr, respectively, to ^{78}Ni yielded persistence of the proton and neutron shell gaps with ~ 5.1 MeV (see fig. 5) and ~ 3.5 MeV, respectively [6]. This is in agreement with experimental evidence on the persistence of $\nu 0g^2_{9/2}$ seniority isomerism from $N = 42$ (^{70}Ni) to $N = 48$ (^{78}Zn, ^{76}Ni) [37,38,39,40] and the $N = 50$ shell strength in Ge isotopes [36]. The inferred ^{78}Ni shell gaps along with the recently determined empirical $T = 1$ interaction and single particle (hole) energies for the $N = 50$ isotones and Ni isotopes [42] provide a new bench mark for tuning the monopole interaction in the ^{48}Ca to ^{78}Ni model space. In the lower right panel of fig. 7 the experimental $E2$ strengths $B(E2; 8^+ \rightarrow 6^+)$ in the Ni isotopes beyond $N = 40$ are compared to recent shell model results in the full $(0f_{5/2}, 1p, 0g_{9/2})$ model space [42]. The shell model accounts very well for the observed $E2$ strength and the disappearance of the $I^\pi = 8^+$ isomerism in ^{72,74}Ni [43,44], which is intimately connected to the low $I^\pi = 2^+$ excitation energies [9,43], *i.e.* a strongly bound $\nu g^2_{9/2}; I^\pi = 2^+$ two-body matrix element. As a consequence in the $\nu g^n_{9/2}$ multiplet for $n = 4, 6$ the seniority $v = 4$ are more strongly bound which opens a new $8^+, v = 2 \rightarrow 6^+, v = 4$ $\Delta v = 2$ decay channel with a large $E2$ strength as shown in fig. 7, lower right panel. The microscopic origin of the low $I^\pi = 2^+$ excitation energy, which is experimentally verified in $^{70-76}$Ni [37,44,40], can be ascribed to a reduction of the $\pi 0f$ SO splitting and the $Z = 28$ shell gap due to the strong tensor force $\nu g_{9/2}\pi f$ monopole as discussed in sect. 3 and fig. 5 [4]. The deviation of the $E2$ trend in the heavy Ni isotopes from their $N = 50$ valence mirror nuclei will be discussed in sect. 4.2.

4.2 Valence mirrors and break-down of the seniority scheme

In fig. 7 experimental and shell model $E2$ transition strengths for the $N > 40$ even Ni isotopes are compared to their $g^n_{9/2}$ valence mirror $N = 50$ isotones. Shell model calculations were performed with a new empirical interaction in the full $T = 1$ $(0f_{5/2}, 1p, 0g_{9/2})$ neutron respective proton model space [42]. The conclusions from the comparison in fig. 7 can be summarised as follows:

- The $B(E2)$ values for the yrast states do not show any mirror symmetry in the $n = 4, 6$ midshell nuclei. Nevertheless the $v = 2, 4$ states in the Ni isotopes have good seniority.
- The agreement for $\Delta v = 0$ transitions of yrast states is excellent.
- The $B(E2; 2^+ \rightarrow 0^+)$ in ^{70}Ni is largely underestimated in the shell model [45]. This is further evidence that the $Z = 28$ shell gap is soft against proton core excitation due to a monopole driven shell quenching (see sect. 4.1). As the $N = 50$ isotones have different neutron shell structure the valence mirror ^{92}Mo is affected only marginally by core excitation (upper left panel in fig. 7).
- The $B(E2; 4^+ \rightarrow 2^+)$ in the $n = 4, 6$ $N = 50$ isotones ^{94}Ru and ^{96}Pd cannot be reproduced in the $T = 1$ shell model approach [46]. It can be shown that this is a general feature of all calculations in a pure proton space [47]. On the other hand the large scale shell model as described in sect. 2 without further modification can describe these transitions if up to $4p4h$ excitations across $N = 50$ are included (dashed line in fig. 7). The core excitations mix proton and neutrons and hence cause the breakdown of the seniority scheme by mixing $v = 2, 4$ configurations in the $I^\pi = 4^+$ state [47].

5 Shell structure along the r-path towards N ≫ Z

The success of the concept of monopole driven shell structure especially for the partially quenched $N = 50$ shell

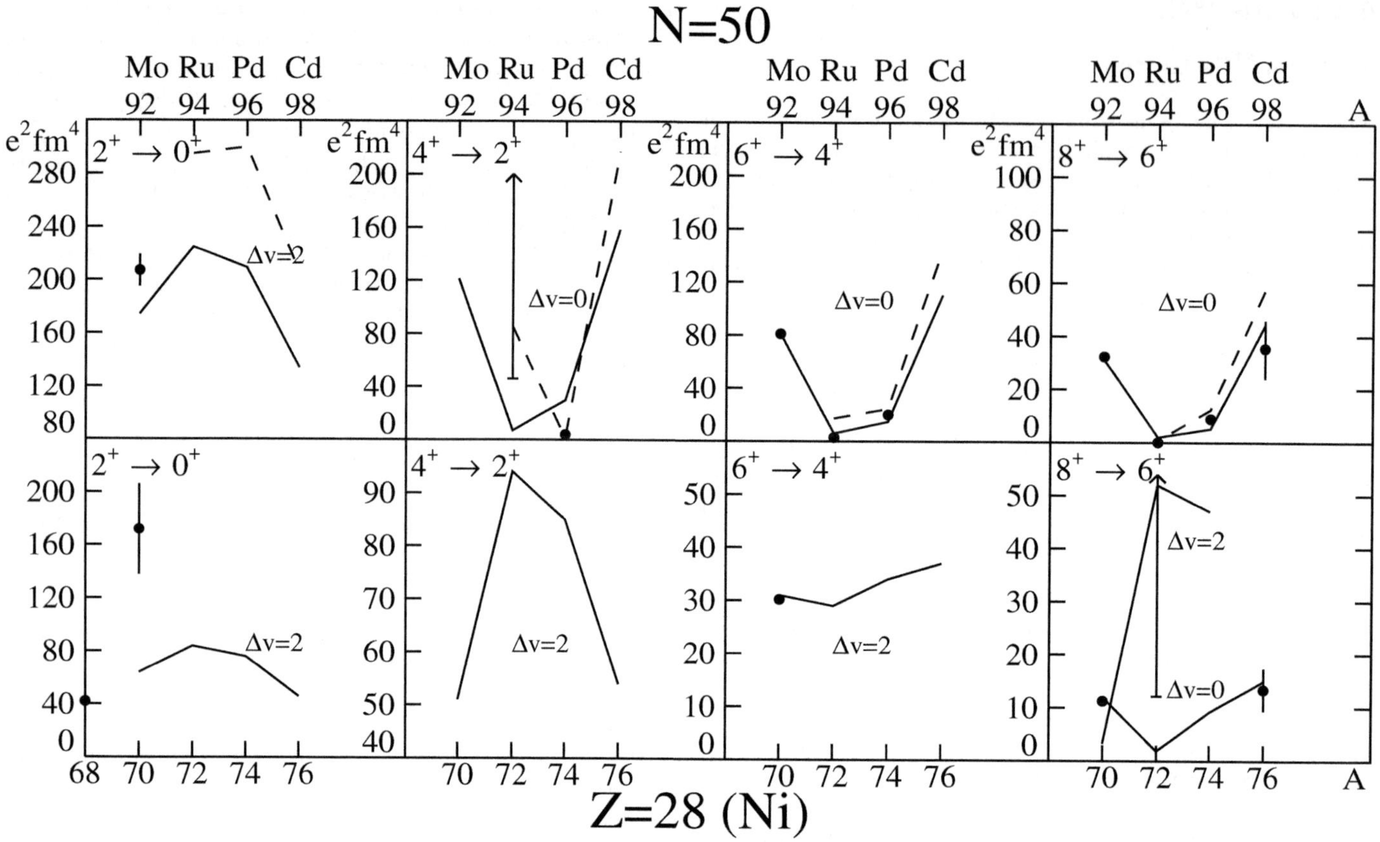

Fig. 7. $E2$ transition strengths within the $g_{9/2}^n$ configuration in even Ni isotopes (lower panel) and the $N = 50$ valence mirror partners (upper panel).

at ^{78}Ni, raises the question whether this could provide a possible scenario to understand the r-path abundance deficiency trough below the $A \simeq 130$ peak in astrophysical network calculations [48]. Quenching of the $N = 82$ shell due to a softening of the neutron potential as described in the introduction [2] has been invoked to explain this abundance deficiency [48] and experimental evidence for a reduced shell gap for $N = 82$, $Z \leq 50$ has been presented [49,50]. While the nuclear structure origin of the astrophysics problem is still controversial, it might be appropriate to also look into alternative structure scenarios. In essence a reduced $N = 82$ shell gap causes increased excitation of neutrons into orbitals beyond $N = 82$ leading eventually to deformation. As a consequence the β-decay halflives at the previous waiting points become shorter due to larger Q_β values while they are increased for smaller neutron numbers due to the delayed filling of the $\nu 0g_{7/2}$ subshell which is the key orbital for the $\nu 0g_{7/2} \to \nu 0g_{9/2}$ allowed Gamow-Teller (GT) transition.

The relevant r-path nuclei are found below ^{132}Sn at $Z \leq 50$ with the single neutron states playing the key rôle. The evolution of the neutron hole states is governed by the same $\pi 0g_{9/2}\nu j$ interaction [24] as for the neutron particles along $N = 50$ as shown in fig. 4 (MHJ) except for a renormalisation due to the different shell model core, which in the simplest case is an $A^{-1/3}$ scaling. In the left panel of fig. 8 the evolution of the $N = 51$ neutron levels according to eq. (2) is shown with this interaction. The right panel shows the evolution of the $N = 81$ neutron hole states with an interaction as determined for a ^{132}Sn core [6,24]. Starting points are the experimental values adopted for ^{132}Sn [8,9]. It should be noted that eq. (2) holds only for closed j' shells, *i.e.* in the example of fig. 8 for the points, provided the proton shell gap is preserved, too. In between due to configuration mixing the trend may deviate from the lines drawn to guide the eye. The exact progression can be inferred from a full shell model calculation (see fig. 4 for the $N = 51$ case). This does not exclude a shell gap reduction due to cross shell excitations when moving away from a doubly-magic nucleus along a semi-magic chain of nuclei. Note that, *e.g.*, from ^{100}Sn to ^{94}Ru this amounts to a ~ 2 MeV reduction.

To validate this extrapolation it has to be proven that the $Z = 40$ gap is preserved for ^{122}Zr. In the upper panel of fig. 9 the evolution of the $Z = 40$ gap and the adjacent single proton levels from $N = 50$ to 82 is shown. The realistic interaction based on the CD Bonn potential is identical with the one denoted by MHJ in fig. 4, derived for a ^{88}Sr core [24]. The results according to eq. (2) are drawn by dashed lines. It is obvious that the experimental points known until $N = 64$ are not well reproduced. The interaction fails in two details: i) it reverses the $\nu 0g_{7/2}$-$1d_{5/2}$

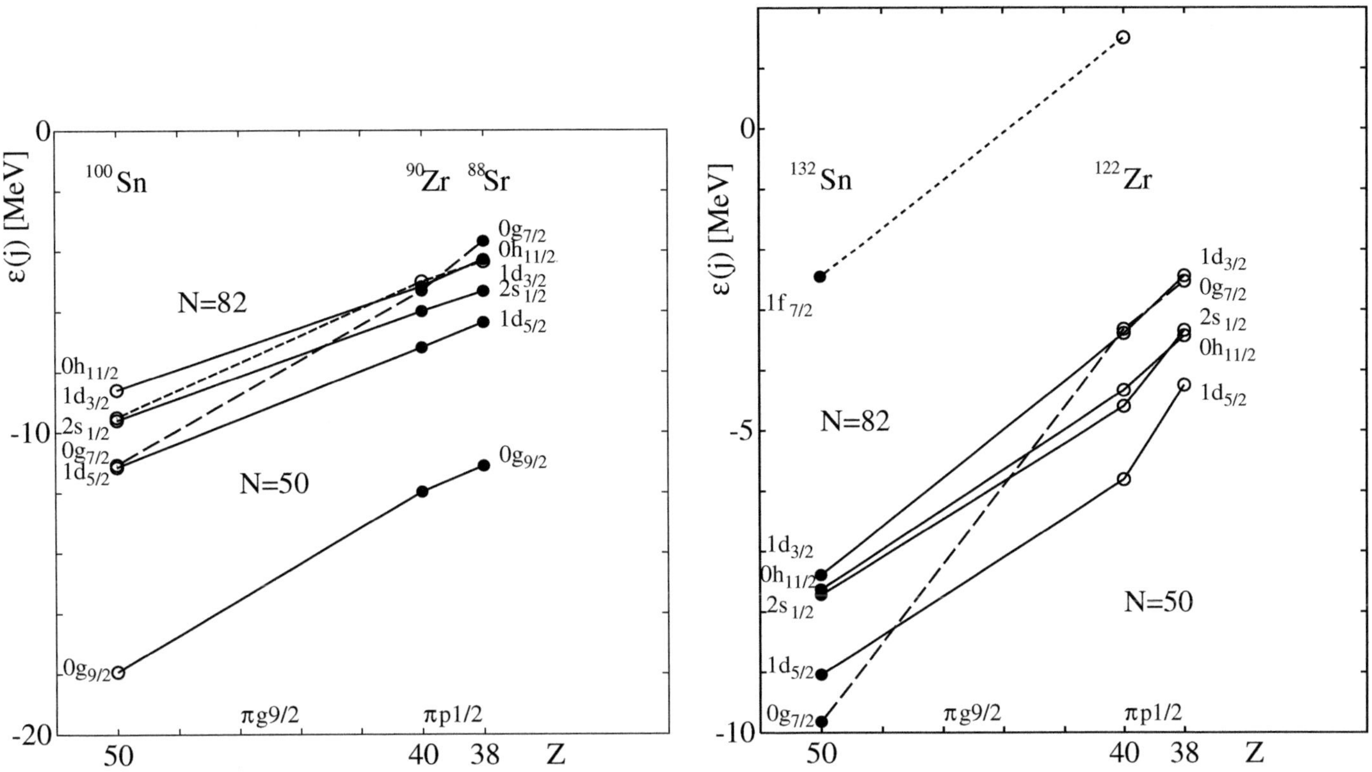

Fig. 8. Shell evolution along $N = 50$ (left) and $N = 82$ (right) isotones with the (MHJ) realistic interaction for ^{88}Sr and ^{100}Sn core, respectively. The interactions for the $\nu g_{9/2}$ orbit below $N = 50$ is from an empirical fit [7,23] and for the $\nu f_{7/2}$ orbit above $N = 82$ from extrapolation from the ^{208}Pb region [6].

sequence in ^{101}Sn (see fig. 4); ii) it calculates the excitation energy of the $I^\pi = 1/2^-$ isomer in ^{103}In much too high relative to the $I^\pi = 9/2^+$ ground state [51]. These excitation energies are shown in the lower panel of fig. 9. Both deficiencies can be cured by tuning the $\pi g_{9/2}$-$\nu g_{7/2}$ and $\pi p_{1/2}$-$\nu d_{5/2}$ monopoles. The results for the shell gap are shown by full lines in the upper panel and by a full line circle in the lower panel next to the experimental ^{103}In point (SM). Due to lacking experimental information the monopoles involving the $\pi p_{3/2}, f_{5/2}$ were not corrected. The features of the $Z = 40$ shell extrapolation are summarised as follows:

- In spite of the qualitative character of the extrapolation by using eq. (2) instead of an involved shell model calculation, the experimental shell gap is well reproduced. Especially the stabilisation from ^{90}Zr ($N = 50$) to ^{96}Zr ($N = 56$) and the following quenching towards the strongly deformed region $^{100-104}$Zr is accounted for.
- Beyond $N = 64$ upon filling of the $\nu 2s_{1/2}, 0h_{11/2}, 1d_{3/2}$ orbitals the gap is widening again to reach ~ 4.35 MeV at $N = 82$, which has to be compared to 3.198 MeV and 4.036 MeV at $N = 50$ and 56, respectively. The $N = 82$ (^{122}Zr) value is an upper limit, as the interaction derived for ^{88}Sr was not scaled by $A^{-1/3}$ which yields a 10% reduction. The weakness of this extrapolation is the lack of any experimental verification.
- Exploiting the above-mentioned sensitivity of the excitation energy of the $I^\pi = 1/2^-$ isomer in ^{103}In to the tuning of the interaction, in the lower panel of fig. 9 these excitation energies are shown for all In isotopes. A clear correlation of this energy with the width of the shell gap is seen up to $N = 64$. Beyond the minimum at $N \sim 64$ the values are rising again as predicted by the shell gap extrapolation. The trend is stopped, however, at $N = 78$ which casts some doubt on the predicted persistence of the $Z = 40$ shell gap. It should be noted though that the last three points are results of β-endpoint mass measurements. Recent mass measurements using ion traps in neutron-rich Sr-Ru isotopes revealed up to ~ 1 MeV discrepancies as compared to β-endpoint results [52].
- We note in passing that the $Z = 40$ shell gap extrapolation including the trend in the In probe does not show any sign of double-magicity for the $Z = 40$, $N = 70$ nucleus ^{110}Zr.

In conclusion the persistence of the $Z = 40$ gap at $N = 82$ is still an open question. So a possible quenching of the $N = 82$ gap may find a simple explanation in the lack of the $Z = 40$ gap, which would make ^{122}Zr a mid-shell nucleus with reduced neutron gap due to *ph* excitations across the shell.

An alternative scenario for filling the abundance trough below $A \simeq 130$ is based on the steep upsloping of the $\nu g_{7/2}$ level from the deepest in the shell at ^{132}Sn to the Fermi surface at ^{122}Zr as shown in fig. 8 right panel. The allowed GT transition is delayed as the $\nu 0g_{7/2}$ starts to be filled only about 12 nucleons below $N = 82$ thus

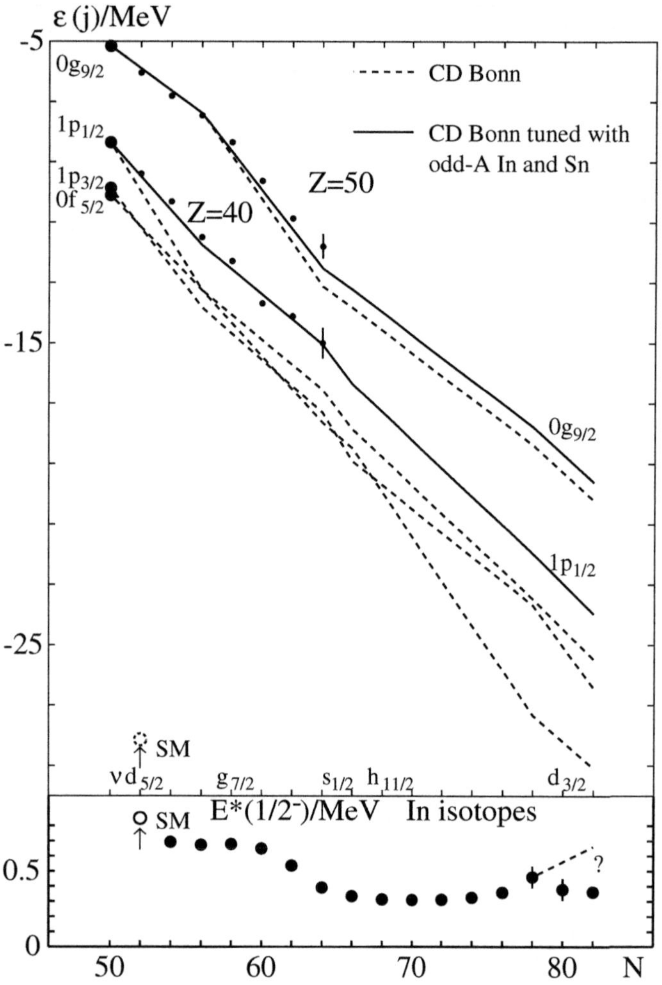

Fig. 9. Evolution of the $Z = 40$ shell gap (top panel) and the splitting of the $I^\pi = 9/2^+$ g.s. and $I^\pi = 1/2^-$ isomer in In isotopes.

increasing β-decay halflives in the region $A < 130$. On the other hand a filled $\nu 0g_{7/2}$ orbit at $N \sim 82$ at the Fermi surface causes large effective Q_β values, which decreases the halflives in this region. As a consequence the abundance peak intensities will be shifted to lighter masses. This scenario, however, neglects the influence of neutron separation energies, which in case of a quenched shell are inherently included.

6 Summary and conclusions

It has been shown that isomer decay spectroscopy close to magic nuclei provides a very sensitive probe of residual interactions and single particle energies employed in shell model calculations. The strong proton-neutron interaction in identical orbitals at $N = Z$ gives rise to spin-gap isomers exhibiting exotic decay modes such as p and 2p decay. An indispensable prerequisite for sound predictions are readily available large scale shell model codes along with realistic interactions that in their monopole part are well adjusted to experimental single particle energies. This does not necessarily hamper the predictive power of shell model calculations as the tuning can be done in regions accessible to detailed spectroscopy (see sects. 2 and 5). Monopole driven shell evolution can account for many aspects of structural changes on the pathway from proton-rich $N \sim Z$ nuclei (^{100}Sn) to the neutron-rich $N \gg Z$ (^{78}Ni) region. The concept has been shown to account for the new shell closures established in light nuclei, shell reordering and evolution of spin-orbit splitting in medium-heavy nuclei, and it provides an alternative access to the structure of $N \gg Z$ r-path nuclei.

The authors enjoyed illuminating discussions with F. Nowacki and T. Otsuka and are grateful for communication of unpublished data by A. Lisetskiy and O. Sorlin.

References

1. A. Bohr, B.R. Mottelson, *Nuclear Structure* (World Scientific, Singapore 1998).
2. J. Dobaczewski *et al.*, Phys. Rev. Lett. **72**, 981 (1994).
3. T. Otsuka *et al.*, Phys. Rev. Lett. **87**, 0852502 (2002).
4. T. Otsuka *et al.*, Acta Phys. Pol. B **36**, 1213 (2005); Phys. Rev. Lett. **95**, 232502 (2005).
5. H. Grawe, Acta Phys. Pol. B **34**, 2267 (2003).
6. H. Grawe *et al.*, Eur. Phys. J. A **25**, s01, 357 (2005).
7. H. Grawe *et al.*, Physica Scripta T **56**, 71 (1995).
8. H. Grawe, M. Lewitowicz, Nucl. Phys. A **693**, 116 (2001).
9. H. Grawe, Springer Lect. Notes Phys. **651**, 33 (2004).
10. A. Blazhev *et al.*, Phys. Rev. C **69**, 064304 (2004).
11. C. Plettner *et al.*, Nucl. Phys. A **733**, 20 (2004).
12. R. Grzywacz *et al.*, Phys. Rev. C **55**, 1126 (1997).
13. M. Górska *et al.*, Z. Physik A **353**, 233 (1995).
14. E. Nolte, H. Hicks, Phys. Lett. B **97**, 55 (1980).
15. J. Döring *et al.*, GSI Ann. Rep. (2003) and to be published.
16. J. Döring *et al.*, Phys. Rev. C **68**, 034306 (2003).
17. K. Ogawa, Phys. Rev. C **28**, 958 (1983).
18. M. Górska *et al.*, *Proceedings of the 8th International Spring Seminar on Nuclear Physics, Key Topics in Nuclear Structure, Paestum, Italy, 2004*, edited by A. Covello (World Scientific, Singapore, 2005) p. 229.
19. M. Górska *et al.*, Phys. Rev. Lett. **79**, 2415 (1997).
20. I. Mukha *et al.*, Phys. Rev. C **70**, 044311 (2004).
21. I. Mukha *et al.*, Phys. Rev. Lett. **95**, 022501 (2005).
22. I. Mukha *et al.*, Nature **439**, 298 (2006).
23. M. Górska *et al.*, ENPE99, AIP Conf. Proc. **495**, 217 (1999).
24. M. Hjorth-Jensen *et al.*, Phys. Rep. **261**, 125 (1995) and private communication.
25. T. Otsuka *et al.*, Eur. Phys. J. A **15**, 151 (2002).
26. T. Motobayashi *et al.*, Phys. Lett. B **346**, 9 (1995).
27. M. Hanawald *et al.*, Phys. Rev. Lett. **82**, 1391 (1999).
28. J.P. Schiffer *et al.*, Phys. Rev. Lett. **92**, 162501 (2004).
29. O. Sorlin *et al.*, Phys. Rev. Lett. **88**, 092501 (2002).
30. K.H. Langanke *et al.*, Phys. Rev. C **67**, 044314 (2003).
31. M. Sawicka *et al.*, Eur. Phys. J. A **16**, 51 (2002).
32. A. Bürger *et al.*, Phys. Lett. B **622**, 29 (2005).
33. D.-C. Dinca *et al.*, Phys. Rev. C **71**, 041302 (2005).
34. R.V.F. Janssens, *et al.*, Phys. Lett. B **546**, 55 (2002).
35. K.-L. Kratz *et al.*, Phys. Rev. C **38**, 278 (1988).
36. Y.H. Zhang *et al.*, Phys. Rev. C **70**, 024301 (2004).

37. R. Grzywacz *et al.*, Phys. Rev. Lett. **81**, 766 (1998).
38. J.M. Daugas *et al.*, Phys. Lett. B **476** 213 (2000).
39. M. Sawicka *et al.*, Eur. Phys. J. A **20**, 109 (2004).
40. R. Grzywacz, Eur. Phys. J. A **25**, s01, 89 (2005).
41. S. Franchoo *et al.*, Phys. Rev. C **64**, 054308 (2001).
42. A. Lisetskiy *et al.*, Eur. Phys. J. A **25**, s01, 95 (2005).
43. H. Grawe *et al.*, Nucl. Phys. A **704**, 211c (2002).
44. M. Sawicka *et al.*, Phys. Rev. C **68**, 044304 (2003).
45. O. Perru *et al.*, submitted to Phys. Rev. Lett.
46. H. Mach *et al.*, *Proceedings of the International Symposium A New Era of Nuclear Structure Physics, Niigata, Japan 2003*, edited by Y. Suzuki, S. Ohya, M. Matsuo, T. Ohtsubo (World Scientific, Singapore, 2004) p. 277.
47. H. Mach *et al.*, to be published.
48. B. Pfeiffer *et al.*, Nucl. Phys. A **693**, 282 (2001).
49. I. Dillmann *et al.*, Phys. Rev. Lett. **91**, 162503 (2003).
50. T. Kautzsch *et al.*, Eur. Phys. J. A **9**, 201 (2000).
51. O. Kavatsyuk *et al.*, Eur. Phys. J. A **25**, 211 (2005).
52. J. Äystö, these proceedings.

Eur. Phys. J. A **27**, s01, 269–276 (2006)
DOI: 10.1140/epja/i2006-08-041-6

EPJ A direct
electronic only

Relation between proton and neutron asymptotic normalization coefficients for light mirror nuclei and its relevance for nuclear astrophysics

N.K. Timofeyuk[1,a], P. Descouvemont[2], and R.C. Johnson[1]

[1] Department of Physics, School of Electronics and Physical Sciences, University of Surrey, Guildford, Surrey GU2 7XH, UK
[2] Physique Nucléaire Théorique et Physique Mathématique, CP 229, Université Libre de Bruxelles, B-1050 Brussels, Belgium

Received: 20 June 2005 /
Published online: 15 March 2006 –

Abstract. It has been realised recently that charge symmetry of the nucleon-nucleon interaction leads to a certain relation between Asymptotic Normalization Coefficients (ANCs) in mirror-conjugated one-nucleon overlap integrals. This relation can be approximated by a simple analytical formula that involves mirror neutron and proton separation energies, the core charge and the range of the strong nucleon-core interaction. We perform detailed microscopic multi-channel cluster model calculations and compare their predictions to the simple analytical formula as well as to calculations within a single-particle model in which mirror symmetry in potential wells and spectroscopic factors are assumed. The validity of the latter assumptions is verified on the basis of microscopic cluster model calculations. For mirror pairs in which one of the states is above the proton decay threshold, a link exists between the proton partial width and the ANC of the mirror neutron. This link is given by an approximate analytical formula similar to that for a bound-bound mirror pair. We compare predictions of this formula to the results of microscopic cluster model calculations. Mirror symmetry in ANCs can be used to predict cross sections for proton capture at stellar energies using neutron ANCs measured with stable or "less radioactive" beams.

PACS. 21.60.Gx Cluster models – 21.10.Jx Spectroscopic factors – 27.20.+n $6 \leq A \leq 19$ – 27.30.+t $20 \leq A \leq 38$

1 Introduction

Over the last 15 years, the nuclear physics community has shown a growing interest in Asymptotic Normalization Coefficients (ANCs). The study of these coefficients, both theoretically and experimentally, is mostly motivated by their application to nuclear astrophysics.

The one-nucleon ANC determines the magnitude of the large distance behaviour of the overlap integral between the bound state wave functions of nuclei A and $A-1$. Such overlaps enter the amplitude for non-resonant nucleon capture reactions. If the capture occurs outside the nuclear interior, as often happens at very low stellar energies, then the overall normalization of its cross sections as well as of the astrophysical S-factors, is determined by the squared ANC [1]. Since the same ANCs play a crucial role in other peripheral processes such as transfer reactions, they can be measured in laboratories and used to predict non-resonant capture processes at low stellar energies [1].

[a] e-mail: n.timofeyuk@surrey.ac.uk

It has been suggested recently in ref. [2] that the ANCs of two mirror overlap integrals should be related if the charge symmetry of nucleon-nucleon (NN) interactions is valid. It has been shown there that mirror ANCs can be linked by an approximate analytical expression which contains only nucleon separation energies, charges of the product nuclei and the range of the strong interaction between the last nucleon and the core. This link can be used to predict cross sections for non-resonant proton capture if mirror neutron ANCs are known. The latter can be determined using direct reactions in experiments with stable beams. Such experiments are less difficult and more accurate than ones involving radioactive beams which are necessary to determine the proton ANCs.

According to ref. [2], if one of the mirror nuclear states is a low-lying narrow proton resonance, and its mirror analog is particle-stable, then a link should exist between the width of the proton decaying state and the neutron ANC of its mirror analog. This link is given by an approximate analytical formula which is similar to that for the case of bound mirror pairs.

A proper understanding of the link between the width of a proton resonance and the neutron ANC of its mirror analog can be important for predicting the resonant proton capture rates for a particular class of resonant reactions at stellar energies. This class includes reactions that proceed via very narrow isolated states for which the proton width Γ_p is either comparable to or much less than its γ-decay width Γ_γ. Such resonances can be found in the neutron-deficient region of the *sd* and *pf* shells (for example, some levels in ^{25}Si, ^{27}P, ^{33}Ar, ^{36}K and ^{43,46}V) and their study is important for understanding nucleosynthesis in the *rp* process. For the resonances mentioned above Γ_p can be less than 1 eV. Direct measurements of such tiny widths using proton elastic scattering are impossible. Proton transfer reactions can be used instead. They provide spectroscopic factors which are combined with the single-particle widths to get the necessary partial proton widths. However, uncertainties in Γ_p extracted through theoretical analysis of these reactions (for example, using the distorted-wave formalism) are about 50% [3]. These uncertainties arise due to problems in the theoretical analysis of stripping reactions to the continuum and the determination of the single-particle proton widths. The determination of proton widths based on the link to ANCs of their mirror particle-stable analogs can provide better accuracy since it avoids the uncertainties mentioned above.

In the present paper, we compare three different calculations of the ratio between mirror ANCs of bound mirror pairs and two calculations for the ratio between the proton widths and neutron ANCs of their mirror analogs. The first calculation uses the analytical formulae of ref. [2]. The second one (applied here only for bound mirror pairs) is based on the idea of mirror symmetry of single-particle potential wells and of spectroscopic factors. The third method uses the microscopic calculations within a multichannel cluster model. These three approaches are described in sect. 2, 3 and 4. The comparison between them is given in sect. 5 and 6 and discussed in sect. 7.

2 Analytical formula

2.1 Bound mirror pairs

The ANC C_{lj} for the one-nucleon virtual decay $A \to (A-1)+N$ is defined via the tail of the overlap integral $I_{lj}(r)$

$$I_{lj}(r) = \langle \chi_{\frac{1}{2}\tau}\left[\left[Y_l(\hat{\boldsymbol{r}}) \otimes \chi_{\frac{1}{2}}\right]_j \otimes \Psi^{J_{A-1}}\right]_{J_A} | \Psi^{J_A}\rangle \qquad (1)$$

between the many-body wave functions Ψ^{J_A} and $\Psi^{J_{A-1}}$ of nuclei A and $A-1$. Here l is the orbital momentum, j is the total relative angular momentum between $A-1$ and N, τ is the isospin projection and $\chi_{\frac{1}{2}\tau}$ is the isospin wave function of nucleon N, and r is the distance between N and the center-of-mass of $A-1$. Asymptotically, this overlap behaves as

$$\sqrt{A}\, I_{lj}(r) \approx C_{lj} \frac{W_{-\eta,l+1/2}(2\kappa r)}{r}, \qquad r \to \infty, \qquad (2)$$

where $\kappa = (2\mu\epsilon/\hbar^2)^{1/2}$, ϵ is the one-nucleon separation energy, $\eta = Z_{A-1}Z_N e^2 \mu/\hbar^2\kappa$, μ is the reduced mass for the $(A-1)+N$ system and W is the Whittaker function.

For the $\langle {}^{A-1}_{N-1}Z \otimes n | {}^{A}_{N}Z\rangle$ overlap, where the separated nucleon is a neutron, the ANC C_{lj} can be represented by the matrix element [4,5,6]:

$$C_{lj} = -i^{-l} \frac{2\mu\sqrt{A}}{\hbar^2} \times$$
$$\langle \chi_{\frac{1}{2}\tau}\left[\left[j_l(i\kappa r) Y_l(\hat{\boldsymbol{r}}) \otimes \chi_{\frac{1}{2}}\right]_j \otimes \Psi^{J_{A-1}}\right]_{J_A} \| \hat{V}^{nucl} \| \Psi^{J_A}\rangle, \quad (3)$$

where $j_l(i\kappa r)$ is the spherical Bessel function,

$$\hat{V}^{nucl} = \sum_{i=1}^{A-1} V_{NN}(|\boldsymbol{r}_i - \boldsymbol{r}_A|), \qquad (4)$$

V_{NN} is the strong two-body NN potential and $\boldsymbol{r}_i$ is the radius-vector of the i-th nucleon.

For the mirror overlap $\langle {}^{A-1}_{Z-1}N \otimes p | {}^{A}_{Z}N\rangle$, where the separated nucleon is a proton, the expression for the ANC C_{lj} can be obtained by considering the inhomogeneous coupled system of differential equations

$$\left(\epsilon_p + \hat{T}_l^{rel} + \sum_{n'l'j'} V^{coul}_{nlj,n'l'j'}(r)\right) I_{n'l'j'}(r)$$
$$= -\langle \chi_{\frac{1}{2}\tau}\left[\left[Y_l(\hat{\boldsymbol{r}}) \otimes \chi_{\frac{1}{2}}\right]_j \otimes \Psi^{J_{A-1}}\right]_{J_A} \| \hat{V}^{nucl} \| \Psi^{J_A}\rangle, \quad (5)$$

which is easily obtained from the simultaneous consideration of the Schrödinger equations for nuclei A and $A-1$ and expanding the wave function of A into complete set of eigenfunctions $\Psi_n^{J_{A-1}}$ of the Hamiltonian H_{A-1} for the core $A-1$. In eq. (5) $V^{coul}_{nlj,n'l'j'}(r)$ is the matrix element of the Coulomb interaction between the last proton and the protons of the core $A-1$:

$$V^{coul}_{nlj,n'l'j'}(r) = \left\langle \left[\left[Y_l(\hat{\boldsymbol{r}}) \otimes \chi_{\frac{1}{2}}\right]_j \otimes \Psi_n^{J_{A-1}}\right]_{J_A}\right.$$
$$\times \left\| \sum_{i=1}^{A-1} \frac{e_i e_A}{r_{iA}} \right\| \left.\left[\left[Y_{l'}(\hat{\boldsymbol{r}}) \otimes \chi_{\frac{1}{2}}\right]_{j'} \otimes \Psi_{n'}^{J'_{A-1}}\right]_{J_A}\right\rangle. \qquad (6)$$

Here n denotes an excited state of $A-1$, e_i is the charge of i-th nucleon and $r_{iA} = |\boldsymbol{r}_i - \boldsymbol{r}_A|$.

If non-diagonal Coulomb couplings in eq. (5) can be neglected and the diagonal Coulomb potential is replaced by a function $V^{coul}_{mod}(r)$ which is a constant $V^{coul}_{mod}(r) = \epsilon_n - \epsilon_p$ inside the Coulomb radius, and $(Z-1)e^2/r$ outside it, then the proton ANC is given by the same expression as eq. (3) but in which the Bessel function is replaced by the solution $\varphi_l^{mod}(r)$ of the Schrödinger equation with the potential $V^{coul}_{mod}(r)$. Since the main contribution to the ANC comes from internal nuclear region, we need to know $\varphi_l^{mod}(r)$ only at $r < R_N$ (R_N is the radius of the nuclear interior). In this region it is given by the expression [2]

$$\varphi_l^{mod}(r) = \frac{F_l(i\kappa_p R_N)}{\kappa_p R_N j_l(i\kappa_n R_N)} j_l(i\kappa_n r), \qquad r \le R_N, \qquad (7)$$

in which $i\kappa_p$ and $i\kappa_n$ are determined by the proton and neutron separation energies ϵ_p and ϵ_n, and F_l is the regular Coulomb wave function at imaginary momentum $i\kappa$.

If the charge symmetry of NN interactions is valid, then the wave functions of the mirror pairs ${}^A_ZN$ - ${}^A_NZ$ and ${}^{A-1}_{N-1}Z$ - ${}^{A-1}_{Z-1}N$ should be approximately the same in the nuclear interior and the ratio

$$\mathcal{R} = \left(\frac{C_p}{C_n}\right)^2, \tag{8}$$

where C_p and C_n are proton and neutron ANCs for mirror nucleon decays, should be approximated by the square of the normalization coefficient in function $\varphi_l^{mod}(r)$:

$$\mathcal{R} \approx \mathcal{R}_0 \equiv \left| \frac{F_l(i\kappa_p R_N)}{\kappa_p R_N\, j_l(i\kappa_n R_N)} \right|^2. \tag{9}$$

2.2 Bound-unbound mirror pairs

It has been shown in ref. [7] that the partial width Γ_p of a narrow proton resonance is related to the single-particle ANC b_p of the Gamow function describing the proton motion in the resonance state times the spectroscopic factor S_p: $\Gamma_p = (\hbar^2\kappa_p/\mu)S_p b_p^2$. One can show that $S_l^{1/2} b_p$ can be represented by an integral containing the wave functions of nuclei A and $A-1$ and the interaction potential between the proton and $A-1$ in the same way as for bound nuclear states. Assuming that the Gamow function in the internal region of the resonance is the same as the internal wave function of its mirror stable analog and repeating the reasoning of the previous section, we get for

$$\mathcal{R}_\Gamma = \Gamma_p / C_n^2, \tag{10}$$

an approximate expression [2]

$$\mathcal{R}_\Gamma \approx \mathcal{R}_0^{res} = \frac{\hbar^2\kappa_p}{\mu} \left| \frac{F_l(\kappa_p R_N)}{\kappa_p R_N\, j_l(i\kappa_n R_N)} \right|^2. \tag{11}$$

In eq. (10) C_n is the neutron ANC of the proton mirror bound analog. We would like to stress that the width Γ_p entering eq. (11) is a residue in the S-matrix pole at the energy of the proton resonance and not the width for the cross sections of resonant reactions. However, for narrow resonances the difference between these two definitions of the width is small.

3 Single-particle model

According to the analytical formula, the ratio of mirror ANCs should depend only on nucleon separation energies and should be independent of the NN potentials. We checked this property for the case of the two-body model. We considered a family of Woods-Saxon potentials that give some chosen neutron separation energy ϵ_n, and some chosen proton separation energy ϵ_p when the Coulomb potential of a uniformly charged sphere was added. This was achieved by simultaneously varying both the depth and the radius of the Woods-Saxon potential at fixed diffusenesses. The actual numerical values of ϵ_n and ϵ_p were

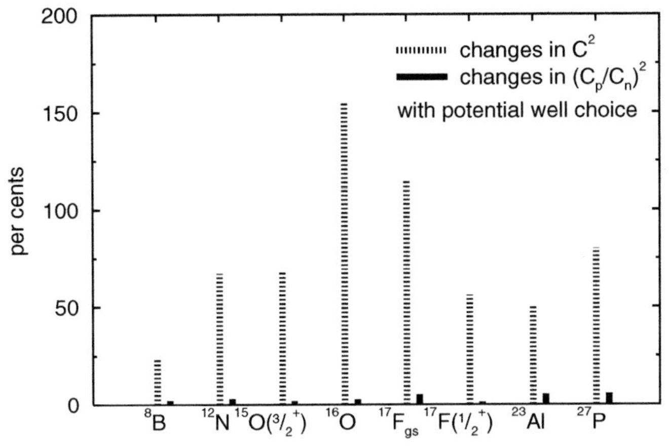

Fig. 1. Changes in ANCs squared and in ratio of proton to neutron ANC squared $(C_p/C_n)^2$ with choice of two-body nuclear potential well for a range of nuclei.

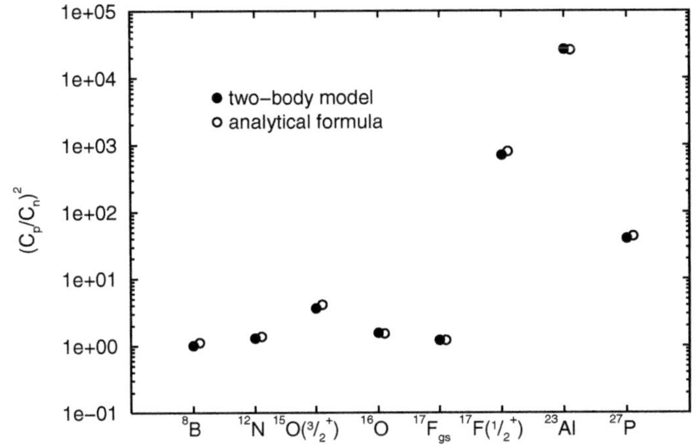

Fig. 2. Ratio of proton to neutron ANC squared $(C_p/C_n)^2$ calculated in the two-body potential model and using the analytical formula (9) for a range of nuclei.

the same as neutron and proton separation energies in the mirror pairs ^{8}Li-^{8}B, ^{12}B-^{12}N, ^{15}N($\frac{3}{2}^+$)-^{15}O($\frac{3}{2}^+$), ^{17}O$_{g.s.}$-^{17}F$_{g.s.}$, ^{17}O($\frac{1}{2}^+$)-^{17}F($\frac{1}{2}^+$), ^{23}Ne-^{23}Al, ^{27}Mg-^{27}P and in the nucleus ^{16}O.

For different potentials from the same family, the neutron and proton ANC values changed significantly but in such a way that their ratio was roughly the same. To illustrate this, we have presented in fig. 1 the changes in ANCs squared C^2 as thick vertical dashed lines and the changes in $(C_p/C_n)^2$ as vertical solid lines. While C^2 changes by 25 to 155%, the changes in $(C_p/C_n)^2$ do not exceed 3%.

The weak sensitivity of the ratio of mirror ANCs to the nuclear potentials suggests an alternative empirical way to determine this ratio. If we assume that mirror neutron and proton single-particle wells are exactly the same and that the spectroscopic factors S_p and S_n are equal for mirror pairs, then the ratio $\mathcal{R}$ can be approximated by the single-particle ratio $\mathcal{R}_{s.p.}$

$$\mathcal{R} \approx \mathcal{R}_{s.p.} \equiv (b_p^{c.s.}/b_n^{c.s.})^2, \tag{12}$$

where the single-particle ANCs $b_p^{c.s.}$ and $b_n^{c.s.}$ are calculated numerically for exactly the same nuclear potential well. Unlike $\mathcal{R}_0$, $\mathcal{R}_{s.p.}$ takes into account the differences in internal wave functions of mirror nuclei due to the Coulomb interaction.

In fig. 2 we compare the ratio $\mathcal{R}_{s.p.}$ with the analytical estimate $\mathcal{R}_0$. One can see that $\mathcal{R}_0$ reproduces the general trend in $\mathcal{R}_{s.p.}$ well. The difference between them is about 2–6% for relatively large proton separation energies but can reach 10–20% for ^{8}B, ^{17}F$(\frac{1}{2}^+)$ and ^{15}O$(\frac{3}{2}^+)$ where this energy becomes very small.

4 Microscopic cluster model

To understand the validity of simple approximations (9), (11) and (12), their predictions should be compared to the numerical calculations using theoretical structure models. One of the models, the best adapted for ANC calculations, is a microscopic cluster model.

The multi-channel cluster wave function for a nucleus A consisting of a core $A-1$ and a nucleon N can be represented as follows:

$$\Psi^{J_A M_A} = \sum_{lSJ_{A-1}\omega} \mathcal{A}[\chi_{\frac{1}{2}\tau}[g^{J_{A-1}}_{\omega lS}(\boldsymbol{r})\otimes[\Psi^{J_{A-1}}_{\omega}\otimes\chi_{\frac{1}{2}}]_S]_{J_A M_A}], \tag{13}$$

where $\mathcal{A} = A^{-\frac{1}{2}}(1-\sum_{i=1}^{A-1} P_{i,A})$ and the operator $P_{i,A}$ permutes spatial and spin-isospin coordinates of the i-th and A-th nucleons. In this work, $\Psi^{J_{A-1}}_{\omega}$ is a wave function of nucleus $A-1$ with the angular momentum J_{A-1} defined either in the translation-invariant harmonic-oscillator shell model, or in a multicluster model. The quantum number ω labels states with the same angular momentum J_{A-1} and S is the channel spin. Transition from the lS coupling scheme to the lj coupling scheme can be done using standard techniques.

The relative wave function $g^{J_{A-1}}_{\omega lS}(\boldsymbol{r}) = g^{J_{A-1}}_{\omega lS}(r)\,Y_{lm}(\hat{r})$ is determined using the R-matrix method. In this method, as explained in detail in ref. [8], the Bloch-Shrödinger equation is solved for the wave function $\Psi^{J_A M_A}$, which allows the correct asymptotic behaviour for the relative wave function $g^{J_{A-1}}_{\omega lS}$ to be obtained. For the states that are stable with respect to particle decay this behaviour is

$$g^{J_{A-1}}_{\omega lS}(r) \approx C^{J_{A-1}}_{\omega lS}\frac{W_{-\eta,l+1/2}(2\kappa r)}{r} \tag{14}$$

and for particle-unstable states

$$g^{J_{A-1}}_{\omega lS}(r) \approx A_{\omega}\frac{\delta_{\omega\nu}I_{\nu}(\kappa_{\nu}r) - U^{J_{A-1}}_{\omega\nu}O_{\nu}(\kappa_{\nu}r)}{\kappa_{\omega}v_{\nu}^{1/2}}. \tag{15}$$

Here, I_ν and O_ν are the ingoing and outgoing Coulomb functions, v_ν is the velocity in the channel ν and U is the collision matrix. The resonance width is determined by assuming a Breit-Wigner shape for the collision matrix near an isolated resonance. This width is the residue at the pole of the R-matrix and its ratio to the ANC squared of the mirror neutron can be compared to the approximation (11).

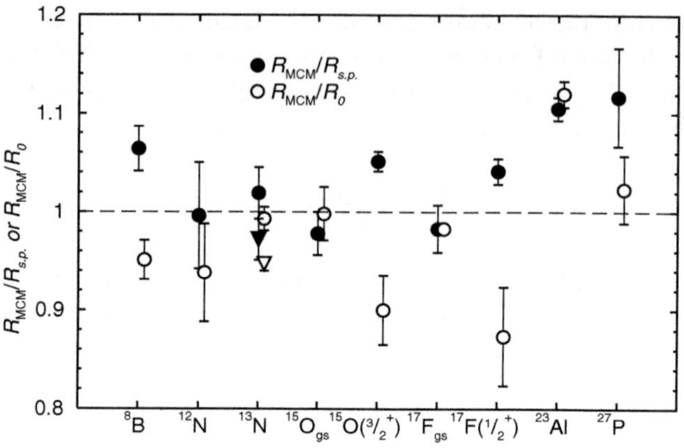

Fig. 3. Ratio $\mathcal{R}_{MCM}/\mathcal{R}_0$ and $\mathcal{R}_{MCM}/\mathcal{R}_{s.p.}$. For the ^{13}C-^{13}N mirror pair the circles represent the calculations within the four-cluster model and the downward triangles represent the calculations in the two-cluster model for various nuclei.

5 Bound mirror pairs

In this section we calculate mirror one-body overlap integrals for several $0p$ and sd shell nuclei within the multi-channel cluster model and explore their properties such as ANCs, spectroscopic factors and single-particle ANCs. The residual nucleus is always taken in its ground state. We use the best adapted effective NN interactions for such calculations, namely, the Volkov potential V2 [9] and the Minnesota (MN) potential [10]. The two-body spin-orbit force [11] and the Coulomb interaction are also included. More details of the calculations can be found in ref. [12] and references therein.

Each of V2 and MN have one adjustable parameter that gives the strength of the odd NN potentials V_{11} and V_{33}. This parameter is usually fitted to reproduce the experimental separation energy for neutron or proton. Such a procedure is crucial for theoretical calculations of ANCs. However, in most cases the same choice of this parameter for mirror states does not reproduce both neutron and proton separation energies. Therefore, we use slightly different interactions in mirror nuclei to reproduce simultaneously the separations energies for neutrons and protons. This simulates charge symmetry breaking of the effective NN interactions that should be a consequence of the charge symmetry breaking in realistic NN interactions.

5.1 Mirror symmetry in ANCs

In this section we compare the ratio $\mathcal{R}_{MCM}$ of mirror ANCs squared obtained in the microscopic cluster model with two different approximations, $\mathcal{R}_0$ and $\mathcal{R}_{s.p.}$. The ratios $\mathcal{R}_{MCM}$ calculated with two different NN potentials differed by no more than 4%. So, we use for $\mathcal{R}_{MCM}$ the result averaged over two potentials V2 and MN. The calculated ratio $\mathcal{R}_{MCM}/\mathcal{R}_0$ and $\mathcal{R}_{MCM}/\mathcal{R}_{s.p.}$ are shown in fig. 3 for several mirror pairs. The error bars in this figure are due to averaging $\mathcal{R}_{MCM}$ over the two NN potentials and because of uncertainties in $\mathcal{R}_0$ due to the choice of R_N

and the uncertainties in $\mathcal{R}_{s.p.}$ due to the residual dependence on the nucleon-core potential. Where two different values of the channel spin are possible, we take the sum of the squared ANCs in these channels for each of mirror nuclei and construct their ratio. In nuclear astrophysics the sum of the squared ANCs is often needed rather than their individual values in channels with different spin.

For the mirror pair ^{13}C-^{13}N, two models were used: a two-cluster model (downward triangles) and a four cluster model (circles). The slightly different ratios $\mathcal{R}_{MCM}$ obtained in these models are due to the different amount of charge symmetry breaking required in each of these models.

According to fig. 3, the deviation of $\mathcal{R}_{MCM}$ from the analytical value $\mathcal{R}_0$ does not exceed 7% for most cases except for the two *s*-wave mirror states with one node ^{15}N($\frac{3}{2}^+$)-^{15}O($\frac{3}{2}^+$) and ^{17}O($\frac{1}{2}^+$)-^{17}F($\frac{1}{2}^+$) and for the mirror pair ^{23}Ne-^{23}Al. The deviations in these cases are 10%, 13% and 12% respectively. In most cases $\mathcal{R}_{MCM}$ is smaller than $\mathcal{R}_0$ but larger than $\mathcal{R}_{s.p.}$. The deviation of $\mathcal{R}_{s.p.}$ from $\mathcal{R}_{MCM}$ is not more than 6% except for the ^{23}Ne-^{23}Al and ^{27}Mg-^{27}P mirror pairs where these deviations are 10% and 12%, respectively.

We have found that the average $\mathcal{R}_{av}$ of two different approximations $\mathcal{R}_0$ and $\mathcal{R}_{g.s.}$ is in reasonably good agreement with $\mathcal{R}_{MCM}$. For all cases except ^{23}Ne-^{23}Al the difference between $\mathcal{R}_{av}$ and $\mathcal{R}_{MCM}$ does not exceed 6%. Therefore, in the absence of detailed microscopic calculations $\mathcal{R}_{av}$ can be a good choice for using mirror symmetry in ANCs to predict proton ANCs from mirror neutron ones and vice versa.

The 10–12% difference between $\mathcal{R}_0$ (or $\mathcal{R}_{s.p.}$) and $\mathcal{R}_{MCM}$ for the mirror pair ^{23}Ne-^{23}Al arises due to strong core excitation effects. This deviation occurs in multichannel calculations which include many excited states in the ^{22}Ne and ^{22}Mg cores. When all the core excitations are removed, the calculated value of $\mathcal{R}_{MCM}$ decreases and agrees with $\mathcal{R}_{s.p.}$ and $\mathcal{R}_0$ within 2% [12].

5.2 Mirror symmetry in spectroscopic factors

The spectroscopic factor S_{lj} is defined as

$$S_{lj} = A \int_0^\infty \mathrm{d}r\, r^2 (I_{lj}(r))^2, \qquad (16)$$

and we obtain them by numerical integration of the overlap integrals squared, calculated in the MCM. We use slightly different odd NN interactions in each mirror state in order to reproduce the experimental separation energies both for neutrons and protons. However, these interactions do not differ much and, therefore, the difference in mirror wave functions in the nuclear interior should arise because of the charge symmetry breaking due to the Coulomb interactions. Since the latter are smaller then strong interactions and because the main contribution to the spectroscopic factor comes from nuclear interior, one expects that the spectroscopic factors in mirror states to be almost equal.

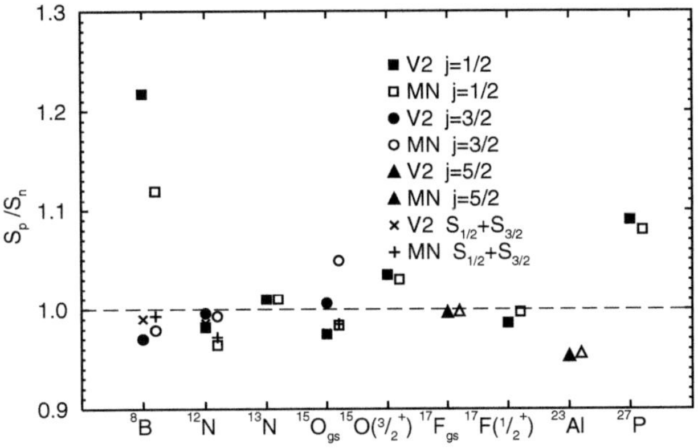

Fig. 4. Ratio of proton to neutron spectroscopic factors for various nuclei.

The ratio S_p/S_n, where S_p is the proton spectroscopic factor and S_n is the spectroscopic factors for its mirror analog, is shown in fig. 4 both for the V2 and the MN potentials. In this figure, the results for ^{13}N-^{13}C are shown only for the four-cluster model. The two-cluster model predicts very similar values for S_p/S_n for this mirror pair.

One can see that for the mirror pairs ^{13}N-^{13}C and ^{17}F-^{17}O with well developed single-particle structure, S_p/S_n is very close to one. For most other cases the deviation of S_p/S_n from one is no more than 4% for both the NN potentials. The strongest deviation is obtained for the small $j = 1/2$ components of the $\langle ^8\mathrm{B}|^7\mathrm{Be}\rangle$ and $\langle ^8\mathrm{Li}|^7\mathrm{Li}\rangle$ overlap integrals and it reaches about 20% for V2 and 11% for MN potential, respectively. Such a sensitivity to the NN potential for $j = 1/2$ is explained by the different amounts of charge symmetry breaking required to reproduce the experimental separation energies in ^{8}B and ^{8}Li with V2 and MN. Another strong deviation of S_p/S_n from one occurs for the ^{27}P-^{27}Mg mirror pair where it reaches 8 to 9%. This deviation arises from core excitation effects.

5.3 Mirror symmetry in single-particle ANCs

The overlap integrals $I_{lj}(r)$, divided by the square roots of their spectroscopic factors S_{lj}, are normalised functions of only one degree of freedom. They play the same role as single-particle wave functions generated by some effective local single-particle potential. These functions are characterised by the single-particle ANCs $b_{lj} = C_{lj} S_{lj}^{-1/2}$. Comparison between single-particle ANCs b_{lj} for mirror nuclei may help to understand if mirror symmetry of the effective local single-particle potential wells is valid.

In this section we compare the ratio $\mathcal{R}_b = C_p^2 S_n/(C_n^2 S_p)$ calculated in the microscopic cluster model with charge symmetry breaking and the ratio $\mathcal{R}_{s.p.}$ obtained using the same single-particle potential well which also reproduces experimental separation energies. If the charge symmetry of the local effective single-particle potentials is valid, then the ratio $\mathcal{R}_b/\mathcal{R}_{s.p.}$ should be equal to one.

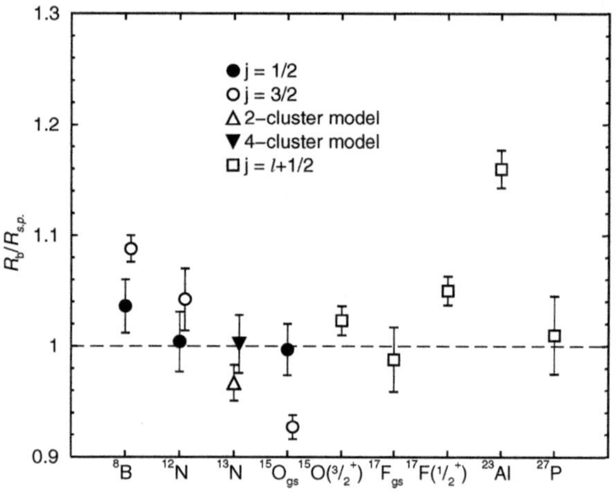

Fig. 5. Ratio $\mathcal{R}_b/\mathcal{R}_{s.p.}$ for various nuclei.

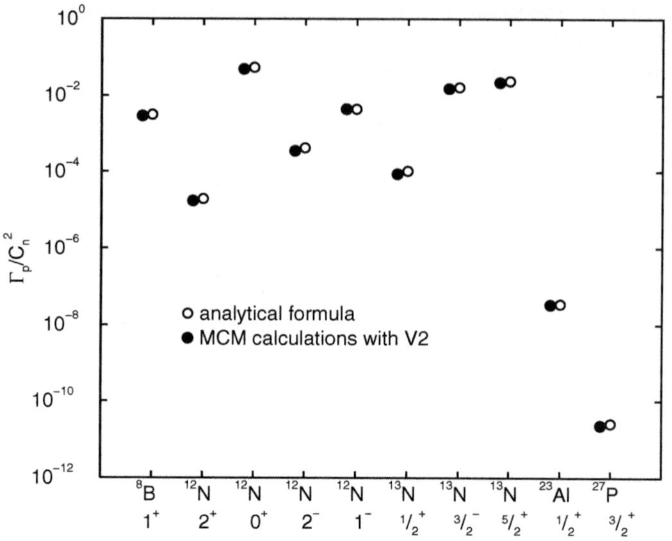

Fig. 6. Ratio of the proton width to the ANC squared of the mirror neutron calculated (given in the units of $\hbar c$) with exactly the same NN interactions in mirror nuclei as compared to the predictions of the analytical formula (11).

In fig. 5 we plot the ratio $\mathcal{R}_b/\mathcal{R}_{s.p.}$. The results of microscopic calculations for $\mathcal{R}_b$ with two different NN potentials are averaged. The error bars in this figure are present due to this averaging and because of uncertainties in $\mathcal{R}_{s.p.}$ due to the residual dependence on nucleon-core potential.

For the mirror pairs of overlap integrals $\langle ^8\mathrm{B}|^7\mathrm{Be}\rangle$-$\langle ^8\mathrm{Li}|^7\mathrm{Li}\rangle$, $\langle ^{12}\mathrm{N}|^{11}\mathrm{C}\rangle$-$\langle ^{12}\mathrm{B}|^{11}\mathrm{B}\rangle$, and $\langle ^{15}\mathrm{O}|^{14}\mathrm{N}\rangle$-$\langle ^{15}\mathrm{N}|^{14}\mathrm{N}\rangle$ with $l = 1$, where two angular momenta j are possible, the ratio $\mathcal{R}_b/\mathcal{R}_{s.p.}$ is different in channels with $j = 3/2$ and $j = 1/2$. The deviation of $\mathcal{R}_b/\mathcal{R}_{s.p.}$ from one for these overlaps correlates with the size of their r.m.s. radii so that the larger is the r.m.s radius, the smaller are the deviations. For the $j = 3/2$ overlaps $\langle ^8\mathrm{B}|^7\mathrm{Be}\rangle$-$\langle ^8\mathrm{Li}|^7\mathrm{Li}\rangle$ and $\langle ^{12}\mathrm{N}|^{11}\mathrm{C}\rangle$-$\langle ^{12}\mathrm{N}|^{11}\mathrm{C}\rangle$ $\langle r^2\rangle^{1/2}_{j=3/2}$ is larger than $\langle r^2\rangle^{1/2}_{j=1/2}$ [12] and the ratio $\mathcal{R}_b/\mathcal{R}_{s.p.}$ deviates from one for $j = 3/2$ overlaps more than in the $j = 1/2$ case. The situation for the $\langle ^{15}\mathrm{O}|^{14}\mathrm{N}\rangle$-$\langle ^{15}\mathrm{N}|^{14}\mathrm{N}\rangle$ mirror overlaps is opposite, the r.m.s. radius for $j = 1/2$ is smaller than that for $j = 3/2$ and $\mathcal{R}_b$ deviates from $\mathcal{R}_{s.p.}$ more for the latter case.

For the $\langle ^{13}\mathrm{N}|^{12}\mathrm{C}\rangle$-$\langle ^{13}\mathrm{C}|^{12}\mathrm{C}\rangle$ pair of mirror overlaps, we have performed the calculations both in four- and two-cluster models. In the first case, the mirror symmetry of local effective potential wells is valid. In the second case, the mirror local effective potentials are not the same due to large charge symmetry breaking required to reproduce the experimental separation energies.

The ratio $\mathcal{R}_b/\mathcal{R}_{s.p.}$ is equal to one within the theoretical uncertainties only for $^{17}\mathrm{F}_{g.s.}$ and $^{27}\mathrm{P}$. It is interesting that significant deviation between $\mathcal{R}_b$ and $\mathcal{R}_{s.p.}$ can be seen for the single-particle nuclear state $^{17}\mathrm{F}(\frac{1}{2}^+)$, in which the mirror symmetry of the mean field is intuitively expected. We believe that the reason for such a deviation is the fact that in $^{17}\mathrm{O}(\frac{1}{2}^+)$ the valence 1s neutron penetrates inside the core more easily than the mirror proton thus more strongly disturbing the mean field. Also a very strong deviation occurs for $^{23}\mathrm{Al}$ which should originate from the strong deformation of the $^{22}\mathrm{Mg}$ core.

6 Bound-unbound mirror pairs

In this section we consider excited mirror states, that lie above proton emission thresholds on the proton-rich side and are bound on the neutron-rich side, for the same $0p$ and sd shell mirror pairs as in the previous section.

First of all we perform the microscopic cluster model calculations with exactly the same NN potentials for each nucleus of a mirror pair, thus imposing charge symmetry of the NN interactions. The adjustable parameters of the NN potentials are chosen to reproduce the experimental energies of proton resonances. We calculate the widths Γ_p for these resonances, the ANCs squared C_n^2 for mirror neutrons and plot their ratio $\mathcal{R}_\Gamma^{MCM}$ in fig. 6 for the V2 potential. We compare this ratio to the prediction $\mathcal{R}_0^{res}$ of the analytical formula (11). As seen in fig. 6, the analytical formula describes very well the general trend in the $\mathcal{R}_\Gamma^{MCM}$ behaviour. The same is true for the MN potential.

To see the differences between $\mathcal{R}_\Gamma^{MCM}$ and $\mathcal{R}_0^{res}$ we have plotted in fig. 7 the ratio $\mathcal{R}_0^{res}/\mathcal{R}_\Gamma^{MCM}$. We calculated this ratio both in the single-channel (no core excitations) and the multi-channel (including several core excitations) cluster model for two different NN potentials, V2 and MN. The results are plotted in fig. 7. One can see from this figure that for $0p$ shell nuclei the results obtained with and without taking core excitations into account differ by not more than 6%, except for $^{12}\mathrm{N}(0^+)$, where this difference is about 10%. Core excitations become more important for nuclei in the middle of the sd shell. For $^{23}\mathrm{Ne}$-$^{23}\mathrm{Al}$, their influence on $\mathcal{R}_\Gamma^{MCM}$ is about 12–16%. A similar effect is seen for $^{27}\mathrm{Mg}(\frac{3}{2}^+)$-$^{27}\mathrm{P}(\frac{3}{2}^+)$ in the calculations with the V2 potential but for MN this influence is much stronger, about 45%. This happens because with the MN potential the d-wave $^{26}\mathrm{Si}(0^+)$ + p configuration in $^{27}\mathrm{P}(\frac{3}{2}^+)$ becomes three times weaker than

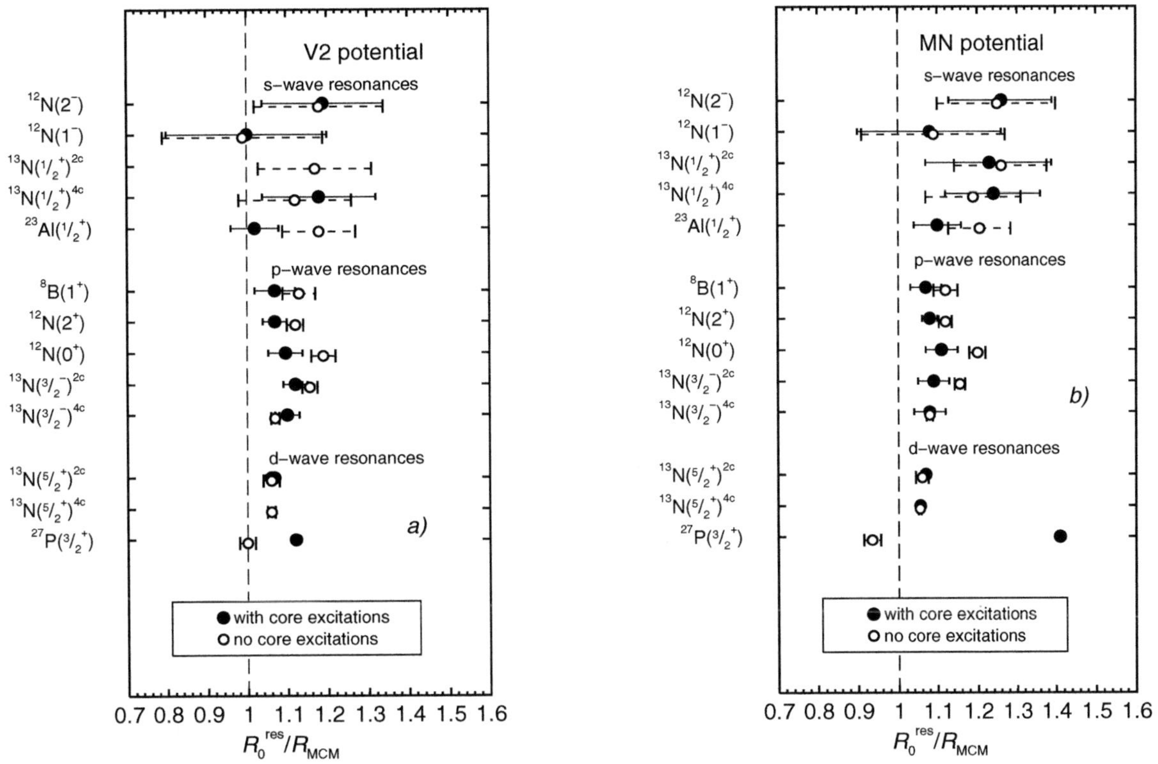

Fig. 7. Ratio between the predictions $\mathcal{R}_0^{res}$ of the analytical formula (11) and those from the microscopic calculations $\mathcal{R}_\Gamma^{MCM}$ for the V2 (*a*) and MN (*b*) potentials with and without taking core excitations into account. Charge symmetry of the *NN* interactions is assumed. Both four-cluster (4*c*) and two-cluster (2*c*) calculations for ^{13}N are shown.

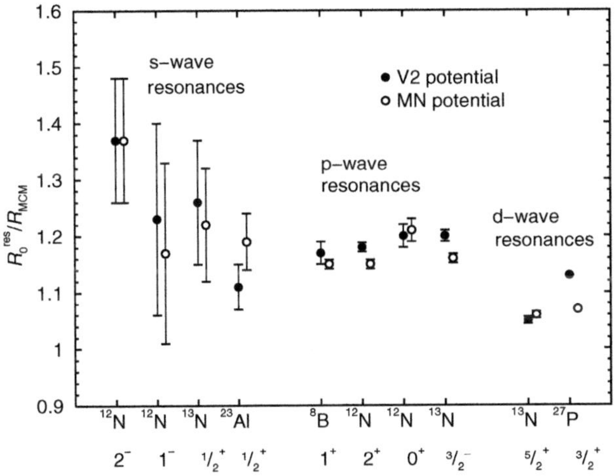

Fig. 8. Ratio between the predictions $\mathcal{R}_0^{res}$ of the analytical formula (11) and the microscopic calculations $\mathcal{R}_\Gamma^{MCM}$ for the V2 and MN potentials. The *NN* potentials are slightly different in mirror nuclei. Core excitations are included. For ^{13}N, the results of the four-cluster calculations are shown.

the *s*-wave ^{26}Si(2^+) + p configuration. In weak configurations the effects of charge symmetry breaking due to the Coulomb interaction are more noticeable. In the particular case of ^{27}Mg$(\frac{3}{2}^+)$-^{27}P$(\frac{3}{2}^+)$, the mirror symmetry breaking in spectroscopic factors, obtained with MN potential, is about 33%, while with V2 the *d*-wave ^{26}Si(0^+) + p configuration dominates and the mirror symmetry breaking for spectroscopic factor of this configuration is only 4%. For other nuclei, the effect of different choices of the *NN* interaction is about 6 to 8%. The average deviation of $\mathcal{R}_0^{res}$ from $\mathcal{R}_{MCM}$ is about 10–12%.

Next, we perform multi-channel calculations in which the experimental energies of proton resonance and neutron separation energies are reproduced. This requires some charge symmetry breaking in the *NN* interactions used. The results of such calculations are presented in fig. 8. The differences in the calculations with V2 and MN potentials do not exceed 8% for all the cases considered here. The largest difference between $\mathcal{R}_0^{res}$ and $\mathcal{R}_\Gamma^{MCM}$ occurs, as expected, for the *s*-wave resonances ^{12}N(2^-), ^{12}N(1^-) and ^{13}N$(\frac{1}{2}^-)$, the widths of which are not small. For the narrow *s*-wave resonance ^{23}Al, $\mathcal{R}_0^{res}$ deviates from $\mathcal{R}_\Gamma^{MCM}$ by about 14%. Similar deviations, of 16-18%, occur for all the *p*-wave resonances. For the *d*-wave resonances ^{13}N$(\frac{5}{2}^+)$ and ^{27}P$(\frac{3}{2}^+)$ this deviation is noticeably smaller, 6% and 10% respectively.

7 Discussion and conclusions

In this paper we have performed calculations of ANCs for mirror one-body overlap integrals within a multi-channel microscopic cluster model. These calculations provide an improved understanding of mirror symmetry in ANCs because the model used takes into account the differences in the internal structure of mirror nuclei due to the Coulomb interaction and the effects that arise due to core excitations, which were ignored in the derivation of the analytical formula (9). Comparison of the results obtained within the microscopic model with the predictions of eq. (9) has confirmed the general trend in the behaviour of ratio of mirror ANCs, given by this formula. This trend is determined only by the separation energies of mirror proton and neutron and the charges of the cores. The difference between the microscopic calculations and the analytical formula is usually less than 7%. However, it may reach, 10 to 12% for the very weakly bound $1s$ nuclei and for nuclei with strongly excited cores.

The microscopic calculations of the ratio of mirror ANCs are also very close to those of the single-particle model, where mirror symmetry in the single-particle potential wells and in the spectroscopic factors is assumed. The difference between such calculations is typically less that 7% except for the sd shell nuclei ^{23}Al and ^{27}P with strongly deformed cores. The average of the single-particle estimate and the prediction of the analytical model is in agreement with the microscopic calculations to within 6%, except for ^{23}Al. Therefore, this average value can be used to predict unknown ANCs from known mirror ones when the microscopic calculations are not available. As far as the mirror symmetry of the single-particle model is concerned, it is not always justified. Our microscopic calculations have shown that spectroscopic factors in mirror states can differ by up to 9% and that mirror symmetry in proton and neutron potential wells is not always present, even for nuclear states with well-pronounced single-particle structure.

Mirror symmetry of ANCs can be used to predict cross sections of proton capture at stellar energies using mirror neutron ANCs. For example, the astrophysical S-factor of the $^7\mathrm{Be}(\mathrm{p},\gamma)^8\mathrm{B}$ reaction can be calculated using the ANCs for the overlap integral $\langle ^8\mathrm{Li}|^7\mathrm{Li}\rangle$. The latter has been measured recently in [13]. Using this experimental value and the predictions for the ratio of the ^{8}B proton ANC to the ^{8}Li neutron ANC from the microscopic cluster calculations, we get $S_{17}(0) = 17.8 \pm 1.7\,\mathrm{eV \cdot b}$ for V2 and $18.2 \pm 1.8\,\mathrm{eV \cdot b}$ for the MN. These results agree well with most measurements based on indirect methods.

The microscopic calculations for bound-unbound mirror states have confirmed that the main trend in the behaviour of the ratio between the proton width and the mirror neutron ANC squared is well reproduced by the analytical formula (11). The difference between the predictions of this formula and the exact microscopic calculations is less than 20% for narrow proton resonances.

The mirror symmetry between the proton width and the mirror neutron ANC can be used to predict unknown widths of very narrow resonances.

It can also be used as a test of the accuracy for experimentally measured ANCs. For example, for the neutron ANC of $^8\mathrm{Li}(1^+)$ measured in [13] and the proton width of $^8\mathrm{B}(1^+)$ from [14] we get $\mathcal{R}_{exp} = (2.29 \pm 0.28) \times 10^{-3}\,\hbar c$. This is significantly larger than the predictions $\mathcal{R}_\Gamma^{MCM} = (1.73 \pm 0.03) \times 10^{-3}\,\hbar c$ of the microscopic model. The proton width of $^8\mathrm{B}(1^+)$ has recently been remeasured in ^{7}Be + p scattering [15], which leads to $\mathcal{R}_{exp} = (1.92 \pm 0.23) \times 10^{-3}\,\hbar c$. The remaining difference between $\mathcal{R}_{exp}$ and $\mathcal{R}_\Gamma^{MCM}$ requires the verification of the accuracy of the determination of the $^8\mathrm{Li}(1^+)$ ANC.

This work has been supported by the UK EPSRC via grant GR/T28577.

References

1. H.M. Xu, C.A. Gagliardi, R.E. Tribble, A.M. Mukhamedzhanov, N.K. Timofeyuk, Phys. Rev. Lett. **73**, 2027 (1994).
2. N.K. Timofeyuk, R.C. Johnson, A.M. Mukhamedzhanov, Phys. Rev. Lett. **91**, 232501 (2003).
3. C. Iliadis, L. Buchmann, P.M. Endt, H. Herndl, M. Wiescher, Phys. Rev. C **53**, 475 (1996).
4. L.D. Blokhintsev, I. Borbely, E.I. Dolinskii, Sov. J. Part. Nucl. **8**, 485 (1977).
5. A.M. Mukhamedzhanov, N.K. Timofeyuk, Sov. J. Nucl. Phys. **51**, 431 (1990), (Yad. Fiz. **51** 679 (1990)).
6. N.K. Timofeyuk, Nucl. Phys. A **632**, 38 (1998).
7. A.M. Mukhamedzhanov, R.E. Tribble, Phys. Rev. C **59**, 3418 (1999).
8. P. Descouvemont, M. Vincke, Phys. Rev. **A** 42, 3835 (1990).
9. A.B. Volkov, Nucl. Phys. **74**, 33 (1965).
10. D.R. Thompson, M. LeMere, Y.C. Tang, Nucl. Phys. A **286**, 53 (1977).
11. D. Baye, N. Pecher, Bull. Sc. Acad. Roy. Belg. **67**, 835 (1981).
12. N.K. Timofeyuk, P. Descouvemont, Phys. Rev. C **71**, 064305 (2005).
13. L. Trache, A. Azhari, F. Carstoiu, H.L. Clark, C.A. Gagliardi, Y.-W. Lui, A.M. Mukhamedzhanov, X. Tang, N. Timofeyuk, R.E. Tribble, Phys. Rev. C **67**, 062801(R) (2003).
14. F. Ajzenberg-Selove, Nucl. Phys. A **490**, 1 (1988).
15. C. Angulo, M. Azzouz, P. Descouvemont, G. Tabacaru, D. Baye, M. Cogneau, M. Couder, T. Davinson, A. Di Pietro, P. Figuera, M. Gaelens, P. Leleux, M. Loiselet, A. Ninane, F. de Oliveira Santos, R.G. Pizzone, G. Ryckewaert, N. de Sereville, F. Vanderbist, Nucl. Phys. A **716**, 211 (2003).

Eur. Phys. J. A **27**, s01, 277–282 (2006)
DOI: 10.1140/epja/i2006-08-042-5

EPJ A direct
electronic only

A simple interpretation of global trends in the lowest levels of p- and sd-shell nuclei

G. Lévai[1,a] and P.O. Hess[2]

[1] Institute of Nuclear Research of the Hungarian Academy of Sciences (ATOMKI), Pf. 51 H-4001 Debrecen, Hungary
[2] Instituto de Ciencias Nucleares, UNAM, Circuito Exterior, C.U., A.P. 70-543, 04510 México D.F., Mexico

Received: 2 July 2005 /
Published online: 23 March 2006 –

Abstract. A simple approach is presented to estimate the position of the first opposite-parity state for p- and sd-shell nuclei. The approach is based on the assumption that the quadrupole-quadrupole interaction determines the energy of the lowest-lying states, and they are a mixture of a few leading $SU(3)$ irreps even in the presence of further symmetry-breaking interactions. The quadrupole-quadrupole interaction together with the indirect effect of the Pauli exclusion principle will then lead to a rich structure in the trends of observables along Z, N and A chains. A comparison with the experimental data is given, with the carbon chain as illustrative example. The findings suggest that the changing shell structure near the neutron drip line might be explained by the appearance of low-lying highly deformed $2\hbar\omega$ states.

PACS. 21.60.Fw Models based on group theory – 21.10.Hw Spin, parity and isobaric spin

1 Introduction

With the evolution of nuclear physics increasingly complex models have been developed in order to describe the structure of the nucleus, this many-body system of nucleons interacting via complicated forces. These models are considered successful if they are able to account for a wide variety of observables with reasonable accuracy. In order to fulfill these requirements it is usually necessary to use numerous parameters and basic assumptions when the models are constructed. Even then many models can be used only in a limited domain of nuclei, and even there their performance is considered really successful only for specific examples. At the same time the complicated machinery of these models often obscures the basic physical picture behind the phenomena they describe.

In light of this one often finds simple approaches, including even some oversimplified ones rather useful to gain insight into nuclear structure and to obtain thumb rules for certain structural properties of nuclei. For example, in [1] a two-level model was introduced which helped to understand pairing properties and the onset of deformation in nuclei. In [2] a simple picture was used to predict shell inversion in ^{11}Be, while in [3] the structure of nuclei is looked upon from the simplest angle and trends in structure can be understood via elementary considerations. Such simple models can be as useful as very sophisticated ones, because they allow for a transparent description of gross structures and shed some light onto usually complicated situations.

In this contribution we present a schematic but systematic approach designed to account for some elementary observables of p- and sd-shell nuclei in a consistent way. These are the excitation energies of the lowest positive- and negative-parity states with special attention to the position of the first state with parity opposite to that of the ground state. This approach rests on the assumption that an extremely simplified (one-parameter) Hamiltonian dominated by the quadrupole-quadrupole interaction and observing the Pauli principle is sufficient to trace qualitatively the position of these states. It is assumed that these states originate from the lowest (*i.e.* the $0\hbar\omega$, $1\hbar\omega$ and $2\hbar\omega$) shells and have dominant contribution only from a few leading $SU(3)$ basis states of the Elliott model [4]. It then follows that even if the $SU(3)$ symmetry is broken, the *relative* energies of these states are not influenced significantly by the mixing.

The rationale of this simple approach is that focusing only on certain basic observables of a large number of nuclei (from ^{4}He to ^{40}Ca) can be at least as useful as more sophisticated models that describe many levels of one (or a few) nucleus. Despite its simplicity, the present approach takes into account the most important factors in this region: the quadrupole-quadrupole interaction that generates deformation and the Pauli principle that selects the allowed ($SU(3)$) configurations. The interplay of these two key elements then generates characteristic structural changes in the trend of the fundamental observables

[a] e-mail: `levai@atomki.hu`

Table 1. Leading $SU(3)$ representations associated with n identical nucleons (protons or neutrons) occupying the i-th shell, and the corresponding permutational symmetry $[f]$. For configurations with $n > 6$ on the sd shell the conjugate representation belonging to $12 - n$ has to be taken. Only configurations appearing up to $2\hbar\omega$ excitations are displayed.

		n	0	1	2	3	4	5	6
		$[f]$	[0]	[1]	[11]	[21]	[22]	[32]	[33]
i	Shell								
1	s		(0,0)	(0,0)	(0,0)				
2	p		(0,0)	(1,0)	(2,0)	(1,1)	(0,2)	(0,1)	(0,0)
3	sd		(0,0)	(2,0)	(4,0)	(4,1)	(4,2)	(5,1)	(6,0), (0,6)
4	pf		(0,0)	(3,0)	(6,0)				
5	sdg		(0,0)	(4,0)					

we investigate, and does so without incorporating further model assumptions concerning *e.g.* the deformation of nuclei, which is an important input for alternative models. Another advantage of the $SU(3)$ scheme is that it is relatively easy to pin down regions where the basic assumptions are expected to break down, and where the results thus have to be accepted with caution. It is also possible to estimate the importance of the neglected interactions (pairing, spin-orbit, etc.)

The aim of this study is not the description of individual nuclei, rather to search for *trends* in the mass, neutron and proton number, which might indicate structural changes. This can be instructive for experiments targeting unexplored regions of the nuclide chart.

2 The approach

In this section we present the basic assumptions of our approach and discuss their importance together with their validity in various domains of p- and sd-shell nuclei [5].

Assumption 1. The lowest-lying states can reasonably be described in terms of the harmonic oscillator shell model of Elliott [4] in the sense that they have dominant contribution only from a few leading $SU(3)$ basis states.

The validity of the Elliott model has been proven for many light nuclei, which means that the physical states can be described reasonably well in terms of the $SU(3)$ basis states of the model. This implies that the $SU(3)$ symmetry is largely respected by the interaction terms considered in the Hamiltonian. The most important term in the Elliott Hamiltonian is the $\mathbf{Q} \cdot \mathbf{Q}$ quadrupole-quadrupole interaction, which is closely related to the second-order Casimir invariant $\mathbf{C}_2$ of the $SU(3)$ group. In fact, in the best examples of the Elliott model (^{20}Ne, ^{24}Mg, etc.) the physical states are almost pure $SU(3)$ states belonging to the leading $SU(3)$ irreducible representation (irrep) (λ, μ), *i.e.* the one for which $\mathbf{C}_2$ has maximal eigenvalue $C_2(\lambda, \mu) = \lambda^2 + \mu^2 + \lambda\mu + 3\lambda + 3\mu$. These leading $SU(3)$ states are also the most deformed ones. While the overall magnitude of λ and μ determine the magnitude of the deformation, their relative size is indicative for the nature of the deformation. In particular, irreps with large λ and small μ represent prolate deformation, while the reverse situation corresponds to oblate shape. When neither numbers are small compared to the other, a triaxial shape occurs.

Obviously, in most cases the $SU(3)$ symmetry is broken, nevertheless, it was found that even when this happens, only $SU(3)$ irreps with similar $C_2(\lambda, \mu)$ get mixed [6], so the average expectation value of the Casimir invariant $\mathbf{C}_2$ is close to the expectation value for the leading $SU(3)$ irrep. If this mechanism holds not only for the $0\hbar\omega$ states, but also for the lowest-lying $1\hbar\omega$ and $2\hbar\omega$ states, then the energy *differences* of the opposite-parity levels are moderately sensitive to the actual symmetry breaking.

This scenario is a reasonable approximation in mid-shell situations, but it might fail close to shell closures, where there are fewer $SU(3)$ irreps and even these have different $C_2(\lambda, \mu)$ expectation value. Furthermore, the breakdown of the $SU(3)$ symmetry is also stronger near the shell closures, *i.e.* where N and Z are close to 2, 8 and 20.

In general the construction of the full $SU(3)$ model space is a difficult task, however, the leading $SU(3)$ irreps can be identified relatively easily. Due to the total antisymmetry of the nuclear wave function, the spatial and the spin-isospin structure of the nucleus is correlated. In practical terms this means that the maximal spatial symmetry (exhibiting itself in maximal deformation) comes with maximal antisymmetry in the spin-isospin sector, so in the lowest-lying configurations the protons and neutrons tend to pair off separately. There is thus a relatively simple recipe which we can use to determine the $SU(3)$ character of the states in question by extracting the $SU(3)$ content of the proton and neutron configurations on each shell (s, p, sd, pf and sdg, the latter two only in excited configurations), and combining them to obtain the largest possible (λ, μ) irrep. Table 1 lists the (λ, μ) quantum numbers associated with proton and neutron configurations possessing maximal symmetry in the spatial sector [4]. We list all configurations on the s and p shells, while for the sd shell we present the (λ, μ) irreps up to particle number $n = 6$, which corresponds to the middle of the shell: for $n > 6$ the appropriate $SU(3)$ irreps are obtained as the conjugates (μ, λ) of the $SU(3)$ irreps belonging to $12 - n < 6$ particles. In table 1 the pf and sdg shells appear with configurations consisting of up to 2 and 1 particles, respectively, because only such systems

Table 2. Possible nucleon configurations and leading $SU(3)$ representations for 0, 1 and $2\hbar\omega$ in the proton and neutron sector for ^{16}C.

	Protons		Neutrons	
$0\hbar\omega$	s^2p^4	(0,2)	$s^2p^6sd^2$	(4,0)
$1\hbar\omega$	$s^2p^3sd^1$	(3,1)	$s^2p^6sd^1pf^1$	(5,0)
	s^1p^5	(0,1)	$s^2p^5sd^3$	(4,2)
$2\hbar\omega$	$s^2p^3pf^1$	(4,1)	$s^2p^6sd^1sdg^1$	(6,0)
	$s^2p^2sd^2$	(6,0)	$s^2p^6pf^2$	(6,0)
	$s^1p^4sd^1$	(2,2)	$s^2p^5sd^2pf^1$	(7,1)
	s^0p^6	(0,0)	$s^2p^4sd^4$	(4,4)
			$s^1p^6sd^3$	(4,1)

can appear when we determine $1\hbar\omega$ and $2\hbar\omega$ excitations of sd-shell nuclei. Table 1 also contains the permutational symmetry $[f]$ associated with the proton and neutron configurations on each major shell, and it is a straightforward task to construct the possible permutational symmetries of both nucleon types and of the whole nuclear state.

In order to determine the leading $SU(3)$ irrep (λ,μ) for configurations without major shell excitation ($0\hbar\omega$) we simply have to take the $(\lambda^i_\pi,\mu^i_\pi)$ and $(\lambda^i_\nu,\mu^i_\nu)$ irreps obtained for protons and neutrons from the i-th shell (s, p, sd, pf, sdg for $i=1,2,3,4,5$, respectively), and combine them to get $\lambda=\sum_{i=1}^5(\lambda^i_\pi+\lambda^i_\nu)$, $\mu=\sum_{i=1}^5(\mu^i_\pi+\mu^i_\nu)$. We illustrate this procedure with the example of the ^{16}C nucleus consisting of $Z=6$ protons and $N=10$ neutrons. Table 2 displays the possible proton and neutron configurations up to two shell excitation quanta, together with the corresponding leading $(\lambda^i_\pi,\mu^i_\pi)$ and $(\lambda^i_\nu,\mu^i_\nu)$ $SU(3)$ irreps. The leading $0\hbar\omega$ is then found to be (4,2).

As can be seen from table 2, there are several configurations for $1\hbar\omega$ depending on whether we excite one nucleon from the valence shell to the next highest shell or from the shell below to the valence shell. We have to calculate the leading irrep from each of these following the recipe given above. The leading $SU(3)$ irrep is found to be (7,1), which originates from the (3,1) proton and (4,0) neutron configuration. It also has to be mentioned that in order to get rid of spurious states, the center-of-mass motion has to be subtracted. In the harmonic-oscillator picture this is done very easily: the $SU(3)$ irreps of the $0\hbar\omega$ model space have to be multiplied by the (1,0) $SU(3)$ irrep representing one excitation quantum in the c.m. motion, and the resulting (λ,μ) states have to be subtracted from the $1\hbar\omega$ model space obtained before [7]. The largest (λ,μ) irrep with multiplicity larger than 0 will then be the leading one for $1\hbar\omega$. It is obvious from table 2 that the (7,1) irrep is not redundant. The $2\hbar\omega$ space can be constructed in a similar way: proton and neutron configurations with altogether two shell excitations have to be considered. Then the leading $SU(3)$ irrep is found to be (10,0), which originates from the $2\hbar\omega$ (6,0) proton and the $0\hbar\omega$ (4,0) neutron configuration.

It has to be mentioned that in certain special situations the simple recipe outlined above might not produce automatically the leading $SU(3)$ state. This is the case in the middle of the sd shell, when 6 identical nucleons can yield the (6,0) and also the (0,6) $SU(3)$ state, as can be seen from table 1. In these ambiguous situations further assumptions have to be made in order to select the configuration that leads to the leading $SU(3)$ irrep of the total wave function. The general rule is that the contribution of the other type of nucleons, as well as that of the same type of nucleons from the p shell have to be taken into consideration: the leading $SU(3)$ irrep is obtained if λ or μ reaches the possible maximal value. This corresponds to a kind of "polarization" effect in the sense that the two types of nucleons attempt to realize similar kind of deformation, *i.e.* prolate or oblate. In fact, this effect can appear in sd^5 and sd^7 configurations too, to which the simple rule would assign (5,1) and (1,5) as in table 1, while the construction of the true leading $SU(3)$ state might require the secondary (2,4) and (4,2) states, with opposite kind of deformation.

Assumption 2. The simplified Hamiltonian

$$H=\hbar\omega\mathbf{N}-\chi\mathbf{C}_2(\lambda,\mu) \tag{1}$$

is sufficient to account for the excitation energy of the lowest few levels.

Here the first term accounts for major shell excitations ($\mathbf{N}=0,1,2$), while $\mathbf{C}_2$ is the second-order Casimir operator of $SU(3)$ mentioned earlier. Since the states are supposed to be composed of several $SU(3)$ basis states the expectation value of $\mathbf{C}_2$ should be the corresponding average of the $C_2(\lambda,\mu)$ eigenvalues. However, based on Assumption 1 we replace this average with the eigenvalue belonging to the leading $SU(3)$ state, which can be determined using the mechanism presented previously. The two coupling constants appearing in (1) are known to be parametrized in terms of A only. In particular, we can take the formula

$$\hbar\omega=45A^{-1/3}-25A^{-2/3}, \tag{2}$$

(in units of MeV), which was deduced from the systematic behaviour for light nuclei near the valley of stability [8]. This value might thus change when the neutron drip line is approached: due to loosely bound neutrons the average $\hbar\omega$ might be lowered. For χ we take

$$\chi=aA^{-\frac{5}{3}}, \tag{3}$$

which is a generally accepted parametrization [9]. We adjusted a, the only parameter appearing in our approach to the χ value calculated for 100 nuclei using the expression (1) and the experimental energy of the first opposite-parity level [5]. The trend of χ and its parametrization in terms of (3) is displayed in fig. 1 using the adjusted $a=30$ MeV value. The deviation for low A might be partly due to the underestimation of $\hbar\omega$: this is indicated by the fact that the adjusted χ is negative in a number of cases, which is the consequence of the fact that $\hbar\omega$ is lower than the energy of the lowest-lying opposite-parity state. The increasing trend of the data points towards $A=40$ and the maximum near $A=16$ might be an indirect effect of the shell closure too, which may lead to deviations from the smooth behaviour of $\hbar\omega$ in (2) and/or χ in (3). This might

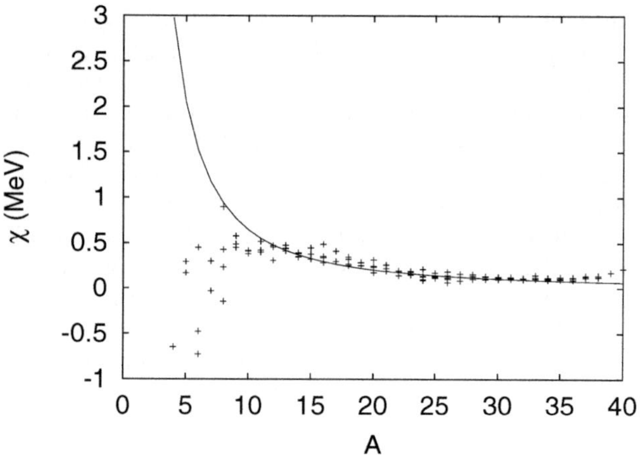

Fig. 1. The parameter χ as a function of the mass number A (full line). The parametrization $aA^{-\frac{5}{3}}$ was chosen with $a = 30\,\text{MeV}$. Crosses indicate χ values calculated back from (1) using the experimental energy of the lowest opposite-parity level.

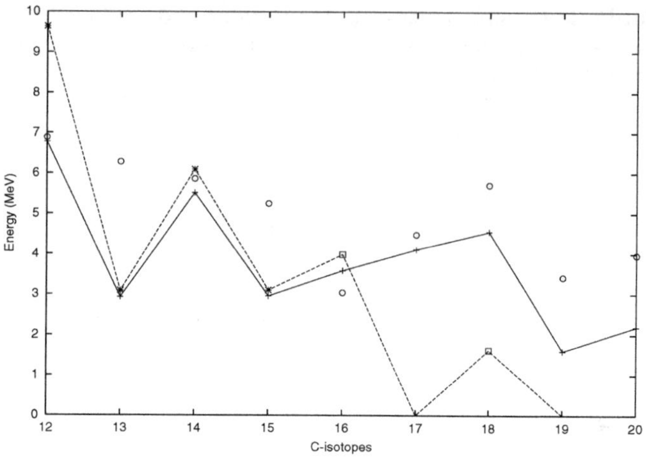

Fig. 2. The energy of the lowest opposite-parity ($1\hbar\omega$) level in the C isotopes, and the lowest expected $2\hbar\omega$ level (open circles). Dashed and full lines connect experimental and theoretical points, respectively. Crosses indicate experimentally well-established opposite-parity states, while open boxes stand for lower limits in energy. The lack of symbol stands for missing data.

be related to the observation made before, *i.e.* that the validity of the $SU(3)$ scheme might break down near shell closures, where the importance of the $\mathbf{Q}\cdot\mathbf{Q}$ interaction is expected to be reduced. In practical terms this means that the $SU(3)$ breaking interactions (*e.g.*, pairing) might mix further $SU(3)$ states with the leading ones, so the replacement of $\langle C_2(\lambda,\mu)\rangle$ with the $C_2(\lambda,\mu)$ eigenvalue of the leading $SU(3)$ irrep might not be a good approximation in this case. In fact, this effect could be compensated by taking larger χ values near the shell closures, which is exactly what fig. 1 indicates in these regions.

In a full scale spectroscopic study further terms should also be considered in (1), however, if we focus only on low-lying levels that have low values for the orbital angular momentum and spin, terms like the spin-orbit interaction, $\mathbf{L}^2$, $\mathbf{L}\cdot\mathbf{S}$ and spin-dependent forces are not expected to contribute significantly to the energy [5].

Before closing this section we return to the major assumptions of our approach, *i.e.* that in order to account for the general trends in the lowest-lying levels of nuclei in this region it is sufficient to incorporate the two most important factors, *i.e.* the quadrupole-quadrupole interaction that generates deformation and the Pauli principle that selects the allowed ($SU(3)$) configurations. The interplay of these two key elements then generates characteristic structural changes in the trend of the fundamental observables we investigate, and does so without incorporating further model assumptions concerning *e.g.* the deformation of nuclei, which is an important input for alternative models. Obviously, the breakdown of $SU(3)$ symmetry and thus the mixing of $SU(3)$ states has strong influence on further observables. This is the case, for example, with the electric quadrupole moment, because its expectation value is calculated from quantities that significantly differ from each other in their magnitude, and in addition, might also have different sign. Nevertheless, Q can be calculated in a straightforward and parameter-free way for any $SU(3)$ state [5], and this can be helpful in determining the character of the nuclear states.

3 Applications

In [5] we presented the results concerning the position of the first opposite-parity state and compared it to experiment, as far as data were available. In total about 180 nuclei in the p and sd shell were considered and of those predictions were made for about 80 nuclei. In general, the agreement was satisfactory. The general trend was well reproduced, although not always in terms of absolute values. As illustration we present our results in tabular and graphical form in table 3 and fig. 2 for the carbon isotope chain. It is seen from fig. 2 that the rich structure of the experimental plot is excellently reproduced by the calculated one and not only in its trend, but also in magnitude. It is remarkable that the odd-even staggering structure is reproduced without any spin-dependent interactions, by considering only terms depending on the orbital structure of the nuclei. This odd-even staggering effect appeared in other isotope chains too, mainly with even value of Z (Be, Ne, Mg, Si, S) [5], and apart from the Mg chain, even its magnitude was reproduced reasonably well. Another characteristic feature reproduced rather successfully was a bump near $N = 8$ in the N, O and F chain, which obviously originates from the shell closure effect.

Deviations from the experimental plot appeared in certain well-defined regions. This was the case for light isotopes $A \leq 10$, where the theoretical curve systematically fell behind the experimental one, leading even to shell inversion in some cases. As we have discussed before, this might be the consequence of the underestimation of $\hbar\omega$ in (2) in this region. We note that shell inversion (*i.e.* $E(1\hbar\omega) < E(0\hbar\omega)$) was reproduced for ^{11}Be too, the only

Table 3. Numerical values of $\hbar\omega$ and χ used in (1) to determine the energies of the lowest $1\hbar\omega$ and $2\hbar\omega$ states ($E_{\rm Th}(1\hbar\omega)$, $E_{\rm Th}(2\hbar\omega)$) with the indicated (λ,μ) quantum numbers for the C isotopes, displayed together with experimental information for the ground and the lowest opposite-parity state.

Nucleus	$\hbar\omega$	χ	(λ,μ)			$J^{\pi}_{\rm g.s.}$	1st opp. p. state		$E_{\rm Th}(1\hbar\omega)$	$E_{\rm Th}(2\hbar\omega)$
			$0\hbar\omega$	$1\hbar\omega$	$2\hbar\omega$		$E_{\rm Exp}$	J^{π}		
^{12}C	14.89	0.477	(0,4)	(3,3)	(6,2)	0^+	9.641	3^-	6.778	6.878
^{13}C	14.62	0.417	(0,3)	(2,4)	(5,3)	$\frac{1}{2}^-$	3.089	$\frac{1}{2}^+$	2.929	6.276
^{14}C	14.37	0.369	(0,2)	(2,3)	(4,4)	0^+	6.093	1^-	5.514	5.863
^{15}C	14.14	0.329	(2,2)	(4,3)	(7,2)	$\frac{1}{2}^+$	3.103	$\frac{1}{2}^-$	2.956	5.254
^{16}C	13.92	0.295	(4,2)	(7,1)	(10,0)	0^+	≥ 3.986	2	3.586	3.037
^{17}C	13.72	0.267	(4,3)	(7,2)	(10,1)	*n.d.*	*n.d.*	*n.d.*	4.111	4.485
^{18}C	13.53	0.243	(4,4)	(7,3)	(10,2)	(0^+)	> 1.62	*n.d.*	4.552	5.707
^{19}C	13.35	0.222	(5,3)	(7,4)	(10,3)	*n.d.*	*n.d.*	*n.d.*	1.600	3.422
^{20}C	13.18	0.204	(0,8)	(8,3)	(10,4)	*n.d.*	*n.d.*	*n.d.*	4.635	3.976

known isotope for which the parity of the ground state differs from that corresponding to a $0\hbar\omega$ configuration. Although this nucleus is close to the $A \leq 10$ region, fig. 1 demonstrates that in this case the low $E_{\rm Th}(1\hbar\omega)$ value is not due to the underestimated $\hbar\omega$: the cross corresponding to ^{11}Be lies almost precisely on the χ curve calculated from (3).

One more systematic deviation appears near $A = 40$, where the $E_{\rm Th}(1\hbar\omega)$ falls behind the energy of the first opposite-parity level. This, again, is obviously due to the shell closure effect which influences the results through an underestimated χ value. This can also be traced down on fig. 1.

The trends described here can be understood from table 3, where the $SU(3)$ states (λ,μ) are displayed for the leading 0, 1 and $2\hbar\omega$ states. It is seen that proceeding along the isotope chain λ and μ rarely change with more than one unit as the shells are filled up with more and more neutrons. There are, however, some discontinuities for 1 and $2\hbar\omega$, *e.g.* at ^{16}C. This is due to the fact that particles excited to higher shells typically contribute to the whole system with $SU(3)$ configurations corresponding to prolate deformation (see table 2). This shows up in a low $E_{\rm Th}(2\hbar\omega)$ level too, as can be seen in fig. 2. It is a general trend in other isotopes too that the $0\hbar\omega$ state can be prolate, oblate or triaxial type, while the excited states tend to proceed towards the prolate direction with relatively large deformation.

We mention here that the ^{16}C nucleus is known to exhibit some unusual features. It was found, for example, that there is an extremely weak electric quadrupole transition from the first 2^+ state at $E_x = 1.766$ MeV to the ground state [10], furthermore, the observations were compatible with a large deformation, especially for protons. In our scheme the leading $0\hbar\omega$ and $2\hbar\omega$ states belong to the $(\lambda,\mu) = (4,2)$ and $(10,0)$ $SU(3)$ irreps, which certainly have rather different structure, leading to a strong hindrance in the electromagnetic transitions between them. The $2\hbar\omega$ (10,0) states also have strongly prolate nature; in particular, they are constructed from an s^2, p^2, sd^2 proton structure, having (6,0) $SU(3)$ character, and an s^2, p^6, sd^2 neutron structure, having (4,0) (see table 2). This is clearly compatible with the experimental findings. The simple scheme yields the energy of the lowest (10,0) state at $E_x = 3.04$ MeV, which is also rather close to the experimental energy of the 2_1^+ state. In summary, we assume that even if the $SU(3)$ symmetry is broken and the states do not have pure $SU(3)$ character, the ground state of ^{16}C is dominated by $0\hbar\omega$, while the 2_1^+ state has dominantly $2\hbar\omega$ structure. This also means that there should be a 0_2^+ state with dominantly $2\hbar\omega$, and a 2_2^+ state with dominantly $0\hbar\omega$ character in the vicinity of these levels. The states at $E_x = 3.027$ MeV and 3.987 MeV might be candidates for these.

The example of the ^{16}C nucleus shows that discontinuities in some basic observables (*e.g.* quadrupole moment, $B(E2)$, deformation) in the ground-state region of nuclei can be the result of the presence of highly deformed low-lying $2\hbar\omega$ states. This mechanism eventually leads to the smearing out of the energy gap and the apparent disappearance of the shell structure. This scenario can also occur near the closure of the $N = 20$ (or $Z = 20$) shell, since the $0\hbar\omega$ configuration has small deformation there, while exciting nucleons to the pf and sdg shells may change the deformation to the strongly prolate direction.

There is one more interesting finding found in some nuclei that can be illustrated with the example of a carbon isotope. In particular, there is a strong discontinuity in the $0\hbar\omega$ (λ,μ) state at ^{20}C: here the protons correspond to the usual p^2 (0,2) configuration, while the neutrons are assigned to sd^6 (0,6) or (6,0) (see table 1). Due to the "polarization" effect discussed previously the lowest energy arises if the (0,6) neutron state is taken, leading to the overall (0,8) state (see table 3). The sd^6 configurations typically appear at nuclei with N, $Z = 14$, and can lead to underestimated $E_{\rm Th}(1\hbar\omega)$ or $E_{\rm Th}(2\hbar\omega)$ energies in a handful of isotopes near the middle of the sd shell [5].

Before closing this section, we mention that calculations for the quadrupole moments for about sixty nuclei with known $Q_{\rm Exp}$ or $|Q_{\rm Exp}|$ showed that the picture in which a few dominant $SU(3)$ irreps get mixed in the lowest-lying states is consistent with the experimental data [5]. While $Q_{\rm Th}$ calculated using the leading $0\hbar\omega$ $SU(3)$ state agrees remarkably well with $Q_{\rm Exp}$ for nuclei that are known have pure $SU(3)$ character in their ground state (^{20}Ne, ^{24}Mg, ^{19}F, etc.), it represents an upper limit in the case of most other nuclei, indicating that some other leading $SU(3)$ states mix with the leading one significantly. In some cases $Q_{\rm Exp}$ was also close to $Q_{\rm Th}$ calculated with the leading $2\hbar\omega$ state indicating that such states might play important role even in the ground-state region.

4 Conclusions

We presented the basic ideas of a simple thumb rule approach designed to estimate the position of the first opposite-parity ($1\hbar\omega$) state and the first $2\hbar\omega$ state in light nuclei. The full details of the method and the specific results concerning about 180 nuclei are given in [5], while here we discussed in more detail only the C isotope chain to illustrate the main findings. The method is specially directed to experimental physicists and they can use it to have a first estimate where to look for low lying opposite parity states or, even more difficult, where to look at shell inversion without the change in parity, *i.e.* shape isomeric states, which is related to the inversion with the $2\hbar\omega$ shell.

The approach rests on two basic assumptions: 1) the lowest-lying states are composed only of a few leading (λ,μ) $SU(3)$ irreps having comparable $C_2(\lambda,\mu)$ $SU(3)$ eigenvalue; and 2) the terms other than the harmonic and quadrupole-quadrupole interaction do not contribute significantly to the energy. The approach contains only one adjustable parameter (the coupling constant of the quadrupole-quadrupole interaction), furthermore, the deformation of the nuclei follows directly from the calculations, rather than being an input parameter.

The simple rules outlined above imply that the joint action of the quadrupole-quadrupole interaction and the Pauli principle is able to reproduce the rich structure manifested in the energy of the first opposite-parity state, apart from some well-defined regions. The basic assumptions also guarantee that even if the $SU(3)$ symmetry is broken, this does not influence significantly the relative energy of the lowest 0, 1 and $2\hbar\omega$ levels. Furthermore, although other observables are more sensitive to the mixing of $SU(3)$ states, some of these, *e.g.* the quadrupole moment can be indicative of the nature (*e.g.* deformation) of the states.

With the example of the C isotopes and ^{16}C in particular, we demonstrated that some unusual findings usually attributed to the disappearance of the shell structure can be explained by our procedure. Though at $0\hbar\omega$ the $SU(3)$ irrep is small near a closed shell or (0,0) at a closed shell, at $n\hbar\omega$ ($n > 0$) the corresponding $SU(3)$ irrep can represent large deformation at low energy. This result does suggest that the mean field description of nuclei near the neutron drip line can still be valid and the disappearance of the shell structure is only apparent. This effect is the result of the Pauli exclusion principle which allows only small $SU(3)$ irreps for nuclei near a closed shell but results in large $SU(3)$ irreps, and thus a large quadrupole moment and deformation for $2\hbar\omega$ excitations. This is also related to the fact that in light nuclei small changes in the size of the $SU(3)$ irreps imply large changes in deformation, which is a less drastic effect for heavier nuclei.

This work was supported by the CONACyT-MTA and CSIC-MTA exchange programs, CONACyT, DGAPA (Grant No. IN119002) and by the OTKA (Grant No. T37502).

References

1. H.J. Lipkin, N. Meschkov, S. Glick, Nucl. Phys. A **62**, 118 (1965).
2. I. Talmi, I. Unna, Phys. Rev. Lett. **4**, 469 (1960).
3. R.F. Casten, *Nuclear Structure from a Simple Perspective* (Oxford University Press, Oxford, 1990).
4. J.P. Elliott, Proc. R. Soc. London, Ser. A **245**, 128; 562 (1958).
5. P.O. Hess, G. Lévai, Int. J. Mod. Phys. E **14**, 845 (2005).
6. J.G. Hirsch, P.O. Hess, L. Hernández, C. Vargas, T. Beuschel, J.P. Draayer, Rev. Mex. Fís. **45**, (S2) 86 (1999); C.E. Vargas, J.G. Hirsch, J.P. Draayer, Nucl. Phys. A **691**, 409 (2001); **697**, 655 (2002).
7. J.P. Elliott, T.H.K. Skyrme, Proc. R. Soc. London, Ser. A **232**, 561 (1955); D.M. Brink, G.F. Nash, Nucl. Phys. A **40**, 608 (1963); K.T. Hecht, Nucl. Phys. A **170**, 34 (1971).
8. J. Blomqvist, A. Molinari, Nucl. Phys. A **106**, 545 (1968).
9. M. Dufour, A.P. Zuker, Phys. Rev. C **54**, 1641 (1996).
10. N. Imai *et al.*, Phys. Rev. Lett. **92**, 62501 (2004).

Eur. Phys. J. A **27**, s01, 283–288 (2006)
DOI: 10.1140/epja/i2006-08-043-4

EPJ A direct
electronic only

Exploring the Nα + 3n light nuclei via the (^{7}Li, ^{7}Be) reaction

C. Nociforo[1,a], F. Cappuzzello[1], A. Cunsolo[1,2], A. Foti[2,3], S.E.A. Orrigo[1], J.S. Winfield[1], M. Cavallaro[1,2], S. Fortier[4], D. Beaumel[4], and H. Lenske[5]

[1] INFN-Laboratori Nazionali del Sud, Via S. Sofia 62, 95123 Catania, Italy
[2] Dipartimento di Fisica e Astronomia, Università di Catania, Via S. Sofia 64, 95123 Catania, Italy
[3] INFN-Sezione di Catania, Via S. Sofia 64, 95123 Catania, Italy
[4] Institut de Physique Nuclèaire, 91406 Orsay Cedex, France
[5] Institut für Theoretische Physik, Universität Giessen, Heinrich-Buff-Ring 16, 35392 Giessen, Germany

Received: 18 June 2005 /
Published online: 29 March 2006 –

Abstract. Experimental signatures of the dynamical correlations of a core with a single-particle neutron have been found in light neutron-rich nuclei investigated via the (^{7}Li,^{7}Be) charge-exchange reactions at around 8 MeV/u. Of particular astrophysical relevance are low-lying narrow resonances ($\Gamma \sim 200$ keV FWHM) BSEC (Bound States Embedded in the Continuum). Because of their long lifetime BSEC states are likely to effect the capture rates in any scenario for nucleosynthesis in neutron-rich environment. They have been observed in the continuum of ^{11}Be and ^{15}C nuclei. A microscopic nuclear structure model based on QRPA theory, which takes into account Dynamical Core Polarisation (DCP) correlations, gives a suitable description of these resonances as well as single-particle states of the studied systems. In this context, high-energy narrow structures populated in nuclei having an integer number of α-particles plus three neutrons are good BSEC candidates and can be systematically investigated.

PACS. 21.10.Pc Single-particle levels and strength functions – 21.60.-n Nuclear structure models and methods – 25.70.Kk Charge-exchange reactions – 27.20.+n $6 \leq A \leq 19$

1 Introduction

During the last decade the investigation of the structure of unstable nuclei has become the major activity of nuclear structure physics, leading to the discovery of a variety of new phenomena, from halos and neutron skins to shell quenching and the appearances of new magic numbers. From a theoretical point of view the explanation of new phenomena represents a precious benchmark strongly supporting the development of refined theories based on the microscopic description of nuclei. Also, the studies of these phenomena have provided an interesting cross-relation to astrophysics because such investigations are most suitable for exploring under laboratory conditions reactions which otherwise occur only in stellar and supernovae environments. Their precise understanding is of importance to comprehend nucleosynthesis in both stellar and primordial processes.

The nuclei considered in this paper are located in the mass region of the CNO cycle. Although the nuclei investigated here are not directly involved in the CNO cycle, we hope to bring attention to a general aspect of nuclear continuum dynamics, namely the existence of extremely sharp resonances in the low-energy continuum. These Bound States Embedded into the Continuum (BSEC) are long-lived states resulting from configuration mixing between simple single-particle states attached to the (inert) ground state of the host system and more complex core-excited configurations which can only decay via the coupling to a continuum of unbound states attached to the core ground state.

Far from the line of β-stability the standard representation of nuclei, based on the assumption of mean-field dynamics, turns out to be often inadequate. The reason is the energy scale of the single-particle states given by the separation energy S_n, which is reduced to a few hundred keV, value close to the typical matrix element of the residual nuclear interaction. Hence, in loosely bound systems a strong competition of mean-field and correlation dynamics is expected even at low excitation energy. In the case of ^{11}Be nucleus, for instance, this is supported by the observations of a considerable amount of ^{10}Be(2$^+$) core excited configuration in the ground state, which was explored via transfer [1] as well as breakup reactions [2].

[a] *Present address*: GSI, Darmstadt, Germany; e-mail: c.nociforo@gsi.de

Experimental signatures of the Dynamical Core Polarisation (DCP) correlations above particle threshold are the BSEC [3,4], which were observed in the inelastic scattering of stable nuclei. The BSEC are quasibound resonant states characterized by a peculiar line shape. They are interpreted as configurations where a nucleon is re-scattered by a core, which can be easily excited into a multi-particle configuration with total energy above particle threshold but neither of the involved particles by itself is in a continuum state. Since such a core excited configuration is degenerate with the single particle continuum attached to the ground-state the decay will take place only through the relatively weak residual interaction. Hence, a nucleon trapped in such a configuration remains for long time attached to the core although the total energy of the direct process would be enough to remove it immediately.

Spectroscopic studies dealing with the weakly bound ^{11}Be and ^{15}C nuclei, explored via the (^{7}Li,^{7}Be) charge-exchange (CEX) reactions at around 8 MeV/u, have shown the appearance of narrow resonances ($\Gamma \sim 200$ keV FWHM) far from the neutron emission threshold, identified most likely as BSEC [5,6]. In sect. 2 the experimental results are summarized. As pointed out in refs. [5,6], the coupling between the valence neutron and the excited states of the core is needed in order to get a correct description also of the observed continuum shape. For this reason, a microscopic nuclear structure model based on the QRPA theory, called DCP model [7,8,9] has been applied to the studied nuclear systems. A brief review of it is given in sect. 3. The obtained results are then discussed and compared to the experimental data.

Nuclei having a $N\alpha$+3n like configuration (where N is an integer number) are good candidates to observe BSEC (see sect. 4). Clear indications may be provided by high-resolution studies of the continuum structures for such nuclear systems. In the low-mass region experimental data obtained via the (^{7}Li,^{7}Be) CEX reaction at $\sim$ 8 MeV/u are already available for the ^{7}He and ^{19}O nuclei, thus giving a systematic overview of these phenomena as a function of asymmetry and mass. Some conclusions are discussed in sect. 5.

2 The (^{7}Li, ^{7}Be) CEX reaction

The (^{7}Li, ^{7}Be) CEX reaction has been extensively studied in the past mainly because it gives useful information on the response function of nuclei to the isovector component of the effective nucleon-nucleon interaction [10, 11]. Since such reactions transform a target proton into a neutron, they have a natural relation to β decay processes. Indeed, charge exchange reactions allow to study, under well-defined conditions, transitions of β decay-type in mass regions which would be otherwise inaccessible. Such data are of direct relevance also for astrophysical purposes, especially for understanding weak interactions in short-lived neutron-rich nuclei far off stability. It is also known to be a useful spectroscopic tool at intermediate energies [12] where the one-step direct mechanism is considered to be dominant. In addition, depending on the system chosen and the q-matching condition [13], this reaction is an interesting spectroscopic probe at Tandem energies as well [14,15,16,17]. In ref. [17] the competition between the direct exchange of isospin degree of freedom and the two-step transfer mechanisms was faced for the ^{11}B(^{7}Li, ^{7}Be)^{11}Be reaction. Due to the high selectivity of the (^{7}Li, ^{7}Be) reaction, the analogy between the β-decay probabilities and the CEX cross sections measured at forward angles for Gamow-Teller transitions, proved for the (p, n) or (n, p) reactions [18,19], is maintained (within 20% accuracy) despite the complications coming from the presence of the heavy ions involved in the reaction [17]. In the case where the (^{7}Li, ^{7}Be$_{g.s.}$) and the (^{7}Li, ^{7}Be$_{0.43}$) transitions are experimentally resolved (by means of high resolution or γ-ray coincidence measurements), the observable $G = \sigma_1/(\sigma_0 + \sigma_1)$, where σ_0 and σ_1 are the corresponding cross sections measured for the two cited transitions, gives a strong indication on how the reaction mechanism proceeds [10,11]. As a consequence of such an analysis, it seems that in the (^{7}Li, ^{7}Be) reaction involving ^{11}Be or ^{15}C the two-step components are hidden also due to the structure properties of these unstable nuclei.

2.1 Experimental results

During the last few years, several light stable and unstable nuclei, the latter having a $N\alpha$ + 3n like configuration (where N is an integer number) have been investigated with the same probe (^{7}Li, ^{7}Be) in a systematic way. All the experiments have been performed with a ^{7}Li^{+++} beam at around 8 MeV/u provided by the Tandem Van de Graaff accelerator at the IPN-Orsay with different targets ($\sim$ 100 μg/cm^2 thickness), both solid and gas ones. In all the cases, the ^{7}Be ejectiles were detected in the focal plane detector of the Split-Pole magnetic spectrometer with a resolution as good as 50 keV. The focal plane detector was a position- and angle-sensitive proportional gas counter followed by a stopping plastic scintillator. Measurements at forward angles including 0° were done to get angular distribution of the cross sections. A ΔE-E telescope of silicon detectors was mounted at around θ_{lab} = 30° and used as a monitor for normalization procedures. Additional runs with ^{12}C target such as other elements normally present as target impurities (or in the windows of the gas target) were performed in order to estimate the source of background in the spectra. The background due to the p(^{7}Li, ^{7}Be)n reaction was always present below θ_{lab} = 7° due to the large cross section ($\sim$ 100 mb/sr).

Some interesting results have been obtained in the case of ^{11}Be (S_n = 0.504 MeV) and ^{15}C (S_n = 1.218 MeV) nuclei. Examples of final spectra detected in the case of ^{11}B(^{7}Li, ^{7}Be)^{11}Be and ^{15}N(^{7}Li, ^{7}Be)^{15}C reactions at around 8 MeV/u are shown in figs. 1 and 2, at θ_{lab} = 10.5° and θ_{lab} = 14°, respectively. The experimental results show that the response of the two nuclei to the same probe is quite similar. Single-particle states are populated at low excitation energy ($\sim$ 100 μb/sr). At high excitation energy, where the DCP regime is expected to be dominant, BSEC have been identified. In particular, in the ^{11}Be

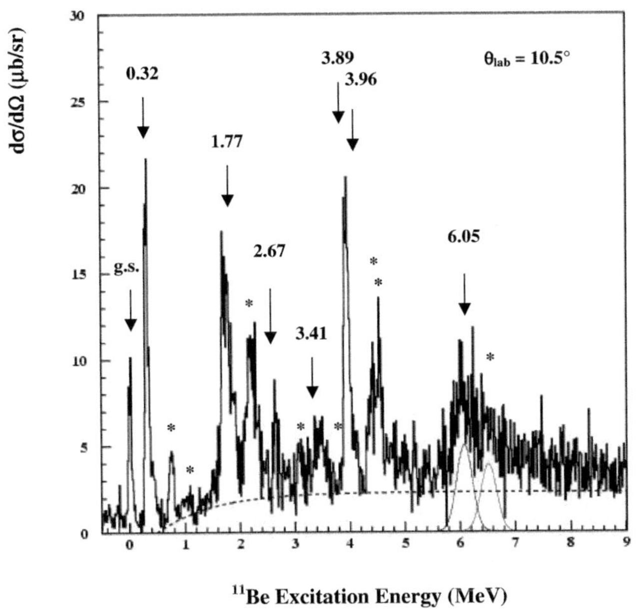

Fig. 1. Final excitation energy spectrum at $\theta_{lab} = 10.5°$ for the ^{11}B(^{7}Li, ^{7}Be)^{11}Be reaction at around 8 MeV/u. The fitted structure represents the state at 6.05 MeV. The peaks marked with an asterisk are associated to the excitation of ^{7}Be at 0.43 MeV. The dashed line is the non-resonant ^{11}B(^{7}Li, ^{7}Be)^{10}Be + n three-body phase space.

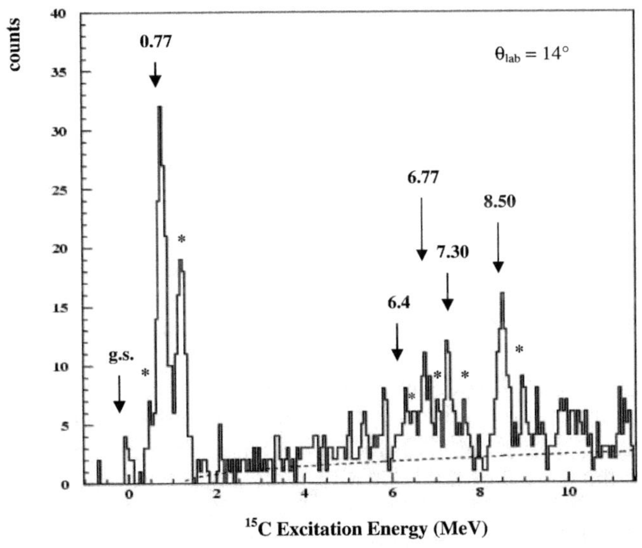

Fig. 2. Final excitation energy spectrum at $\theta_{lab} = 14°$ for the ^{15}N(^{7}Li, ^{7}Be)^{15}C reaction at around 8 MeV/u. The peaks indicated with arrows are the populated ones, those marked with an asterisk are associated to the excitation of ^{7}Be at 0.43 MeV. The dashed line is the non-resonant ^{15}N(^{7}Li, ^{7}Be)^{14}C + n three-body phase space.

spectrum the structure at 6.05 MeV ($\Gamma = 320 \pm 40$ keV) and in the ^{15}C spectrum the narrow peak at 8.50 MeV ($\Gamma \leq 140$ keV). Peaks marked by an asterisk refer to transitions in which the ^{7}Be is emitted in the 0.43 MeV first (bound) excited state, and are well resolved due to the high resolution given by the Tandem beam and magnetic spectrometer combination. A non-resonant three-body phase space due to the projectile breakup was calculated analytically for the exit channel like in ref. [20], transformed in the proper excitation energy scale and normalised for the most backward angle measurement ($\theta_{lab} = 18°$). In both spectra it is plotted. The ^{11}Be state at 6.05 MeV was previously unknown. For the ^{15}C case, narrow structures at around 7–8 MeV were observed before in the less selective ^{9}Be(^{7}Li, p)^{15}C reaction performed at 20 MeV [21]. There, the attempts of spin assignment for these states led to high spin values, to account for the observed hindrance for neutron emission.

Our data were analyzed in a QRPA many-body approach using the formalism of ref. [22]. Special features of these exotic systems that must be taken into account are described in details in refs. [5,17]. Calculations of microscopic QRPA transition densities, assuming configurations where the valence neutron is coupled to an inert core, reproduce well in both cases the level structure below 2 MeV [5,6], but are not able to explain the strong fragmentation of the strength at higher excitation energy.

A comparison with DWBA calculations, which was possible only for the ^{11}Be(g.s., $1/2^+$), ^{11}Be($0.32, 1/2^-$) and ^{11}Be($1.77, 5/2^+$) states [17], and for the ^{15}C(g.s., $1/2^+$) and ^{15}C($0.77, 5/2^+$) ones [23], shows that the corresponding angular distributions are quantitatively well reproduced without any scaling factor and without the necessity to introduce two-steps contributions. The unnatural-parity state transitions, both in the projectile and in the target, account for most of the observed cross sections. The distribution at around 0° of the G ratio as a function of the excitation energies are also well described in the DWBA framework.

In addition, a quite asymmetric line shape has been observed in the spectra of ^{15}C at several angles in correspondence of the 8.5 MeV resonance (see fig. 2), explained as the result of the interference between the BSEC and the three-body continuum [24,25]. Such an observation is an important experimental signature of the existence of BSEC.

3 Nuclear structure model

Single-particle configurations of the odd-mass nuclei with respect to an even-even vibrating core nucleus have been investigated theoretically [26]. In ref. [27], correlations in the ^{11}Be nucleus were calculated as corrections to the Hartree-Fock potential due to the coupling of single-particle states to RPA collective one-phonon states of the ^{10}Be core. The result, in case of the ^{11}Be ground state, leads to a strong admixture ($\sim 20\%$) of a $d_{5/2}$ neutron coupled to the ^{10}Be(2^+) core, in agreement with experimental results (see [2] for a review). In these calculations a non microscopic effective interaction was used. Recently, coupling of single particle motion with core vibrations has been further investigated as correlations always in the ^{11}Be case [28], in a similar way as we did.

Our DCP model is a microscopic version of the quasiparticle-core coupling (QPC) model originally proposed by Bohr and Mottelson [29], and adopted in the past for stable nuclei. The ground state of the even-even nucleus (core) is supposed to be known as the vacuum state for the quasiparticle operators. The wave functions of the odd-mass system can be written as

$$|jm, E_\lambda\rangle = \sum_n z_{nj}|nljm\rangle + \sum_{j'J_c} z_{j'J_c}|(j'J_c)jm\rangle, \quad (1)$$

where $|nljm\rangle$ is the one-quasiparticle (1QP) state with orbital and total angular momentum l and j, parity $\pi_j = (-1)^l$ and radial quantum number $n \geq 1$. The second term of eq. (1) corresponds to the core excited components (3QP) with a 1QP j' and one-particle-one-hole (two-quasiparticle (2QP)) core excitations J_c coupled to have total angular momentum j.

The effective Hamiltonian of the odd-mass system, is separated into H_0, which is diagonal in the 1QP and 2QP states, and the residual interaction V_{13}, which couples 1QP to 3QP configurations. The residual interaction used is derived from the M3Y-G matrix interaction [30].

The main feature of the DCP model is the derivation of an effective Schrödinger equation for the single-particle part coupled to core excited configurations, leading to an energy dependent contribution in the single-particle self-energy operator affecting the separation energy and the wave function [31]. The originality of the method consists in expanding the single-particle component of the wave function of the odd-mass nucleus into a set of unperturbed wave function of fixed orbital and total angular momentum (l, j) but different radial quantum number n ("major shell mixing"). It is due to the fact that the nucleon is rescattered by the core, which can have a very large energy. Then, the high-lying core excitations enter off-the-energy-shell into the mass operator of low-lying excitations of the odd-mass system. Since the parity of the eigenstate is fixed by π_j, the core excited configurations are restricted by the parity selection rule $\pi_j = \pi_{j'}\pi_{J_c}$. The single particle strengths $|z|^2$ are distributed over the whole spectrum, but in the most cases a fraction is found in an interval of several MeV around the eigenvalue E_λ. The BSEC are directly related to those solutions which lead to a wave function in which one or several core excited states carry the main part of the strength. However, since also the 1QP strength is non-vanishing these states can still be excited in one-step transfer process.

3.1 Results of the calculations

In order to give an interpretation of the continuum states populated in the (^{7}Li, ^{7}Be) CEX reactions at about 8 MeV/u DCP calculations were done for ^{15}C [32,33] and recently also for ^{11}Be [34].

Being a microscopic approach we first describe the ground state of the even-even core ^{10}Be and ^{14}C nuclei by Hartree-Fock-Bogolyubov (HFB) theory. The proton and neutron single-particle energies and wave functions have been calculated assuming an isoscalar potential Wood-Saxon (WS) shaped, for the central part, and the radial derivative of a WS, for the spin-orbit part. The depth parameters of such potentials have been chosen in order to reproduce the HFB single-particle energies of the core.

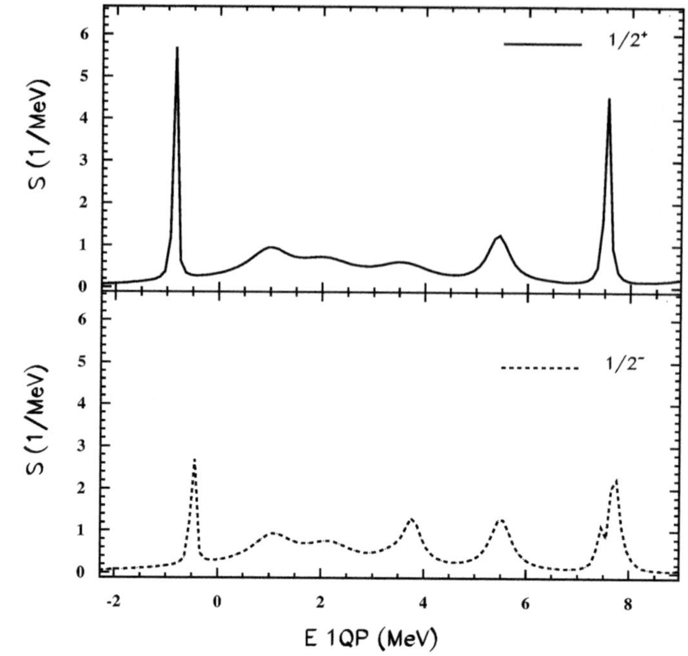

Fig. 3. Calculated single-particle strength of ^{11}Be for $j = 1/2^+$ (solid line) and $j = 1/2^-$ (dashed line). See the text for details.

Especially the QRPA strength functions of ^{10}Be(2^+) and ^{14}C(1^-) are quantitatively well described [34]. The interaction is kept the same for all the studied cases.

In fig. 3 the single-particle strength functions $s_{1/2}$ and $p_{1/2}$ calculated for the ^{11}Be nucleus are plotted as a function of the 1QP energy. The DCP results of the calculated $s_{1/2}$ and $d_{5/2}$ single-particle strengths of ^{15}C are presented in fig. 4. In all the cases, the contribution coming from the natural parity states of the core up to $J_c = 3$ have been included. Even if the absolute energy values are not quantitatively well reproduced, the level inversion between $2s_{1/2}$ and $1p_{1/2}$ in ^{11}Be is obtained and their energy difference calculated is around 0.390 MeV. The energy difference between the $2s_{1/2}$ and $1d_{5/2}$ in ^{15}C is underestimated, being 0.110 MeV, but, also here, the experimental level ordering is correctly reproduced. Fragmentation of the strength above 3 MeV and 7 MeV is present in ^{11}Be and ^{15}C calculations, respectively, demonstrating that a detailed microscopic description of the core coupling is very important. Strong overlap between 1QP and 3QP states appears above 8 MeV in the ^{15}C case, with the dominance of 1^-, 2^+ and 3^- ^{14}C exited states. In the ^{11}Be case, contributions of the ^{10}Be(2^+) are mainly found at around 3–4 MeV. According to our theoretical results, we do not support the necessity to assume in our experimental data configurations where the neutron occupies an orbital with an high orbital angular momentum.

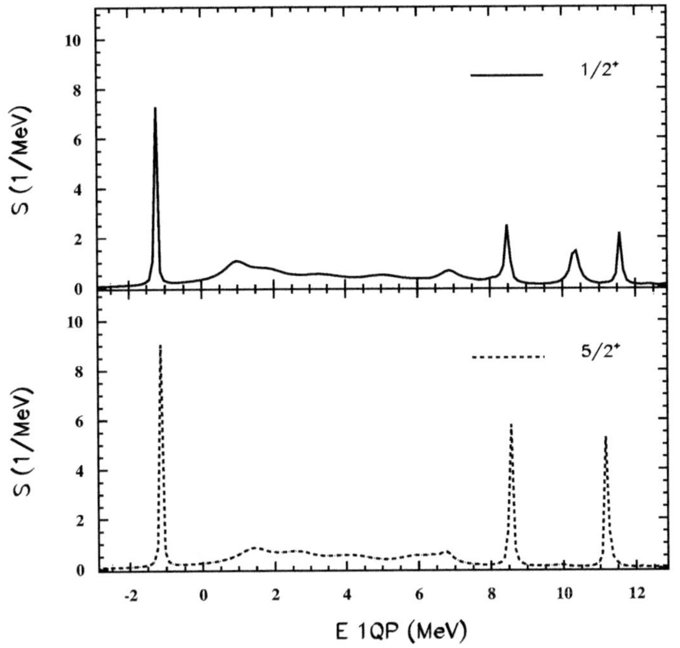

Fig. 4. Calculated single-particle strength of ^{15}C for $j = 1/2^+$ (solid line) and $j = 5/2^+$ (dashed line). See the text for details.

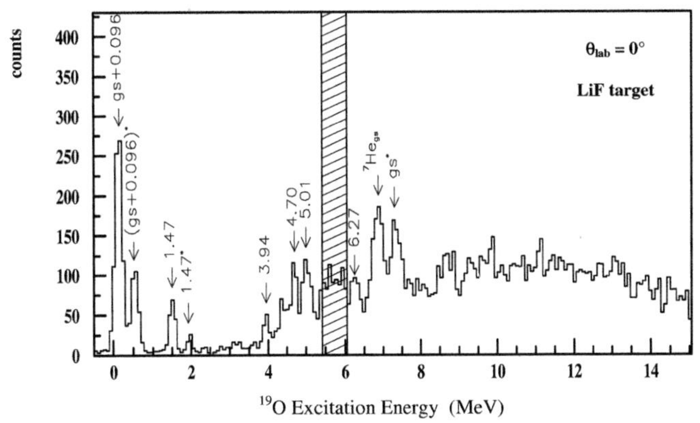

Fig. 5. Excitation energy spectrum at 0° for the ^{19}F(^{7}Li, ^{7}Be)^{19} reaction at around 8 MeV/u. The ground state of ^{7}He, which is present due to the Li impurities in the target, is also indicated. The dashed area represents a mask indicating a region not safe for identification due to the background. The states marked with an asterisk are associated to the excitation of ^{7}Be at 0.43 MeV.

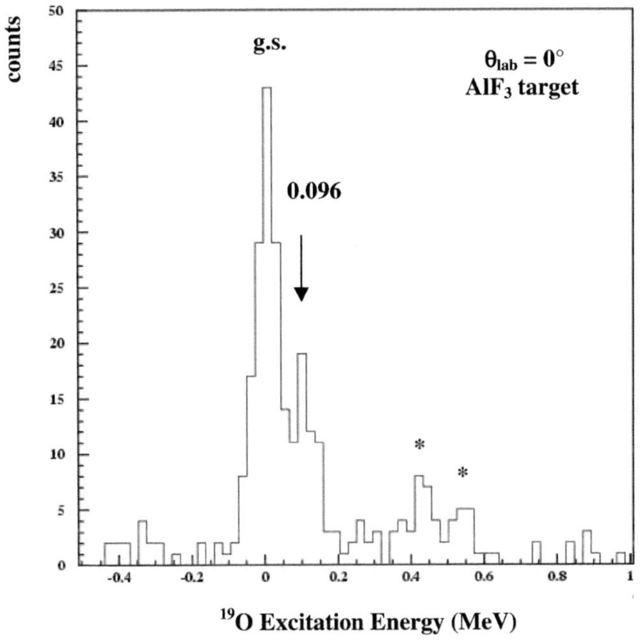

Fig. 6. Detail of the excitation energy spectrum at 0° obtained for the ^{19}F(^{7}Li, ^{7}Be)^{19}O reaction at around 8 MeV/u with an AlF$_3$ target. The states marked with an asterisk are associated to the excitation of ^{7}Be at 0.43 MeV.

Spectroscopic factors in agreement with the values provided experimentally have been obtained. Core excited state configurations like $d_{5/2}\otimes^{10}$Be(2^+) are present in the wave function of the ground state of ^{11}Be at 18%. The main component for the ground state of ^{15}C is represented by the $s_{1/2}\otimes^{14}$C(0^+) configuration.

4 Other explored Nα + 3n systems

In a more general context, the ^{11}Be and ^{15}C nuclei, can be imagined as systems where a hard core made of an integer number of alpha particles ($N\alpha$) is coupled to three neutrons. Assuming that single-particle excitations of the α-clusters are rather unlikely (hard core) at low excitation energy, the few-body dynamics in the nuclear medium can be studied. Since the pairing is expected to play an important role in the three-neutron phase space, a large contribution to it is represented by one valence nucleon coupled to the core, which is made softer by the remaining nucleon pair. Phenomena directly associated to that may be experimentally observed by studying such systems. In this context, experiments involving systems having $N = 1$ and $N = 4$, *i.e.* ^{7}He and ^{19}O, have already been done.

Concerning ^{7}He, the ^{7}Li(^{7}Li, ^{7}Be)^{7}He reaction was performed at about 8 MeV/u using a special target in order to minimized the oxidation of the Li contained in it. The ground state, which is particle instable with respect to the decay into ^{6}He + n by 0.44 MeV and has a width of 0.160 MeV, was easily identified. Another resonant state was found at around 2.90 MeV, supposed to be the same populated in other transfer reaction experiments at low and intermediate energies [35]. No trace of the low resonance identified at around 1.2 MeV [36] was found. The presence of ^{12}C impurities in the target in the 0° and 4° spectra did not allow to extract further information.

The ^{19}O nucleus represents an interesting case in order to find similarities with the system having $N = 2$ and $N = 3$. The populated states in the reaction ^{19}F(^{7}Li, ^{7}Be)^{19}O were the ground and those at 0.096, 1.47, 3.94, 5.01 and 6.27 MeV. In figs. 5 and 6 they are indicated by arrows. The bin size of the 0° spectrum of fig. 5 is compressed so that the ground and 0.096 MeV states are not resolved. The ground state of the ^{7}He, present due to the

Li impurities of the target, is also indicated. Due to higher neutron separation energy ($S_n = 3.957\,\mathrm{MeV}$) in comparison to the ^{11}Be and ^{15}C ones, the interesting region to explore here is at higher excitation energy. Moreover, the presence of the background due to Li and other impurities in the LiF target, makes it difficult to have clean access to an excitation energy interval greater than 8 MeV. The ground and the level at 0.096 MeV of ^{19}O nucleus, are visibly well separated in fig. 6 where a detail of the spectrum measured at 0°, obtained using an AlF_3 target, is shown.

5 Conclusion

Exploration of excited states in light neutron-rich nuclei is a rich source of information about nuclear structure. We also hope to have indicated a special aspect of nuclear continuum dynamics which might be of relevance for reaction rates and lifetimes of neutron-rich nuclei in astrophysical processes. Excited states in ^{11}Be, ^{15}C, ^{7}He and ^{19}O have been investigated via the (^{7}Li, ^{7}Be) at very similar bombarding energy ($\sim 8\,\mathrm{MeV}/u$). The obtained high-energy resolution has been crucial for that purpose. A dominant observed feature in the ^{15}C and ^{11}Be cases is the strong fragmentation of strength in BSEC at high excitation energy. The use of refined microscopic theories, such as the DCP model, has been fundamental in order to interpret them as single-particle states where the neutron is coupled to core excited states. The experiments and their analysis discussed here are considered as a first step into the largely unexplored regime of continuum dynamics and BSEC states in weakly bound nuclei. The results deserve further work, especially also in order to clarify the role of such states in astrophysical processes.

Particularly interesting for this phenomenology is the exploration of $N\alpha + 3$n like nuclei with $N > 4$. They still have not been explored since the cross sections at this low bombarding energies were expected to go down, making the experiments quite difficult. On the other hand, the study of these heavier systems is very important in order to learn about the influence of mass and asymmetry parameters on these phenomena. The advent of the MAGNEX spectrometer [37] at the LNS-INFN in Catania will allow to go on with the systematics by studying less favorable cases, fully exploiting the large solid angle. This device will make it possible to perform measurements with increasing statistics and much less number of angle settings, preserving a good energy resolution.

References

1. J.S. Winfield *et al.*, Nucl. Phys. A **683**, 48 (2001).
2. R. Palit *et al.*, Phys. Rev. C **68**, 034318 (2004).
3. G. Baur, H. Lenske, Nucl. Phys. A **282**, 201 (1977).
4. H. Fuchs *et al.*, Nucl. Phys. A **343**, 133 (1980).
5. F. Cappuzzello *et al.*, Phys. Lett. B **516**, 21 (2001).
6. F. Cappuzzello *et al.*, Europhys. Lett. **75(6)**, 766 (2004).
7. F.J. Eckle *et al.*, Nucl. Phys. A **506**, 199 (1990).
8. P. von Neumann-Cosel *et al.*, Nucl. Phys. A **516**, 385 (1990).
9. H. Lenske, J. Phys. G **24**, 1429 (1998).
10. S.M. Austin *et al.*, Phys. Rev. Lett. **44**, 972 (1980).
11. S. Nakayama *et al.*, Phys. Rev. Lett. **67**, 1082 (1991).
12. S. Nakayama *et al.*, Phys. Lett. B **246**, 342 (1990); Nucl. Phys. A **507**, 515 (1990).
13. W. von Oertzen, Nucl. Phys. A **482**, 357c (2004).
14. J. Cook *et al.*, Phys. Rev. C **30**, 1538 (1984).
15. A. Etchegoyen *et al.*, Phys. Rev. C **38**, 2124 (1988).
16. J. Jänecke *et al.*, Phys. Rev. C **54**, 1070 (1996).
17. F. Cappuzzello *et al.*, Nucl. Phys. A **739**, 30 (2004).
18. W.P. Alford, B.M. Spicer, *Advances in Nuclear Physics* edited by J.W. Negel, E. Vogt, Vol. **24** (Plenum, New York, 1998).
19. F. Osterfeld, Rev. Mod. Phys. **64**, 491 (1992).
20. G. Ohlsen, Nucl. Instrum. Methods **37**, 240 (1965).
21. J.D. Garrett *et al.*, Phys. Rev. C **10**, 1730 (1974).
22. F.T. Baker *et al.*, Phys. Rep. **289**, 235 (2001).
23. S.E.A. Orrigo *et al.*, *Proceedings of the 10th International Conference on Nuclear Reaction Mechanism, Villa Monastero, Varenna, 9-13 June 2003*, published in Suppl. J. Milan Univ. (Ed. Ricerca Scientifica ed Educazione Permanente, Milano, 2003) p. 147.
24. S.E.A. Orrigo, PhD Thesis, Università di Catania (2004).
25. S.E.A. Orrigo *et al.*, to be published in Phys. Lett. B.
26. G.F. Bertsch *et al.*, Rev. Mod. Phys. **55**, 287 (1983).
27. N. Vinh Mau, Nucl. Phys. A **592**, 33 (1995).
28. G. Gori *et al.*, Phys. Rev. C **69**, 041302(R) (2004).
29. A. Bohr, B. Mottelson, *Nuclear Structure*, Vols. **1** and **2** (Benjamin, New York, 1969 and 1970).
30. F. Hofmann, H. Lenske, Phys. Rev. C **57**, 2281 (1998).
31. H. Lenske, Prog. Part. Nucl. Phys. A **693**, 616 (2001).
32. C. Nociforo, PhD Thesis, Università di Catania (2001).
33. C. Nociforo *et al.*, Acta Phys. Pol. B **34**, 2387 (2003).
34. C. Nociforo, H. Lenske, to be submitted.
35. F. Ajzenberg-Selove, Nucl. Phys. A **708**, 3 (2002).
36. M. Meister *et al.*, Phys. Rev. Lett. **88**, 102501 (2002).
37. A. Cunsolo *et al.*, Nucl. Instrum. Methods **484**, 56 (2002).

Eur. Phys. J. A **27**, s01, 289–294 (2006)
DOI: 10.1140/epja/i2006-08-044-3

EPJ A direct
electronic only

Equation of state of strongly interacting matter in compact stars

A. Lavagno[1,2] and G. Pagliara[1,3,a]

[1] Dipartimento di Fisica, Politecnico di Torino, 10129 Torino, Italy
[2] INFN, Sezione di Torino, 10125 Torino, Italy
[3] INFN, Sezione di Ferrara, 44100 Ferrara, Italy

Received: 17 June 2005 /
Published online: 21 March 2006 –

Abstract. We study the equation of state of strongly interacting matter at large densities and vanishing temperature. The hadronic matter equation of state is computed in a relativistic mean-field model and the quark matter equation of state is computed using a NJL-type model which takes into account the possibility of formation of the gapless color flavor locked phase. We focus in particular on the possible phase transition from hadronic matter to quark matter using both Maxwell and Gibbs constructions. We finally discuss the relevance of the equation of state in the context of compact stars and we propose some astrophysical signatures of the presence of quark matter in compact stars.

PACS. 26.60.+c Nuclear matter aspects of neutron stars – 25.75.Nq Quark deconfinement, quark-gluon plasma production, and phase transitions

1 Introduction

The equation of state of strongly interacting matter at densities below the saturation density of nuclear matter $\rho_0 = 0.16\,\mathrm{fm}^{-3}$ is relatively well known [1,2] due to the large amount of experimental data on nuclear physics available. At larger densities there are many uncertainties due to the lack of experimental data; the saturation of nuclear force makes, in fact, the compression of nuclear matter at larger densities quite difficult. In ultra relativistic heavy-ion collision experiments the baryon densities can reach values of a few times ρ_0 at temperatures of $\sim$ 150–200 MeV. The only "natural laboratories" in which matter is compressed to densities up to ten times ρ_0 are compact stars. For these reasons, the study of these stellar objects can shed light on the equation of state of strongly interacting matter at extreme conditions. In particular, it has been extensively studied the possibility of formation, in the core of a compact star, of "exotic" particles as the hyperons or meson condensates of pions or kaons, or finally, it has been suggested that a phase transition from hadronic matter to quark matter, in which the quarks composing the baryons are deconfined, can occur. Concerning quark matter, in particular, recent studies on the QCD phase diagram at finite densities and temperatures have revealed the existence of a rich structure of the phase diagram with several possible phases in which a color superconducting state can be formed [3]. The fundamental phenomenological problem in the study of compact stars is to investigate how the measurable properties of these stellar objects as masses, radii, thermal evolution, periods of rotation etc., depend on the equation of state of matter. In principle, it is therefore possible to put constraints on the theory of the QCD phase diagram from astrophysical observations and measurements on compact stars. In this paper we will first analyze the equation of state of strongly interacting matter focusing in particular on the phase transition from hadronic matter to quark matter, eventually in its color superconducting phase and we will propose some signatures of the presence of quark matter in a compact star.

[a] e-mail: pagliara@fe.infn.it

2 Equation of state of compact stars matter

2.1 Hadronic matter

Concerning hadronic matter, we use the relativistic field-theoretical approach to the nuclear equation of state [4]. In this theory the interactions between hadrons are described by the exchange of three mesons, the scalar field σ, the vector field ω and the isovector field ρ. The Lagrangian of the model has five free parameters which are fixed imposing that the model reproduces five measured quantities of the nuclear matter, *i.e.* the saturation density, the binding energy per nucleon, the incompressibility, the effective mass of nucleons and the symmetry energy. This model can easily incorporate all the particles of the baryon octet [5,6] and in particular baryons containing

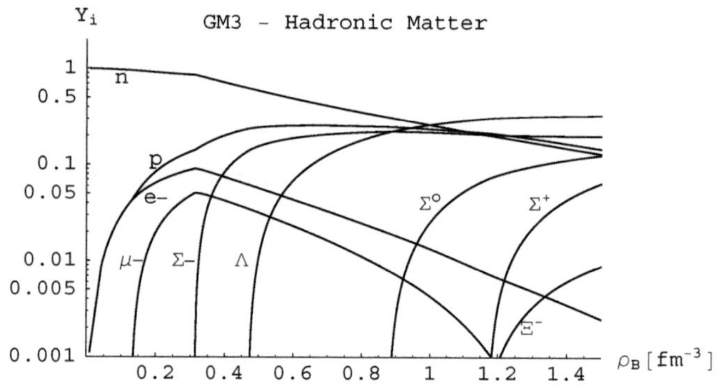

Fig. 1. Particle abundances as functions of the total baryon density.

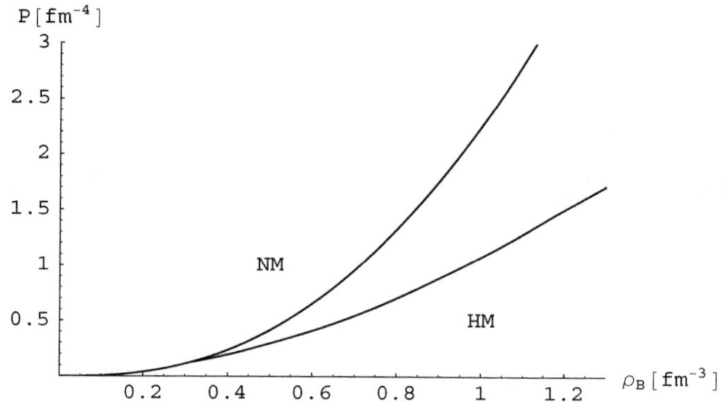

Fig. 2. Pressure as a function of baryon density for nucleonic matter (NM) and hadronic matter (HM).

strangeness, the hyperons. At $T = 0$, in the mean-field approximation, the thermodynamic potential Ω per unit volume can be written as

$$\Omega = -\frac{1}{3\pi^2}\sum_B \int_0^{k_{FB}} \mathrm{d}k \frac{k^4}{E_B^\star(k)} + \frac{1}{2}m_\sigma^2\sigma^2 + \frac{1}{3}a\sigma^3 + \frac{1}{4}b\sigma^4 - \frac{1}{2}m_\omega^2\omega^2 - \frac{1}{2}m_\rho^2\rho^2, \qquad (1)$$

where the $\sum_B$ runs over the eight baryon species, $E_B^\star(k) = \sqrt{k^2 + M_B^{\star 2}}$ and the baryon effective masses are $M_B^\star = M_B - g_\sigma\sigma$. The effective chemical potentials ν_B are given in terms of the thermodynamic chemical potentials μ_B and of the vector meson fields as follows:

$$\nu_B = \mu_B - g_\omega\omega - t_{3B}g_\rho\rho, \qquad (2)$$

where t_{3B} is the isospin 3-component for baryon B and the relation to the Fermi momentum k_{FB} is provided by $\nu_B = \sqrt{k_{FB}^2 + M_B^{\star 2}}$. The isoscalar and isovector meson fields (σ, ω and ρ) are obtained as a solution of the field equations in the mean-field approximation [6].

The equation of state of compact star matter must satisfy the beta equilibrium and charge neutrality conditions. The former allows to express the chemical potentials of all the particles as linear combinations of the baryonic and electric charge chemical potential (μ_B and μ_C):

$$\mu_i = b_i\mu_B + c_i\mu_C, \qquad (3)$$

where b_i is the baryon number of the particle and c_i is its electric charge in unit of the electron charge.

The condition of charge neutrality, taking into account also the densities of leptons, reads

$$0 = \rho_p + \rho_{\Sigma^+} - \rho_{\Sigma^-} - \rho_{\Xi^-} - \rho_e - \rho_\mu, \qquad (4)$$

where ρ_i indicate the number densities of the various particles. With all these conditions we can calculate the hadronic-matter equation of state as a function of only one chemical potential, the baryon chemical potential μ_B.

In fig. 1 the particle abundances as functions of the total baryon density are displayed. Notice in particular how the density of neutrons start to decrease as neutral hyperons start to form. In fig. 2 the pressures as a function of baryon density for nuclear matter and hadronic matter are compared. As it can be seen the presence of hyperons makes the equation of state softer.

The model described so far will be used for densities of matter of the order of or larger than the saturation density ρ_0. For lower densities we will use the Negele-Vautherin and the Baym-Pethick-Sutherland equations of state [1,2].

2.2 Quark matter

In high-density hadronic matter, baryons are forced to stay so close to one another that they would overlap, the constituent quarks are shared by neighboring baryons and should eventually become mobile over a distance larger than the typical size of one baryon. This means that quarks become deconfined and that at large densities they are the real degrees of freedom of strongly interacting matter instead of baryons. The process of deconfinement and the equation of state of quark matter can in principle be described by QCD. However in the energy scale involved in a compact star, QCD is non-perturbative and therefore simple phenomenological models are usually adopted to describe quark matter as the MIT bag model [7], the NJL model [8] or the Color Dielectric Model [9].

2.2.1 Unpaired quark matter

The simplest model to describe quark matter is the MIT bag model. In this model, quark matter is described as a gas of free quarks with massless up and down quarks and strange quarks having a "current mass" variable between 80–200 MeV. All the "non-perturbative physics of QCD" is simulated by the bag constant B which represents the pressure of the vacuum. The thermodynamic potential reads

$$\Omega = \frac{3}{\pi^2}\left(\int_0^{k_{Fu}} dk k^2(k-\mu_u) + \int_0^{k_{Fd}} dk k^2(k-\mu_d) + \int_0^{k_{Fs}} dk k^2(\sqrt{k^2+m_s^2}-\mu_s)\right) + B, \quad (5)$$

where

$$k_{Fu} = \mu_u = \frac{1}{3}\mu_B + \frac{2}{3}\mu_C, \quad (6)$$

$$k_{Fd} = \mu_d = \frac{1}{3}\mu_B - \frac{1}{3}\mu_C, \quad (7)$$

$$k_{Fs} = \sqrt{\mu_s^2 - m_s^2}, \quad (8)$$

$$\mu_s = \mu_d, \quad (9)$$

and B is the bag constant. Beta stability is included in eqs. (6-9) and charge neutrality is imposed by the equation

$$0 = \frac{2}{3}\rho_u - \frac{1}{3}\rho_d - \frac{1}{3}\rho_s - \rho_e - \rho_\mu . \quad (10)$$

As the hadronic equation of state, the unpaired quark matter equation of state has only one independent variable which can be chosen to be the baryon chemical potential μ_B.

2.2.2 Color superconducting quark matter

Quark matter is actually a gas of interacting fermions, the interaction being mediated by the exchange of gluons as described by QCD. Considering just the one-gluon exchange potential, there is a channel of interaction between quarks which is attractive and which corresponds with the two incoming quarks to be in the $\bar{3}$ channel. From the results of the Bardeen-Cooper-Schrieffer theory of superconductivity, it is known that if in a Fermi gas there is an arbitrary weak attractive potential, the Fermi surface becomes unstable with respect to the formation of a condensate of Cooper pairs. Therefore, as it happens in metals at low temperature, also in quark matter an instability with respect to the formation of Cooper pairs between quarks does develop, originating the phenomenon of the so-called color superconductivity. The order parameter characterizing this phase is the value of the diquark condensate or, in other words, the color superconducting gap Δ. The possibility of the existence of this phase of QCD was first shown at asymptotic densities, *i.e.* in a perturbative regime [10,11]. The same results have been extrapolated to lower densities using models of quark matter as for example the NJL model [12]. Both schemes leads to a value of $\Delta \sim 100$ MeV. Such a high value indicates that, at variance with the superconductivity in metals, color superconductivity is very robust because in the only place of the universe in which it can appear, the core of compact stars, the temperature is very low (few keV). This observation has encouraged further studies in this direction, see ref. [3] for an exhaustive review.

It is not yet understood which type of color superconducting phase can appear in a compact star. It is widely accepted that the Color-Flavor Locking phase (CFL) is the real ground state of QCD at asymptotically large densities. In this phase, up, down, and strange quarks are present and all of them are involved in the formation of Cooper pairs. As a consequence, all the quarks have a common Fermi level and therefore charge neutrality and beta stability are automatically satisfied without the presence of leptons [13]. The CFL phase can form only if the mass of the strange quark m_s is small with respect to the superconducting gap and the chemical potential. All these considerations are valid at high density. At lower densities, there is still uncertainty about the presence of CFL phase and in particular about the transition from the CFL phase to the hadronic phase. At intermediate values of m_s, it is in general difficult to involve strange quarks in BCS pairing due to their Fermi momentum, which is lower than that of up and down quarks. Recently, it has been shown that the CFL phase can form only if the ratio $m_s^2/\mu \lesssim 2\Delta_{CFL}$ [14]. At larger values of m_s^2/μ, but not too large values of m_s, the most energetically favored phase is the so-called gapless CFL (gCFL) phase instead of the 2SC phase or unpaired quark matter [15,16,17]. The gCFL phase has the same symmetries as the CFL phase but there are two gapless quark modes and a nonzero electron density and therefore it can have very different transport properties respect to the CFL phase.

To compute the equation of state of the (g)CFL phase we adopt the NJL-like formalism of refs. [14,15,16] in which the thermodynamic potential per unit volume can be written as

$$\Omega = -\frac{1}{4\pi^2}\int \mathrm{d}p\, p^2 \sum_j |\epsilon_j(p)|\rho_j(p) + \frac{1}{G}(\Delta_1^2+\Delta_2^2+\Delta_3^2) - \frac{\mu_e^4}{12\pi^2}, \quad (11)$$

where Δ_1, Δ_2, Δ_3 are the superconducting gaps characterizing the gCFL phase (which reduce to a single gap in the CFL phase), G is the strength of the diquark coupling, $\epsilon_j(p)$ are the dispersion relations of quarks as in ref. [15] and μ_e is the electron chemical potential. Following the approximations used in refs. [14,15], the effect of m_s is introduced as a shift $-m_s^2/2\mu$ in the chemical potential for the strange quarks and the contributions of antiparticles is neglected. In eq. (11) we have introduced the quasiparticle probabilities

$$\rho_j(p) = \frac{1}{2}\left(1 - \frac{\tilde{\epsilon}_j(p)}{\epsilon_j(p)}\right), \quad (12)$$

where $\tilde{\epsilon}_j(p)$ are the dispersion relations with vanishing gaps. To assure the convergence of the integral in eq. (11), a form factor $f = (\Lambda^2/(p^2+\Lambda^2))^2$, which multiplies the gaps, is introduced in the dispersion relations $\epsilon_j(p)$. The form factor was fixed to mimic the effects of the asymptotic freedom of QCD [12] and the parameter Λ was fixed at a value of 800 MeV.

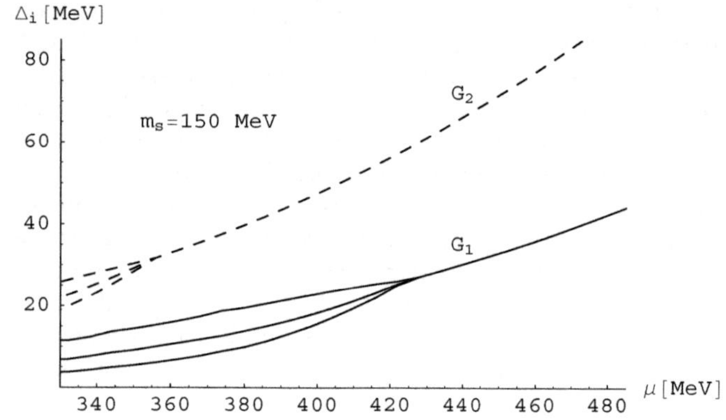

Fig. 3. Gap parameters as a function of the quark chemical potential for two different values of diquark coupling G_1 and G_2 and for a fixed value of the strange quark mass $m_s = 150\,\mathrm{MeV}$. The larger the value of G, the larger the window of the chemical potential in which the CFL phase occurs.

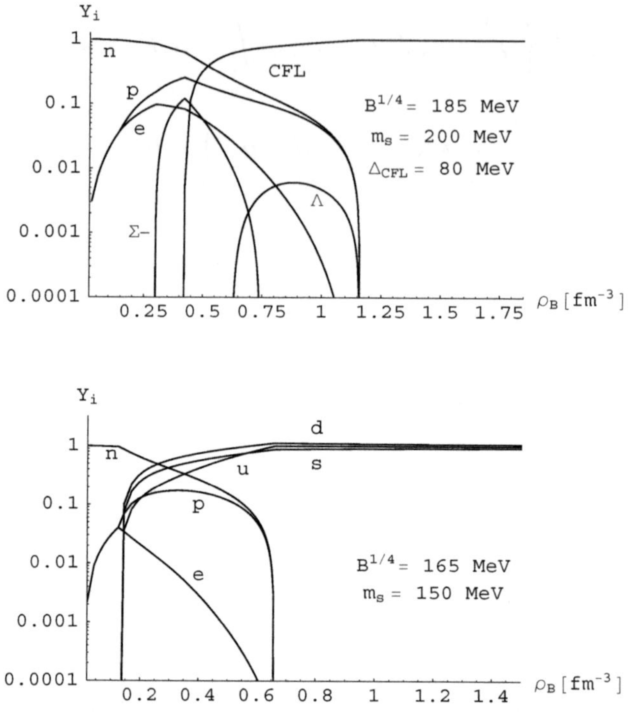

Fig. 4. Particle abundances as functions of the total baryon density. The upper panel corresponds to the case of the hyperon-quark mixed phase and the lower panel to the nucleon-quark mixed phase.

As in the other equations of state discussed so far we must impose the chemical equilibrium between the different quark species. The chemical equilibrium conditions (which also include β-stability) allow to express the chemical potential μ_{cf} (c and f are the indexes of color and flavor, respectively) of each quark as a function of the quark (baryonic) chemical potential μ, the electron chemical potential μ_e and the two chemical potentials, μ_3 and μ_8, associated to the $U(1) \times U(1)$ subgroup of the color gauge group (see ref. [15] for details). The color and electric charge neutrality are imposed by the following three equations:

$$\frac{\partial \Omega}{\mu_3} = 0, \qquad \frac{\partial \Omega}{\mu_8} = 0, \qquad \frac{\partial \Omega}{\mu_e} = 0. \tag{13}$$

Moreover, the thermodynamic potential must be minimized with respect to the gap parameters and therefore we have to impose the three additional conditions:

$$\frac{\partial \Omega}{\Delta_1} = 0, \qquad \frac{\partial \Omega}{\Delta_2} = 0, \qquad \frac{\partial \Omega}{\Delta_3} = 0. \tag{14}$$

The above equations allow us to compute the thermodynamic potential and all the thermodynamic variables as a function of the quark chemical potential only. In fig. 3 the gap parameters are displayed as functions of the quark chemical potential μ for two different values of the diquark coupling. G_1 and G_2 correspond, respectively, to values of $\Delta_{CFL} \sim 40$ and $\Delta_{CFL} \sim 100\,\mathrm{MeV}$ at $\mu = 500\,\mathrm{MeV}$ and $m_s = 150\,\mathrm{MeV}$. It is interesting to observe that the window in which the gCFL phase appears depends noticeably on the value of the diquark coupling and, in particular, it decreases with G (see fig. 3). This confirms the general argument for which the transition from gCFL to CFL occurs when $m_s^2/\mu \simeq 2\Delta_{CFL}$.

2.3 Phase transitions

Although there is no theoretical evidence, it is believed that the hadronic matter-quark matter phase transition is a first order phase transition. Maxwell construction is the usual tool to connect two phases: the phase transition occurs at constant pressure with a discontinuity of the number density. In the context of compact stars, it has been shown that the two phases can be connected by an intermediate window of mixed phase [6]. In the calculation of the possible mixed phases, hadronic-unpaired quark phase or hadronic-CFL phase, we use the Gibbs construction in which the equations of mechanical, thermal and chemical equilibrium are simultaneously imposed. The conservation of the baryon number and the electrical neutrality are imposed as global conditions. Due to the existence of two conserved charges in the matter of a compact star the pressure need not to be constant in the mixed phase which is crucial for the stability of the star.

In fig. 4 the abundances of various particles as functions of the baryon density are shown for two different equations of state. In the upper panel, we show a mixed phase of hadronic matter and CFL quark matter where the superconducting gap is fixed at a value of 80 MeV. In the lower panel, we show a mixed phase of nuclear matter and unpaired quark matter.

The hypothesis of a direct transition from hadronic matter to the CFL phase is too simple at the light of recent results on the QCD phase diagram. Actually, a scenario in which an intermediate window of unpaired quark

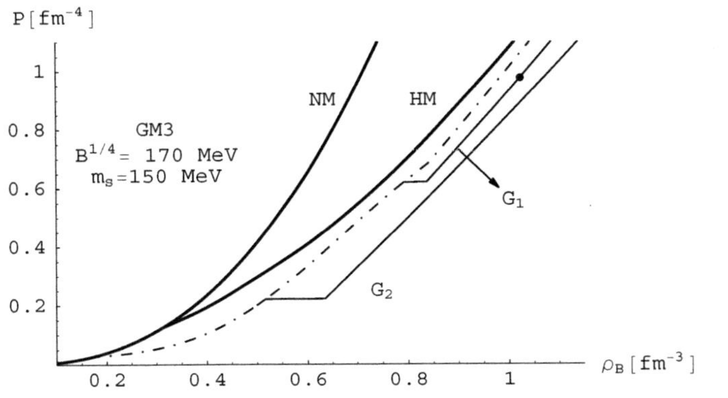

Fig. 5. Pressure as a function of baryon density for the scenario in which a first phase transition from hadronic matter to quark matter occurs via a mixed phase and then a second phase transition (here computed using the Maxwell construction) occurs from this mixed phase to the gCFL phase. The dot-dashed line is related to the hadron-unpaired quark mixed phase. The thin solid lines represent the phase transitions from the mixed phase to the gCFL phase for the two diquark couplings. Notice that in the case of G_1 a gCFL phase window is still present.

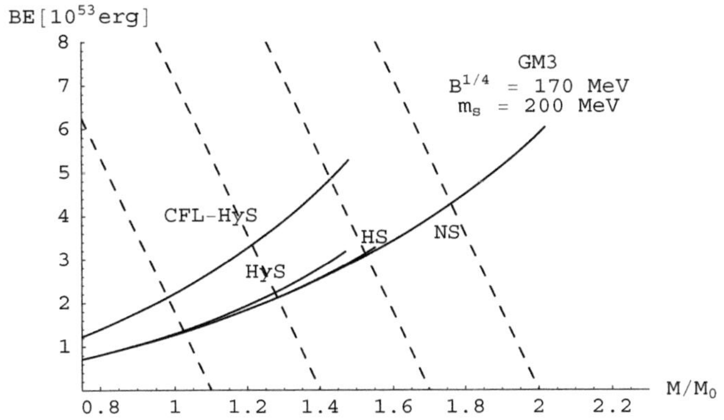

Fig. 6. The binding energies for neutron stars (NS), hadronic stars (HS), hybrid stars HyS (without CFL phase) and CFL hybrid stars (CFL-HyS) are shown as functions of the gravitational mass. The dashed line are the lines of constant baryonic mass.

matter or 2SC quark matter is more plausible. In this case the equation of state of strongly interacting matter has two first-order phase transitions: a first one from hadronic matter to unpaired quark matter (or 2SC) and a second one from unpaired quark matter (or 2SC) to the (g)CFL phase (see refs. [18,19,20]). This scenario is shown in fig. 5 where the pressure as a function of the baryon density is shown for different equations of state. The first phase transition occurs between nuclear matter and unpaired quark matter (thick and dot-dashed lines) and the second phase transition (here computed using Maxwell construction for simplicity) occurs between the nuclear-quark mixed phase and the (g)CFL phase (dot-dashed and thin solid lines). The couplings G_1 and G_2 are the same of fig. 3 and the dot on the curve labeled with G_1 represents the onset of the CFL phase. Notice that for large diquark coupling (G_2) the transition from the mixed phase to the (g)CFL phase occurs above the onset of the gCFL-CFL phase and for small diquark coupling (G_1) a window of the gCFL phase is instead present. Concerning the hyperons, within this choice of parameters the transition from hadronic matter to gCFL quark matter occurs before reaching the threshold of the formation of hyperons. If we use larger values of the bag constant, the first critical density may be larger than the threshold of the formation of hyperons. In that case, however, the phase transition would involve the CFL phase directly.

3 Signatures of quark matter in compact stars

Several signatures of the presence of quark matter in compact stars have been proposed in the literature. The most extensively discussed is the mass-radius relation of compact stars which should allow the existence of very compact stellar objects (radius less than 10 km) if quark matter is present [21]. Other interesting ways to study the composition of a compact star come from the analysis of the cooling of the stars [22] or the stability with respect to the r-modes [23]. In both cases, the weak-decay channels involving the strange quark can strongly affect the transport properties of the matter of the compact star. Recently, it has been shown that also from gravitational wave signals from isolated compact stars, we can obtain important informations on the structure and composition of a star [24,25]. Finally, the formation of quark matter in a compact star can have a role also in the most violent explosive phenomena of the Universe, *i.e.* supernovae and gamma-ray-bursts [26]. Here, as an example, we will discuss only the effect of the formation of quark matter during a supernova explosion.

Let us first introduce the expression of the baryonic mass of a star:

$$M_B = m_n \int_0^R \mathrm{d}r \frac{4\pi r^2}{(1-2m(r)/r)^{1/2}}\, \rho_B(r), \qquad (15)$$

where m_n is the neutron mass, $m(r)$ is the mass in a sphere of radius r, ρ_B is the total baryon density. The energy released in a supernova explosion is the difference between the gravitational mass of the core of the progenitor star M_c and the gravitational mass of the final compact star. The energy released corresponds also to the binding energy of the star. For practical purposes we can assume that $M_c \sim M_B$. Therefore, the energy released can be approximated as the difference $M_B - M$. Almost all of this energy is released in neutrinos. In fig. 6, we display the binding energy for neutron stars, hadronic stars and hybrid stars (without CFL) and CFL-hybrid stars, as functions of the gravitational mass. For a fixed value of the baryonic mass, which as already stated roughly corresponds to the mass of the collapsing core, we can calculate the energy released in the collapse. Typical values of this energy are of the order of 10^{53} ergs. It is evident from fig. 6 that, for a fixed baryonic mass (dashed lines), the binding energy of a CFL-hybrid star can be a few times that of a neutron star and therefore the corresponding energy released is larger. Figure 6, suggests, moreover, that, if in the future new supernova events will be detected, it will be possible from the signal of the emitted neutrinos to obtain information on the equation of state of the matter in compact stars. For instance, if the energy released in neutrinos is of the order of or larger than $\sim 4 \times 10^{53}$ erg, we can exclude hybrid or hadronic stars. A similar energy may be released in the formation of a neutron star but this would require a baryon mass for the core $\sim 2M_{\odot}$ what seems difficult in the light of recent supernovae simulations [27]. In this case only a CFL (CFL-hybrid) star can be a plausible candidate. This conclusion agrees with the well-known results that the stars containing quark matter are more bounded that hadronic stars.

4 Conclusions

In this paper, we studied the equation of state of strongly interacting matter at large densities and vanishing temperature. We analyzed in particular the possibility of a phase transition from hadronic matter and quark matter. We included in our calculation the recent results on the QCD phase diagram indicating two first-order phase transitions: a first one from hadronic matter to unpaired quark matter and a second one from unpaired quark matter to the color flavor locked phase. These theoretical results can be tested studying compact stars which are the only places in the universe in which the density of matter can reach values up to ten times the nuclear matter saturation density. We listed the different signatures of the presence of quark matter in a compact star and we discussed, in particular, a scenario in which during a supernova explosion deconfinement of quarks is realized in the newly born compact star.

References

1. G. Baym, C. Pethick, P. Sutherland, Astrophys. J. **170**, 299 (1971).
2. J.W. Negele, D. Vautherin, Nucl. Phys. A **207**, 298 (1973).
3. K. Rajagopal, F. Wilczek, hep-ph/0011333 (2000).
4. B.D. Serot, J.D. Walecka, Phys. Lett. B **87**, 172 (1979).
5. N.K. Glendenning, S.A. Moszkowski, Phys. Rev. Lett. **67**, 2414 (1991).
6. N. Glendenning, *Compact Stars* (Springer-Verlag, 1997).
7. G. Baym, S.A. Chin, Phys. Lett. B **62**, 241 (1976).
8. K. Schertler, S. Leupold, J. Schaffner-Bielich, Phys. Rev. C **60**, 025801 (1999).
9. A. Drago, U. Tambini, M. Hjorth-Jensen, Phys. Lett. B **380**, 13 (1996).
10. B.C. Barrois, Nucl. Phys. B **129**, 390 (1977).
11. D. Bailin, A. Love, Phys. Rep. **107**, 325 (1984).
12. M.G. Alford, K. Rajagopal, F. Wilczek, Nucl. Phys. B **537**, 443 (1999).
13. K. Rajagopal, F. Wilczek, Phys. Rev. Lett. **86**, 3492 (2001).
14. M. Alford, C. Kouvaris, K. Rajagopal, Phys. Rev. Lett. **92**, 222001 (2004).
15. M. Alford, C. Kouvaris, K. Rajagopal, Phys. Rev. D **71**, 054009 (2005).
16. K. Fukushima, C. Kouvaris, K. Rajagopal, Phys. Rev. D **71**, 034002 (2005).
17. S.B. Ruster, I.A. Shovkovy, D.H. Rischke, Nucl. Phys. A **743**, 127 (2004).
18. S.B. Ruster, V. Werth, M. Buballa, I.A. Shovkovy, D.H. Rischke, hep-ph/0503184 (2005).
19. D. Blaschke, S. Fredriksson, H. Grigorian, A.M. Oztas, F. Sandin, hep-ph/0503194 (2005).
20. A. Lavagno, G. Pagliara, nucl-th/0504066 (2005).
21. A. Drago, A. Lavagno, G. Pagliara, Phys. Rev. D **69**, 057505 (2004).
22. H. Grigorian, D. Blaschke, D. Voskresensky, Phys. Rev. C **71**, 045801 (2005).
23. A. Drago, A. Lavagno, G. Pagliara, Phys. Rev. D **71**, 103004 (2005).
24. N. Andersson, D.I. Jones, K.D. Kokkotas, Mon. Not. R. Astron. Soc. **337**, 1224 (2002).
25. A. Drago, G. Pagliara, Z. Berezhiani, gr-qc/0405145 (2004).
26. Z. Berezhiani *et al.*, Astrophys. J. **586**, 1250 (2003).
27. S.E. Woosley, A. Heger, T.A. Weaver, Rev. Mod. Phys. **74**, 1015 (2002).

Eur. Phys. J. A **27**, s01, 295–300 (2006)
DOI: 10.1140/epja/i2006-08-045-2

EPJ A direct
electronic only

Clustering in light nuclei in fragmentation above 1 A GeV

N.P. Andreeva[1], D.A. Artemenkov[2], V. Bradnova[2], M.M. Chernyavsky[3], A.Sh. Gaitinov[1], N.A. Kachalova[2], S.P. Kharlamov[3], A.D. Kovalenko[2], M. Haiduc[4], S.G. Gerasimov[3], L.A. Goncharova[3], V.G. Larionova[3†], A.I. Malakhov[2], A.A. Moiseenko[5], G.I. Orlova[3], N.G. Peresadko[3], N.G. Polukhina[3], P.A. Rukoyatkin[2], V.V. Rusakova[2], V.R. Sarkisyan[5], T.V. Shchedrina[2], E. Stan[2,4], R. Stanoeva[2,6], I. Tsakov[6], S. Vokál[2,7], A. Vokálová[2], P.I. Zarubin[2,a], and I.G. Zarubina[2]

[1] Institute for Physics and Technology, Almaty, Republic of Kazakhstan
[2] Joint Institute for Nuclear Research, Dubna, Russia
[3] Lebedev Institute of Physics, Russian Academy of Sciences, Moscow, Russia
[4] Institute of Space Sciences, Magurele, Romania
[5] Yerevan Physics Institute, Yerevan, Armenia
[6] Institute for Nuclear Research and Nuclear Energy, Sofia, Bulgaria
[7] P.J. Šafărik University, Košice, Slovak Republic

Received: 20 June 2005 /
Published online: 14 March 2006 –

Abstract. The relativistic invariant approach is applied to analyzing the 3.3 A GeV ^{22}Ne fragmentation in a nuclear track emulsion. New results on few-body dissociations have been obtained from the emulsion exposures to 2.1 A GeV ^{14}N and 1.2 A GeV ^{9}Be nuclei. It can be asserted that the use of the invariant approach is an effective means of obtaining conclusions about the behavior of systems involving a few He nuclei at a relative energy close to 1 MeV per nucleon. The first observations of fragmentation of 1.2 A GeV ^{8}B and ^{9}C nuclei in emulsion are described. The presented results allow one to justify the development of few-body aspects of nuclear astrophysics.

PACS. 21.45.+v Few-body systems – 23.60.+e α decay – 25.10.+s Nuclear reactions involving few-nucleon systems

1 Introduction

Interactions in few-body systems consisting of more than two ^{1,2}H and ^{3,4}He nuclei can contribute to a nucleosynthesis pattern. A macroscopic medium composed of the lightest nuclei having energy of the nucleosynthesis scale can possess properties analogous to those of dilute quantum gases of atomic physics. In this sense, few-body fusions imply a phase transition to "drops" of a quantum liquid, that is, to heavier nuclei. Fusions can proceed via the states corresponding to low-lying cluster excitations in forming nuclei.

On a microscopic level the phase transition can proceed through the production of quasi-stable and loosely bound quantum states. Among candidates for such states one can consider the dilute α-particle Bose condensate [1] as well as radioactive and unbound nuclei along the proton drip line. At first glance, exploration of few-body transitions in laboratory conditions seems to be impossible. Nevertheless, such processes can indirectly be explored in the inverse processes of relativistic nucleus breakups in a nuclear track emulsion by selecting excitations close to the few-body decay threshold.

Experimental data on the charged topology for the final states of a number of light nuclei have been described in [2,3,4,5,6]. More specific studies were performed for the leading channels like ^{12}C $\to 3\alpha$ [7], ^{16}O $\to 4\alpha$ [8,9], ^{6}Li $\to d\alpha$ [10,11], ^{7}Li $\to t\alpha$ [12], ^{10}B $\to d\alpha\alpha$ [13], and ^{7}Be $\to {}^{3}$Heα [14]. A collection of appropriate reaction images can be found in [15,16].

In the present paper the behaviour of relativistic systems consisting of several H and He nuclei will be described in terms of invariant variables of a 4-velocity space as suggested in [17]. The invariant presentation makes it possible to extract qualitatively new information about few-cluster systems from fragmentation of relativistic nuclei in peripheral interactions. An invariant approach is applied to the existing data on 3.3 A GeV ^{22}Ne interactions in a nuclear track emulsion, as well as to new data for 2.1 A GeV ^{14}N and 1.2 A GeV ^{9}Be nuclei extracted from a portion of a recently exposed emulsion. The first observations of the fragment topology for neutron-deficient

[a] e-mail: zarubin@lhe.jinr.ru
[†] Deceased.

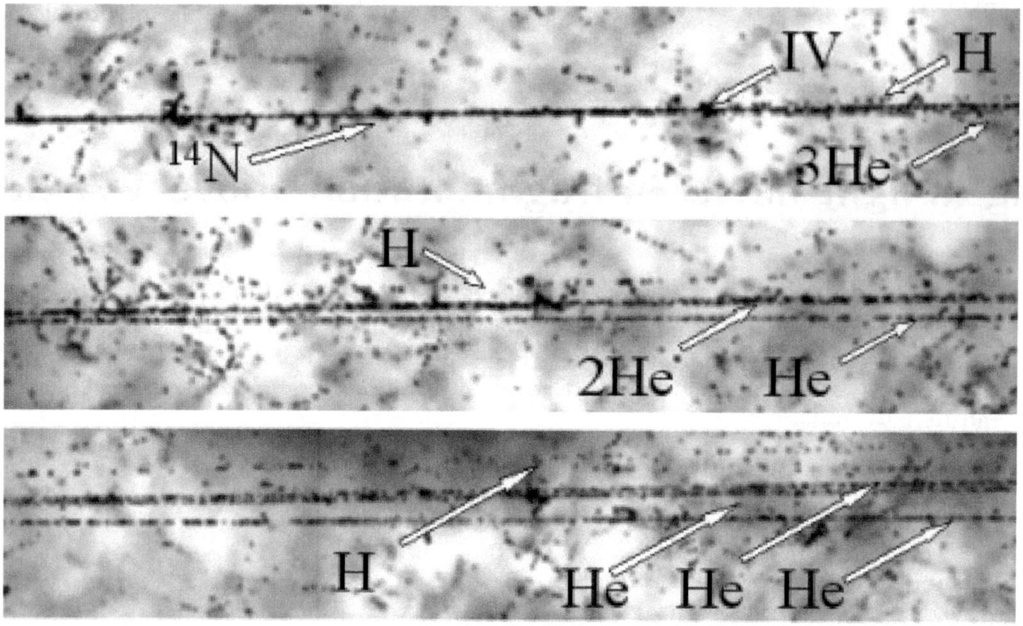

Fig. 1. Example of peripheral interaction of a 2.1 A GeV ^{14}N nucleus in a nuclear track emulsion ("white" star). The interaction vertex (indicated as IV) and nuclear fragment tracks (H and He) in a narrow angular cone are seen on the upper microphotograph. Following the direction of the fragment jet, it is possible to distinguish 1 singly and 3 doubly charged fragments on the middle and bottom microphotograph.

^{8}B and ^{9}C nuclei in emulsion are described in this report. Last emulsion exposures were performed at the JINR Nuclotron in the years 2002-4 [18].

2 Nuclear fragment jets

The relativistic projectile fragmentation results in the production of a fragment jet which can be defined by invariant variables characterizing relative motion:

$$b_{ik} = -\left(\frac{P_i}{m_i} - \frac{P_k}{m_k}\right)^2, \qquad (1)$$

with $P_{i(k)}$ and $m_{i(k)}$ being the 4-momenta and the masses of the i or k fragments. Following [17], one can suggest that a jet is composed of the nuclear fragments having relative motion within the non-relativistic range $10^{-4} < b_{ik} < 10^{-2}$. The lower limit corresponds to the ground-state decay $^8\mathrm{Be} \to 2\alpha$, while the upper one to the boundary of low-energy nuclear interactions. The expression of the data via the relativistic invariant variable b_{ik} makes it possible to compare the target and projectile fragmentation in a common form. Figure 1 shows the microphotograph of a special example of a projectile fragment jet — the "white" star as introduced in [3]. It corresponds to the case of a relativistic nucleus dissociation accompanied by neither a target fragment nor meson production.

The variable characterizing the excitation of a fragment jet as a whole is an invariant mass M^* defined as

$$M^{*2} = (\Sigma P_j)^2 = \Sigma(P_i \cdot P_k). \qquad (2)$$

The system excitation can be characterized also by

$$Q = M^* - M \qquad (3)$$

with M being the mass of the ground state of the nucleus corresponding to the charge and weight of the fragment system. The variable Q corresponds to the excitation energy of the system of fragments in their cms. A useful option is

$$Q' = \frac{(M^* - M')}{A}, \qquad (4)$$

with M' being the sum of fragment masses and A the total atomic weight. The normalized variable Q' characterizes a mean kinetic energy of fragments per nucleon in their cms. Precision of the experimental b_{ik} and Q values is driven to a decisive degree by the angular resolution in the determination of unit vectors defining the direction of the fragment emission.

Due to excellent spatial resolution (about 0.5 μm) the emulsion technique is known to be most adequate for the observation and angular measurements of projectile fragments down to a total breakup of relativistic nuclei. Nevertheless, it has restrictions on the determination of the 4-momentum components of fragments. Firstly, the fragment spatial momentum in the projectile fragmentation cone is suggested to be equal within a few percent error to the primary nucleus value when normalized to the nucleon numbers. Secondly, by multiple scattering measurements it is possible to identify the mass only for relativistic H isotopes and much harder for He ones. Normally, the α-particle mass is taken for the mass of doubly charged fragments in a narrow fragmentation cone. Both assumptions are proven to be reasonable for light stable nuclei.

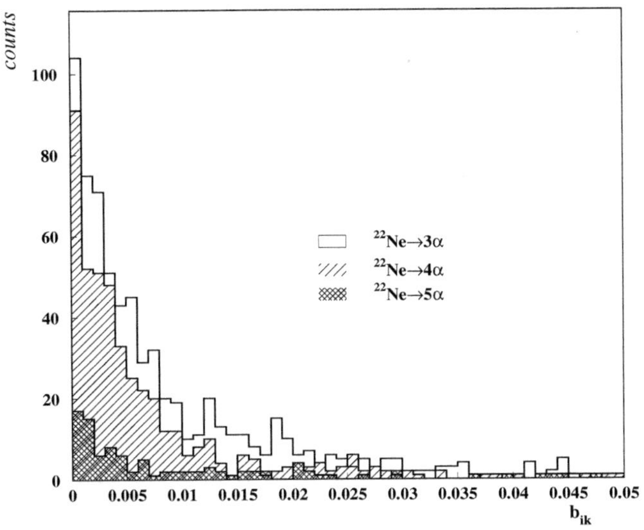

Fig. 2. Distribution of α-particle pairs *vs.* the relative variable b_{ik} (1) for the fragmentation modes $^{22}\mathrm{Ne} \to n\alpha$.

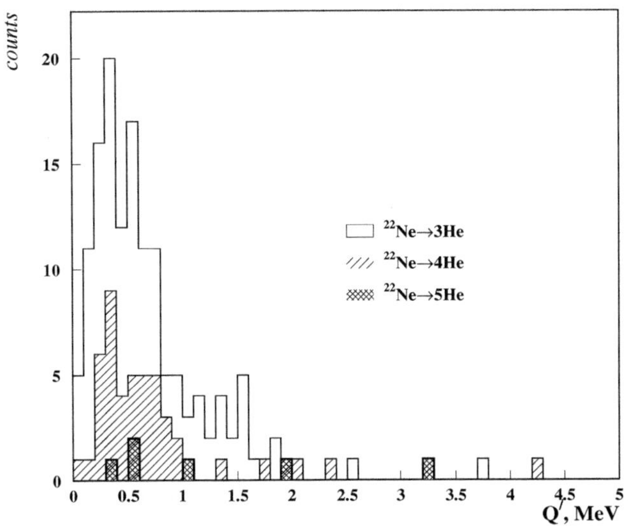

Fig. 3. Distribution of α-particle pairs *vs.* Q' (4) for the fragmentation modes $^{22}\mathrm{Ne} \to n\alpha$.

3 Fragmentation of ^{22}Ne nuclei

A nuclear state analogous to the dilute Bose gas can be revealed in the formation of $n\alpha$ particle ensembles possessing quantum coherence near the production threshold. The predicted property of these systems is a narrow velocity distribution in the cms [1]. Originating from relativistic nuclei, they can appear as narrow $n\alpha$ jets in the forward cone defined by the nucleonic Fermi motion. The determination of the cms for each event is rather ambiguous while analysis of jets in the b_{ik} space enables one to explore $n\alpha$ particle systems in a universal way.

At our disposal there are data on 4100 events from 3.3 A GeV ^{22}Ne nucleus interactions with emulsion nuclei (presented in [2]) which contain the classification of secondary tracks by ionization and angles. The key feature for the ^{22}Ne fragmentation consists in a suppression of binary splittings into medium charge fragments with respect to He and H cluster formation. The increase of a fragmentation degree is revealed in a growth of the α particle multiplicity. Thus, transition to the $n\alpha$ particle states having a high level density predominates over the binary splittings occurring at lower energy thresholds.

In the present analysis, the doubly charged particles found in a forward 6° cone were classified as relativistic α particles. Figure 2 shows the b_{ik} distribution (1) for the fragmentation channel $^{22}\mathrm{Ne} \to n\alpha$ for n equal to 3 (240 events), 4 (79 events), and 5 (10 events) which is rather narrow. The distribution "tails" appear to be due to the ^{4}He diffractive scattering or ^{3}He formation proceeding at higher momentum transfers. The events, satisfying the non-relativistic criterion $b_{ik} < 10^{-2}$ for each α-particle pair were selected for n equal to 3 (141 events), 4 (47 events), and 5 (6 events).

Figure 3 presents the normalized Q' distribution (4) for them. Being considered as estimates of the mean kinetic energy per nucleon in the center-of-mass of the $n\alpha$ system, the Q values does not exceed the Coulomb barrier values. Thus, in spite of the high $n\alpha$ multiplicity, the $n\alpha$ jets are seen to remain rather "cold" and consimilar. Among 10 $^{22}\mathrm{Ne} \to 5\alpha$ events there were found 3 "white" stars. Of these, in 2 "golden" events α particle tracks are contained even within a 1° cone. For these two events the value of Q' is estimated to be as low as 400 and 600 keV per nucleon. The detection of such "ultracold" 5α states is a serious argument in favor of the reality of the phase transition of α clusterized nuclei to the dilute Bose gas of α-particles. It gives a special motivation to explore lighter $n\alpha$ systems produced as potential "building blocks" of the dilute α-particle Bose gas.

Table 1. Charge-topology distribution of the "white" stars originated from the dissociation of 2.1 A GeV ^{14}N nuclei.

Z_f	6	5	5	4	3	3	–	–
N_1	1	–	2	1	4	2	3	1
N_2	–	1	–	1	–	1	2	3
N_{ws}	6	2	3	1	1	1	1	10

4 Fragmentation of ^{14}N nuclei

We are presently engaged in accumulating statistics on the interactions of 2.1 A GeV ^{14}N nuclei in emulsion impacted on "white" star searches. Twenty-five "white" stars have already been found among 540 inelastic events by scanning over primary tracks. Such a systematic scanning allows one to estimate relative probabilities of various fragmentation modes. The secondary tracks of "white" stars were selected to be concentrated in a forward 8° cone.

Table 1 shows the number of the found "white" stars N_{ws} composed of a single heavy fragment having charge Z_f and of N_1 singly and N_2 doubly charged fragments. The predominant role of the 4-prong mode He + H among

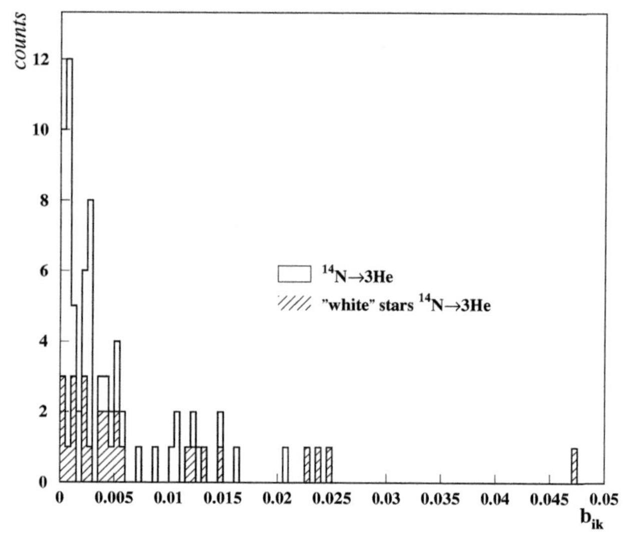

Fig. 4. Distribution of α-particle pairs *vs.* the relative variable b_{ik} (1) for the fragmentation mode $^{14}N \rightarrow 3\alpha$.

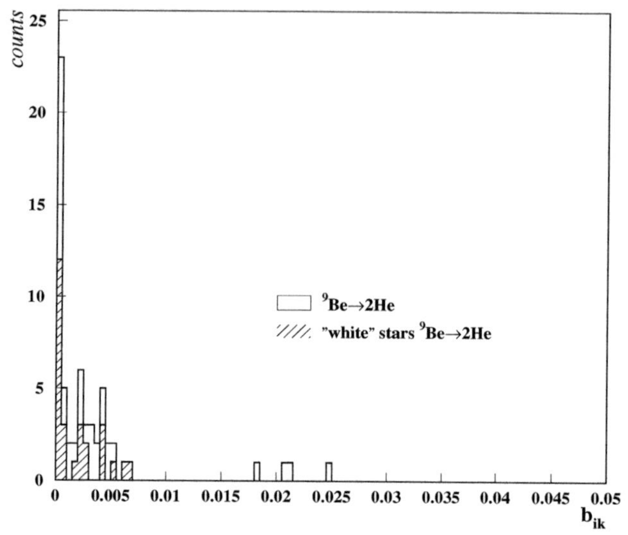

Fig. 6. Distribution of α-particle pairs *vs.* the relative variable b_{ik} (1) for the fragmentation mode $^{9}Be \rightarrow 2\alpha$.

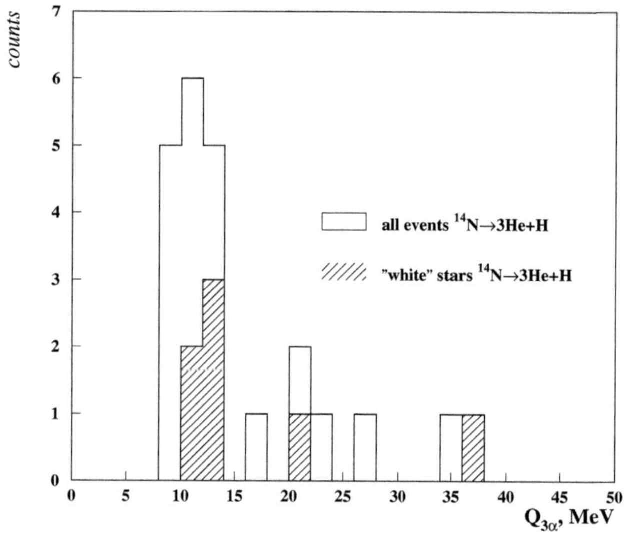

Fig. 5. Distribution of α-particle triplets *vs.* $Q_{3\alpha}$ (3) for the fragmentation mode $^{14}N \rightarrow 3\alpha + H$.

the "white" stars is clearly seen. It implies that the exploration of the 3α systems originated in ^{14}N fragmentation is rather promising.

Figure 4 allows one to compare the b_{ik} distribution for the "white" 3α stars to the case of 3He + H events, where a prohibition on a target fragmentation is lifted of. In both cases, the criterion of a non-relativistic character of fragment interactions is satisfied.

Figure 5 shows the $Q_{3\alpha}$ distribution in which the excitation energy is counted out from the ^{12}C nucleus mass. It can be concluded that the major fraction of entries is concentrated within a range of 10 to 14 MeV covering the known ^{12}C nucleus levels. Softening of the selection conditions for 3He + H events, under which the target fragment formation is allowable, does not result in a shift of the position of the 3α excitation peak. This circumstance points out the universality of the mechanism of population of 3α particle states. Besides, one can readily estimate that the normalized values $Q'_{3\alpha}$ are of the same magnitude as in fig. 3. Another conclusion is that the contribution of the α-^{8}Be state in the 3α configuration does not exceed a 10% level. This topic awaits for higher statistics allowing a reliable ^{8}Be identification.

5 Fragmentation of ^{9}Be nuclei

The relativistic ^{9}Be nucleus fragmentation is an attractive source for ^{8}Be generation due to the absence of a combinatorial background. The ^{8}Be nucleus reveals itself in the formation of α-particle pairs having an extremely small opening angle of the order of a few 10^{-3} rad in the range of a few GeV. The estimation of the ^{8}Be production probability will make it possible to clear up the interrelation between n-^{8}Be and α-n-α excitation modes which are important in understanding the ^{9}Be structure and fragmentation of heavier nuclei.

A secondary ^{9}Be beam was produced through fragmentation of the primary 1.2 *A* GeV ^{10}B beam. In scanning emulsion layers exposed to ^{9}Be nuclei, about 200 interactions are detected with He pair produced in a forward 8° cone. As in the previously considered cases, the b_{ik} distribution for 50 measured events, which is shown in fig. 6, confirms the non-relativistic behavior of the relative motion of the produced α-particles. In just the same way as in the case of the ^{14}N nuclei, softening of the criterion of selection of the 2He pairs does not lead to changes of the distribution shape.

Figure 7 shows the $Q_{2\alpha}$ distribution (3) allowing one to estimate the excitation scale. There is an event concentration below 1 MeV which is relevant for the ^{8}Be ground-state decay. Besides, one can resolve a bump at around

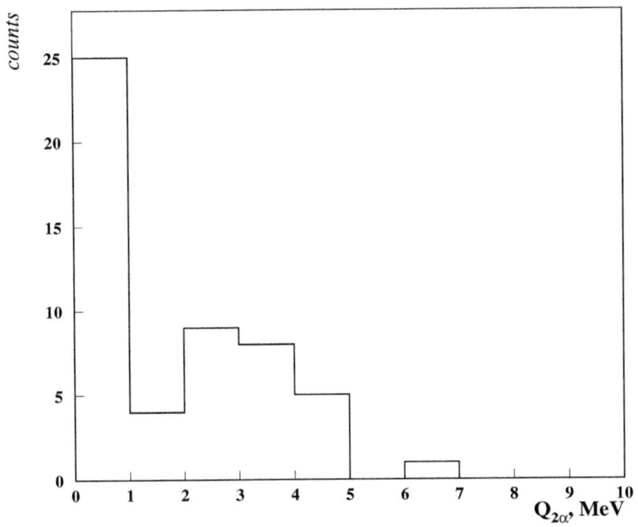

Fig. 7. Distribution of α-particle pairs *vs.* $Q_{2\alpha}$ (3) for the fragmentation mode ^{9}Be $\rightarrow 2\alpha$.

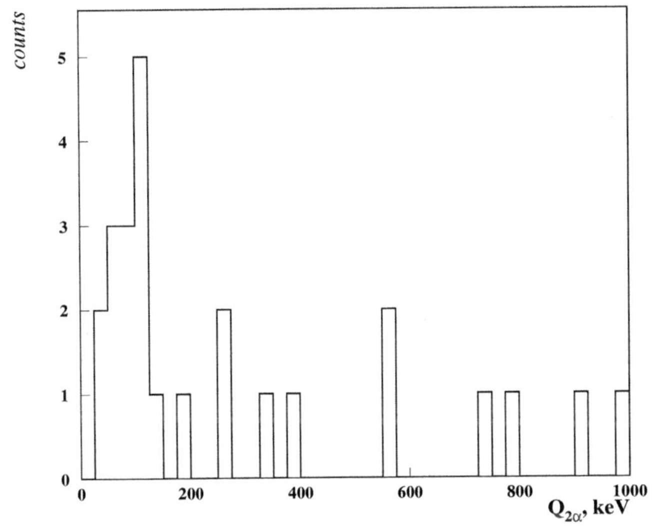

Fig. 8. Distribution of α-particle pairs *vs.* $Q_{2\alpha}$ (3) for the fragmentation mode ^{9}Be $\rightarrow 2\alpha$ zoomed between 0–1000 keV.

3 MeV corresponding to the ^{8}Be decay from the first excited state 2^+. A zoomed part of this distribution near zero is presented in fig. 8. A clear peak is seen as a concentration of 14 events around the mean value $Q_{2\alpha}$ equal to 88 keV which is close to the decay energy of the ^{8}Be ground state. Thus, the achieved identification of ^{8}Be production allows one to justify the spectroscopy of $n\alpha$ decays from the lowest decay energy.

6 Charged topology of ^{8}B fragmentation

In order to expose an emulsion to ^{8}B nuclei, use was made of the fragmentation of 1.2 *A* GeV ^{10}B nuclei. In this case, the absence of ^{9}B nuclei among secondary fragments

Table 2. Charge-topology distribution of the "white" stars originated from the dissociation of 2.1 *A* GeV ^{8}B nuclei.

Z_f	4	3	–	–	–
N_2	–	–	2	1	–
N_1	1	2	1	3	5
N_{ws}	15	1	9	11	3

turned out to be good luck, resulting in a clear separation of the primary and secondary beams by their magnetic rigidity. When scanning emulsions, this fact was indirectly confirmed by the absence of "white" stars with a charge topology He + H. They could be produced by background ^{6}Li having the same magnetic rigidity as ^{10}B nuclei. A 15% admixture of ^{7}Be nuclei was eliminated according to the charge topology of the found "white" stars. The most intensive background presented by ^{3}He nuclei was visually separated.

By scanning over the incoming tracks, a total of 39 "white" stars, in which the charge in the 15°-cone is equal to 5, have been found. Their distribution by the charge modes is shown in table 2 in the same manner as in table 1. The significance of the ^{8}B modes can be compared with the topology of "white" stars produced by the 1 *A* GeV ^{10}B nuclei [13]. The fraction of the 3-prong stars ^{10}B $\rightarrow$ 2He + ^{1,2}H was established to be equal to 80% with 40% deuteron clustering. The probability of the 2-prong mode ^{10}B $\rightarrow$ ^{9}Be + ^{1}H was found to be equal only to 3%. Such a strong difference can be explained by the lower value of the ^{2}H binding energy.

As is shown in table 2, an obvious distinction of the ^{8}B case consists in a high yield of the 2-prong mode ^{8}B $\rightarrow$ ^{7}Be+^{1}H. This feature is due to the weak ^{1}H binding. Thus, one can conclude that the loosely bound ^{8}B nucleus manifests its structure already in the charge-topology. Further it is planned to increase statistics, identify the H and He isotopes, and reconstruct emission angles.

Obtaining ^{10,11}B beams at the JINR Nuclotron makes it possible to form ^{10,11}C secondary beams by the use of charge-exchange reactions analogous to the ^{7}Li $\rightarrow$ ^{7}Be process [15]. This method is optimal for the emulsion technique where the simplicity of identification of incoming nuclei rather than their intensity is of importance. The existence and the cross sections of such processes will be established in a separate experiment. The suggested emulsion exposures will allow one to explore the 3-prong modes ^{10,11}C $\rightarrow$ 3He analogous to the case ^{12}C $\rightarrow 3\alpha$ [7]. Clustering in ^{12}C $\rightarrow 3\alpha$ reflects the ternary α process. Study of the 3He clustering in ^{10,11}C fragmentation would serve as a basis for studying the possible role of the 3He fusion process in nucleosynthesis, that is, in media with a mixed composition of He isotopes.

7 Charged topology of ^{9}C fragmentation

Unfortunately, it is impossible to use the approach based on a charge-exchange reaction for the formation of a ^{9}C nucleus beam. An emulsion was exposed to a secondary

Table 3. Charge-topology distribution of the "white" stars originated from the dissociation of 2.1 A GeV ^{9}C nuclei.

Z_f	5	4	–	–	–
N_1	1	2	–	2	4
N_2	–	–	3	2	1
N_{ws}	1	2	3	7	5

beam produced in fragmentation of the 2.1 A GeV ^{12}C nuclei and having a ^{9}C magnetic rigidity. A search was made for the events in which the total charge of tracks in a forward 15° cone is equal to 6. The presently found 17 events are distributed as shown in table 3. The 3He mode, which dominates the ^{12}C "white" stars, is seen to be suppressed. Besides, there is an indication that multiple dissociation channels predominate over possible candidates to the p+^{8}B and p+p+^{7}Be modes. Verification of this observation makes a farther increase in statistics to be particularly intriguing.

The cases, which might be interpreted as 3^{3}He, are of especial importance, since they point to highly-lying cluster excitations associated with a strong nucleon rearrangement to produce the 3^{3}He system. Like the process $^{12}\mathrm{C} \to 3\alpha$, this dissociation can be considered as a visible reflection of the inverse process of a ternary N_{ws}He fusion. It can provide a significantly higher energy output followed by ^{4}He pair production. Search for a ternary ^{3}He process appears to be a major goal of further accumulation of statistics.

8 Conclusions

The invariant approach is applied to analyzing the relativistic fragmentation of ^{22}Ne, ^{14}N and ^{9}Be nuclei having a significant difference in the primary energy. It is shown that doubly charged fragments having relative b_{ik} within the range $b_{ik} < 10^{-2}$ form well-separated $n\alpha$ jets. It corresponds to the relative motion of α-particles with relative kinetic energy of the order of 1 MeV per nucleon in the jet center-of-mass system.

New experimental observations are reported from the emulsion exposures to ^{14}N and ^{9}Be nuclei with energy above 1 A GeV. Being applied to the fragmentation of these nuclei the invariant analysis is shown to be a promising means to study excited states of simple α-particle systems. The internal energy of a system involving He fragments can be estimated in an invariant form down to the ^{8}Be nucleus decays.

The pattern of the relativistic fragmentation becomes more complete in the case of proton excess in the explored nucleus. It is shown that nuclear track emulsions provide unique possibilities to explore few-body decays of ^{8}B and ^{9,10,11}C nuclei. The paper describes the start of this work. The invariant approach applied for the stable nuclei will be of special benefit in the case of the neutron-deficient nuclei.

In spite of statistical restrictions, nuclear track emulsions ensure the initial stage of investigations in an unbiased way and enable one to develop scenarios for dedicated experiments [16]. Our experimental observations concerning few-body aspects of nuclear physics can be described in the relativistic invariant form allowing one to enlarge nuclear physics grounds of the nucleosynthesis pattern.

The work was supported by the Russian Foundation for Basic Research (Grants nos. 96-1596423, 02-02-164-12a, 03-02-16134, 03-02-17079, 04-02-16593, 04-02-17151), the Agency for Science of the Ministry for Education of the Slovak Republic and the Slovak Academy of Sciences (Grants VEGA 1/9036/02 and 1/2007/05) and grants from the JINR Plenipotentiaries of Bulgaria, Czech Republic, Slovak Republic, and Romania during 2002-5.

References

1. P. Schuck, H. Horiuchi, G. Roepke, A. Tohsaki, C.R. Phys. **4**, 537 (2003).
2. A. El-Naghy *et al.*, J. Phys. G **14**, 1125 (1988).
3. G. Baroni *et al.*, Nucl. Phys. A **516**, 673 (1990).
4. G. Baroni *et al.*, Nucl. Phys. A **540**, 646 (1992).
5. M.A. Jilany, Phys. Rev. **70**, 014901 (2004).
6. N.P. Andreeva *et al.*, Phys. At. Nucl. **68**, 455 (2005).
7. V.V. Belaga *et al.*, Phys. At. Nucl. **58**, 1905 (1995).
8. N.P. Andreeva *et al.*, Phys. At. Nucl. **59**, 106 (1996).
9. V.V. Glagolev *et al.*, Eur. Phys. J. A **11**, 285(2004).
10. F.G. Lepekhin *et al.*, Eur. Phys. J. A **1**, 137(1998).
11. M.I. Adamovich *et al.*, Phys. At. Nucl. **62**, 1378 (1999).
12. M.I. Adamovich *et al.*, J. Phys. G **30**, 1479 (2004).
13. M.I. Adamovich *et al.*, Phys. At. Nucl. **67**, 514 (2004).
14. V. Bradnova *et al.*, Nucl. Phys. A **734**, E92 (2004).
15. V. Bradnova *et al.*, Acta Phys. Slovaca **54**, 351 (2004).
16. Web site of the BECQUEREL Project, `http://becquerel.jinr.ru`.
17. A.M. Baldin, L.A. Didenko, Fortsch. Phys. **38**, 261 (1990).
18. A.I. Malakhov, Nucl. Phys. A **734**, 82, (2004).

Eur. Phys. J. A **27**, s01, 301–306 (2006)
DOI: 10.1140/epja/i2006-08-046-1

EPJ A direct
electronic only

Effects of the particle-number projection on the isovector pairing energy

N.H. Allal[1,2,a], M. Fellah[1,2], M.R. Oudih[1], and N. Benhamouda[1]

[1] Laboratoire de Physique Theorique, Faculte de Physique, USTHB, BP 32 El-Alia, 16111 Bab-Ezzouar, Alger, Algeria
[2] Centre de Recherche Nucleaire d'Alger, COMENA, BP 399 Alger-Gare, Alger, Algeria

Received: 7 June 2005 /
Published online: 13 March 2006 –

Abstract. The usual neutron-proton BCS wave function is simultaneously projected on both the good neutron and proton numbers using a discrete projection operator. The projected energy of the system is deduced as a limit of rapidly convergent sequence. It is numerically studied for the $N = Z$ nuclei of which "experimental" pairing gaps may be deduced from the experimental odd-even mass differences. It then appears that the particle-number fluctuation effect is even more important than in the case of pairing between like-particles.

PACS. 21.60.-n Nuclear structure models and methods – 21.30.Fe Forces in hadronic systems and effective interactions

1 Introduction

Renewed interest in the study of the neutron-proton (np) pairing correlations occurred recently (cf., *e.g.*, [1,2,3,4,5,6,7,8,9,10,11,12,13,14,15,16,17,18,19,20,21]) due to the development of the radioactive beam facilities that made possible the experimental study of medium mass nuclei such as $N \simeq Z$. The study of these proton-rich nuclei is not only interesting from the nuclear structure point of view, but it is also important in the astrophysical context. Indeed, it is believed that this kind of nuclei are synthesized in the rapid-proton capture process (rp-process) under astrophysical conditions [22,23]. Moreover, most of these nuclei are not accessible in experiments and therefore many nuclear astrophysics calculations crucially depend on accurate theoretical prediction for the nuclear properties. In $N \simeq Z$ nuclei, the neutron and proton Fermi levels are close to each other and therefore the np pairing correlations are expected to play a significant role in their structure.

On the other hand, one of the most exciting subjects of the modern nuclear physics is the double-beta decay, because one expects an answer to the question whether the neutrino is a Majorana or a Dirac particle (cf., *e.g.*, [24,25,26,27]). Indeed, neutrinoless double-beta decay proceeds only when neutrinos are massive Majorana particles, hence its observation would resolve the question [28]. It has been shown that the inclusion of the np pairing influences the neutrinoless double-beta decay rates significantly [9].

Most often, pairing is treated within a generalized BCS treatment [1,2,3,4,5,6,7,8]. However, the shortcomings of this approach are well known: the most important of them is the number symmetry breaking that may imply serious errors in the evaluation of several physical observables. The usual techniques used in order to remedy this shortcoming are those already used for the study of pairing correlations between like-particles, *i.e.* the Quasiparticle Random Phase Approximation (QRPA) (cf., *e.g.*, [9,10,11,12,13,14,15,16,17]), the Lipkin-Nogami (LN) method [18] or the Generator Coordinate Method (GCM) [19]. However, in these methods, the particle-number symmetry is only approximately restored. Recently, an exact particle-number projection method in the isovector case has been proposed [29]. It is based on a discrete form of the projection operator that allows one to derive an explicit form of the projected wave-function. The method was applied within the Richardson schematic model [30]. The aim of the present work is to study numerically the particle-number fluctuations effects on the energy in a realistic case, *i.e.* for the $N = Z$ nuclei of which the "experimental" pairing gaps Δ_p, Δ_n and Δ_{np} may be deduced from the experimental odd-even mass differences [3]. The single-particle energies used are those of a Woods-Saxon deformed mean field. The structure of this paper is as follows: the BCS theory is briefly recalled in sect. 2. The projection method is presented in sect. 3. The numerical results are given and discussed in sect. 4.

[a] e-mail: allaln@yahoo.com

2 BCS theory

In the second quantization and isotopic spin formalism, a system of mass A constituted by N neutrons and Z protons is described by the Hamiltonian [1]

$$\hat{H}=\sum_{\nu t}\varepsilon_{\nu t}a^+_{\nu t}a_{\nu t}-\frac{1}{4}\sum_{\nu\mu>0,t_1t_2}G_{t_1t_2}a^+_{\nu t_1}a^+_{\tilde{\nu}t_2}a_{\tilde{\mu}t_2}a_{\mu t_1}\,, \quad (1)$$

where:

- t is the subscript corresponding to the isotopic spin component ($t=n,p$).
- $a^+_{\nu t}$ and $a_{\nu t}$ respectively represent the creation and annihilation operators of the particle in the state $|\nu t\rangle$, of energy $\varepsilon_{\nu t}$; and $a^+_{\tilde{\nu}t}$ and $a_{\tilde{\nu}t}$ those of its time-reverse $|\tilde{\nu}t\rangle$, that has the same energy.
- $G_{t_1t_2}$ characterizes the pairing strength and is supposed independent of the levels.

The neutrons and protons are supposed to occupy the same energy levels. The standard procedure is to use the generalized Bogoliubov-Valatin transformation approach [1,2,3,8,31] where the quasi-particle operators are given by

$$\begin{pmatrix}\alpha^+_{\nu 1}\\ \alpha^+_{\nu 2}\\ \alpha_{\tilde{\nu}1}\\ \alpha_{\tilde{\nu}2}\end{pmatrix}=\begin{pmatrix}U_{\nu 1p} & U_{\nu 1n} & V_{\nu 1p} & V_{\nu 1n}\\ U_{\nu 2p} & U_{\nu 2n} & V_{\nu 2p} & V_{\nu 2n}\\ -V_{\nu 1p} & -V_{\nu 1n} & U_{\nu 1p} & U_{\nu 1n}\\ -V_{\nu 2p} & -V_{\nu 2n} & U_{\nu 2p} & U_{\nu 2n}\end{pmatrix}\begin{pmatrix}a^+_{\nu p}\\ a^+_{\nu n}\\ a_{\tilde{\nu}p}\\ a_{\tilde{\nu}n}\end{pmatrix}. \quad (2)$$

The indices 1, 2 reveal the existence of two kinds of quasi-particles.

The BCS state $|\Psi\rangle$ is obtained by eliminating all the quasiparticles in the actual vacuum, *i.e.*

$$|\Psi\rangle\propto\prod_{\nu>0}\alpha_{\nu 1}\alpha_{\tilde{\nu}1}\alpha_{\nu 2}\alpha_{\tilde{\nu}2}|0\rangle\,. \quad (3)$$

Using the transformation (2), one obtains, after normalization,

$$|\Psi\rangle=\prod_{\nu>0}|\Psi_\nu\rangle \quad (4)$$

with

$$|\Psi_\nu\rangle=\Big[B^\nu_1a^+_{\tilde{\nu}p}a^+_{\nu p}a^+_{\tilde{\nu}n}a^+_{\nu n}+B^\nu_2a^+_{\tilde{\nu}p}a^+_{\nu p}+B^\nu_3a^+_{\tilde{\nu}n}a^+_{\nu n}$$
$$+B^\nu_4\left(a^+_{\tilde{\nu}p}a^+_{\nu n}+a^+_{\tilde{\nu}n}a^+_{\nu p}\right)+B^\nu_5\Big]|0\rangle\,, \quad (5)$$

and where the B^ν_i factors depend on the $U_{\nu\tau t}$ and $V_{\nu\tau t}$ ($t=n,p$; $\tau=1,2$) coefficients.

The pairing gap parameters are defined by

$$\Delta_{tt'}=\frac{G_{tt'}}{2}\sum_{\nu>0}\langle\Psi|a_{\tilde{\nu}t}a_{\nu t'}|\Psi\rangle\,,\qquad t,t'=n,p\,. \quad (6)$$

The particle-number fluctuation will be measured by the quantity

$$\Delta N_{BCS}=\frac{1}{A}\sqrt{\langle\Psi|\hat{N}^2|\Psi\rangle-\langle\Psi|\hat{N}|\Psi\rangle^2}\,, \quad (7)$$

where $\hat{N}$ is the particle-number operator.

The BCS energy is defined by $E^{np}_{BCS}=\langle\Psi|\hat{H}|\Psi\rangle$, that is

$$E^{np}_{BCS}=\sum_{\nu>0}E_\nu+\sum_{\nu\neq\mu}E_{\nu\mu}\,, \quad (8)$$

where E_ν and $E_{\nu\mu}$ are evaluated in the particle-number representation:

$$E_\nu=(B^\nu_1)^2\left[2\varepsilon_{\nu n}+2\varepsilon_{\nu p}-\frac{G_{nn}}{4}-\frac{G_{pp}}{4}-\frac{G_{np}}{2}\right]$$
$$+(B^\nu_2)^2\left[2\varepsilon_{\nu p}-\frac{G_{pp}}{4}\right]+(B^\nu_3)^2\left[2\varepsilon_{\nu n}-\frac{G_{nn}}{4}\right]$$
$$+2(B^\nu_4)^2\left[\varepsilon_{\nu n}+\varepsilon_{\nu p}-\frac{G_{np}}{4}\right] \quad (9)$$

and

$$E_{\nu\mu}=-\frac{G_{nn}}{4}(B^\nu_1B^\nu_2+B^\nu_3B^\nu_5)(B^\mu_1B^\mu_2+B^\mu_3B^\mu_5)$$
$$-\frac{G_{pp}}{4}(B^\nu_1B^\nu_3+B^\nu_2B^\nu_5)(B^\mu_1B^\mu_3+B^\mu_2B^\mu_5)$$
$$-\frac{G_{np}}{2}B^\nu_4(B^\nu_1-B^\nu_5)B^\mu_4(B^\mu_1-B^\mu_5)\,. \quad (10)$$

3 Projection

3.1 Projected wave function

In the case of pairing between identical particles, the projector that allows one to obtain the state of an even system with $2P_t$ particles (where P_t refers to the neutron or proton pair number), from the BCS vacuum, is given by [32]

$$\hat{\wp}_t=\frac{1}{2(n+1)}\left\{\sum_{k=0}^{n+1}\xi_kz_k^{-P_t}\prod_\nu\left[1+(\sqrt{z_k}-1)\,a^+_{\nu t}a_{\nu t}\right]+\text{c.c.}\right\}, \quad (11)$$

where

$$\xi_k=\begin{cases}1/2, & \text{if } k=0 \text{ or } k=n+1,\\ 1, & \text{if } 1\le k\le n.\end{cases} \quad (12)$$

$z_k=\exp(i\frac{k\pi}{n+1})$, n is a non-zero integer and c.c. means the complex conjugate with respect to z_k. As has been shown in refs. [33] and [34], as soon as the inequality $2(n+1)>\max(P_t,\Omega-P_t)$ is satisfied, the projected state coincides with the P_t pairs of the particles component.

In the np pairing case, the projector that allows one to obtain the state that has both the good proton and neutron numbers (and hence the good isospin) is then of the form

$$\hat{\wp}=\prod_{t=n,p}\hat{\wp}_t. \quad (13)$$

The projected wave function is then

$$|\Psi_{nn'}\rangle=C_{nn'}\hat{\wp}|\Psi\rangle\,, \quad (14)$$

where $C_{nn'}$ is a normalization factor. That is

$$|\Psi_{nn'}\rangle = C_{nn'}\left[\sum_{k=0}^{n+1}\sum_{k'=0}^{n'+1}\xi_k\xi_{k'}\left(z_k^{-P_n}z_{k'}^{-P_p}|\Psi(z_k,z_{k'})\rangle + \overline{z}_k^{-P_n}z_{k'}^{-P_p}|\Psi(\overline{z}_k,z_{k'})\rangle + \text{c.c.}\right)\right], \quad (15)$$

where $\overline{z}_k$ is the complex conjugate of z_k, and

$$|\Psi(z_k,z_{k'})\rangle = \prod_{\nu>0}|\Psi_\nu(z_k,z_{k'})\rangle \quad (16)$$

with

$$|\Psi_\nu(z_k,z_{k'})\rangle = \Big[B_1^\nu z_k z_{k'} a_{\tilde{\nu}p}^+ a_{\nu p}^+ a_{\tilde{\nu}n}^+ a_{\nu n}^+ + B_2^\nu z_{k'} a_{\tilde{\nu}p}^+ a_{\nu p}^+ + B_3^\nu z_k a_{\tilde{\nu}n}^+ a_{\nu n}^+ + B_4^\nu\sqrt{z_k z_{k'}}\left(a_{\tilde{\nu}p}^+ a_{\nu n}^+ + a_{\tilde{\nu}n}^+ a_{\nu p}^+\right) + B_5^\nu\Big]|0\rangle. \quad (17)$$

It appears that the expressions (5) and (17) are formally similar. The effect of the projection in these expressions consists in a renormalization of the B_i^ν factors.

The integers n and n', respectively, measure the extraction degree of the neutron and proton false components of $|\Psi\rangle$. As soon as the condition

$$\begin{aligned} 2(n+1) &> \mathrm{Max}(P_n, \Omega - P_n),\\ 2(n'+1) &> \mathrm{Max}(P_p, \Omega - P_p), \end{aligned} \quad (18)$$

where 2Ω is the total degeneracy of pairs, is satisfied, the state (15) coincides with the physical component (*i.e.* with N neutrons and Z protons).

We have thus obtained an explicit expression of the projected wave function. The latter allows one to deduce the expectation value of any physical observable.

The particle-number fluctuations will be measured by

$$\Delta N_{nn'} = \frac{1}{A}\sqrt{\langle\Psi_{nn'}|\hat{N}^2|\Psi_{nn'}\rangle - \langle\Psi_{nn'}|\hat{N}|\Psi_{nn'}\rangle^2}. \quad (19)$$

3.2 Energy

Any operator $\hat{O}$ that conserves the particle-number symmetry satisfies the property

$$\langle\Psi_{nn'}|\hat{O}|\Psi_{nn'}\rangle = 4(n+1)(n'+1)C_{nn'}\langle\Psi_{nn'}|\hat{O}|\Psi\rangle. \quad (20)$$

Using (20) the energy of the system is then given by

$$E_{nn'} = 4(n+1)(n'+1)C_{nn'}^2 \times\left[\sum_{k=0}^{n+1}\sum_{k'=0}^{n'+1}\xi_k\xi_{k'}\left(z_k^{-P_n}z_{k'}^{-P_p}E(z_k,z_{k'}) + \overline{z}_k^{-P_n}z_{k'}^{-P_p}E(\overline{z}_k,z_{k'}) + \text{c.c.}\right)\right], \quad (21)$$

where

$$E(z_k,z_{k'}) = \sum_{\nu>0}E_\nu(z_k,z_{k'})\prod_{j\neq\nu}A_j(z_k,z_{k'}) + \sum_{\nu\neq\mu}E_{\nu\mu}(z_k,z_{k'})\prod_{j\neq\nu\mu}A_j(z_k,z_{k'}) \quad (22)$$

with

$$A_j(z_k,z_{k'}) = \langle\Psi_j(z_k,z_{k'})|\Psi_j\rangle, \quad (23)$$

$$\begin{aligned} E_\nu(z_k,z_{k'}) = {}& z_k z_{k'}(B_1^\nu)^2\left(2\varepsilon_{\nu n} + 2\varepsilon_{\nu p} - \frac{G_{nn}}{4} - \frac{G_{pp}}{4} - \frac{G_{np}}{2}\right)\\ &+ z_{k'}(B_2^\nu)^2\left(2\varepsilon_{\nu p} - \frac{G_{pp}}{4}\right) + z_k(B_3^\nu)^2\left(2\varepsilon_{\nu n} - \frac{G_{nn}}{4}\right)\\ &+ 2\sqrt{z_k z_{k'}}(B_4^\nu)^2\left(\varepsilon_{\nu n} + \varepsilon_{\nu p} - \frac{G_{np}}{4}\right) \end{aligned} \quad (24)$$

and

$$\begin{aligned} E_{\nu\mu}(z_k,z_{k'}) = {}& -\frac{G_{nn}}{4}z_k\left(z_{k'}B_1^\nu B_2^\nu + B_3^\nu B_5^\nu\right)\left(z_{k'}B_1^\mu B_2^\mu + B_3^\mu B_5^\mu\right)\\ &- \frac{G_{pp}}{4}z_{k'}\left(z_k B_1^\nu B_3^\nu + B_2^\nu B_5^\nu\right)\left(z_k B_1^\mu B_3^\mu + B_2^\mu B_5^\mu\right)\\ &- \frac{G_{np}}{2}\sqrt{z_k z_{k'}}B_4^\nu B_4^\mu\left(\sqrt{z_k z_{k'}}B_1^\nu - B_5^\nu\right)\\ &\times\left(\sqrt{z_k z_{k'}}B_1^\mu - B_5^\mu\right). \end{aligned} \quad (25)$$

Here again, the expressions (9) (respectively (10)) and (24) (respectively (25)) are formally similar and the particle-number projection leads to a renormalization of the B_i^ν factors.

4 Numerical results. Discussion

The previously described method is applied within the framework of a Woods-Saxon deformed mean field with a maximum number of major shells $N_{\max} = 12$. The ground-state deformation parameters used in the present work are those of Moller and Nix [35].

4.1 Pairing strength choice

The choice of the G_{np} pairing strength is still an open question. Several propositions have been formulated, let us cite among others:

- Chen and Goswami [36], that have arbitrarily chosen the form
$$G_{np} = G_{nn} + \frac{6}{A}.$$
- Satula and Wyss [18], for their part, justify their choice,
$$G_{np} = \frac{1}{2}(G_{nn} + G_{pp}),$$
by assuming, using arguments based on the spin invariance, that, on the $N \simeq Z$ line, $G_{nn(pp)} \simeq G_{np}$.

Table 1. Variation of the overlap $\langle\Psi|\Psi_{nn'}\rangle$, the particle-number fluctuations $\Delta N_{nn'}$ and the projected energy $E_{nn'}$ (MeV) as a function of the extraction degrees of the false components n and n' for the nucleus ^{36}Ar.

n	n'	$\langle\Psi\|\Psi_{nn'}\rangle$ (%)	$\Delta N_{nn'}$	$E_{nn'}$(MeV)
0	0	50	0.09	−649.12
0	1	41	0.06	−653.95
0	2	41	0.06	−654.36
0	3	41	0.06	−654.36
0	4	41	0.06	−654.36
0	5	41	0.06	−654.36
0	6	41	0.06	−654.36
0	7	41	0.06	−654.36
1	0	42	0.06	−654.50
1	1	34	0.02	−659.47
1	2	34	0.01	−659.89
1	3	34	0.01	−659.89
1	4	34	0.01	−659.89
1	5	34	0.01	−659.89
1	6	34	0.01	−659.89
1	7	34	0.01	−659.89
2	0	41	0.06	−654.70
2	1	34	0.02	−659.68
2	2	34	0.00	−660.09
2	3	34	0.00	−660.10
2	4	34	0.00	−660.10
2	5	34	0.00	−660.10
2	6	34	0.00	−660.10
2	7	34	0.00	−660.10
3	0	41	0.06	−654.70
3	1	34	0.02	−659.68
3	2	34	0.00	−660.09
3	3	34	0.00	−660.10
3	4	34	0.00	−660.10
3	5	34	0.00	−660.10
3	6	34	0.00	−660.10
3	7	34	0.00	−660.10

n	n'	$\langle\Psi\|\Psi_{nn'}\rangle$ (%)	$\Delta N_{nn'}$	$E_{nn'}$(MeV)
4	0	41	0.06	−654.70
4	1	34	0.02	−659.68
4	2	34	0.00	−660.09
4	3	34	0.00	−660.10
4	4	34	0.00	−660.10
4	5	34	0.00	−660.10
4	6	34	0.00	−660.10
4	7	34	0.00	−660.10
5	0	41	0.06	−654.70
5	1	34	0.02	−659.68
5	2	34	0.00	−660.09
5	3	34	0.00	−660.09
5	4	34	0.00	−660.09
5	5	34	0.00	−660.10
5	6	34	0.00	−660.10
5	7	34	0.00	−660.10
6	0	41	0.06	−654.70
6	1	34	0.02	−659.68
6	2	34	0.00	−660.09
6	3	34	0.00	−660.09
6	4	34	0.00	−660.09
6	5	34	0.00	−660.09
6	6	34	0.00	−660.09
6	7	34	0.00	−660.09
7	0	41	0.06	−654.70
7	1	34	0.02	−659.68
7	2	34	0.00	−660.09
7	3	34	0.00	−660.09
7	4	34	0.00	−660.09
7	5	34	0.00	−660.09
7	6	34	0.00	−660.09
7	7	34	0.00	−660.09

– Chasman [20] sets for the same reason

$$G_{np} = \frac{1}{2}G_{nn} = \frac{1}{2}G_{pp}\,.$$

– Civitarese *et al.* propose either the form [1]

$$G_{np} = \frac{C(Z)}{A}\,,$$

where $C(Z)$ is a constant that varies as a function of the considered element, or the form [2]

$$G_{np} = C\frac{16}{A+56}\,,$$

where C is a constant.

The latter expression has also been used by Szpikowski [37].

In the present work, we restrict ourselves to $N = Z$ even-even nuclei of which the "experimental" pairing gap may be deduced from the experimental masses [3] (*i.e.* with $18 \leq Z \leq 32$). The pairing strengths G_{pp}, G_{nn} and G_{np} are then chosen such as to exactly reproduce the "experimental" values of Δ_{pp}, Δ_{nn} and Δ_{np}. For the study of $N = Z$ nuclei of higher masses (and of which "experimental" pairing gaps are not known), one should first have to establish by a fit an expression of the pairing strengths from the previously obtained values, and then to extrapolate this expression.

4.2 Convergence of the method

We have reported in table 1, the overlap $\langle\Psi|\Psi_{nn'}\rangle$, the particle-number fluctuations $\Delta N_{nn'}$ as well as the projected energy $E_{nn'}$(MeV) values as a function of the extraction degrees of the false components n and n' ($0 \leq n, n' \leq 7$) for the nucleus ^{36}Ar chosen as an example. The values obtained using the BCS theory are respectively $\Delta N_{BCS} = 0.09$ and $E_{BCS} = -648.78$ MeV. From table 1, one can conclude that

i) The convergence of the method is very fast. Indeed, following expression (18), the convergence should be theoretically reached as soon as $n, n' > 218$, since $\Omega = 455$. But it clearly appears that the results are stable for $n, n' > 6$ for all the evaluated quantities. This fact proves the efficiency of the projection method.

Table 2. Overlap between the BCS and projected wave functions, without inclusion of the np pairing correlations for the proton (first row) and neutron (second row) systems; and with inclusion of the np pairing correlations (third row). Difference between the projected and BCS energies (MeV) without (fourth row) and with (fifth row) inclusion of the np pairing correlations.

Nucleus	^{32}S	^{36}Ar	^{40}Ca	^{44}Ti	^{48}Cr	^{52}Fe	^{56}Ni	^{60}Zn	^{64}Ge
$\langle\Psi\|\Psi_n\rangle_p$ (%)	59	58	62	56	56	58	56	56	55
$\langle\Psi\|\Psi_n\rangle_n$ (%)	59	59	63	57	57	59	56	57	54
$\langle\Psi\|\Psi_{nn'}\rangle$ (%)	36	34	39	34	32	35	32	35	31
δE (MeV)	−4.17	−4.86	−7.28	−4.18	−4.12	−4.83	−4.40	−3.50	−4.26
δE^{np} (MeV)	−9.59	−11.31	−17.72	−11.57	−8.72	−11.00	−8.63	−7.84	−9.13

ii) The particle-number projection is indispensable since the overlap $\langle\Psi|\Psi_{nn'}\rangle$ between the BCS and projected wave function is only 34% (after convergence). It is thus very far from the theoretical overlap.
iii) The particle-number fluctuations are efficiently eliminated since the quantity ΔN that was 0.09 with the BCS theory vanishes after projection.
iv) The energy value is significantly reduced with regard to the BCS one (the difference is more than 11 MeV).

4.3 Effects of the particle-number fluctuations on the system energy

In order to evaluate the particle-number fluctuations effect on the system energy, we have calculated the difference between the projected and BCS energies,

$$\delta E^{np} = E_{nn'} - E_{BCS}^{np}, \tag{26}$$

for the ground state of the previously cited nuclei. The obtained values are compared in table 2 to those of the same quantity, when only the pairing between like-particles is taken into account, *i.e.*

$$\delta E = E_n - E_{BCS}, \tag{27}$$

where E_n is the energy evaluated using the SBCS particle-number projection method [33,34,38].

It then appears that the particle-number fluctuations effects are even more important when the np pairing is considered. Indeed, the δE value is on average −4.62 MeV, whereas that of δE^{np} is −10.61 MeV.

We have also reported in the same table the overlap between the BCS and projected wave functions without ($\langle\Psi|\Psi_n\rangle_p$ and $\langle\Psi|\Psi_n\rangle_n$ for the proton and neutron systems, respectively) and with ($\langle\Psi|\Psi_{nn'}\rangle$) inclusion of the np pairing correlations. Here again, it appears that the particle-number fluctuations are even more important when the np correlations are included. Indeed, the average value of $\langle\Psi|\Psi_n\rangle_p$ or $\langle\Psi|\Psi_n\rangle_n$ is 57%, whereas that of $\langle\Psi|\Psi_{nn'}\rangle$ is only 34%.

These facts show the necessity of both the inclusion of the np pairing correlations and the particle-number projection in $N \simeq Z$ nuclei. It will be the case in the astrophysical context and specially for the study of the rp-process. Indeed, as underlined in the introduction, this study is not only hindered by the uncertainty of the astrophysical conditions but also by the lack of experimental information on the nuclei along this process. The theoretical predictions have thus to be particularly rigorous.

Finally, it is worth noticing that the present particle-number projection method could be easily generalized to the excited states and hence used for the double-beta decay study.

References

1. O. Civitarese, M. Reboiro, Phys. Rev. C **56**, 1179 (1997).
2. O. Civitarese, M. Reboiro, P. Vogel, Phys. Rev. C **56**, 1840 (1997).
3. F. Simkovic, Ch.C. Moustakidis, L. Pacearescu, A. Faessler, Phys. Rev. C **68**, 054319 (2003).
4. J. Engel, S. Pittel, M. Stoitsov, P. Vogel, J. Dukelsky, Phys. Rev. C **55**, 1781 (1997).
5. A.L. Goodman, Phys. Rev. C **60**, 014311 (1999).
6. D.R. Bes, O. Civitarese, E.E. Moqueda, N.N. Scoccola, Phys. Rev. C **61**, 024315 (2000).
7. D. Mokhtari, N.H. Allal, M. Fellah, Heavy Ion Phys. **19**, 187 (2004).
8. A. Goodman, Adv. Nucl. Phys. **11**, 263 (1979).
9. G. Pantis, F. Simkovic, J.D. Vergados, A. Faessler, Phys. Rev. C **53**, 695 (1996).
10. J. Dobes, S. Pittel, Phys. Rev. C **57**, 688 (1998).
11. O. Civitarese, F. Montani, M. Reboiro, Phys. Rev. C **60**, 24305 (1999).
12. A. Bobyk, W.A. Kaminski, P. Zareba, Eur. Phys. J. A **5**, 385 (1999).
13. S.G. Frauendorf, J.A. Sheikh, Nucl. Phys. A **645**, 509 (1999).
14. D.S. Delion, J. Dukelsky, P. Schuck, E.J.V. de Passos, F. Krmpotic, Phys. Rev. C **62**, 44311 (2000).
15. A.A. Raduta, L. Pacearescu, V. Baran, P. Sarriguren, E. Moya de Guerra, Nucl. Phys. A **675**, 503 (2000).
16. P. Sarriguren, E. Moya de Guerra, R. Alvarez-Rodriguez, Nucl. Phys. A **716**, 230 (2003).
17. N. Paar, T. Niksic, D. Vretenar, P. Ring, Phys. Rev. C **69**, 054303 (2004).
18. W. Satula, R. Wyss, Nucl. Phys. A **676**, 120 (2000).
19. M. Kyotoku, H.T. Chen, Phys. Rev. C **36**, 1144 (1987).
20. R.R. Chasman, Phys. Lett. B **524**, 81 (2002).
21. Perlinska, S.G. Rohozinski, J. Dobaczewski, W. Nazarewicz, Phys. Rev. C **69**, 014316 (2004).
22. Y. Sun, Eur. Phy. J. A **20**, 133 (2004).
23. K. Kaneko, M. Hasegawa, Phys. Rev. C **72**, 031302 (2005).
24. A.A. Raduta, O. Haug, F. Simkovic, A. Faessler, J. Phys. G **27**, 2429 (2001).
25. J. Suhonen, M. Aunola, Nucl. Phys. A **723**, 271 (2003).

26. H.V. Klapdor-Kleingrothaus, A. Dietz, I.V. Krivosheina, Ch. Dorr, C. Tomei, Phys. Lett. B **578**, 54 (2004).
27. R. Arnold *et al.*, Phys. Rev. Lett. **95**, 182302 (2005).
28. N. Gupta, H.S. Mani, J. Phys. G **31**, 599 (2005).
29. N.H. Allal, M. Fellah, M.R. Oudih, N. Benhamouda, *Exotic Nuclei, EXON2004, Proceedings of the International Symposium, Peterhof, 2004*, edited by Yu. E. Penionzhkevich, E.A. Cherepanov (World Scientific, Singapore, 2005).
30. R.W. Richardson, N. Sherman, Nucl. Phys. **52**, 253 (1964).
31. A. Goswami, Nucl. Phys. **60**, 228 (1964).
32. M.R. Oudih, M. Fellah, N.H. Allal, Int. J. Mod. Phys. E **12**, 109 (2003).
33. V.N. Fomenko, J. Phys. A **3**, 8 (1970).
34. M. Fellah, T.F. Hammann, Nuovo Cimento A **30**, 239 (1975).
35. P. Moller, J.R. Nix, W.D. Myers, W.J. Swiatecki, At. Data Nucl. Data Tables **59**, 185 (1995).
36. H. Chen, A. Goswami, Phys. Lett. B **24**, 257 (1967).
37. S. Szpikowski, Acta Phys. Pol. B **31**, 443 (2000).
38. M. Fellah, T.F. Hammann, D.E. Medjadi, Phys. Rev. C **8**, 1585 (1973).

7 Rare-Ion-Beam Facilities and Experiments

Eur. Phys. J. A **27**, s01, 309–314 (2006)
DOI: 10.1140/epja/i2006-08-047-0

EPJ A direct
electronic only

Study of the N = 28 shell closure in the Ar isotopic chain

A SPIRAL experiment for nuclear astrophysics

L. Gaudefroy[1,a], O. Sorlin[2], D. Beaumel[1], Y. Blumenfeld[1], Z. Dombrádi[3], S. Fortier[1], S. Franchoo[1], M. Gélin[2], J. Gibelin[1], S. Grévy[2], F. Hammache[1], F. Ibrahim[1], K. Kemper[4], K.L. Kratz[5], S.M. Lukyanov[6], C. Monrozeau[1], L. Nalpas[7], F. Nowacki[8], A.N. Ostrowski[5], Yu.-E. Penionzhkevich[6], E. Pollacco[7], P. Roussel-Chomaz[2], E. Rich[1], J.A. Scarpaci[1], M.G. St. Laurent[2], T. Rauscher[9], D. Sohler[3], M. Stanoiu[1], E. Tryggestad[1], and D. Verney[1]

[1] IPN, IN2P3-CNRS, F-91406 Orsay Cedex, France
[2] GANIL, BP 55027, F-14076 Caen Cedex 5, France
[3] Institute of Nuclear Research, H-4001 Debrecen, Pf. 51, Hungary
[4] Department of Physics, Florida State University, Tallahassee, FL 32306, USA
[5] Institut für Kernchemie, Universität Mainz, D-55128 Mainz, Germany
[6] FLNR/JINR, 141980 Dubna, Moscow Region, Russia
[7] CEA-Saclay, DAPNIA-SPhN, F-91191 Gif-sur-Yvette Cedex, France
[8] IReS, Univiversité Louis Pasteur, BP 28, F-67037 Strasbourg Cedex, France
[9] Departement für Physik und Astronomie, Universität Basel, Switzerland

Received: 17 June 2005 /
Published online: 15 March 2006 –

Abstract. The structure of the neutron-rich nucleus ^{47}Ar has been investigated through the d(^{46}Ar,^{47}Ar)p transfer reaction. The radioactive beam of ^{46}Ar at $10\,A \cdot$MeV was provided by the SPIRAL facility at GANIL. The protons corresponding to a neutron pick-up on bound or unbound states mechanism in ^{47}Ar nuclei were detected at backward angles by the position-sensitive Si array-detector MUST. The transfer-like ejectiles were detected in the SPEG spectrometer. Level scheme, spin assignments and spectroscopic factors have been deduced for ^{47}Ar and compared to shell model predictions. They suggest a slight erosion of the $N = 28$ shell gap from the weakening of the spin-orbit interaction arising from the f and p orbitals. These spectroscopic information are subsequently used to infer (n, γ) reaction rates in the Ar isotopic chain to understand the origin of the ^{48}Ca/^{46}Ca abnormal isotopic ratio observed in certain inclusions of meteorites.

PACS. 21.10.Hw Spin, parity, and isobaric spin – 21.10.Jx Spectroscopic factors – 25.60.Je Transfer reactions – 26.30.+k Nucleosynthesis in novae, supernovae and other explosive environments

1 Introduction

The role of the spin-orbit interaction is essential to create the magic numbers as $N = 28$, revealed for instance in the doubly magic nucleus ^{48}Ca. This interaction lowers the $f_{7/2}$ neutron orbit just into the middle of the gap between the sd and fp oscillator shells, resulting in a neutron magic number at $N = 28$. The ordering of the neutron orbitals around $N = 28$ ($f_{7/2}$, $p_{3/2}$, $p_{1/2}$ and $f_{5/2}$) exhibits 2 pairs of spin-orbit partners originating from the f and p states. A differential change in the size of these two pairs of orbitals could enhance or reduce the $N = 28$ gap. A strong reduction of the f spin-orbit splitting, as compared to the p one, would erode the $N = 28$ gap. This would create particle-hole (ph) excitations between orbitals of the same oscillator shell $f_{7/2}$ and $p_{3/2}$ which are strongly connected by quadrupole interactions. Thus, even a small amount of excitations across $N = 28$ may lead to permanent quadrupole deformation. Therefore, the emergence of deformed nuclei with $N = 28$ is intimately related to the strength of the spin-orbit interaction.

Recent experimental data suggest an erosion of the $N = 28$ shell in very neutron-rich nuclei. According to β-decay [1] and Coulomb-excitation [2] experiments, quadru- pole ground-state deformation develops at $Z = 16$ four protons below the doubly magic ^{48}Ca. Study of the neutron-rich $^{40-44}$S using the in-beam γ-spectroscopy technique [3] suggests deformed ground states for ^{40,42}S and a mixed deformed configuration for ^{44}S in accordance with both the mean-field [4,5,6,7,8] and the recent large-scale shell model calculations [9]. A 0_2^+ level (of mainly 2p2h origin) was tentatively assigned at an energy of 2.7 MeV in ^{46}Ar [10], as compared to 4.28 MeV in ^{48}Ca. The recent finding of a 0_2^+ at 1.365 keV in ^{44}S shows that

[a] e-mail: gaudefroy@ganil.fr

the decrease of the 0_2^+ level has continued with the removal of protons in the *sd* shell. The large $\rho^2(E0 : 0_2^+ \to 0_1^+)$ points towards the existence of two strongly interacting states with very different shapes in ^{44}S. This witnesses that the $N = 28$ gap has been overcome by *ph* excitations to generate permanent quadrupole deformation.

To understand the erosion of the $N = 28$ shell gap south to ^{48}Ca, and its probable link to a reduced spin-orbit splitting, it is essential to see the evolution of the $f_{7/2}$, $p_{3/2}$, $p_{1/2}$ and $f_{5/2}$ orbitals in the ^{46}Ar nucleus. The tensor monopole interaction [11] could be responsible for this reduction. It acts in the present case between the protons in the *sd* shell (mainly in the $d_{3/2}$ orbital) and the neutrons in the two $f_{7/2}$-$f_{5/2}$ and $p_{3/2}$- $p_{1/2}$ spin-orbit partners. As protons are removed from the $d_{3/2}$ orbital, the spin-orbit splittings are weakened. To see if such an effect is observed in the $N = 28$ region we should compare the neutron single-particle energies in the ^{49}Ca and ^{47}Ar isotones. In ^{49}Ca the size of the spin-orbit splitting between the p orbits is of about 2 MeV, and the main component of the $f_{5/2}$ strength lies at 3.9 MeV excitation energy (referenced to the $p_{3/2}$ ground state (g.s.)). The present goal of the d(^{46}Ar,^{47}Ar)p transfer reaction is to obtain similar information in ^{47}Ar. To achieve this goal, we have used the newly operating SPIRAL facility to produce a radioactive beam of ^{46}Ar.

The study of the $N = 28$ shell closure is expected to shed some light on the hitherto mysterious anomalies observed in certain meteorites, as the large ^{48}Ca/^{46}Ca ratio. The present study aims to deduce radiative neutron-capture rates in the Ar isotopic chain from the spectroscopic information obtained through the (d, p) transfer reaction. Astrophysical consequences are subsequently discussed.

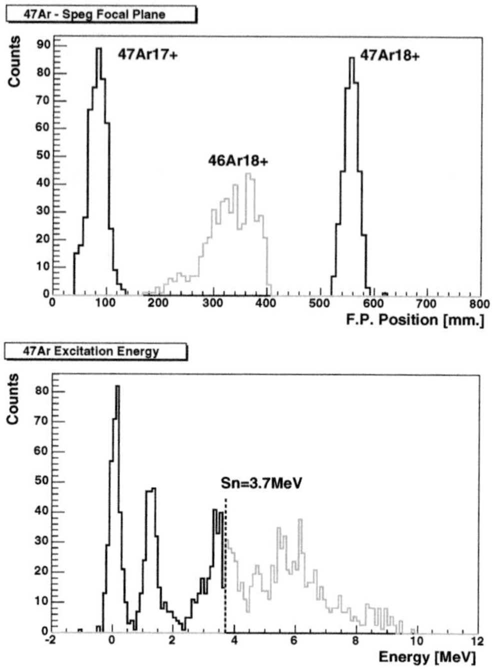

Fig. 1. Top: Position of the nuclei transmitted in the focal plane of SPEG for the ^{46}Ar (d, p) ^{47}Ar reaction. The grey line indicates transfer above the neutron separation energy S_n. The bottom part shows the corresponding energy spectrum.

2 Experimental methods

The transfer reaction d(^{46}Ar,^{47}Ar)p has been performed at GANIL with a radioactive beam of ^{46}Ar produced by the SPIRAL facility. This beam has been produced through the projectile fragmentation of a 66 $A \cdot$ MeV ^{48}Ca primary beam of about 200 pnA intensity in a thick carbon target located at the underground production cave of SPIRAL. The isotopes of interest were produced into the target which was heated at 2000 K in order to favor their extraction. They were subsequently ionized by an ECR source to the charge 9^+ and accelerated by the cyclotron CIME up to the energy of 10 $A \cdot$ MeV. The beam intensity of ^{46}Ar was $2 \cdot 10^4\,\mathrm{s}^{-1}$, without isobaric contamination. In addition to this, a stable beam of ^{40}Ar has been produced by the same devices at similar energy in order to validate our analysis of the pick-up (d, p) reaction with a known case studied in direct kinematics [12]. The total number of ^{40}Ar and ^{46}Ar nuclei which crossed the target was $2 \cdot 10^9$ and $4 \cdot 10^9$, respectively.

Neutron pick-up reactions (d, p) were induced by a 380 μg $\cdot$ cm^{-2} thick CD_2 target. The tracking of the secondary beams was achieved by a position-sensitive gas-filled detector CATS [13] located 11 cm downstream the target. We have reconstructed the impact points of the nuclei on the target with an accuracy of 1.0 mm. Protons were detected at backward angles (between 110 and 170 degrees) using the 8 position-sensitive MUST telescopes [14]. By measuring the energies (from 0.3 to 6 MeV) and angles of the protons, the two-body pick-up reaction can be characterized. The beam-like transfer products were selected by the SPEG [15] spectrometer and identified at its dispersive focal plane through their position (see fig. 1 top), energy loss and time-of-flight information. In order to obtain the energy spectrum of ^{47}Ar (^{46}Ar), we have used the measured proton energy and angle in the relativistic kinematics formula.

The experimental proton angular distributions have been compared to those obtained with DWBA calculations using the DWUCK4 code [16] to deduce the ℓ value and spectroscopic factor (SF) of each identified level (fig. 2). Several optical potentials have been used to see their influence on the SF values. It was found that, depending on the potential, the SF vary within about 20% around a mean value. We have considered two different potentials for both the entrance and exit channels, and used the four combinations between them to describe the (d, p) reaction. The (d+^{46}Ar) channel was described by using the deuteron global optical potential parametrization of W.W. Daehnick et al. [17] (denoted as D1 in the following) or the adiabatical deuteron potential of G.L. Wales and R.C. Johnson [18] (D2) which takes into account the effect of the deuteron break-up. The deuteron potential is described with a combination of neutron and proton potentials (CH89 [19]) for the parametrization of the break-up.

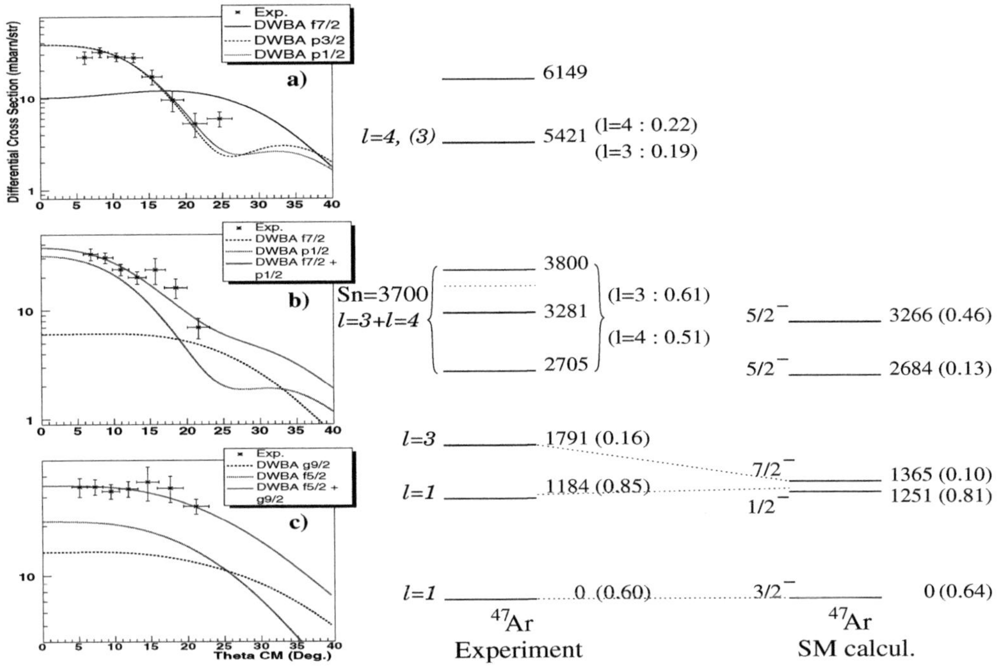

Fig. 2. Left: Angular distributions for the g.s. **a)**, 1st + 2nd excited states **b)** and the structure around 3.5 MeV (3 states) **c)** in ^{47}Ar are shown in comparison to DWBA calculations assuming ℓ=1, 3 or 4 distributions. Right: Experimental level scheme in ^{47}Ar obtained from the present work compared to shell model calculations using the sdpf interaction [25]. Calculated and experimental spectroscopic factors are indicated in parentheses. Only calculated levels with spectroscopic factors greater than 0.1 are presented.

The exit channel (p+^{47}Ar) was described both by the potential CH89 (denoted as P1) and by the parametrization of C.M. Perey and F.G. Perey [20] (P2).

3 Experimental results

The energy spectrum of ^{47}Ar is shown in fig. 1 bottom. Prior to the present work only the mass excess (Δm) and beta-decay half-life of ^{47}Ar were known. However, the use of tabulated value $\Delta m = -25.9(1)$ MeV for ^{47}Ar [21] would shift by 600 keV the peak corresponding to the $p_{3/2}$ ground state (g.s.). Since the method to deduce the excitation spectrum has been proven to be successful already to deduce the whole excitation spectra of ^{41,45}Ar [22], we would suggest a new mass-excess value of $\Delta m = -25.3(2)$ MeV for ^{47}Ar to bring back the g.s. level to zero excitation energy as shown in fig. 1. This new mass-excess leads to a neutron separation energy (S_n) in ^{47}Ar of 3.7(2) MeV instead of 4.25(14) MeV. The full width at half maximum of the g.s. peak is 420 keV. With this energy resolution, an unfolding procedure is often required to separate several peaks which lie at nearby energies. Two Gaussian distributions centered at 1.1(2) and 1.8(2) MeV are used to reproduce the observed structure between 0.8 and 2.2 MeV. In the same manner, three Gaussians centered at 2.7(2), 3.3(2) and 3.8(2) MeV are required to reproduce the broad structure around 3 MeV. As in the case of ^{45}Ar [22], well-defined structures are still visible in ^{47}Ar above the neutron separation energy value S_n. This indicates that at this energy, the level scheme cannot be considered as a continuum of states. We note however that the spectrum cannot be exploited above 6 MeV in ^{47}Ar because it is obtained with protons whose energies become too low to be detected at the most backward angles of the MUST detector.

Table 1. Spectroscopic factors obtained in the data analysis of the ^{46}Ar(d, p)^{47}Ar reaction with combinations of the four optical potentials D1, D2, P1 and P2 described in the text.

E^* (keV)	D1P1	D1P2	D2P1	D2P2
0 ($p_{3/2}$)	0.60	0.53	0.65	0.57
1184 ($p_{1/2}$)	0.93	0.81	0.84	0.77
1791 ($f_{7/2}$)	0.17	0.13	0.20	0.18
2705-3800	$\ell = 3 : 0.57$	0.57	0.66	0.66
(ℓ = 3,4)	$\ell = 4 : 0.51$	0.44	0.58	0.53
5421	$\ell = 3 : 0.17$	0.17	0.19	0.20
(ℓ = 3,4)	$\ell = 4 : 0.21$	0.20	0.23	0.22

Angular distributions have been deduced for all states but the one at 6.1(2) MeV. The spectroscopic factors listed in table 1 have been extracted using the procedure described in section 2. The g.s. exhibits an angular distribution which is typical of an $\ell = 1$ state. This is in agreement with the $p_{3/2}$ assignment obtained with the Shell Model (SM) calculations using the ANTOINE code [23,24] with the *sdpf* interaction [25] (see fig. 2). The spectroscopic factors of the $p_{3/2}$ is 0.6 in ^{47}Ar as compared to 0.84 in the isotone ^{49}Ca. Two components of orbital momenta $\ell = 1$ and $\ell = 3$ have to been used to fit the states at 1.1(2) MeV and 1.8(2) MeV. We tentatively assign the configurations $p_{1/2}$ and $f_{7/2}$ to the states at 1.1(2) MeV and 1.8(2) MeV, respectively. The spectroscopic factor of the $p_{1/2}$ state is close to unity, indicating its quasi-pure single neutron configuration added to the ^{46}Ar core. The intruder state $7/2^-$ built on a $(\nu p_{3/2})^2(\nu f_{7/2})^{-1}$ configuration has a spectroscopic factor of 0.16 in ^{47}Ar. The experimental energies and spectroscopic factors of the g.s. and first excited states compare very well with SM calculations. The SF value of the intruder configuration also agrees very well, but the level lies about 400 keV above the calculated value. This indicates that SM calculations either slightly underestimate the size of the $N = 28$ gap or overestimate pairing/correlation energies gained by promoting a neutron in the $p_{3/2}$ orbital in ^{47}Ar. The present result will help for better understanding the onset of collectivity in the $Z = 16$ nuclei through intruder excitations. It is also interesting to compare the spin-orbit splitting between the $p_{1/2}$ and $p_{3/2}$ states which is reduced from about 2 MeV in ^{49}Ca to 1.2 MeV in ^{47}Ar, as two protons are removed from the quasi-degenerated proton $d_{3/2}$ and $s_{1/2}$ orbitals. The origin of this decrease of spin-orbit splitting with the removal of $d_{3/2}$ protons could be ascribed to the tensor monopole force [11] which acts in opposite sign between the two orbitals $p_{1/2}$ and $p_{3/2}$.

The fit of the experimental angular distribution located around 3 MeV excitation energy requires the presence of at least three states, and an admixture of $\ell = 3$ and/or $\ell = 4$ components as shown in table 1. This is in nice agreement with the SM calculations. Another solution of the fit leads to a much higher $\ell = 4$ SF of 0.76. However this value is very unlikely as the SF of the $g_{9/2}$ orbital is weaker than 0.2 in ^{49}Ca and never exceeds 0.6 at that excitation energy in the $N = 29$ isotones. From the present assumption, it is found that the components of the $f_{5/2}$ orbital are lowered by about 500 keV between ^{49}Ca and ^{47}Ar. This could be an effect of the reduction of the spin-orbit splitting between the $f_{7/2}$ and $f_{5/2}$ orbitals, which is evidenced through the lowering of the $f_{5/2}$ orbital with respect to the $p_{3/2}$ g.s.

We have used the same interaction, *sdpf*, to calculate the level scheme of ^{49}Ar (fig. 3), which is currently not reachable via (d, p) reaction as the intensity of ^{48}Ar beam is too weak. Given the good agreement between calculated and experimental levels in ^{45,47}Ar (see ref. [22] for ^{45}Ar), we expect to obtain reliable predictions for the case of ^{49}Ar whose astrophysical interest will be emphasized in the next section. This nucleus exhibits two low-energy $p_{3/2}$ and $p_{1/2}$ states with SF of 0.23 and 0.77, respectively. In ^{48}Ar$_{30}$ the $p_{3/2}$ is in principle half-filled. Therefore a SF of 0.23 carries about 50% of the $p_{3/2}$ total strength. The first excited state carries almost all the $p_{1/2}$ strength. It is shifted down by about 1 MeV as compared to the ^{47}Ar because of the pairing energy gain of the 2 holes in the $p_{3/2}$ orbital when promoting a neutron in the $p_{1/2}$ shell.

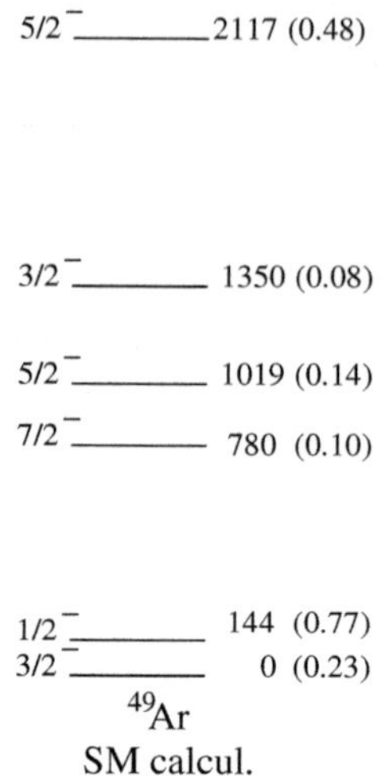

Fig. 3. Calculated level scheme and spectroscopic factors (in parenthesis) of ^{49}Ar.

4 Determination of (n, γ) rates from (d, p) reaction

The experimental results obtained on ^{45,47}Ar nuclei from (d, p) reactions are used in the following to determine the neutron capture rates (n, γ) on ^{44,46}Ar. Similar procedure is applied to determine ^{48}Ar(n, γ) capture rate using the calculated structure of ^{49}Ar. This will bring a wealth of data to be used for a possible explanation of the abnormally high abundance ratio ^{48}Ca/^{46}Ca observed in certain refractory inclusions of meteorites. We have applied a procedure similar to that described in ref. [26] for the case of ^{48}Ca to calculate the neutron capture cross section in the Ar isotopic chain. It was pointed out in ^{48}Ca that 95% of the cross section is of Direct Capture (DC) origin, which was ascribed to the intrinsic structure of the ^{49}Ca g.s. which exhibits a low angular momentum and a large spectroscopic factor. The contribution of resonant capture to unbound states was negligible since the level density at the S_n value is low. Predictions in the Ar isotopic chain around $N = 28$ depend strongly on the nuclear structure of ^{46}Ar nucleus. In particular, the neutron-capture rates depend on the SF of the low-energy p states, the S_n value and the level density above the latter. These three points are examined in the following.

Taking into account the spin-conservation rules in the DC reaction, the neutron is mainly captured into $\ell = 1$ bound states through the $E1$ operator without centrifugal barrier as the transferred angular momentum is $\ell_n = 0$ (s-wave capture). The ratio between s-wave (final state with $\ell = 1$) and d-wave (final state with $\ell = 3$) direct neutron-capture rate is approximately 10^4 in the present nuclei [27]. Therefore the s-wave DC to the g.s. and first excited states that have both large spectroscopic factors and $\ell = 1$ values dominates over all other contributions. Moreover, in an s-wave capture, the velocity dependence of the cross section is known to be $1/v$, and the Maxwell-average neutron capture rate (MACS $= Na\langle\sigma v\rangle$, with Na

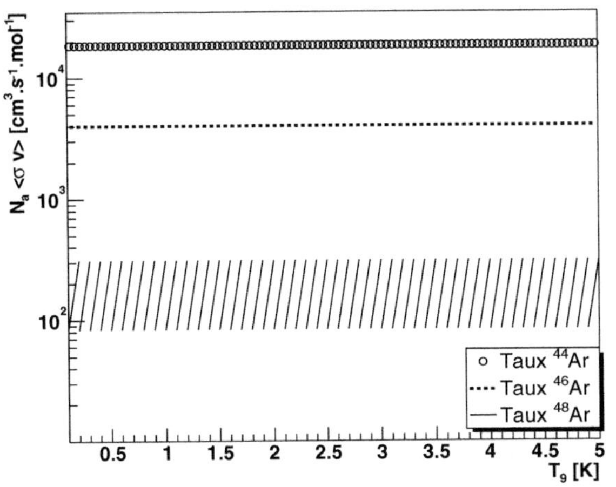

Fig. 4. From top to bottom curves: neutron capture rates on ^{44,46,48}Ar deduced from our results (for ^{44,46}Ar) and from SM calculations (for ^{48}Ar). The dashed region witnesses the large uncertainty on the S_n value of ^{49}Ar.

the Avogadro number) is not expected to show energy dependence.

The neutron separation energy S_n in the final nuclei is also a key ingredient to determine the direct neutron-capture rate. It defines the Q-value of the (n,γ) reaction to the bound states in the ^{45,47,49}Ar nuclei. Owing to the known mass-excesses of ^{44,45,46}Ar and the newly determined one of ^{47}Ar, we can determine the reaction Q-values for the ^{44,46}Ar neutron captures with sufficient precision. In the case of ^{49}Ar, only extrapolated value exists [21]. The large uncertainty of about 800 keV on the Q-value (2.5(8) MeV for a capture to the g.s.) implies a large uncertainty in the calculation of the neutron-capture rate on ^{48}Ar, as shown in fig. 4.

Well-defined energy peaks are still present above the neutron emission threshold in both ^{45}Ar and ^{47}Ar nuclei. This means that the statistical treatment of neutron captures using the Hauser-Feshbach formalism is no longer valid for these light nuclei, close to magic numbers. Instead, individual resonances should be considered. As the centrifugal barrier will strongly hinder neutron-captures on orbitals of high angular momenta, we have to search for resonances with low-ℓ values in the vicinity of the neutron separation energy. As mentioned above, no state with $\ell \leq 1$ has been found, making the resonant capture a negligible process for these nuclei.

The direct neutron capture rates on ^{44,46}Ar have been calculated using the procedure described in ref. [26] and the experimental spectroscopic information presented here and in ref. [22]. As the DC process mainly occurs at the surface of the nucleus, the choice of an appropriate nuclear density distribution is important. These have been obtained from mean-field calculations using HFB formalism in a spherical geometry [28]. These distributions reproduce remarkably well the measured root mean square radii of ^{44,46}Ar [29]. The neutron capture rate on ^{48}Ar was deduced from the calculated structure of ^{49}Ar, presented in the previous section. The calculated MACS are shown in fig. 4. A MACS ratio of about 5 occurs between $A = 44$ and $A = 46$, and a more pronounced one (10 to 50 depending on the S_n value in ^{49}Ar) between $A = 46$ and $A = 48$. Mass measurements of the neutron-rich ^{48,49}Ar isotopes are of great interest to infer the ^{48}Ar(n,γ) rate with better accuracy. From our results summarized in fig. 4, it arises that nuclear structure (*i.e.* the presence of low angular momenta states — $\ell = 1$ — with closed to one SF and high binding energies) speed-up the neutron capture at $A = 46$ (as compared to $A = 48$), despite the presence of the $N = 28$ shell closure.

5 Astrophysical implications

About two decades ago, G.J. Wasserburg and his group at Caltech identified correlated isotopic anomalies for the neutron-rich ^{48}Ca, ^{50}Ti and ^{54}Cr isotopes in peculiar refractory inclusions of the Allende meteorite [30,31]. As an example, the ^{48}Ca/^{46}Ca ratio was found to be 250, a factor of 5 larger than in the Solar System. It was concluded that these highly unusual isotopic compositions witness late-stage nucleosynthesis processes which preceded the formation of the solar nebula. However, astrophysical models existing at that time encountered severe difficulties when trying to reproduce these observed anomalies, in particular those in the EK-1-4-1 inclusions. Since that time major progresses have been made in particular on mass measurements, β-decay lifetimes of unstable nuclei and neutron-capture cross sections on stable and unstable nuclei (present work). We now gather all information to see which stellar conditions could account for these observations.

A plausible astrophysical scenario to account for the overabundance of ^{48}Ca is a weak rapid neutron-capture process [32,27]. There the neutron-rich stable ^{46,48}Ca isotopes are produced during a neutron-capture and β-decay process. The main contribution to the production of these Ca isotopes is provided by the β-decay of their progenitor isobars in the Ar isotopic chain [27]. This was traced back from the fact that, in the $Z < 18$ chains, the measured β-decay lifetimes of ^{44}S and ^{45}Cl are shorter than the neutron-capture rates at the $N = 28$ shell closure. Consequently, the matter flow in the S and Cl chains is depleted by β-decay to the upper Z isotopes before reaching masses $A = 46$ or 48. It was noted earlier that this feature arises from the erosion of the $N = 28$ shell gap in the S and Cl isotopic chains [1]. Thus the main progenitors of ^{46,48}Ca are produced either directly in the Ar chain or from the β-decay of $Z < 18$ nuclei which subsequently could capture neutrons in the Ar chain. Therefore the determination of neutron-capture rates in the Ar isotopes is important. Competition between the neutron capture and β-decay reactions is reflected by the reaction mean times. The mean neutron capture time t_n can be expressed as $t_n = 1/(d_n\langle\sigma v\rangle)$. Knowing the β-decay half lives and neutron capture rates in the nuclei of interest, we can deduce an approximative value of the neutron density, d_n, which could account for the large observed isotopic ratio ^{48}Ca/^{46}Ca $= 250$ in the weak r-process

Table 2. Neutron capture and β decay mean times for $A = 44$, 46 and 48 in the argon chain. The neutron capture mean times are deduced from the present work. The mean rate value of ^{49}Ar has been taken to deduce its neutron capture mean time.

Isotope	$Na\langle\sigma v\rangle$	t_n	$T^{\beta}_{1/2}$
^{44}Ar	$6.19\ 10^6$	$2.8\ 10^{-5}$ s	11.9 min
^{46}Ar	$4.04\ 10^3$	45 ms	7.8 s
^{48}Ar	$2.03\ 10^2$	1 s	0.45 s

temperature range ($T_9 = 0.1$ to $\simeq 5$). Such a neutron density would give rise to a neutron capture mean time, $t_n^{^{46}\mathrm{Ar}}$, on ^{46}Ar which favors the capture reaction in comparison to the β-decay one. This can be expressed by the relation $1-\exp(\frac{-\ln 2\cdot t_n^{^{46}\mathrm{Ar}}}{7.8}) = 1/250$ which leads to a $t_n^{^{46}\mathrm{Ar}}$ value of about 45 ms implying a neutron density of about $3\cdot 10^{21}\ \mathrm{cm}^{-3}$. Table 2 presents the neutron capture mean times corresponding to that neutron density, as well as the β-decay half lives.

At $A = 48$, calculated t_n becomes longer than $T^{\beta}_{1/2}$. Consequently the neutron capture is halted in the Ar chain at $A = 48$, accumulating substantial amount of ^{48}Ca. Conversely, few depletion occurs through β-decay at $A = 46$ as the lifetime of ^{46}Ar is longer than neutron capture time. These two features can account for explaining the observed high ^{48}Ca/^{46}Ca ratio. More realistic nucleosynthesis network calculations are in progress in order to confirm or otherwise the naive picture drawn.

6 Conclusions

Spectroscopic information has been obtained in $^{47}\mathrm{Ar}_{29}$ using the d(^{46}Ar,^{47}Ar)p transfer reaction at GANIL/SPIRAL. We have used the position-sensitive MUST detector for the protons and the SPEG spectrometer to select and identify the transfer-like nuclei ^{47}Ar. Energies and angular distributions of hitherto unknown levels in ^{47}Ar have been determined for the first time. A new value of the mass-excess of ^{47}Ar, $\Delta m = -25.3(2)$ MeV, is proposed. The two components of the spin-orbit splitting $p_{3/2}$ and $p_{1/2}$ have been identified at low excitation energies in ^{47}Ar. The approximate location of the $f_{5/2}$ strength has been determined too. These data point to a weakening of the p and f spin-orbit splittings, the latter being responsible for the reduction of the $N = 28$ shell gap. The intruder configuration $2p1h$ across the $N = 28$ gap has been observed at 1.8(2) MeV. Its energy will constrain the size of the $N = 28$ gap in ^{47}Ar and the amount of correlations. Comparison between experimental results and shell model calculations shows a very good agreement, except for the location of the intruder state whose calculated energy is 400 keV lower. Combining the present information and the one obtained on $^{45}\mathrm{Ar}_{27}$ [22], we deduce that the magicity at the $N = 28$ shell closure is preserved at ^{46}Ar. The stellar neutron capture rates ^{44,46}Ar(n,γ)^{45,47}Ar have been deduced using our data obtained from the (d,p) reaction. A nearby extrapolation led to the estimation of ^{48}Ar(n,γ)^{49}Ar rate. It is shown that nuclear structure does matter for a correct determination of these capture rates. A fast(slow) neutron-capture rate is found for ^{46}Ar(^{48}Ar). In a weak r-process scenario, with neutron density of about $3\cdot 10^{21}\ \mathrm{cm}^{-3}$, the matter flow in the Ar chain is slightly depleted at $A = 46$ and accumulated at $A = 48$. This would result, after β-decay of the unstable Ar isotopes to an overproduction of the stable ^{48}Ca as compared to ^{46}Ca, as is observed in certain refractory inclusions of the Allende meteorite.

References

1. S. Grévy *et al.*, Phys. Lett. B **594**, 252 (2004).
2. T. Glasmacher *et al.*, Phys. Lett. B **395**, 163 (1997).
3. D. Sohler *et al.*, Phys. Rev. C **66**, 054302 (2002).
4. T.R. Werner *et al.*, Nucl. Phys. A **597**, 327 (1996).
5. P.-G. Reinhardt *et al.*, Phys. Rev. C **60**, 014316 (1999).
6. S. Péru *et al.*, Eur. Phys. J. A **49**, 35 (2000).
7. R. Rodriguez-Guzman *et al.*, Phys. Rev. C **65**, 024304 (2002).
8. G.A. Lalazissis *et al.*, Phys. Rev. C **60**, 014310 (1999).
9. E. Caurier *et al.*, Eur. Phys. J. A **15**, 145 (2002).
10. Z. Dombrádi *et al.*, Nucl. Phys. A **727**, 195 (2003).
11. T. Otsuka *et al.*, *Proceedings of the XXXIX Zakopane School of Physics*, Acta Phys. Pol. B **36**, 1213 (2005).
12. S. Sen *et al.*, Nucl. Phys. A **250**, 45 (1975).
13. S. Ottini-Hustache *et al.*, Nucl. Instrum. Methods A **431**, 476 (1991).
14. Y. Blumenfeld *et al.*, Nucl. Instrum. Methods A **421**, 471 (1999).
15. L. Bianchi *et al.*, Nucl. Instrum. Methods A **276**, 509 (1989).
16. P.D. Kunz, computer code DWUCK4, Colorado University, unpublished.
17. W.W. Daehnick *et al.*, Phys. Rev. C **21**, 2253 (1980).
18. G.L. Wales, R.C. Johnson, Nucl. Phys. A **274**, 168 (1976).
19. R.L. Varner *et al.*, Phys. Rep. **201**, 57 (1991).
20. C.M. Perey, F.G. Perey, At. Data Nucl. Data Tables **17**, 1 (1991).
21. G. Audi *et al.*, Nucl. Phys. A **729**, 3 (2003).
22. L. Gaudefroy *et al.*, J. Phys. G, to be published.
23. E. Caurier, Shell Model code ANTOINE, IReS, Strasbourg 1989-2002.
24. E. Caurier, F. Nowacki, Acta Phys. Pol. B **30**, 705 (1999).
25. S. Nummela *et al.*, Phys. Rev. C **63**, 044316 (2001).
26. E. Kraussmann *et al.*, Phys. Rev. C **53**, 469 (1996).
27. O. Sorlin *et al.*, C.R. Phys. **4**, 541 (2003).
28. M. Girod, private communication.
29. A. Klein *et al.*, Nucl. Phys. A **607**, 1 (1996).
30. T. Lee *et al.*, Astrophys. J. **220**, L21 (1978).
31. F.R. Niederer *et al.*, Astrophys. J. **240**, L73 (1980).
32. K.L. Kratz *et al.*, Mem. Soc. Astron. Ital. **2**, 453 (2001).

Eur. Phys. J. A **27**, s01, 315–320 (2006)
DOI: 10.1140/epja/i2006-08-048-y

EPJ A direct
electronic only

Status of the TRIUMF annular chamber for the tracking and identification of charged particles (TACTIC)

G. Ruprecht[1,a], D. Gigliotti[1], P. Amaudruz[1], L. Buchmann[1], S.P. Fox[2], B.R. Fulton[2], T. Kirchner[1], A.M. Laird[2], P.D. Mumby-Croft[2], R. Openshaw[1], M.M. Pavan[1], J. Pearson[1], G. Sheffer[1], and P. Walden[1]

[1] TRIUMF, 4004 Wesbrook Mall, Vancouver, BC, V6T 2A3, Canada
[2] University of York, Heslington, York, YO10 5DD, UK

Received: 1 July 2005 /
Published online: 15 March 2006 –

Abstract. TACTIC (TACTIC web site: `http://tactic.triumf.ca`) is a new detector for low-energy nuclear reactions currently under development at TRIUMF. The cylindrical ionization chamber allows three-dimensional reconstruction of particle tracks by means of a two-dimensional anode array combined with a TOF measurement of the drift electrons. In addition, the integrated charge for each pulse provides information about the energy loss of the particle and therefore allows a better identification of the nuclear species producing the track. The geometry of TACTIC covers a large angular range permitting the measurement of differential cross-sections over a large solid angle. It will be ideal for investigations of nuclear processes pertinent to the field of nuclear astrophysics.

PACS. 25.55.-e ^{3}H-, ^{3}He-, and ^{4}He-induced reactions – 29.40.Cs Gas-filled counters: ionization chambers, proportional, and avalanche counters – 29.40.Gx Tracking and position-sensitive detectors

1 Introduction

With the advent of radioactive ion beam (RIB) facilities many reactions have to be measured in inverse kinematics. For capture reactions, the cone of the ejected heavy ions is usually sufficiently small so that a recoil separator can be used for detecting 100% of the ions. For reactions with two or more heavy ejectiles, the cone is larger and a detector array like TUDA [1] is the better option. However, a fraction of the angular range is lost for small angles in order to let the beam through, and for larger angles where the ejectiles do not reach the detector. When using gas targets, if the energy of the ejectiles is low they cannot penetrate the gas and/or the exit window and also lose energy in the dead layer of the detector.

Another approach to cover a large forward angular range is to use an ionization chamber. The problem here is that the target and detection region are not separated, resulting in a large background and poor statistics. High segmentation is needed in order to collect as many track points as possible and this would require several amplifying gas cells as well as amplification and digitalization electronics.

Our approach is to employ Gas Electron Multiplier (GEM) [2,3] foils for the first stage of amplification inside the chamber. This considerably reduces the complexity of a cylindrical chamber design which in turn makes a separation of target and detection region more feasible.

[a] e-mail: `ruprecht@triumf.ca`

2 The TACTIC chamber

The TRIUMF Annular Chamber for the Tracking and Identification of Charged Particles (TACTIC) is a combined cylindrical ionization (IC)/time projection chamber (TPC) detector where the gas target along the central axis can be "windowless" to ejectiles (*i.e.* the target and detector gases are the same), or "windowed" (*i.e.* a thin window separates disparate target and detector gases). In either case field-defining cathode wires delimit the target region. Using this method, the beginning part of the track within the target region cannot be "seen" in the drift region, but the vertex point can still be inferred by extrapolating the reconstructed TPC anode hits. The total energy deposited by the stopping ejectiles is also measured by the accumulated charge on the anodes. Furthermore, using the vertex reconstruction, the energy loss of the beam in the target allows the simultaneous measurement of cross-sections and angular distributions over a range of energies.

The GEM foil [2,3] acts as a preamplifier inside the chamber wall providing low-noise signals requiring only one further stage of amplification. The signals are digitized using a multi-channel VME-based flash ADC board, thus minimizing the amount of electronics required.

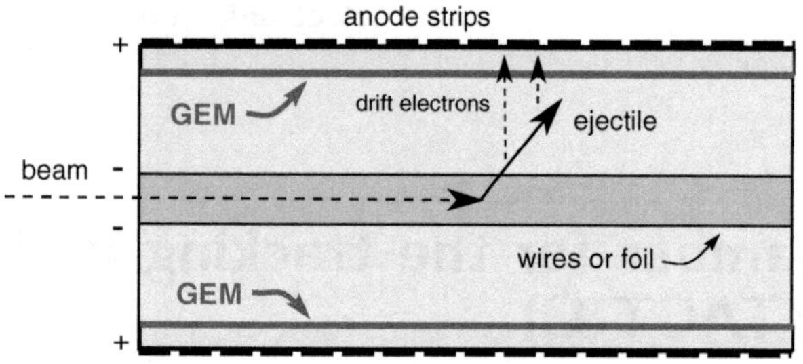

Fig. 1. Schematic side view of the proposed TACTIC detector.

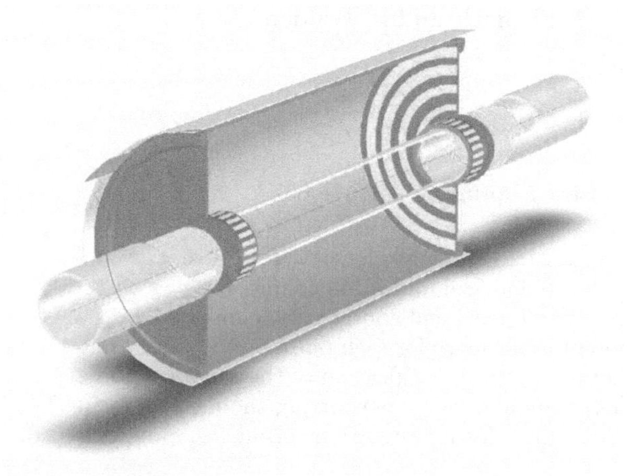

Fig. 2. Three-dimensional cut-away view of the proposed TACTIC detector.

A schematic view of the chamber is shown in fig. 1. Target and drift region are separated by wires or a foil (depending on the experiment) acting as the cathode for the drift electrons. The ejectiles move into the drift region where they produce electrons along the track which drift slowly (compared to the flight time of the ejectile) towards the GEM surrounding the entire drift cylinder. These primary electrons create an electron avalanche in the high electric field inside the GEM, producing a signal gain of 10–100. The avalanche electrons are collected by the nearby anode pads on the chamber wall. Like a TPC, the time differences of the signals between the pads gives information about the trajectory, whereas like an IC the collected charge is a measure of the ejectile energy loss. The trajectory together with the released charge along the track allows a unique identification of the nuclei ejected, which is essential for experiments involving ejectiles with similar energy but different charges or masses.

In the three-dimensional view, fig. 2, the support structure for the cathode wires (or target foil) can be seen. There are also biased rings at the end caps to straighten the drift field (see also sect. 5). The anode pads (not shown) are sub-divided azimuthally. This is necessary for high counting rates caused either by elastically scattered nuclei or decaying beam particles but may be also required by special experiments like reactions with polarized beam or with several heavy-ion ejectiles. The pads are etched onto flexible PCB which also holds the electrical connectors. The chamber will be constructed as two half cylinders allowing the target tube, including the separation wires or foil, to be removed or replaced. The chamber is placed in a detector-gas filled outer box that connects to the beam tube.

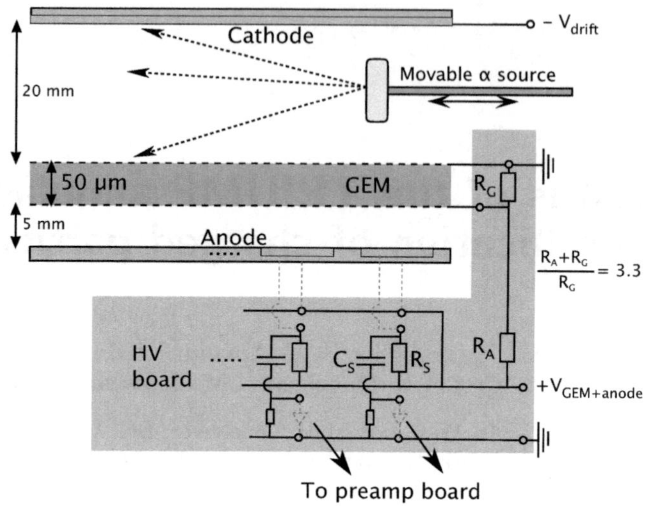

Fig. 3. Schematic view of the planar TACTIC test chamber investigating the properties of the GEM and testing particle tracking.

3 The TACTIC test chamber.

For the proposed first experiment, ^{8}Li(α, n)^{11}B, the energy of the recoils for the interesting part of the excitation curve is too small to penetrate a foil separating the detection region from the drift region, so a set of wires will replace the foil. This requires the target gas (primary helium) to be working as detection gas as well. In order to investigate the dependence of the GEM gain on the helium mixture and gas pressure, and to determine the local resolution of the particle track points, tests have been performed on a planar test chamber (see fig. 3). A movable 5486 keV alpha particle source was mounted perpendicular to the strips. At lower pressures a 16 μm Mylar foil was mounted in front of the source to reduce the alpha particle range.

The test chamber has a drift volume of about $20 \times 20 \times 2\,\text{cm}^3$ and an active GEM area of $8 \times 8\,\text{cm}^2$, covering 16 active anode strips, each 5 mm in width. The anodes are under positive high voltage and decoupled by capacitors. The signals are amplified by a single 16-channel preamplifier board. Analogue electronics and CAMAC ADCs and TDCs have been used to process the signals. We measured the change of the GEM gain for different mixtures of Ar/CO_2 as well as He/CO_2. The CO_2 quenching gas fraction was adjusted by a gas handling system (GHS). The highest amplification was achieved with a 90%/10% He/CO_2 ratio. The He/CO_2 mixture was comparatively better than an Ar/CO_2 mixture which is often used as a detector gas (see fig. 4).

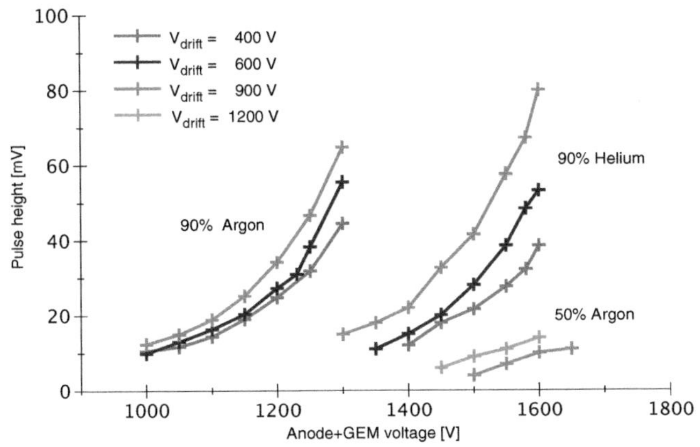

Fig. 4. Pulse heights *vs.* GEM+Anode voltage (the GEM voltage is a factor of 3.3 smaller).

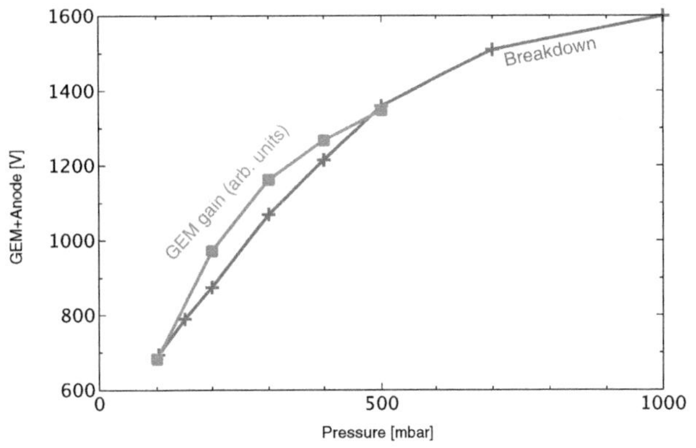

Fig. 6. The GEM trip voltage, and the relative gain at a constant GEM voltage, are shown as a function of gas pressure. Both nearly scale with the square root of pressure.

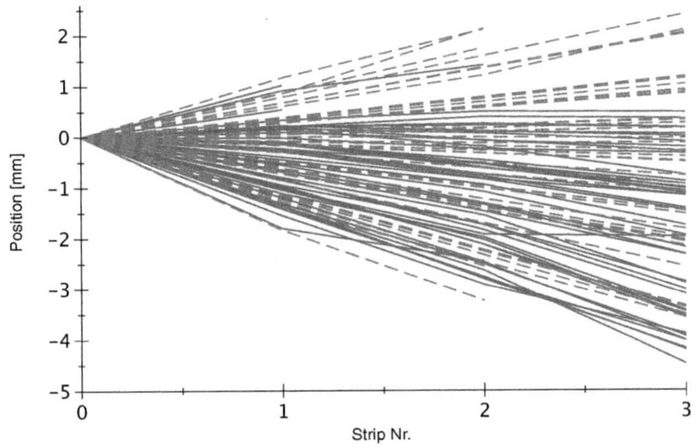

Fig. 5. (Colour on-line) Electron drift times with respect to the first strip, converted to distances assuming a drift velocity of 12 mm/μs. The electrons are released along the tracks of the α's. The dashed lines stem from a GEANT4 simulation, the solid purple lines from the measurement.

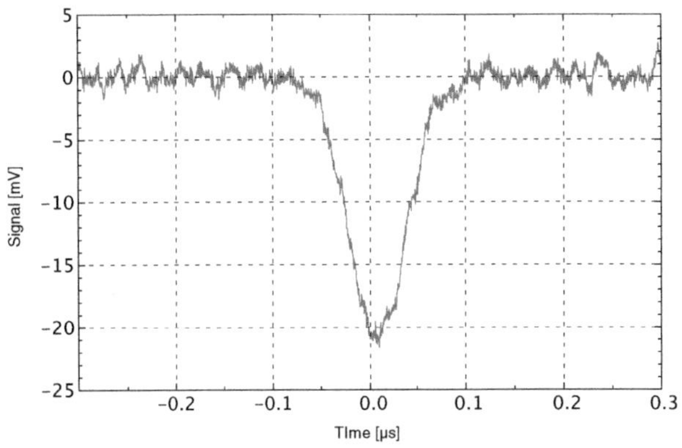

Fig. 7. Amplified signal of $\approx 5\,\mathrm{MeV}$ α particles at 500 mbar.

By measuring the time differences between the signals on 4 consecutive strips we have been able to gain a projected image of the particle tracks, as can be seen in fig. 5. One observes good agreement between measurement and simulation. The source was mounted 5 cm before the first strip. The first strip signal was used as the trigger, so all track times are with respect to the first strip.

In a second phase, the gas pressure was varied down to 100 mbar while keeping the He/CO_2 mixture at a constant ratio (below 100 mbar the oxygen contamination becomes too high and quenches the signals) and the flow rate at 200 cm^3/min. This enabled the dependence of the GEM gain on pressure and applied voltage to be determined. The source was not collimated, therefore particles were emitted in all directions producing a broad energy spectrum at each anode. The relative gain was found by comparing the measured spectra with the results of a GEANT4 [4] simulation. While the shape of the gain vs. GEM voltage curves remains nearly the same, the gain at a constant GEM voltage scales approximately with the square root of pressure, as does the breakdown voltage (see fig. 6). This is a useful result for the design of the cylindrical chamber. A typical signal is shown in fig. 7. Despite noise in the laboratory environment and the very weak initial ionization, the signals look very clear and promising.

4 GEANT4 simulation

A change of the gas pressure affects the initial ionization per anode strip, the track length and straggling of the ejectiles, the energy loss and straggling of the projectiles in the target region, as well as the gain of the GEM. To study the mutual dependence of all these parameters a Monte-Carlo simulation of all relevant processes taking part in the chamber was performed using the GEANT4 framework [4]. The most important questions to be answered from the simulation are

- How accurately can the reaction vertex point be reconstructed from the anode signals? This is important because the interaction energy must be well known for measurements below the Coulomb threshold.
- What is the angular resolution?
- What is the energy resolution?
- When the emission of low-energetic gammas is expected, what attenuation can be expected at the gamma detectors placed outside the chamber?

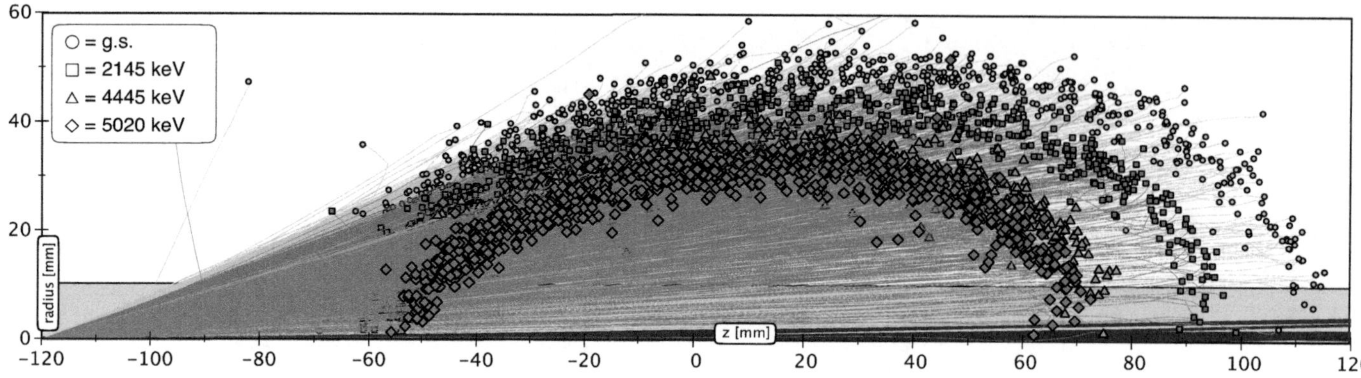

Fig. 8. (Colour on-line) Trajectories and end-point distribution calculated with GEANT4 of the ^{11}B ejectiles for the ground state and the first three excited states. The simulated detector is filled with a 90/10 He/CO_2 gas mixture at room temperature and a pressure of 250 mbar. The target region ranges from $r = 0$–10 mm (red background). The 9 MeV ^{8}Li beam comes from the left-hand side and hits the target at $z = -120$ mm (thick blue lines). Only the ejectiles that stem from 8.9–9.0 MeV beam energies are shown.

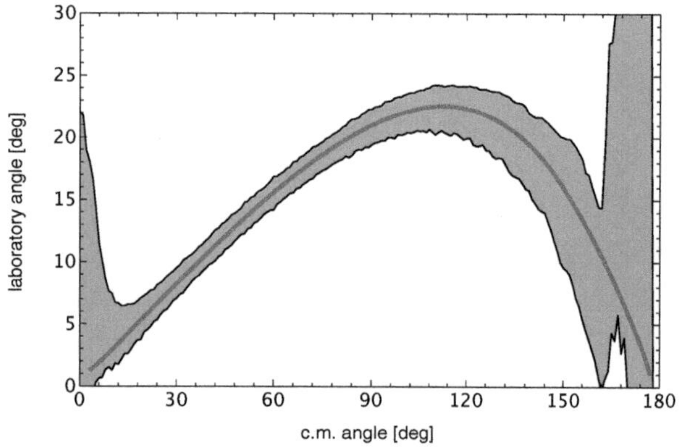

Fig. 9. Uncertainty ($\pm\sigma$ range calculated for 10000 events) of the ^{11}B laboratory angles, reconstructed using 48 anode rings, each 5 mm in width, emitted at different c.m. angles as expected from the GEANT4 simulation for the ^{11}B ground state. Overlayed is the theoretical curve.

GEANT4 is tailored for high-energy physics, and support for low-energy nuclear reactions is limited. There are currently problems with the energy loss of ions at low energies and more work in this area is needed, but some preliminary results are presented here.

The geometry is defined by the maximum of the z- and r-projected ejectile range, where z is the coordinate along beam axis and r the radial distance. The simulated tracks of the ejectiles are shown in fig. 8 for $E_{\text{beam}} = 9$ MeV and 4 different ^{11}B states [1]. Regarding the endpoints, for small laboratory angles and high energies the ^{11}B levels can be better resolved than for lower energies corresponding to a backwards emission in the centre-of-mass (c.m.) frame, or larger angles. However, combined with the measured ejectile energies, there is some improvement possible.

The angular resolution using 48 anode rings can be seen in fig. 9. The diagram shows the standard deviation around the true angle in the c.m. system. For a measured laboratory angle of 20° there is a large uncertainty for the c.m. angle (between 70° and 150°). This calculation is based on a simple linear least squares fit of the radial distances expected to be measured with the anodes (the error of the drift time is not currently included in the simulation). The uncertainty comes from the z-resolution as well as from the beam and ejectile straggling. With a more sophisticated analysis (*e.g.* giving smaller radii more weight) an improvement is possible. Again, no energy signals have been taken into account which can further refine the analysis.

The simulation is a crucial contribution for the detector design as well as for later experimental analysis. Further refinements are required to be able to obtain reliable statements about the vertex reconstruction and energy resolution. The estimates presented here have been done without taking the drift time uncertainties into account. The latter can be calculated using GARFIELD [6]. Combining GEANT4 with GARFIELD, a simulation of the anode signal shape can be achieved providing a further piece of information from the particles.

We have implemented the software for low-energy reactions (C++ objects) in a generic manner that can be used for other GEANT4 applications as well. As detectors for nuclear reactions become more complex, this will be a helpful contribution to the GEANT4 repository that other nuclear physics groups could use for detector simulation or analysis.

5 Drift field and beam induced electrons

Electrons induced directly by the beam could potentially drift from the target region to the detection region where they produce unwanted signals. To avoid this, a second cage of wires held at a slightly more positive voltage will be inserted a few mm inside the cathode wire cage, fully encompassing the expected incident beam spread. A GARFIELD [6] calculation has shown that electrons released in the target region will be collected at the inner cage and not drift to the detection region (see fig. 10).

[1] For the energies involved here the stopping powers and ranges are in agreement with the results from SRIM [5].

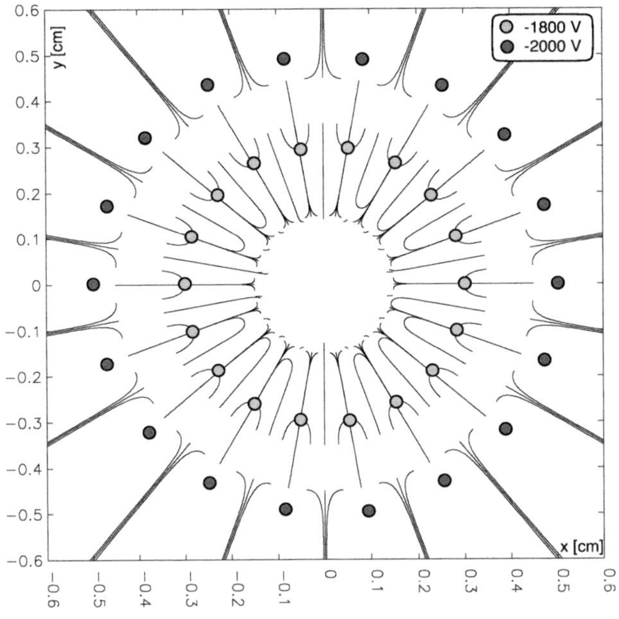

Fig. 10. GARFIELD simulation of the electron drift in the target region. The circles mark the wires, and the thin black lines the paths of the electrons released at different positions. Only electrons close to the cathode wire cage reach escape the target region and enter the detection volume.

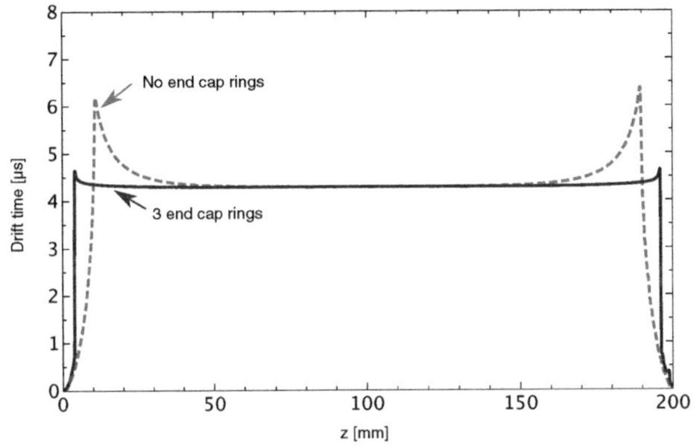

Fig. 11. Drift time of electrons released at -500 V biased cathode wires to the GEM *vs.* the z-axis, with and without voltage-shaping rings at the end caps at 10 mm radius. The radii of the cathode cage and of the GEM cylinder are 10 mm and 50 mm, respectively.

An accurate measurement of the drift time demands a uniform electric field within TACTIC. Since the enclosure outside the end caps is electrically grounded, the drift field is distorted close to the end caps. Therefore, the field near the endcaps must be shaped by appropriately biased rings (see fig. 2). Figure 11 shows the field uniformity improvement by using three rings. The electric field has been calculated using FEMLAB [7] and the total drift time was obtained by integrating along the field lines, assuming a linear approximation for the drift velocity given in [8] for a 90/10 mixture of Ar/CO_2.

6 Data acquisition

We will use flash ADCs for the data acquisition —a new technique that is becoming more popular for nuclear physics applications. The entire anode signal is sampled and can be stored for later analysis. For the TACTIC prototype we will use a 48-channel VME board with a sampling rate of 40–70 MHz and a resolution of 10 bits per sample. However, due to the large data rate it is desirable to extract time and charge information immediately proceeding the digitalization. The firmware running on the VME board offers a large number of configuration possibilities based on custom analysis algorithms. This feature is required because the pulse shapes differ depending on the direction of the recoil.

The raw charge and time data provided by the flash ADC board will be stored and pre-analysed using TRIUMF's MIDAS [9] system. DAQ systems for nuclear physics usually lack the ability to display particle tracks, so a more sophisticated analysis system, ROOT [10], will be used. There also exists a new MIDAS/ROOT interface, ROME [11], which is still under development at PSI, Switzerland. We are currently testing the suitability of these tools for this kind of particle tracking experiment.

7 Applications

The first experiment planned to be measured with TACTIC is the $^{8}\mathrm{Li}(\alpha,\mathrm{n})^{11}\mathrm{B}$ reaction which plays a role in r-process nucleosynthesis [12]. Including light elements in a scenario of neutrino-driven wind reactions with light nuclei can change the synthesis of heavier elements by an order of magnitude. One important reaction chain is $\alpha(\mathrm{t},\gamma)^{7}\mathrm{Li}(\mathrm{n},\gamma)^{8}\mathrm{Li}(\alpha,\mathrm{n})^{11}\mathrm{B}$. The total cross-section at 9 MeV ^{8}Li impact energy is about 400 mb but only a fraction ($\approx 20\%$) goes into the ^{11}B ground state, while there are up to 8 excited states involved.

In a recently published measurement [13], an ion chamber filled with a 90/10 He/CO_2 gas mixture and 2×64 flash ADCs for the readout enabled a three-dimensional reconstruction of the tracks. The resulting energy spectra allowed a rough separation of the ^{11}B levels. In contrast, TACTIC uses a GEM which, in principle, allows for much better local resolution and therefore better tracking. The layout of the TACTIC prototype will be designed to enclose the ^{11}B tracks for ^{8}Li impact energies between 1.2 and 9.0 MeV.

Another experiment is the measurement of the ^{7}Be+p elastic scattering cross-section at low energies. The angular distributions give information on the phase shifts for different angular momentum and spin combinations. There are 16 phase shifts involved if only s, p, and d-waves are taken into account, therefore a high angular resolution is required for this experiment. The contradicting results when compared with the mirror reaction $^{7}\mathrm{Li}+\mathrm{n}$ [14, 15,16] could be resolved by an accurate angular distribution measurement using TACTIC. A better understanding of the resonance structure will help in extrapolating the $^{7}\mathrm{Be}(\mathrm{p},\gamma)^{8}\mathrm{B}$ cross-section to low energies —an important reaction for the high-energy neutrino production in the Sun.

Another interesting application is the measurement of nuclear reactions with three or more heavy-ion ejectiles. The more complex phase space requires a large angular range with a good energy resolution to be measured for two or more ejectiles in coincidence. This is hard to achieve with passive detectors like silicon counters, while in a TACTIC-like detector a high fraction of the ejectiles can be tracked simultaneously.

8 Conclusions

A cylindrical ionization chamber layout in combination with GEM foils for time projection chamber-like tracking is a promising detector configuration for low-energy nuclear reactions, in particular for measurements in inverse kinematics. The GEM is easy to handle and there are fewer restrictions as opposed to gas cell pads. Moreover, it works very well with a He/CO_2 gas mixture which opens the possibility of ^{4}He and ^{3}He induced RIB reactions at low beam energies.

The suppression of the ionization electrons created by the beam reduces the event rate by orders of magnitude and makes TACTIC suitable for high-current RIB facilities like ISAC/TRIUMF. The first measurement —the reaction $^8\mathrm{Li}(\alpha,\mathrm{n})^{11}\mathrm{B}$— will show the actual resolving power of the TACTIC prototype and the maximum counting rate that can be achieved.

The capability of simultaneous particle tracking qualifies TACTIC for reactions with three or more heavy-ion ejectiles. For this, however, a higher azimuthal segmentation of the anodes is required but it remains an interesting application for the future.

References

1. TUDA web site: `http://tuda.triumf.ca`.
2. F.Sauli, A. Sharma, Annu. Rev. Nucl. Part. Sci. **49**, 341 (1999).
3. A. Sharma, F. Pauli, Nucl. Instrum. Methods A **350**, 470 (1994).
4. GEANT4 web site: `http://cern.ch/geant4`.
5. SRIM web site: `http://www.srim.org`.
6. GARFIELD web site: `http://garfield.web.cern.ch`.
7. FEMLAB web site: `http://www.comsol.com/products/femlab`.
8. A. Peisert, F. Sauli, *Drift and Diffusion of Electrons in Gases*, CERN, 1984.
9. MIDAS web site: `http://midas.triumf.ca`.
10. ROOT web site: `http://root.cern.ch`.
11. ROME web site: `http://midas.psi.ch/rome`.
12. M. Terasawa, K. Sumiyoshi, T. Kajino, J. Mathews, I. Tanihata, Astrophys. J. **562**, 470 (2001).
13. Hashimoto *et al.*, Nuc. Phys. A **764**, 330c (2004).
14. C. Angulo *et al.*, Nucl. Phys. A **716**, 211 (2003).
15. G.V. Rogachev *et al.*, Phys. Rev. C **64**, 061601 (2001).
16. F. Barker, A.M. Mukhamezhanov, Nucl. Phys. A **673**, 526 (2000).

Eur. Phys. J. A **27**, s01, 321–324 (2006)
DOI: 10.1140/epja/i2006-08-049-x

EPJ A direct
electronic only

Testing of the RIKEN-ATOMKI CsI(Tl) array in the study of ^{22,23}O nuclear structure

Z. Elekes[1,2,a], Zs. Dombrádi[1], S. Bishop[2], Zs. Fülöp[1], J. Gibelin[3], T. Gomi[2], Y. Hashimoto[4], N. Imai[2], N. Iwasa[5], H. Iwasaki[6], G. Kalinka[1], Y. Kondo[4], A.A. Korsheninnikov[2,7], K. Kurita[8], M. Kurokawa[2], N. Matsui[4], T. Motobayashi[2], T. Nakamura[4], T. Nakao[6], E.Yu. Nikolskii[2,7], T.K. Ohnishi[2], T. Okumura[4], S. Ota[9], A. Perera[2], A. Saito[6], H. Sakurai[6], Y. Satou[4], D. Sohler[1], T. Sumikama[2], D. Suzuki[8], M. Suzuki[8], H. Takeda[6], S. Takeuchi[2], Y. Togano[8], and Y. Yanagisawa[2]

[1] Institute of Nuclear Research of the Hungarian Academy of Sciences, P.O. Box 51, Debrecen, H-4001, Hungary
[2] The Institute of Physical and Chemical Research, 2-1 Hirosawa, Wako, Saitama 351-0198, Japan
[3] Institut de Physique Nucléaire, 15 rue Georges Clemenceau, 91406 Orsay, France
[4] Tokyo Institute of Technology, 2-12-1 Ookayama, Meguro-ku, Tokyo 152-8550, Japan
[5] Tohoku University, Sendai, Miyagi 980-8578, Japan
[6] University of Tokyo, Tokyo 113-0033, Japan
[7] Kurchatov Institute, Kurchatov sq. 1, 123182, Moscow, Russia
[8] Rikkyo University, 3-34-1 Nishi-Ikebukuro, Toshima, Tokyo 171-8501, Japan
[9] Kyoto University, Kyoto 606-8501, Japan

Received: 27 July 2005 /
Published online: 15 March 2006 –

Abstract. The paper reports on the test of the newly developed RIKEN-ATOMKI CsI(Tl) array. It is demonstrated that high-quality detectors with excellent light collection efficiency and narrow resolution distribution could be produced. The setup has been commissioned in the study of the reaction of ^{22}O + CD_2. In the present paper, we summarize the results mainly on the (d, d′) channel. From the cross section for the transition between the ground state and the first 2^+ state, we could deduce the "matter" deformation parameter to be $\beta_M = 0.23 \pm 0.02$ by distorted wave analysis. Comparing this data with previous measurements it can be concluded that ^{22}O isotope has moderate and similar neutron and proton deformations.

PACS. 25.70.Hi Transfer reactions – 27.30.+t $20 \leq A \leq 38$ – 29.30.Kv X- and gamma-ray spectroscopy – 29.30.Ep Charged-particle spectroscopy

1 Introduction

The existing and forthcoming radioactive beam facilities open a wide range of possibilities to study exotic nuclei and nuclear processes which play a crucial role in nucleosynthesis. To study direct nuclear reactions of astrophysical and nuclear structure interest in inverse kinematics a light ion spectrometer has been constructed in RIKEN to be applied at the RIBF accelerator facility that is under construction [1].

2 Characterization of the CsI(Tl) detectors

For charged particle detection, CsI(Tl) is an excellent material of choice: it has high light yield; the wavelength of the emitted light (~ 550 nm) matches the sensitivity of silicon photodiodes well; the decay time of the light emission is a function of the particle type, enabling thereby particle identification. Since CsI(Tl) is only lightly hygroscopic, light reflector wrapping need not to be humidity tight. The only serious drawback of CsI(Tl) is the existence of a long (~ 8 μs) time constant light component. Despite the good light yield, highly efficient and uniform collection of scintillation light from all over the detector volume is mandatory from the point of view of good energy resolution and particle identification alike [2]. This is especially important for low energies where the signal/noise ratio is limited by the inherent leakage current noise of the photodiode. While the necessary thickness of the scintillator is determined by the highest energy of the particles to be stopped in it, the lateral dimensions are dependent on factors like the degree of granularity needed, the size of the photodiode, etc. In our case the detectors are 55 mm thick, making them able to stop charged particles up to 120 MeV/amu, and their cross section is 16 × 16 mm^2 except for the last 5 mm, where they are tapered to match the 10 × 10 mm^2

[a] e-mail: elekes@atomki.hu

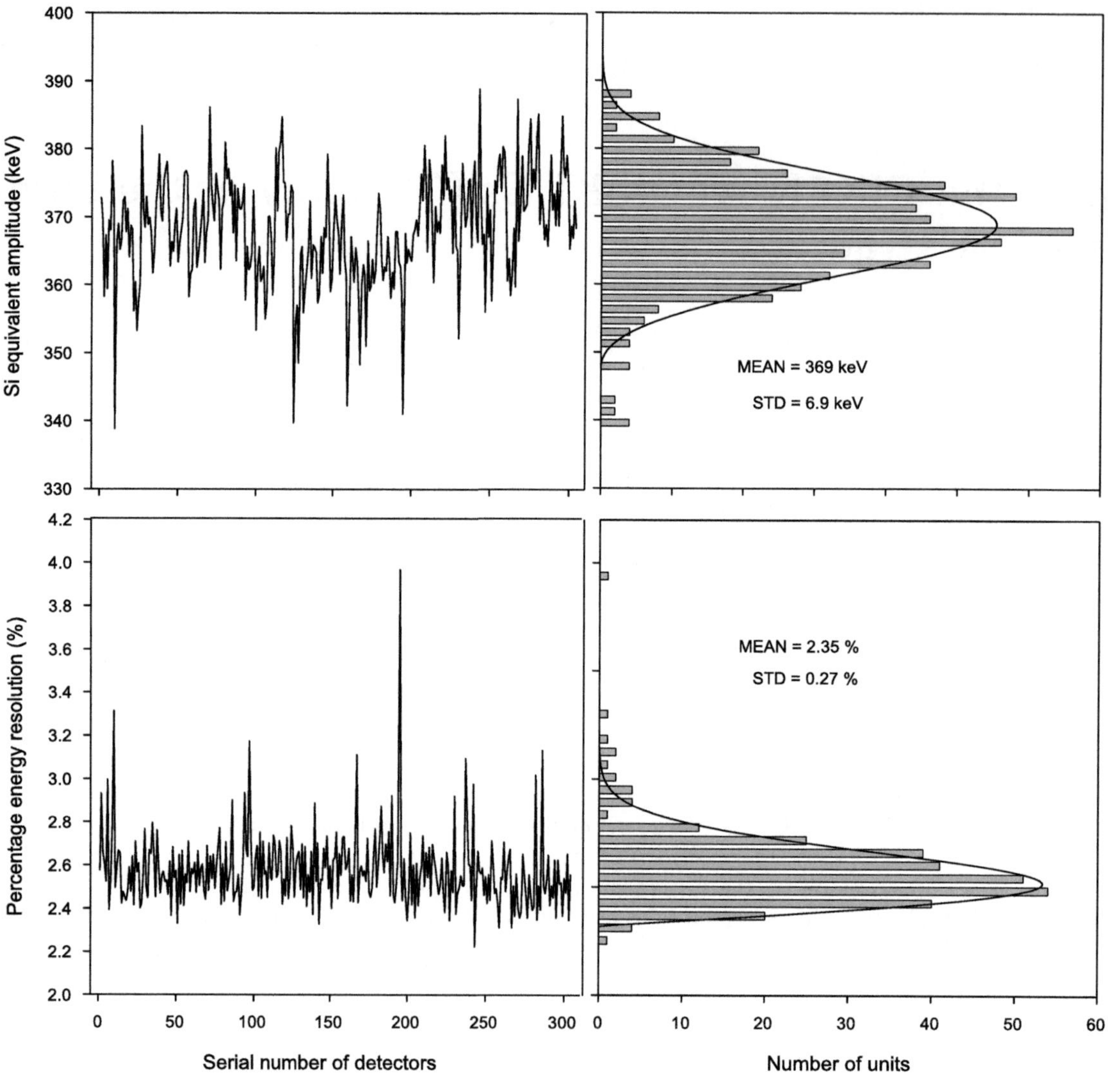

Fig. 1. Distribution of the light collection efficiency and resolution of the detectors.

cross section photodiodes. The light collection efficiency and its uniformity has been optimized through measurements using numerous combinations of different surface conditions and of various wrapping materials for the sides and front face, and also by modeling with Monte Carlo calculations [3]. The best solution for our purpose [4] is front face polished, side surfaces specially depolished and covered with two layers of 60 μm thick VM2000 mirror film [5]. For the front face, this mirror film or thin aluminized Mylar foil can be used, depending on the type and energy of the particles to be detected. Similar solutions in the third generation of DIAMANT detector system [6] in conjunction with Euroball, or in the GLAST space detector system [7], also justifies the choice of this mirror foil as a promising new alternative in scintillation detection to the presently overwhelmingly applied diffuse reflectors.

The quality of the applied technology can be assessed from fig. 1, which summarizes the performance test results obtained with 5.5 MeV alpha particles for all the 312 detector units completed, in the form of amplitude and resolution distribution graphs. The light collection is so well reproducible that the amplitude distribution is only slightly wider (FWHM = 3.2%), than the width of an individual spectral peak (2.55%). This means, that practically no gain matching in the array is necessary. During this test all detectors were covered with a thin Aluminized Mylar front reflector. By replacing it with a VM2000 film, which poses no problem in the detection of high-energy light particles, the amplitudes can be increased by $\approx 20\%$, significantly improving thereby the energy resolution. The average value of the low-energy background continuum for 5.5 MeV alphas, not shown in the figure, is less than 3%. It is worth mentioning that despite of the much larger scintillator crystals ($16 \times 16 \times 55\,\mathrm{mm}^3$ *vs.* $14.5 \times 14.5 \times 3\,\mathrm{mm}^3$) of this system the energy resolution values are only slightly lower (2.5% *vs.* 2.1%) than that of the DIAMANT system [6]. Please note, that besides charged particles, CsI(Tl) is a sensitive and high-performance gamma ray detector. For gammas, the light yield is ≈ 30 photon/keV with $< 0.3\%$ nonuniformity along the crystal length, whereas the energy resolution for 511 keV is $< 10\%$.

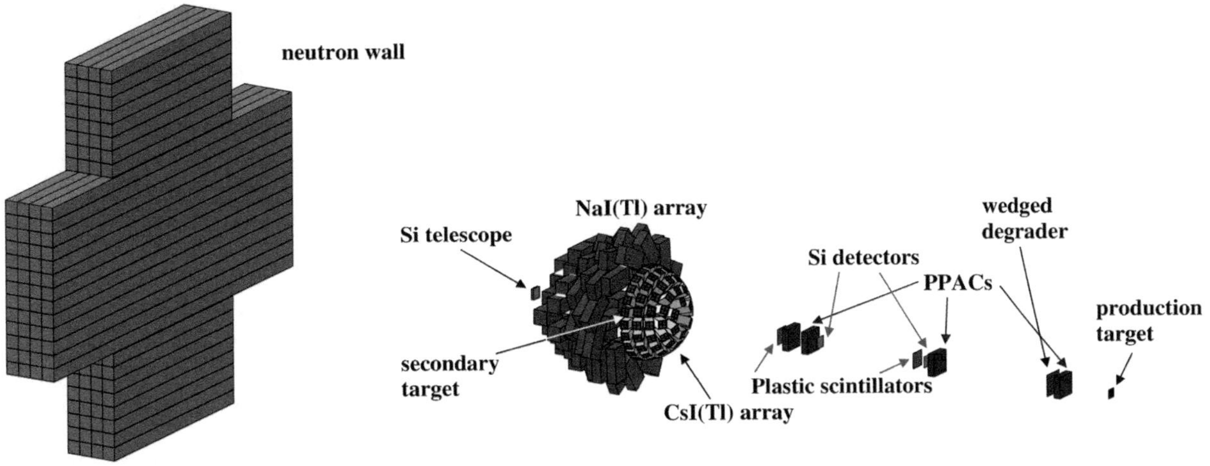

Fig. 2. Schematic view of the experimental setup.

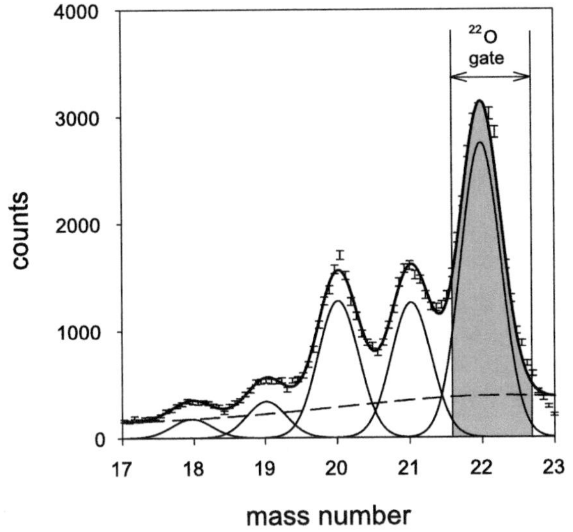

Fig. 3. Separation of oxygen isotopes using ΔE-E information in the silicon telescope. The bold solid line is a sum of 5 Gaussian functions and a polynomial background. The individual Gaussians and the background function are also plotted with thin solid lines.

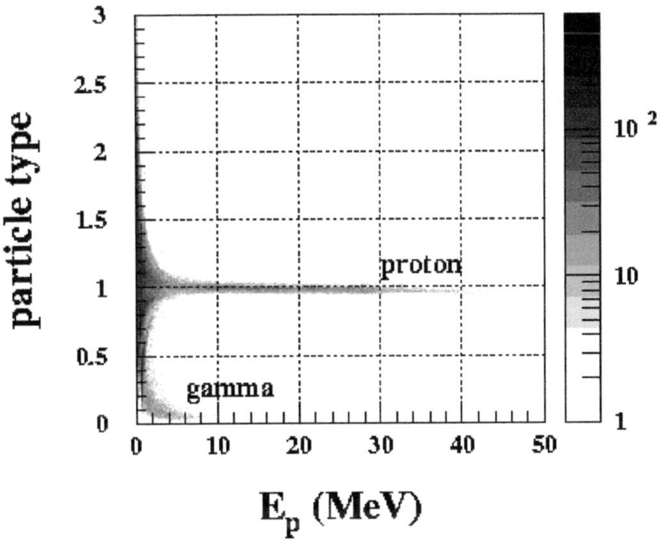

Fig. 4. Particle identification performed by the CsI(Tl) array in coincidence with the incident ^{22}O beam.

3 The commissioning experiment

In the commissioning experiment, a 94 A·MeV energy primary beam of ^{40}Ar with 60 pnA intensity hit a ^{9}Be production target of 0.3 cm thickness. The schematic view of the experimental setup is shown in fig. 2. The reaction products were momentum- and mass-analyzed by the RIPS fragment separator [8]. The secondary beam mainly included neutron-rich ^{25}Ne and ^{22}O nuclei. The RIPS was operated at 6% momentum acceptance. The total intensity was approximately 1500 cps having an average ^{22}O intensity of 600 cps. The identification of incident beam species was performed by energy loss and time of flight. The separation of ^{22}O particles was complete. Two plastic scintillators of 1 mm thickness were placed at the first and second focal planes (F2 and F3) to measure the TOF. Silicon detectors with thickness of 0.5 mm were inserted at F2 and F3 for energy loss determination. The secondary beam was transmitted to a secondary CD_2 target of 30 mg/cm^2 at the final focus of RIPS. The reaction occurred at an energy of 34 A·MeV. The position of the incident particles was determined by two PPACs placed at F3 upstream of the target. The scattered particles were detected and identified by a 2×2 matrix silicon telescope placed 96 cm downstream of the target. The telescope consisted of four layers with thicknesses of 0.5, 0.5, 2 and 2 mm. The first two layers were made of stripped detectors measuring the x and y positions of the fragments. On the basis of ΔE-E information, separation was carried out among the different oxygen isotopes which is demonstrated in fig. 3 where the linearized mass spectrum of oxygen nuclei is shown for one segment of the telescope. The protons emitted backward in the reaction were detected by 156 CsI(Tl)

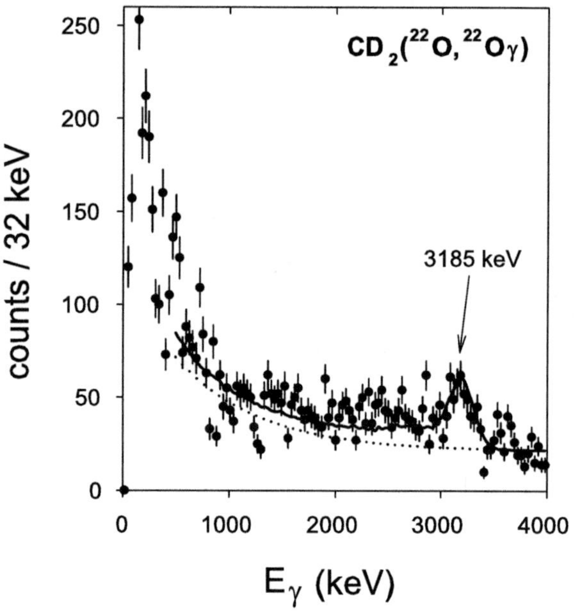

Fig. 5. Doppler-corrected spectra of γ-rays emerging from $^{22}O + CD_2$ reaction. The solid line is the final fit including the spectrum curve from GEANT4 simulation and an additional smooth polynomial background plotted as a separate dotted line.

scintillator crystals read out by photodiodes. The particle identification quality is presented in fig. 4 where the gamma rays and protons are well separated from each other down to 1–2 MeV energy of protons. 80 NaI(Tl) scintillator detectors also surrounded the target to detect de-exciting γ-rays emitted by the inelastically scattered nuclei. The intrinsic energy resolution of the array was 10% (FWHM) for a 662 keV γ-ray energy. The neutrons coming from the produced excited ^{23}O nuclei were detected by a neutron wall consisting of four layers of plastic scintillators placed at 2.5 m downstream of the target.

4 Results and discussion

In fig. 5 the Doppler-corrected γ-ray spectra for ^{22}O nucleus is presented, which is produced by putting an additional gate on the time spectra of the NaI(Tl) detectors selecting the prompt events and subtracting the random coincidences. By fitting the spectrum with a Gaussian function and smooth exponential background, first, the position of the single peak was determined at 3185(15) keV. The quoted uncertainty of the peak position is the square root of the sum of the squared uncertainties including two main errors namely the statistical one and the one due to Doppler correction. The above energy for ^{22}O is in a good agreement with the value 3199(8) keV determined earlier [9]. After the peak position has been determined it was fed into the detector simulation software GEANT4 [10] and the resultant response curve plus a smooth polynomial background was used to analyze the experimental spectrum and determine the cross section for $^{22}O + {}^{2}H$ reaction to be $\sigma(0_1^+ \to 2_1^+) = 19 \pm 3$ mb. From a distorted-wave analysis, we derived the "matter" deformation length (δ_M). In the calculation, the standard collective form factors were applied and the parameter set in [11] was employed for the optical potential. The "matter" deformation length deduced in this way is $\delta_M = 0.77 \pm 0.07$ fm which corresponds to a moderate mass deformation of $\beta_M = 0.23 \pm 0.02$. We can compare this result with the data from $^{22}O + {}^{197}Au$ reaction [12] where the sensitivity of the probe for proton and neutron distributions is different from that of our case. In the cited work, the proton deformation (β_p) was derived to be between 0.2 and 0.24. This means that the neutron deformation of ^{22}O is very close to that of the proton one taking into account the mass deformation determined in the present study. This result is in contrast with the expectations that the increasing neutron number may lead to a stronger neutron decoupling. In reality, the $M_n/M_p \sim \beta_n/\beta_p$ ratios are 2, 3 and 1 for ^{18}O, ^{20}O and ^{22}O, respectively. The increasing trend is stopped by the $N = 14$ subshell closure, which was indicated already by the high energy of the 2_1^+ state as well as by the small value of the $B(E2; 0_1^+ \to 2_1^+)$. The subshell closure makes both the proton and neutron distributions nearly spherical in ^{22}O.

We would like to thank the RIKEN Ring Cyclotron staff for their assistance during the experiment. One of the authors (Z.E.) is grateful for the JSPS Fellowship Program in RIKEN and thanks the support from OTKA F60348. The European authors thank the kind hospitality and support from RIKEN. The present work was partly supported by the Grant-in-Aid for Scientific Research (No. 1520417) by the Ministry of Education, Culture, Sports, Science and Technology and by OTKA T38404, T42733 and T46901.

References

1. T. Motobayashi *et al.*, Nucl. Instrum. Methods A **204**, 736 (2003).
2. J. Gál *et al.*, Nucl. Instrum. Methods A **366**, 120 (1995).
3. E. Frlez *et al.*, Comp. Phys. Comm. **134**, 110 (2001).
4. Z. Elekes *et al.*, Nucl. Phys. A **719**, 316C (2003).
5. M.F. Weber *et al.*, Science **287**, 2451 (2000).
6. G. Kalinka *et al.*, ATOMKI Annu. Rep. **64** (2001).
7. `http://glast.stanford.edu`.
8. T. Kubo *et al.*, Nucl. Instrum. Methods B **70**, 309 (1992).
9. M. Stanoiu *et al.*, Phys. Rev. C **69**, 034312 (2004).
10. S. Agostinelli *et al.*, Nucl. Instrum. Methods A **506**, 250 (2003).
11. R.D. Cooper *et al.*, Nucl. Phys. A **218**, 249 (1974).
12. P.G. Thirolf *et al.*, Phys. Lett. B **485**, 16 (2000).

8 Perspectives of Nuclear Physics and Astrophysics

Eur. Phys. J. A **27**, s01, 327–332 (2006)
DOI: 10.1140/epja/i2006-08-050-5

EPJ A direct
electronic only

Nuclear astrophysics at the east drip line

S. Kubono[1,a], T. Teranishi[2], M. Notani[3], H. Yamaguchi[1], A. Saito[1], J.J. He[1], Y. Wakabayashi[1,2], H. Fujikawa[1], G. Amadio[1], H. Baba[4], T. Fukuchi[5], S. Shimoura[1], S. Michimasa[4], S. Nishimura[4], M. Nishimura[2], Y. Gono[4], A. Odahara[6], S. Kato[7], J.Y. Moon[8], J.H. Lee[8], Y.K. Kwon[8], C.S. Lee[8], K.I. Hahn[9], Zs. Fülöp[10], V. Guimarães[11], and R. Lichtenthaler[11]

1 Center for Nuclear Study (CNS), University of Tokyo, Wako Branch at RIKEN, Hirosawa 2-1, Wako, Saitama, 351-0198 Japan
2 Department of Physics, Kyushu University, Fukuoka, 812-8581 Japan
3 Physics Division, Argonne National Laboratory, IL 60439, USA
4 RIKEN, Saitama, 351-0198 Japan
5 Rikkyo University, Toshima, Tokyo, 171-0021 Japan
6 Nishi-Nippon Institute of Technology, Fukuoka, 800-0394 Japan
7 Department of Physics, Yamagata University, Yamagata, 999-8560 Japan
8 Department of Physics, Chung-Ang University, Seoul, 156-756 Korea
9 Ewha Womens' University, Seoul, 120-750 Korea
10 Institute of Nuclear Research (ATOMKI), Debrecen, H-4001 Hungary
11 Departmento de Fysica Nuclear, Universidade de São Paulo, São Paulo, Brazil

Received: 28 June 2005 /
Published online: 23 March 2006 –

Abstract. In the first half of the paper, the nuclear astrophysics activities in Japan, especially in experimental studies are briefly overviewed. A variety of beams have been developed and used for nuclear astrophysics experiments in Japan. The activities include the RI beam facilities at low energies by the in-flight method at the Center for Nuclear Study (CNS), University of Tokyo and by the ISOL-based method at the JAERI tandem facility, and the RI beam facility at intermediate energies at RIKEN. Other activities include a study of the $^{12}C(\alpha,\gamma)^{16}O$ reaction exclusively at the tandem accelerator at the Kyushu University, and studies at the neutron facility at Tokyo Institute of Technology and at the photon facility at AIST (Sanso-ken). Research opportunities in the future at RIBF, J-PARC, and SPRING8 are also discussed. A discussion on the research activities at CNS has been specifically extended in the latter half, including various possibilities in collaboration at the RI beam factory at RIKEN.

PACS. 25.70.Ef Resonances – 25.60.-t Reactions induced by unstable nuclei – 26.30.+k Nucleosynthesis in novae, supernovae and other explosive environments – 29.25.Rm Sources of radioactive nuclei

1 Introduction

The nuclear astrophysics activity in Japan is partially stimulated by the high activities in astronomy in Japan. The Kamiokande detector successfully observed the supernova neutrinos in 1987 for the first time, and the large-scale, high-resolution optical telescope SUBARU has been operational since 1997 in Hawaii. The successful operation of X-ray observatories in Japan is another element. Recently, the radio-observatory activities at Nobeyama have been decided to extend to the ALMA project, that is the next generation of the radio observatories, based on the US-Japan-Europe collaboration. Of course, the nuclear astrophysics activity inversely has stimulated, for instance, astronomical observations of r-process elements as well as the s-process elements in very metal-poor stars, and also observations of isotopic ratios rather than elements. For instance, a recent SUBARU observation has succeeded in determining the isotopic ratios for the element Eu.

Experimental efforts in nuclear astrophysics have been expanded very rapidly in the last two decades since the introduction of RI beams in nuclear physics. One of the major reasons is that the nuclear reactions involved in explosive phenomena in the universe can be directly investigated with RI beams at very low energies. Along the development of nuclear astrophysics several useful methods have been invented for the field. These include indirect methods such as the Coulomb dissociation method, the asymptotic normalization coefficient (ANC) method, and so on. These developments are summarized in ref. [1]. The

[a] e-mail: kubono@cns.s.u-tokyo.ac.jp

Table 1. Accelerator facilities for nuclear astrophysics in Japan.

Beam	Facility	Affiliation	Subjects
Stable beams	Tandem	Kyushu University	$^{12}C(\alpha,\gamma)^{16}O$
RI beams (low energies, in-flight)	CRIB	Univ. of Tokyo	Primordial NS, rp-process
RI beams (intermediate energies, in-flight)	RIPS	RIKEN	Coul. Dissoc., ANC
RI beams (low energies, ISOL based)	TRIAC	KEK-JAERI	primordial NS, SN-NS
Neutron beams	Pelletron	TIT	s-process, prim. NS
Photon beams	e-ring	Sanso-ken(AIST)	p-nuclei, s-process

early stage of the development was initiated also in Japan that can be seen in refs. [2,3,4,5,6,7,8]. The research activity of nuclear astrophysics is now widely accepted as an important subfield of nuclear physics. It can be easily understood by looking at session names in most large conferences, and also at research propagandas in proposals for large-scale facilities. One never misses subjects related to nuclear astrophysics.

As is well known, there are two ways for RI beam production. In-flight RI beam separation is the most popular way adopted and realized, but ISOL-based method is more powerful in a sense for nuclear astrophysics. These two type facilities are available in Japan. The former method has been adopted at RIKEN at intermediate energies, and at the Center for Nuclear Study (CNS), University of Tokyo at low energies, whereas the latter one is adopted at the KEK-JAERI collaboration. A neutron source is available at the pelletron facility at the Tokyo Institute of Technology. Photon beams, available at AIST (Sanso-ken), were introduced for nuclear astrophysics experiments. On the other hand, a pure, high intensity ^{12}C beam is also realized at low energies at the Tandem facility of the Kyushu University.

Table 1 summarizes the major beam facilities now available for nuclear astrophysics in Japan. Details will be described for each facility in sect. 2. There are some other facilities relevant to nuclear astrophysics. Nuclear reaction mechanisms at very low energies have been investigated at Tohoku University, and some stellar reactions are under investigation at the Research Center of Nuclear Physics, Osaka University by an indirect method. We specifically discuss the detail of the low-energy in-flight RI beam separator CRIB [9] at CNS in sect. 3, and the experimental results in sects. 4 and 5. A short summary is given in sect. 6.

2 Experimental facilities in japan

There are two-type RI beam facilities known, and both of them are available in Japan, as mentioned above. In-flight type RI beam separators include RIPS at RIKEN at intermediate energies and CRIB at CNS. In-flight separation mostly uses inverse kinematics to obtain the kinematical focusing effect. Unstable nuclei can be produced with heavy-ion induced reactions, separated in-flight and focused at the double-achromatic focal plane. A detail of RIPS will be explained in the contribution by Togano to this symposium. The most typical activities for nuclear astrophysics with RIPS are the investigation of stellar reactions using the Coulomb dissociation method. The low-energy separator CRIB will be discussed in detail in the next section. The RI Beam Factory (RIBF) at RIKEN, which is under construction, will begin delivering RI beams in 2006 or 2007, which will enable us to investigate the pathway of the r-process for the cosmochronology, first generation stars, and the supernova mechanism. The detail of the RIBF project may be found at `http://www.rarf.riken.go.jp/index-e.html`.

The ISOL-based facility, called TRIAC, very recently has been established at the Tandem laboratory in JAERI, Tokai by a collaboration of JAERI and KEK. Most part was transferred from the E-arena of the old INS (Institute for Nuclear Study, University of Tokyo). They are going to place their emphasis on acceleration of fission products for nuclear physics as well as for nuclear astrophysics. The facility may be operational for routine use from the fall of 2005. They also have a plan to increase the RI beam energies by the existing super-conducting Linac up to 8 MeV/u for ions of $q/A \geq 1/4$. A detailed introduction to the facility can be found at `http://triac.kek.jp/en/`.

Neutron capture reactions play a crucial role specifically for heavy element synthesis. The neutron facility at the Tokyo Institute of Technology has been running for more than ten years for nuclear astrophysics. The group headed by Nagai introduced large-volume NaI crystals to measure directly the capture-gamma rays, which made the measurements more reliable than the activation method. They have found an important contribution of p-wave under a certain condition even at very low energies, indicating breakdown of the $1/v$ rule. A proposal of an extensive neutron facility plan for nuclear astrophysics has been approved in the J-PARC project, the large hadron project by the joint venture of KEK and JAERI, although the proposal is not funded yet.

Another interesting beam for nuclear astrophysics is the real photon beam, which can be obtained by backscattering of the laser beam from stored electron beams. A beautiful experiment was demonstrated by Utsunomiya, which can be explained in detail in this volume of the proceedings. A new-generation photon beam facility is under preparation at Spring8 in Harima, Japan.

The last facility is the Tandem accelerator facility at Kyushu University. They have constructed an extensive recoil separator exclusively for the measurement of the stellar $^{12}C(\alpha,\gamma)^{16}O$ reaction at the He burning temperature region. They have succeeded to modify the Tandem

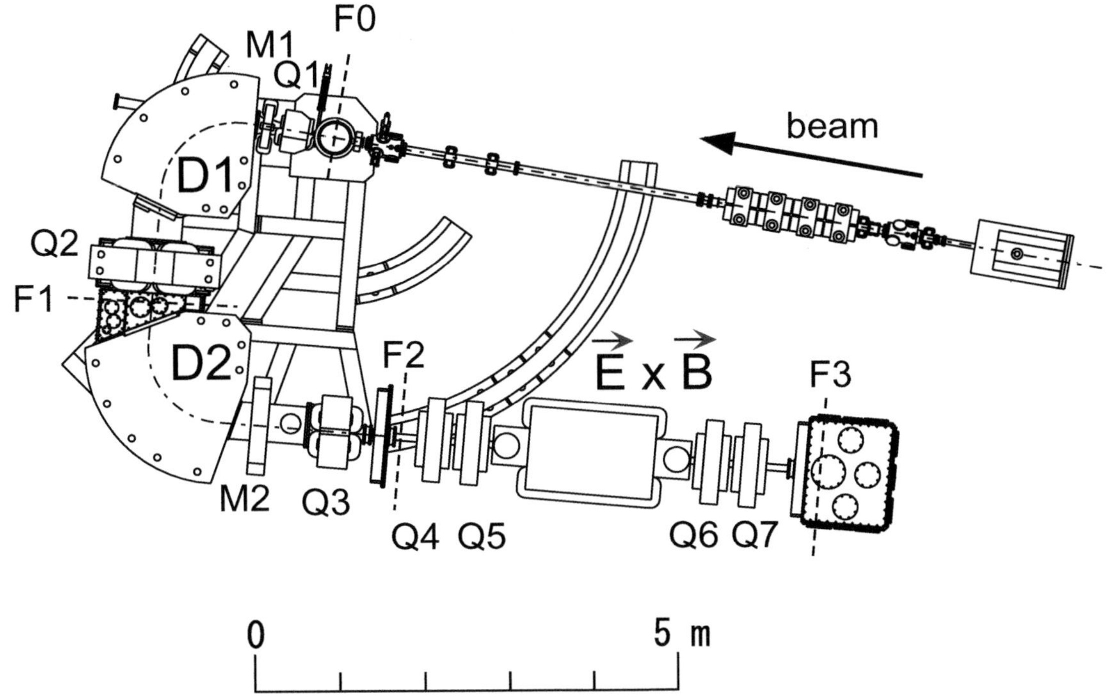

Fig. 1. Plane view of CRIB. Primary low-energy heavy-ion beams are provided from the AVF cyclotron of RIKEN.

Table 2. Intensities of RI beams below 10 AMeV obtained at CRIB.

RI beam	Primary beam	Reaction	Intensity (pps)	Purity (%)
^{7}Be	^{7}Li	^{1}H(^{7}Li, ^{7}Be)	1×10^{6}	90
^{8}Li	^{8}Li	^{2}H(^{7}Li, ^{8}Li)	1×10^{6}	100
^{10}C	^{10}B	^{1}H(^{10}B, ^{10}C)	1.6×10^{5}	90
^{14}O	^{14}N	^{1}H(^{14}N, ^{14}O)	1.6×10^{6}	90

operation for high intensities at very low energies. Currently, they are fighting to attain the beam suppression factor of 10^{-19} by improving the whole system by one order of magnitude.

As we overviewed above, we have a variety of opportunities for experimental nuclear astrophysics in Japan as well as great possibilities in the years to come.

3 The CRIB project at CNS

The CNS shut down their own cyclotron in 2000, and moved to the RIKEN campus, and immediately initiated a joint venture with RIKEN at the RIKEN Accelerator Research Facility (RARF). CRIB [10] is one of the major facilities that CNS introduced to RARF. In order to maximize the capability, we set a CNS-RIKEN joint project, AVF-Up Grade Project, under which we had decided to establish an extensive low-energy RI beam separator. Figure 1 displays the plane view of CRIB which consists of a double-achromatic magnetic section and a Wien Filter section. The configuration is F0-QMDQ-F1-DMQ-F2-QQ-WF-QQ-F3, where F*i* is the *i*-th focal plane of the optics, D a dipole magnet, Q a quadrupole magnet, and M a multipole magnet. One may use a degrader at F1, which is a momentum dispersive focal plane, for a better separation of the RI beam of interest.

The Wien filter section gives a capability of better particle separation and also provides some interesting features for other studies. The velocity separation section has 1.5 m-long electric parallel plates that have the maximum voltages of ±200 kV for a gap of 8 cm, giving 50 kV/cm. The maximum velocity dispersion designed was about 0.8 cm/%. The separation capability was verified with an ^{14}O beam produced from the ^{1}H(^{14}N, ^{14}O) reaction at F0. It gave almost 100% purity at F3. Another favorable feature of the RI beam from the filter is the beam quality. Under some condition, one may need only the Wien filter without the degrader, to obtain a small beam spot since the major factor for the RI beam size is due to the straggling at the degrader.

For example, an ^{14}O beam of 1.6×10^{6} pps was obtained from the ^{1}H(^{14}N, ^{14}O) reaction with a moderate beam intensity of 200 pnA of the primary beam at F0. We have succeeded to eliminate most contamination by the Wien filter. The RI beam size at F3 is still as large as 7×7 mm^{2} without degrader. Since this can be improved by correcting for the higher order aberration of the beam optics, a new sextupole magnet was installed. This work is in progress. The RI beam intensities (table 2) are limited

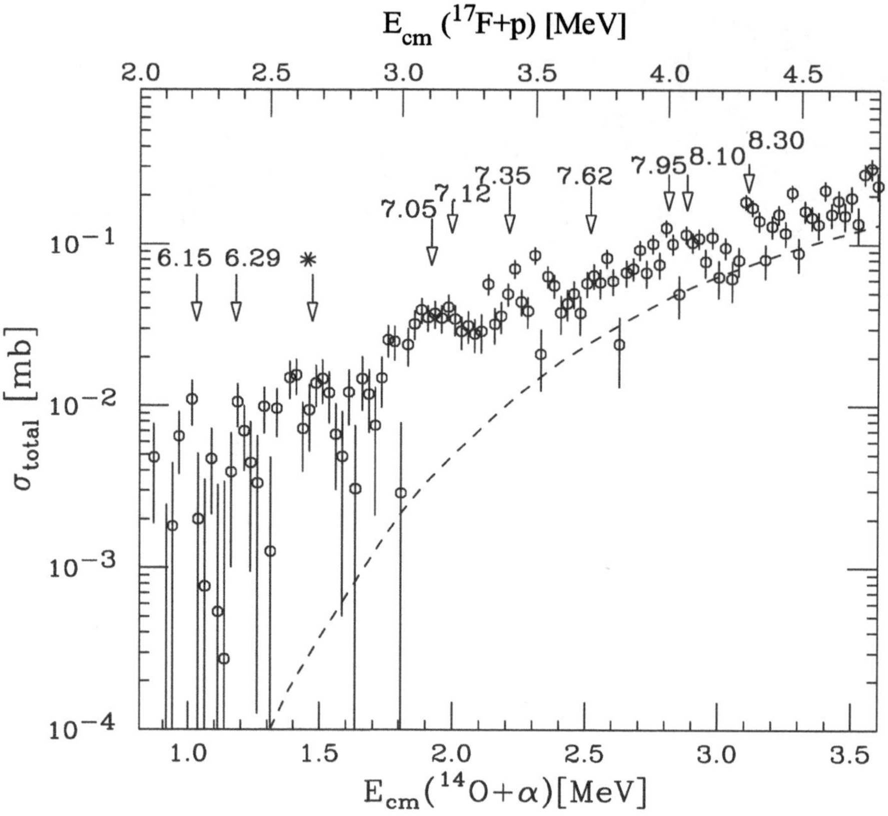

Fig. 2. Proton spectrum from the $^{14}O(\alpha, p)^{17}F$ reaction measured at $\Theta_{Lab} = 0°$. The asterisk indicates a transition to the 0.495 MeV first excited state in ^{17}F.

by two factors at this moment, the primary beam intensities and the production target. An intensive program is under way for the first part by installing the Hyper ECR source, and by upgrading many parts of the AVF cyclotron for the acceleration efficiency. This includes successful installation of a flat-top acceleration mode, a new deflector, new DC power supplies, etc. The accelerator energy factor K is now about 80, which was about 45 in practice before. As for the second part, a cooled gas target system is being developed that has a higher target density and can stand high beam currents. These efforts are in progress so that in the near future we may begin to study the stellar (p, γ) reaction directly with CRIB. This project should be a nice demonstration that a small accelerator facility can make a great contribution for RI beam physics.

4 The direct measurement of the $^{14}O(\alpha, p)^{17}F$ reaction

A series of reaction studies with low-energy RI beams has been undertaken at CRIB, specifically, for the onset and the early stage of the explosive hydrogen burning (rp) process.

The high-temperature (high-T) rp-process may typically take place in an X-ray burst, which is considered to be an event on the surface of a neutron star with accretion of hydrogen from the companion star in the main sequence phase. Here, one of the most critical stellar reactions is $^{14}O(\alpha, p)^{17}F$ for the ignition of the high-T rp-process. There are many experiments performed by indirect methods [11,12] for the problem, but not by the direct method with an ^{14}O beam. Two experiments were made previously using the time-reverse reaction $^{17}F(p, \alpha)^{14}O$. Only some transitions through resonances above $E_{cm} = 3$ MeV in ^{18}Ne were reported [13,14]. Note that the most critical energy region is around E_{cm} = 1–2 MeV for the present problem. This reaction has been successfully investigated for the first time using a high intensity ^{14}O beam from CRIB.

A low-energy ^{14}O beam was produced by the $^{1}H(^{14}N, ^{14}O)$ reaction at 8.4 MeV/u, and separated. The intensity and the purity of the beam was 1.6×10^6 pps and 85%, respectively, at F2. The momentum spread of the beam was defined to 1% by setting an aperture at F1, the momentum dispersive focal plane. The secondary target of He was cooled down to about 30 K, so that the target length was shortened roughly by a factor of 10, which made the present experiment possible. This is described in detail in ref. [15].

For the measurement of the $^{14}O(\alpha, p)^{17}F$ cross section, we applied the thick-target method [1], which has been developed in the last decade for low-energy RIB experiments and applied for proton elastic scattering experiments. This method was successfully used in the present case. Figure 2 displays a proton spectrum measured at 0° with a

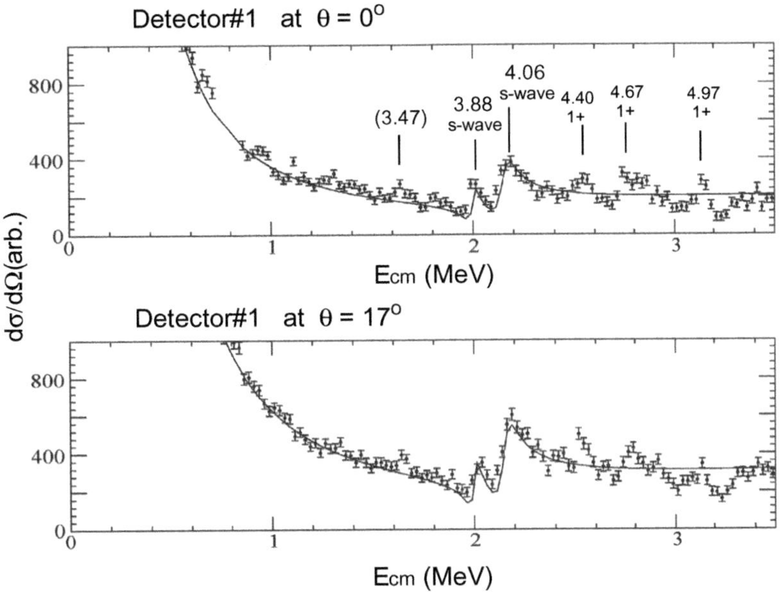

Fig. 3. Elastic scattering of ^{23}Mg+p measured at $\Theta_{Lab} = 0°$ and 17°. The solid line is an *R*-matrix fit. Possible states are indicated by the excitation energies in ^{24}Al.

silicon counter telescope. Several peaks are clearly seen that correspond to the (α, p) reaction mostly leading to the ground state in ^{17}F. The transitions through the 6.15 and 6.29 MeV states in ^{18}Ne were seen for the first time. These transitions were considered to be the main contributions to the stellar reactions under the X-ray burst condition [11]. The cross sections are roughly the same as predicted in ref. [11], thus confirming the primary importance of the two contributions. The transitions through the states at 7–8 MeV are also clearly observed. The peak around 6.5 MeV is considered to be the transition through the state at 7.1 MeV in ^{18}Ne decaying to the first excited state at 0.495 MeV in ^{17}F. Since there is no state of large proton width in the ^{17}F+p scattering [14] and the states in this energy region in ^{18}Ne cannot have a large α width as it is so close to the α threshold, the peak around 6.5 MeV cannot be explained by a state in ^{18}Ne. This implies that the transition through the 7.1 MeV state in ^{18}Ne increases the reaction rate roughly by 50%. Note that the reaction study with the time-reverse reaction cannot access this process.

5 Search for proton resonances relevant to the early stage of the rp-process

The mechanism of the early stage of the rp-process is of great interest. Previously, we studied by indirect methods the excited states near and above the proton threshold in the proton-rich nuclei, relevant to the early stage of the rp-process [16], where many new states were identified. However, the reaction rates are not determined yet because the resonance properties are not known. Thus, we have started to investigate the properties of these proton resonances by the direct method. So far, we studied the proton resonant scattering of ^{21}Na+p, ^{22}Mg+p, ^{23}Mg+p, ^{25}Al+p and ^{26}Si+p as well as ^{24}Mg+p for testing the thick target method [1] with the present experimental setup. The present data of ^{24}Mg+p are very well reproduced by the *R*-matrix calculation [17] with known resonance parameters, confirming the validity of the method. Preliminary results on ^{25}Al+p and ^{26}Si+p are presented elsewhere [18].

Figure 3 displays the proton excitation functions of ^{23}Mg+p. This is the first experiment to investigate ^{24}Al by proton resonance scattering. No resonance parameters were known before for the states in ^{24}Al [19]. We can see clearly two resonances at 3.88 and 4.06 MeV, and they are fitted well by *s*-wave resonances and are in agreement with previously known states at 3.885 and 4.059 MeV [19]. We can also see three resonances that probably correspond to the 1^+ states known by the beta decay study of ^{24}Si [20]. Detailed analysis is in progress.

6 Summary

Experimental facilities for nuclear astrophysics in Japan are briefly overviewed together with their research activities. Most of the interesting beams are available and the future scope is also bright for this field. These include high-energy RI beams of very short-lived, very neutron-rich nuclei at RIKEN, extensive photon beams at Spring8, and high-intensity neutron beams at J-PARC.

In contrast to these grand scale facilities, small machine facilities also can make great contributions to nuclear astrophysics. Our extensive low-energy RI beam facility, CRIB, demonstrates such feasibility. We may investigate stellar reactions of the rp-process very efficiently at CRIB. We have shown that an in-flight RI beam production method at low energies is very useful, and has a possibility to obtain RI beam intensity of the order of 10^8 pps for light nuclides near the line of stability. The Wien filter gives a better condition for the property of RI beams because one does not always have to use the degrader that deteriorates the RI beam quality considerably. The present work demonstrates that even small accelerator laboratories can devote to extensive RI beam programs including nuclear astrophysics.

References

1. S. Kubono, Nucl. Phys. A **693**, 221 (2001) and references therein.
2. S. Kubono, M. Ishihara, T. Nomura, *The Proceedings of the International Symposium on Heavy Ion Physics and Nuclear Astrophysical Problems* (World Scientific Publishing Company, Singapore, 1989).
3. S. Kubono, T. Kajino, *The Proceeding of the International Workshop on Unstable Nuclei in Astrophysics* (World Scientific Publishing Company, Singapore, 1992).
4. S. Kubono, T. Kajino, *The Proceedings of the International Symposium on Origin and Evolution of the Elements* (World Scientific Publishing Company, Singapore, 1993).
5. T. Kajino, S. Kubono, Y. Yoshii, *The Proceedings of the International Symposium on Origin of Matter and Evolution of Galaxies 1996* (World Scientific Publishing Company, Singapore, 1996).
6. S. Kubono, T. Kajino, K. Nomoto, I. Tanihata, *The Proceedings of the International Symposium on Origin of Matter and Evolution of Galaxies 97* (World Scientific Publishing Company, Singapore, 1999).
7. T. Kajino, S. Kubono, K. Nomoto, I. Tanihata, *The Proceedings of the International Symposium on Origin of Matter and Evolution of Galaxies 2000* (World Scientific Publishing Company, Singapore, 2003).
8. M. Terasawa, S. Kubono, T. Kishida, T. Kajino, T. Motobayashi, K. Nomoto, *The Proceedings of the International Symposium on Origin of Matter and Evolution of Galaxies 2004* (World Scientific Publishing Company, Singapore, 2004).
9. S. Kubono *et al.*, Eur. Phys. J. A **13**, 217 (2002).
10. Y. Yanagisawa, S. Kubono *et al.*, Nucl. Instrum. Methods A **539**, 74 (2005).
11. K.I. Hahn *et al.*, Phys. Rev. C **54**, 1999 (1996).
12. I.S. Park *et al.*, Phys. Rev. C **59**, 1182 (1999).
13. B. Harss *et al.*, Phys. Rev. Lett. **82**, 3964 (1999).
14. J.C. Blackmon *et al.*, Nucl. Phys. A **688**, 142c (2001).
15. M. Notani, S. Kubono, T. Teranishi *et al.*, Nucl. Phys. A **738**, 411 (2004).
16. S. Kubono, Prog. Theor. Phys. **96**, 275 (1996).
17. J.M. Blatt, L.C. Biedenharn, Rev. Mod. Phys. **24**, 258 (1952).
18. J.Y. Moon *et al.*, Nucl. Phys. A **758**, 158c (2005).
19. S. Kubono, T. Kajino, S. Kato, Nucl. Phys. A **588**, 521 (1995).
20. V. Banerjee *et al.*, Phys. Rev. C **67**, 024307 (2002).

Eur. Phys. J. A **27**, s01, 333–335 (2006)
DOI: 10.1140/epja/i2006-08-051-4

EPJ A direct
electronic only

Radiative electron capture —A tool to detect He^{++} in space

D. Chmielewska[a] and Z. Sujkowski

The Andrzej Soltan Institute for Nuclear Studies, 05-400 Otwock - Świerk, Poland

Received: 13 July 2005 /
Published online: 23 March 2006 –

Abstract. Large clouds of totally ionised helium and hydrogen might exist in the intergalactic space, invisible for the usual optical observations. These clouds, bombarded with medium energy electrons, should generate photon radiation in the X-ray region. The radiation is associated with the Radiative Electron Capture, REC, by He^{++} ions and should be observable with space born detectors. The photon spectra calculated for a range of temperatures of the electron spectra as well as several column densities of the plasma clouds are discussed.

PACS. 25.40.Lw Radiative capture – 98.62.Ra Intergalactic matter; quasar absorption and emission-line systems; Lyman forest

1 Introduction

One of the most intriguing questions of to-day astrophysics is the mass balance of the Universe. The observed baryonic mass is only about $4 \pm 1\%$ of the total mass.

The nature of the "cold dark matter" [$(29 \pm 4)\%$] and of the "dark energy" [$(67 \pm 6)\%$], *i.e.* of the unobserved missing mass is largely unknown [1]. Various hypotheses are being put forward, such as, *e.g.*, the SUSY particles or the disappearance of massive particles in extra dimensions [2].

Fascinating as this discussion is, it requires precise book-keeping of the known, ordinary baryonic matter spread throughout the Universe. A possible contribution to this baryonic component, which has so far been largely neglected, may be due to the diffuse clouds of completely ionized hydrogen and helium in the intergalactic medium, IGM. This matter, presumably of primordial origin, escapes observation by the usual optical methods.

Clouds of neutral hydrogen have been detected in the earth-bound observatories as the so-called "Lyman forest" in the quasar's absorption spectra in a large range of z. The $^2H/^1H > 4 \times 10^{-5}$ isotope ratios have been determined. The column densities of about $5 \times 10^{17}\,cm^{-2}$ been deduced for the neutral ^1H [3]. Much larger quantities of the hydrogen plasma are presumably associated with these clouds.

The first evidence of the existence of the helium plasma clouds stems from the observation of the red shifted 304 Å absorption line in the light of the quasar Q0302-003 [4,5,6]. This line is characteristic for absorption by singly ionized helium, He^+. No lines expected for the neutral He ($\lambda_0 = 584$ Å) have been seen. These facts can be considered as an indirect evidence of the existence of substantial amounts of completely ionized hydrogen and helium plasma scattered in the IGM. The total mass of the baryonic matter distributed there may be far from insignificant.

Besides contributing to the total mass of the baryonic matter the clouds of the primordial hydrogen and helium in the IGM carry precious information relevant for the theories of Big-Bang nucleosynthesis (BBN) and for the cosmic chemical evolution. Information of this kind might be particularly valuable in the light of the recent high-precision measurements of the Cosmic Microwave Background —CMB [7].

The principle of the observation is illustrated in fig. 1. The He^+ clouds situated between a quasar and the observer display various red shifts. The corresponding z-values are a measure of the distances. To quote [4] "the intergalactic space appears to be peppered with tenuous clouds of possibly primordial gas that have not yet condensed into galaxies".

The observation of the He^+ line, combined with the non-observation of the neutral He lines, are not sufficient to determine the total amount of gas or gas-plasma in the clouds. They give some model dependent limits. To quote [4] further, they show "the tip of the iceberg in terms of the total baryonic mass present". It is thus obvious that any independent observable shedding light on the hidden part of this mass would be of considerable value. One such observable could be the photon spectrum in the X-ray region due to the radiative capture of fast electrons by the He^{++} ions. The present work describes the principle of such observation. The experiment verifying the

[a] e-mail: danka@ipj.gov.pl

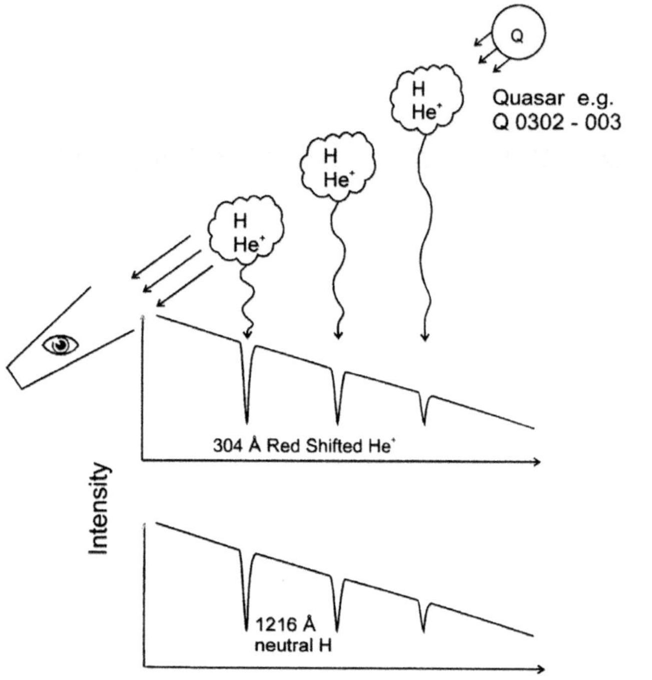

Fig. 1. The white spectrum light from a distant quasar is absorbed in clouds of ionised gas. The absorption lines are red-shifted depending on the distance of the clouds, giving rise to the "Lyman-α forest".

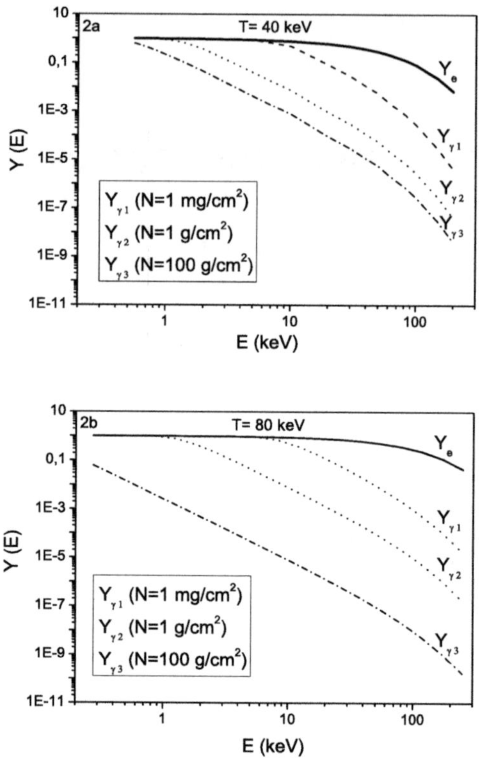

Fig. 2. The REC photon spectra due to the electron capture in the plasma clouds for electrons with $T = 40\,\mathrm{keV}$ (top) and $T = 80\,\mathrm{keV}$ (bottom) and for the column densities varying between $1\,\mathrm{mg/cm^2}$ and $100\,\mathrm{g/cm^2}$ (or between 1.5×10^{21} and 1.5×10^{26} atoms/cm^2).

theoretical cross-sections for the radiative capture process in the difficult to reach relativistic region is described separately [8].

2 Expected photon spectra due to Radiative Electron Capture in the plasma clouds

A general description of the Radiative Electron Capture process for fast relativistic electrons can be found, *e.g.*, in [9] (see also [10,11,12]).

The process corresponds to the time reversed photoelectric effect for atoms in partly ionized states:

$$Z^{q^+} + \hbar\nu \Longrightarrow Z^{(q+1)^+} + e^-, \qquad (1)$$

where Z is the atomic number and q is the charge of the ion.

In the case of He^{++} ions the inverse effect occurs for singly ionized helium, He^+. The cross-section for the REC effect, σ_{REC}, depends strongly on the atomic number of the capturing ion Z as well as on the velocity of the electrons, v_e.

Crudely

$$\sigma_{REC} \sim \frac{Z^{4.5}}{v^5}. \qquad (2)$$

The plasma clouds in the IGM are subject to continuous bombardment by fluxes of photons and fast electrons. The electron energy spectra are characterized by a power law with temperatures, T, ranging typically between few tens and few hundreds keV (*e.g.*, [13]).

The plasma clouds at sufficient column density are opaque to these electron fluxes, especially so at low electron energies. Due to the REC effect they convert the electron spectra to photon spectra of practically the same energy, cross-section weighted at each energy value:

$$Y_\gamma(E_\gamma) = Y_e(E_e)\left(1 - e^{-N\sigma_{REC}}\right), \qquad (3)$$

where $E\gamma = E_e$, $Y_e(E_e)$ is the number of electrons of energy E_e and N is the column number density of the cloud. The column number density is defined as the total number of particles per cm^2 in the column of the length equal the size of the cloud at a given section.

Figure 2 shows the photon spectra calculated for several temperatures and column densities. The background spectra of electrons, indicated schematically in fig. 2, are taken from the BATSE catalogue [13]. The REC spectra are calculated under the assumption of a spherical plasma cloud bombarded isotropically from all directions. This averages out the otherwise strongly anisotropic REC emission.

Only some educated guesses can be made about the N (plasma) values for He^{++} and H^+. The range of the values used in fig. 2 reflects in an exaggerated way the uncertainties in these estimates.

The relative contribution of H^{++} and H^{+} to the REC spectra can be obtained from the estimate

$$[N(\mathrm{He}) + N(\mathrm{He}^{+}) + N(\mathrm{He}^{++})] \approx 0.08[N(\mathrm{H}) + N(\mathrm{H}^{+})]$$

and from the approximate relationship $\sigma_{REC}(\mathrm{He}) \approx 2^{4.5}\sigma_{REC}(\mathrm{H})$ for any given E_e (see formula (2)).

Hence helium and hydrogen contribute about 2/3 and 1/3 to the total photon spectrum, respectively.

3 Summary and conclusions

It is shown that helium + hydrogen plasma clouds in the IGM can act as effective converters of electron-to-photon spectra in the observationally attractive energy range of a few to a few hundred keV.

This offers a possibility of obtaining an observational, quantitative information on the completely ionized hydrogen and helium plasma clouds in the Inter-Galactic Medium.

References

1. E. Sheldon, Acta Phys. Pol. B **3**, 243 (2002).
2. K. Ichiki *et al.*, Phys. Rev. D **68**, 083518 (2003).
3. A. Songaila *et al.*, Nature **385**, 137 (1997) and references therein.
4. P. Jakobsen *et al.*, Nature **370**, 35 (1994).
5. A. Songaila *et al.*, Nature **375**, 124 (1995).
6. A.F. Davidsen *et al.*, Nature **380**, 47 (1996).
7. B.D. Fields *et al.*, Phys. Rev. D **66**, 010001 (2002) and these proceedings.
8. A. Gójska *et al.*, these proceedings.
9. J. Eichler, Phys. Rep. **193**, 165 (1993).
10. Th. Stöhlker, habilitation thesis, GSI Darmstadt (1999).
11. Th. Stöhlker *et al.*, Phys. Rev. A **58**, 2043 (1998).
12. Z. Sujkowski, Nucl. Phys. A **719**, 266c (2003).
13. J.C. Ling *et al.*, Astron. J. Suppl. Ser. **127**, 79 (2000).

Eur. Phys. J. A **27**, s01, 337–342 (2006)
DOI: 10.1140/epja/i2006-08-052-3

EPJ A direct
electronic only

AMS —A powerful tool for probing nucleosynthesis via long-lived radionuclides

A. Wallner[1,a], R. Golser[1], W. Kutschera[1], A. Priller[1], P. Steier[1], and C. Vockenhuber[1,2]

[1] VERA-Laboratory, Institut für Isotopenforschung und Kernphysik, Universität Wien, Währinger Strasse 17, A-1090 Wien, Austria
[2] TRIUMF Laboratory, 4004 Wesbrook Mall, Vancouver, BC, V6T 2A3, Canada

Received: 3 July 2005 /
Published online: 23 March 2006 –

Abstract. Well-established data on production-rates of long-lived radionuclides are important for the understanding and calculation of various nucleosynthesis processes. However, lack of information exists for a list of nuclides as pointed out by nuclear-data requests. In addition, the search for supernova (SN)-produced radionuclides will give an improved insight into explosive scenarios. Accelerator mass spectrometry (AMS) represents a technique, which is capable to quantify such long-lived radionuclides using mass spectrometric methods. The potential of AMS is presented here as a powerful tool for probing nucleosynthesis. Applications of AMS are exemplified for a few specific cases: the detection of extraterrestrial radioactivity on Earth in terrestrial archives as a signature of nearby SN explosions, and the measurement of cross-sections, as an important ingredient for stellar as well as nuclear model calculations.

PACS. 07.75.+h Mass spectrometers – 26.20.+f Hydrostatic stellar nucleosynthesis – 26.30.+k Nucleosynthesis in novae, supernovae and other explosive environments – 26.35.+c Big Bang nucleosynthesis

1 Introduction

Nuclear astrophysics continues to be a rich and exciting research field. Pushed by the public interest, nuclear astrophysics is called to provide answers to various outstanding questions. In order to meet these inquiries, astrophysical research is a growing field [1]. Within this astrophysical context, nuclear-data activities are one basic instrument to help to answer such questions [2,3]. Nuclear data are needed for a better understanding of spectacular events such as the Big-Bang, star evolution and supernova (SN) explosions, but also to gain insight into the formation of our solar system. Nucleosynthesis gives us one of the very important fingerprints of nature, the elemental abundances. In order to understand this signature, the physics of nuclear reactions have to be well understood. Network calculations of element production in various scenarios have shown a great progress in the last years and they allow for reproducing a lot of the abundance distribution of the elements. However, we are still at a very early stage of understanding the physics leading to the well-known isotopic pattern of our Solar System. The success of astrophysical models depends strongly on the accuracy of nuclear data. These ingredients have to be provided from experimental investigations. One essential key to this request is the precise knowledge of reaction and production cross-sections. Such experimental input helps to verify nuclear model calculations and gives us a deeper understanding of the nuclear forces responsible for the nuclear interactions. This huge nuclear-data need pushes the activities of experimental facilities to investigate astrophysical questionnaires. Here we want to discuss the technique of accelerator mass spectrometry (AMS) as a valuable tool for probing nucleosynthesis. In the following sections the potential of AMS for quantifying long-lived radionuclides will be highlighted.

[a] e-mail: `anton.wallner@univie.ac.at`

2 Accelerator mass spectrometry at the Vienna Environmental Research Accelerator facility

Well-established data on production-rates of long-lived radionuclides are important for the various nucleosynthesis processes and they are highly desired. Lack of information on cross-section data exists for a number of nuclides as pointed out by nuclear-data requests. Especially long-lived radionuclides have often been inaccessible to decay counting techniques, *e.g.* because of low activity or an unfavorable decay scheme. In such cases the technique of AMS may provide valuable contributions to our understanding of celestial evolution and nucleosynthesis.

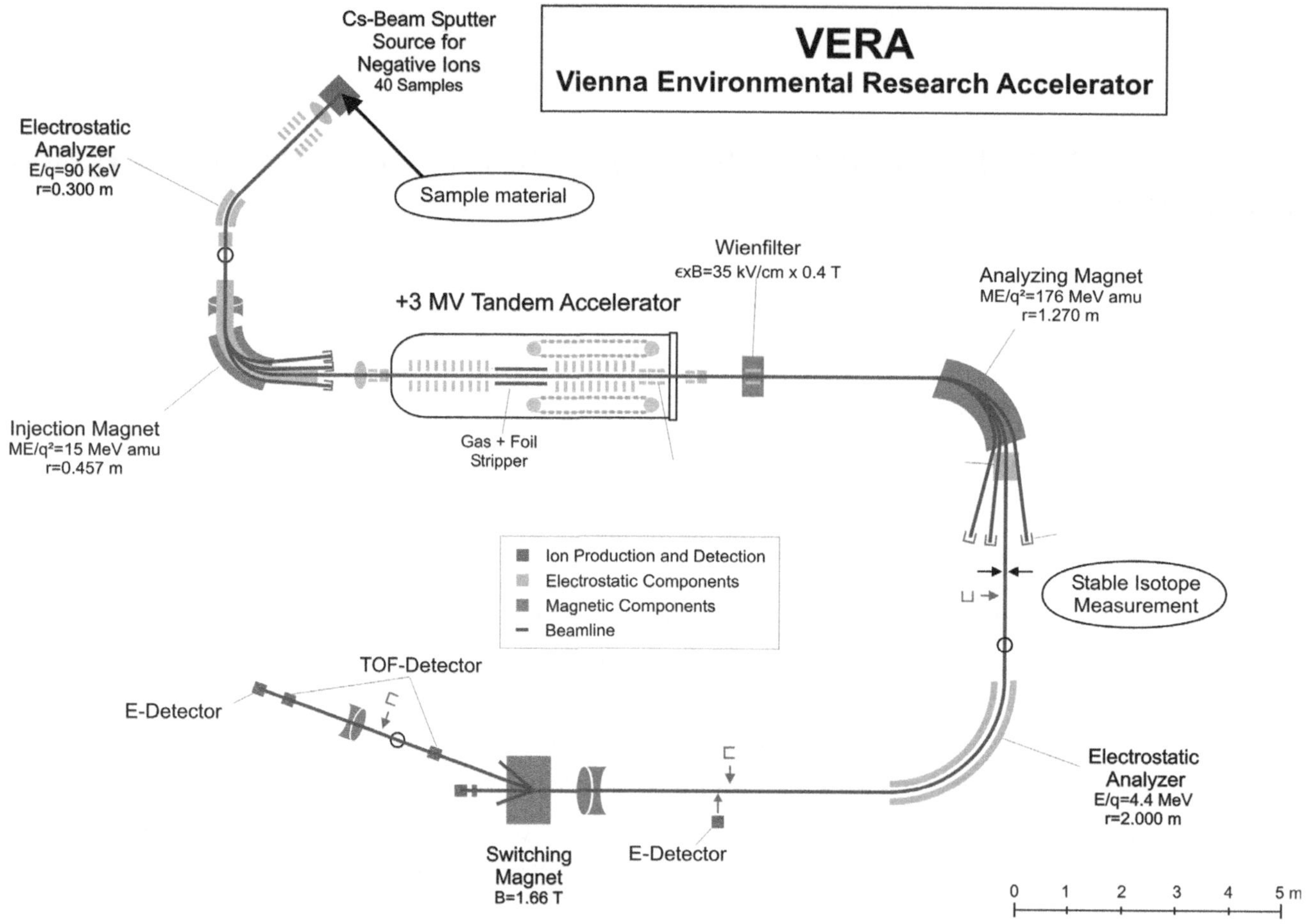

Fig. 1. Schematic layout of the VERA facility: Stable ions are measured with Faraday cups positioned after the injection magnet (low-energy section, negative ions) and after the analyzing magnet (high-energy section, positive ions). The rare isotope count-rates are obtained using energy-sensitive particle detectors. The detection system of the heavy-ion beamline (which is also used for ^{41}Ca-detection) consists of the time-of-flight (TOF) detectors and an energy (E) detector.

Accelerator mass spectrometry (AMS) represents an analytical method for the detection of various nuclides through direct atom-counting. This mass spectrometric technique offers the potential for quantifying isotope abundances down to levels of 10^{-15} and below. AMS is applied for the detection of radionuclides with half-lives between a few years and up to hundred million years. In combination with very low sample masses needed, AMS offers a tremendously higher sensitivity compared to the decay counting method, which is a consequence of the long half-life of those radionuclides.

In the following the method of AMS is summarized with focus on a typical AMS facility as represented by the Vienna Environmental Research Accelerator (VERA). VERA is based on a 3-MV pelletron tandem. It represents a state-of-the-art AMS facility (fig. 1), which provides the ability for quantifying nuclides over the whole mass range [4,5,6]. Some of the measured radioisotopes are, *e.g.*, ^{10}Be, ^{14}C, ^{26}Al, ^{36}Cl, ^{41}Ca, ^{129}I, ^{182}Hf, ^{236}U and ^{244}Pu. Since it offers highest sensitivity, AMS at VERA is predestinated for the measurement of minute concentrations of such isotopes. At VERA, AMS is used for quantifying long-lived radionuclides within a wide range of applications —from archaeology via climate research to astrophysics (for details see, *e.g.*, [7,8]).

AMS uses negative ion sputter sources: Solid sample material is inserted into an ion source. At VERA a sample-wheel containing positions for 40 samples can be loaded into the ion source. The sample material is sputtered using a Cs beam focused onto the sample. In most cases the sample material has to be converted for the AMS measurement into a suitable chemical form. AMS itself represents a "sample-destructive" technique. Typical samples masses are a few mg of material. The range of isotope ratios measured with AMS is of the order of 10^{-10} to 10^{-15}. It is also this dynamic range, which makes AMS a flexible and attractive technique. The sensitivity of AMS is exemplified using the following typical values: Assume 10 mg sample material (*e.g.*, Ca) is available for a measurement (which will give several sputter targets for the AMS measurement). If we combine a typical isotope ratio of 10^{-12} (or 10^{-14}) and the sample mass, we obtain a total number of rare isotopes of about 10^8 (10^6) atoms. With an assumed half-life of 100 000 years ($\lambda = 2\cdot10^{-13}\,\mathrm{s}^{-1}$), these numbers correspond to activities of 20 μBq (or 0.2 μBq), far below any chance to be quantified by activity measurements.

In AMS the typical measurement procedure is the following: In a cesium sputter source, negatively charged ions are produced and pre-accelerated before they pass a low-energy mass spectrometer, which analyzes a specific mass. The ions are further injected into a tandem accelerator (at VERA we use up to +3.5 MV terminal voltage). Any molecules that might contribute to a molecular interference are completely destroyed in the terminal stripper of the accelerator. Due to this stripping process, only atomic, mostly positively charged ions leave the tandem accelerator. A specific charge state is selected by a second, high-energy 90° analyzing magnet for further transport to the detector. At VERA using the 3-MV tandem, the particles have energies between 10 and 25 MeV. Stable isotopes will be measured as currents using Faraday cups while the radionuclides are counted directly with a particle detector. Rejection of isotopic interferences can be achieved with additional filters, like a Wien-filter, an electrostatic analyzer, and a time-of-flight system. Further reduction of any isobaric and isotopic interference is also achieved by means of specific energy-loss techniques, used either in front of the detection system or as a part of the particle detector (see also sect. 4, Ca-measurements with VERA). A final particle detector delivers the energy of the incoming ions while they are fully stopped. The detection beam-line at VERA is described in [6]; typical parameters for the various nuclides measured at VERA are listed in [5,9].

AMS is a mass spectrometric technique. Basically, it measures isotope count rates for different isotopes. To this end, sequentially, stable ion currents are measured with Faraday cups, positioned at the low-energy and high-energy sides of the AMS beamline (fig. 1). These current measurements are sandwiched by counting the rare isotope with the particle detector. With this raw data, *i.e.* countrate and particle current, an isotope ratio is calculated. However, possible long-term drifts of the particle transmission along the beamline have to be monitored. For quality control, the transmission is regularly monitored by means of standards with well-known isotope ratios. Another ion may also mimic a true event in the detector. Contamination from chemistry or in the ion source may induce an additional "true" signal. In addition, isobaric interferences have to be controlled. Note, that impurities in the sample after chemical preparation are typically on the ppm-level, whereas the rare isotope content is another 6 to 9 orders of magnitude lower. In order to quantify or check the background level, blank samples are therefore measured, too.

The overall efficiency (*i.e.* fraction of particles detected to that inserted into the ion source), which includes the efficiency for producing negative ions, stripping yield, transmission through the beam line and detector efficiency, depends strongly on the isotope under investigation. For carbon measurements up to several percent can be obtained at VERA but in other cases one has to deal with an overall efficiency as low as 10^{-5} [9].

3 Applications of AMS related to astrophysics

3.1 Search for supernova-produced radionuclides on Earth

Ellis, Fields and Schramm 1996 [10] and Fields [11] pointed out that there might be a chance for finding long-lived radionuclides in terrestrial archives, which were originally produced in a supernova (SN). Several candidates had been identified by [10], among them ^{26}Al, ^{53}Mn, ^{60}Fe, ^{146}Sm, ^{182}Hf, and ^{244}Pu which are produced in sufficient amounts to be in principle detectable on Earth. Such radioisotopes, which have been ejected into the interstellar medium, may be picked up by our Solar System when it passes through such a region and those nuclides will finally become incorporated into terrestrial archives, like sediments or ice cores. SN-rates in our stellar vicinity, are expected to be rather common. In a rough estimation Fields [11] calculates a rate of the order of a few SN events per million years within a distance of 100 parsec (pc). From that follows, that proper radionuclides should have half-lives also in the million-year range, far too short-lived for surviving from the formation of the Solar System. Radionuclides with those half-lives are best measured using the technique of AMS. It is crucial to exclude contamination of these SN-produced radionuclides from other natural sources (*e.g.*, cosmic ray production, natural fission).

Indeed, with the use of AMS, a strong peak of ^{60}Fe ($T_{1/2}$ = 1.5 Ma) was found recently in a deep-sea manganese crust profile [12], which confirms the presumption of Ellis *et al.*. This unusual isotope signal is interpreted as live radioactivity deposited on Earth originating from a supernova explosion three million years ago, and at a distance of several tens of pc [12,13]. Hence, they are eagerly sought fingerprints of recent, explosive nucleosynthesis. However, this signal needs confirmation from other radionuclides. Possible other candidates are the r-process nuclides ^{182}Hf (8.9 Ma), ^{244}Pu (80.6 Ma), and ^{247}Cm (16 Ma). In some cases (*e.g.*, ^{244}Pu), man-made contributions have to be carefully evaluated.

The extremely low isotopic abundance of such radionuclides asks for a very sensitive detection method. Accelerator mass spectrometry (AMS) is currently the only technique capable to measure such ultra-low isotope ratios. At VERA, techniques for measuring the isotopes ^{182}Hf and ^{244}Pu have been developed [34,14,15]. Whereas the detection of SN-produced ^{182}Hf still suffers from interference from the stable isobar ^{182}W and needs further technological developments, in the case of transuranium elements no stable isobar exists. Therefore, *e.g.* ^{244}Pu detection has to deal with the rejection of isotopic background only. The mass-selective low-energy injection system at VERA [16] and the dedicated low-background heavy-ion beamline in combination with the highly reproducible conditions make VERA a favorable system for measuring those isotopes. For the case of ^{244}Pu measurements, no isotopic interferences from other isotopes are found at VERA. This setup has also been proven to be very valuable for measuring the various other Pu isotopes in environmental samples [17]. In those cases, the man-made content of Pu is investigated which allows for example to reconstruct the neutron history of that environment. However, the isotopic signature of these other Pu isotopes (^{239,240,241,242}Pu) is a measure of the man-made ^{244}Pu, too, because this information can be used to extrapolate the anthropogenic content of those Pu isotopes to the expected anthropogenic ^{244}Pu concentration in the sample material. Using the 14 MV tandem accelerator in Munich, C. Wallner *et al.* found a ratio for ^{244}Pu/^{239}Pu of 10^{-3} [18]. In a first measurement of a deep-sea manganese crust they found one single count of ^{244}Pu in a background-free measurement. Since no ^{239}Pu was detected, no ^{244}Pu counts from anthropogenic sources are expected. A second, independent measurement to quantify ^{244}Pu in deep-sea sediment has been performed by M. Paul *et al.* [19]. They also found one count of ^{244}Pu in their sample, however this signal is compatible with their expectations of anthropogenic origin using also the ^{239}Pu content of that sample.

In ref. [17] typical spectra from Pu measurements of environmental samples measured at VERA are shown. Best discrimination is obtained using the information of 2-dimensional spectra which comprise time-of-flight signals and energy signals. It is demonstrated that the high-energy mass separator is capable to suppress ions from neighboring masses completely which is important for detecting ^{244}Pu. To summarize, using this setup, VERA has proven to be capable to quantify supernova-produced ^{244}Pu in terrestrial archives. If successful, the long half-life of ^{244}Pu (80 Ma) gives the possibility to trace the mean abundance of ^{244}Pu in the interstellar medium [18,19,20], which may contain ejected material from several SN explosions.

3.2 Measurement of production cross-sections

The observation of live radionuclides provides direct evidence for ongoing nucleosynthesis. Radionuclides, like ^{26}Al [21] and ^{44}Ti, can be traced in the sky as live radioactivity. Such mappings identify active areas in our Galaxy. The relevant stellar scenarios are an essential key for understanding celestial evolution. At VERA cross-section measurements have been performed, *e.g.* for the quantification of ^{26}Al [22,23]. The background level for ^{26}Al detection at VERA is found to be at $6 \cdot 10^{-16}$. This extremely high sensitivity enables us also to apply AMS for the measurement of the ^{25}Mg(p, γ)^{26}Al strength functions at astrophysical interesting energies [24]. To this end, Mg targets have been irradiated with protons. After the irradiation the produced ^{26}Al is extracted for the successive AMS measurement [25,26]. Because of the very small cross-sections a very limited number of ^{26}Al atoms is produced, which requires both, a high overall efficiency of the measurement and very low background interference.

Another prominent example for applying AMS is the neutron capture cross section of ^{62}Ni producing the radionuclide ^{63}Ni. This reaction plays an important role in the control of the flow path of the slow neutron-capture nucleosynthesis process [27,28]. Its cross-section was identified as a key value to describe correctly the isotopic pattern around ^{62}Ni [29,30]. AMS has also been used to determine the production of ^{44}Ti via the ^{40}Ca(α, γ) reaction [19]. Apart from these examples, a lot of additional measurements have been performed or are being performed using AMS. At VERA a measurement program is underway for studying neutron capture reactions for various isotopes, too.

4 Ca-measurements with VERA

Another AMS nuclide is ^{41}Ca. This radioisotope has a half-life of 104 000 years (see, *e.g.*, [31]). The long half-life together with its direct electron-capture decay to the ground state of ^{41}K makes this nuclide difficult to detect via decay counting. The measurement of this nuclide is of interest in astrophysics for a quantification of the ^{40}Ca(n, γ) cross section at stellar energies, which is underway at different laboratories. In the following some recent developments at VERA for the detection of the long-lived radionuclide ^{41}Ca are presented.

Natural ^{41}Ca/Ca isotope ratios are of the order of 10^{-14} to 10^{-15}. Such low ratios have been measured at other AMS laboratories using CaH_2 samples for AMS. CaH_2 needs a very elaborate sample preparation and sample handling. This chemical form was necessary because the stable isobar ^{41}K does not form stable negative $^{41}KH_3^-$ -ions. Thus, using CaH_2, effective isobar suppression is achieved and isotope ratios of the order of 10^{-15} can be quantified using AMS [32,33]. However, in cases where the ^{41}Ca content is not as low as in natural concentrations, *i.e.* for isotope ratios of the order of 10^{-13} or higher, an alternative chemical form can be used. Calcium fluoride samples, CaF_2, have been proven to be a suitable material for such applications with much simpler chemical preparation.

Here we want to present first results on ^{41}Ca detection with a 3 MV tandem accelerator using CaF_2. To this end a new isobar discrimination technique, the so-called deltaTOF technique [34,35] has been applied. Note, ^{41}Ca is not a typical nuclide for small accelerators since isobaric interferences at this mass range are usually not sufficiently resolved by standard AMS-techniques having particle energies below 1 MeV/amu. The measurement of CaF_2 samples at VERA has become possible by the availability of so-called silicon nitride (SiN) foils, which offer a very homogenous foil thickness. The method applied is the following: mass selected ions pass through sufficiently thick degrader foils and loose energy, depending on their atomic number Z. Thus isobars, which have initially the same velocity, have different velocities after the foil which is measured using a high resolution time-of-flight setup. The homogeneity of those SiN-foils yields resolvable energy discrimination above the inherent energy straggling.

We could utilize a series of ^{41}Ca standards, which were kindly supplied by K. Nishiizumi [36] and G. Korschinek. They span a range of isotope ratios from $1.2 \cdot 10^{-10}$ and $5.9 \cdot 10^{-13}$. These standards allow us to optimize our AMS-setup and to investigate the limits of a 3 MV tandem accelerator for quantifying ^{41}Ca using CaF_2 samples.

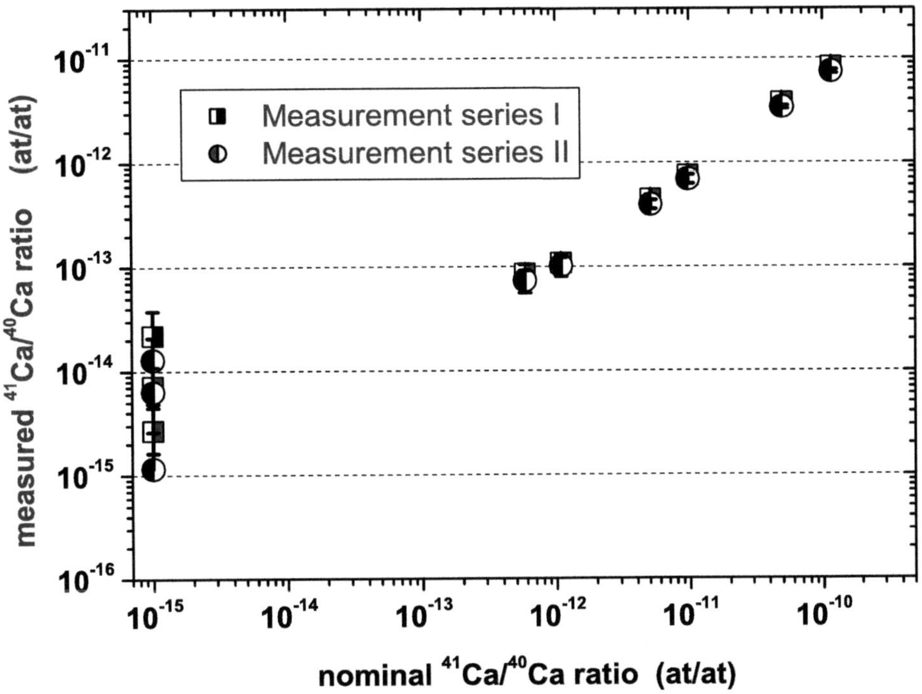

Fig. 2. Measurement of a ^{41}Ca-standard dilution series. The nominal ratios represent the quoted $^{41}Ca/^{40}Ca$ isotope ratios, the measured $^{41}Ca/^{40}Ca$ ratio is the raw uncorrected ratio of the ^{41}Ca count-rate and the ^{41}Ca particle current. Blank samples with expected negligible ^{41}Ca content were assigned to a nominal ratio of $1 \cdot 10^{-15}$.

The measurement procedure applied at VERA is the following: in the cesium sputter source, negatively charged $^{41}CaF_3^-$-ions are produced and injected into the tandem accelerator, which is set to a voltage of 3.3 MV. After the tandem accelerator, $^{41}Ca^{4+}$ ions, having energies of about 15 MeV, are identified with the detector setup described above. However, the detection efficiency drops with the additional scattering when the beam is passing the thick SiN foils. To this end, the flight path has been reduced from 2.2 m as used for heavy ion measurements (see fig. 1) to 0.65 m. The time resolution is still sufficient since energy-loss straggling is the main contribution to the width of the peaks. Besides the time-of-flight information, utilizing an ionization chamber at the very end of the beam-line, the total residual energy of the incoming ions is measured, too.

Six different Ca-standards have been measured applying this technique. In fig. 2 the measured $^{41}Ca/^{40}Ca$ isotope ratios are plotted *versus* their nominal values. Measured data represent the absolute ratios, uncorrected for background events. The absolute value of these data reflects the beam losses between the position of the current measurement and the particle detection system. Losses are mainly due to the beam scattering in the thick SiN foils. The data should follow a straight line if background interference is of no concern. Blank samples are plotted at the left part of the figure assigning them a nominal ratio of 10^{-15}. The background value found for such samples during these measurements was about 3 orders of magnitude lower than the value of the highest Ca standard with a quoted ratio $^{41}Ca/^{40}Ca$ of $1.16 \cdot 10^{-10}$ [36], which results in an normalized value of about $1 \cdot 10^{-13}$ for that blank samples. The uncertainty of these blank samples enters into the final values and dominates already the error of the standard material with the (lowest) nominal value of $5.9 \cdot 10^{-13}$ (see error bars). Different symbols in the plot depict the use of SiN degrader foils of different thickness: measurement series I was carried out using 1000 nm thick SiN foils and measurement series II using a 1650 nm thick foil. This figure clearly shows that $^{41}Ca/^{40}Ca$ isotope ratios of a few 10^{-13} can be quantified with this setup. In a later measurement series the K-background showed a somewhat enhanced count rate, which was up to a factor of 20 higher than for the previous measurements. A possible reason for this higher background rate might be the preceding cleaning of the ion source which was combined with the replacement of some source parts. Those new parts are suspicious to K-contamination. A systematic investigation of possible sources for this enhanced background is presently underway.

5 Summary

It has been shown —for the first time— that a 3 MV tandem accelerator is capable to measure $^{41}Ca/^{40}Ca$ ratios down to a few 10^{-13}. The reproducibility of those measurements was found to be between 5 to 8 %. Applying the deltaTOF technique allows the use of CaF_2 samples despite potassium will form negative ions and therewith isobars of ^{41}Ca will be produced, too. For samples with ratios on the order of or below 10^{-12} a moderate ^{41}K background count-rate will become desirable because it may dominate the uncertainty of the Ca data.

Our recent investigations on ^{41}Ca measurements at VERA opens now the possibility for quantitative Ca measurements. As a first project, the determination of the cross section for the ^{40}Ca(n, γ)^{41}Ca reaction at astrophysical interesting neutron energies (25 keV) is in progress.

We thank F. Käppeler and M. Paul for including VERA into the ^{40}Ca(n, γ) measurement program. We gratefully acknowledge K. Nishiizumi and G. Korschinek for leave some Ca standard material for our measurement program.

References

1. L. Buchmann, M. Comyn, J. Thomson (Editors) *Proceedings of the 8th International Symposium on Nuclei in the Cosmos Vancouver, British Columbia, Canada, 19-23 July, 2004*, Nucl. Phys. **758**, 3-859 (2005).
2. F.-K. Thielemann, D. Argast, F. Brachwitz, J.L. Fisker, C. Fröhlich, R. Hirschi, E. Kolbe, D. Mocelj, T. Rauscher, Nucl. Phys. A **751**, 301 (2005).
3. M. Wiescher, Nucl. Phys. A **751**, 285 (2005).
4. W. Kutschera, P. Collon, H. Friedmann, R. Golser, P. Hille, A. Priller, W. Rom, P. Steier, S. Tagesen, A. Wallner, E. Wild, G. Winkler, Nucl. Instrum. Methods B **123**, 47 (1997).
5. P. Steier, R. Golser, W. Kutschera, A. Priller, C. Vockenhuber, S. Winkler, Nucl. Instrum. Methods B **223-224**, 67 (2004).
6. C. Vockenhuber, I. Ahmad, R. Golser, W. Kutschera, V. Liechtenstein, A. Priller, P. Steier, S. Winkler, Int. J. Mass Spectrom. **223**, 713 (2003).
7. W. Kutschera, Nucl. Phys. A **752**, 645 (2005).
8. W. Kutschera, Int. J. Mass Spectrom. **242**, 145 (2005).
9. A. Wallner, R. Golser, W. Kutschera, A. Priller, P. Steier, C. Vockenhuber, H. Vonach, T. Faestermann, K. Knie, G. Korschinek, R.C. Haight *et al.* (Editors), AIP Conf. Proc. **769**, 621 (2005).
10. J. Ellis, B.D. Fields, D.N. Schramm, Astrophys. J. **470**, 1227 (1996).
11. B.D. Fields, New Astron. Rev. **48**, 119 (2004).
12. K. Knie, G. Korschinek, T. Faestermann, E.A. Dorfi, G. Rugel, A. Wallner, Phys. Rev. Lett. **93**, 171103 (2004).
13. B.D. Fields, K.A. Hochmuth, J. Ellis, Astrophys. J. **621**, 902 (2005).
14. C. Vockenhuber, C. Feldstein, M. Paul, N. Trubnikov, M. Bichler, R. Golser, W. Kutschera, A. Priller, P. Steier, S. Winkler, New Astron. Rev. **48**, 161 (2004).
15. S. Winkler, I. Ahmad, R. Golser, W. Kutschera, K.A. Orlandini, M. Paul, A. Priller, P. Steier, C. Vockenhuber, New Astron. Rev. **48**, 151 (2004).
16. P. Steier, R. Golser, W. Kutschera, V. Liechtenstein, A. Priller, A. Valenta, C. Vockenhuber, Nucl. Instrum. Methods B **188**, 283 (2002).
17. E. Hrnecek, P. Steier, A. Wallner, Appl. Radiat. Isotopes **63**, 633 (2005).
18. C. Wallner, T. Faestermann, U. Gerstmann, K. Knie, G. Korschinek, C. Lierse, G. Rugel, New Astron. Rev. **48**, 145 (2004).
19. M. Paul, A. Valenta, I. Ahmad, D. Berkovits, C. Bordeanu, S. Ghelberg, Y. Hashimoto, A. Hershkowitz, S. Jiang, T. Nakanishi, K. Sakamoto, Nucl. Phys. A **719**, C29 (2003).
20. C. Wallner, T. Faestermann, U. Gerstmann, W. Hillebrandt, K. Knie, G. Korschinek, C. Lierse, C. Pomar, G. Rugel, Nucl. Instrum. Methods B **172**, 333 (2000).
21. W. Prantzos, R. Diehl, Phys. Rep. **267**, 1 (1996).
22. A. Wallner, S.V. Chuvaev, A.A. Filatenkov, Y. Ikeda, W. Kutschera, G. Mertens, A. Priller, W. Rochow, P. Steier, H. Vonach, Eur. Phys. J. A **17**, 285 (2003).
23. A. Wallner, Y. Ikeda, W. Kutschera, A. Priller, P. Steier, H. Vonach, E. Wild, Nucl. Instrum. Methods B **172**, 382 (2000).
24. A. Arazi, T. Faestermann, J.O. Fernández Niello, K. Knie, G. Korschinek, M. Poutivtsev, E. Richter, G. Rugel, A. Wallner, submitted to Phys. Rev. C (2006).
25. A. Arazi, T. Faestermann, J.O. Fernández Niello, D. Frischke, K. Knie, G. Korschinek, H.J. Maier, E. Richter, G. Rugel, A. Wallner, Nucl. Instrum. Methods B **223-224**, 259 (2004).
26. A. Arazi, T. Faestermann, J.O. Fernández Niello, K. Knie, G. Korschinek, E. Richter, G. Rugel, A. Wallner, New Astron. Rev. **46**, 525 (2002).
27. H. Nassar *et al.*, Phys. Rev. Lett. **94**, 092504 (2005).
28. A. Tomyo *et al.*, Astrophys. J. **623**, L153 (2005).
29. T. Rauscher, K.H. Guber, Phys. Rev. C **66**, 028802 (2002); T. Rauscher, K.H. Guber, Phys. Rev. C **71**, 059903(E) (2005).
30. M. Paul *et al.*, Nucl. Phys. A **718**, 239c (2003); H. Nassar *et al.*, Nucl. Phys. **758**, 411 (2005).
31. M. Paul, I. Ahmad, W. Kutschera, Z. Phys. A **340**, 249 (1991); W. Kutschera, I. Ahmad, M. Paul, Radiocarbon **34**, 436 (1992).
32. W. Kutschera, I. Ahmad, P.J. Billquist, B.G. Glasgola, K. Furer, R.C. Pardo, M. Paul, K.E. Rehm, P.J. Slota jr., R.E. Taylor, J.L. Yntema, Radiocarbon **31**, 311 (1989).
33. D. Fink, R. Middleton, J. Klein, P. Sharma, Nucl. Instrum. Methods B **47**, 79 (1990); D. Fink, J. Klein, R. Middleton, Nucl. Instrum. Methods B **52**, 572 (1990).
34. P. Steier, R. Golser, V. Liechtenstein, W. Kutschera, A. Priller, C. Vockenhuber, A. Wallner, Nucl. Instrum. Methods B **240**, 445 (2005).
35. C. Vockenhuber, R. Golser, W. Kutschera, A. Priller, P. Steier, K. Vorderwinkler, A. Wallner, Nucl. Instrum. Methods B **240**, 490 (2005).
36. K. Nishiizumi, M.W. Caffee, D.J. DePaolo, Nucl. Instrum. Methods B **172**, 399 (2000).

Author index

GPSR Compliance

The European Union's (EU) General Product Safety Regulation (GPSR) is a set of rules that requires consumer products to be safe and our obligations to ensure this.

If you have any concerns about our products, you can contact us on ProductSafety@springernature.com

In case Publisher is established outside the EU, the EU authorized representative is:

Springer Nature Customer Service Center GmbH
Europaplatz 3
69115 Heidelberg, Germany

Batch number: 10762213

Printed by Printforce, the Netherlands